ABHIJIT CHATTERJEE
Aug. 25, 1979

# CALCULUS WITH ANALYTIC GEOMETRY

# CALCULUS

**ABRAHAM SPITZBART**
*University of Wisconsin*
*Milwaukee*

*with Analytic Geometry*

**Scott, Foresman and Company • Glenview, Illinois**
*Dallas, Tex. • Oakland, N.J. • Palo Alto, Cal.*
*Tucker, Ga. • Brighton, England*

# CALCULUS WITH ANALYTIC GEOMETRY
*Abraham Spitzbart*

Library of Congress Catalog Card Number 73-83284
ISBN: 0-673-07907-4
AMS 1970 Subject Classification: 98A20

Copyright © 1975, 1969 Scott, Foresman and Company.
Portions of this text are adapted from *Analytic Geometry,* Copyright © 1969
  Scott, Foresman and Company.
Philippines Copyright 1975 Scott, Foresman and Company.
All Rights Reserved.
Printed in the United States of America.

Regional offices of Scott, Foresman and Company are located in
Dallas, Texas; Glenview, Illinois; Oakland, New Jersey; Palo Alto,
California; Tucker, Georgia, and Brighton, England.

# Preface

This book is designed for a course in analytic geometry and calculus which traditionally occupies a three-semester sequence or the equivalent, but it can be adapted to shorter courses. A thorough preparation in these subjects is provided, for application in fields of study which use mathematics, as well as for further study in mathematics itself. A knowledge of college algebra and trigonometry is presupposed.

Throughout the book an effort is made to balance the ideas of formal proof, a need for proof, and an intuitive approach. A number of proofs are included to emphasize the logical reasoning that underlies the formal development of mathematics in general and of calculus in particular. No effort is made to prove each statement. Instead, in many cases the emphasis is on the "need for proof," the idea that a proof is needed for a statement that cannot be accepted as "obvious," even though the proof is, in fact, omitted. Basic concepts often fall in this category, ones that are easy to state and to understand but difficult to prove, such as fundamental properties of continuous functions. At times a statement is accepted without proof, especially if a plausible argument can be given for its validity.

A persistent goal in the presentation has been the smooth progression from one topic to another, with constant attention to the relation of the topic under consideration to what has preceded and what is to come.

Vectors are introduced in the first chapter and are used frequently in subsequent chapters. In the development of analytic geometry considerable attention is given to graphing, with emphasis on showing the essential features of a graph quickly. Much of the work in calculus requires just such rapid sketching.

The point of view in the study of limits, introduced in Chapter 3, can be varied from one that is largely intuitive to one which is more rigorous. The derivative is based on the concept of rate of change of a function, and is then developed to a point where applications can be meaningfully considered. Integration is introduced as an inverse of differentiation, and this is followed by a discussion of the definite integral; the connection between the two concepts is given careful attention. Applications of integration, both geometric and physical, are treated in a manner which blends considerations of rigor and an intuitive approach. An earlier introduction of integration can be achieved, if desired, by omitting or deferring parts of earlier chapters.

Chapters 18–23 form a body of material in three-dimensional analytic geometry, calculus of several variables, and infinite series. The treatment of linear equations in Chapter 19 can serve as an introduction to a general study of linear systems, by providing a good knowledge of special cases in advance of a more general and abstract study. Chapter 23 on Infinite Series can, if desired, be studied at any point after Chapter 15.

Section numbers indicate the chapter and section. Theorems, definitions, equations, and figures are numbered according to the section in which they appear. A reference to an exercise without the designation of its exercise group means the next set of exercises following that reference. In other cases an exercise is referred to by both number and exercise group.

Some sections are starred to indicate that the topics may be omitted without disturbing the development of the subject. Starred sections are accompanied by starred exercise groups. On the other hand, there may also be starred exercises in both starred and unstarred exercise groups, to suggest that they are somewhat more difficult or that they make use of less familiar material from earlier courses.

A few numerical tables appear in the Appendix. However, it may be desirable for the student to have access to more extensive tables, such as appear in many standard compilations. The Appendix also contains a short table of integrals and a brief review of trigonometry.

Answers to odd-numbered exercises appear at the back of the book. Answers to even-numbered exercises are available in a separate manual.

I should like to express my appreciation to the staff of Scott, Foresman and Company for its assistance and cooperation in the preparation of this book, and to the reviewers of the manuscript, who helped sharpen the presentation.

*A. Spitzbart*

# Contents

**Chapter 1  FUNDAMENTAL CONCEPTS**
    1.1  The real number system  *1*
    1.2  Inequalities  *3*
    1.3  Absolute value  *7*
    1.4  Coordinate lines. Intervals  *10*
    1.5  Rectangular coordinate system  *12*
    1.6  Directed line segments. Projections  *13*
    1.7  Vectors  *15*
    1.8  Base vectors  *18*
    1.9  Geometric interpretation of vectors  *20*

**Chapter 2  LENGTH AND DIRECTION. THE STRAIGHT LINE**
    2.1  The distance formula; the midpoint formula  *25*
    2.2  Scalar or dot product of vectors  *29*
    2.3  Slope of a line  *33*
    2.4  Parallel and perpendicular lines  *34*
    2.5  Angles between lines  *37*
    2.6  Relations. Graph of an equation  *40*
    2.7  The equation $Ax + By + C = 0$  *43*
    2.8  Forms of the equation of a straight line  *44*
    2.9  Intersections of straight lines  *47*
    2.10  Linear inequalities in $x$ and $y$  *49*

**Chapter 3  FUNCTIONS AND LIMITS**
    3.1  Functions  *53*
    3.2  Combinations of functions  *56*
    3.3  Introduction to limits  *59*
    3.4  The limit concept  *63*
    3.5  Properties of limits  *66*
    3.6  Continuity  *70*
    3.7  Some consequences of continuity  *72*

**Chapter 4 THE DERIVATIVE**
    4.1  Introduction   *76*
    4.2  The derivative   *76*
    4.3  Tangent to a curve   *81*
    4.4  Differentiation formulas   *85*
    4.5  Derivative of a composite function. The Chain Rule   *88*
    4.6  Derivative of a product and of a quotient   *93*
    4.7  Implicit differentiation   *97*
    4.8  Derivatives of higher order   *100*
    4.9  Differentials   *103*
    4.10  Interpretation of differentials   *108*

**Chapter 5 APPLICATIONS OF DERIVATIVES**
    5.1  Introduction   *112*
    5.2  Direction of a curve   *112*
    5.3  Increasing and decreasing functions   *115*
    5.4  Maxima and minima   *119*
    5.5  Maximum and minimum in an interval   *125*
    5.6  Applications of maxima and minima   *127*
    5.7  Extremal problems with auxiliary conditions   *128*
    5.8  Concavity. Point of inflection   *132*
    5.9  *Extended criterion for an extremum or point of inflection   *137*
    5.10  Rate of change   *139*
    5.11  Linear motion   *141*
    5.12  Related rates   *143*

**Chapter 6 CURVE SKETCHING**
    6.1  Introduction   *148*
    6.2  Additional criteria   *150*
    6.3  Behavior of functions for large $x$   *155*
    6.4  Horizontal and vertical asymptotes   *159*
    6.5  Vertical tangents   *163*
    6.6  The graph of $y^2 = R(x)$   *167*
    6.7  *Asymptotic curves   *171*

**Chapter 7 EQUATIONS OF SECOND DEGREE**
    7.1  Introduction   *175*
    7.2  Equation of a locus   *175*
    7.3  The equation of a circle   *179*
    7.4  *Circles determined by various conditions   *184*
    7.5  *Intersections involving circles   *186*
    7.6  The parabola   *190*
    7.7  Translation of axes   *194*
    7.8  The ellipse   *198*
    7.9  The hyperbola   *204*
    7.10  *Asymptotes of the hyperbola   *210*
    7.11  A word on standard forms   *212*

**Chapter 8  EQUATIONS OF SECOND DEGREE,** *Continued*
    8.1  Introduction  *214*
    8.2  Rotation of axes  *214*
    8.3  The general second-degree equation  *218*
    8.4  Identification of conics  *222*
    8.5  Focus-directrix property of the conics  *225*
    8.6  *Tangents of the conics  *228*
    8.7  *Optical properties of the conics  *229*

**Chapter 9  INTEGRATION**
    9.1  Introduction  *234*
    9.2  The Mean-Value Theorem  *234*
    9.3  The indefinite integral  *238*
    9.4  Use of the chain rule in integration  *242*
    9.5  Change of variable  *246*
    9.6  Introduction to the definite integral  *248*
    9.7  Summation  *249*
    9.8  An area problem  *253*
    9.9  The definite integral  *257*
    9.10  Fundamental Theorem of Integral Calculus  *262*
    9.11  Change of variable in a definite integral  *266*
    9.12  Area of a plane region  *268*

**Chapter 10  TRIGONOMETRIC FUNCTIONS AND THEIR INVERSES**
    10.1  Introduction  *275*
    10.2  Fundamental trigonometric limits  *276*
    10.3  Derivatives of trigonometric functions  *279*
    10.4  Integrals of trigonometric functions  *282*
    10.5  Inverse functions  *285*
    10.6  Inverse trigonometric functions  *288*
    10.7  Derivatives of inverse trigonometric functions  *291*
    10.8  Integrals leading to inverse trigonometric functions  *294*

**Chapter 11  LOGARITHMIC AND EXPONENTIAL FUNCTIONS**
    11.1  Introduction  *297*
    11.2  The logarithm function  *297*
    11.3  The exponential function  *299*
    11.4  Graphs of logarithmic and exponential functions  *304*
    11.5  Differentiation of logarithmic and exponential functions  *308*
    11.6  Integrals leading to logarithmic and exponential functions  *311*
    11.7  The hyperbolic functions  *314*
    11.8  Integrals of hyperbolic functions  *318*
    11.9  The inverse hyperbolic functions  *319*
    11.10  Derivatives of inverse hyperbolic functions  *321*

**Chapter 12  GEOMETRIC APPLICATIONS OF THE DEFINITE INTEGRAL**
- 12.1  Introduction   *324*
- 12.2  Volume of a solid with known cross section   *325*
- 12.3  Volume of a solid of revolution   *327*
- 12.4  Volume of a solid of revolution by cylindrical shells   *332*
- 12.5  Arc length   *336*
- 12.6  Area of a surface of revolution   *339*

**Chapter 13  PHYSICAL APPLICATIONS OF THE DEFINITE INTEGRAL**
- 13.1  Introduction   *345*
- 13.2  Work   *345*
- 13.3  Fluid pressure   *349*
- 13.4  Moments in a plane. Center of mass   *352*
- 13.5  Center of mass of a lamina   *356*
- 13.6  *Center of mass of a composite body   *361*
- 13.7  Centroids of other figures   *364*
- 13.8  Moment of inertia   *368*

**Chapter 14  METHODS OF INTEGRATION**
- 14.1  Formal integration   *373*
- 14.2  Powers and products of sine and cosine   *376*
- 14.3  Powers of tangent and secant   *379*
- 14.4  Trigonometric substitution   *381*
- 14.5  Integrals containing complete quadratic functions   *385*
- 14.6  Partial fractions. Case 1.   *388*
- 14.7  Partial fractions. Case 2.   *391*
- 14.8  Integration by parts   *394*
- 14.9  *Reduction formulas   *399*
- 14.10 Rational functions of $\sin \theta$ and $\cos \theta$   *401*

**Chapter 15  INDETERMINATE FORMS. IMPROPER INTEGRALS**
- 15.1  Introduction   *406*
- 15.2  Indeterminate forms   *406*
- 15.3  L'Hospital's Rule   *408*
- 15.4  Indeterminate forms of exponential type   *413*
- 15.5  Improper integrals   *415*

**Chapter 16  PARAMETRIC EQUATIONS. VECTOR FUNCTIONS**
- 16.1  Equations in parametric form   *421*
- 16.2  Elimination of a parameter. Slope   *424*
- 16.3  The use of a parameter   *428*
- 16.4  Arc length. Area of a surface of revolution   *433*
- 16.5  Vector functions   *437*
- 16.6  Velocity and acceleration   *441*
- 16.7  Curvature   *446*
- 16.8  Arc length as parameter   *449*

## Chapter 17  POLAR COORDINATES
17.1  The polar coordinate system  *452*
17.2  Equations of curves in polar form  *455*
17.3  Graphing polar equations  *457*
17.4  *Rotation of a curve  *460*
17.5  Conics in polar form  *462*
17.6  Intersections of curves in polar coordinates  *466*
17.7  Direction of a curve in polar coordinates  *469*
17.8  Area of a region in polar coordinates  *472*
17.9  Arc length; area of a surface of revolution  *476*
17.10  *Velocity and acceleration in polar coordinates  *479*

## Chapter 18  VECTORS IN THREE DIMENSIONS
18.1  The rectangular coordinate system  *483*
18.2  Three-dimensional vectors  *484*
18.3  Geometrical interpretation of vectors; distance formula  *485*
18.4  *The midpoint formula; point of division formulas  *489*
18.5  Scalar product; directions in space  *491*
18.6  Direction angles  *494*
18.7  Vector product or cross product  *497*
18.8  *Triple product  *501*

## Chapter 19  LINES AND PLANES
19.1  Direction numbers  *504*
19.2  The plane  *507*
19.3  The straight line  *514*
19.4  Intersection of two planes  *517*
19.5  Intersections of lines and planes  *521*
19.6  *Normal form of the equation of a plane  *525*
19.7  *Intersection of three planes  *529*
19.8  *Specialized distance formulas  *531*

## Chapter 20  SURFACES AND CURVES
20.1  Introduction  *534*
20.2  Sketching a surface  *534*
20.3  Cylindrical surfaces  *537*
20.4  Surfaces of revolution  *539*
20.5  Quadric surfaces  *542*
20.6  *Ruled surfaces  *549*
20.7  Curves in space  *552*
20.8  Vector functions. Parametric representation of a curve  *554*
20.9  Unit tangent and unit normal vectors  *559*
20.10  Velocity and acceleration. Curvature. The moving trihedral  *561*
20.11  Cylindrical and spherical coordinates  *566*

**Chapter 21  PARTIAL DIFFERENTIATION**
- 21.1 Functions of several variables  *570*
- 21.2 Directional derivative. Partial derivative  *574*
- 21.3 Partial derivatives of higher order  *579*
- 21.4 The Chain Rule  *583*
- 21.5 Use of the Chain Rule  *588*
- 21.6 Tangent planes  *590*
- 21.7 Directional derivative and gradient  *593*
- 21.8 Total differential  *597*
- 21.9 Use of the total differential  *600*
- 21.10 More on higher order derivatives  *604*
- 21.11 Extrema of functions of two variables  *606*
- 21.12 Exact differential  *612*
- 21.13 *Line integrals  *616*
- 21.14 *Independence of path  *624*

**Chapter 22  MULTIPLE INTEGRATION**
- 22.1 Introduction  *629*
- 22.2 Double integrals  *630*
- 22.3 Evaluation of double integrals. Iterated integrals  *634*
- 22.4 Double integrals in polar coordinates  *637*
- 22.5 Area by double integrals  *640*
- 22.6 Volume by double integrals  *643*
- 22.7 Physical applications of double integrals  *647*
- 22.8 Moment of inertia  *650*
- 22.9 Triple or volume integral  *654*
- 22.10 Triple integrals in cylindrical and spherical coordinates  *660*
- 22.11 Area of a curved surface  *665*

**Chapter 23  INFINITE SERIES**
- 23.1 Sequences  *671*
- 23.2 The limit of a sequence  *674*
- 23.3 Infinite series  *677*
- 23.4 The integral test for series  *682*
- 23.5 Comparison tests for series  *686*
- 23.6 *An extended comparison test  *688*
- 23.7 The ratio test for series  *690*
- 23.8 Testing a series of positive terms  *692*
- 23.9 Alternating series  *693*
- 23.10 Absolute convergence of series  *695*
- 23.11 Power series  *698*
- 23.12 Sums of power series  *702*
- 23.13 Taylor series  *707*
- 23.14 The binomial series  *711*
- 23.15 Use of the remainder in Taylor's formula  *714*
- 23.16 Computation with power series  *716*
- 23.17 Finding series from known series  *719*
- 23.18 Indeterminate forms via infinite series  *723*
- 23.19 Taylor's formula for a function of two variables  *724*

**TRIGONOMETRY REVIEW AND FORMULAS**  *728*

**TABLES**
1. Squares and Square Roots  *732*
2. Common Logarithms  *733*
3. Natural Logarithms  *735*
4. Powers of $e$  *735*
5. Trigonometric Values for Angles in Degrees  *736*
6. Trigonometric Values for Angles in Radians  *737*
7. Table of Integrals  *738*

**ANSWERS TO ODD-NUMBERED EXERCISES**  *740*

**INDEX**  *766*

# 1 Fundamental Concepts

## 1.1 THE REAL NUMBER SYSTEM

The material to be covered in this book will deal exclusively with real numbers. You are, to be sure, familiar with many properties of real numbers, but there are many others which are not explicitly stated in introductory courses. While it is not possible to list all the properties, it is possible to give a basic list from which all properties can be derived. In this section we present such a basic list of properties of the real numbers in the form of a set of axioms. After listing these axioms we shall discuss briefly some of their implications.

We shall use the word "set" in various connections. A *set* is a collection of objects, defined in such a way that it may be determined whether any object is in the set or not. Each object in a set is called an *element* or *member* of the set. We may then speak of a set of numbers, or a set of points, or a set of straight lines, and so on. Other terms, like "subset," will be defined as they occur.

In the context of the above discussion we now define the *real numbers* as a set whose elements satisfy the axioms listed below. Each axiom is stated in the form of a property which is assumed to hold. The letters $a$, $b$, $c$ denote elements of the set, or real numbers.

All properties of the real numbers can be derived from these axioms, including properties of subtraction and division, which are defined in terms of addition and multiplication. These include all the familiar properties of arithmetic and algebra. It is not feasible to go into a complete discussion of these properties here.

## Axioms for the real numbers

1. *Closure under addition and multiplication.* For any real numbers $a$ and $b$ there is a *unique sum*, written as $a + b$, and a *unique product*, written as $a \times b$, or $a \cdot b$, or $ab$.

2. The *commutative properties* of addition and multiplication, respectively, hold; that is, if $a$ and $b$ are any real numbers, then

$$a + b = b + a; \quad ab = ba.$$

3. The *associative properties* of addition and multiplication, respectively, hold; that is, if $a$, $b$, and $c$ are any real numbers, then

$$a + (b + c) = (a + b) + c = a + b + c;$$
$$a(bc) = (ab)c = abc.$$

4. An *additive identity*, or *zero*, exists; that is, there exists a real number 0 such that for any real number $a$,

$$a + 0 = 0 + a = a.$$

A *multiplicative identity*, or *unity*, exists; that is, there exists a real number $1 \neq 0$ such that for any real number $a$,

$$a \cdot 1 = 1 \cdot a = a.$$

5. *Additive inverse property.* For any real number $a$ there exists a real number $b$, also written as $-a$, such that

$$a + b = b + a = 0, \quad \text{or} \quad a + (-a) = (-a) + a = 0.$$

*Multiplicative inverse property.* For any real number $a \neq 0$ there exists a real number $c$, also written as $a^{-1}$, or $1/a$, such that

$$ac = ca = 1 \quad \text{or} \quad aa^{-1} = a^{-1}a = 1 \quad \text{or} \quad a \cdot \frac{1}{a} = \frac{1}{a} \cdot a = 1.$$

6. The *distributive property of multiplication with respect to addition* holds; that is, if $a$, $b$, and $c$ are any real numbers, then

$$a(b + c) = ab + ac \quad \text{and} \quad (b + c)a = ba + ca.$$

7. *Order property.* There is a relation $<$ such that

   (a) One and only one of the following holds:

   $$a < b, \quad a = b, \quad b < a.$$

   (b) If $a < b$ and $b < c$, then $a < c$.
   (c) If $a < b$ and $c$ is any real number, then $a + c < b + c$.
   (d) If $0 < c$ and $a < b$, then $ac < bc$.

8. The *least upper bound property* holds; that is, any nonempty set of real numbers which has an upper bound must have a least upper bound. (Appropriate definitions appear in the following discussion.)

The presence of the ordinary positive integers within the set of real numbers is somewhat concealed. There is an inkling of them in the existence of the multiplicative unit, 1. It would seem that repeated additions of this unit would lead to the positive integers. This is basically the case, but we shall not develop the logical definition of the positive integers based on the axioms of the real numbers. Once the positive integers are defined, the negative integers are easily defined by the axiom which provides the additive inverse. The number "zero" exists by Axiom 4 and is included among the integers, but zero is neither positive nor negative. The number zero has special properties in addition to that of the additive identity. We mention in particular the property that if $ab = 0$, then at least one of the numbers $a$ and $b$ is zero; this property can be proved.

A *subset* of the real numbers (that is, a part of the set of real numbers), which includes the integers, is the set of *rational numbers*, numbers which may be expressed as quotients of integers (division by zero always being excluded). Whereas integers have the property that for each integer there is a next larger one, rational numbers have the property that between any two of them there always exists another rational number, and in fact infinitely many.

As closely packed as the rational numbers are, they do not exhaust the set of real numbers. In a certain sense, which we shall not enter upon here, there are more real numbers which are not rational than there are rational numbers. The nonrational numbers, called the *irrational numbers*, include numbers involving radicals, like $\sqrt{2}$, $\sqrt[3]{5}$, and other familiar numbers such as $\pi$ and $e$. (The number $e$ is used in the study of natural logarithms.)

The *least upper bound axiom* expresses the so-called *completeness property* of the real numbers, of which there are many forms. It states that if all the numbers of a nonempty set are less than or equal to some number (an *upper bound* of the set), then there exists a smallest number with this property (the *least upper bound* of the set). One application of the least upper bound axiom concerns the proof of existence of certain real numbers. For example, it is true (the proof may be known to you) that no rational number $r$ exists such that $r^2 = 2$. Now consider the set of all positive rational numbers $x$ such that $x^2 < 2$. It can be shown that this set is not empty (it contains 1), and it has an upper bound, 2, for example. Then by the least upper bound axiom this set has a least upper bound $u$, for which it can be shown that $u^2 = 2$. This proves that the square root of 2 exists as a real number.

In general, the least upper bound axiom insures that the gaps in the set of rational numbers are filled by other real numbers.

In closing this brief discussion, we mention that the order axiom leads to the study of inequalities, which we will now discuss.

## 1.2 INEQUALITIES

It is assumed that you know the properties of the arithmetic operations with real numbers. The order axiom of real numbers (Axiom 7) permits us to define and work with inequalities. You will recall that the symbol $<$ is read "is less than"; we also use the symbol $>$ to mean "is greater than."

We introduce here a notation for logical implication. If $A$ and $B$ represent statements, then

$A \Rightarrow B$ means "$A$ implies $B$"; that is, if $A$ holds then $B$ holds;

$A \Leftarrow B$ means "$B$ implies $A$"; that is, if $B$ holds then $A$ holds;

$A \Leftrightarrow B$ means "$A$ is equivalent to $B$"; that is, $A$ implies $B$ and $B$ implies $A$.

**DEFINITION 1.2.1.** $\quad a > b \Leftrightarrow b < a.$

**DEFINITION 1.2.2.** *If $a$ is a real number such that $0 < a$ (or $a > 0$), then $a$ is called positive. If $a < 0$, then $a$ is called negative.*

From Axiom 7(a) and this definition it follows that for a real number $a$, one and only one of the following holds: $a$ is positive; $a$ is zero; or $a$ is negative.

**THEOREM 1.2.1.** $\quad 0 < 1.$

*Proof.* By Axiom 4, $1 \neq 0$. Suppose $1 < 0$. Then by Axiom 7(c),

$$1 + (-1) < 0 + (-1), \quad \text{or} \quad 0 < -1.$$

But then we have by Axiom 7(d),

$$0 \cdot (-1) < (-1) \cdot (-1) \quad \text{or} \quad 0 < 1,$$

which contradicts the supposition $1 < 0$. Hence we must have $0 < 1$.

**THEOREM 1.2.2.** $\quad a > 0 \Leftrightarrow -a < 0.$

*Proof.* Suppose $a > 0$. If we add $-a$ to both members [Axiom 7(c) and Definition 1.2.1], we have

$$a + (-a) > 0 + (-a).$$

But $a + (-a) = 0$ and $0 + (-a) = -a$; we then get $0 > -a$, or $-a < 0$. The converse is proved in a similar manner.

**THEOREM 1.2.3.** (a) $\quad a < b \Leftrightarrow a - b < 0.$

(b) $\quad a > b \Leftrightarrow a - b > 0.$

*Proof of* (a). If $a < b$, then $a + (-b) < b + (-b)$, by Axiom 7(c). But $a + (-b) = a - b$ and $b + (-b) = 0$. Hence $a - b < 0$.

For the converse, if $a - b < 0$, then $(a - b) + b < 0 + b$. But

$$(a - b) + b = a + [(-b) + b] = a, \quad \text{and} \quad 0 + b = b.$$

Hence $a < b$, and the proof of (a) is complete. The proof of (b) is similar.

**THEOREM 1.2.4.** $\quad a < b \Rightarrow a - c < b - c.$

*Proof.* This follows at once from Axiom 7(c), since subtraction of $c$ is the same as addition of $-c$.

The counterpart of Axiom 7(d), when $c < 0$, is contained in the following statement.

**THEOREM 1.2.5.** $\qquad a < b$ and $c < 0 \Rightarrow ac > bc$.

*Proof.* $c < 0 \Rightarrow 0 < -c$. Hence, by Axiom 7(d), $a(-c) < b(-c)$, and $-ac < -bc$. Then, adding $ac + bc$ to both members of the last relation, we have

$$-ac + ac + bc < -bc + ac + bc.$$

This reduces to $bc < ac$, so that $ac > bc$, as was to be shown.

Corresponding properties hold when $a > b$, with entirely similar proofs. We collect the appropriate properties in the following statement.

**THEOREM 1.2.6.** (a) *For any real number $c$,*

$$a < b \Rightarrow a \pm c < b \pm c, \qquad a > b \Rightarrow a \pm c > b \pm c,$$

*where the upper signs in each are taken together, as are the lower signs.*

(b) *If $c > 0$, then $a < b \Rightarrow ac < bc$; $a > b \Rightarrow ac > bc$.*
(c) *If $c < 0$, then $a < b \Rightarrow ac > bc$; $a > b \Rightarrow ac < bc$.*

An *inequality* expresses the relation "less than" or "greater than" between two quantities. We refer to the quantity on the left (right) of the inequality symbol as the left (right) side or left (right) *member* of the inequality.

Theorem 1.2.6 then states in words that (1) both sides of an inequality may be increased or decreased by equal quantities, and may be multiplied or divided by equal *positive* quantities (excluding division by zero) without changing the sense of the inequality; (2) both sides of an inequality may be multiplied or divided by equal *negative* quantities (excluding division by zero), if the inequality sign is reversed. (*Note:* Division by $c \neq 0$ is equivalent to multiplication by $c^{-1}$, or $1/c$.)

A relation may involve a combined inequality and equality sign. Thus $a \leq b$ means that either $a < b$ or $a = b$. It is not difficult to determine the statements corresponding to Theorem 1.2.6 when such combined signs are permitted.

A *continued inequality* such as $a < b < c$ means that $b$ is simultaneously greater than $a$ and less than $c$; the corresponding meanings for statements such as $a \leq b < c$ or $a \leq b \leq c$ or $a < b \leq c$ should be clear.

We are now in a position to solve a variety of inequalities. A *solution* of an inequality in the undetermined quantity $x$ is any value of $x$ for which the inequality is true; the *complete solution*, or *solution set*, of such an inequality consists of the collection of *all* such values of $x$.

***Example 1.2.1.*** Solve the inequality $2x + 3 < 3x - 4$.

*Solution.* By Theorem 1.2.6(a) we may subtract $3x + 3$ from both sides, to obtain $-x < -7$. By (c) of the same theorem, we may multiply both sides by $-1$, reversing the inequality sign at the same time, and we obtain $x > 7$ as the solution, that is, the set of all numbers such that $x > 7$.

More formally, we have shown that if $2x + 3 < 3x - 4$, then it must be true that $x > 7$. To establish the solution, we must show that if $x > 7$, then $2x + 3 < 3x - 4$. This is indeed the case, since each step that was used can be reversed.

In the following two examples we use some familiar properties of real numbers concerning the sign of a product of numbers.

***Example 1.2.2.*** Solve the inequality $2x^2 - 3x - 2 > 0$.

*Solution.* We may write the inequality as

$$(2x + 1)(x - 2) > 0.$$

For a product of two quantities to be positive, either both must be positive or both must be negative. The two possibilities are:

(a) $2x + 1 > 0$ and $x - 2 > 0$;
(b) $2x + 1 < 0$ and $x - 2 < 0$.

To satisfy both parts of (a), we must have $x > 2$. To satisfy both parts of (b), we must have $x < -1/2$. The steps used are reversible. The complete solution is then written as

$$x < -1/2, \quad x > 2;$$

this means the set consisting of all numbers which are less than $-1/2$, as well as all numbers which are greater than 2.

***Example 1.2.3.*** Solve the inequality $(2x - 1)(x + 3)(4x - 7) > 0$.

*Solution.* For the product to be positive, either all three factors must be positive, or any two of them must be negative and the third one positive. The tentative possibilities are:

(a) $2x - 1 > 0$ and $x + 3 > 0$ and $4x - 7 > 0$;
(b) $2x - 1 > 0$ and $x + 3 < 0$ and $4x - 7 < 0$;
(c) $2x - 1 < 0$ and $x + 3 < 0$ and $4x - 7 > 0$;
(d) $2x - 1 < 0$ and $x + 3 > 0$ and $4x - 7 < 0$.

The third part of (a) requires $x > 7/4$, and such values also satisfy the first two parts. The inequalities in (b) are inconsistent and are satisfied by no value of $x$; this is also true for the inequalities in (c). The inequalities in (d) are all satisfied by any number $x$ such that $-3 < x < 1/2$. The complete solution is written as

$$x > 7/4, \quad -3 < x < 1/2,$$

which means the set consisting of those numbers which are greater than 7/4, and those numbers which are between $-3$ and $1/2$.

## 1.3 ABSOLUTE VALUE

We now turn to the concept of *absolute value* which plays a very important part in the study of mathematics.

**DEFINITION 1.3.1.** *The* **numerical value,** *or* **absolute value,** *of $x$, denoted by $|x|$, is defined as follows*

$$|x| = \begin{cases} x & \text{if } x \geq 0 \\ -x & \text{if } x < 0. \end{cases}$$

Accordingly, *an absolute value is never negative.* Thus $|2| = 2$, and $|-2| = -(-2) = 2$.

Some very basic properties of absolute values are expressed in the following.

**THEOREM 1.3.1.** (a) $|xy| = |x||y|$

(b) $\left|\dfrac{x}{y}\right| = \dfrac{|x|}{|y|}, \quad y \neq 0$

(c) $\sqrt{x^2} = |x|$.

*Proof.* (a) If $x = 0$ or $y = 0$ the statement clearly holds. Otherwise, the possibilities are

$x > 0, y > 0: |xy| = xy, |x| = x, |y| = y;$ hence $|xy| = |x||y|;$
$x < 0, y < 0: |xy| = xy, |x| = -x, |y| = -y;$ again $|xy| = |x||y|;$
$x < 0, y > 0: |xy| = -xy, |x| = -x, |y| = y; |xy| = |x||y|;$
$x > 0, y < 0: |xy| = -xy, |x| = x, |y| = -y; |xy| = |x||y|.$

In each case the statement holds.
The proof of (b) is entirely similar.

(c) We know that if $a \geq 0$, the value of $\sqrt{a}$ is positive or zero, and that $\sqrt{x^2} = \pm x$. We must then choose the sign so that $\sqrt{x^2}$ is positive or zero. Hence we choose the plus sign if $x \geq 0$, and the minus sign if $x < 0$; the value of $\sqrt{x^2}$ is then the same as that of $|x|$ in the above definition.

**THEOREM 1.3.2.** *If $a > 0$, then*

(a) $|x| < a \Leftrightarrow -a < x < a \Leftrightarrow x^2 < a^2$

(b) $|x| > a \Leftrightarrow (x < -a$ or $x > a) \Leftrightarrow x^2 > a^2.$

**8**   FUNDAMENTAL CONCEPTS                                           **1.3**

*Proof.*   We shall prove (a) by showing that
$$|x| < a \Rightarrow -a < x < a \Rightarrow x^2 < a^2 \Rightarrow |x| < a.$$
Suppose $|x| < a$. If $0 \leq x$, then $|x| = x$ and $0 \leq x < a$. If $x < 0$, then $|x| = -x$, so that $0 < -x < a$ and $0 > x > -a$. In either case, $-a < x < a$.

Now suppose $-a < x < a$. If $x \geq 0$, then $x < a \Rightarrow x^2 \leq ax < a^2$. If $x < 0$, then $-x < a \Rightarrow (-x)^2 < a(-x) < a^2$. In either case, $x^2 < a^2$.

Now suppose that $x^2 < a^2$; we must show $|x| < a$. Suppose $|x| \geq a$. Then $|x|^2 \geq a|x| \geq a^2$ or $|x|^2 \geq a^2$. But $|x|^2 = x^2$. Hence $x^2 \geq a^2$, and we have a contradiction. Therefore, $|x| < a$, and the proof of (a) is complete.

The proof of (b) is similar and is called for in the exercises.

*Example 1.3.1.*   (a) $|x - 1| < 2 \Leftrightarrow -2 < x - 1 < 2 \Leftrightarrow -1 < x < 3$

(b) $|2x - 3| < 1 \Leftrightarrow -1 < 2x - 3 < 1 \Leftrightarrow 2 < 2x < 4 \Leftrightarrow 1 < x < 2$

(c) $|3x - 1| \geq 4 \Leftrightarrow (3x - 1 \leq -4$ or $3x - 1 \geq 4) \Leftrightarrow (x \leq -1$ or $x \geq 5/3)$.

*Example 1.3.2.*   Solve the inequality $2x^2 - 4x - 3 < 0$.

*Solution.*   The left member does not factor; a method of *completing the square* may be used, whereby $b^2/4$ is added to $x^2 + bx$, to give $(x + b/2)^2$. Adding 3 to both sides and then dividing by 2, we obtain
$$x^2 - 2x < \tfrac{3}{2}.$$
Now adding 1 to both members of the latter inequality, we have
$$x^2 - 2x + 1 < \tfrac{5}{2} \quad \text{or} \quad (x - 1)^2 < \tfrac{5}{2}.$$
By Theorem 1.3.2(a), with $x - 1$ in place of $x$, we have $|x - 1| < \tfrac{1}{2}\sqrt{10}$; then
$$-\tfrac{1}{2}\sqrt{10} < x - 1 < \tfrac{1}{2}\sqrt{10} \quad \text{or} \quad 1 - \tfrac{1}{2}\sqrt{10} < x < 1 + \tfrac{1}{2}\sqrt{10}.$$

An additional theorem is of paramount importance in the use of inequalities in the study of calculus. It is known as the *triangle inequality*.

**THEOREM 1.3.3.**         $|a + b| \leq |a| + |b|.$

*Proof.*   Since either $a = |a|$ or $a = -|a|$, we may always write
$$-|a| \leq a \leq |a|;$$
similarly,
$$-|b| \leq b \leq |b|.$$
Combining these, we have
$$-(|a| + |b|) \leq a + b \leq |a| + |b|.$$
Therefore, by Theorem 1.3.2(a), adapted to include equality, we get the desired inequality in the statement of the present theorem.

Some consequences of Theorem 1.3.3 appear in the exercises.

# EXERCISE GROUP 1.1

In each of Exercises 1–22 determine the values of $x$ which satisfy the given inequality.

1. $3x - 2 < 5$
2. $3x - 5 < 10$
3. $-3x - 5 < 2$
4. $-7x - 5 > 4$
5. $2(x + 3) > -1$
6. $x - 2 < 3x$
7. $2(x + 3) > 3(x + 4)$
8. $3(2 - x) > 4(x - 2)$
9. $x(x + 1) < x^2 + 3x + 2$
10. $2(2x - 1)(x - 3) < 4x(x - 2)$
11. $x^2 + x - 6 < 0$
12. $x^2 - 5x + 6 > 0$
13. $3x^2 - 2x - 8 > 0$
14. $6x^2 - 7x - 20 < 0$
15. $2x^2 < x + 10$
16. $3x + 2 < 5x^2$
17. $(2x - 1)(x + 2)(3x - 7) < 0$
18. $(x + 2)(x - 1)(x + 3) > 0$
19. $(x + 1)^2(x - 2) > 0$
20. $(3x + 2)^2(x - 1) < 0$
21. $\dfrac{(2x + 5)(3x + 2)}{7x - 3} < 0$
22. $\dfrac{2x - 3}{(x + 4)(x - 2)} > 0$

In Exercises 23–28 find the values of $x$ for which the expression is real. (*Note:* If $a$ is any real number, then $a^2 \geq 0$.)

23. $\sqrt{2x - 1}$
24. $\sqrt{3 - 4x}$
25. $\sqrt{x^2 - 1}$
26. $\sqrt{x^2 + x - 2}$
27. $\sqrt{\dfrac{x - 1}{x}}$
28. $\sqrt{\dfrac{x - 2}{1 - 2x}}$

In each of Exercises 29–40 write an equivalent inequality (or inequalities) for $x$ without absolute value symbols.

29. $|x| < 4$
30. $|x| < 5$
31. $|x| > 5$
32. $|x| > a^2$
33. $|x + 1| < 3$
34. $|2x - 3| < \frac{1}{3}$
35. $|4 - 2x| > \frac{1}{2}$
36. $|3 - x| > b$
37. $2 < |x - 3| < 4$
38. $3 < |2x - 3| < 5$
39. $0 < |x + 2| < 2$
40. $0 < |2x - 1| < 4$

In each of Exercises 41–46 determine the values of $x$ which satisfy the given inequality.

41. $2x^2 - 3x - 7 > 0$
42. $2x^2 - 3x - 7 < 0$
43. $2x^2 < 5(x + 1)$
44. $3(x + 1) < 5x^2$
45. $2x^2 + 5 > 7x + 1$
46. $x^2 + 2 < 2\sqrt{3}x$

47. Prove part (b) of Theorem 1.3.2.
48. Prove: $|a - b| \leq |a| + |b|$.
49. Prove: $|a + b| \geq |a| - |b|$.
50. Prove: $|a - b| \geq |a| - |b|$.
51. Prove: $|a + b| \geq ||a| - |b||$.

## 1.4 COORDINATE LINES. INTERVALS

Making use of the familiar concept of length on a straight line, together with direction on the line, we may represent the real numbers geometrically. On a straight line which extends indefinitely in both directions, a point $O$ is chosen as *origin*. One of the two possible directions on the line is chosen as the *positive* direction; the opposite direction then is *negative*. A unit of length is chosen on the line. A positive number $a$ is then represented by a point on the line in the positive direction from $O$ at a distance of $a$ units from $O$; the negative number $(-a)$ is represented by a point on the opposite side of $O$, the same distance from $O$. The line thus described is called a *coordinate line*. *To each point on the line there corresponds one real number, and to each real number there corresponds one point on the line.*

If $A$ is the point on a coordinate line representing the real number $a$, then $a$ is called the *linear coordinate* of $A$. We shall often refer simply to the point $a$, to mean more explicitly the point whose linear coordinate is $a$. For example, "the point $-2$" is to be interpreted as the point whose linear coordinate is $-2$.

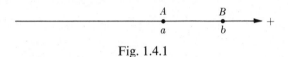

Fig. 1.4.1

Consider a coordinate line, as in Fig. 1.4.1, in which the positive direction is to the right. If $A$ and $B$ are points with coordinates $a$ and $b$ respectively, then $a < b$ if and only if $A$ is to the left of $B$.

The inequality $a < x < b$ is satisfied by all real numbers which are simultaneously greater than $a$ and less than $b$, that is, by those numbers which are between $a$ and $b$. Such a set of numbers is called an *open interval of real numbers*, and is designated as $(a, b)$. The corresponding points on the line form an *open interval of points*. The set of numbers satisfying the inequality $a \leq x \leq b$ forms a *closed interval*, and is designated as $[a, b]$. If the set of numbers includes only one of the end numbers, the interval is called either *half-open* or *half-closed*; such an interval may be denoted by $[a, b)$ or $(a, b]$, depending on whether $a$ or $b$ is included.

We may use this notation for intervals to describe the solution of inequalities. For example, $-2 < x < 3$ becomes $(-2, 3)$. In order to describe an inequality such as $x > 3$ in this manner, we use $(3, \infty)$; while $x < 3$ becomes $(-\infty, 3)$. The symbol $\infty$ for "infinity" is used solely for the purpose of describing the interval, and is assigned no meaning by itself for the present. The solution of the inequality in Example 1.2.3,

$$(2x - 1)(x + 3)(4x - 7) > 0,$$

becomes $(-3, 1/2)$ and $(7/4, \infty)$ in our notation for intervals; the geometric representation of these intervals appears in Fig. 1.4.2, where the use of parentheses indicates exclusion of an endpoint. To indicate inclusion of an endpoint, the scheme of Fig. 1.4.3, which represents the interval $-3 \leq x \leq 1/2$, or $[-3, 1/2]$, is clear, with brackets indicating inclusion.

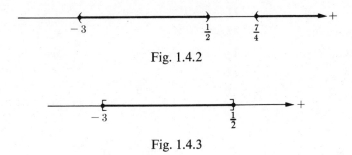

Fig. 1.4.2

Fig. 1.4.3

This interval notation is summarized as follows:

| Interval | Notation |
|---|---|
| $a \leq x \leq b$ | $[a, b]$ |
| $a \leq x < b$ | $[a, b)$ |
| $a < x \leq b$ | $(a, b]$ |
| $a < x < b$ | $(a, b)$ |
| $a < x$ | $(a, \infty)$ |
| $a \leq x$ | $[a, \infty)$ |
| $x < a$ | $(-\infty, a)$ |
| $x \leq a$ | $(-\infty, a]$ |

**Example 1.4.1.** Express in interval notation the solution of the inequality

$$\frac{2x+3}{x-2} > 0.$$

*Solution.* By properties of real numbers, the numerator and denominator of a fraction must be both negative or both positive for a fraction to be positive. Hence we must have

either $\quad 2x + 3 < 0 \quad$ and $\quad x - 2 < 0,$

or $\quad 2x + 3 > 0 \quad$ and $\quad x - 2 > 0.$

The first alternative gives $x < -3/2$, while the second gives $x > 2$. The solution set consists of the intervals $(-\infty, -3/2)$ and $(2, \infty)$.

Note the usage of the words "or" and "and" in Example 1.4.1. The word "or" indicates alternatives, while "and" implies simultaneity. For example, one of the alternatives was both $2x + 3 < 0$ *and* $x - 2 < 0$; these inequalities *together* implied $x < -3/2$.

## EXERCISE GROUP 1.2

Express the set of numbers satisfying the inequality in each of the following exercises in interval notation. Sketch each solution set on a coordinate line.

1. $-4 < x < -1$
2. $-3 \leq x < 0$
3. $1 < x \leq 4$
4. $x > 2$
5. $x < 2$
6. $|x| > 1$
7. $|x| < 1.4$
8. $|x - 1| \leq 2$
9. $|x + 2| < 2$
10. $|x - 3| \leq a, a > 0$
11. $|x + 4| > b, b > 0$
12. $|2x + 3| \leq 2$
13. $|3x - 1| > 1$
14. $1 < |x| < 2$
15. $2 \leq |x - 1| \leq 4$
16. $1 < |5x - 4| < 2$
17. $0 < |3x + 5| \leq 1$
18. $2x^2 + 3x - 9 < 0$
19. $3x^2 + 5x - 1 < 0$
20. $x^2 - x - 1 \geq 0$
21. $\dfrac{x - 1}{x - 2} < 0$
22. $\dfrac{2x + 1}{x - 3} > 0$
23. $\dfrac{x^2}{2x - 1} > 0$
24. $\dfrac{(2x + 4)(x - 1)}{x + 1} < 0$
25. $\dfrac{2x - 5}{(2x - 3)(3x - 2)} > 0$

## 1.5 RECTANGULAR COORDINATE SYSTEM

It is assumed that the student is familiar with the *rectangular Cartesian coordinate system*. He may then draw any figure, as needed, for the following brief account. Two coordinate lines, called *axes*, are drawn in a plane so that they are perpendicular to each other, with the origin on each at the point of intersection. The usual position is such that one axis can be designated as horizontal, the other as vertical. The point of intersection is then called the *origin* of the coordinate system; the horizontal axis is commonly called the $x$ axis; and the vertical axis is the $y$ axis. It is usual practice to have the positive direction to the right on the $x$ axis, and upward on the $y$ axis.

If $P$ is any point in the plane, let $M$ be the foot of the perpendicular from $P$ to the $x$ axis, and $N$ the foot of the perpendicular from $P$ to the $y$ axis. If $x$ and $y$ are the linear coordinates of the points $M$ and $N$, respectively, the point $P$ is designated by the number pair $(x, y)$. The number $x$ is called the *abscissa*, or $x$ *coordinate*, of $P$, and $y$ is the *ordinate*, or $y$ *coordinate*, of $P$; while $x$ and $y$ together are simply the *rectangular coordinates* of $P$.

The correspondence between points in the plane and pairs of real numbers is such that for each point there is exactly one pair of real numbers, and for each pair of real numbers there is exactly one point.

The coordinate axes separate the plane into four *quadrants*, with each one characterized by the signs of the coordinates of a point in it. The quadrant in which both $x > 0$ and $y > 0$ is designated as $Q_1$; then $Q_2$, $Q_3$, and $Q_4$ are the quadrants met successively in a counterclockwise rotation about the origin. *Points on the coordinate axes are not in any quadrant.*

## 1.6 DIRECTED LINE SEGMENTS. PROJECTIONS

If $A$ and $B$ are two points on a coordinate line with positive direction to the right, we define the *directed line segment* from $A$ to $B$ as the line segment $AB$, with direction from $A$ to $B$, and denote it by $\overline{AB}$. We also use $\overline{AB}$ for the *directed length* of $AB$, which means the length of $AB$ if $B$ is to the right of $A$, and the negative of the length of $AB$ if $B$ is to the left of $A$. In Fig. 1.6.1, $\overline{AB}$ is positive, and $\overline{BA}$ is negative. We have, for any positions of $A$ and $B$,

$$\overline{AB} = -\overline{BA}.$$

Fig. 1.6.1

For any three points $A$, $B$, and $C$ on a coordinate line, we have in general

(1.6.1) $$\overline{AB} + \overline{BC} = \overline{AC}.$$

To establish this, we may consider all possible cases for the relative positions of $A$, $B$, and $C$. For example, one possible case is represented in Fig. 1.6.2. We have in this case

$$\overline{AB} = \overline{AC} + \overline{CB},$$

or

$$\overline{AB} - \overline{CB} = \overline{AC}.$$

But $\overline{CB} = -\overline{BC}$, and we obtain

$$\overline{AB} + \overline{BC} = \overline{AC}.$$

Fig. 1.6.2

Other cases are treated similarly. *Relation (1.6.1) also holds if any of the points coincide.*

Let $A$ and $B$ be points on a coordinate line with coordinates $a$ and $b$ respectively. Then, regardless of the positions of $A$ and $B$, we have

$$\overline{AB} = \overline{AO} + \overline{OB} = \overline{OB} - \overline{OA}.$$

But $\overline{OA} = a$ and $\overline{OB} = b$; hence

(1.6.2) $$\overline{AB} = b - a.$$

Now, let $P_1(x_1, y_1)$ and $P_2(x_2, y_2)$ be two points in a coordinate plane.

The *projection* of $P_1$ on the $x$ axis is the point $A$ which is the foot of the perpendicular from $P_1$ to the $x$ axis, and has the linear coordinate $x_1$ (Fig. 1.6.3). Similarly, the *projection* of $P_2$ on the $x$ axis is the point $B$ with coordinate $x_2$.

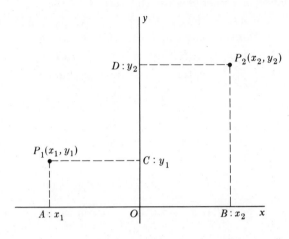

Fig. 1.6.3

The projection of the directed line segment $\overline{P_1P_2}$ on the $x$ axis is then defined as the directed line segment $\overline{AB}$, and we have

(1.6.3) $$\text{proj}_x \overline{P_1P_2} = \overline{AB} = x_2 - x_1.$$

We have, similarly,

(1.6.4) $$\text{proj}_y \overline{P_1P_2} = \overline{CD} = y_2 - y_1.$$

**Example 1.6.1.** Find $\text{proj}_x \overline{AB}$ and $\text{proj}_y \overline{AB}$ for the points $A(2, -4)$ and $B(-3, -7)$.

*Solution.* Since $A$ is the initial point of the directed line segment, we may take $x_1 = 2$, $y_1 = -4$, $x_2 = -3$, $y_2 = -7$. By the above relations, we have

$$\text{proj}_x \overline{AB} = x_2 - x_1 = -3 - 2 = -5,$$
$$\text{proj}_y \overline{AB} = y_2 - y_1 = -7 - (-4) = -3.$$

You may draw a figure to obtain a visual impression.

It is easy to see that the projection of a directed line segment on any straight line parallel to the $x$ axis is equal to the projection on the $x$ axis, if the positive direction on this line is the same as on the $x$ axis. A similar statement applies to the $y$ axis.

*In general, the projection of a point on any straight line is the foot of the perpendicular drawn from the point to the line.*

***Example 1.6.2.*** Given a point $P_1(-2, 3)$, find the coordinates of a point $P_2$ if $\text{proj}_x \overline{P_1P_2} = 3$, $\text{proj}_y \overline{P_1P_2} = -7$.

*Solution.* Formulas (1.6.3) and (1.6.4) may be used directly, but the following procedure may be helpful. Since $\text{proj}_x \overline{P_1P_2} = 3$, the point $P_2$ is 3 units *to the right* of $P_1$, and hence $x_2 = -2 + 3 = 1$. Similarly, $P_2$ is 7 units *below* $P_1$, and $y_2 = 3 - 7 = -4$. The point $P_2$ is $(1, -4)$.

**EXERCISE GROUP 1.3**

1. Describe the set of points whose abscissas are $-3$; whose ordinates are 2.
2. Describe the set of points whose abscissas are $a$; whose ordinates are $b$.
3. Describe the set of points which are equidistant from $(2, -3)$ and $(2, 6)$.
4. Describe the set of points which are equidistant from $(-1, 4)$ and $(5, 4)$.
5. Three vertices of a parallelogram are $(-3, 2)$, $(1, 0)$, and $(0, 3)$. Locate the fourth vertex. Is there more than one solution?
6. Three vertices of a rectangle are $(-8, -3)$, $(5, -3)$, and $(-8, 2)$. Locate the fourth vertex. Is there more than one solution?
7. The ends of the base of an isosceles triangle are at $(3, 0)$, $(8, 0)$, and the altitude to the base is of length 4. Find the coordinates of any point which might be the third vertex.
8. Two vertices of an equilateral triangle are $(1, 3)$ and $(7, 3)$. Find the coordinates of any point which might be the third vertex.
9. The legs of a right triangle are on the coordinate axes, and the hypotenuse is of length $a$. If $(x, y)$ are the coordinates of the midpoint of the hypotenuse, show that $4x^2 + 4y^2 = a^2$.

In Exercises 10–15 find the projections on the $x$ axis and on the $y$ axis of the directed line segment $\overline{P_1P_2}$ where $P_1$ is the first given point, and $P_2$ the second.

10. $(2, 3), (5, 7)$
11. $(2, -3), (4, -6)$
12. $(7, -2), (-2, -7)$
13. $(-4, 8), (6, -6)$
14. $(0, 0), (2, -3)$
15. $(-4, 3), (0, 0)$

In each of the following verify that $\text{proj } \overline{AB} + \text{proj } \overline{BC} = \text{proj } \overline{AC}$, where the projections are either on the $x$ axis or on the $y$ axis. Draw a figure for each.

16. $A(-9, 0), B(2, 4), C(-2, -4)$
17. $A(2, -6), B(0, 0), C(8, 4)$
18. $A(2, -1), B(4, 3), C(-3, -1)$
19. $A(-2, -5), B(3, -7), C(1, 3)$
20. $A(0, 3), B(0, -7), C(-3, 5)$
21. $A(-3, 0), B(4, 0), C(0, 8)$

## 1.7 VECTORS

In this section we define and develop vectors algebraically; a geometric interpretation is given in Section 1.9. Vectors are mathematical entities, defined in terms of real numbers, which have many applications, some of which we consider in the work to follow. Vectors are also an illustration of a mathematical system, with a structure based on a suitable set of definitions.

**DEFINITION 1.7.1.** *A* **vector** *a is an ordered pair of real numbers*

$$\mathbf{a} = (a_1, a_2),$$

*subject to the following:*

1. *If* $\mathbf{a} = (a_1, a_2)$ *and* $\mathbf{b} = (b_1, b_2)$, *then*

$$\mathbf{a} = \mathbf{b} \Leftrightarrow a_1 = b_1 \quad \text{and} \quad a_2 = b_2.$$

2. *The* **sum** *of the vectors* **a** *and* **b** *is*

$$\mathbf{a} + \mathbf{b} = (a_1 + b_1, a_2 + b_2).$$

3. *If c is any real number, then*

$$c\mathbf{a} = (ca_1, ca_2).$$

The word "ordered" in the definition signifies that, for example, (2, 3) and (3, 2) are distinct pairs. We may now define the following additional terms.

**DEFINITION 1.7.2.** 1. *The* **negative** *of the vector* $\mathbf{a} = (a_1, a_2)$ *is*

$$-\mathbf{a} = (-1)\mathbf{a} = (-a_1, -a_2).$$

2. *The* **difference** *of* $\mathbf{a} = (a_1, a_2)$ *and* $\mathbf{b} = (b_1, b_2)$ *is*

$$\mathbf{a} - \mathbf{b} = \mathbf{a} + (-\mathbf{b}) = (a_1 - b_1, a_2 - b_2).$$

3. *The* **magnitude** *of* **a** *is*

$$|\mathbf{a}| = \sqrt{a_1^2 + a_2^2}.$$

4. **a** *is a* **unit vector** *if* $|\mathbf{a}| = 1$.

5. *The* **zero vector** *is* $\mathbf{0} = (0, 0)$.

Some immediate consequences follow from this definition and the known properties of real numbers. We shall state these, and call for the proofs in the exercises.

**THEOREM 1.7.1.** *Addition of vectors is commutative and associative.*

$$\mathbf{a} + \mathbf{b} = \mathbf{b} + \mathbf{a}; \quad (\mathbf{a} + \mathbf{b}) + \mathbf{c} = \mathbf{a} + (\mathbf{b} + \mathbf{c}).$$

**THEOREM 1.7.2.** *Multiplication of a vector by a real number is distributive with respect to addition.*

$$c(\mathbf{a} + \mathbf{b}) = c\mathbf{a} + c\mathbf{b}.$$

**THEOREM 1.7.3.** $\mathbf{a} = \mathbf{0} \Leftrightarrow |\mathbf{a}| = 0.$

**Example 1.7.1.** Let $\mathbf{a} = (2, 3)$, $\mathbf{b} = (1, -2)$. Then

$$2\mathbf{a} = 2(2, 3) = (4, 6);$$
$$\mathbf{a} + \mathbf{b} = (2 + 1, 3 + [-2]) = (3, +1);$$
$$\mathbf{a} - \mathbf{b} = (2 - 1, 3 - [-2]) = (1, 5);$$
$$2\mathbf{a} - 3\mathbf{b} = (2 \cdot 2 - 3 \cdot 1, 2 \cdot 3 - 3[-2]) = (1, 12);$$
$$|\mathbf{b}| = \sqrt{1^2 + (-2)^2} = \sqrt{5}.$$

**DEFINITION 1.7.3.** *To* **normalize** *the vector* $\mathbf{a}$, *means to find a vector* $\mathbf{u} = c\mathbf{a}$ *for c a real number, such that* $|\mathbf{u}| = 1$.

**Example 1.7.2.** To normalize the vector $\mathbf{a} = (2, -3)$, we must have $|c\mathbf{a}| = 1$. Hence $c|\mathbf{a}| = 1$, if $c > 0$ (see Exercise 21), and $c = 1/|\mathbf{a}| = 1/\sqrt{13}$. The normalized vector is then

$$\mathbf{u} = \frac{1}{\sqrt{13}} \mathbf{a} = \frac{1}{\sqrt{13}} (2, -3) = \left( \frac{2}{\sqrt{13}}, -\frac{3}{\sqrt{13}} \right).$$

Note that $\mathbf{v} = -\mathbf{u} = (-2/\sqrt{13}, 3/\sqrt{13})$ is also a normalized vector.

As is customary, we have used the same notation for magnitude of a vector as for absolute value of a real number. This causes no difficulty, since it is always clear which is meant.

**EXERCISE GROUP 1.4**

The questions in Exercises 1–14 apply to the vectors $\mathbf{a} = (2, -5)$, $\mathbf{b} = (-3, 2)$, $\mathbf{c} = (5, 3)$.

1. Find (a) $\mathbf{a} + \mathbf{b}$, (b) $\mathbf{a} - \mathbf{b}$, (c) $\mathbf{a} + 2\mathbf{b}$
2. Find (a) $\mathbf{a} + \mathbf{c}$, (b) $\mathbf{a} - \mathbf{c}$, (c) $2\mathbf{a} + \mathbf{c}$
3. Find (a) $|\mathbf{a}|$, (b) $|\mathbf{b}|$, (c) $|\mathbf{a} + \mathbf{b}|$
4. Find (a) $|\mathbf{c}|$, (b) $|\mathbf{a} + 2\mathbf{c}|$, (c) $|\mathbf{a} - \mathbf{c}|$
5. Find (a) $\mathbf{a} - \mathbf{b} + 2\mathbf{c}$, (b) $|\mathbf{a} - \mathbf{b} + 2\mathbf{c}|$
6. Find (a) $2\mathbf{a} + 3\mathbf{b} - \mathbf{c}$, (b) $|2\mathbf{a} + 3\mathbf{b} - \mathbf{c}|$
7. Normalize (a) $\mathbf{a}$, (b) $\mathbf{b}$, (c) $\mathbf{a} + \mathbf{b}$
8. Normalize (a) $\mathbf{a} + \mathbf{c}$, (b) $\mathbf{a} - \mathbf{c}$, (c) $\mathbf{a} - \mathbf{b} + 2\mathbf{c}$
9. Find $\mathbf{u}$ if $\mathbf{u} + \mathbf{a} = 0$
10. Find $\mathbf{u}$ if $\mathbf{u} + \mathbf{a} = \mathbf{b} - \mathbf{c}$
11. Find $\mathbf{u}$ if $2\mathbf{u} = \mathbf{a} + \mathbf{c}$
12. Find $\mathbf{u}$ if $3\mathbf{u} = 2\mathbf{a} - \mathbf{b}$
13. Find $\mathbf{u}$ if $\mathbf{a} + 2\mathbf{u} = 3\mathbf{b}$
14. Find $\mathbf{u}$ if $\mathbf{a} + 2\mathbf{b} + 3\mathbf{c} + 4\mathbf{u} = 0$

15. Find (a) $|(a, a+1)|$, (b) $|(a-1, a+1)|$
16. Find (a) $|(a+1, \sqrt{a})|$, (b) $|(\sqrt{a}+1, \sqrt{a}-1)|$
17. Prove the commutative property of vector addition.
18. Prove the associative property of vector addition.
19. Prove the distributive property of multiplication of a vector by a real number.
20. Show that $\mathbf{a} + (-\mathbf{a}) = \mathbf{0}$.
21. Show that $|c\mathbf{a}| = c|\mathbf{a}|$ if $c > 0$. What is $|c\mathbf{a}|$ if $c < 0$?
22. Prove that $\mathbf{a} = \mathbf{0} \Leftrightarrow |\mathbf{a}| = 0$.
23. Show that $\mathbf{a}/|\mathbf{a}|$ gives a normalization of $\mathbf{a}$, if $\mathbf{a} \neq \mathbf{0}$.

## 1.8 BASE VECTORS

We define two *base vectors* $\mathbf{i}$ and $\mathbf{j}$ as follows:

$$\mathbf{i} = (1, 0), \qquad \mathbf{j} = (0, 1).$$

It is easily seen that $\mathbf{i}$ and $\mathbf{j}$ are unit vectors, that is, $|\mathbf{i}| = |\mathbf{j}| = 1$. Furthermore, these vectors serve as a base for all vectors in the sense of the following theorem.

**THEOREM 1.8.1.** *Any vector* $\mathbf{a} = (a_1, a_2)$ *may be expressed uniquely in the form* $\mathbf{a} = a_1\mathbf{i} + a_2\mathbf{j}$.

*Proof.* The reasons for the following steps are easily supplied

$$a_1\mathbf{i} + a_2\mathbf{j} = a_1(1, 0) + a_2(0, 1) = (a_1, 0) + (0, a_2) = (a_1, a_2).$$

These steps prove the validity of the representation. Uniqueness follows from Theorem 1.8.2 or Exercise 17 (or it may be shown directly).

We also say that $\mathbf{a}$ is expressed as a *linear combination* of $\mathbf{i}$ and $\mathbf{j}$, that is, a real multiple $a_1$ of $\mathbf{i}$ plus a real multiple $a_2$ of $\mathbf{j}$. The numbers $a_1$ and $a_2$ are called the *components* of the vector $\mathbf{a}$.

In terms of the representation of a vector as in Theorem 1.8.1, called the *base vector representation*, the operations with vectors follow in large part the ordinary rules of algebra. For example,

1. $(2\mathbf{i} - 3\mathbf{j}) + (-\mathbf{i} + 5\mathbf{j}) = (2-1)\mathbf{i} + (-3+5)\mathbf{j} = \mathbf{i} + 2\mathbf{j};$
2. $3(3\mathbf{i} - 4\mathbf{j}) = 9\mathbf{i} - 12\mathbf{j}.$

(See also Section 2.3.)

Theorem 1.8.1 is a special case of the following more general result.

**THEOREM 1.8.2.** *If* $\mathbf{a} \neq \mathbf{0}$, $\mathbf{b} \neq \mathbf{0}$, *and* $\mathbf{a} \neq k\mathbf{b}$ *for any number* $k$, *then any vector* $\mathbf{d}$ *may be expressed uniquely in the form*

(1.8.1) $\qquad \mathbf{d} = u\mathbf{a} + v\mathbf{b}$ *for some numbers* $u$ *and* $v$.

## 1.8 BASE VECTORS

*In words*, **d** *is a linear combination of* **a** *and* **b**.

*Proof.* Let us write
$$\mathbf{a} = (a_1, a_2), \quad \mathbf{b} = (b_1, b_2), \quad \mathbf{d} = (d_1, d_2).$$

We must then have
$$(d_1, d_2) = u(a_1, a_2) + v(b_1, b_2) = (ua_1 + vb_1, ua_2 + vb_2).$$

It follows from Definition 1.7.1 that

(1.8.2) $$d_1 = ua_1 + vb_1, \quad d_2 = ua_2 + vb_2$$

These equations are linear in $u$ and $v$. If $a_1 b_2 - a_2 b_1 \neq 0$, these equations have a unique solution for $u$ and $v$, and (1.8.1) is established.

We now show that $a_1 b_2 - a_2 b_1 = 0$ cannot hold. Suppose that it does hold. Since $\mathbf{b} \neq \mathbf{0}$, at least one of $b_1$ and $b_2$ is not zero. Suppose $b_1 \neq 0$. Then $a_1 = kb_1$ for some number $k$. Then, since $a_1 b_2 - a_2 b_1 = 0$,

$$kb_1 b_2 = a_2 b_1, \quad \text{and} \quad a_2 = kb_2.$$

Hence $(a_1, a_2) = (kb_1, kb_2) = k(b_1, b_2)$, which contradicts an original hypothesis. Thus $a_1 b_2 - a_2 b_1 \neq 0$, and the proof is complete.

***Example 1.8.1.*** Express the vector $(2, 3)$ as a linear combination of the vectors $(-1, 2)$ and $(4, 3)$.

*Solution.* By (1.8.2), we must have
$$2 = -u + 4v, \quad 3 = 2u + 3v.$$

This pair of equations has the solution $u = 6/11$, $v = 7/11$. Hence
$$(2, 3) = \tfrac{6}{11}(-1, 2) + \tfrac{7}{11}(4, 3).$$

**EXERCISE GROUP 1.5**

Given $\mathbf{a} = 2\mathbf{i} - 5\mathbf{j}$ and $\mathbf{b} = 4\mathbf{i} + 7\mathbf{j}$, express the vector in each of Exercises 1–8 in the base vector representation.

1. $-3\mathbf{a}$
2. $\mathbf{a} - 2\mathbf{b}$
3. $-2\mathbf{a} + 4\mathbf{b}$
4. $9(\mathbf{a} + \mathbf{b})$
5. $3(3\mathbf{a} + \mathbf{b})$
6. $-2(4\mathbf{a} - 3\mathbf{b})$
7. $\mathbf{a}/|\mathbf{a}|$
8. $\mathbf{b}/|\mathbf{b}|$

In each of Exercises 9–16 express the given vector in the form $u\mathbf{a} + v\mathbf{b}$, if $\mathbf{a} = (2, 3)$ and $\mathbf{b} = (-3, 2)$.

9. $(5, 1)$
10. $(1, 8)$
11. $(8, -1)$
12. $(-5, 12)$
13. $(1, 1)$
14. $(2, 0)$
15. $(0, 3)$
16. $(-2, 1)$

17. Prove directly that under the hypotheses of Theorem 1.8.2 a vector **d** can be expressed as $\mathbf{d} = u\mathbf{a} + v\mathbf{b}$ in only *one* way; that is, if $u_1\mathbf{a} + v_1\mathbf{b} = u_2\mathbf{a} + v_2\mathbf{b}$, then $u_1 = u_2$ and $v_1 = v_2$. This proves the uniqueness stated in that theorem.

## 1.9 GEOMETRIC INTERPRETATION OF VECTORS

Any two parallel lines are said to have the same *direction*. Thus, when we refer to the direction of a line in a plane we may think of any line of a set of parallel lines. In addition, we may speak of two opposite *senses* on a line, depending on the way in which the line is traced.

A directed line segment, as defined in Section 1.6, then has a direction, a sense, and initial and terminal points. Let us consider the set of directed line segments which are obtainable from a given directed line segment by moving it parallel to itself. This set is completely determined by specifying the directed lengths of its horizontal and vertical projections (Section 1.6). Given a vector $\mathbf{a} = (a_1, a_2)$, any directed line segment whose $x$ and $y$ projections are $a_1$ and $a_2$, respectively, is called a *geometric representation* of the vector **a**; more loosely, we shall also refer to it as the vector **a**. In this sense, the geometric representation of a vector has a length, a direction, and a sense, but its position in the plane is unspecified. As a result of this discussion, any directed line segments with the same length, direction, and sense may be labeled as the same vector. A vector is often drawn with an arrow tip to indicate the sense of the vector, as in Fig. 1.9.1 and Fig. 1.9.2.

We now interpret geometrically some of the definitions of Section 1.7. The vector $c\mathbf{a}$ has length $|c|$ times that of **a**, is parallel to **a**, and has the same sense as **a** if $c > 0$ and the opposite sense if $c < 0$.

The *sum* $\mathbf{a} + \mathbf{b}$ is a vector formed from **a** and **b** according to either the *triangle law* or the *parallelogram law*. According to the triangle law, $\mathbf{a} + \mathbf{b}$ is obtained by starting a vector **b** at the terminal point of **a**, and then joining the initial point of **a** to the terminal point of **b**. In Fig. 1.9.1 this process gives the vector $\overrightarrow{OP}$. It is clear from the figure that

$$\text{proj}_x \overrightarrow{OP} = a_1 + b_1, \qquad \text{proj}_y \overrightarrow{OP} = a_2 + b_2.$$

Hence $\overrightarrow{OP}$ is the vector $\overrightarrow{OP} = (a_1 + b_1, a_2 + b_2)$, and the vector yielded by the triangle law is the sum $\mathbf{a} + \mathbf{b}$ as given by Definition 1.7.1.

According to the parallelogram law, **a** and **b** are drawn with the same initial point $O$ (Figure 1.9.2), a parallelogram is constructed with sides **a** and **b**, and the diagonal $\overrightarrow{OP}$ of the parallelogram which starts at $O$ gives the sum $\mathbf{a} + \mathbf{b}$. It is clear from the figure that $\overrightarrow{RP} = \mathbf{b}$. Hence $\overrightarrow{OP}$ is also the sum of **a** and **b** by the triangle law. Thus the same vector is obtained by the parallelogram law and the triangle law. If two vectors are given geometrically, their sum is independent of any coordinate system.

The other diagonal of the parallelogram of Fig. 1.9.2, taken as $\overrightarrow{QR}$, represents the difference $\mathbf{a} - \mathbf{b}$. For, by the triangle law applied to $\triangle OQR$, we have

$$\overrightarrow{OQ} + \overrightarrow{QR} = \overrightarrow{OR}, \quad \text{or} \quad \mathbf{b} + \overrightarrow{QR} = \mathbf{a},$$

and therefore $\overrightarrow{QR} = \mathbf{a} - \mathbf{b}$.

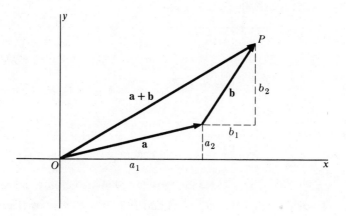

Fig. 1.9.1

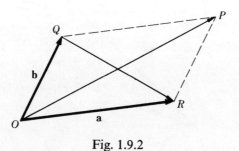

Fig. 1.9.2

According to Fig. 1.9.3, the vector $\mathbf{a}$ may be written as

$$\mathbf{a} = \overrightarrow{PR} + \overrightarrow{RQ} = a_1\mathbf{i} + a_2\mathbf{j},$$

and the figure may be taken as an interpretation of the base vector representation of $\mathbf{a}$ in terms of $\mathbf{i}$ and $\mathbf{j}$. Applying the Theorem of Pythagoras to $\triangle PQR$, we may also write

$$|\mathbf{a}|^2 = a_1^2 + a_2^2 = (\text{proj}_x \mathbf{a})^2 + (\text{proj}_y \mathbf{a})^2.$$

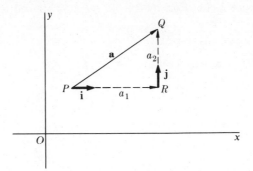

Fig. 1.9.3

If $P$ and $Q$ have coordinates $(x_1, y_1)$ and $(x_2, y_2)$ respectively, then $a_1 = x_2 - x_1$, $a_2 = y_2 - y_1$, and the vector $\overrightarrow{PQ}$ becomes

(1.9.1) $$\overrightarrow{PQ} = (x_2 - x_1, y_2 - y_1).$$

For example, given the points $A(2, -3)$ and $B(3, 7)$, we have, by (1.9.1),

$$\overrightarrow{BA} = (2 - 3, -3 - 7) = (-1, -10).$$

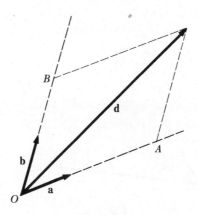

Fig. 1.9.4

Now let **a** and **b**, shown in Fig. 1.9.4, be two vectors as in Theorem 1.8.2. By the parallelogram law,

$$\mathbf{d} = \overrightarrow{OA} + \overrightarrow{OB}.$$

But $\overrightarrow{OA} = u\mathbf{a}$ for some number $u$, and $\overrightarrow{OB} = v\mathbf{b}$ for some number $v$. Then

$$\mathbf{d} = u\mathbf{a} + v\mathbf{b},$$

as in Theorem 1.8.2. In this way the meaning of the statement that **d** can be expressed as a sum of multiples of **a** and **b** is illustrated, and in particular, the multipliers $u$ and $v$ in this case are

$$u = \frac{|\overrightarrow{OA}|}{|\mathbf{a}|}, \quad v = \frac{|\overrightarrow{OB}|}{|\mathbf{b}|}.$$

In other cases, $u$ or $v$ or both may be negative.

**Example 1.9.1.** In any triangle $ABC$ (draw a figure) we have

$$\overrightarrow{AB} + \overrightarrow{BC} = \overrightarrow{AC}, \quad \overrightarrow{AC} + \overrightarrow{CA} = \mathbf{0}.$$

Hence $\overrightarrow{AB} + \overrightarrow{BC} + \overrightarrow{CA} = \mathbf{0}$.

**Example 1.9.2.** Let $M$ be the midpoint of the diagonal $QR$ in the parallelogram of Fig. 1.9.2. Express the vector $\overrightarrow{OM}$ in terms of **a** and **b**.

*Solution.* By the above interpretation of the diagonals of a parallelogram, we have $\overrightarrow{QR} = \mathbf{a} - \mathbf{b}$. If $M$ is the midpoint of $QR$, it follows that

$$\overrightarrow{QM} = \frac{1}{2}\overrightarrow{QR} = \frac{1}{2}(\mathbf{a} - \mathbf{b}),$$

and hence

$$\overrightarrow{OM} = \overrightarrow{OQ} + \overrightarrow{QM} = \mathbf{b} + \frac{1}{2}(\mathbf{a} - \mathbf{b}) = \frac{\mathbf{a} + \mathbf{b}}{2}.$$

**EXERCISE GROUP 1.6**

A figure should be drawn for each of the following.

1. Given the points $A(3, 4)$, $B(2, -7)$, and $C(-4, 3)$, find the vectors $\overrightarrow{AB}$, $\overrightarrow{BC}$, and $\overrightarrow{CA}$, and verify that their sum is zero.

2. Given the points $A(2, 3)$, $B(-4, 2)$, and $C(5, -1)$:
   (a) Find the vector sum $\overrightarrow{OA} + \overrightarrow{OB} + \overrightarrow{OC}$.
   (b) Find the vector sum $\overrightarrow{PA} + \overrightarrow{PB} + \overrightarrow{PC}$ for the point $P(3, 4)$.

3. Given the quadrilateral with vertices $A(1, 0)$, $B(2, -3)$, $C(4, -7)$, and $D(0, -3)$:
   (a) Verify directly that $\overrightarrow{AB} + \overrightarrow{BC} + \overrightarrow{CD} + \overrightarrow{DA} = \mathbf{0}$.
   (b) Can you draw a conclusion for an arbitrary quadrilateral?
   (c) Can you draw a conclusion for an arbitrary polygon?

4. Given the points $A(1, 3)$, $B(5, -7)$, $C(5, 9)$, and $D(-3, 7)$, find the vector sum $\overrightarrow{QA} + \overrightarrow{QB} + \overrightarrow{QC} + \overrightarrow{QD}$ for the point $Q(2, 3)$.

5. Given the points $P(4, 7)$ and $Q(2, -5)$:
   (a) Find the vector $\overrightarrow{PQ}$.
   (b) If $M$ is the midpoint of $PQ$, find the vector $\overrightarrow{PM}$.
   (c) Find the coordinates of the midpoint $M$.

6. In $\triangle ABC$ let $\overrightarrow{AB} = \mathbf{a}$, $\overrightarrow{BC} = \mathbf{b}$. Express each of the following vectors in terms of $\mathbf{a}$ and $\mathbf{b}$:
   (a) the median from $A$, that is, the vector drawn from $A$ to the midpoint of $BC$.
   (b) the median from $B$.
   (c) the median from $C$.

7. In parallelogram $ABCD$ let $\overrightarrow{AB} = \mathbf{a}$, $\overrightarrow{AD} = \mathbf{b}$. Express each of the following vectors in terms of $\mathbf{a}$ and $\mathbf{b}$:
   (a) the vector from $A$ to the midpoint of $BC$.
   (b) the vector from $A$ to the midpoint of $CD$.
   (c) the vector from the midpoint of $BC$ to the midpoint of $CD$.

8. (a) Show geometrically that
$$|\mathbf{a} + \mathbf{b}| \leq |\mathbf{a}| + |\mathbf{b}|.$$
   (b) Write out the equivalent inequality in terms of the components of $\mathbf{a}$ and $\mathbf{b}$. The inequality you obtain is a special case of the so-called *Minkowski inequality* (important in more advanced mathematics).

9. Show geometrically that $|\mathbf{a} + \mathbf{b}| \geq |\mathbf{a}| - |\mathbf{b}|$. Assume that $|\mathbf{a}| > |\mathbf{b}|$.

10. Illustrate geometrically the associative property of vector addition:
$$(\mathbf{a} + \mathbf{b}) + \mathbf{c} = \mathbf{a} + (\mathbf{b} + \mathbf{c}).$$

11. In $\triangle ABC$, let $P$ and $Q$ be the midpoints of $AB$ and $BC$ respectively. Prove that $\overrightarrow{PQ} = \frac{1}{2}\overrightarrow{AC}$. Can you relate this to a theorem of Plane Geometry?

12. Prove by vector methods that the diagonals of a parallelogram bisect each other.

13. Prove by vector methods that the midpoints of the sides of any quadrilateral are the vertices of a parallelogram.

# 2 Length and Direction. The Straight Line

## 2.1 THE DISTANCE FORMULA; THE MIDPOINT FORMULA

In this chapter we consider basic matters concerning length and direction as they relate to straight lines. In later chapters these ideas are extended to more general curves.

The distance between two points $P_1(x_1, y_1)$ and $P_2(x_2, y_2)$ in a coordinate plane is defined as the length of the line segment $P_1P_2$, or the magnitude of the vector $\overrightarrow{P_1P_2}$ (or of $\overrightarrow{P_2P_1}$). We may then denote this distance by $|\overrightarrow{P_1P_2}|$ or $|\overrightarrow{P_1P_2}|$, or even by $|P_1P_2|$. Since

$$\overrightarrow{P_1P_2} = (x_2 - x_1, y_2 - y_1),$$

we have, by Definition 1.7.2,

(2.1.1) $\qquad |P_1P_2| = |\overrightarrow{P_1P_2}| = \sqrt{(x_2 - x_1)^2 + (y_2 - y_1)^2}.$

Formula (2.1.1) is called, simply, the *distance formula*.

It should be noted that (2.1.1) is also valid when $P_1P_2$ is parallel to a coordinate axis. For example, if $x_1 = x_2$, then

$$|P_1P_2| = \sqrt{(y_2 - y_1)^2} = |y_2 - y_1|.$$

Furthermore, the order in which the points are taken is immaterial in the application of the distance formula.

**Example 2.1.1.** The distance between the points $(-2, 3)$ and $(4, 7)$ is, by (2.1.1),

$$\sqrt{[4-(-2)]^2 + (7-3)^2} = \sqrt{36 + 16} = \sqrt{52} = 2\sqrt{13}.$$

**Example 2.1.2.** The distance between the fixed point $(3, -4)$ and the variable point $(x, y)$ is, by (2.1.1),

$$\sqrt{(x-3)^2 + (y+4)^2}.$$

**Example 2.1.3.** Show that the triangle with vertices $A(-3, 2)$, $B(1, 1)$, and $C(-4, -2)$ is an isosceles right triangle.

*Solution.* We have

$$|AB| = \sqrt{(1+3)^2 + (1-2)^2} = \sqrt{17},$$
$$|AC| = \sqrt{(-4+3)^2 + (-2-2)^2} = \sqrt{17}.$$

Hence $|AB| = |AC|$, and the triangle is isosceles. Furthermore,

$$|BC| = \sqrt{(-4-1)^2 + (-2-1)^2} = \sqrt{34}.$$

Since $|BC|^2 = |AB|^2 + |AC|^2$, the Pythagorean relation holds and $ABC$ is a right triangle, with the right angle at $A$.

Let $P(x_m, y_m)$ be the midpoint of the line segment joining $P_1(x_1, y_1)$ and $P_2(x_2, y_2)$. Then $\overrightarrow{P_1P} = \overrightarrow{PP_2}$ (Fig. 2.1.1) or, in component form,

$$(x_m - x_1, y_m - y_1) = (x_2 - x_m, y_2 - y_m).$$

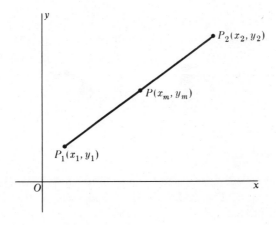

Fig. 2.1.1

Using Definition 1.7.1, for equality of vectors, we get

$$x_m - x_1 = x_2 - x_m, \qquad y_m - y_1 = y_2 - y_m,$$

from which we find

(2.1.2) $\qquad x_m = \tfrac{1}{2}(x_1 + x_2), \qquad y_m = \tfrac{1}{2}(y_1 + y_2).$

Formula (2.1.2) is the *midpoint formula*.

Hence *the midpoint of the line segment joining $P_1(x_1, y_1)$ and $P_2(x_2, y_2)$ has the coordinates $[\tfrac{1}{2}(x_1 + x_2), \tfrac{1}{2}(y_1 + y_2)]$*.

**Example 2.1.4.** The midpoint of the line segment joining the points $(-1, 7)$ and $(4, 9)$ has the coordinates

$$x_m = \tfrac{1}{2}(-1 + 4) = \tfrac{3}{2}, \qquad y_m = \tfrac{1}{2}(7 + 9) = 8.$$

**Example 2.1.5.** Show that the midpoint of the line segment joining $A(-5, 0)$ and $B(0, 3)$ is equidistant from $A$, $B$, and the origin (see Fig. 2.1.2).

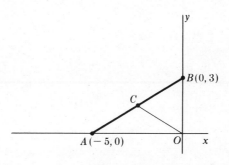

Fig. 2.1.2

*Solution.* By (2.1.2) the midpoint of $AB$ is $C(-5/2, 3/2)$. It is sufficient to show that $|AC| = |OC|$ (why?) or $|AC|^2 = |OC|^2$. By the distance formula we have

$$|AC|^2 = (-5/2 + 5)^2 + (3/2)^2 = 17/2,$$
$$|OC|^2 = (-5/2)^2 + (3/2)^2 = 17/2.$$

Hence $|AC|^2 = |OC|^2$, and the desired property is established.

**EXERCISE GROUP 2.1**

In Exercises 1–4, $A$, $B$, and $C$ are vertices of a triangle. Find (a) the length of side $AC$, (b) the midpoint of $AC$, (c) the length of the median to $AC$.

1. $A(2, 4)$, $B(4, 5)$, $C(-1, 3)$
2. $A(-3, 4)$, $B(0, -3)$, $C(7, 2)$
3. $A(0, 0)$, $B(3, 5)$, $C(5, 3)$
4. $A(-7, -2)$, $B(-2, -3)$, $C(-4, -1)$

In Exercises 5–10 show that the triangle with the given vertices is a right triangle.

5. $(-1, -4), (5, 2), (3, 4)$  
6. $(2, 5), (5, 2), (10, 7)$  
7. $(3/2, 1), (-1, 5/2), (4, 31/6)$  
8. $(-2, -3), (19, 4), (-1, -6)$  
9. $(2, 4), (6, 7), (4 + \sqrt{6}, 6)$  
10. $(4, \sqrt{11}), (-1, -4), (3, 4)$  

11. Show that the triangle with vertices $(7, -2), (-3, -8)$, and $(8, -15)$ is isosceles.

12. Show that the triangle with vertices $(0, 2), (-4, 0)$, and $(8, -19)$ is isosceles.

13. Show that the triangle with vertices $(1 + 5\sqrt{3}, 5 + 5\sqrt{3}), (6, 0)$, and $(-4, 10)$ is equilateral.

14. Given the point $P(x, y)$, find (a) the distance between $P$ and the origin, (b) the distance between $P$ and $(2, -1)$, (c) the midpoint of the line segment joining $P$ and $(2, -1)$.

In Exercises 15–18 show that the midpoints of $AB$, $BC$, $CD$, and $AD$ are the vertices of a parallelogram.

15. $A(1, 2), B(4, 7), C(5, 1), D(1, 0)$
16. $A(2, -1), B(1, -2), C(-2, 4), D(1, 1)$
17. $A(0, 2), B(2, 4), C(6, -1), D(0, -2)$
18. $A(-1, -1), B(3, -2), C(-2, 2), D(5, 3)$

In Exercise 19–24 show by the use of distances that the given points are *collinear*; that is, they lie on a straight line.

19. $(1, 2), (-3, -4), (2, 3.5)$
20. $(-1, 2), (0.5, 0), (5, -6)$
21. $(1, 1), (6, 8), (3, 3.8)$
22. $(9, 1), (1, -3), (8, 0.5)$
23. $(\sqrt{2}, 1), (\sqrt{8}, 3), (-\sqrt{2}, -3)$
24. $(12, -4), (5, -5), (-0.6, -5.8)$

25. The same method by which the midpoint formula was obtained may be used to find the points of trisection of the line segment joining $P_1(x_1, y_1)$ and $P_2(x_2, y_2)$. Show that this method yields, as points of trisection,

$$\left(\frac{2x_1 + x_2}{3}, \frac{2y_1 + y_2}{3}\right) \text{ and } \left(\frac{x_1 + 2x_2}{3}, \frac{y_1 + 2y_2}{3}\right).$$

26. Use the results of Exercise 25 to find the points of trisection of the line segment joining the given points.
    (a) $(2, 3)$ and $(-1, 4)$  (b) $(3, -1)$ and $(7, 5)$

★ 27. If $P(x, y)$ is a point on the line through the points $P_1(x_1, y_1)$ and $P_2(x_2, y_2)$ such that $r_2 \overrightarrow{P_1P} = r_1 \overrightarrow{PP_2}$, show that

$$x = \frac{r_2 x_1 + r_1 x_2}{r_1 + r_2}, \quad y = \frac{r_2 y_1 + r_1 y_2}{r_1 + r_2}.$$

## 2.2 SCALAR OR DOT PRODUCT OF VECTORS

Two kinds of multiplication play an important part in the study of vectors. For two-dimensional vectors, in which we are presently interested, only one of these is appropriate. The *scalar product* or *dot product* $\mathbf{a} \cdot \mathbf{b}$ of two vectors $\mathbf{a}$ and $\mathbf{b}$ is defined as follows. (Vector product is defined in Section 18.7.)

**DEFINITION 2.2.1.** *If* $\mathbf{a} = (a_1, a_2)$ *and* $\mathbf{b} = (b_1, b_2)$ *then the* **dot product** *or* **scalar product** *of* $\mathbf{a}$ *and* $\mathbf{b}$ *is*

$$\mathbf{a} \cdot \mathbf{b} = a_1 b_1 + a_2 b_2.$$

It is clear that scalar product yields a real number. Except when $\mathbf{a} = \mathbf{b}$ the scalar product is always written with the dot between the two vectors. If $\mathbf{a} = \mathbf{b}$ we have

$$\mathbf{a} \cdot \mathbf{a} = \mathbf{a}^2 = |\mathbf{a}|^2 = a_1^2 + a_2^2.$$

*Example 2.2.1.* Let $\mathbf{a} = (2, 3)$ and $\mathbf{b} = (1, -2)$. Then

$$\mathbf{a} \cdot \mathbf{b} = 2 \cdot 1 + 3 \cdot (-2) = -4;$$
$$\mathbf{a}^2 = 2^2 + 3^2 = 13;\ \mathbf{b}^2 = 1^2 + (-2)^2 = 5.$$

**THEOREM 2.2.1.** *Scalar product is commutative,*

$$\mathbf{a} \cdot \mathbf{b} = \mathbf{b} \cdot \mathbf{a},$$

*and distributive,*

$$\mathbf{a} \cdot (\mathbf{b} + \mathbf{c}) = \mathbf{a} \cdot \mathbf{b} + \mathbf{a} \cdot \mathbf{c}.$$

The proof is left to the exercises. There is no associative property for scalar product (see Exercise 26).

Recalling the base vectors $\mathbf{i} = (1, 0)$, $\mathbf{j} = (0, 1)$ [Section 1.8], we have

(2.2.1)
$$\mathbf{i} \cdot \mathbf{i} = \mathbf{i}^2 = \mathbf{j} \cdot \mathbf{j} = \mathbf{j}^2 = 1,$$
$$\mathbf{i} \cdot \mathbf{j} = \mathbf{j} \cdot \mathbf{i} = 0.$$

If we now write

$$\mathbf{a} = a_1 \mathbf{i} + a_2 \mathbf{j}, \qquad \mathbf{b} = b_1 \mathbf{i} + b_2 \mathbf{j},$$

it is seen that $\mathbf{a} \cdot \mathbf{b}$ may be obtained from the form

$$\mathbf{a} \cdot \mathbf{b} = (a_1 \mathbf{i} + a_2 \mathbf{j}) \cdot (b_1 \mathbf{i} + b_2 \mathbf{j})$$

by using the ordinary rules of multiplication in conjunction with (2.2.1).

*Example 2.2.2.* $(2\mathbf{i} - 3\mathbf{j}) \cdot (-\mathbf{i} + 5\mathbf{j}) = -2\mathbf{i}^2 + 10\mathbf{i} \cdot \mathbf{j} + 3\mathbf{j} \cdot \mathbf{i} - 15\mathbf{j}^2$
$$= -2 + 0 + 0 - 15 = -17.$$

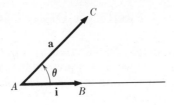

Fig. 2.2.1.

Let $\mathbf{a} = (a_1, a_2)$ be any nonzero vector (Fig. 2.2.1). Through the initial point $A$ of $\mathbf{a}$ draw a vector $\mathbf{i}$ (that is, with the sense of the positive $x$ axis). Let $\theta$ be an angle with $\overrightarrow{AB}$ as initial side and $\overrightarrow{AC}$ as terminal side. Then $\theta$ defines the *direction* of $\mathbf{a}$ in the plane. It then follows from the properties of sine and cosine that

(2.2.2) $\qquad a_1 = |\mathbf{a}|\cos\theta, \qquad a_2 = |\mathbf{a}|\sin\theta.$

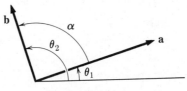

Fig. 2.2.2.

Now let $\mathbf{a}$ and $\mathbf{b}$ be two vectors with directions $\theta_1$ and $\theta_2$. We may represent them by two directed line segments with the same initial point (Fig. 2.2.2). The *angle $\alpha$ between $\mathbf{a}$ and $\mathbf{b}$* is defined as the smallest nonnegative (that is, positive or zero) angle whose initial and terminal lines are along these directed line segments. We then have $0 \leq \alpha \leq \pi$, and

$$\alpha = \pm(\theta_2 - \theta_1) + 2k\pi \quad \text{for some integer } k.$$

Note that $\cos\alpha = \cos(\theta_2 - \theta_1)$ in any case.

Using (2.2.2) and corresponding equations for $b_1$ and $b_2$, we have

$$\begin{aligned}
\mathbf{a} \cdot \mathbf{b} &= a_1 b_1 + a_2 b_2 \\
&= |\mathbf{a}||\mathbf{b}|(\cos\theta_1 \cos\theta_2 + \sin\theta_1 \sin\theta_2) \\
&= |\mathbf{a}||\mathbf{b}|\cos(\theta_2 - \theta_1) \\
&= |\mathbf{a}||\mathbf{b}|\cos\alpha.
\end{aligned}$$

We have derived the formula

(2.2.3) $\qquad \mathbf{a} \cdot \mathbf{b} = |\mathbf{a}||\mathbf{b}|\cos\alpha,$

where $\alpha$ is the angle between $\mathbf{a}$ and $\mathbf{b}$. Formula (2.2.3) may be used to find $\alpha$; it may also be considered as a geometric interpretation of the scalar product of **a** and **b**.

*Example 2.2.3.* Find the angle between the vectors $2\mathbf{i} + 3\mathbf{j}$ and $\mathbf{i} - \mathbf{j}$.

*Solution.* Let $\mathbf{a} = 2\mathbf{i} + 3\mathbf{j}$, $\mathbf{b} = \mathbf{i} - \mathbf{j}$. We find

$$\mathbf{a} \cdot \mathbf{b} = 2 \cdot 1 + 3(-1) = -1, \quad |\mathbf{a}| = \sqrt{13}, \quad |\mathbf{b}| = \sqrt{2}.$$

Hence, if $\alpha$ is the angle between **a** and **b**, we have by (2.2.3),

$$\cos \alpha = \frac{\mathbf{a} \cdot \mathbf{b}}{|\mathbf{a}||\mathbf{b}|} = \frac{-1}{\sqrt{13}\sqrt{2}} = -\frac{1}{\sqrt{26}}.$$

Then $\alpha$ is the angle between $\pi/2$ and $\pi$ such that $\cos \alpha = -1/\sqrt{26}$, or $\alpha = \cos^{-1}(-1/\sqrt{26})$.

**THEOREM 2.2.2.** *Two nonzero vectors **a** and **b** are perpendicular if and only if $\mathbf{a} \cdot \mathbf{b} = 0$. More briefly,*

$$\mathbf{a} \perp \mathbf{b} \Leftrightarrow \mathbf{a} \cdot \mathbf{b} = 0.$$

*Proof.* $a \perp b \Leftrightarrow \alpha = \pi/2$. If $\alpha = \pi/2$, then $\cos \alpha = 0$ and $\mathbf{a} \cdot \mathbf{b} = 0$. On the other hand if $\mathbf{a} \cdot \mathbf{b} = 0$ then, since $|\mathbf{a}| \neq 0$ and $|\mathbf{b}| \neq 0$, we see from (2.2.3) that $\cos \alpha = 0$ and $\mathbf{a} \perp \mathbf{b}$.

*Example 2.2.4.* Prove that the points $M(4, 3)$, $N(1, 2)$, and $P(4, -7)$ are vertices of a right triangle.

*Proof.* We find

$$\overrightarrow{MN} = (-3, -1), \quad \overrightarrow{NP} = (3, -9).$$

Therefore, $\overrightarrow{MN} \cdot \overrightarrow{NP} = (-3) \cdot 3 + (-1)(-9) = 0$. It follows by Theorem 2.2.2 that $\overrightarrow{MN} \perp \overrightarrow{NP}$, and thus $MNP$ is a right triangle with the right angle at $N$.

As another, less direct application of Theorem 2.2.2 we shall prove a theorem of Plane Geometry by the use of vector methods.

*Example 2.2.5.* Prove that the altitudes of a triangle meet in a point.

*Solution.* Let the vectors **a**, **b**, and **c** represent the sides of the triangle $PQR$, directed as in Fig. 2.2.3. Then

(2.2.4) $$\mathbf{a} + \mathbf{b} + \mathbf{c} = \mathbf{0}.$$

The following steps are also valid when one of the angles is obtuse. $KQ$ and $LP$ are two of the altitudes, intersecting at $M$. We shall show that $MR \perp PQ$.

From $\triangle QMR$ and $\triangle PMR$, respectively, we see that

$$\overrightarrow{MQ} = \overrightarrow{MR} - \mathbf{b}, \quad \overrightarrow{MP} = \overrightarrow{MR} + \mathbf{c}.$$

Now, $MQ \perp RP$ and $MP \perp QR$. Hence

$$\mathbf{c} \cdot \overrightarrow{MQ} = \mathbf{c} \cdot (\overrightarrow{MR} - \mathbf{b}) = \mathbf{c} \cdot \overrightarrow{MR} - \mathbf{b} \cdot \mathbf{c} = 0,$$
$$\mathbf{b} \cdot \overrightarrow{MP} = \mathbf{b} \cdot (\overrightarrow{MR} + \mathbf{c}) = \mathbf{b} \cdot \overrightarrow{MR} + \mathbf{b} \cdot \mathbf{c} = 0.$$

Hence $\mathbf{b} \cdot \mathbf{c} = \mathbf{c} \cdot \overrightarrow{MR} = -\mathbf{b} \cdot \overrightarrow{MR}$, so that $(\mathbf{b} + \mathbf{c}) \cdot \overrightarrow{MR} = 0$. Since $\mathbf{b} + \mathbf{c} = -\mathbf{a}$ from (2.2.4), it follows that $\mathbf{a} \cdot \overrightarrow{MR} = 0$. Thus $MR \perp PQ$, $MR$ is an altitude, and the three altitudes intersect in the point $M$.

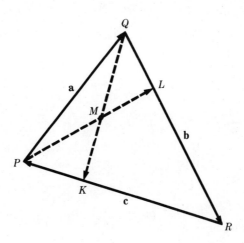

Fig. 2.2.3

## EXERCISE GROUP 2.2

1. Evaluate (a) $\mathbf{i} \cdot (\mathbf{i} + 2\mathbf{j})$, (b) $(2\mathbf{i} + \mathbf{j}) \cdot (\mathbf{i} - 2\mathbf{j})$, (c) $(3\mathbf{i} + 4\mathbf{j})^2$
2. Evaluate (a) $\mathbf{j} \cdot (2\mathbf{i} + 3\mathbf{j})$, (b) $(-\mathbf{i} + 3\mathbf{j}) \cdot (2\mathbf{i} + \mathbf{j})$, (c) $(2\mathbf{i} - 5\mathbf{j})^2$

Find the indicated value in Exercises 3–10 if $\mathbf{a} = (2, 3)$, $\mathbf{b} = (-2, 4)$ and $\mathbf{c} = (-1, -2)$.

3. (a) $\mathbf{a} \cdot \mathbf{b}$, (b) $\mathbf{a}^2$
4. (a) $\mathbf{b} \cdot \mathbf{c}$, (b) $\mathbf{b}^2$
5. $\mathbf{a} \cdot (\mathbf{b} + \mathbf{c})$
6. $\mathbf{a} \cdot (2\mathbf{b} - \mathbf{c})$
7. $(\mathbf{a} - \mathbf{b})^2$
8. $(\mathbf{b} - 2\mathbf{c})^2$
9. $(\mathbf{a} - \mathbf{b}) \cdot (\mathbf{b} + 2\mathbf{c})$
10. $(\mathbf{a} - \mathbf{b} + \mathbf{c}) \cdot (2\mathbf{a} - \mathbf{c})$

Find the angle between the two *vectors* in each of Exercises 11–18.

11. $\mathbf{i} + \mathbf{j}$ and $\mathbf{j}$
12. $2\mathbf{i} + \mathbf{j}$ and $\mathbf{i} + 2\mathbf{j}$
13. $-\mathbf{i} + 2\mathbf{j}$ and $3\mathbf{i} + \mathbf{j}$
14. $\mathbf{i} - \sqrt{3}\mathbf{j}$ and $\sqrt{3}\mathbf{i} - \mathbf{j}$
15. $(-2, 3)$ and $(4, 2)$
16. $(1, -1)$ and $(0, 1)$
17. $(4, 7)$ and $(-7, 4)$
18. $(2, 3)$ and $(3, 2)$

19. If **a** is any vector, prove that $\mathbf{a} = (\mathbf{a} \cdot \mathbf{i})\mathbf{i} + (\mathbf{a} \cdot \mathbf{j})\mathbf{j}$.
20. Given the points $A(-1, 2)$ and $P(x, y)$, express in nonvector form the condition (a) that $\overrightarrow{OA}$ be perpendicular to $\overrightarrow{AP}$, (b) that $\overrightarrow{OP}$ be perpendicular to $\overrightarrow{AP}$.
21. Given the points $A(-2, 0)$, $B(2, 0)$, and $P(x, y)$, express in nonvector form the condition that $\overrightarrow{AP}$ be perpendicular to $\overrightarrow{BP}$.
22. Given the points $A(1, 4)$, $B(-2, 3)$, and $P(x, y)$, express in nonvector form the condition that $\overrightarrow{AP}$ be perpendicular to $\overrightarrow{BP}$.
23. Prove the commutativity of scalar product: $\mathbf{a} \cdot \mathbf{b} = \mathbf{b} \cdot \mathbf{a}$.
24. Prove the distributivity of scalar product: $\mathbf{a} \cdot (\mathbf{b} + \mathbf{c}) = \mathbf{a} \cdot \mathbf{b} + \mathbf{a} \cdot \mathbf{c}$.
25. Prove that $\mathbf{a} \cdot \mathbf{a} = |\mathbf{a}|^2$.
26. Explain why $\mathbf{a} \cdot (\mathbf{b} \cdot \mathbf{c})$ has no meaning.
27. Prove by vector methods that the diagonals of a rhombus are perpendicular.
28. Let $ABCD$ be a quadrilateral with $|AB| = |AD|$, $|BC| = |CD|$. Prove by vector methods that $AC$ is perpendicular to $BD$.
* 29. Derive the Law of Cosines for a triangle as follows. In a triangle $ABC$ let $\mathbf{a} = \overrightarrow{AB}$, $\mathbf{b} = \overrightarrow{AC}$. Then $\overrightarrow{CB} = \mathbf{a} - \mathbf{b}$, by Section 1.9, and

$$|\overrightarrow{CB}|^2 = |\mathbf{a} - \mathbf{b}|^2 = (\mathbf{a} - \mathbf{b}) \cdot (\mathbf{a} - \mathbf{b}).$$

Expand the right member and interpret the result, using (2.2.3).

## 2.3 SLOPE OF A LINE

Let $P_1(x_1, y_1)$ and $P_2(x_2, y_2)$ be two points in a coordinate plane (Fig. 2.3.1). Then $\overrightarrow{P_1 P_2} = -\overrightarrow{P_2 P_1}$, and, specifically,

$$\overrightarrow{P_1 P_2} = (x_2 - x_1, y_2 - y_1)$$
$$\overrightarrow{P_2 P_1} = (x_1 - x_2, y_1 - y_2).$$

Let us assume $x_2 \neq x_1$. Then for both vectors the ratio of the $y$ component to the $x$ component is the same. Thus the ratio

(2.3.1) $$m = \frac{y_2 - y_1}{x_2 - x_1}, \quad x_2 \neq x_1,$$

has the same value whichever sense on the line is used, and *this value may be used to describe the direction of the line without regard to the sense.*

The ratio $m$ in (2.3.1) is called the *slope* of the line which contains the points $(x_1, y_1)$ and $(x_2, y_2)$ if $x_2 \neq x_1$. It is independent of the points chosen on the line.

If $x_2 = x_1$, the points lie on a line parallel to the $y$ axis, and *the line has no slope*. (A line is also considered as parallel to itself.)

For a line parallel to the $x$ axis, which includes the $x$ axis itself, we have $y_1 = y_2$ and hence $m = 0$.

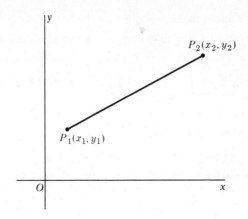

Fig. 2.3.1

When a line is not parallel to the $y$ axis, *the slope of the line expresses the rate of change in the ordinate with respect to the change in the abscissa.*

If $m > 0$, then $y_2 - y_1$ and $x_2 - x_1$ have the same sign, and the line rises as $x$ increases or falls as $x$ decreases. If $m < 0$, then $y_2 - y_1$ and $x_2 - x_1$ have opposite signs, and the line falls as $x$ increases or rises as $x$ decreases.

*Example 2.3.1.* To find the slope of the line joining the points (2, 5) and (−3, 7) we use (2.3.1), it being immaterial which of the points is taken as $(x_1, y_1)$. We get

$$m = \frac{7-5}{-3-2} = -\frac{2}{5}, \quad \text{or} \quad m = \frac{5-7}{2-(-3)} = -\frac{2}{5},$$

the same value in each case. We may say that the line falls 2 units for every 5 unit increase in $x$, or rises 2 units for every 5 unit decrease in $x$ (or any proportionate amounts).

Since $y_2 - y_1 = m(x_2 - x_1)$, the vector $\overrightarrow{P_1P_2}$ in Fig. 2.3.1 may be written as

$$\overrightarrow{P_1P_2} = [x_2 - x_1, m(x_2 - x_1)] = (x_2 - x_1)(1, m).$$

If $x_2 - x_1 \neq 0$, it follows that the vector $(1, m)$, and hence also the vector $(-1, -m)$, is parallel to $\overrightarrow{P_1P_2}$. We may state this result as follows.

**THEOREM 2.3.1.** *If $m$ is the slope of a straight line, the vectors $(1, m)$ and $(-1, -m)$ are parallel to the line.*

## 2.4 PARALLEL AND PERPENDICULAR LINES

Lines parallel to the $y$ axis have no slope. Let $L_1$ and $L_2$ be two *parallel lines* which are not parallel to the $y$ axis. Since the geometric representation of a vector

(Section 1.9) may lie on any one of a set of parallel lines, a vector **a** on $L_1$ is also a vector **a** on $L_2$. Hence *the parallel lines $L_1$ and $L_2$ have the same slope.*

Now let $L_1$ with slope $m_1$ and $L_2$ with slope $m_2$ be two *perpendicular lines* with neither one parallel to a coordinate axis. By Theorem 2.3.1 we may say that

$$\mathbf{t}_1 = (1, m_1), \quad \mathbf{t}_2 = (1, m_2)$$

are vectors on $L_1$ and $L_2$, respectively. By Section 2.2,

$$\mathbf{t}_1 \perp \mathbf{t}_2 \Leftrightarrow \mathbf{t}_1 \cdot \mathbf{t}_2 = 0.$$

The latter condition becomes, by Definition 2.2.1,

(2.4.1) $$1 + m_1 m_2 = 0,$$

which is the *condition for perpendicularity of $L_1$ and $L_2$*. We collect these results in the following.

**THEOREM 2.4.1.** *Let $L_1$ and $L_2$ be two lines with slopes $m_1$ and $m_2$, respectively. Then*
1. $L_1 \| L_2 \Leftrightarrow m_1 = m_2$;
2. $L_1 \perp L_2 \Leftrightarrow m_1 m_2 = -1.$

The perpendicularity condition $m_1 m_2 = -1$ is also expressed in the statement that $m_1$ and $m_2$ are *negative reciprocals*, and we may also write

$$m_1 = -1/m_2 \quad \text{or} \quad m_2 = -1/m_1.$$

If one or both lines are parallel to the $y$ axis, then one or both lines fail to have a slope, and parallelism or perpendicularity of the lines is easily determined.

*Example 2.4.1.* Show that the quadrilateral with vertices $A(-1, 4)$, $B(2, 5)$, $C(5, 2)$, and $D(2, 1)$ is a parallelogram.

*Solution.* This type of problem can be handled by the use of the distance formula. However, we wish to use directions here. You may wish to draw a figure.

We shall write $m(AB)$, for example, to denote the slope of the straight line containing $A$ and $B$. By the slope formula (2.3.1) we find

$$m(AB) = \tfrac{1}{3}, \quad m(BC) = -1, \quad m(CD) = \tfrac{1}{3}, \quad m(DA) = -1.$$

Hence

$$m(AB) = m(CD), \quad m(BC) = m(DA).$$

And, by Theorem 2.4.1, $AB$ is parallel to $CD$, and $BC$ is parallel to $DA$. It follows that $ABCD$ is a parallelogram.

*Example 2.4.2.* Show that the triangle with vertices $A(3, -4)$, $B(7, 2)$, and $C(12, -10)$ is a right triangle.

*Solution.* We dispense with a figure. We find

$$m(AB) = \tfrac{3}{2}, \quad m(BC) = -\tfrac{12}{5}, \quad m(CA) = -\tfrac{2}{3}.$$

Hence

$$m(AB)m(CA) = -1;$$

thus $AB$ is perpendicular to $CA$ by Theorem 2.4.1, and there is a right angle at $A$.

*Example 2.4.3.* Show that the points $A(-1, 1)$, $B(0, 3)$, and $C(3, 9)$ are collinear (lie on a straight line).

*Solution.* It is sufficient to show that any two lines formed by the three points have the same slope. We find

$$m(AB) = \frac{3-1}{0+1} = 2, \quad m(AC) = \frac{9-1}{3+1} = 2.$$

Since $m(AB) = m(AC)$, the given points are collinear.

## EXERCISE GROUP 2.3

In each of Exercises 1–8 find the slope of the line joining the two given points.

1. $(-1, 3), (2, 4)$
2. $(7, 3), (3, -5)$
3. $(4, 3), (0, 0)$
4. $(-1, -1), (9, -8)$
5. $(1, 3), (x, y)$
6. $(0, 0), (x, y)$
7. $(a, 0), (x, y)$
8. $(0, b), (x, y)$

In each of Exercises 9–12 show by use of slopes that the three points are collinear; that is, they lie on a straight line.

9. $(3, 8), (1, 5), (-1, 2)$
10. $(3, 3), (7, 8), (-9, -12)$
11. $(-1, 3), (-1/3, 4), (5/3, 7)$
12. $(3, 2/3), (-2, -8/3), (6, 8/3)$

In Exercises 13–18 show, by use of slopes, that the triangle with the given vertices is a right triangle.

13. $(-1, -4), (5, 2), (3, 4)$
14. $(2, 5), (5, 2), (10, 7)$
15. $(3/2, 1), (-1, 5/2), (4, 31/6)$
16. $(-2, -3), (19, 4), (-1, -6)$
17. $(2, 4), (6, 7), (4 + \sqrt{6}, 6)$
18. $(4, \sqrt{11}), (-1, -4), (3, 4)$

19. Find $a$ such that the line through the points $(-7, -9)$ and $(8, -2)$, and the line through the points $(4, -1)$ and $(a, 2)$ are (a) parallel, (b) perpendicular.

20. Find $b$ such that the line through the points $(11, 4)$ and $(2, -2)$, and the line through the points $(7, 3)$ and $(4, b)$ are (a) parallel, (b) perpendicular.

21. Find $b$ such that the line through the points $(2, 3)$ and $(4, b)$ has slope $1/3$.

22. Find $a$ such that the line through the points $(-1, 4)$ and $(a, -2)$ has slope $-1/3$.

**23.** Find $a$ such that the points $(2, 3)$, $(4, 1)$, and $(a, -1)$ are collinear.

**24.** Find $b$ such that the points $(2, b)$, $(-1, 3)$, and $(4, -1)$ are collinear.

## 2.5 ANGLES BETWEEN LINES

From Fig. 2.5.1 it may be seen that either of the two angles $\alpha_1$ and $\alpha_2$ may be considered as an angle between the two lines shown. Each angle may be restricted to the interval $0 \leq \alpha \leq \pi$. For many purposes it does not matter which angle is used, since the angles are supplementary,

(2.5.1) $$\alpha_1 + \alpha_2 = \pi.$$

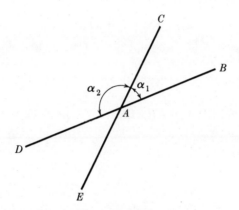

Fig. 2.5.1

On the other hand, if vectors lying on the two lines are used, the angle between them is explicit. Using the notation $\measuredangle\,(\mathbf{a}, \mathbf{b})$ to denote the angle between the two vectors $\mathbf{a}$ and $\mathbf{b}$, we have, referring to Fig. 2.5.1,

$$\measuredangle\,(\overrightarrow{AB}, \overrightarrow{AC}) = \measuredangle\,(\overrightarrow{AD}, \overrightarrow{AE}) = \alpha_1,$$
$$\measuredangle\,(\overrightarrow{AC}, \overrightarrow{AD}) = \measuredangle\,(\overrightarrow{AE}, \overrightarrow{AB}) = \alpha_2.$$

We recall that a formula for the angle $\measuredangle\,(\mathbf{a}, \mathbf{b})$ between the vectors $\mathbf{a}$ and $\mathbf{b}$ is available. From (2.2.3) we have

(2.5.2) $$\cos \measuredangle\,(\mathbf{a}, \mathbf{b}) = \frac{\mathbf{a} \cdot \mathbf{b}}{|\mathbf{a}|\,|\mathbf{b}|}.$$

Let us now apply this to two lines $L_1$ and $L_2$ with slopes $m_1$ and $m_2$, respectively. We may use $(1, m_1)$ or $(-1, -m_1)$ as a vector on $L_1$, and $(1, m_2)$

or $(-1, -m_2)$ as a vector on $L_2$ (Theorem 2.3.1). Choosing these vectors as

$$\mathbf{a} = (1, m_1), \quad \mathbf{b} = (1, m_2),$$

we get, by (2.5.2),

(2.5.3) $$\cos \measuredangle (\mathbf{a}, \mathbf{b}) = \frac{1 + m_1 m_2}{\sqrt{1 + m_1^2} \sqrt{1 + m_2^2}}.$$

*Formula* (2.5.3) *may then be used to find an angle between two lines with slopes* $m_1$ *and* $m_2$. If the other possible choices for $\mathbf{a}$ and $\mathbf{b}$ were used, $\cos \measuredangle (\mathbf{a}, \mathbf{b})$ would be the same as in (2.5.3) or the negative of that expression, which is consistent with (2.5.1).

*Example 2.5.1.* Find an angle between two lines whose slopes are $-1$ and $-\frac{1}{2}$.

*Solution.* If $\alpha_1$ is such an angle, we find at once, by (2.5.3),

$$\cos \alpha_1 = \frac{1 + (-1)(-\frac{1}{2})}{\sqrt{1 + 1} \sqrt{1 + 1/4}} = \frac{3}{\sqrt{10}}.$$

The other possible angle $\alpha_2$ is such that $\cos \alpha_2 = -3/\sqrt{10}$.

*Example 2.5.2.* Find the angles (or their cosines) of the triangle with vertices $A(1, 2)$, $B(9, 1)$, and $C(6, 7)$.

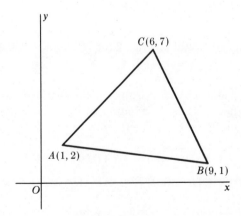

Fig. 2.5.2

*Solution.* For each angle of the triangle (Fig. 2.5.2) we may use the appropriate vector determined by the coordinates, and formula (2.5.2). Thus for angle $A$ we use vectors $(5, 5)$ and $(8, -1)$ and find

$$\cos A = \frac{40 - 5}{\sqrt{50}\sqrt{65}} = \frac{7}{\sqrt{130}};$$

for angle $B$ we use vectors $(-8, 1)$ and $(-3, 6)$ and find

$$\cos B = \frac{24 + 6}{\sqrt{65}\sqrt{45}} = \frac{2}{\sqrt{13}};$$

for angle $C$ we use vectors $(3, -6)$ and $(-5, -5)$ and find

$$\cos C = \frac{-15 + 30}{\sqrt{45}\sqrt{50}} = \frac{1}{\sqrt{10}}.$$

To do the same problem using slopes we find

$$m(CA) = 1, \quad m(AB) = -1/8, \quad m(BC) = -2;$$

we then use the vectors

$(1, 1)$ and $(1, -1/8)$ for angle $A$,
$(-1, 1/8)$ and $(-1, 2)$ for angle $B$,
$(1, -2)$ and $(-1, -1)$ for angle $C$.

It is easily verified that the same values of the cosines are obtained.

**EXERCISE GROUP 2.4**

In Exercises 1–4 find the smallest positive angle between two lines whose slopes are given.

1. $3/4; 4/3$
2. $1; -1/2$
3. $0; -1/3$
4. $-5; 0$

In Exercises 5–8 find an angle between the line determined by $A$ and $B$ and the line determined by $A$ and $C$.

5. $A(1, 2), B(2, 3), C(3, -1)$
6. $A(-1, 3), B(-2, -4), C(5, -2)$
7. $A(1, 0), B(3, 1), C(4, -3)$
8. $A(-2, 3), B(0, 2), C(7, 5)$

In Exercises 9–12 find the angles of the triangle with the given vertices.

9. $(-1, -3), (1, 4), (2, -1)$
10. $(5, 6), (-8, 2), (4, -5)$
11. $(-3, 0), (0, 3), (-4, -6)$
12. $(3, 9), (4, -1), (-3, 5)$

13. Find $a$ such that an angle between the lines determined by the points $(1, -3), (4, 2)$, and $(2, -1), (a, 3)$ is $45°$.

14. Find $b$ such that an angle between the lines determined by the points $(4, 0), (2, -1)$, and $(1, b), (3, 0)$ is $30°$.

15. If $\alpha$ is an angle between two lines with slopes $m_1$ and $m_2$, use (2.5.3) and trigonometric identities to show that

(a) $\sin \alpha = \pm \dfrac{m_1 - m_2}{\sqrt{1 + m_1^2}\sqrt{1 + m_2^2}}$, (b) $\tan \alpha = \pm \dfrac{m_1 - m_2}{1 + m_1 m_2}$ if $1 + m_1 m_2 \neq 0$.

In Exercises 16–19 find the smallest positive angle between the lines whose slopes are given, by using the tangent formula of Exercise 15.

16. 3; −2
17. 4; −3
18. 1/3; 1/2
19. −3/2; 1/5

## 2.6 RELATIONS. GRAPH OF AN EQUATION

Ordered pairs of numbers are used in many places in mathematics. For example, we defined a vector as any ordered pair of real numbers. In other cases, one or both of the numbers of the pairs under consideration may be restricted in some manner. Whatever the purpose for which ordered pairs may be used, the ordered pairs themselves may be taken as the elements of a set.

Let us define a *relation* as any set of *ordered pairs of numbers* $(x, y)$, specified in some manner. The significance of the word "ordered" in the definition is again indicated in the distinction between (2, 3) and (3, 2), for example. In some cases the complete relation can be enumerated, as in

(2.6.1) $\qquad R: \{(2, 3), (-1, 3), (2, 0), (3, 5)\}.$

Here the relation $R$ consists of the four pairs listed. In other cases, where the pairs of numbers cannot be enumerated, a relation can be described in some other explicit manner. For example,

(2.6.2) $\qquad R: \{(0, y) \mid y \text{ is any real number}\},$

which means the set of number pairs $(0, y)$, where $y$ is any real number, defines a relation; and

(2.6.3) $\qquad R: \{(x, 1) \mid x \text{ is any real number}\},$

also defines a relation.

The *graph of a relation* $R$ is the set of points in a coordinate plane which represent the number pairs of $R$, the first number of any pair being taken as the abscissa of the corresponding point, and the second number as the ordinate. The graph of (2.6.1) then consists of four points in the plane, the graph of (2.6.2) is the $y$ axis, and the graph of (2.6.3) is a straight line, parallel to the $x$ axis and one unit above it.

We are particularly interested at this point in relations which are defined by equations containing two letters which represent *variables*, that is, letters which may be assigned different number values. Consider the equation

$$2x - 3y = 7;$$

this may be treated as a condition which restricts the values of $x$ and $y$ so as to satisfy the equation. Any pair of values of $x$ and $y$ which satisfies the equation is called a *solution* of the equation. Thus $x = 2$, $y = -1$ is a solution; $x = 0$, $y = -7/3$ is a solution; and so is $x = 7/2$, $y = 0$. There are, in fact, infinitely many solutions. The set of all such solutions is a relation, and the graph of this relation is defined as the *graph of the equation*. We may express the relation in the form

(2.6.4) $$\{(x, y) | 2x - 3y = 7\},$$

which is read as "the set of number pairs $(x, y)$ such that $2x - 3y = 7$." When graphs are involved, we shall speak interchangeably of sets of number pairs or sets of points.

In different words, *the graph of an equation consists of those points, and only those points, whose coordinates satisfy the equation.*

Later in the book methods of graphing equations will be considered which will minimize the amount of plotting of points which is required for this purpose. Here we shall rely principally on plotting a sufficient number of points to permit us to draw a curve connecting them.

*Example 2.6.1.* Graph the relation in (2.6.4).

*Solution.* We construct a table of values for the equation

(2.6.5) $$2x - 3y = 7$$

by assigning values to $x$ and computing the corresponding values of $y$:

| $x$ | $-1$ | $0$ | $1$ | $2$ |
|---|---|---|---|---|
| $y$ | $-3$ | $-\frac{7}{3}$ | $-\frac{5}{3}$ | $-1$ |

The points obtained are then plotted in a coordinate system (Fig. 2.6.1) and joined by a smooth curve. The graph of the relation (2.6.4), as well as of the equation (2.6.5), is apparently a straight line. (It is, in fact, a straight line.)

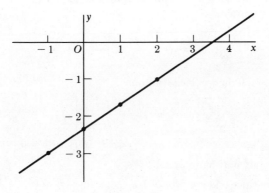

Fig. 2.6.1

*Example 2.6.2.* Graph the equation: $x = y^2 - 2y$.

*Solution.* It is easier to assign values to $y$ and solve for $x$. In this way we get the table:

| $x$ | 3 | 0 | $-1$ | 0 | 3 | $\frac{5}{4}$ | $\frac{5}{4}$ |
|---|---|---|---|---|---|---|---|
| $y$ | $-1$ | 0 | 1 | 2 | 3 | $-\frac{1}{2}$ | $\frac{5}{2}$ |

The corresponding points, and graph, are shown in Fig. 2.6.2. If a gap between points leaves doubt as to how they should be joined, additional intermediate points may be obtained. We may add to the table of points (5/4, −1/2) and (5/4, 5/2).

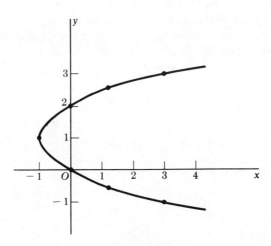

Fig. 2.6.2

In the above examples the graphs continue indefinitely, beyond the ends of the curve actually shown.

### EXERCISE GROUP 2.5

In Exercises 1–11 graph the specified relation.

1. $\{(x, y) | x = 0\}$
2. $\{(x, y) | y = -1\}$
3. $\{(x, y) | x + y = 0\}$
4. $\{(x, y) | y = x\}$
5. $\{(x, y) | y^2 = 1\}$
6. $\{(x, y) | y^2 - y = 0\}$
7. $\{(x, y) | x$ and $y$ are positive integers$\}$
8. $\{(x, y) | y$ is a positive odd multiple of 1/2$\}$
9. $\{(x, y) | x$ is a positive integer and $y = x\}$
10. $\{(x, y) | y = 1$, if $x \geq 0; y = -1$, if $x < 0\}$
11. $\{(x, y) | y = x$, if $x \geq 0; y = -x$, if $x < 0\}$

Graph each of the following equations, by use of a table of values.

12. $4x + y - 2 = 0$
13. $x - 3y + 1 = 0$

14. $2x - 4y = 5$
15. $y = x^2$
16. $2y = x^2 + 3x$
17. $2x^2 - x - y = 0$
18. $x + 4y^2 + 4y = 0$

19. $x^2 + y^2 = 4$
20. $2x^2 + y^2 = 4$
21. $y = |x + 2|$
22. $y = |x - 2|$
23. $y = |x| + 2$

## 2.7 THE EQUATION $Ax + By + C = 0$

In Example 2.6.1 and the exercises of the preceding section various equations of the form

$$Ax + By + C = 0$$

were graphed. In each case a straight line was obtained. Let us show that this holds in general. We use as a criterion for a straight line the fact that *a line segment joining any two points is in the same direction.*

**THEOREM 2.7.1.** (a) *The graph of the equation*

(2.7.1)    $Ax + By + C = 0,$    *A and B not both zero,*

*is a straight line.*
(b) *Any straight line has an equation of the form* (2.7.1).

*Proof.* (a) Let $(x_1, y_1)$ be a point on (2.7.1), (that is, on its graph). Then

(2.7.2)    $Ax_1 + By_1 + C = 0.$

Subtracting (2.7.2) from (2.7.1), we may write

(2.7.3)    $A(x - x_1) + B(y - y_1) = 0.$

We may consider $(A, B)$ as a fixed, nonzero vector, and $(x - x_1, y - y_1)$ also as a vector. Then by the criterion for perpendicularity of two vectors, we see from (2.7.3) that $(A, B) \perp (x - x_1, y - y_1)$. Hence the segment joining any point $(x, y)$ on (2.7.1) to the point $(x_1, y_1)$ has the same direction, perpendicular to $(A, B)$, and the graph of (2.7.1) is a straight line.

(b) Let $P_1(x_1, y_1)$ be a fixed point in a given straight line and let $P(x, y)$ be an arbitrary point. Then $PP_1$ has a fixed direction. Let $(A, B)$ be a vector perpendicular to this direction. We must have (2.7.3) which may be written as

$$Ax + By - Ax_1 - By_1 = 0.$$

This is of the form (2.7.1) with $C = -Ax_1 - By_1$, and the proof is complete.

Equation (2.7.1) is called the *general form* of the equation of a straight line. Because of its identification with a straight line, Equation (2.7.1) is also called a *linear equation in x and y.*

If $B = 0$, the vector $(A, 0)$ is parallel to the $x$ axis, and the line (2.7.1) is parallel to the $y$ axis. Since $A \neq 0$ in this case, the equation of such a line becomes $x = -C/A$, or $x = $ constant.

If $B \neq 0$, (2.7.3) gives

(2.7.4) $$\frac{y - y_1}{x - x_1} = -\frac{A}{B}, \qquad x \neq x_1.$$

Hence *the slope of the line $Ax + By + C = 0$ is $m = -A/B$*.

*Example 2.7.1.* The equation $2x - 3y = 7$ of Example 2.6.1 is of the form (2.7.1), and therefore represents a straight line. It has slope 2/3 since $A = 2$, $B = -3$. We may draw the graph by finding any point on it, say $(2, -1)$, and drawing through this point a line which rises 2 units when $x$ advances 3 units. The graph appears in Fig. 2.6.1.

## 2.8 FORMS OF THE EQUATION OF A STRAIGHT LINE

We consider here various forms of the equation of a straight line. Each of them will exhibit two characteristics that determine a straight line. Algebraically, each will involve two independent constants in some way. There are three constants in the general form $Ax + By + C = 0$, but only two of them are independent, since we may divide both sides of the equation by any of the three constants which is not zero, and then consider as independent constants the ratios which appear.

*Point-slope form.* Let $m$ be the slope of a line (not parallel to the $y$ axis), and let $(x_1, y_1)$ be any point on the line. Then from (2.7.4) we may write

(2.8.1) $$y - y_1 = m(x - x_1).$$

This is the *point-slope* form of the equation of a straight line, which exhibits, in the way it is written, the slope of the line, and the coordinates of a point on it. Note carefully that (2.7.4) does not represent the straight line since it has no meaning when $x = x_1$; however, (2.8.1) is satisfied also for $x = x_1$, $y = y_1$.

*Example 2.8.1.* Find an equation of the line which passes through the points $(2, -3)$ and $(5, 2)$.

*Solution.* We may use (2.8.1). We find $m = (2 + 3)/(5 - 2) = 5/3$. Then using the point $(2, -3)$ in (2.8.1) we have

$$y + 3 = \tfrac{5}{3}(x - 2),$$

as an equation of the straight line. In general form (2.7.1) we get

$$5x - 3y - 19 = 0.$$

It is easily verified that the use of the point $(5, 2)$ in (2.8.1) will lead to the same result.

*Slope-intercept form.* If a straight line crosses the $y$ axis at the point $(0, b)$, the number $b$ is called the *$y$ intercept* of the line. Let $m$ be the slope of the line. Then by (2.8.1), we have $y - b = m(x - 0)$, or

(2.8.2) $$y = mx + b.$$

This is the *slope-intercept form* of the equation of a straight line, which exhibits the slope and $y$ intercept of the line. It is simply the general form (2.7.1) solved for $y$, as

$$y = -\frac{A}{B}x - \frac{C}{B}.$$

We have $m = -A/B$, as in (2.7.4), and $b = -C/B$.

**Example 2.8.2.** Write the equation $-2x - 5y + 2 = 0$ in slope-intercept form.

*Solution.* Solving the equation for $y$, we have the desired form,

$$y = -\tfrac{2}{5}x + \tfrac{2}{5};$$

it follows that $m = -2/5$, $b = 2/5$.

*Intercept form.* If a straight line crosses the $x$ axis at the point $(a, 0)$, the number $a$ is called the *$x$ intercept* of the line. If the line crosses both axes, at $(a, 0)$ and $(0, b)$, then $m = -b/a$, and by (2.8.2),

$$y = -\frac{b}{a}x + b, \qquad a \neq 0.$$

This may be written as

(2.8.3) $$\frac{x}{a} + \frac{y}{b} = 1, \qquad a \neq 0, b \neq 0.$$

This is the *intercept form* of the equation of a straight line, also called the *two-intercept form*. It exhibits directly the (nonzero) $x$ and $y$ intercepts of the line.

**Example 2.8.3.** Write the equation $-2x - 5y + 2 = 0$ in intercept form.

*Solution.* If we set $y = 0$, we find $a = 1$, and with $x = 0$, we find $b = 2/5$. Hence the intercept form is

$$\frac{x}{1} + \frac{y}{2/5} = 1.$$

This may also be obtained directly from the given equation by obtaining, in turn,

$$2x + 5y = 2, \qquad x + \tfrac{5}{2}y = 1, \qquad \frac{x}{1} + \frac{y}{2/5} = 1.$$

We know from Section 2.4 that two parallel lines have the same slope, while two perpendicular lines have slopes which are negative reciprocals (if the slopes exist). We may use these criteria as in the following example. (See also Exercises 32 and 33.)

*Example 2.8.4.* Find an equation of a line passing through the point $(-5, 2)$ which is (a) parallel to, (b) perpendicular to the line $4x + 3y - 7 = 0$.

*Solution.* The slope of the given line is $-4/3$. Any parallel line must therefore have slope $-4/3$, and any perpendicular line must have slope $3/4$. Using the point-slope form in each case we then have for:
(a) the parallel line

$$y - 2 = -\tfrac{4}{3}(x + 5) \quad \text{or} \quad 4x + 3y + 14 = 0;$$

(b) the perpendicular line

$$y - 2 = \tfrac{3}{4}(x + 5) \quad \text{or} \quad 3x - 4y + 23 = 0.$$

### EXERCISE GROUP 2.6

In each of Exercises 1–6 find an equation of the straight line passing through the given points.

1. $(2, 7), (5, -2)$
2. $(2, 0), (-2, -\tfrac{5}{2})$
3. $(2, 0), (0, -3)$
4. $(1, 0), (0, 1)$
5. $(a, 0), (0, 2a)$
6. $(x_1, y_1), (x_2, y_2)$

In each of Exercises 7–12 find an equation of the straight line passing through the given point, and with the given slope.

7. $(0, -1)$, slope $\tfrac{1}{2}$
8. $(0, 3)$, slope $2$
9. $(-3, 2)$, slope $-2$
10. $(4, -2)$, slope $\tfrac{2}{3}$
11. $(a, 0)$, slope $m$
12. $(0, -b)$, slope $m$

In each of Exercises 13–20 find an equation of the straight line which satisfies the given conditions.

13. Passes through the point $(-3, 5)$ and is parallel to the line $7x - 2y = 4$.
14. Passes through the point $(5, -2)$ and is parallel to the line $14x + 3y + 7 = 0$.
15. Passes through the point $(-3, 5)$ and is perpendicular to the line $7x - 2y = 4$.
16. Passes through the point $(5, -2)$ and is perpendicular to the line $14x + 3y + 7 = 0$.
17. Passes through the point $(4, \tfrac{1}{2})$ and has $y$ intercept $2$.
18. Passes through the point $(4, \tfrac{1}{2})$ and has $x$ intercept $2$.
19. Passes through the point $(2, 3)$ and with $x$ intercept twice its $y$ intercept.
20. Passes through the point $(2, 5)$, with the sum of the $x$ and $y$ intercepts equal to zero.

Exercises 21–29 apply to the triangle with vertices $A(2, -3)$, $B(4, 5)$, and $C(-2, -1)$. In each find an equation of the specified line.

21. The median to $AB$
22. The median to $AC$
23. The median to $BC$
24. The altitude to $AB$
25. The altitude to $AC$
26. The altitude to $BC$
27. The perpendicular bisector of $AB$.
28. The perpendicular bisector of $AC$.
29. The perpendicular bisector of $BC$.
30. Show that for any value of $k$ the line $2x - 3y = k$ is perpendicular to the line $6x + 4y - 3 = 0$.
31. Show that the lines $A_1 x + B_1 y + C_1 = 0$, $A_2 x + B_2 y + C_2 = 0$, with all constants different from zero,
    (a) are parallel if and only if $A_1/A_2 = B_1/B_2$,
    (b) are coincident if and only if $A_1/A_2 = B_1/B_2 = C_1/C_2$,
    (c) are perpendicular if and only if $A_1 A_2 + B_1 B_2 = 0$.
32. Show that an equation of the line through $(x_1, y_1)$ and parallel to $ax + by = c$ is $ax + by = ax_1 + by_1$.
33. Show that an equation of the line through $(x_1, y_1)$ and perpendicular to $ax + by = c$ is $bx - ay = bx_1 - ay_1$.
★ 34. Show that the (perpendicular) distance between the line $L$: $Ax + By + C = 0$ and a point $P_0(x_0, y_0)$ is given by

$$d = \frac{|Ax_0 + By_0 + C|}{\sqrt{A^2 + B^2}}.$$

[Suggestion: Choose any point $P_1(x_1, y_1)$ on $L$. Now $\mathbf{N} = (A, B)$ is a vector perpendicular to $L$. Show that $d = |\mathbf{N} \cdot \overrightarrow{P_1 P_0}|/|\mathbf{N}|$. Then show that $\mathbf{N} \cdot \overrightarrow{P_1 P_0} = Ax_0 + By_0 + C$, and obtain the desired formula.]

## 2.9 INTERSECTIONS OF STRAIGHT LINES

Two straight lines in a plane either (a) are coincident, (b) are parallel and distinct, or (c) have exactly one point of intersection. Let equations of the lines be

(2.9.1) $\qquad A_1 x + B_1 y + C_1 = 0, \qquad A_2 x + B_2 y + C_2 = 0.$

In case (a) the equations have all solutions in common (infinitely many in number), are called *dependent*, and (Exercise 31 of Exercise Group 2.6)

(2.9.2) $\qquad\qquad \dfrac{A_1}{A_2} = \dfrac{B_1}{B_2} = \dfrac{C_1}{C_2}.$

If a coefficient of an equation in (2.9.1) is zero, Equation (2.9.2) is to **mean** simply that the corresponding coefficient in the other one is also zero. For example, if $B_1 = 0$, then $B_2 = 0$ and (2.9.2) becomes

$$\frac{A_1}{A_2} = \frac{C_1}{C_2}, \qquad B_1 = B_2 = 0.$$

In case (b) the equations (2.9.1) have no common solution and are called *inconsistent*. In this case

$$\frac{A_1}{A_2} = \frac{B_1}{B_2} \neq \frac{C_1}{C_2}.$$

If $C_2 = 0$ here, then $C_1 \neq 0$, and so on.

In case (c) *the equations of the lines have one solution in common, the coordinates of the point of intersection of the lines.*

**Example 2.9.1.** Find the point of intersection of the lines

$$4x + 2y + 1 = 0, \qquad 5x - 3y - 7 = 0.$$

*Solution.* We may eliminate $y$ by adding three times the first equation to twice the second one. This gives $22x - 11 = 0$, or $x = 1/2$. Substituting this value into either of the given equations, we then find $y = -3/2$. The point of intersection of the lines is $(1/2, -3/2)$.

**Example 2.9.2.** (a) The equations

$$2x - 3y = 2, \qquad 4x - 6y = 4$$

are dependent, and represent the same straight line.

(b) The equations

$$2x - 3y = 2, \qquad 4x - 6y = 2$$

are inconsistent; that is, they have no common solution, and represent distinct, parallel straight lines.

**EXERCISE GROUP 2.7**

In Exercises 1–9 find the intersection of the given pair of lines.

1. $2x - 3y + 5 = 0$
   $3x + 2y = 12$
2. $3x + 5y + 1 = 0$
   $y - 2x = 5$
3. $2x - 3y = 1$
   $30x - 42y = 17$
4. $5x + 7y = 1$
   $2x - 3y = 12$
5. $4x - 3y = 3$
   $-7x + 4y = 1$
6. $\dfrac{3x}{2} + 1 = -\dfrac{4y}{3}$
   $8x + 3y = 7$
7. $\frac{2}{3}x - \frac{3}{4}y = \frac{3}{2}$
   $10x + 21y = 1$
8. $3x + 5y - 9 = 0$
   $5y - \frac{7}{2}x + 4 = 0$
9. $4y + 6x + 9 = 0$
   $4x + 6y + 1 = 0$

In each of Exercises 10–13 show that the three lines are concurrent, that is, meet in a point.

10. $2x - 3y - 3 = 0$, $x + 5y - 8 = 0$, $4x - y - 11 = 0$
11. $x - 5y + 10 = 0$, $7x - 2y + 1 = 0$, $2x + y - 3 = 0$
12. $2x + y + 3 + 0$, $x - 2y + 1 = 0$, $4x + 7y + 7 = 0$
13. $2x + 3y - 1 = 0$, $8x + 7y - 11 = 0$, $x + 4y + 3 = 0$
14. For the triangle with vertices $(-3, -5)$, $(1, 1)$, and $(5, -1)$:
    (a) find the intersection of the medians
    (b) find the intersection of the altitudes
    (c) find the intersection of the perpendicular bisectors
* 15. Given the parallelogram with vertices $(0, 0)$, $(a, 0)$, $(b, c)$, and $(b - a, c)$, show that the lines drawn from the origin to the midpoints of the opposite sides trisect a diagonal.
* 16. (a) If $P(x, y)$ is a point on the line segment joining $P_0(x_0, y_0)$ and $P_1(x_1, y_1)$, show that for some number $t$, $0 \le t \le 1$, we must have

$$x = x_0 + t(x_1 - x_0), \qquad y = y_0 + t(y_1 - y_0).$$

[Hint: We must have $\overrightarrow{P_0P} = t\,\overrightarrow{P_0P_1}$.]

(b) Show that the line segment $P_0P_1$ of (a) has a point in common with the line $Ax + By + C = 0$, if and only if for some number $t$, $0 \le t \le 1$,

$$(1 - t)(Ax_0 + By_0 + C) + t(Ax_1 + By_1 + C) = 0.$$

* 17. Deduce from (b) of Exercise 16 that $P_0(x_0, y_0)$ and $P_1(x_1, y_1)$ are on the same side of the line $Ax + By + C = 0$, if and only if $Ax_0 + By_0 + C$ and $Ax_1 + By_1 + C$ have the same sign, and are on opposite sides, if and only if these expressions have different signs.

## 2.10 LINEAR INEQUALITIES IN $x$ AND $y$

It can be shown (it follows for example from Exercise 17 of Exercise Group 2.7) that the straight line

(2.10.1) $$Ax + By + C = 0$$

divides the points of the plane which are not on the line into two parts or regions, in one of which $Ax + By + C > 0$ and in the other $Ax + By + C < 0$. Thus in order to determine which inequality applies for *all* the points in one of these regions, it is sufficient to test a single point.

*Example 2.10.1.* The straight line $3x + 4y - 12 = 0$ is shown in Fig. 2.10.1. One of the regions into which this line divides the plane contains the origin. Substituting $x = 0$, $y = 0$ in the expression $3x + 4y - 12$ we find

$$3 \cdot 0 + 4 \cdot 0 - 12 < 0.$$

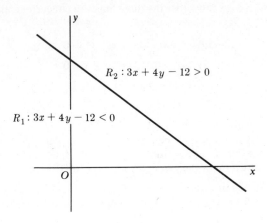

Fig. 2.10.1

Hence the inequality $3x + 4y - 12 < 0$ is satisfied by *every* point in the region containing the origin, labeled $R_1$ in the figure, and $3x + 4y - 12 > 0$ for every point in $R_2$.

Let us write (2.10.1) as $f = 0$, where $f = Ax + By + C$. We may say that $f < 0$ for every point of one of the half-planes determined by the line $f = 0$, and $f > 0$ for every point of the other half-plane. In other words, the line $f = 0$ divides the plane into two regions, in one of which $f < 0$ and in the other of which $f > 0$.

If two lines

$$f_1 = A_1 x + B_1 y + C_1 = 0, \qquad f_2 = A_2 x + B_2 y + C_2 = 0,$$

intersect, *four* distinct regions are formed. Each of these regions may be characterized by the signs of $f_1$ and $f_2$ for points in it. For example, in one region we have $f_1 > 0, f_2 > 0$; in a second region we have $f_1 > 0, f_2 < 0$; and so on. An illustration of this is the separation of the plane into quadrants by the coordinate axes.

We shall consider, by means of an example, a more general case where three lines are involved.

**Example 2.10.2.** Show on a graph the region of the plane for which the following inequalities hold:

$$f_1 = 2x + y - 5 < 0, \qquad f_2 = x - y + 6 > 0, \qquad f_3 = 2x + 6y - 9 > 0.$$

*Solution.* Figure 2.10.2 shows the regions of the plane formed by the lines $f_1 = 0, f_2 = 0, f_3 = 0$, and the combinations of signs of $f_1, f_2$ and $f_3$ in each of these regions. We shall describe a systematic procedure for arriving at this figure.

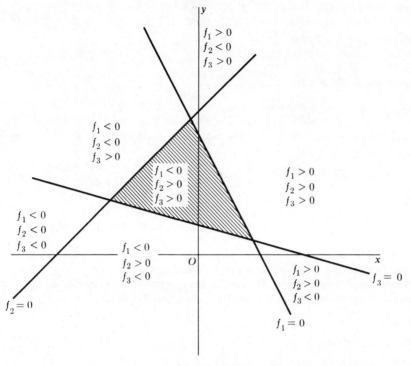

Fig. 2.10.2

The three lines $f_1 = 0$, $f_2 = 0$, and $f_3 = 0$ are drawn and labeled. We fix our attention on $f_1$. For $x = 0$, $y = 0$ we find $f_1 < 0$. Hence all the regions on the same side of $f_1 = 0$ as the origin are designated as $f_1 < 0$ (four of them), and the regions on the other side of $f_1 = 0$ are designated as $f_1 > 0$ (three of them).

Similarly, for $x = 0$, $y = 0$ we find $f_2 > 0$. Hence the regions on the same side of $f_2 = 0$ as the origin are designated as $f_2 > 0$, and those on the other side of $f_2 = 0$ are designated as $f_2 < 0$. The same is done for $f_3$. In this way Fig. 2.10.2 was obtained.

The region for which $f_1 < 0$, $f_2 > 0$, $f_3 > 0$ may now be identified, and is shaded in the figure.

The use of the origin in Example 2.10.2 was a matter of convenience. Any other point may be used for the same purpose. If a line passes through the origin, some convenient point other than the origin will be chosen for this purpose.

The kind of problem considered in Example 2.10.2 is met often in certain applications of mathematics, frequently with many more than three inequalities. The set of points, if any, for which a group of such (linear) inequalities holds, must have the following property: *if the points $P_1$ and $P_2$ are in the set, then every point of the line segment $P_1P_2$ must lie in the set* (see Exercises 14 and 15).

A set of points with this property is called *convex*. The interior of a triangle is a convex set (for example, the shaded region in Fig. 2.10.2). In fact, each region in Fig. 2.10.2 is convex.

**EXERCISE GROUP 2.8**

In each of Exercises 1–4 two expressions of the form $f = Ax + By + C$ are given. Each equation $f_i = 0$ determines a straight line. Draw the lines, and designate for each region formed by these lines, the signs of the expressions $f_i$. In each case shade the region, if any, for which both $f_1 < 0$ and $f_2 > 0$.

1. $f_1 = 2x - y + 5,\quad f_2 = 3x + 2y - 3$
2. $f_1 = x - y - 1,\quad f_2 = x - y + 2$
3. $f_1 = x - 3y + 1,\quad f_2 = x + 2y$
4. $f_1 = 2x - 3y,\quad f_2 = 4x + y$

In each of Exercises 5–12 three expressions of the form $f = Ax + By + C$ are given. Each equation $f_i = 0$ determines a straight line. Draw the lines, and designate for each region formed by these lines, the signs of the expressions $f_i$, as in Fig. 2.10.2. In each case shade the region, if any, for which $f_1 < 0, f_2 > 0, f_3 < 0$.

5. $f_1 = 8x - 3y - 7,\quad f_2 = 2x + y - 7,\quad f_3 = 2x - 5y + 11$
6. $f_1 = x + y,\quad f_2 = x - y,\quad f_3 = 2x - y$
7. $f_1 = 2x - y,\quad f_2 = 2x - y - 1,\quad f_3 = 2x - y + 2$
8. $f_1 = 4x - 3y + 4,\quad f_2 = 3x + 4y - 5,\quad f_3 = x + y - 2$
9. $f_1 = 4x - y + 2,\quad f_2 = 5x - 9y - 4,\quad f_3 = 7x + 3y + 2$
10. $f_1 = 2 - x + y,\quad f_2 = 3x - 2y,\quad f_3 = 6x - y + 7$
11. $f_1 = 2x + 5,\quad f_2 = 9 + x - 2y,\quad f_3 = 3y - 7$
12. $f_1 = 3x - 5y - 9,\quad f_2 = 4x + 5y + 7,\quad f_3 = x - 2y$

★ 13. In Fig. 2.10.2 for Example 2.10.2, it is seen that there is no region for which $f_1 > 0, f_2 < 0, f_3 < 0$. Show *analytically* (without the use of a figure) that these inequalities cannot be satisfied simultaneously.

14. If $R_1$ and $R_2$ are convex regions which overlap, show that the set $R$, consisting of those points common to $R_1$ and $R_2$, is also convex.

15. Draw a closed curve such that the points interior to the curve do not form a convex set.

# 3 Functions and Limits

## 3.1 FUNCTIONS

The concept of *function* is basic in much of mathematics. In broad terms it concerns a correspondence between two classes of objects which may be of any types whatsoever. Many experiences of everyday living reflect the idea of correspondence, or relationship, as we wish to use it. In a given day the temperature bears a definite relation to the time. The yield of an acre of wheat clearly depends on the amount of rainfall. The speed of an airplane is certainly dependent on the amount of fuel fed to the engines. The strength of a wooden beam corresponds to the thickness of the beam. Such examples can be multiplied indefinitely. Each one of them exhibits correspondence between two specific quantities.

The concept can be extended to include a correspondence among three or more quantities. Thus the yield of an acre of wheat depends on temperature as well as rainfall. The speed of an airplane depends on its altitude as well as the fuel consumed. In large part, however, we shall be concerned with relations or correspondences between sets of numbers.

The definition of *function* as used in mathematics is similar to that of relation (Section 2.6), but with an important distinction. In fact, a function is a special type of relation.

**DEFINITION 3.1.1.** *A **function** is a set of number pairs $(a, b)$ such that no distinct number pairs have the same first number $a$. The set of all first numbers $a$ is called the **domain** of the function. The set of all second numbers $b$ is called the **range** of the function.*

*Example 3.1.1.* The number pairs

$$(2, 3), (-1, 4), (3, -5), (4, 3)$$

define a *function* with the set of numbers $2, -1, 3, 4$ as the *domain* and the set of numbers $3, 4, -5$ as the *range*. Note that a number in the range may correspond to more than one number in the domain. The given number pairs also define a relation.

The set of number pairs

$$(1, 3), (2, 4), (1, 7), (-2, 0)$$

defines a relation, but not a function, since both 3 and 7 correspond to the same number, 1.

The second number of any of the pairs in a function is called the *value of the function* corresponding to the first number of the pair. If the function is denoted by $f$, the value of the function corresponding to any number $x$ in the domain is written as $f(x)$.

If $f$ denotes the function in Example 3.1.1, we have

$$f(2) = 3, \quad f(-1) = 4, \quad f(3) = -5, \quad f(4) = 3.$$

The value of a function may be given by a mathematical expression. If we write, for example,

$$f(x) = 2x - 3,$$

then for any value of $x$ the value of the function is $2x - 3$. In particular,

$$f(3) = 2 \cdot 3 - 3 = 3, \quad f(-4) = 2(-4) - 3 = -11,$$

and so on. The function may be defined as the set of number pairs $(x, 2x - 3)$, where the domain in which $x$ lies is the set of real numbers. We may also say, a bit more loosely, that $2x - 3$ is a function of $x$.

In general, if $f(x)$ is the value of a function corresponding to $x$, we also say that $f(x)$ is a function of $x$, for $x$ in some stated or implied domain. In the latter case, *the domain is usually taken to consist of all real numbers for which the values $f(x)$ exist*. (See Exercises 3–8.)

The graph of a function $f$ is the set of points in the coordinate plane given by $(x, f(x))$ for $x$ in the domain of $f$. It then follows that *the graph of the function $f$ is the graph of the relation*

$$\{(x, y) \mid y = f(x), x \text{ in } A, \text{ the domain}\}.$$

The graph of the function $f$ as thus defined is also called the graph of the equation $y = f(x)$ for $x$ in $A$.

**Example 3.1.2.** The graph of the function $2x - 3$ is the graph of the equation $y = 2x - 3$. It is therefore a straight line with slope 2 and $y$ intercept $-3$ (slope-intercept form).

The function $f(x) = mx + b$, $m$ and $b$ constant, is called a *linear function*, because its graph, or the graph of the equation $y = mx + b$, is a straight line, as in Example 3.1.2.

While it is true that many, perhaps most, of the functions which will concern us are defined by the use of explicit mathematical expressions, this feature is not inherent in the function concept. We give some illustrations.

**Example 3.1.3.** Let $f$ be defined in the interval $[0, 1]$ so that

$$f(x) = \begin{cases} 1 \text{ if } x \text{ is rational,} \\ 0 \text{ if } x \text{ is irrational.} \end{cases}$$

The function $f$ as thus defined is perfectly acceptable; its graph cannot be drawn in the usual way.

**Example 3.1.4.** Let $g$ be defined for $x \geq 0$ so that

$$g(x) = 10 \text{ if } 0 \leq x \leq 1,$$
$$g(x) = 20 \text{ if } 1 < x \leq 2,$$
$$g(x) = 30 \text{ if } 2 < x \leq 3,$$
$$\vdots$$

This function is sometimes called the *postage function*, and its graph is shown in Fig. 3.1.1.

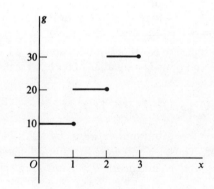

Fig. 3.1.1

## EXERCISE GROUP 3.1

1. (a) Draw the graph of the function $f$: $[(1, 3), (2, 2), (3, -1), (5, 2)]$.
   (b) Find $f(2)$ and $f(5)$.   (c) Find $x$ if $f(x) = -1$.

2. (a) Draw the graph of the function $g : [(1, 2), (2, 3), (3, 4), (4, 5)]$.
   (b) Find $g(1)$ and $g(3)$.   (c) Find $x$ if $g(x) = 3$.

In Exercises 3–8 determine the domain of the function represented by the given expression.

3. $\dfrac{x-2}{x}$

4. $\dfrac{x+1}{2x+3}$

5. $\dfrac{x-1}{x^2+1}$

6. $\dfrac{1}{\sqrt{4-x}}$

7. $\dfrac{1}{\sqrt{4-x^2}}$

8. $\dfrac{2x}{x^2+2x-3}$

In Exercises 9–16 the given number pairs define a function $f$ with domain consisting of all real numbers for which the value of $f$ is defined. In each case draw the graph of $f$.

9. $(x, 2x)$

10. $(x, -3x)$

11. $(x, -2x-1)$

12. $(x, -x+5)$

13. $(x, 3x^2)$

14. $(x, x^2-x)$

15. $(x, x^2-3x+6)$

16. $(x, 2x^2-5)$

17. Find a function (in terms of $x$) which represents the distance from any point on the curve $y = x^2$ to (a) the origin, (b) the point $(-2, 3)$.

18. Find a function (in terms of $x$) which represents the slope of the line joining any point of the curve $y = (x-1)/(x+1)$ to (a) the origin, (b) the point $(4, -3)$.

19. Let $P$ be any point on the curve $y = f(x)$. Find an equation of the *locus* of the midpoint of $OP$, that is, of the set of midpoints $OP$ for all points $P$.

20. Let $P$ be any point on the curve $y = f(x)$, let $M$ be the projection of $P$ on the $x$ axis, and let $N$ be the projection of $P$ on the $y$ axis. (a) Find an equation of the locus of the midpoint of $MP$. (b) Find an equation of the locus of the midpoint of $NP$. (c) Find an equation of the locus of the midpoint of $MN$.

21. Let $y = f(x)$ and $y = g(x)$ be equations of two curves. If a line parallel to the $y$ axis intersects them in $P$ and $Q$, find an equation of the locus of the midpoint of $PQ$.

## 3.2 COMBINATIONS OF FUNCTIONS

Functions may be combined in familiar ways, by addition, subtraction, multiplication, and division. Thus, if $f$ and $g$ are functions defined in the same domain $A$, the *sum*, *difference*, *product*, and *quotient* of $f$ and $g$ are defined, respectively, as

$$f \pm g : \{(x, y) \,|\, y = f(x) \pm g(x), x \text{ in } A\};$$
$$fg : \{(x, y) \,|\, y = f(x)g(x), x \text{ in } A\};$$
$$f/g : \{(x, y) \,|\, y = f(x)/g(x), x \text{ in } A \text{ and } g(x) \neq 0\}.$$

Thus the domain of $f \pm g$ and of $fg$ is the set $A$ of numbers common to the domains of $f$ and $g$ (see Section 3.1), while the domain of $f/g$ is that set less those numbers, if any, for which $g(x) = 0$.

*Example 3.2.1.* Given the functions
$$f: [(1, 3), (2, 2), (3, -1), (5, 2)],$$
$$g: [(1, -1), (2, 4), (3, 2), (5, -1)],$$
we have
$$f + g: [(1, 2), (2, 6), (3, 1), (5, 1)];$$
$$f/g: [(1, -3), (2, 1/2), (3, -1/2), (5, -2)].$$

*Example 3.2.2.* If $f(x) = x^2 - 1$, $g(x) = 2x + 3$, we have
$$f(x) - g(x) = x^2 - 1 - (2x + 3) = x^2 - 2x - 4, \quad \text{all real } x;$$
$$f(x)/g(x) = \frac{x^2 - 1}{2x + 3}, \quad x \neq -3/2.$$

An important way of combining functions is to use, for members of the domain of one function, values of a second function. Specifically, the *composite* of $f$ by $g$, written as $f \circ g$, is defined as

$$f \circ g : \{(x, y) \,|\, y = f[g(x)], \quad x \text{ in the domain of } g, g(x) \text{ in the domain of } f\}.$$

Thus *the domain of $f \circ g$ is the set of numbers $x$ such that, at the same time, $x$ is in the domain of $g$ and $g(x)$ is in the domain of $f$.*

*Example 3.2.3.* Determine $f[g(x)]$ and $g[f(x)]$ and their domains, if
$$f(x) = \frac{1}{x - 3}, \quad g(x) = \sqrt{x}.$$

*Solution.* For $f[g(x)]$ we must have $x \geq 0$ in order for $x$ to be in the domain of $g$, and also $g(x) \neq 3$ in order for $g(x)$ to be in the domain of $f$. Thus the domain of $f \circ g$ consists of all numbers $x$ such that $x \geq 0$, $x \neq 9$. Then we have

$$f[g(x)] = \frac{1}{g(x) - 3} = \frac{1}{\sqrt{x} - 3}, \quad x \geq 0, x \neq 9.$$

For $g[f(x)]$ we must have $x \neq 3$ for $x$ to be in the domain of $f$, and $f(x) \geq 0$ for $f(x)$ to be in the domain of $g$. Hence the domain of $g \circ f$ is the set of numbers greater than 3, and

$$g[f(x)] = \sqrt{f(x)} = \frac{1}{\sqrt{x - 3}}, \quad x > 3.$$

We now illustrate a use of composite functions which is of much importance in later work.

*Example 3.2.4.* Express the function $\sqrt{x^2 + 1}$ as a composite $f[g(x)]$ of two functions $f$ and $g$.

*Solution.* Let $f(x) = \sqrt{x}$, $g(x) = x^2 + 1$. Then
$$f[g(x)] = \sqrt{g(x)} = \sqrt{x^2 + 1}.$$
Thus $f$ and $g$ as given are a solution.

Note that the solution to a problem such as Example 3.2.4 is *not unique*. We may also write
$$\sqrt{x^2 + 1} = \sqrt[4]{(x^2 + 1)^2}.$$
Hence $f(x) = \sqrt[4]{x}$, $g(x) = (x^2 + 1)^2$ is another solution (and so is $f(x) = x$, $g(x) = \sqrt{x^2 + 1}$). The goal in such a problem is usually to find "simplest" functions.

**EXERCISE GROUP 3.2**

1. Define the functions $f$ and $g$ as follows:
$$f: [(0, 2), (1, 3), (2, 4), (3, 5)],$$
$$g: [(0, -1), (1, 1), (2, -5), (3, 6)].$$
   Give the functions whose values are
   (a) $2f(x)$, (b) $f(x) + g(x)$, (c) $f(x) - g(x)$, (d) $[f(x)]^2$

In Exercises 2–7 the functions $f$ and $g$ are defined by the number pairs
$$f: (x, x^2 + x), \qquad g: (x, x - 1),$$
where the domain of each function is the set of real numbers. Find the indicated values.

2. $f[g(1)]$   3. $g[f(2)]$   4. $f[g(a)]$   5. $g[f(a)]$

6. (a) $g(-x)$, (b) $g(x + h)$, (c) $\dfrac{1}{h}[g(x + h) - g(x)]$

7. (a) $f(-x)$, (b) $f(x + h)$, (c) $\dfrac{1}{h}[f(x + h) - f(x)]$

In Exercises 8–13 determine (a) the function $f[g(x)]$ and its domain, (b) the function $g[f(x)]$ and its domain.

8. $f: [(1, 3), (2, 2), (3, -1), (5, 2)]$;  $g: [(1, 2), (2, 3), (3, 4), (4, 5)]$

9. $f: [(2, 3), (-1, 4), (3, -5), (4, 3)]$;  $g: [(1, 3), (2, 4), (3, -1), (-2, 0)]$

10. $f(x) = 2x - 3$,  $g(x) = \sqrt{x}$

11. $f(x) = \dfrac{1}{x}$,  $g(x) = \dfrac{1}{x + 1}$

12. $f(x) = x^2$,  $g(x) = \sqrt{x}$

13. $f(x) = x^3$,  $g(x) = \sqrt[3]{x}$

14. If $f(x) = 2x - 3$, find $f(2x), f(x^2), f(2x - 3), f[(x - 1)(x + 1)]$

**15.** If $g(x) = \dfrac{x+3}{x-2}$, find $g(x+2)$, $g(x^2+1)$, $g[(x-1)/(x+1)]$

**16.** If $f$ is the function defined in Example 3.1.3, find $f(f)$.

**17.** If $g$ is the function defined in Example 3.1.4, find $g[g(1)]$ and $g[g(2)]$.

In each of the following express the given function as a composite function $f[g(x)]$. (See Example 3.2.4 and the comment following it.)

**18.** $(x+2)^2$

**19.** $(2x^2+3x-1)^{17}$

**20.** $3(2x)^3 - (2x)^2 + 4x - 1$

**21.** $\dfrac{1}{x^2 + x}$

**22.** $\sqrt{x-1}$

**23.** $\sqrt[3]{\dfrac{x-1}{x+1}}$

## 3.3 INTRODUCTION TO LIMITS

Perhaps the most fundamental question concerning a function is simply the value of the function for a specified number in the domain. More specifically, is a number $a$ in the domain of a function $f$, and, if so, what is $f(a)$? A deeper question concerns the behavior of a function $f(x)$ for values of $x$ near a specified number $a$ whether or not $a$ is in the domain, apart from the value of the function at that specified number. This question leads directly to the *limit concept* which is at the heart of the study of calculus and which is amplified in what follows. We discuss this concept in a nonrigorous way at this time, leaving the more rigorous development to later sections.

For example, consider the function

$$g(x) = 3x - 2.$$

We have $g(2) = 4$. In addition, it appears to be a reasonable statement that the closer $x$ is to 2, the closer the value $g(x)$ is to 4. It is the latter behavior which we describe by writing

$$\lim_{x \to 2} g(x) = 4;$$

we read this as "the limit of $g(x)$ as $x$ approaches 2 is (or equals) 4." Similarly, we have

$$\lim_{x \to -3} g(x) = -11, \qquad \lim_{x \to a} g(x) = 3a - 2.$$

In the same vein let us examine the function

$$\frac{x}{x^2 + 4}$$

near the value $x = 3$. As $x$ approaches 3, the numerator gets closer to 3, the denominator gets closer to $3^2 + 4 = 13$, and the quotient of $x$ by $x^2 + 4$ gets closer to 3/13. We then have

$$\lim_{x \to 3} \frac{x}{x^2 + 4} = \frac{3}{13}.$$

We note that 3/13 is also the value of the function at $x = 3$.

In fact, it will be seen later that for the functions defined by the usual mathematical expressions a limit can be obtained, for the most part, by evaluating the function at the specified number. We shall use this observation for purposes of illustration in this introduction to limits, with the reservation that, to be rigorous, more precise statements with proofs are necessary.

An example where a limit does not exist is helpful here. Consider the function

$$f(x) = \frac{x}{|x|}, \quad x \neq 0.$$

Since $|x| = x$ if $x > 0$, and $|x| = -x$ if $x < 0$, we have

$$f(x) = 1 \text{ if } x > 0, \quad f(x) = -1 \text{ if } x < 0.$$

There is no point of the graph for $x = 0$. For $x$ close to zero, $f(x) = 1$ for some values and $-1$ for some values (Fig. 3.3.1). There can be no *single* number that the values of $f(x)$ are close to for $x$ near zero. Hence the limit of $f(x)$ as $x$ approaches zero does not exist. It should be clear, however, that $\lim_{x \to a} f(x)$ does exist if $a \neq 0$.

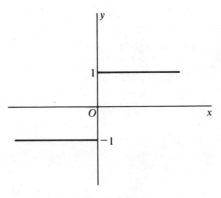

Fig. 3.3.1

We now consider the function

$$f(x) = \frac{x - 2}{x^2 - 4}, \quad x \neq \pm 2.$$

This function is undefined at $x = 2$, since the denominator vanishes (is zero) there. (The function is undefined also at $x = -2$, but we are here concerned only with its behavior near $x = 2$.) However, if $x$ is near 2, say $x > 0$ but not equal to 2, we may divide numerator and denominator by $x - 2 \neq 0$, and write

$$f(x) = \frac{1}{x + 2}, \quad x > 0 \text{ and } x \neq 2.$$

Therefore, if $x \neq 2$ and $x$ is near 2, the value of $f(x)$ is given by $1/(x+2)$, and the closer $x$ is to 2 the closer $f(x)$ is to $1/4$. We have

$$\lim_{x \to 2} \frac{x-2}{x^2-4} = \frac{1}{4}.$$

The example just completed should be studied carefully. The function was *not defined* for $x = 2$; however, for $x$ near 2 it could be represented by a different expression, and the limit could be found.

We emphasize that, in general, the existence of $\lim_{x \to a} f(x)$ is not affected by the existence of $f(a)$, or by the value of $f(a)$ if it exists. If the limit happens to equal $f(a)$, special considerations apply (Section 3.6).

**Example 3.3.1.** Evaluate: $\lim_{h \to 0} \dfrac{(2+h)^3 - 2^3}{h}$.

*Solution.* As $h$ gets close to zero, the numerator and denominator both get close to zero, and there is no indication of any number that the fraction approaches. We rewrite the given function as

(3.3.1) $\quad \dfrac{(2+h)^3 - 2^3}{h} = \dfrac{8 + 12h + 6h^2 + h^3 - 8}{h} = 12 + 6h + h^2, \quad h \neq 0.$

Since the desired limit depends only on values of the function for values of $h$ near zero and in no way on what happens at $h = 0$, and since (3.3.1) holds for $h \neq 0$, we have

$$\lim_{h \to 0} \frac{(2+h)^3 - 2^3}{h} = \lim_{h \to 0} (12 + 6h + h^2).$$

As $h \to 0$, it appears that the value of $12 + 6h + h^2$ approaches 12, and we have

$$\lim_{h \to 0} \frac{(2+h)^3 - 2^3}{h} = 12.$$

**Example 3.3.2.** Evaluate: $\lim_{x \to 4} \dfrac{\sqrt{x} - 2}{x - 4}$.

*Solution.* We write

$$\frac{\sqrt{x} - 2}{x - 4} = \frac{\sqrt{x} - 2}{(\sqrt{x} + 2)(\sqrt{x} - 2)} = \frac{1}{\sqrt{x} + 2}, \quad x > 0, x \neq 4.$$

Therefore,

$$\lim_{x \to 4} \frac{\sqrt{x} - 2}{x - 4} = \lim_{x \to 4} \frac{1}{\sqrt{x} + 2} = \frac{1}{2 + 2} = \frac{1}{4}.$$

We changed the form of the given function in order to be able to evaluate the limit. We emphasize again that the limit does not depend on what occurs at $x = 4$.

**Example 3.3.3.** Evaluate: $\lim\limits_{h \to 0} \dfrac{\sqrt{x+h} - \sqrt{x}}{h}$.

*Solution.* We write

$$\frac{\sqrt{x+h} - \sqrt{x}}{h} = \frac{\sqrt{x+h} - \sqrt{x}}{h} \cdot \frac{\sqrt{x+h} + \sqrt{x}}{\sqrt{x+h} + \sqrt{x}} = \frac{1}{\sqrt{x+h} + \sqrt{x}}.$$

Hence

$$\lim_{h \to 0} \frac{\sqrt{x+h} - \sqrt{x}}{h} = \lim_{h \to 0} \frac{1}{\sqrt{x+h} + \sqrt{x}} = \frac{1}{\sqrt{x} + \sqrt{x}} = \frac{1}{2\sqrt{x}}, \quad x > 0.$$

We note that because the original form of the function gave no indication of the limit, a change in form was effected, which consisted of *rationalizing the numerator*. We also note that the limit obtained applies for any value $x > 0$, and may itself be considered as a function of $x$.

### EXERCISE GROUP 3.3

Evaluate the limits in Exercises 1–12

1. $\lim\limits_{x \to -2} (x^2 + 3)$

2. $\lim\limits_{x \to 3} (2x^2 - 7)$

3. $\lim\limits_{x \to 2} \dfrac{x^2 + 1}{x^2 - 1}$

4. $\lim\limits_{x \to -2} \dfrac{2x^2 - 3}{x^2 - 7}$

5. $\lim\limits_{x \to 4} \dfrac{\sqrt{x}}{\sqrt{x+4}}$

6. $\lim\limits_{z \to 9} \dfrac{\sqrt{z}+2}{\sqrt{z}-2}$

7. $\lim\limits_{x \to \sqrt{3}} \dfrac{x^4 - 2x^2 + 4}{x^2 + 3}$

8. $\lim\limits_{x \to -\sqrt{2}} \dfrac{3x^2 - 2}{x^4 + 4}$

9. $\lim\limits_{x \to 16} \dfrac{x^{1/4} + x^{1/2}}{x + x^{1/2}}$

10. $\lim\limits_{x \to 27} \dfrac{x^{1/3} - 3}{x^{1/3} + 4}$

11. $\lim\limits_{y \to 4} \dfrac{y^{1/2} + y - 6}{y^{1/2} - 3}$

12. $\lim\limits_{s \to 64} \dfrac{s^{1/2} - s^{1/3}}{s^{1/2} + s^{1/3}}$

Evaluate the limits in Exercises 13–20 by first making, as needed, a change in form.

13. $\lim\limits_{x \to 1} \dfrac{x - 1}{x^2 - 1}$

14. $\lim\limits_{x \to 2} \dfrac{x - 2}{x^3 - 8}$

15. $\lim\limits_{x \to 2} \dfrac{x^2 - 4}{x^3 - 8}$

16. $\lim\limits_{y \to 9} \dfrac{y - 9}{y^{1/2} - 3}$

17. $\lim\limits_{x \to -64} \dfrac{x^{1/3} + 4}{x + 64}$

18. $\lim\limits_{h \to 0} \dfrac{(2+h)^2 - 2^2}{h}$

19. $\lim\limits_{x \to 2} \dfrac{3x^2 - 10x + 8}{x^3 - 2x^2 - 2x + 4}$

20. $\lim\limits_{x \to 1/2} \dfrac{2x^2 - 3x + 1}{2x^2 + x - 1}$

In each of the following evaluate $\lim_{h \to 0} [f(x+h) - f(x)]/h$.

21. $f(x) = 3x - 4$
22. $f(x) = x^2 - 2$
23. $f(x) = 2x^2 - x$
24. $f(x) = 1/x$
25. $f(x) = \sqrt{x}$
26. $f(x) = 1/\sqrt{x}$
27. $f(x) = x^3$
28. $f(x) = 2x^3 - x$
29. $f(x) = \dfrac{x-1}{x+1}$

## 3.4 THE LIMIT CONCEPT

We collect the ideas of the preceding section in the following somewhat intuitive statement:

(3.4.1) $\quad \lim_{x \to a} f(x) = L$ means that the values $f(x)$ can be made to lie as close to $L$ as we choose by restricting $x$ to be close enough, but not equal, to $a$.

The existence of a limit, and its value, are properties of a function "in the small," a phrase which is intended to signify that only values of the function for numbers in the domain near a given value are involved. The meanings of "near" and "close" are vague, however; no matter how close numbers are to a given number, we must be concerned with numbers that are still closer. A mathematically precise treatment cannot be based on a statement which is that loose. For this reason the following formal definition of limit is given.

**DEFINITION 3.4.1.** *Let $a$ lie in an open interval which, except possibly for $a$ itself, is in the domain of a function $f$. Then*

$$\lim_{x \to a} f(x) = L,$$

*if, given any number $\epsilon > 0$, a number $\delta$ exists such that for $x$ in the domain of $f$,*

(3.4.2) $\quad |f(x) - L| < \epsilon \quad \text{whenever} \quad 0 < |x - a| < \delta.$

The statement (3.4.1) may be considered as a less precise formulation of this definition. In either form, certain important features are clear.

1. It is the closeness to the limit $L$ which is prescribed; this then restricts the values of $x$.
2. The question of existence, and the value, of a limit depend in no way on the value of the function at $a$ itself.

The closeness of $f(x)$ to $L$ is measured by $|f(x) - L|$, and the fact that this is prescribed corresponds to the fact that $\epsilon$ is given. The fact that the values of $x$ are then restricted corresponds to the fact that $\delta$ of the definition is determined, or shown to exist, after $\epsilon$ is given. Note that the way in which $\epsilon$ determines the closeness of $f(x)$ to $L$, and $\delta$ determines the closeness of $x$ to $a$, is an appropriate inequality in each case. It should also be noted that, if a suitable value of $\delta$ has been found for a given $\epsilon$, *any smaller (positive) number also may serve as a value for $\delta$*. The value of $\delta$ depends in some manner on $\epsilon$, and is often written as $\delta_\epsilon$ or $\delta(\epsilon)$, but, as just noted, $\delta$ is not uniquely determined when $\epsilon$ is given.

It is clear from the discussion that the interest is in smaller and smaller values of $\epsilon$, and that in general, the smaller $\epsilon$ is, the smaller $\delta$ will be. However, it is not necessary to include any of this in the definition.

We do not intend to give a complete treatment of limits; in particular, results will often be stated without proof. However, it is instructive to see how some limits can be established directly from the definition.

***Example 3.4.1.*** Prove: (a) $\lim_{x \to a} k = k$, if $k$ is a constant,

(b) $\lim_{x \to a} x = a$.

*Solution.* (a) Apply Definition 3.4.1. Given $\epsilon > 0$, $\delta$ may be *any* positive number, since (3.4.2) becomes the trivial statement, $|k - k| < \epsilon$ whenever $0 < |x - a| < \delta$.

(b) Apply Definition 3.4.1. Given $\epsilon > 0$, we may choose $\delta = \epsilon$, and (3.4.2) becomes the true statement, $|x - a| < \epsilon$ whenever $0 < |x - a| < \epsilon$.

***Example 3.4.2.*** Prove: $\lim_{x \to 4} (2x - 5) = 3$.

*Solution.* We must show that given $\varepsilon > 0$ we may find $\delta$ so that

$$0 < |x - 4| < \delta \quad \text{implies} \quad |(2x - 5) - 3| < \epsilon.$$

Now

$$|(2x - 5) - 3| = |2x - 8| = 2|x - 4|.$$

In order to have this quantity less than $\epsilon$ it is sufficient to have $|x - 4| < \epsilon/2$; that is, if $|x - 4| < \epsilon/2$, then $2|x - 4| < \epsilon$. Hence we may take $\delta = \epsilon/2$, and the proof is complete.

For some limits the details are less obvious.

***Example 3.4.3.*** Show that $\lim_{x \to 1/2} 1/x = 2$.

*Solution.* We must show that, given $\epsilon > 0$, we may find $\delta$ such that

$$0 < \left|x - \frac{1}{2}\right| < \delta \quad \text{implies} \quad \left|\frac{1}{x} - 2\right| < \epsilon.$$

Now,

(3.4.3) $$\left|\frac{1}{x} - 2\right| = \frac{2|x - 1/2|}{|x|}.$$

We may require initially that $0 < \delta \leq 1/4$. Then if $|x - 1/2| < 1/4$, it is certainly true that $x - 1/2 > -1/4$ and $x > 1/4$, so that $1/|x| < 4$. With this initial restriction on $x$, it follows from (3.4.3) that

$$\left|\frac{1}{x} - 2\right| < 8\left|x - \frac{1}{2}\right|.$$

From this we see that if $|x - 1/2| < \epsilon/8$, then $|1/x - 2| < \epsilon$. We thus take $\delta = \epsilon/8$ if this turns out to be less than $1/4$, and otherwise we take $\delta = 1/4$. Briefly, we may take $\delta$ as the smaller of $1/4$ and $\epsilon/8$, and we may state that $0 < |x - 1/2| <$ (smaller of $\epsilon$ and $1/4$) $\Rightarrow |1/x - 2| < \epsilon$. This completes the proof. (*Note*: the number $1/4$ is of no particular significance; any positive number less than $1/2$ could have been used, but the value of $\delta$ might need to be changed accordingly.)

Properties of limits will be considered in the next section. Examples 3.4.2 and 3.4.3 can be generalized, and furnish particular limits that may be used in applying the general properties. The following theorem provides another particular limit.

**THEOREM 3.4.1.** *If $n$ is a positive integer, then*

$$\lim_{x \to a} x^{1/n} = a^{1/n}.$$

*If $a < 0$ this is meant to apply only if $n$ is odd.*

*Proof.* To illustrate the method, we give a proof for $n = 2$, $a > 0$. The method of proof is motivated by the relation

(3.4.4) $$x^{1/2} - a^{1/2} = \frac{x - a}{x^{1/2} + a^{1/2}}, \quad a > 0.$$

Given $\epsilon > 0$ choose $\delta_1 = a/2$. Then $0 < |x - a| < \delta_1$ implies $x > a/2$, so that

$$x^{1/2} > (a/2)^{1/2} \quad \text{and} \quad x^{1/2} + a^{1/2} > (a/2)^{1/2} + a^{1/2} = N > 0.$$

It follows from (3.4.4) that

(3.4.5) $$|x^{1/2} - a^{1/2}| < \frac{|x - a|}{N}.$$

Now let $\delta_2 = \epsilon N$. Then if (3.4.5) holds, and if also $0 < |x - a| < \delta_2$, we have

(3.4.6) $$|x^{1/2} - a^{1/2}| < \frac{\epsilon N}{N} = \epsilon.$$

Hence, if $\delta$ is chosen as the smaller of $\delta_1$ and $\delta_2$, both (3.4.5) and (3.4.6) hold, and the proof is complete for the special case. We shall not carry out the general proof, but the method of proof just given can be extended to apply in general (Exercises 25, 26).

***Example 3.4.4.*** Prove the falsity of the statement

$$\lim_{x \to 2} (2x - 3) = 3.$$

*Solution.* It is sufficient to show that for even one value of $\epsilon$, no value $\delta$ can be found such that (3.4.2) holds. Let us choose $\epsilon = 1$. Then $|f(x) - 3| < \epsilon$ becomes $|2x - 3 - 3| < 1$, from which we get

$$5/2 < x < 7/2.$$

Now if (3.4.2) holds for any $\delta$, it must also hold for any smaller $\delta$. However, no value of $\delta$ smaller than $1/2$ can work. For, if $|x - 2| < 1/2$, then certainly $x < 5/2$, and we cannot satisfy the requirement $x > 5/2$. Thus Definition 3.4.1 cannot be satisfied, and the above limit statement is false. (See Exercise 27.)

It is noteworthy that the definition of limit, as well as the examples, involve a specific number $L$ which is in question. Then the attempt is made to show that the specific number either satisfies the definition or does not. Clearly, we cannot test all possible numbers in order to determine whether a limit exists. For this reason we shall develop certain properties of limits, as well as certain specific limits, which will enable us to find the limit of a function in many cases of practical interest. We may be guided, in finding a prospective limit, by the intuitive statement (3.4.1) and by the discussion of Section 3.3. We may then attempt to prove that it is, in fact, a limit by applying Definition 3.4.1, or, as is more likely to be the case, by applying the properties to be developed.

## 3.5 PROPERTIES OF LIMITS

The examples of the preceding section illustrate methods of working with limits directly from the definition. However, these methods are rarely used for the purpose of finding specific limits. Generally, they are used to prove general properties of limits, properties which may then be applied to particular functions. We present some of the more useful of these properties in the following theorem, and prove one of them to illustrate the methods of proof.

**THEOREM 3.5.1.** *If* $\lim_{x \to a} f(x) = A$ *and* $\lim_{x \to a} g(x) = B$, *then*

1. $\lim_{x \to a} [kf(x)] = k \lim_{x \to a} f(x) = kA$, *for any constant* $k$;

2. $\lim_{x \to a} [f(x) \pm g(x)] = \lim_{x \to a} f(x) \pm \lim_{x \to a} g(x) = A \pm B$;

3. $\lim_{x \to a} [f(x)g(x)] = \lim_{x \to a} f(x) \cdot \lim_{x \to a} g(x) = AB$;

4. $\lim_{x \to a} \dfrac{f(x)}{g(x)} = \dfrac{\lim_{x \to a} f(x)}{\lim_{x \to a} g(x)} = \dfrac{A}{B}$, *if* $B \neq 0$.

*Proof of Part 3.* We must show that given $\epsilon > 0$ we may find $\delta$ such that

$$|f(x)g(x) - AB| < \epsilon \quad \text{whenever} \quad 0 < |x - a| < \delta.$$

We may write

$$f(x)g(x) - AB = f(x)g(x) - Ag(x) + Ag(x) - AB$$
$$= g(x)[f(x) - A] + A[g(x) - B].$$

The method will be to show that each term on the right can be made small. By properties of inequalities, starting with the triangle inequality, we obtain

$$|f(x)g(x) - AB| \leq |g(x)[f(x) - A]| + |A[g(x) - B]|,$$

and hence

(3.5.1) $\quad |f(x)g(x) - AB| \leq |g(x)||f(x) - A| + |A||g(x) - B|.$

Since $\lim_{x \to a} g(x) = B$, we may find $\delta_1$ such that

$$|g(x) - B| < 1 \quad \text{whenever} \quad 0 < |x - a| < \delta_1;$$

from this it follows that, whenever $0 < |x - a| < \delta_1$,

(3.5.2) $\quad |g(x)| = |[g(x) - B] + B| \leq |g(x) - B| + |B| < 1 + |B|.$

Since $\lim_{x \to a} f(x) = A$, we may also find, for the given $\epsilon$, a number $\delta_2$ such that

(3.5.3) $\quad |f(x) - A| < \dfrac{\epsilon}{2(1 + |B|)} \quad \text{whenever} \quad 0 < |x - a| < \delta_2.$

We may also find $\delta_3$ such that

(3.5.4) $\quad |g(x) - B| < \dfrac{\epsilon}{2(|A| + 1)} \quad \text{whenever} \quad 0 < |x - a| < \delta_3.$

If we now choose $\delta$ as the smallest of $\delta_1$, $\delta_2$, and $\delta_3$, then (3.5.2), (3.5.3), and (3.5.4) all apply, and (3.5.1) gives

$$|f(x)g(x) - AB| < (1 + |B|) \cdot \frac{\epsilon}{2(1 + |B|)} + |A| \cdot \frac{\epsilon}{2(|A| + 1)} = \frac{\epsilon}{2} + \frac{\epsilon}{2} = \epsilon$$

whenever $0 < |x - a| < \delta$. This completes the proof.

The properties in Theorem 3.5.1 clearly extend to sums and products of more than two functions. (A formal proof can be given with the use of mathematical induction.)

**Example 3.5.1.** Prove: $\lim_{x \to a} (mx + b) = ma + b$, $m$ and $b$ constant.

*Solution.* We have (supply the reasons!)

$$\lim_{x \to a} (mx + b) = \lim_{x \to a} (mx) + \lim_{x \to a} b$$
$$= m \lim_{x \to a} x + b = ma + b.$$

**Example 3.5.2.** Prove: $\lim\limits_{x \to a} 1/x = 1/a$ if $a \neq 0$.

*Solution.* We have
$$\lim_{x \to a} 1/x = \lim_{x \to a} 1 / \lim_{x \to a} x = 1/a.$$

**Example 3.5.3.** Prove: $\lim\limits_{x \to a} x^n = a^n$, $n$ a positive integer.

*Solution.* Apply the limit theorem for a product of several functions to the case of $n$ equal functions $f(x) = x$.

**Example 3.5.4.** Find $\lim\limits_{x \to -2} (2x^3 - 3x^2 + x + 5)$.

*Solution.*
$$\lim_{x \to -2} (2x^3 - 3x^2 + x + 5) = 2 \lim_{x \to -2} x^3 - 3 \lim_{x \to -2} x^2 + \lim_{x \to -2} x + \lim_{x \to -2} 5$$
$$= 2(-8) - 3(4) + (-2) + 5$$
$$= -25.$$

**Example 3.5.5.** Find $\lim\limits_{x \to 3} \dfrac{4x^2 - 3}{2x^3 + 1}$.

*Solution.* We obtain $\lim\limits_{x \to 3} (4x^2 - 3) = 33$, $\lim\limits_{x \to 3} (2x^3 + 1) = 55$. The limit theorem for a quotient applies, and we have
$$\lim_{x \to 3} \frac{4x^2 - 3}{2x^3 + 1} = \frac{33}{55} = \frac{3}{5}.$$

**Example 3.5.6.** Let us reconsider the following limit of Section 3.3:
$$\lim_{x \to 2} \frac{x - 2}{x^2 - 4}.$$

Since $\lim\limits_{x \to 2} (x^2 - 4) = 0$, we may not apply the limit theorem for a quotient. However, the method applied in Section 3.3 is valid; we have
$$\lim_{x \to 2} \frac{x - 2}{x^2 - 4} = \lim_{x \to 2} \frac{1}{x + 2} = \frac{1}{4}.$$

For later use we present here a consequence of Part 2 of Theorem 3.5.1.

**THEOREM 3.5.2.** *Let* $\lim\limits_{x \to a} f(x) = A$; *then for some function* $\epsilon(x)$ *we may write*
$$f(x) = A + \epsilon(x), \quad \text{where} \quad \lim_{x \to a} \epsilon(x) = 0.$$

*Proof.* We define $\epsilon(x)$ as $\epsilon(x) = f(x) - A$. We need to show that $\lim\limits_{x \to a} \epsilon(x) = 0$.

We have
$$\lim_{x \to a} \epsilon(x) = \lim_{x \to a} [f(x) - A] = \lim_{x \to a} f(x) - A = A - A = 0,$$
and the proof is complete.

We include here the following special case of Theorem 3.6.2, proved in Section 3.6.

**THEOREM 3.5.3.** *If* $\lim_{x \to a} f(x) = A \geq 0$, *and* $n$ *is a positive integer, then*
$$\lim_{x \to a} [f(x)]^{1/n} = A^{1/n}.$$
*The same result also holds if* $A < 0$ *and* $n$ *is odd.*

### EXERCISE GROUP 3.4

Evaluate each of the following limits by use of properties of limits.

1. $\lim\limits_{x \to -2} \dfrac{x^2 + 4x - 2}{x + 3}$

2. $\lim\limits_{x \to 2} \dfrac{2x - 4}{x^2 - 3x}$

3. $\lim\limits_{x \to -1/3} (x + 1/x)$

4. $\lim\limits_{x \to 8} (x^{1/3} - x^{-1/3})$

5. $\lim\limits_{x \to 9} (\sqrt{x} + 1/\sqrt{x})$

6. $\lim\limits_{x \to 4} \dfrac{2x - x^{1/2} + 3}{x + 3x^{1/2} - 2}$

7. $\lim\limits_{x \to -2} \sqrt{x + 4}$

8. $\lim\limits_{x \to -3} \sqrt[3]{2x - 2}$

9. $\lim\limits_{x \to 4} \dfrac{\sqrt{x} + 3}{\sqrt{x} - 3}$

10. $\lim\limits_{x \to 3} \dfrac{x - \sqrt{4 - x}}{x + \sqrt{5 - x}}$

11. $\lim\limits_{x \to 2} \dfrac{x^{1/2} - 2^{1/2}}{x - 2}$

12. $\lim\limits_{h \to 0} \dfrac{h}{\sqrt{2} - \sqrt{(h+1)^2 + 1}}$

13. $\lim\limits_{y \to -3} \dfrac{y + 3}{y^3 - 3y + 9}$

14. $\lim\limits_{h \to 0} \dfrac{h}{1 - \sqrt[3]{h + 1}}$

In Exercises 15–22 prove the statement by direct use of the definition of limit.

15. $\lim\limits_{x \to 2} (3x + 1) = 7$

16. $\lim\limits_{x \to -1} (3 - x) = 4$

17. $\lim\limits_{x \to -2} (2x + 5) = 1$

18. $\lim\limits_{x \to 0} (4x - 3) = -3$

19. $\lim\limits_{x \to 2} x^2 = 4$

20. $\lim\limits_{x \to 3} (x^2 - 3) = 6$

21. $\lim\limits_{x \to 1} (x^2 - x) = 0$

22. $\lim\limits_{x \to -2} (x^2 + 2x) = 0$

23. Prove the uniqueness of a limit; that is, if
$$\lim_{x \to a} f(x) = A \quad \text{and} \quad \lim_{x \to a} f(x) = B,$$
then $A = B$.

**24.** Let $g(x) = f(x) - L$. If $\lim\limits_{x \to a} f(x) = L$, prove that $\lim\limits_{x \to a} g(x) = 0$.

**25.** Prove Theorem 3.4.1, if $n = 3$ and $a > 0$.

**★ 26.** Prove Theorem 3.4.1, if $n$ is any positive integer and $a > 0$.

**★ 27.** Show that $|x - 2| < 1/2 \Rightarrow |(2x - 3) - 3| < 3$. This means that for the function $f(x) = 2x - 3$ of Example 3.4.4, given $\epsilon = 3$, we have found $\delta$ $(=1/2)$ such that $|x - 2| < \delta \Rightarrow |f(x) - 3| < \epsilon$. However, the fact that we have found a suitable $\delta$ for *one* value of $\epsilon$ is of no significance, since for $\lim\limits_{x \to 2} f(x) = 3$ to hold we must be able to do it for *every* $\epsilon$, and in that example we showed that this could not be done.

## 3.6 CONTINUITY

In most of the examples of the preceding section the functions were such that the value of the limit as $x$ approached a number $a$ was the same as the value of the function at $a$. Such behavior is not automatic for functions; it illustrates a property known as *continuity* which is possessed by some functions, although not necessarily at each number in the domain.

**DEFINITION 3.6.1.** *A function is continuous at a value $a$, if*

$$\lim_{x \to a} f(x) = f(a).$$

*A function is continuous in an interval, if it is continuous at each number in the interval.*

According to the definition, continuity of a function $f$ at a value $a$ implies the following:

1. $f(a)$ is defined ($a$ is in the domain of $f$);
2. $\lim\limits_{x \to a} f(x)$ exists (as a real number);
3. The values in 1. and 2. are equal.

The absence of any of these conditions means that the function is not continuous at $a$, and the function is said to be *discontinuous* at $a$.

According to the examples of the preceding section, the linear function $mx + b$ is continuous at all values, and the function $1/x$ is continuous at all values except $x = 0$.

**Example 3.6.1.** Determine all numbers for which the function $(x - 3)/(x + 1)$ is continuous.

*Solution.* The function is not defined at $-1$, and hence is discontinuous there. For any other number we have, by the properties of limits of Section 3.5,

$$\lim_{x \to a} \frac{x - 3}{x + 1} = \frac{a - 3}{a + 1}, \quad a \neq -1.$$

Since this limit is also the value of the function at $a$, *the function is continuous* for all $x$ such that $x \neq -1$.

***Example 3.6.2.*** By properties of limits the function

$$f(x) = \frac{x-2}{x^2-4}$$

is continuous if $x \neq \pm 2$. For $x = 2$ or $x = -2$, the function is undefined, *hence discontinuous*. The discontinuity at $x = 2$ is particularly noteworthy since, as seen in Example 3.5.7, $\lim_{x \to 2} f(x)$ does exist.

We now define a new function $F(x)$ as follows

$$F(x) = \begin{cases} \dfrac{x-2}{x^2-4} & \text{if } x \neq 2, \\ 1/4 & \text{if } x = 2. \end{cases}$$

This is a different function from $f(x)$ above, since $F(2)$ is defined, whereas $f(2)$ is not defined. We still have $F(x)$ discontinuous at $x = -2$. However, we now have

$$\lim_{x \to 2} F(x) = 1/4 = F(2),$$

since the limit on the left does not depend on anything that occurs at $x = 2$, and for other values of $x$ the function $F(x)$ has the same values as $f(x)$. Hence $F(x)$ is continuous at $x = 2$, and we may say that $F(x)$ is continuous for all $x \neq -2$.

The function $f(x)$ of Example 3.6.2 is said to have a *removable discontinuity* at $x = 2$. We shall graph such functions later. For the present we merely state that the graph of $f(x)$ has a gap at $x = 2$; while in the graph of $F(x)$ this gap is filled.

While the functions with which we shall be concerned will in general be continuous, except possibly for isolated values of $x$, continuity should be considered as a restrictive property of functions. In fact, a function may be discontinuous at all numbers in an interval which is in the domain of the function. (An example is the function of Example 3.1.3, although we shall not prove this.)

By Example 3.5.3 it follows that $x^n$, where $n$ is a positive integer, is continuous for all values. It then follows from the properties of limits that any polynomial

$$P(x) = a_0 x^n + a_1 x^{n-1} + \cdots + a_n, \qquad a_0 \neq 0,$$

is continuous for all values.

The following statement about continuous functions is a direct consequence of Theorem 3.5.1.

**THEOREM 3.6.1.** *If $f(x)$ and $g(x)$ are continuous at $x = a$, then $f(x) \pm g(x)$ and $f(x)g(x)$ are continuous at $x = a$; and $f(x)/g(x)$ is continuous at $x = a$ if $g(a) \neq 0$.*

We define a *rational function* $R(x)$ as a function which may be expressed as the quotient of two polynomials,

$$R(x) = \frac{P(x)}{Q(x)}, \qquad Q(x) \neq 0,$$

where $P(x)$ and $Q(x)$ are polynomials in $x$. It then follows that *a rational function is continuous except where* $Q(x) = 0$. If $Q(a) = 0$, there may be a removable discontinuity at $x = a$, provided also $P(a) = 0$ (see Example 3.6.2).

With Theorem 3.4.1 we may state that if $n$ is a positive integer, then $x^{1/n}$ is continuous for $x = a \geq 0$; if $n$ is odd, then $x^{1/n}$ is continuous also for $x = a < 0$.

We conclude this section with a result concerning composite functions.

**THEOREM 3.6.2.** *If* $\lim_{x \to a} g(x) = b$, *and if* $f$ *is continuous at* $b$, *then*

$$\lim_{x \to a} f[g(x)] = f(b).$$

*Proof.* Given $\epsilon > 0$, there exists $\epsilon_1 > 0$ (by the continuity of $f$ at $b$), such that

(3.6.1) $\qquad |f(g(x)) - f(b)| < \epsilon$ when $|g(x) - b| < \epsilon_1$.

For that $\epsilon_1$ there exists $\delta > 0$ such that

(3.6.2) $\qquad |g(x) - b| < \epsilon_1$ when $0 < |x - a| < \delta$.

By combining (3.6.1) and (3.6.2), we see that when $\epsilon > 0$ is given, we may find $\delta > 0$ such that

$$|f(g(x)) - f(b)| < \epsilon \quad \text{when} \quad 0 < |x - a| < \delta,$$

and the proof is complete.

As a consequence of the above theorem, if also $b = g(a)$, so that $g$ is continuous at $a$, then $f[g(x)]$ is continuous at $a$. In other words, *a continuous function of a continuous function is continuous*.

## 3.7 SOME CONSEQUENCES OF CONTINUITY

Because of the fundamental nature of the property of continuity, the proofs of some theorems relating to this concept are rather abstract. We dispense with the proofs for the following important properties.

**THEOREM 3.7.1.** *If* $f$ *is continuous in the closed interval* $[a, b]$, *there exist numbers* $x_1$ *and* $x_2$ *in* $[a, b]$ *such that*

$$f(x_1) \geq f(x), \qquad a \leq x \leq b,$$
$$f(x_2) \leq f(x), \qquad a \leq x \leq b.$$

*In other words, $f(x_1)$ is a maximum value of $f(x)$ for $x$ in $[a, b]$ and $f(x_2)$ is a minimum value of $f(x)$ for $x$ in $[a, b]$.*

We also say that *f takes on its maximum and its minimum in the interval* [a, b].

Various aspects of this result are illustrated in Fig. 3.7.1. In part (a) the graph is shown for the function

$$f(x) = \begin{cases} x^2, & 0 \le x < 1 \\ 1/2, & x = 1. \end{cases}$$

This function is discontinuous at $x = 1/2$, and the theorem does not apply. There is a minimum at $x = 0$, but no maximum.

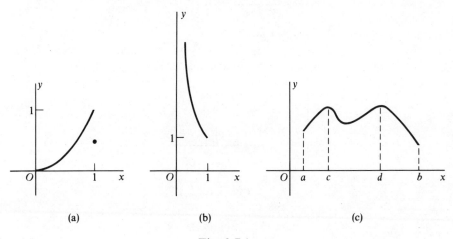

Fig. 3.7.1

In Fig. 3.7.1(b) the function shown is $1/x$, $0 < x \le 1$. The interval $(0, 1]$ is not closed, and the theorem does not apply. The function has a minimum at $x = 1$, but no maximum.

For the function shown in Fig. 3.7.1(c) the hypotheses of Theorem 3.7.1 hold. Hence there is a maximum, which occurs at both $x = c$ and $x = d$; and there is a minimum, which occurs at $x = b$.

If the hypotheses are not satisfied, Theorem 3.7.1 simply does not apply. Thus, if a function has a discontinuity, or if the interval of continuity is not closed, there may yet be a maximum or a minimum or both, but there need not be.

**THEOREM 3.7.2. Intermediate Value Theorem.** *Let f be a function continuous in* [a, b], *and suppose that* $f(a) < f(b)$. *If K is a number between* $f(a)$ *and* $f(b)$, *that is,*

$$f(a) < K < f(b),$$

*there is at least one number* $x_1$ *in* (a, b) *such that* $f(x_1) = K$. *A similar result holds if* $f(a) > f(b)$.

We may interpret this theorem by reference to Fig. 3.7.2. The theorem implies that any horizontal line, drawn between $P$ and $Q$, must intersect the graph of $f$. This seems so apparent as to be trivial. However, it is not trivial, and the proof (not given here) is far from trivial.

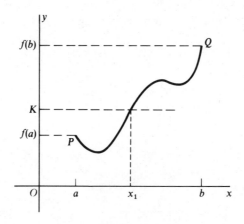

Fig. 3.7.2

### EXERCISE GROUP 3.5

Use the definition of continuity to show directly that each of the functions in Exercises 1–8 is continuous for all real values of $x$.

1. $3x - 5$
2. $4x + 3$
3. $x^2 + 3x$
4. $2x^3 - x + 1$
5. $\dfrac{x-1}{x^2+1}$
6. $\dfrac{x^2 - 2x}{2x^2 + 3}$
7. $\dfrac{x^2 - 1}{x^2 + x + 1}$
8. $\dfrac{1}{2x^2 - x + 1}$

In Exercises 9–20 determine the values of $x$ for which the given function is continuous.

9. $\dfrac{x}{x-1}$
10. $\dfrac{x^2}{2x+1}$
11. $\dfrac{x^3}{x^2 - x - 2}$
12. $\dfrac{4x - 5}{3x^2 - 2x - 1}$
13. $\dfrac{x+2}{x^2 - 4}$
14. $\dfrac{x^2 - 5x + 6}{x^2 + x - 12}$
15. $\dfrac{(x+2)(x-4)}{x(3x-2)}$
16. $\dfrac{2x - 7}{(x-3)(2x-3)}$

17. $\dfrac{1}{\sqrt[3]{x}}$

18. $\dfrac{1}{1+\sqrt[3]{x}}$

19. $\dfrac{2x}{\sqrt[3]{x+1}}$

20. $\dfrac{4x^2}{\sqrt[3]{x}+x}$

21. Find the maximum and minimum of $f(x)$ in the interval [0, 1] if
$$f(x) = x \text{ in } [0, 1/2], \qquad f(x) = 1 - x \text{ in } [1/2, 1].$$

22. Find the maximum and minimum of $f(x)$ in the interval [0, 2] if
$$f(x) = 1 - x^2 \text{ in } [0, 1/2), \qquad f(x) = (5x + 2)/6 \text{ in } [1/2, 2].$$

23. Illustrate by a graph a function which has a discontinuity in some closed interval but which nevertheless has both a maximum and a minimum in that interval.

24. Show how the Intermediate Value Theorem is used in the following argument. To solve the equation $f(x) = 0$, suppose that we have found numbers $a$ and $b$ such that
$$f(a) < 0, \qquad f(b) > 0.$$
Then a solution $x = c$ exists between $a$ and $b$.

25. Restate the definition of continuity, Definition 3.6.1, in terms of $\epsilon$ and $\delta$, that is, in a form similar to the definition of limit.

# 4  The Derivative

## 4.1  INTRODUCTION

The ideas which lie at the heart of calculus were developed over a period of many years, prior to the so-called invention of the subject. Its development is still in progress. Perhaps the single feature which distinguishes calculus from "pre-calculus" mathematics is the extensive use of the limit concept in formulating the subject matter. The limit concept has now been studied (in Chapter 3), and we are in a position to proceed with the introduction of a basic concern of calculus, *the derivative of a function*.

The derivative of a function, sometimes also called the *derived function*, is defined in a very concise manner as a certain limit, and the ramifications and applications of the derivative will lead to many interesting and important developments.

## 4.2  THE DERIVATIVE

Let $f$ be a function defined in an interval $(a, b)$, and let $x$ be a specified number in this interval. The difference

$$f(x + h) - f(x)$$

represents the change in $f$ as $x$ increases by an amount $h$. Here $h$ may be positive or negative, provided only that $x + h$ remains in the interval. The quotient

(4.2.1) $$\frac{f(x + h) - f(x)}{h}$$

is the ratio of the change in $f$ to the change in $x$. It is called the *relative rate of change* or *average rate of change of $f$ in the interval* $[x, x + h]$. As $h$ varies, the rate of change varies, in general.

***Example 4.2.1.*** Determine the average rate of change of $f(x) = x^2$ as $x$ increases from 2 by the amounts $h = 0.1, 0.01, -0.1, -0.01$.

*Solution.* We have $f(2) = 4$. With $h = 0.1$ we have $x + h = 2.1$. Then $f(2.1) = 4.41$, and the average rate of change is

$$\frac{4.41 - 4}{0.1} = \frac{0.41}{0.1} = 4.1.$$

In this way we find

| $h$ | average rate of change of $f$ |
| --- | --- |
| 0.1 | 4.1 |
| 0.01 | $(4.0401 - 4)/0.01 = 4.01$ |
| $-0.1$ | $(3.61 - 4)/(-0.1) = 3.9$ |
| $-0.01$ | $(3.9601 - 4)/(-0.01) = 3.99$ |

A more general method is to calculate the ratio (4.2.1) before substituting any specific value. We find, for $f(x) = x^2$,

$$f(x + h) - f(x) = (x + h)^2 - x^2 = x^2 + 2hx + h^2 - x^2 = 2hx + h^2.$$

If now $h \neq 0$, we obtain

$$\frac{f(x + h) - f(x)}{h} = \frac{2hx + h^2}{h} = 2x + h.$$

Substituting $x = 2$, and for $h$ the various given values, we obtain the same results as before.

We are particularly interested in what happens to the ratio (4.2.1), for a fixed number $x$ in the domain of $f$, as $h$ gets small in absolute value. As $|h|$ becomes smaller, the value of the ratio in general changes. The ratio *may* approach a limit as $h \to 0$. Such a limit, if it exists, is called the *instantaneous rate of change of $f$ at $x$*, or, more simply, the *rate of change of $f$ at $x$*.

In Example 4.2.1 the average rate of change of the function $x^2$ in the interval $[x, x + h]$ was found to be $2x + h$. Letting $h \to 0$, we see that the rate of change for any value of $x$ is $2x$. In particular, at $x = 2$ the rate of change of the function is 4.

The ideas leading to the rate of change are the basis of the study of calculus. The rate of change of a function $f$ at a value $x$ is, in fact, the *derivative* of $f$ at $x$, for which we give a formal definition. For the present we use the notation $f'(x)$ to denote this derivative.

**DEFINITION 4.2.1.** *If $f$ is a function defined in an interval $[a, b]$, the derivative of $f$ at a value $x$ in $(a, b)$ is defined as*

$$f'(x) = \lim_{h \to 0} \frac{f(x+h) - f(x)}{h},$$

*if the limit exists. It is understood that $x + h$ is always in $[a, b]$.*

If a function $f$ has a derivative at $x$, $f$ is said to be *differentiable* at $x$. If the limit in Definition 4.2.1 does not exist, then $f$ does not have a derivative, and is not differentiable, at $x$.

It must be understood at the outset that in Definition 4.2.1 *the number $x$ is fixed in value*. The function whose limit is in question represents the average rate of change of $f$ in $[x, x + h]$, and *the variable in this function is $h$*. Because the definition involves a limit as $h \to 0$, the number $h = 0$ is not involved (see Definition 3.4.1), and in the use of Definition 4.2.1 the restriction $h \neq 0$ is always present.

***Example 4.2.2.*** Find $f'(x)$ for $f(x) = 2x^2 - 3x$.

**Solution.** We have

$$\frac{f(x+h) - f(x)}{h} = \frac{[2(x+h)^2 - 3(x+h)] - [2x^2 - 3x]}{h}$$

$$= \frac{4hx - 3h + 2h^2}{h} = 4x - 3 + 2h.$$

Hence

$$f'(x) = \lim_{h \to 0} (4x - 3 + 2h) = 4x - 3.$$

***Example 4.2.3.*** Find the derivative of $1/\sqrt{x}$, $x > 0$.

**Solution.** Letting $f(x) = 1/\sqrt{x}$, we have

$$\frac{f(x+h) - f(x)}{h} = \frac{1}{h}\left[\frac{1}{\sqrt{x+h}} - \frac{1}{\sqrt{x}}\right] = \frac{\sqrt{x} - \sqrt{x+h}}{h\sqrt{x+h}\sqrt{x}}$$

$$= \frac{\sqrt{x} - \sqrt{x+h}}{h\sqrt{x+h}\sqrt{x}} \cdot \frac{\sqrt{x} + \sqrt{x+h}}{\sqrt{x} + \sqrt{x+h}} = -\frac{1}{\sqrt{x+h}\sqrt{x}(\sqrt{x} + \sqrt{x+h})}$$

Hence

$$f'(x) = \lim_{h \to 0} \left[-\frac{1}{\sqrt{x+h}\sqrt{x}(\sqrt{x} + \sqrt{x+h})}\right] = -\frac{1}{\sqrt{x}\sqrt{x}(\sqrt{x} + \sqrt{x})}$$

$$= -\frac{1}{2x\sqrt{x}}.$$

The notation $f'(a)$ is used for the value at $x = a$ of the derivative of $f$. We may then write

$$f'(a) = \lim_{h \to 0} \frac{f(a + h) - f(a)}{h}.$$

**Example 4.2.4.** Find $f'(2)$ for $f(x) = x^3$.

*Solution.*

$$\frac{f(2 + h) - f(2)}{h} = \frac{(2 + h)^3 - 2^3}{h}.$$

The limit of this expression as $h \to 0$ was found in Example 3.3.1 to be 12. Hence $f'(2) = 12$.

An alternate method is to find $f'(x)$ for arbitrary $x$ and then to evaluate the derivative at $x = 2$. Thus

$$\frac{f(x + h) - f(x)}{h} = \frac{(x + h)^3 - x^3}{h} = 3x^2 + 3hx + h^2.$$

Then $f'(x) = 3x^2$, and $f'(2) = 12$.

The following *alternate definition of derivative* is sometimes useful:

(4.2.2) $$f'(a) = \lim_{x \to a} \frac{f(x) - f(a)}{x - a}.$$

It follows directly from Definition 4.2.1 by replacing $x$ by $a$ and $h$ by $x - a$.

**Example 4.2.5.** Let us examine the differentiability of the function $f(x) = |x|$. We consider three cases:

(i) $a > 0$. If $x$ is close enough to $a$, then $x > 0$, $|x| = x$, $|a| = a$, and we have, using (4.2.2),

$$f'(a) = \lim_{x \to a} \frac{x - a}{x - a} = 1.$$

(ii) $a < 0$. If $x$ is close enough to $a$, then $x < 0$, $|x| = -x$, $|a| = -a$, and we have

$$f'(a) = \lim_{x \to a} \frac{-x + a}{x - a} = -1.$$

(iii) $a = 0$. Then $f(a) = 0$, and

$$\frac{f(x) - f(a)}{x - a} = \frac{|x|}{x} = \begin{cases} 1 & \text{if } x > 0 \\ -1 & \text{if } x < 0. \end{cases}$$

Hence there can be no limit as $x \to 0$ and no derivative of $f$ at 0.

Thus $f'(x) = 1$ if $x > 0$, $f'(x) = -1$ if $x < 0$, and there is no derivative at $x = 0$.

Other notations for derivative are in common use, and we shall often use them interchangeably. In particular, $Df(x)$ is also used; in this notation we may write, for example, $Dx^3 = 3x^2$ (see Example 4.2.4).

Note that in general *the derivative of a function f is also a function*, whose domain consists of those numbers in the domain of $f$ for which $f$ is differentiable. The *derived function* may then be referred to as $f'$ or $Df$.

There is a very definite, although one-sided, relationship between differentiability and continuity, expressed in the following result.

**THEOREM 4.2.1.** *If $f'(a)$ exists, then $f$ is continuous at $x = a$.*

*Proof.* Since $f'(a)$ exists we may write (see Theorem 3.5.2, and the alternate definition of derivative (4.2.2)), if $x \neq a$,

$$\frac{f(x) - f(a)}{x - a} = f'(a) + r(x), \quad \text{where} \quad \lim_{x \to a} r(x) = 0.$$

Then

$$f(x) = f(a) + (x - a)[f'(a) + r(x)].$$

It follows that $\lim_{x \to a} f(x) = f(a)$, and $f$ is continuous at $a$.

The converse of this theorem is false. A function may be continuous at $x = a$ without being differentiable there. For example, the function $|x|$ has no derivative at $x = 0$ (Example 4.2.5), but it is continuous at $x = 0$. We noted earlier that continuity was, in a sense, a restrictive property of functions; differentiability is even more restrictive.

## EXERCISE GROUP 4.1

In each of Exercises 1–12 find $f'(x)$ for the given function $f(x)$ by direct use of the definition of derivative.

1. $5x - 3$
2. $4 - 3x$
3. $x^2 - 3$
4. $2x^2 - x$
5. $\dfrac{3}{x}$
6. $\dfrac{1}{3 - x}$
7. $\dfrac{x}{x - 1}$
8. $\dfrac{1}{2x - 3}$
9. $\sqrt{x}$
10. $\sqrt{1 - 2x}$
11. $\dfrac{1}{1 + \sqrt{x}}$
12. $\dfrac{1}{\sqrt{2x}}$

In Exercises 13–18 find the value of the derivative of the given function at the specified value $x = a$.

13. $2x + 1$, $a = 2$
14. $2x^2$, $a = 3$
15. $2x^2 + 3x$, $a = -1$
16. $x^3/3$, $a = 2$
17. $\dfrac{2}{x}$, $a = 2$
18. $\dfrac{1}{x-3}$, $a = 1$

Each of the following limit expressions could arise in obtaining the derivative of some function by direct use of the definition of derivative. Find that derivative. (The purpose is *not* to evaluate the limit.)

*Example.* Given the expression $\lim\limits_{x \to -3} \dfrac{(x^2 + x) - 6}{x + 3}$, we recognize that it arises as the derivative of $x^2 + x$ at $x = -3$, since if $f(x) = x^2 + x$, we have $f(-3) = 6$. The answer is given as $f'(-3)$ for $f(x) = x^2 + x$.

19. $\lim\limits_{x \to 2} \dfrac{x^2 - 4}{x - 2}$
20. $\lim\limits_{h \to 0} \dfrac{(1 + h)^3 - 1}{h}$
21. $\lim\limits_{h \to 0} \dfrac{\sqrt{4 + h} - 2}{h}$
22. $\lim\limits_{h \to 0} \dfrac{(x + h)^2 - x^2}{h}$
23. $\lim\limits_{x \to 1} \dfrac{x^3 - 1}{x - 1}$
24. $\lim\limits_{h \to 0} \dfrac{(2 + h)^2 - 4}{h}$
25. $\lim\limits_{x \to -2} \dfrac{x^3 - 3x + 2}{x + 2}$
26. $\lim\limits_{x \to 3} \dfrac{x^2 + 2x - 15}{x - 3}$

## 4.3 TANGENT TO A CURVE

From Section 4.2 it is clear that derivatives are applicable to the study of rate of change; later, we shall develop this idea further. We wish to apply the derivative here to the graphs of equations and functions.

## 82  THE DERIVATIVE  4.3

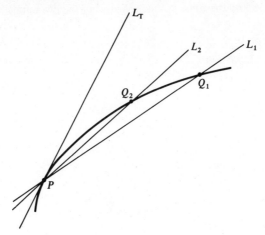

Fig. 4.3.1

In Fig. 4.3.1 each of the lines $L_1$ and $L_2$ intersects the curve in two points; whereas $L_T$ intersects the curve in one point. However, the latter point may also be considered as a limiting position of the second point of intersection as the line is rotated about $P$ to approach the line $L_T$. Line $L_T$ is called a *tangent* or *tangent line* to the curve at $P$.

**DEFINITION 4.3.1.**  *A line $L_T$ is tangent to a curve at a point $P$, if it is the limiting position of a line $L$ through $P$ and a second point $Q$ of the curve as the point $Q$ approaches $P$ along the curve. In other words, $L_T$ is a tangent line if the slope of $L_T$ is the limit of the slope of $L$ as $Q$ approaches $P$ along the curve.*

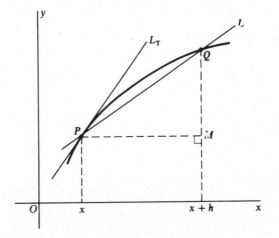

Fig. 4.3.2

Let $y = f(x)$ represent an equation of a curve (Fig. 4.3.2), and let $P(x, f(x))$ be a point on the curve. A second point $Q$ may be represented by $Q(x + h, f(x + h))$. The slope of secant line $PQ$, written as $m(PQ)$, is given by the ratio of the directed line segments $\overline{MQ}$ and $\overline{PM}$, and we have, since

$$\overline{MQ} = f(x + h) - f(x), \quad \overline{MP} = h,$$

the relation

$$m(PQ) = \frac{\overline{MQ}}{\overline{MP}} = \frac{f(x + h) - f(x)}{h}.$$

Now let $Q \to P$ along the curve. Then $h \to 0$, and if $f'(x)$ exists, we have

$$\lim_{Q \to P} m(PQ) = \lim_{h \to 0} \frac{f(x + h) - f(x)}{h} = f'(x).$$

Thus the curve has a tangent line at $P$, with slope given by $f'(x)$.

The direction of a curve at a point is characterized by the direction of the tangent line (if it exists) to the curve at the point; statements relating to directions of curves become statements relating to tangent lines.

**DEFINITION 4.3.2.** *The* **slope of a curve** *at a point is the slope of the tangent line to the curve at that point.*

**DEFINITION 4.3.3.** *An* **angle between two curves** *at a point of intersection is an angle between tangent lines to the curves at that point.*

***Example 4.3.1.*** Find the slope of the curve $y = x^2$ at the points for which $x = 0, 2, -1/2$.

*Solution.* The slope of the curve at any point is given by $Dx^2$. Using the definition of derivative, we find for the slope $m(x)$,

$$m(x) = Dx^2 = 2x.$$

Hence $m(0) = 0$, $m(2) = 4$, $m(-1/2) = -1$.

***Example 4.3.2.*** Find (a) an equation of the tangent to the curve $y = 2x^2 - 3x$ at the point (3, 9), and (b) an equation of the normal to the curve at that point.

*Solution.* (a) The derivative $D(2x^2 - 3x)$ was found in Example 4.2.2. Hence $m(x) = 4x - 3$ and $m(3) = 9$. The point-slope form of the equation of a straight line may be used, and an equation of the tangent is

$$y - 9 = 9(x - 3) \quad \text{or} \quad 9x - y = 18.$$

(b) The normal to a curve at a point is the line perpendicular to the tangent at that point. The normal at (3, 9) thus has slope $-1/9$ (the negative reciprocal of the slope of the curve), and an equation of the normal is

$$y - 9 = -\tfrac{1}{9}(x - 3) \quad \text{or} \quad x + 9y = 84.$$

**Example 4.3.3.** Find an angle between the curves $y = x^2$ and $y = 2x^2 - 3x$ of Examples 4.3.1 and 4.3.2 at the point of intersection $P(3, 9)$.

*Solution.* The slopes of the curves were obtained in the cited examples as

$$m_1(x) = 2x, \qquad m_2(x) = 4x - 3.$$

At $P$ we have then

$$m_1(3) = 6, \qquad m_2(3) = 9.$$

It is convenient to use the result of Exercise 15 of Exercise Group 2.4. Thus if $\alpha$ is an angle between the curves at $(3, 9)$, we find

$$\tan \alpha = \frac{m_2 - m_1}{1 + m_1 m_2} = \frac{9 - 6}{1 + 54} = \frac{3}{55}.$$

Hence the curves intersect at the point $(3, 9)$ at an angle (approximately $3°$) whose tangent is $3/55$.

Other problems relating to slopes of curves or to angles between curves can be treated by finding derivatives of appropriate functions, and appear in the exercises.

## EXERCISE GROUP 4.2

In each of Exercises 1–8 find (a) an equation of the tangent to the given curve at the specified point, (b) an equation of the normal to the curve at the point.

1. $y = x^2 - 3, (-1, -2)$
2. $y = 2x^2 - x, (2, 6)$
3. $y = 1/x, (2, 1/2)$
4. $y = 1/(3 - x), (1, 1/2)$
5. $y = \sqrt{x}$, at $x = 4$
6. $y = 1/\sqrt{x}$, at $x = 4$
7. $y = 1/x^2$, at $x = 1$
8. $y = x^2$, at $x = 0$

In Exercises 9–14 find the points at which the curve has the specified slope.

9. $y = x^2$, slope 1
10. $y = 2x^2 + x$, slope 0
11. $y = x^3 - 4x^2$, slope 3
12. $y = x + \frac{1}{x}$, slope 0
13. $y = 1/x$, slope 0
14. $y = \sqrt{x}$, slope 1/2

15. Show that the line $y = 5x$ is tangent to the curve $y = x^2 + x + 4$.
16. The curves $y = x^2 - 1/2$, $y = 1/(2x)$ intersect at the point $(1, 1/2)$. Show that the tangents to the curves at their point of intersection are perpendicular.
17. Find the angle between the tangents to the curves $y = x^2$, $y = 2 - x^2$ at each point of intersection.
18. Show that the slope of the curve $y = x^3 + 4x - 3$ is never less than 4.

**19.** Find an angle between the tangents to the curves $y = x^3 - x^2$ and $y = 2x^2 - 2x$ at each point of intersection.

**20.** Find equations of those tangent lines to the curve $y = x^2$ which pass through the point $(5, 9)$.

## 4.4 DIFFERENTIATION FORMULAS

In Section 4.2 derivatives were found by direct application of the definition. However, it is desirable to develop formulas of differentiation that permit us to avoid, where feasible, the direct use of this definition. We shall develop some of these and then use them. It is well to realize, even at this stage, that it will be necessary from time to time to revert to the use of the definition of derivative whenever a differentiation cannot be performed with previously established formulas.

**THEOREM 4.4.1.** *If $f(x) = c$, $c$ constant, then*

$$f'(x) = 0;$$

*that is, the derivative of a constant function is identically zero.*

*Proof.* We have

$$\frac{f(x+h) - f(x)}{h} = \frac{c - c}{h} = 0, \quad h \neq 0.$$

To complete the proof, we obtain

$$f'(x) = \lim_{h \to 0} 0 = 0.$$

**THEOREM 4.4.2.** *If $f(x) = x^n$, $n$ a positive integer, then*

$$f'(x) = nx^{n-1}.$$

*Proof.* We have, with $h \neq 0$,

$$\frac{f(x+h) - f(x)}{h} = \frac{(x+h)^n - x^n}{h}.$$

We factor the numerator on the right by use of the formula

$$u^n - v^n = (u - v)(u^{n-1} + u^{n-2}v + \cdots + v^{n-1}).$$

Then

$$\frac{f(x+h) - f(x)}{h} = \frac{1}{h}[(x+h) - x][(x+h)^{n-1} + (x+h)^{n-2}x + \cdots + x^{n-1}]$$

$$= (x+h)^{n-1} + (x+h)^{n-2}x + \cdots + (x+h)x^{n-2} + x^{n-1}, \ (n \text{ terms}).$$

Each of the $n$ terms in this sum approaches $x^{n-1}$ as $h \to 0$. Hence the limit of the sum is $nx^{n-1}$, and we have

$$f'(x) = nx^{n-1},$$

and the theorem is established.

Note that Theorem 4.4.2 applies for $n = 1$, so that $Dx = 1$. Other simple examples are

$$Dx^3 = 3x^2, \quad Dx^7 = 7x^6, \quad Dx^{43} = 43x^{42}.$$

We may classify each differentiation formula we develop as of *specific* type or *general* type. By a specific formula of differentiation, we mean a formula for the derivative of a particular function. The two preceding theorems provide specific formulas. The formula in Theorem 4.4.2 is sometimes called the *power formula* or *power rule*; it will be shown later that this formula holds, in fact, for any real exponent and not only for positive integral exponents. Surprisingly, the power formula and Theorem 4.4.1 are the only specific formulas we shall obtain until the derivatives of trigonometric, logarithmic, and exponential functions are developed in Chapters 10 and 11. However, we can differentiate very many functions by using, in addition, the general formulas of differentiation; the latter are formulas of differentiation which involve *combinations of functions*. We proceed to develop some general formulas, and then show how they are used. The proofs make use of the definition of derivative.

**THEOREM 4.4.3.** *If $f$ is a differentiable function and $F(x) = cf(x)$, $c$ constant, then*

$$F'(x) = cf'(x).$$

*Proof.* We have, with $h \neq 0$,

$$\frac{F(x+h) - F(x)}{h} = \frac{cf(x+h) - cf(x)}{h} = c\frac{f(x+h) - f(x)}{h}.$$

As $h \to 0$ the fraction on the right approaches $f'(x)$ by the definition of derivative, and we have

$$F'(x) = \lim_{h \to 0} c\frac{f(x+h) - f(x)}{h} = cf'(x).$$

***Example 4.4.1.*** Find $f'(x)$, if $f(x) = 4x^7$.

*Solution.* We have

$$\begin{aligned} f'(x) &= 4Dx^7 & \text{[by Theorem 4.4.3]} \\ &= 4(7x^6) & \text{[by Theorem 4.4.2]} \\ &= 28x^6. \end{aligned}$$

With the addition of the following theorem we shall be in a position to differentiate any polynomial.

**THEOREM 4.4.4.** *If $f$ and $g$ are differentiable in an interval $(a, b)$, then so is $F(x) = f(x) + g(x)$, and*

$$F'(x) = f'(x) + g'(x);$$

*in brief, the derivative of a sum is equal to the sum of the derivatives.*

*Proof.* We have

$$\frac{F(x+h) - F(x)}{h} = \frac{[f(x+h) + g(x+h)] - [f(x) + g(x)]}{h}$$

$$= \frac{f(x+h) - f(x)}{h} + \frac{g(x+h) - g(x)}{h},$$

and

$$F'(x) = \lim_{h \to 0} \frac{f(x+h) - f(x)}{h} + \lim_{h \to 0} \frac{g(x+h) - g(x)}{h} = f'(x) + g'(x).$$

The following consequences of the above theorems are easily shown; the functions involved are assumed differentiable.

**COROLLARY.** 1. $D[f(x) - g(x)] = Df(x) - Dg(x)$.
2. $D[f_1(x) + f_2(x) + \cdots + f_n(x)] = Df_1(x) + Df_2(x) + \cdots + Df(x)$.

With use of the above we can now differentiate any polynomial; we state the result explicitly.

**THEOREM 4.4.5.** *If $P_n(x) = a_n x^n + a_{n-1} x^{n-1} + \cdots + a_1 x + a_0$, with $n$ a positive integer, $a_i$ constant, $i = 0, 1, \ldots, n$, and $a_n \neq 0$, then $P_n(x)$ is differentiable, and*

$$P'_n(x) = n a_n x^{n-1} + (n-1) a_{n-1} x^{n-2} + \cdots + a_1.$$

*Example 4.4.2.* By Theorem 4.4.5, or a combination of earlier theorems, we have

$$D(3x^8 + 2x^6 - x^3 + 4x - 7) = 24x^7 + 12x^5 - 3x^2 + 4.$$

Although the power formula has thus far been proved only for a positive integral exponent (in Theorem 4.4.2), *we are assuming, for purposes of illustration, its validity for any real exponent*—the proofs will come later.

*Example 4.4.3.* Differentiate $3x^4 - x^{1/2} - 2x^{-3}$.

*Solution.* We have

$$D(3x^4 - x^{1/2} - 2x^{-3}) = D(3x^4) - Dx^{1/2} - D(2x^{-3})$$

$$= 3(4x^3) - (1/2)x^{1/2 - 1} - 2(-3)x^{-3 - 1}$$

$$= 12x^3 - x^{-1/2}/2 + 6x^{-4}.$$

**Example 4.4.4.** Find $f'(x)$ for $f(x) = (2x^2 - 3x)(2x + 1)$.

*Solution.* At this time we are obliged to expand the product. Thus
$$f(x) = 4x^3 - 4x^2 - 3x,$$
and we get
$$f'(x) = 12x^2 - 8x - 3.$$

**EXERCISE GROUP 4.3**

In each of Exercises 1–14 find the derivative of the given function. (You may assume the power rule of differentiation for any rational exponent.)

1. $2x^3 - 3x$
2. $\sqrt[3]{x^2}$
3. $1/x^3$
4. $\tfrac{4}{3}x^{3/2}$
5. $2x - 4x^3 + 7$
6. $\tfrac{2}{3}(3x^4 - x^3 + 6x)$
7. $\sqrt{x} + x^2 - 3x$

8. $(x - 1)(x + 2)$
9. $(x^{1/2} - 1)(x^{1/2} + 1)$
10. $(x - 1)(x - 2)(x - 3)$
11. $x^2(x - 1)$
12. $(x^2 + 1)(x^3 + x - 2)$
13. $\dfrac{x^2 - 2x - 3}{x + 1}$
14. $\dfrac{x^4 - 1}{x^2 - 1}$

15. If $f(x) = 2x^3 - 4x^2 + 7x - 8$, find $f'(2), f'(-1/2)$.
16. If $f(x) = 2x^{1/2} - 3x$, find $f'(1), f'(4)$.
17. Let $y = 4x^3 - 13x^2 + 12x - 2$ be an equation of a curve.
    (a) Find the slopes at $x = 0$ and $x = 1$.
    (b) For what values of $x$ is the slope zero?
    (c) For what values of $x$ does the slope equal $-2$?
18. Let $y = x + \sqrt{x}$ be an equation of a curve.
    (a) Find an equation of the tangent to the curve at $x = 4$.
    (b) Find an equation of the normal to the curve at $x = 1$.
    (c) Does the curve have slope zero at any point?
    (d) Does the curve have slope 1 at any point?
    (e) Does the curve have slope 5/2 at any point?
19. Carry out the steps in the proof of Theorem 4.4.5 for $n = 4$.
20. Derive the formula $Dx^n = nx^{n-1}$, $n$ a positive integer, by using the alternate definition of derivative.
21. Prove the corollary to Theorem 4.4.4.

## 4.5 DERIVATIVE OF A COMPOSITE FUNCTION. THE CHAIN RULE

In Section 3.2 we defined and discussed the idea of a composite function. The formula for the derivative of such a function is one of the most important and useful differentiation formulas of general type. Such a formula enables us to obtain, for example, the derivative of $\sqrt{1-x^2}$ by using the derivatives of $\sqrt{x}$ and $1-x^2$. After stating the theorem we shall discuss it before giving a formal proof.

**THEOREM 4.5.1.** *Let $g'(x)$ exist for $a < x < b$, and let $f'(x)$ exist for $x$ in the range of $g$. If we set $F(x) = f[g(x)]$, then*

(4.5.1) $$F'(x) = f'[g(x)]g'(x).$$

*Discussion.* The first derivative on the right warrants closer inspection. The letter $f$ is a name given to a certain function, which denotes a set of ordered pairs of numbers $(a, b)$ for $a$ and $b$ in specified sets. The symbol $f(x)$ denotes the value of the second number of that pair which has $x$ as its first member; thus $f(x) = f(u) = f(g)$ if $x = u = g$. The use of a particular letter is a matter of convenience or choice. The derivative function $f'$ or $Df$ is the function whose value for any number is defined in Definition 4.2.1. Then $f'(x)$ is the value of this function at $x$, $f'(u)$ is the value at $u$, and $f'(g)$ is the value at $g$. Accordingly, what is meant by $f'[g(x)]$ or $Df[g(x)]$ is simply the value of $f'$ or $Df$ evaluated at $g(x)$. To illustrate this idea, suppose $f(x) = x^2$. Then $f'(x) = 2x$, $f'(u) = 2u$, or

$$f'[g(x)] = 2g(x).$$

Now suppose that we also have $g(x) = x^2 - 1$. Then

$$f'[g(x)] = 2g(x) = 2(x^2 - 1).$$

We also have $g'(x) = 2x$. Then, according to (4.5.1), we obtain, if $F(x) = f[g(x)] = (x^2 - 1)^2$,

(4.5.2) $$F'(x) = [2(x^2 - 1)](2x) = 4x(x^2 - 1).$$

Since we may also write $F(x) = x^4 - 2x^2 + 1$, we may also obtain by direct differentiation,

$$F'(x) = D(x^4 - 2x^2 + 1) = 4x^3 - 4x = 4x(x^2 - 1),$$

the same result as in (4.5.2).

Formula (4.5.1) is usually called the *chain rule* of differentiation. Its use in general applications will appear from time to time. We illustrate now the more specific type of application.

**Example 4.5.1.** Find $f'(x)$ for $f(x) = (x^2 - 3x)^{14}$.

*Solution.* We have the choice of expanding $f(x)$ by the binomial formula, and then differentiating as a polynomial. However, we should like to avoid this, if possible, and the chain rule makes it possible. Let $g(x) = x^2 - 3x$ and $h(x) = x^{14}$. Then $f(x) = h[g(x)]$, and, by the chain rule,

$$f'(x) = h'[g(x)] \cdot g'(x) = 14(x^2 - 3x)^{13}(2x - 3).$$

We may describe the technique involved as follows. Differentiate $(x^2 - 3x)^{14}$ as though $x^2 - 3x$ were a single variable, to give $14(x^2 - 3x)^{13}$; then multiply by $(2x - 3)$, the derivative of the function which was considered as a single variable.

It is convenient here to introduce a modified notation for a derivative. The statement $Dy^3 = 3y^2$ is correct. On the other hand, suppose $y$ is a differentiable function, and we want the derivative of $f(x) = [y(x)]^3$. We agree to write the derivative as $D_x y^3$, where *the subscript denotes the variable in the composite function*. With this notation the chain rule may be written as

$$D_x f[g(x)] = Df[g(x)]g'(x).$$

***Example 4.5.2.*** If $y$ is a differentiable function of $x$, find $D_x y^3$.

*Solution.* We differentiate $y^3$ as though $y$ were the variable of differentiation to get $3y^2$. We must now multiply by $D_x y$, which we assume exists, and we have

$$D_x y^3 = D_y y^3 \cdot D_x y = 3y^2 D_x y.$$

***Example 4.5.3.*** Assume $\tan x$ is differentiable, and let us write

$$D \tan x = s(x).$$

Find the derivative of $\tan(x^3)$.

*Solution.* We differentiate $\tan(x^3)$ as though $x^3$ were a single variable, and get $s(x^3)$. We now multiply by $Dx^3$ and find as the result

$$D_x \tan(x^3) = 3x^2 s(x^3).$$

The explicit function $s(x)$ will be obtained in Chapter 10.

***Example 4.5.4.*** Differentiate $(x^2 - 1)^{20}$.

*Solution.* We may write

$$(x^2 - 1)^{20} = f[g(x)], \quad \text{where} \quad f(x) = x^{20}, g(x) = x^2 - 1.$$

Hence, by the chain rule,

$$\begin{aligned} D(x^2 - 1)^{20} &= f'[g(x)] \cdot g'(x) \\ &= 20[g(x)]^{19} \cdot 2x = 40x(x^2 - 1)^{19}. \end{aligned}$$

***Example 4.5.5.*** Differentiate $\sqrt{1 - x}$.

*Solution.* We have $\sqrt{1 - x} = f[g(x)]$, where $f(x) = \sqrt{x}, g(x) = 1 - x$. Hence

$$D\sqrt{1 - x} = \frac{1}{2\sqrt{g(x)}} \cdot g'(x) = -\frac{1}{2\sqrt{1 - x}}.$$

With the use of the chain rule we may express the *power rule of differentia-*

*tion* in general form as follows. If $F(x) = [f(x)]^n$, then

(4.5.3) $$F'(x) = n[f(x)]^{n-1}f'(x).$$

Formula (4.5.3) could have been used directly in Examples 4.5.4 and 4.5.5. Additional examples will now be given.

**Example 4.5.6.** Given that $y$ is a differentiable function of $x$, obtain the derivative of the function $y^{1/3}$.

*Solution.* Assuming the power rule for rational exponents, we find

$$D_x y^{1/3} = \tfrac{1}{3} y^{-2/3} D_x y.$$

This is as far as we can proceed without knowing $D_x y$.

For purposes of illustration we shall assume that the function $\sin x$ is differentiable, and shall call its derivative $p(x)$, so that

(4.5.4) $$D(\sin x) = p(x).$$

**Example 4.5.7.** Find $D \sin 2x$.

*Solution.* We have

$$\sin 2x = f[g(x)], \quad \text{where} \quad f(x) = \sin x, \, g(x) = 2x.$$

Hence

$$D \sin 2x = Df[g(x)] \cdot D[g(x)] = p(2x) \cdot 2 = 2p(2x).$$

**Example 4.5.8.** Find $f'(x)$ if $f(x) = \sin^3 2x$.

*Solution.* By (4.5.3) we get [since $\sin^3 2x = (\sin 2x)^3$]

$$f'(x) = 3 \sin^2 2x \cdot D \sin 2x$$
$$= 3 \sin^2 2x \cdot 2p(2x)$$
$$= 6 \sin^2 2x \cdot p(2x).$$

We proceed to the proof of Theorem 4.5.1.

*Proof of Theorem 4.5.1.* By the definition of derivative we have

$$F'(x) = \lim_{h \to 0} \frac{F(x+h) - F(x)}{h} = \lim_{h \to 0} \frac{f[g(x+h)] - f[g(x)]}{h}.$$

Since $f'(u)$ exists (the letter $u$ is used for convenience), we have by Theorem 3.5.2,

$$\frac{f(u+k) - f(u)}{k} = f'(u) + r(k), \quad k \neq 0, \quad \lim_{k \to 0} r(k) = 0,$$

from which we obtain

(4.5.5) $$f(u+k) - f(u) = k[f'(u) + r(k)].$$

So far this equation holds if $k \neq 0$; however, it also holds when $k = 0$, if $r(0)$ is defined. We now define $r(0) = 0$, and we have $\lim_{k \to 0} r(k) = r(0) = 0$, so that $r(k)$ as now defined is continuous at $k = 0$. In (4.5.5) we now let $u = g(x), k = g(x+h) - g(x)$. Then $u + k = g(x+h)$. With these values inserted we divide (4.5.5) by $h \neq 0$, and we have

$$(4.5.6) \quad \frac{f[g(x+h)] - f[g(x)]}{h} = \frac{g(x+h) - g(x)}{h} [f'[g(x)] + r(k)].$$

We now let $h \to 0$. Since $g$ has a derivative at $x$, we know that $g$ is continuous at $x$, and therefore

$$\lim_{h \to 0} k = \lim_{h \to 0} [g(x+h) - g(x)] = 0.$$

Hence $r(k)$ approaches zero, $f'[g(x)] + r(k)$ approaches $f'[g(x)]$, and

$$[g(x+h) - g(x)]/h$$

approaches $g'(x)$. Thus as $h \to 0$ in (4.5.6), we get

$$F'(x) = g'(x) \cdot f'[g(x)],$$

and the proof of the chain rule is complete.

*Discussion of the proof.* If $h \neq 0$ and $g(x+h) - g(x) \neq 0$, we may write

$$\frac{f[g(x+h)] - f[g(x)]}{h} = \frac{f[g(x+h)] - f[g(x)]}{g(x+h) - g(x)} \cdot \frac{g(x+h) - g(x)}{h},$$

from which it appears to follow easily that, as $h \to 0$,

$$F'(x) = f'[g(x)] \cdot g'(x).$$

However, as $h \to 0$ we do *not* know that $g(x+h) - g(x)$ is always different from zero, even though $h \neq 0$; and we cannot draw this conclusion directly. It is precisely this realization that necessitates the type of proof given.


### EXERCISE GROUP 4.4

In Exercises 1–15 assume that the general power rule, (4.5.3), holds for any rational exponent. Find the derivative of the function with respect to the variable which appears ($a$ is constant).

1. $(x^2 - 1)^9$
2. $(u - 1)^{14}$
3. $(2w - 3)^8$
4. $\sqrt{1-x}$
5. $\sqrt{x^2-1}$
6. $(x - 2)^{2/3}$
7. $(1 - x^2)^{-1/2}$
8. $(x^3 + 3x)^{1/3}$
9. $(v^4 - 1)^{2/5}$
10. $(y^2 - y - 1)^{-1/2}$
11. $(s - s^2)^{4/3}$
12. $(2x + 7)^{3/2}$
13. $(a^{1/2} - x^{1/2})^2$
14. $(a^{2/3} - x^{2/3})^{3/2}$
15. $(x^{1/2} - x^{-1/2})^3$

In Exercises 16–21 we assume that $\sin x$ is a differentiable function, and we write $D \sin x = p(x)$. Express the derivative of each function in terms of $p(x)$.

16. $\sin 4x$
17. $\sin(1-x)$
18. $\sin^2 x$
19. $\sqrt{\sin 2x}$
20. $\sin^2(x/2)$
21. $\sqrt{1 - \sin^2 x}$

* 22. Assuming any needed differentiability, find a formula for the derivative $D_x f[u(v(x))]$ in terms of the derivatives of $f$, $u$, and $v$.

## 4.6 DERIVATIVE OF A PRODUCT AND OF A QUOTIENT

With the formulas to be presented here we complete the present development of differentiation formulas of general type.

**THEOREM 4.6.1.** *If $F(x) = f(x)g(x)$, with $f(x)$ and $g(x)$ differentiable, then $F(x)$ is differentiable, and $F'(x) = f(x)g'(x) + g(x)f'(x)$.*

*Proof.* We have

$$\frac{F(x+h) - F(x)}{h} = \frac{f(x+h)g(x+h) - f(x)g(x)}{h}$$

$$= \frac{f(x+h)[g(x+h) - g(x)] + g(x)[f(x+h) - f(x)]}{h}$$

$$= f(x+h) \cdot \frac{g(x+h) - g(x)}{h} + g(x) \cdot \frac{f(x+h) - f(x)}{h}.$$

We now let $h \to 0$. We have $\lim_{h \to 0} f(x+h) = f(x)$, since $f$ is continuous by Theorem 4.2.1. The fractions on the right approach the derivatives of $f(x)$ and $g(x)$. Hence the limit of the left member exists, and the result is established. The second line above was obtained by subtracting and adding $f(x+h)g(x)$ in the numerator of the preceding line.

**THEOREM 4.6.2.** *If $G(x) = f(x)/g(x)$, with $f(x)$ and $g(x)$ differentiable, then $G(x)$ is differentiable, and*

$$G'(x) = \frac{g(x)f'(x) - f(x)g'(x)}{[g(x)]^2}.$$

*Proof.* Since $g(x) \neq 0$ by hypothesis, and $g$ is continuous, we may choose $\delta$ so that $g(x+h) \neq 0$, $|h| < \delta$. For $h$ in this interval, and with $h \neq 0$, we have

$$\frac{G(x+h)-G(x)}{h} = \frac{1}{h}\left[\frac{f(x+h)}{g(x+h)} - \frac{f(x)}{g(x)}\right] = \frac{1}{h}\left[\frac{f(x+h)g(x) - f(x)g(x+h)}{g(x+h)g(x)}\right]$$

$$= \frac{1}{h}\left[\frac{g(x)[f(x+h)-f(x)] - f(x)[g(x+h)-g(x)]}{g(x+h)g(x)}\right]$$

$$= \frac{1}{g(x+h)g(x)}\left[g(x) \cdot \frac{f(x+h)-f(x)}{h} - f(x) \cdot \frac{g(x+h)-g(x)}{h}\right].$$

As $h \to 0$, we have $g(x+h) \to g(x)$, the fractions within the brackets approach $f'(x)$ and $g'(x)$, respectively, and the result is established.

The formulas in Theorems 4.6.1. and 4.6.2. are known as the *product formula* and *quotient formula of differentiation*, respectively. Each of these may be stated in words.

The derivative of a product of two functions is equal to the product of the first times the derivative of the second plus the second times the derivative of the first.

The derivative of a fraction is the denominator times the derivative of the numerator minus the numerator times the derivative of the denominator, all divided by the square of the denominator.

**Example 4.6.1.** Differentiate $f(x) = (x^2 - 1)^{12}(x+3)^7$.

*Solution.* The function can be expanded, and differentiated as a polynomial, but this is impractical. We may apply the product formula and obtain

$$f'(x) = (x^2-1)^{12} D(x+3)^7 + (x+3)^7 D(x^2-1)^{12}$$
$$= (x^2-1)^{12} \cdot 7(x+3)^6 + (x+3)^7 \cdot 12(x^2-1)^{11} \cdot 2x$$
$$= (x^2-1)^{11}(x+3)^6[7(x^2-1) + 24x(x+3)]$$
$$= (x^2-1)^{11}(x+3)^6(31x^2 + 72x - 7).$$

Note the separate use of the chain rule in finding $D(x+3)^7$ and $D(x^2-1)^{12}$.

**Example 4.6.2.** Differentiate: $f(x) = \dfrac{x}{\sqrt{x^2+1}}$.

*Solution.* Using the quotient formula, we find

$$f'(x) = \frac{\sqrt{x^2+1}\, Dx - xD\sqrt{x^2+1}}{x^2+1}$$

$$= \frac{\sqrt{x^2+1} - x \cdot \dfrac{2x}{2\sqrt{x^2+1}}}{x^2+1}$$

$$= \frac{(x^2+1) - x^2}{(x^2+1)\sqrt{x^2+1}} = \frac{1}{(x^2+1)^{3/2}}.$$

We stated earlier that the *power formula* applies to any real exponent, but the proof had been given only for a positive integral exponent. We now use the quotient formula to extend the proof to an exponent which is a negative integer. Let $n = -m$, where $m$ is a *positive* integer. Then

$$Dx^n = Dx^{-m} = D\left(\frac{1}{x^m}\right) = \frac{x^m D1 - 1 \cdot Dx^m}{x^{2m}}$$

$$= \frac{-mx^{m-1}}{x^{2m}} = -mx^{-m-1} = nx^{n-1}.$$

Thus the power formula has been shown to hold also when $n$ is a negative integer.

Now let $n = 1/q$, with $q$ a positive integer. Starting with the formula

$$u^q - v^q = (u - v)(u^{q-1} + u^{q-2}v + \cdots + v^{q-1}),$$

and letting $u = x^{1/q}$, $v = a^{1/q}$, we obtain

(4.6.1) $$\frac{x^{1/q} - a^{1/q}}{x - a} = \frac{1}{x^{1-1/q} + x^{1-2/q}a^{1/q} + \cdots + a^{1-1/q}}.$$

The limit of the left member as $x \to a$ is $f'(a)$ for $f(x) = x^{1/q}$. As $x \to a$ each of the $q$ terms in the denominator of the fraction on the right of (4.6.1) approaches $a^{1-1/q}$, and hence

$$f'(a) = \frac{1}{qa^{1-1/q}} = \frac{1}{q}a^{1/q-1} = na^{n-1}.$$

The above is valid, if $a > 0$, or if $a < 0$ and $q$ is odd. This proves the power formula for an exponent which is the reciprocal of a positive integer.

Finally, if $n = p/q$, where $p$ and $q$ are integers with no common factor (other than 1), and $q > 0$, we have

$$x^n = x^{p/q} = (x^{1/q})^p.$$

We may now apply the chain rule, using the forms of the power rule already established, and conclude that *the power rule of differentiation is valid for a rational exponent.*

**EXERCISE GROUP 4.5**

In Exercises 1–34 find the derivative of the function with respect to the variable which appears ($a$ is constant).

1. $t^2(2 + 3t)^4$
2. $(x + 3)^2(x - 5)^2$
3. $(2x + 1)^2(x - 2)^4$
4. $x^2\sqrt{x^2 - 1}$
5. $x^{2/3}\sqrt[3]{x^2 - 4}$
6. $(x - 2)^2(x + 1)$
7. $\dfrac{x}{1 - x}$
8. $\dfrac{x}{x^4 + 1}$

9. $\dfrac{x^2-4}{x^2+4}$

10. $\dfrac{1-x}{1+x}$

11. $\dfrac{2x+3}{(x+3)^2}$

12. $\dfrac{(x-2)^2}{(x-3)^3}$

13. $\dfrac{2-x^2}{1+x^2}$

14. $\dfrac{(x^2+1)^3}{(x^3+1)^2}$

15. $(1+x)\sqrt{2+x}$

16. $(x-1)\left(\dfrac{1}{\sqrt{x}}+\sqrt{x}\right)$

17. $\dfrac{1+\sqrt{x}}{1-\sqrt{x}}$

18. $\dfrac{t}{\sqrt{at^2+1}}$

19. $\dfrac{x^2}{(1+x)^3}$

20. $\dfrac{(2x+1)^2}{(x+1)^2}$

21. $\sqrt{\dfrac{1-x}{1+x}}$

22. $\dfrac{v^2-1}{\sqrt{v^2+1}}$

23. $\sqrt{\dfrac{u}{u^2+1}}$

24. $\dfrac{x}{\sqrt{x^2-1}}$

25. $\dfrac{x^2}{1+\sqrt{x}}$

26. $\sqrt{\dfrac{1-x^2}{1+x^2}}$

27. $\sqrt[3]{\dfrac{1+x}{1-x}}$

28. $\sqrt[3]{\dfrac{1-x}{1+x}}$

29. $\dfrac{4+x^2}{\sqrt{4-x^2}}$

30. $\dfrac{x^2}{1+\sqrt{1-x^2}}$

31. $\dfrac{x^2}{2-\sqrt{4-x^2}}$

32. $\dfrac{\sqrt{x}}{\sqrt{x^2-a^2}}$

33. $\dfrac{\sqrt{1+x^2}-\sqrt{1-x^2}}{\sqrt{1+x^2}+\sqrt{1-x^2}}$

34. $\dfrac{\sqrt{a+x}-\sqrt{a-x}}{\sqrt{a+x}+\sqrt{a-x}}$

In Exercises 35–43 assume that $y = y(x)$ is a differentiable function of $x$, with derivative $y'$. Then find the derivative of the specified function in terms of $x$, $y$, and $y'$.

35. $xy$

36. $xy^3$

37. $x^2y^{1/2}$

38. $x/y$

39. $\dfrac{x}{x+y}$

40. $\dfrac{x-y}{x+y}$

41. $2xy+y^2$

42. $\dfrac{1}{x^2+y^2}$

43. $\dfrac{xy}{x^2+y}$

In the following assume that $\sin x$ is differentiable, with $D \sin x = p(x)$. Find the derivative of each function in terms of $p(x)$.

44. $x \sin x$

45. $x^2 \sin x$

46. $x \sin 2x$

47. $\dfrac{x}{\sin x}$

48. $\sin x \sin 2x$

49. $x \sin^2 x$

50. (a) Use the definition of derivative to show directly that

$$D[g(x)]^{-1} = D[1/g(x)] = -g'(x)/[g(x)]^2.$$

(b) Use the result of (a) and the product formula applied to the product $f(x)[g(x)]^{-1}$, to derive the quotient formula.

## 4.7 IMPLICIT DIFFERENTIATION

The function concept, defined in Section 3.1, is quite general and may be used with a domain (or range) which is other than a set of numbers. In particular, the concept applies also in the case where the domain consists of a set of *pairs* of real numbers.

**DEFINITION 4.7.1.** *Let $A_1$ be a set of ordered pairs of real numbers, $p$, and $A_2$ a set of real numbers, $z$. A set of ordered pairs $(p, z)$, where $p$ is in $A_1$ and $z$ is in $A_2$, is a **function** $f$, if no two distinct numbers $z$ correspond to the same element $p$. If each $p$ in $A_1$ is used at least once, $A_1$ is the **domain** of $f$; if each $z$ in $A_2$ is used at least once, $A_2$ is the **range** of $f$.*

We may interpret $A_1$ geometrically as a set of points in the $xy$ plane, and $z$ as the *value* of $f$ for the point $(x, y)$. We then write $z = f(x, y)$. We often refer, somewhat loosely, to the function $f(x, y)$, and understand as the *domain* the set of points $(x, y)$ at which $f$ has a value. Such a function is also called a *function of two variables*.

***Example 4.7.1.*** (i) If $f(x, y) = 1 + xy + y^2$, the domain consists of *all* pairs of real numbers $(x, y)$. We have, for example,

$$f(2, 3) = 1 + 2 \cdot 3 + 3^2 = 16; \qquad f(\sqrt{3}, 1) = 1 + \sqrt{3} \cdot 1 + 1^2 = 2 + \sqrt{3}.$$

(ii) If $f(x, y) = (x - y)/(x + y)$, the domain consists of those pairs of real numbers such that $x + y \neq 0$, which correspond to all points in the $xy$ plane *not* on the line $x + y = 0$.

Let us examine the question of *solving the equation*

(4.7.1) $\qquad\qquad\qquad f(x, y) = 0,$

that is, of finding pairs of numbers $(x, y)$ for which (4.7.1) holds. For example, if $f(x, y) = 1 + xy + y^2$, we have

(4.7.2) $$1 + xy + y^2 = 0.$$

We may treat this as a quadratic equation in $y$, and solve as

$$y_1 = \frac{-x + \sqrt{x^2 - 4}}{2}, \quad y_2 = \frac{-x - \sqrt{x^2 - 4}}{2}.$$

Each of these equations defines a function of $x$. Each of these functions, which may be written as $y_1(x)$ and $y_2(x)$, satisfies (4.7.2), so that

$$f(x, y_1(x)) = 0 \quad \text{and} \quad f(x, y_2(x)) = 0.$$

We may also treat (4.7.2) as a linear equation in $x$, and solve it as

$$x = -\frac{y^2 + 1}{y}, \quad y \neq 0;$$

this defines $x$ as a function $x(y)$ of $y$, and we have

$$f(x(y), y) = 0.$$

In general it can be shown that, under fairly broad conditions imposed on a function $f(x, y)$, functions $y(x)$ and $x(y)$ exist such that

(4.7.3) $$f(x, y(x)) = 0, \quad f(x(y), y) = 0.$$

The equation $f(x, y) = 0$ is then said to *define $y(x)$ implicitly as a function of $x$, and $x(y)$ implicitly as a function of $y$*. As the illustration above suggests, an equation in $x$ and $y$ may define more than one function $y(x)$; similarly, more than one function $x(y)$ may be defined.

We have seen that functions defined by (4.7.1) can sometimes be found explicitly by solving the equation; it is clear then that a function defined implicitly differs from a function defined explicitly only in the way it is defined. More generally, an equation (4.7.1) may define a function implicitly even when the function cannot be obtained explicitly, as in the case of the equation

$$x^5 + xy + y^5 = 0.$$

It can also be shown that under rather general conditions functions $y(x)$ and $x(y)$ defined by (4.7.1) are differentiable in suitable domains. We assume this to be the case for the functions we consider, and shall show by means of examples how derivatives may be found.

**Example 4.7.2.** Find $D_x y$ for a function $y(x)$ defined by the equation

(4.7.4) $$1 + xy + y^2 = 0.$$

*Solution.* When $y(x)$ is substituted for $y$ in the left member of (4.7.4), the resulting function of $x$ is identically zero. Hence its derivative is zero, and

$$D_x(1 + xy + y^2) = 0.$$

Carrying out the differentiation by the method of Section 4.5, we have

$$xD_xy + y + 2yD_xy = 0, \quad (x + 2y)D_xy = -y,$$

and we find

$$D_xy = -\frac{y}{x + 2y}.$$

**Example 4.7.3.** Find $D_yx$ for a function $x(y)$ defined by (4.7.4).

*Solution.* If $x(y)$ is substituted for $x$ in (4.7.4), an identity in $y$ results, the derivative of the left member is zero, and

$$D_y(1 + xy + y^2) = 0.$$

Hence

$$x + yD_yx + 2y = 0, \quad D_yx = -\frac{x + 2y}{y}.$$

*Note 1.* The function $x(y)$ as defined in Example 4.7.3 was obtained explicitly earlier as

(4.7.5) $$x = -\frac{y^2 + 1}{y}.$$

This can be differentiated to give $D_yx = (1 - y^2)/y^2$. It can be verified directly that this is equivalent to the result of Example 4.7.3, if, in the latter, $x$ is replaced by its explicit value (4.7.5).

*Note 2.* The derivatives $D_xy$ and $D_yx$ in Examples 4.7.2 and 4.7.3 are reciprocals. This can be proved generally. (See Theorem 10.5.2.)

*Note 3.* A characteristic of implicit differentiation is that the expression for a derivative usually contains both $x$ and $y$.

### EXERCISE GROUP 4.6

In Exercises 1–12 find $D_xy$ by implicit differentiation, assuming that the equation defines a function $y = y(x)$ which is differentiable.

1. $y^2 = 8ax$
2. $x^2 + y^2 = 4$
3. $x^2 - y^2 = 7$
4. $2x^2 + 3y^2 = 6$
5. $x^{1/2} + y^{1/2} = 4$
6. $x^{2/3} + y^{2/3} = 7$
7. $x^3 + y^3 - 3xy = 0$
8. $x^2 + 2xy + y^2 - 3x = 7$
9. $y^2 = x^2 - x^3$
10. $x^2 + y^2 = 3x^2y^2$
11. $x^2 = \frac{y^2}{y^2 - 1}$
12. $x^2 + y^2 = 4\frac{y^2}{x^2}$

In Exercises 13–20 find an equation of the tangent to the curve at the specified point.

13. $x^2 + y^2 = 5$, $(1, -2)$
14. $x^2 - y^2 = 8$, $(-3, 1)$
15. $x^{1/2} + y^{1/2} = 3$, $(4, 1)$
16. $x^{2/3} + y^{2/3} = 5$, $(1, -8)$
17. $x^3 + y^3 - 6xy = 0$, $(3, 3)$
18. $y^2 - 4x + 3 = 0$, $(3, 3)$
19. $x^2 - 3xy + 2y^2 - 7x + 19 = 0$, $(3, 1)$
20. $2x^2 + 3xy - 3y^2 + 2x - 3y + 50 = 0$, $(-2, 3)$

* 21. Consider any tangent to the curve $x^{1/2} + y^{1/2} = a^{1/2}$. If the tangent intersects the $x$ axis in the point $A$ and the $y$ axis in $B$, prove that $OA + OB$ is constant, equal to $a$.
* 22. Prove that the length of the segment between the coordinate axes, of any tangent to the curve $x^{2/3} + y^{2/3} = a^{2/3}$, is always $a$.
23. *Assuming* differentiability of $x^n$ if $n$ is rational, derive the power formula for $n = p/q$ by using the fact that $y = x^{p/q}$ satisfies the equation $y^q = x^p$.

## 4.8 DERIVATIVES OF HIGHER ORDER

The derivative $f'(x)$ of a function $f(x)$ is itself a function of $x$, its domain consisting of the values of $x$ in the domain of $f$ for which $f(x)$ is differentiable. We may ask whether $f'(x)$ has a derivative. If $f'(x)$ is differentiable, its derivative is called the *second derivative* of $f(x)$, and is denoted by $f''(x)$. Thus

$$f''(x) = [f'(x)]'.$$

In another notation the second derivative is

$$D^2 f(x) = D[Df(x)].$$

Derivatives of higher order than the second are also defined, and may be expressed similarly. For the third derivative we have (see Exercise 29)

$$f'''(x) = [f''(x)]' = [f'(x)]'';$$
$$D^3 f(x) = D[D^2 f(x)] = D^2[Df(x)].$$

Fourth and higher order derivatives may also be expressed in each notation.

*Example 4.8.1.* Find the derivatives of *all* orders of the function
$$f(x) = 3x^3 - 2x^2 + 7x - 9.$$

*Solution.* Applying the usual formulas of differentiation, we find

$$f'(x) = 9x^2 - 4x + 7, \quad f''(x) = 18x - 4, \quad f'''(x) = 18,$$
$$f^{(4)}(x) = f^{(5)}(x) = \cdots = 0.$$

We see that every derivative of order more than 3 is identically zero; we express this as $f^{(n)}(x) = 0$, $n \geq 4$. The superscript in parentheses indicates an order of derivative, not an exponent.

*Example 4.8.2.* Find the derivatives of all orders of $f(x) = x^{-1}$.

*Solution.* Using the power formula, we have

$$f'(x) = -x^{-2}, \quad f''(x) = 2x^{-3}, \quad f'''(x) = -2 \cdot 3 x^{-4},$$
$$f^{(4)}(x) = 2 \cdot 3 \cdot 4 x^{-5}.$$

By inspection we can write a general formula for which a formal proof can be given by mathematical induction:

$$f^{(n)}(x) = (-1)^n n! x^{-n-1}.$$

The symbol $n!$ (read *n factorial*) is the product of all the positive integers from 1 to $n$, inclusive, that is,

$$n! = 1 \cdot 2 \cdot 3 \cdots n.$$

**Example 4.8.3.** Find $f'(x)$ and $f''(x)$ for $f(x) = x/\sqrt{2x+1}$.

*Solution.* Using the quotient formula, we find

$$f'(x) = \frac{\sqrt{2x+1} - \dfrac{x \cdot 2}{2\sqrt{2x+1}}}{2x+1} = \frac{x+1}{(2x+1)^{3/2}},$$

$$f''(x) = \frac{(2x+1)^{3/2} - (x+1)(3/2)(2x+1)^{1/2} \cdot 2}{(2x+1)^3} = -\frac{x+2}{(2x+1)^{5/2}}.$$

The determination of higher order derivatives by use of implicit differentiation involves no new methods and will be illustrated.

In Example 4.7.2 the derivative of a function $y(x)$ defined by the equation $1 + xy + y^2 = 0$ was determined as

(4.8.1) $$D_x y = -\frac{y}{x+2y}.$$

To find $D_x^2 y$ we differentiate both members of (4.8.1) with respect to $x$.

$$D_x^2 y = -\frac{(x+2y)D_x y - y(1 + 2D_x y)}{(x+2y)^2} = \frac{y - xD_x y}{(x+2y)^2}.$$

We now substitute for $D_x y$ its value from (4.8.1), and we obtain

$$D_x^2 y = \frac{y + x \cdot \dfrac{y}{x+2y}}{(x+2y)^3} = \frac{2xy + 2y^2}{(x+2y)^3}.$$

We may simplify the expression for $D_x^2 y$ by making use of the given equation which defines the function $y$. The numerator of $D_x^2 y$ is equal to $2xy + 2y^2$, and from the given equation this is equal to $-2$. Hence we obtain the simpler form

$$D_x^2 y = \frac{-2}{(x+2y)^3}.$$

**Example 4.8.4.** Find $y'$ and $y''$ for a function $y(x)$ defined by the equation

(4.8.2) $$x^3 - 3xy + y^3 = 0.$$

*Solution.* Setting the derivative, with respect to $x$, of the left member of (4.8.2) equal to zero, we have

$$3x^2 - 3(xy' + y) + 3y^2 y' = 0,$$

which is solved for $y'$ as

(4.8.3)
$$y' = \frac{y - x^2}{y^2 - x}.$$

Then

(4.8.4)
$$y'' = \frac{(y^2 - x)(y' - 2x) - (y - x^2)(2yy' - 1)}{(y^2 - x)^2}.$$

Using (4.8.3), we find

$$y' - 2x = \frac{y - 2xy^2 + x^2}{y^2 - x}, \qquad 2yy' - 1 = \frac{y^2 - 2x^2y + x}{y^2 - x},$$

and substitution of these in (4.8.4) gives

$$y'' = \frac{(y^2 - x)(y - 2xy^2 + x^2) - (y - x^2)(y^2 - 2x^2y + x)}{(y^2 - x)^3}$$

$$= -\frac{2xy(x^3 - 3xy + y^3 + 1)}{(y^2 - x)^3}.$$

Since $x$ and $y$ satisfy (4.8.2), the value of $y''$ reduces to

$$y'' = -\frac{2xy}{(y^2 - x)^3}.$$

The examples just completed also illustrate the manner in which the defining equation can sometimes be used to simplify a derivative obtained by implicit differentiation.

### EXERCISE GROUP 4.7

In Exercises 1–4 find the derivatives of all orders of the given function.

1. $f(x) = 2x^4 - x^3 + 3x^2 + 2x - 3$
2. $f(x) = x^5 - 1$
3. $f(x) = (x^2 - 2)(x^2 + 4)$
4. $f(x) = ax^3 + bx^2 + cx + d$

In Exercises 5–10 find the second derivative of the given function.

5. $\dfrac{x}{\sqrt{x+1}}$

6. $\dfrac{\sqrt{x-1}}{x^2}$

7. $\dfrac{x+1}{x-1}$

8. $\dfrac{\sqrt[3]{x+1}}{x^2}$

9. $\sqrt{\dfrac{x+1}{x-1}}$

10. $\dfrac{x^2}{\sqrt{x^2+1}}$

In Exercises 11–16 find a general expression for the $n$th derivative of the function in terms of $n$.

11. $\dfrac{1}{x-1}$

12. $\sqrt{x}$

13. $\dfrac{1}{\sqrt{x}}$

14. $\dfrac{x+1}{x-1}$

15. $\sqrt[3]{x}$

16. $\dfrac{1}{\sqrt[3]{x}}$

In each of Exercises 17–22 find $y''$ for a function $y = y(x)$ defined by the given equation.

17. $x^2 + y^2 = 4$

18. $x^2 - 2y^2 = 7$

19. $4x^2 + 3y^2 = 2$

20. $x^2 - 2xy - y^2 = 1$

21. $2x^2y + 4x + y^2 = 7$

22. $4x^2 + 2y^3 = 9$

In each of the following show that the given function satisfies the equation which follows it.

23. $y = -x^2 + 4x^{-1}$; $\quad x^2y'' - 2y = 0$

24. $y = 2x^{1/2} - 7x^3$; $\quad 2x^2y'' - 5xy' + 3y = 0$

25. $y = 3x/2 + 3x^{-1} - 4x^2$; $\quad x^3y''' + x^2y'' - 2xy' + 2y = 0$

26. $y = 3x^{-2} + x^4$; $\quad x^2y'' - 6y = 6x^4$

27. $y = x^3 + x - 2$; $\quad x^2y'' - 3xy' + 3y = -6$

28. $y = (2x+1)^2 - (2x+1)^{-3}$; $\quad (2x+1)^2y'' + 4(2x+1)y' - 24y = 0$

29. (a) If $f'''$ is defined as $(f'')'$, show that it is also equal to $(f')''$.

(b) If $f^{(4)}$ is defined as $(f''')'$, show that
$$f^{(4)}(x) = [f''(x)]'' = [f'(x)]'''.$$

★ 30. If $F(x) = f[g(x)]$, show that
$$F''(x) = f'[g(x)]g''(x) + f''[g(x)][g'(x)]^2.$$

## 4.9 DIFFERENTIALS

We present here the concept of the *differential*, which is closely related to that of the derivative. The use of differentials is not essential in any connection, but it can be very helpful in some places, particularly in working with integrals, which are introduced in Chapter 9. Formulas for working with differentials parallel those for derivatives.

Let $f$ be a differentiable function, and let $dx$ denote a variable which may take on any real value. We define a function $df$ of the two variables $x$ and $dx$ by the equation

(4.9.1) $$df(x, dx) = f'(x)\, dx.$$

The function $df$ is called the *differential* of the function $f$.

If we apply (4.9.1) to the function $f(x) = x$, we obtain, since $f'(x) = 1$, the equation $df = dx$. Thus the variable $dx$ may be called the *differential* of the function $x$.

We obtain immediately from (4.9.1), if $dx \neq 0$,

(4.9.2) $$\frac{df}{dx} = f'(x);$$

that is, *the quotient of the differential of f by the differential of x is equal to the derivative of f with respect to x* (*not* surprisingly, since definitions were chosen to lead to this result).

Suppose now that $x$ is a function of another variable, $x = g(u)$, where $g$ is differentiable. Then we may write

(4.9.3) $$f(x) = f[g(u)] = \phi(u).$$

In terms of $u$ and $du$ we have, using the chain rule to express $\phi'(u)$,

$$d\phi(u, du) = \phi'(u)\, du = f'[g(u)]g'(u)\, du.$$

This becomes, since $x = g(u)$ and $dx(u, du) = g'(u)\, du$,

(4.9.4) $$d\phi(u, du) = f'[g(u)]\, dx(u, du) = f'(x)\, dx(u, du).$$

Regardless of how $dx$ is determined, the right member of (4.9.4) is $df(x, dx)$, by (4.9.1), and we may write

$$d\phi(u, du) = df(x, dx).$$

Thus the function $f$, defined in terms of $x$, and the function $\phi$, defined in terms of $u$, have equal differentials, provided that $x$ is a function of $u(x = g(u))$ and $dx$ and $du$ are related by this function ($dx = g'(u)du$).

It follows that the differential relations are valid without reference to what the independent variable is. Thus, if $y = f(x)$, we have

(4.9.5) $$dy = f'(x)\, dx$$

whether or not $x$ depends on another variable. If, in addition, the relation $x = g(u)$ holds, then $dx = g'(u)\, du$, and from (4.9.5) we get

(4.9.6) $$dy = f'(x)g'(u)\, du.$$

Hence, since $dy/du$ is the derivative of $y(=f[g(u)])$ with respect to $u$ and $f'(x) = f'[g(u)]$, we have from (4.9.6)

$$D_u y = f'(x)g'(u) = f'[g(u)]g'(u);$$

*this is a statement of the chain rule of differentiation.*

The relation

$$\frac{dy}{dx} = f'(x),$$

obtained from (4.9.5) *may now be used to define a new notation, $dy/dx$, for the derivative of the function $y = f(x)$.* In this notation for derivative what is significant is that a variable $y$ is functionally dependent on a variable $x$, whether this dependence is given directly, or by virtue of intermediate variables, or implicitly as in Section 4.7.

Let us substitute the following in (4.9.6):

$$f'(x) = \frac{dy}{dx}, \qquad g'(u) = \frac{dx}{du}.$$

We may then write, in terms of differentials,

(4.9.7) $$\frac{dy}{du} = \frac{dy}{dx}\frac{dx}{du}.$$

We know that this is algebraically correct if $du \neq 0$, $dx \neq 0$, since differentials are numbers. Thus (4.9.7) is *valid whether interpreted in terms of differentials, or with quotients of differentials interpreted as derivatives*, and herein lies an advantage in the use of differentials.

**Example 4.9.1.** Let $y = f(x) = x^3$. Then $f'(x) = 3x^2$, and we may express the differential relation in any of the forms

$$df = 3x^2 \, dx, \qquad dy = 3x^2 \, dx, \qquad d(x^3) = 3x^2 \, dx.$$

**Example 4.9.2.** Find $dy$ if $y = \dfrac{x^2 + 1}{x + 1}$.

*Solution.* We have

$$D\left(\frac{x^2 + 1}{x + 1}\right) = \frac{(x+1)2x - (x^2 + 1)}{(x+1)^2} = \frac{x^2 + 2x - 1}{(x+1)^2}.$$

Hence

$$dy = \frac{x^2 + 2x - 1}{(x+1)^2} \, dx.$$

The following example illustrates the process of finding a derivative as the quotient of two differentials.

**Example 4.9.3.** If $y = x^3 - 2x$ we have

$$dy = (3x^2 - 2) \, dx.$$

If also $u = \sqrt{x - 1}$, we have

$$du = \frac{1}{2\sqrt{x - 1}} \, dx.$$

Dividing $dy$ by $du$ we obtain

$$\frac{dy}{du} = 2(3x^2 - 2)\sqrt{x - 1}.$$

Since the given equations define $y$ as a function of $u$, the value obtained for $dy/du$ is the derivative of $y$ respect to $u$ (even though $u$ does not appear explicitly in that expression).

We may obtain the explicit function by setting $x = u^2 + 1$ in the expression for $y$, getting
$$y = (u^2 + 1)^3 - 2(u^2 + 1),$$
and then verify that the derivative $D_u y$, obtained directly, is equal to the value of $dy/du$ obtained above.

The formulas of differentiation in terms of derivatives carry over completely to differentials; some of them are listed:

1. $dc = 0$, if $c$ is a constant;
2. $d(x^n) = nx^{n-1}\, dx$;
3. $d[cf(x)] = c\, df(x)$, if $c$ is a constant;
4. $d[f(x) + g(x)] = df(x) + dg(x)$;
5. $d[f(x)g(x)] = f(x)\, dg(x) + g(x)\, df(x)$;
6. $d\dfrac{f(x)}{g(x)} = \dfrac{g(x)\, df(x) - f(x)\, dg(x)}{g^2(x)}$;
7. $df[g(x)] = f'[g(x)]\, dg(x)$.

The following example illustrates the use of the differential formulas.

**Example 4.9.4.** Find $dy$ if $y = \dfrac{x}{(x+1)^2}$.

*Solution.* We have

$$dy = \frac{(x+1)^2\, dx - x\, d(x+1)^2}{(x+1)^4} \qquad \text{[by 6.]}$$

$$= \frac{(x+1)^2\, dx - x \cdot 2(x+1)\, d(x+1)}{(x+1)^4} \qquad \text{[by 2. and 7.]}$$

$$= \frac{(x+1)^2\, dx - 2x(x+1)\, dx}{(x+1)^4} \qquad \text{[by 4. and 1.]}$$

$$= \frac{1-x}{(x+1)^3}\, dx.$$

Differentials may be very conveniently used to carry out implicit differentiation.

**Example 4.9.5.** Find $dy/dx$ for a function $y = f(x)$ defined by the equation
$$x^2 - xy + 2y^2 = 4.$$

*Solution.* It really does not matter whether we consider that the equation defines $y$ as a function of $x$ or the reverse. In either case, using rules for differentials as given above, we have

$$d(x^2 - xy + 2y^2) = d(4) = 0,$$
$$2x\, dx - x\, dy - y\, dx + 4y\, dy = 0,$$
$$(2x - y)\, dx - (4y - x)\, dy = 0.$$

We solve the last equation for $dy/dx$, and we have

$$\frac{dy}{dx} = \frac{2x - y}{4y - x}.$$

If we want $dx/dy$, the derivative of the function $x = x(y)$ determined by the equation, we obtain

$$\frac{dx}{dy} = \frac{4y - x}{2x - y}.$$

The notation $dy/dx$ also extends to higher derivatives, and becomes

$$\frac{d^2y}{dx^2} = f''(x), \quad \frac{d^3y}{dx^3} = f'''(x), \ldots, \frac{d^ny}{dx^n} = f^{(n)}(x).$$

The symbol for each higher derivative will be considered as an entity and *not* as the quotient of two quantities.

**EXERCISE GROUP 4.8**

Find the differential of the function in Exercises 1–12.

1. $x^3 - 3x^2$
2. $-4x^3 + 2x^2 - 7$
3. $3x^{2/3} - 4x^{1/3}$
4. $\sqrt{2t - 1}$
5. $\sqrt{t^2 + 1}$
6. $\dfrac{t}{t - 1}$
7. $\dfrac{2t + 1}{3t - 1}$
8. $\dfrac{t}{t^2 + 1}$
9. $\dfrac{t}{\sqrt{t + 1}}$
10. $(2u^2 - 1)^5$
11. $\dfrac{x^2 - 1}{x^2 + 1}$
12. $(1 - x^2)^{1/3}$

In Exercises 13–18 assume that $u$ and $v$ are functions of $x$. Find the differential of the function (a) in terms of $du$ and $dv$, and (b) in terms of $dx$.

13. $u^2$
14. $\sqrt{u^2 + 1}$
15. $\sqrt{u^2 + v^2}$
16. $\dfrac{u}{v}$
17. $uv$
18. $\dfrac{1}{u - v}$

Use differentials and implicit differentiation to find the indicated derivative in each of the following, assuming that the appropriate function is defined by the equation.

19. $D_x y\colon x^2 + y^2 = 4$
20. $D_x y\colon x^2 - 2xy - 3y^2 = 4$
21. $D_y x\colon x^2 - 3xy + 2 = 0$
22. $D_x y\colon x^3 - xy + y^2 - 2 = 0$
23. $D_u t\colon tu + t^2 + 2t = 7$
24. $D_u v\colon (u+v)^2 - 2(u+v) = 1$

## 4.10 INTERPRETATION OF DIFFERENTIALS

If we rewrite the definition of derivative of a function $f$, with the independent variable $h$ replaced by $dx$, we have

(4.10.1) $$\lim_{dx \to 0} \frac{f(x + dx) - f(x)}{dx} = f'(x).$$

We now set

(4.10.2) $$\frac{f(x + dx) - f(x)}{dx} - f'(x) = \epsilon.$$

Since $x$ is kept fixed, $\epsilon$ is a function of $dx$, and we may write

$$\lim_{dx \to 0} \epsilon = 0.$$

We obtain, from (4.10.2),

(4.10.3) $$f(x + dx) - f(x) = f'(x)\, dx + \epsilon \cdot dx.$$

The left member of (4.10.3) is the change in $f$ as $x$ increases by $dx$ (which may take on any real value). Denoting this change, called the *increment* of $f$, by $\Delta f$, and using the relation $f'(x)\, dx = df$, we have

(4.10.4) $$\Delta f = df + \epsilon \cdot dx.$$

Since $\epsilon$ is small when $dx$ is small, Equation (4.10.4) states that the difference $\Delta f - df$ is a product of two small quantities; if $df \neq 0$ and $dx$ is small, the difference $\Delta f - df$ is small in comparison with $df$ (or with $\Delta f$). We may state the following.

*For a small increment of $x$, the differential $df$ of a function $f$ may be used as an approximation to the increment $\Delta f$ of the function.*

This principle may be used to compute the approximate value of a function in certain cases.

**Example 4.10.1.** Find the approximate value of $\sqrt[3]{7.98}$.

*Solution.* To do this with the use of differentials, we set up the function $y = x^{1/3}$, and compute $dy$ with $x_0 = 8$, $dx = -0.02$. We obtain $dy$ as

$$dy = \tfrac{1}{3} x^{-2/3}\, dx,$$

and substitute the values of $x_0$ and $dx$; thus

$$dy = \tfrac{1}{3}(8)^{-2/3}(-0.02) = -0.00167.$$

Then $y = y_0 + dy = 2 - 0.00167 = 1.99833$, which is an approximation to $\sqrt[3]{7.98}$. Indeed, the approximation is very good; it happens to be correct in all the figures given.

The principle stated earlier in this section may be adapted to the computation of errors. If $f(x)$ is to be computed, and the value of $x$ may be in error by an amount $dx$, then

*df is an approximation to the error in $f(x)$,*

*$df/f(x)$ is an approximate relative error in $f(x)$,*

*100 $df/f(x)$ is an approximate percentage error in $f(x)$.*

***Example 4.10.2.*** Find an approximate relative error in the volume of a cube, if the length of a side is measured as 6 in., with a possible error of 0.01 in.

*Solution.* The volume $V$ is given as a function of the length $x$ of a side as $V = x^3$. Then

$$\frac{dV}{V} = \frac{3x^2\,dx}{x^3} = \frac{3\,dx}{x}.$$

With $x = 6$ and $dx = 0.01$ as the maximum possible error in $x$, we get

$$\frac{3(0.01)}{6} = 0.005$$

as the maximum possible relative error (it has no units) in the volume.

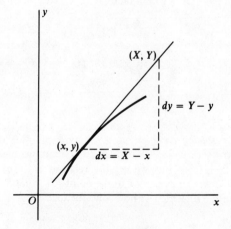

Fig. 4.10.1

A geometric interpretation of the differential of a function may be given as follows. Let $(X, Y)$ be the coordinates of a point on the tangent line to the graph of $y = f(x)$ at the point $(x, y)$, (Fig. 4.10.1). An equation of the tangent line is, by the slope-intercept form,

(4.10.5) $$Y - y = f'(x)(X - x).$$

Note that $x$ and $y$ are fixed, and that $X$ and $Y$ are the variable coordinates of a point on the line. If we set $dx = X - x$ in (4.10.5), we get

$$Y - y = f'(x)\, dx.$$

But $f'(x)\, dx = dy$. Hence $dy = Y - y$, and the differential $dy$ of a function $y = f(x)$ may be interpreted geometrically as the *increase in the ordinate of the tangent line* to the graph at $(x, y)$ when the abscissa increases by $dx$.

### EXERCISE GROUP 4.9

In Exercises 1–6 use differentials to find an approximate value of the given number.

1. $\sqrt{3.98}$
2. $\sqrt{4.02}$
3. $\sqrt[3]{1.03}$
4. $(2.1)^3 - \tfrac{4}{3}(2.1)^{1/2}$
5. $(2.99)^5$
6. $(7.02)^3$

7. The area of an equilateral triangle is computed from the length of a side. Determine the maximum percentage error in the area if a side is measured as $2 \pm 0.1$ in.

8. Approximate the amount of material in a closed cubical box if the inside dimension is 2.5 ft and the thickness is 0.2 in.

9. (a) Find an approximation to the relative change in the area of a sphere due to a change of $h$ in the radius.
   (b) Same as (a) for the volume of a sphere.

10. Find the relative error in the function $x^n$, if the relative error in $x$ is $E$.

Find $\Delta f - df$ for each of the functions in Exercises 11–16, in terms of $h = dx$, in such a form as to exhibit $h^k$, with $k > 1$, as a factor.

11. $f(x) = x^3 - x^2$
12. $f(x) = 2x^2 - 3x - 1$
13. $f(x) = \dfrac{1}{x}$
14. $f(x) = \dfrac{1}{x^2}$
* 15. $f(x) = \dfrac{x}{x+1}$
* 16. $f(x) = \sqrt{x}$

*Newton's Method for solving an equation.* We may apply (4.9.5) at $x = a$, and write $dy = f'(a)\, dx$. We now set $dx = x - a$ and, if $dx$ is small, we have as an approximate relation $dy = \Delta y = y - f(a)$. Thus we get the relation $y - f(a) = f'(a)(x - a)$. Now referring to Fig. 4.10.2, we see that $y = 0$ at the point $P$, so that for the abscissa of $P$ we obtain $-f(a) = f'(a)(x - a)$, which we now solve for $x$ as

$$x = a - \frac{f(a)}{f'(a)}.$$

Thus, if $a$ is close in value to the solution of an equation $f(x) = 0$, we may use the above value of $x$ to obtain a (usually) better approximation to the solution. The process may then be repeated with the new approximation as $a$.

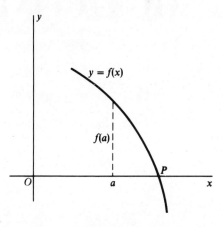

Fig. 4.10.2

* **17.** Apply Newton's Method with $a = 1$ to obtain an approximate solution $a_1$ of the equation $x^2 - 2 = 0$. Then apply the method again with $a = a_1$ to get another approximation $a_2$.

* **18.** Apply Newton's Method once with $a = 1.5$ to obtain an approximate solution of the equation $x^3 - 2 = 0$.

# 5 Applications of Derivatives

## 5.1 INTRODUCTION

The derivative and differentiation have been presented as purely analytic concepts. References to coordinate systems or graphs are not essential for this purpose, although use of these aids can certainly help in fixing the concepts in one's mind. On the other hand, there are many applications of these concepts, in geometry, in the physical sciences, and in the social sciences. We will study some of these applications in the present chapter in order to indicate the nature of the practical uses to which the derivative can be put, and to serve as an effective aid in understanding the ideas themselves. Additional applications will appear often in subsequent chapters.

## 5.2 DIRECTION OF A CURVE

Graphical or geometric applications of the derivative were introduced in Section 4.3. At that stage we defined the slope of a curve at a point as the slope of the tangent line to the curve at that point. If the curve was represented as $y = f(x)$ or as $g(x, y) = 0$, we saw that the slope was given by the value of the derivative $y'$ at a point. The effect of these statements is that questions relating to the direction of a curve become questions relating to slopes; these in turn become questions involving derivatives.

In Section 4.3 we also considered tangents and normals to curves. We are now in a position to consider a wider variety of curves in this way. Furthermore, questions affecting angles of intersection of curves can be answered by working with slopes. We note the appropriate items from the study of straight lines.

1. Two lines, neither one parallel to the $y$ axis, are parallel, if and only if their slopes are equal: $m_1 = m_2$.
2. Two lines, neither one parallel to a coordinate axis, are orthogonal (perpendicular), if and only if their slopes are negative reciprocals: $m_1 m_2 = -1$.
3. An angle $\theta$ between two nonperpendicular lines with slopes $m_1$ and $m_2$ is given by (see Exercise 15 of Exercise Group 2.4)

$$\tan \theta = \frac{m_1 - m_2}{1 + m_1 m_2}, \qquad 1 + m_1 m_2 \neq 0.$$

*Example 5.2.1.* Show that the following curves intersect at right angles: $x^2 + 2y^2 = 4$, $x^2 - y^2 = 1$.

*Solution.* The equations may be solved simultaneously by first eliminating $x^2$, for example. This yields $y^2 = 1$, and then we find $x^2 = 2$. The curves, an ellipse and a hyperbola, intersect in the four points $(\pm\sqrt{2}, \pm 1)$. We use implicit differentiation. For the ellipse (the first curve) we have

$$2x + 4yy' = 0, \qquad y' = -x/2y.$$

For the hyperbola we have

$$2x - 2yy' = 0, \qquad y' = x/y.$$

Consider the point $(\sqrt{2}, 1)$. At this point the slopes are, respectively, $m_1 = -\sqrt{2}/2$, $m_2 = \sqrt{2}$. Hence $m_1 m_2 = -1$, and the curves are orthogonal at this point of intersection. The orthogonality at the other points of intersection follows similarly.

*Example 5.2.2.* Find an angle between the curves

$$y^2 = 2x, \qquad x^2 - xy - 2y^2 + 8 = 0$$

at the point (2, 2).

*Solution.* It is easily verified that (2, 2) is a point of intersection. The derivatives are

$$y' = \frac{1}{y}, \qquad y' = \frac{2x - y}{x + 4y}.$$

At the point (2, 2) the slopes are then $m_1 = 1/2$, $m_2 = 1/5$, and if $\theta$ is an angle between the curves,

$$\tan \theta = \frac{1/2 - 1/5}{1 + (1/2)(1/5)} = \frac{3}{11}, \qquad \theta = 15°, \text{ approx.}$$

**Example 5.2.3.** Show that if a curve of the form $x^2 - y^2 = k \neq 0$ intersects a curve of the form $y^2 + 2xy - x^2 = c \neq 0$, it does so at an angle of 45°.

*Solution.* The slopes of the curves are obtained, respectively, as

$$m_1 = \frac{x}{y}, \quad m_2 = \frac{x-y}{x+y}.$$

We cannot have $x + y = 0$ since then $x^2 - y^2 = 0$. We assume for the moment that $y \neq 0$. If $(x_0, y_0)$ is a point of intersection, and $\theta$ is an angle of intersection, we get

$$\tan \theta = \frac{\dfrac{x_0}{y_0} - \dfrac{x_0 - y_0}{x_0 + y_0}}{1 + \dfrac{x_0}{y_0} \dfrac{x_0 - y_0}{x_0 + y_0}} = \frac{x_0^2 + y_0^2}{x_0^2 + y_0^2} = 1.$$

Hence an angle of intersection is 45°, and the property holds. If $y_0 = 0$ then $x_0 \neq 0$, the tangent to the first curve is parallel to the $y$ axis, the tangent to the second curve has slope 1, and 45° is also an angle between them.

**EXERCISE GROUP 5.1**

In Exercises 1–6 find an equation of the indicated tangent line.

1. Tangent to $x^2 + xy + 1 = 0$ at $(1, -2)$.
2. Tangent to $y = 2x^3 - 3x^2 - 6$ at $x = 2$.
3. Tangent to $y = (x - 1)/(2x - 1)$ at $x = -1$.
4. Tangent to $y = (x^2 - 1)/x^2$ at $x = 1/2$.
5. Tangent to $y = x^2 + 1/x$ at $x = 1$.
6. Tangent to $x^3y + y^3 + x + 1 = 0$ at $(-1, -1)$.
7. Find an equation of the tangent line and an equation of the normal line to the curve $y = x/(x - 1)$ at the origin.
8. Find an equation of the tangent line and an equation of the normal line to the curve $y = x^{3/2} - 3x + x^{1/2} + 1$ at the point with $x = 4$.
9. Find the points on the curve $y = x^3 - 6x^2 + 12x + 2$ at which the tangent is horizontal.
10. Find the points on the curve $y = 4x^3 - 5x^2 - 2x + 4$ at which the tangent is horizontal.
11. Find equations of the tangents to the curve $2x^2 + 3y^2 = 5$ which have slope $2/3$.
12. Find equations of the tangents to the curve $y = x + 1/x$ which have slope $-3$.
13. Show that the curves $y^2 = 4x + 4$ and $y^2 + 6x = 9$ intersect at right angles.
14. Show that the curves $4x^2 + 7y^2 = 28$ and $2x^2 - y^2 = 2$ intersect at right angles.
15. Show that the line $y = 2x + 3$ is tangent to the curve $y = 4x^2 - 6x + 7$.

16. Show that the line $y = 3x - 2$ is tangent to the curve $6x^2 - y^2 = 8$.
17. Show that the curves $y = 4x^2 - 6$ and $y = 3x^2 - 2x - 7$ are tangent to each other.
18. Find the smaller angle at which the curves $4y = -12x^2 + 5$ and $y = 2x^2$ intersect.
19. Show that the slope of the curve $y = (x^4 + 6x^2 - 3)/x$ is never negative.
20. Show that the sum of the slopes of the curves $y = x + 1/x$ and $y = ax$ at any point of intersection is constant.

## 5.3 INCREASING AND DECREASING FUNCTIONS

We have a general idea of what is meant in ordinary usage by the statement that a quantity is increasing or that a quantity is decreasing. For mathematical purposes, however, these concepts must be made precise, and will be defined for a function. It should be kept in mind that the idea of increasing or decreasing function expresses a type of behavior *as the variable increases*. In what follows the student should attempt to relate the precise statements to the intuitive idea. The concepts are defined initially at a specific value, and in this form involve an *instantaneous* type of behavior (Definition 5.3.1). In Definition 5.3.2 the ideas are extended to an interval.

**DEFINITION 5.3.1.** *A function $f$ is (strictly) increasing at $x_0$ if $x_0$ lies in some interval $(a, b)$ in the domain of $f$ such that*

$$f(x) < f(x_0) \text{ when } x \text{ is in } (a, x_0),$$
$$f(x) > f(x_0) \text{ when } x \text{ is in } (x_0, b).$$

*Similarly, $f$ is (strictly) decreasing at $x_0$ if $x_0$ lies in some interval $(a, b)$ in the domain of $f$ such that*

$$f(x) > f(x_0) \text{ when } x \text{ is in } (a, x_0),$$
$$f(x) < f(x_0) \text{ when } x \text{ is in } (x_0, b).$$

In general, $f$ is increasing if $f$ is smaller to the left and larger to the right. A corresponding statement applies to a decreasing function.

For differentiable functions a simple and very useful criterion exists in the sign of the derivative.

**THEOREM 5.3.1.** *If $f'(x_0) > 0$, then $f$ is increasing at $x_0$. If $f'(x_0) < 0$, then $f$ is decreasing at $x_0$.*

*Proof.* We prove the first part; the proof of the second part is similar. If $f'(x_0) > 0$, we have

$$f'(x_0) = \lim_{h \to 0} \frac{f(x_0 + h) - f(x_0)}{h} > 0.$$

The fact that the limit is a positive number tells us that for $|h|$ small enough and not zero the ratio $[f(x_0 + h) - f(x_0)]/h$ is also positive; hence $f(x_0 + h) - f(x_0)$ has the same sign as $h$. Thus

$$f(x_0 + h) - f(x_0) \begin{cases} > 0 \text{ if } h > 0 \\ < 0 \text{ if } h < 0, \end{cases}$$

and $f$ is increasing at $x_0$ according to Definition 5.3.1.

The case $f'(x_0) = 0$ is not covered by Theorem 5.3.1. There are various possibilities for this case; one of them is given in the following result.

**THEOREM 5.3.2.** *Let $f$ be continuous at $x_0$. If $x_0$ is in an interval $(a, b)$ such that $f'(x) > 0$ for $x$ in $(a, x_0)$ and for $x$ in $(x_0, b)$, then $f$ is increasing at $x_0$. If $f'(x) < 0$ for $x$ in $(a, x_0)$ and for $x$ in $(x_0, b)$, then $f$ is decreasing at $x_0$.*

The proof will be postponed until the end of Section 9.2.

Theorem 5.3.2 states that if a function, which is differentiable, except perhaps at $x_0$, and continuous, is increasing at each point of an interval to the left of $x_0$, and at each point of an interval to the right of $x_0$, then it is also increasing at $x_0$. Note that $f'(x_0)$ need not even exist for Theorem 5.3.2, but the theorem still applies if $f'(x_0) = 0$. A further consideration of what may occur if $f'(x_0) = 0$ appears in Section 5.4.

*Example 5.3.1.* Show that $f(x) = (x - 1)^3$ is increasing for every value of $x$.

*Solution.* We find $f'(x) = 3(x - 1)^2$. Hence $f'(x) > 0$ if $x \neq 1$, and by Theorem 5.3.1, $f$ is increasing for any $x \neq 1$. By Theorem 5.3.2, $f$ is also increasing at $x = 1$. Hence $f$ is increasing for all $x$. The graph of $f$ is shown in Fig. 5.3.1.

*Example 5.3.2.* Find the values of $x$ for which the function

$$f(x) = x^3 + 2x^2 - 4x + 8$$

is increasing and the values of $x$ for which it is decreasing.

*Solution.* We find

$$f'(x) = 3x^2 + 4x - 4 = (3x - 2)(x + 2).$$

We solve in turn, the inequalities $f'(x) > 0$ and $f'(x) < 0$ and find:

if $x < -2$ or if $x > 2/3$, $f' > 0$ and $f$ is increasing,

if $-2 < x < 2/3$, $f' < 0$ and $f$ is decreasing.

Thus $f$ is increasing for each $x$ in the intervals $(-\infty, -2)$ and $(2/3, \infty)$, and is decreasing for each $x$ in the interval $(-2, 2/3)$. At $-2$ and at $2/3$ the function is neither increasing nor decreasing. The graph of the function appears in Fig. 5.3.2.

To conclude this section we state a definition and a theorem which extend the idea of increasing and decreasing function to an interval.

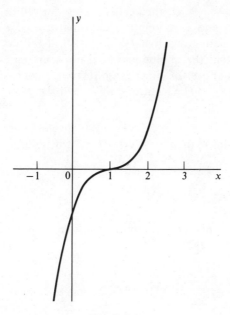

Fig. 5.3.1

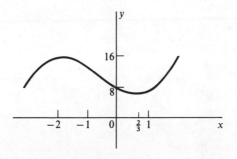

Fig. 5.3.2

**DEFINITION 5.3.2.** *A function is increasing (decreasing) in an interval if it is increasing (decreasing) at every point of the interval.*

The following result clearly relates to one's intuitive concept.

**THEOREM 5.3.3.** *If a continuous function $f$ is increasing in an interval $(a, b)$, then $f(x_2) > f(x_1)$ for any numbers $x_1$ and $x_2$ in $(a, b)$ such that $x_2 > x_1$. A corresponding statement applies to a decreasing function.*

∗ *Proof.* Suppose $f(x_2) \leq f(x_1)$ with $a \leq x_1 < x_2 \leq b$. By Theorem 3.7.1, the function $f$ takes on its minimum value in the interval $[x_1, x_2]$ at some number $c$,

$x_1 \le c \le x_2$. If $x_1 < c \le x_2$ then $f(x) \ge f(c)$ for all $x$ in some interval $[x', c]$, $x_1 < x' < c$, and $f$ cannot be increasing at $c$. If $c = x_1$ then $f(x_1) \le f(x_2)$, and since we supposed that $f(x_2) \le f(x_1)$, it follows that $f(x_1) = f(x_2)$. We can then say that the minimum of $f$ also occurs at $x_2$, and $f$ cannot be increasing at $x_2$ as above. In either case there is a contradiction, the original supposition that $f(x_2) \le f(x_1)$ must be false, and the proof is complete.

The converse of Theorem 5.3.3. is clearly valid, since Definition 5.3.1 is satisfied at each point in $(a, b)$, if $f(x_2) > f(x_1)$ whenever $x_2 > x_1$.

**EXERCISE GROUP 5.2**

Determine the values of $x$ for which the function in each of Exercises 1–10 is (a) increasing, (b) decreasing.

1. $2x^3 + x - 1$
2. $2x^2 + x - 1$
3. $2x^3 - 3x^2 - 12x + 4$
4. $6x^3 + 6x^2 + 5x - 1$
5. $\dfrac{1}{x}$
6. $\dfrac{1}{x^2}$
7. $\dfrac{1}{1 + x^2}$
8. $\dfrac{x}{2 + x^2}$
9. $\dfrac{x}{\sqrt{x - 2}}$
10. $\dfrac{x}{\sqrt{x^2 - 2}}$

Determine the values of $x$ for which the derivative of the function in each of Exercises 11–16 is (a) increasing, (b) decreasing.

11. $x^4 + x^3$
12. $x^3 + 2x^2 - 3x + 4$
13. $\dfrac{1}{x + 1}$
14. $\dfrac{x}{x + 2}$
15. $\sqrt{2x + 1}$
16. $x\sqrt{2x + 1}$

17. Show that the function $x^3 + x^2 + x - 7$ increases for all real $x$.
18. Show that the function $-x^3 + 3x^2 - 6x + 4$ decreases for all real $x$.
19. Show that the function $x/(1 + x)$ increases everywhere in the domain of the function.
20. Show that the function $(x + 2)/(x + 1)$ decreases everywhere in the domain of the function.
21. Show that the derivative of the function $2x^4 - 4x^3 + 3x^2 - 2x - 6$ is everywhere increasing.
22. Show that the derivative of the function $3x^2 + 8x^{3/2}$ increases everywhere in the domain of the function.

## 5.4 MAXIMA AND MINIMA

Closely related to the discussion of the preceding section are the ideas of maximum and minimum of a function. These ideas connote "largest" and "smallest," but in a way that must be made precise for mathematical purposes.

**DEFINITION 5.4.1.** *A function f has a* **maximum** *in an interval $[a, b]$ if there is a number $x_0$ in $[a, b]$ such that*

(5.4.1) $\qquad f(x_0) \geq f(x)$ *for all $x$ in $[a, b]$.*

*A function f has a* **minimum** *in $[a, b]$ if there is a number $x_0$ in $[a, b]$ such that*

(5.4.2.) $\qquad f(x_0) \leq f(x)$ *for all $x$ in $[a, b]$.*

If the inequality in (5.4.1) holds for **all** $x$ in the domain of $f$, then $f(x_0)$ is called the **absolute maximum** of $f$; if the inequality in (5.4.2) holds for **all** $x$ in the domain of $f$, then $f(x_0)$ is called the **absolute minimum** of $f$.

The definitions just given apply to any function. In the case of continuous or differentiable functions more specific criteria apply. A highly important property for continuous functions was given in Theorem 3.7.1, which stated, in brief, that, "a continuous function in a closed interval takes on its maximum and minimum values there."

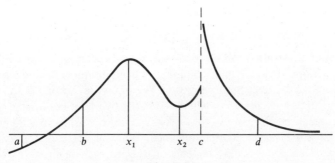

Fig. 5.4.1

Some of the above ideas are illustrated by the function whose graph appears in Fig. 5.4.1. Theorem 3.7.1 applies to the closed interval $[a, b]$, and there is a minimum at $a$ and a maximum at $b$. In the open interval $(a, b)$ there is no maximum or minimum. On the other hand, although the theorem does not apply to the open interval $(b, c)$, there is still a maximum at $x_1$, but no minimum. In the closed interval $[b, d]$ there is a minimum at $b$, but no maximum. In the interval $[c, d]$ there is no maximum, but there is a minimum at $d$; this statement still applies if $c$ is excluded from the interval. Finally, in the semi-infinite interval $[d, \infty)$ there is a maximum, but no minimum.

The following criterion applies in the case of a differentiable function.

**THEOREM 5.4.1.** *If $f(x_0)$ is a maximum or minimum of a function $f$ in an interval $[a, b]$, where $x_0$ is an interior point of the interval, and if $f'(x_0)$ exists, then $f'(x_0) = 0$.*

*Proof.* If $f'(x_0)$ exists and is not zero, then either $f'(x_0) > 0$ or $f'(x_0) < 0$. By Theorem 5.3.1, $f$ is then increasing at $x_0$ or decreasing at $x_0$, and there can be no maximum or minimum at $x_0$.

Theorem 5.4.1 is illustrated in Fig. 5.4.1 for the interval $[b, c]$. There is a maximum at $x_1$ for this interval, and $f'(x_1) = 0$. The theorem does not apply to the interval $[b, c]$ in regard to a minimum, since the minimum is as $b$, which is not an interior point. However, the theorem applies to the minimum of $f$ in $[x_1, c]$; the minimum there occurs at $x_2$, an interior point, and $f'(x_2) = 0$.

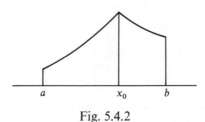

Fig. 5.4.2

A further illustration of some of the above ideas is seen in Fig. 5.4.2. Theorem 3.7.1 applies to the interval $[a, b]$, with a minimum at $a$ and a maximum at $x_0$. Theorem 5.4.1 does not apply, since the minimum occurs at an endpoint, while the maximum occurs at $x_0$ where the derivative does not exist.

Points on a graph such as those for $x_1$ and $x_2$ in Fig. 5.4.1 and for $x_0$ in Fig. 5.4.2 are of particular significance. Whether there is a maximum or minimum at such a point may depend on the interval in question. However, in Fig. 5.4.1 the value $f(x_1)$ is a maximum in some interval containing $x_1$, and $f(x_2)$ is a minimum in some interval containing $x_2$. Similarly, in Fig. 5.4.2 $f(x_0)$ is a maximum in some interval containing $x_0$. We now define this situation precisely.

**DEFINITION 5.4.2.** *The value $f(c)$ is a **relative maximum** of the function $f$, if there is some interval containing $c$ in its interior and lying in the domain of $f$ such that $f(c)$ is the maximum of $f$ in this interval. A **relative minimum** of $f$ is defined similarly. A relative maximum or relative minimum is also called an **extremum** of the function.*

In Fig. 5.4.1 there is a relative maximum at $x_1$ and a relative minimum at $x_2$, with $f'(x_1) = 0$ and $f'(x_2) = 0$. In Fig. 5.4.2 there is a relative maximum at $x_0$, but the derivative does not exist at $x_0$.

By Theorem 5.4.1 *the extrema of a function, where the function is differentiable, occur at values where the derivative is zero*; however, *a zero derivative does not necessarily yield an extremum* (see Fig. 5.4.3). On the other hand, *an extremum may occur even where the derivative does not exist*.

**DEFINITION 5.4.3.** *A number c in the domain of a function f is called a* **critical value** *of f if either $f'(c) = 0$, or the derivative of f does not exist at c.*

Accordingly, *any extrema of a function must occur at the critical values of the function.*

***Example 5.4.1.*** Consider the function $f(x) = x^3/(x^2 - 2)$. We find

$$f'(x) = \frac{x^4 - 6x^2}{(x^2 - 2)^2} = \frac{x^2(x^2 - 6)}{(x^2 - 2)^2}.$$

The critical values are those for which $f'(x) = 0$, or the values for which

$$x^2(x^2 - 6) = 0,$$

and are $x = 0, \pm\sqrt{6}$. These are the values that may yield extrema. The numbers $x = \pm\sqrt{2}$, for which $f'(x)$ does not exist, are not critical values, since they are not in the domain of $f$. (See Fig. 5.4.3.)

We have seen that if an extremum of a function exists, this can happen only at a critical value of the function. We still require a criterion to determine whether a critical value yields an extremum, and if so, whether the extremum is a relative maximum or a relative minimum. We now present such a criterion.

**First derivative test for an extremum.** *Let c be a critical value of a function f at which f is continuous. If c is an interior point of an interval $(a, b)$ such that*

(5.4.3)    $f'(x) > 0$, $x$ in $(a, c)$,   and   $f'(x) < 0$, $x$ in $(c, b)$,

*then $f(c)$ is a relative maximum of f.*

*If c is an interior point of an interval $(a, b)$ such that*

(5.4.4)    $f'(x) < 0$, $x$ in $(a, c)$,   and   $f'(x) > 0$, $x$ in $(c, b)$,

*then $f(c)$ is a relative minimum of f.*

*Proof.* We establish the test for a relative maximum. Suppose that (5.4.3) holds and that $f(c)$ is not a relative maximum; then, for some $x_1 \neq c$ in $(a, b)$

(5.4.5)    $f(x_1) > f(c).$

If $x_1$ is in $(a, c)$ then, since $f$ is continuous, there is, by the Intermediate Value

Theorem, an $x_2$ between $x_1$ and $c$ such that
$$f(x_1) > f(x_2), \qquad a < x_1 < x_2 < c.$$
But this contradicts Theorem 5.3.3, and $x_1$ cannot be in $(a, c)$. In a similar way $x_1$ cannot be in $(c, b)$. Hence (5.4.5) cannot hold, and $f(c)$ is a relative maximum of $f$. The proof of the test for a relative minimum is similar.

The proof just given is simply a formulation of the idea that if $f$ is increasing to the left of $c$ and decreasing to the right of $c$, then there is a relative maximum at $c$.

As a complement to the first derivative test for an extremum, we state the following result, which follows immediately from Theorem 5.3.2.

**THEOREM 5.4.2.** *If $c$ is an interior point of some interval $(a, b)$ such that $f'(x)$ is of the same sign for $a < x < b$, $x \neq c$, then $f$ has no extremum at $c$.*

**Example 5.4.2.** Find any relative maxima and minima of $f(x) = x^3/(x^2 - 2)$.

**Solution.** In Example 5.4.1 we found
$$f'(x) = \frac{x^2(x^2 - 6)}{(x^2 - 2)^2},$$
and the critical values $0$, $\pm\sqrt{6}$. We find also:

i. if $x < -\sqrt{6}$, then $f'(x) > 0$,
ii. if $-\sqrt{6} < x < -\sqrt{2}$, then $f'(x) < 0$,
iii. if $-\sqrt{2} < x < 0$, then $f'(x) < 0$,
iv. if $0 < x < \sqrt{2}$, then $f'(x) < 0$,
v. if $\sqrt{2} < x < \sqrt{6}$, then $f'(x) < 0$,
vi. if $x > \sqrt{6}$, then $f'(x) > 0$.

From i. and ii. there is a relative maximum at $(-\sqrt{6}, -3\sqrt{6}/2)$, while from v. and vi. there is a relative minimum at $(\sqrt{6}, 3\sqrt{6}/2)$. The relations in iii. and iv. were obtained to test the critical value zero, but since the derivative does not change sign in passing through $x = 0$, there is in this case, by Theorem 5.4.2, neither a relative maximum nor minimum. The graph appears in Fig. 5.4.3.

In Example 5.4.2 we see that the intervals that may be used for the first derivative test are determined by the critical values of $f$, and also by the values where $f$ does not exist. In each such interval the sign of the derivative does not change and may be determined by using a particular value. For example, to test the sign of $f'$ in the interval $-\sqrt{6} < x < -\sqrt{2}$, we may choose a convenient value, say $x = -2$. Since $f'(-2) < 0$ it follows that $f'$ is negative for all $x$ in this interval. This property is stated here without proof; it may be used if $f'$ is continuous in the interval.

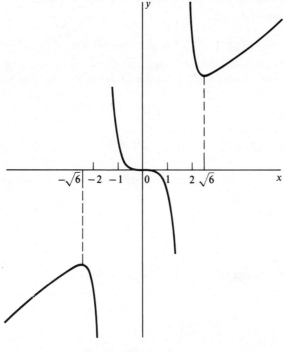

Fig. 5.4.3

There is another useful test for extrema, which utilizes the second derivative.

**Second derivative test for extrema.** *Let $f'(c) = 0$. If $f''(c) > 0$ there is a relative minimum of $f$ at $c$; if $f''(c) < 0$ there is a relative maximum of $f$ at $c$.*

*Proof.* Consider the case where $f''(c) > 0$. Then, since $f''$ is the derivative of $f'$, it follows that $f'(x)$ is increasing at $c$; this means, since $f'(c) = 0$, that

$$f'(x) < 0 \text{ in some interval } (a, c),$$
$$f'(x) > 0 \text{ in some interval } (c, b).$$

It follows by the first derivative test that $f$ has a relative minimum at $c$. The proof for the case $f''(c) < 0$ is similar.

**Example 5.4.3.** Find the extrema of $f(x) = x + 1/x$.

*Solution.* We find

$$f'(x) = 1 - \frac{1}{x^2} = \frac{x^2 - 1}{x^2}, \qquad f''(x) = \frac{2}{x^3}.$$

The critical numbers are $x = \pm 1$. Then we find $f''(-1) < 0$, and there is a relative maximum at $(-1, -2)$; furthermore, $f''(1) > 0$, and thus there is a relative minimum at (1, 2). (Note that $x = 0$ is not a critical number since it is not in the domain of $f$.)

If $f''$ is easily found, or if $f''$ is also needed for some other purpose, the second derivative test is generally used. However, the first derivative test can be used in situations where the second derivative test cannot be used or where it may be inconvenient to do so.

**EXERCISE GROUP 5.3**

In Exercises 1–12 find any relative maxima and minima of the function. Use the first derivative test.

1. $4 - 3x - 2x^2$
2. $2x^3 - x^2 - 4x - 5$
3. $4x^3 + 3x^2 - 6x - 2$
4. $4x^3 - 18x^2 + 27x - 3$
5. $(x + 2)^4$
6. $x^{4/3}$
7. $x\sqrt{1 - x^2}$
8. $x^{1/2} + x^{-1/2}$
9. $\dfrac{x - 1}{\sqrt{x^2 + 1}}$
10. $\dfrac{x - 3}{x^2 - 4x + 7}$
11. $(x - 2)^2(2x + 5)^3$
12. $\dfrac{x - 2}{(2x + 3)^2}$

In Exercises 13–20 find any relative maxima and minima of the function. Use the second derivative test.

13. $x^3 - x$
14. $2x^4 - 4x + 3$
15. $\dfrac{x^2 + 1}{x}$
16. $\dfrac{1 - 4x^3}{x}$
17. $\dfrac{x - 1}{x^2 + 3}$
18. $x - \sqrt{2x^2 + 2}$
19. $2x - 3 + \sqrt{x^2 - 3}$
20. $2x + \sqrt{1 - x}$

In Exercises 21–30 find any relative maxima and minima of the function. Use any appropriate test.

21. $12x - x^3$
22. $x^4 - 4x^3 - 7$
23. $x^4 + 2x^3 - 2$
24. $x^4 - 4x^3 + 4x^2 - 12$
25. $\dfrac{4x + 3}{x^2 + 1}$
26. $\dfrac{x^3}{x^2 - 1}$
27. $x - x^{2/3}$
28. $x^{2/3}(5 - x)$
29. $\sqrt{2x - 1} - \sqrt{3x - 1}$
30. $\dfrac{1}{\sqrt{x - 2}} + \dfrac{1}{\sqrt{4 - x}}$

## 5.5 MAXIMUM AND MINIMUM IN AN INTERVAL

A maximum or minimum of a function in a specified interval may occur either at an interior point, a case which has been treated previously, or at an end point of the interval. The latter case may also be handled by known methods. To determine any maximum or minimum in an interval, we may find any extrema in the interior and then compare with the values of the function at the endpoints.

***Example 5.5.1.*** Find the maximum and minimum of the function

$$f(x) = x^3 - 3x^2 - 2$$

in the interval $[1, 4]$.

*Solution.* We obtain $f'(x) = 3x^2 - 6x = 3x(x - 2)$; the critical values are 0 and 2, of which only 2 is in the interval $[1, 4]$. By the second derivative test we see that this value yields a relative minimum. Since $f(1) > f(2)$ and $f(4) > f(2)$, it follows that $f(2) = -6$ is the minimum of $f$ in the interval. We also find $f(4) > f(1)$; hence $f(4) = 14$ is the maximum in the interval. (See Fig. 5.5.1.)

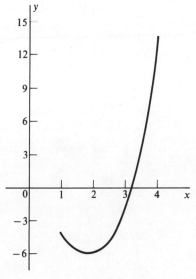

Fig. 5.5.1

We sometimes refer to the behavior of the function in Fig. 5.5.1 at $x = 1$ by saying that there is an *endpoint* maximum. This signifies that an interval in the domain of $f$ has an endpoint at $x = 1$, and for some interval $[1, c]$ the maximum occurs at $x = 1$.

The interval in which a maximum or minimum is to be found may be determined by the domain of the function, as in the following example.

**Example 5.5.2.** Find the absolute maximum and absolute minimum of the function $f(x) = x - x^{1/2}$.

*Solution.* The domain of $f$ is the interval $[0, \infty)$. We find

$$f'(x) = 1 - \frac{1}{2x^{1/2}} = \frac{2x^{1/2} - 1}{2x^{1/2}}.$$

The critical value in the interior of the domain is $x = 1/4$. By the first derivative test there is a relative minimum at $x = 1/4$ with a value of $-1/4$. Since $f(0) = 0 > -1/4$, and since $f$ increases when $x > 1/4$, there is an absolute minimum of $-1/4$ at $x = 1/4$, but there is no absolute maximum. See the graph of $f$ in Fig. 5.5.2. The function has an endpoint maximum at $x = 0$.

Fig. 5.5.2

## EXERCISE GROUP 5.4

In each of the following exercises determine the maximum and minimum of the function in the specified interval.

1. $f(x) = x^3$, $[-3, 1]$
2. $f(x) = x^3 - 3x$, $[-2, 3]$
3. $f(x) = x^2 - 2x + 7$, $[0, 3]$
4. $f(x) = x^4 - 4x$, $[0, 2]$
5. $f(x) = x^{1/3}$, $[-3, 1]$
6. $f(x) = 2x^3 + 3x^2 - 6x$, $[-1, 2]$
7. $f(x) = x + \dfrac{1}{x}$, $[1/2, 2]$
8. $f(x) = x^{2/3} - 2$, $[-2, 1]$
9. $f(x) = (x + 2)\sqrt{x}$, $[0, 4]$
10. $f(x) = (x - 2)\sqrt{x}$, $[0, 4]$
11. $f(x) = \dfrac{x^{1/2}}{3 + x^2}$, $[0, 3]$
12. $f(x) = \dfrac{x^{2/3}}{4 + x^2}$, $[-1, 3]$
13. $f(x) = \begin{cases} x^2, 0 \leq x < 2 \\ 8/x, 2 \leq x \leq 4 \end{cases}$ $[0, 4]$
14. $f(x) = \begin{cases} x + 1/x, 1/2 \leq x < 2 \\ x^2 - 13x/4 + 5, 2 \leq x \leq 3, \end{cases}$ $[1/2, 3]$
15. $f(x) = \begin{cases} x^{2/3}, -1 \leq x < 1 \\ 2 - x, 1 \leq x \leq 3 \end{cases}$ $[-1, 3]$

16. $f(x) = \begin{cases} x + \sqrt{4 - x^2}, & -1 \leq x < 2 \\ -x^2 + 6x - 8, & 2 \leq x \leq 5 \end{cases}$  $[-1, 5]$

## 5.6 APPLICATIONS OF MAXIMA AND MINIMA

Many problems of an applied nature are stated directly in a form which requires finding a maximum or minimum of some quantity. In other problems the connection with a maximum or minimum may be somewhat concealed. Once the type of problem is determined, the quantity to be maximized or minimized is expressed in terms of some variable quantity. Then our previous methods may be used. At times special care is required in interpreting the solution relative to the original problem.

***Example 5.6.1.*** Of all pairs of numbers with a constant difference $k$, find that pair for which the product is the smallest possible.

*Solution.* Any pair of numbers with difference $k$ may be expressed as $x$ and $x + k$. We are to minimize their product, the function

$$f(x) = x(x + k) = x^2 + kx.$$

We find $f'(x) = 2x + k$, the one critical value (for which $f'(x) = 0$) is $-k/2$, and by the second derivative test this yields a minimum of $f(x)$. The desired numbers are $-k/2$ and $k/2$.

For example, if $k = 5$ the numbers are $-5/2$ and $5/2$, and the smallest product is $-25/4$.

In the problems treated here it is important that the function to be maximized or minimized be expressed (directly or indirectly) in terms of a *single* variable. We shall amplify this idea.

***Example 5.6.2.*** A rectangular pen of area 162 sq yd is to be enclosed on three sides. Find the dimensions that require the least amount of material.

*Solution.* The amount of material needed is measured by the three sides indicated in Fig. 5.6.1, that is, by $2x + y$. Here both $x$ and $y$ are variable and are such that $xy = 162$. We must minimize $2x + y$ subject to the condition $xy = 162$.

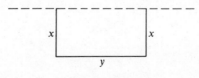

Fig. 5.6.1

Since $y = 162/x$ we must minimize the function

$$f(x) = 2x + \frac{162}{x}.$$

We find

$$f'(x) = 2 - \frac{162}{x^2} = \frac{2x^2 - 162}{x^2},$$

and there are two critical values, $\pm 9$ ($x = 0$ is not in the domain of $f$). Of these only the positive value relates to the physical problem. Then

$$f''(x) = \frac{324}{x^3}, \qquad f''(9) > 0,$$

and there is a minimum value when $x = 9$, $y = 18$.

In Example 5.6.2 the variable $y$ may be considered as an *auxiliary variable*. The variable $x$ could just as well have been considered as the auxiliary variable, in which case a function of $y$ would have been minimized. (What is this function of $y$?) In the following section an alternative method of using an auxiliary variable is presented.

## 5.7 EXTREMAL PROBLEMS WITH AUXILIARY CONDITIONS

It may happen that a quantity to be maximized or minimized is expressed initially in terms of two variables which are related, as in Example 5.6.2. An equation connecting the two variables is called an *auxiliary condition*. We choose one of the variables as independent, and consider the auxiliary condition as defining (perhaps implicitly) the other variable as a function of the independent one. The procedure is amplified in the examples.

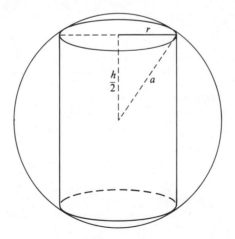

Fig. 5.7.1

**Example 5.7.1.** Determine the right circular cylinder of largest volume which can be inscribed in a given sphere, as in Fig. 5.7.1.

*Solution.* Let $h$ be the altitude and $r$ the radius of the cylinder; let $a$ be the fixed radius of the sphere. The volume to be maximized is given by $\pi r^2 h$, the volume of the cylinder. But $r$ and $h$ are clearly related, by an equation which must be determined. From Fig. 5.7.1 we see that

$$\text{(5.7.1)} \qquad \frac{h^2}{4} + r^2 = a^2.$$

If the value of $h$ as obtained from (5.7.1) is substituted in $\pi r^2 h$, the volume is expressed as a function $V(r)$ of $r$ alone, and this is the function we choose to maximize. We can, however, carry out the process without solving (5.7.1) explicitly. Thus we let

$$V(r) = \pi r^2 h,$$

where $h$ is a function of $r$ defined implicitly by (5.7.1). For any critical value of $V(r)$ we must have

$$\text{(5.7.2)} \qquad V'(r) = \pi(r^2 h' + 2rh) = 0,$$

with primes denoting derivatives with respect to $r$. By implicit differentiation in (5.7.1) with respect to $r$ we have

$$\text{(5.7.3)} \qquad \frac{hh'}{2} + 2r = 0.$$

From (5.7.3) we find $h' = -4r/h$; this, substituted in (5.7.2), gives

$$-\frac{4r^3}{h} + 2rh = 0, \qquad 4r^3 - 2rh^2 = 0, \qquad 2r(2r^2 - h^2) = 0.$$

Hence $r = 0$ or $h = \pm r\sqrt{2}$. Clearly, the value $r = 0$ cannot yield a maximum; while $h = -r\sqrt{2}$ has no meaning for the physical problem. Hence we are limited to a consideration of $h = r\sqrt{2}$, which we shall test in the following manner. If $r$ is small (near zero) or if $r$ is large (near $a$), then $V(r)$ is small (near zero). There must then be a maximum of $V$ for some positive value of $r$, and since there is only one applicable *critical relation*, $h = r\sqrt{2}$, this relation between $h$ and $r$ yields the maximum. The maximum volume occurs for the cylinder for which $h/r = \sqrt{2}$.

The values of $h$ and $r$ in terms of $a$, for the largest inscribed cylinder, may be found by solving (5.7.1) simultaneously with $h = r\sqrt{2}$, and are $r = a\sqrt{6}/3$, $h = 2a\sqrt{3}/3$.

Example 5.7.1 could have been done by solving (5.7.1) explicitly for $h$ and substituting in $\pi r^2 h$. The method used, however, yielded directly the ratio of the altitude and radius of the cylinder; and such a relation is often more significant than explicit values for the quantities!

An additional comment concerns the test for a maximum or minimum. The first and second derivative tests are often laborious in this method. However, an applied extremal problem rarely yields more than one critical relation, and a brief analysis such as that in Example 5.7.1 will determine whether a maximum or minimum exists.

Example 5.6.2 may be done by the use of an auxiliary condition. Referring to that example, we see that the function to be maximized and the auxiliary condition are

$$f(x) = 2x + y, \qquad xy = 162.$$

We then get

$$f'(x) = 2 + y' = 0, \qquad xy' + y = 0.$$

Substituting $y' = -y/x$ from the second equation into the first equation, we have

$$2 - \frac{y}{x} = 0, \qquad y = 2x.$$

The maximum occurs when $y$ and $x$ are in the ratio $y/x = 2$. The values $x = 9$, $y = 18$, as in Example 5.6.2, can then also be obtained.

***Example 5.7.2.*** Find the area of the largest isosceles triangle that can be inscribed in a circle of radius $a$.

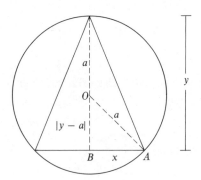

Fig. 5.7.2

*Solution.* One altitude of the isosceles triangle must pass through the center of the circle. Let $y$ be its length, and let $x$ be the length of half the base (Fig. 5.7.2). The length of $OB$ is $y - a$ if $y > a$ and $a - y$ if $y < a$. In either case we obtain from $\triangle OAB$, $x^2 + (y - a)^2 = a^2$, or

(5.7.4) $$x^2 + y^2 - 2ay = 0.$$

This is the auxiliary condition.

The function to be maximized is

$$A = xy.$$

The requirement $dA/dx = 0$ becomes

(5.7.5) $$\frac{dA}{dx} = x\frac{dy}{dx} + y = 0.$$

Differentiation in (5.7.4) with respect to $x$ gives

(5.7.6) $$2x + 2y\frac{dy}{dx} - 2a\frac{dy}{dx} = 0.$$

Equating the values of $dy/dx$ obtained from (5.7.5) and (5.7.6), we have

$$-\frac{y}{x} = \frac{x}{a-y}.$$

This gives $x^2 = y^2 - ay$, which we substitute in (5.7.4), to yield $2y^2 - 3ay = 0$. The solution $y = 0$ cannot furnish a maximum area. Thus the solution $y = 3a/2$ is the desired value. Then from (5.7.4) we find $x = \sqrt{3}\,a/2$. The maximum area is

$$A = \frac{\sqrt{3}\,a}{2} \cdot \frac{3a}{2} = \frac{3\sqrt{3}\,a^2}{4}.$$

## EXERCISE GROUP 5.5

1. A square piece of tin of side 18 in. is made into an open box by cutting out squares at the corners and turning up the edges. Find the dimensions of the largest box that can be formed.
2. Find the relative dimensions of a box with square base and fixed volume, for which the total surface area is least.
3. Do Exercise 2 in the case where the box is open at the top.
4. Two vertical poles are 20 ft and 30 ft high, and are 40 ft apart. Wires are stretched to their tops from a point on the (horizontal) ground directly between their bases. Where should this point be so that the least amount of wire is used?
5. A man in a boat 12 mi from a straight shore wishes to reach a point 10 mi downshore. He can travel on the water at 5 mph, and on land at 13 mph. Where should he land so as to reach his destination in the shortest time?
6. In Exercise 5, where should he land if the point is only 4 mi downshore?
7. Prove that the minimum perimeter for a rectangle of fixed area occurs if the rectangle is square.
8. Find the largest area possible for a rectangle inscribed in a semicircle of radius $a$.
9. A right circular cone is formed from a sector of a circle by pulling the two bounding radii of the sector together. Find that angle of the sector which produces the cone of maximum volume.

10. Find the dimensions of the right circular cylinder of maximum lateral area that can be inscribed in a sphere of radius $a$.

11. Find the dimensions of the right circular cone of maximum volume that can be inscribed in a sphere of radius $a$.

12. Find the dimensions of the right circular cylinder of maximum volume that can be inscribed in a right circular cone of altitude $h$ and radius of base $b$.

13. Find the largest area for a rectangle inscribed in a circle of radius $a$.

14. Find the point on the line $3x + 4y - 2 = 0$ which is closest to the point $(1, 3)$.

15. Find the shortest distance from the point $(x_0, y_0)$ to a point on the line $ax + by + c = 0$.

* 16. Prove that if $(x_0, y_0)$ is the point on a curve which is closest to a point $(a, b)$, then the line determined by $(x_0, y_0)$ and $(a, b)$ is normal to the curve.

17. A line segment of length $a$ is split into two parts on each of which a square is constructed. Find (a) the minimum, (b) the maximum for the sum of the areas of the squares.

18. A line segment of length $a$ is split into two parts on one of which a square is constructed, and on the other of which, as diameter, a semicircle is constructed. Find (a) the minimum, (b) the maximum, for the sum of the areas.

* 19. A man in a boat $a$ mi offshore wishes to reach a point $b$ mi inland and $c$ mi downshore in the least time. He can travel on water at $v_1$ mph and on land at $v_2$ mph. If $\alpha$ is the angle his path in water makes with the shore, and $\beta$ is the angle his path on land makes with the shore, show that $\cos \alpha / \cos \beta = v_1/v_2$.

20. Find the points on the curve $x^2 + xy = 1$ which are closest to the origin.

21. Find the point on the curve $4xy^2 = 1$ in the first quadrant which is closest to the origin.

22. Let $P$ be a point on the curve $y = x^2$ for $0 < x < 1$. Lines are drawn through $P$ parallel to the coordinate axes, forming a rectangle with the $x$ axis and the line $x = 1$. Find the maximum area for such a rectangle.

23. Find the point on the curve $y^2 = 4x + 9$ which is closest to the origin.

24. Find the point on the curve $y^2 = 4x + 4$ which is closest to the origin.

25. At what point of the ellipse $b^2x^2 + a^2y^2 = a^2b^2$ in the first quadrant is the sum of the intercepts of the tangent a minimum?

26. At what point of the ellipse $b^2x^2 + a^2y^2 = a^2b^2$ in the first quadrant is the area of the triangle formed by the tangent and the coordinate axes the smallest?

* 27. Find the length of the longest ladder that can be moved horizontally around a right angle corner, if the two intersecting corridors are of widths 1 ft and 8 ft.

* 28. Do Exercise 27 in the case where if the corridors are of widths $a$ and $b$.

## 5.8 CONCAVITY. POINT OF INFLECTION

There is, in Fig. 5.8.1, a difference in character between the part of the curve to the left of $P$ and the part to the right, a distinction which can be described by the behavior of the slope. As $x$ increases, the slope of the left part of the curve is increasing while the slope of the right part is decreasing. (This applies whether the slope is positive, negative, or zero; "increasing" and "decreasing" mean in the algebraic sense.)

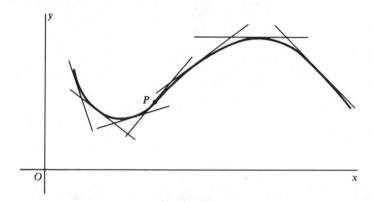

Fig. 5.8.1

**DEFINITION 5.8.1.** *A curve is* **concave up** *at any point at which the slope is increasing, and is* **concave down** *at any point at which the slope is decreasing.*

*A curve is* **concave up in an interval** *if it is concave up at each point in the interval, and is* **concave down in an interval** *if it is concave down at each point in the interval.*

Thus the part of the curve in Fig. 5.8.1 to the left of $P$ is concave up, while the part to the right of $P$ is concave down. We may also say that, in an interval where a curve is concave up, the tangent to the curve rotates in a positive (counterclockwise) direction as $x$ increases; and, in an interval where a curve is concave down, the tangent to the curve rotates in a negative direction.

The following criterion for concavity at a point follows at once from Theorem 5.3.1 and the fact that $f''$ is the derivative of $f'$.

**THEOREM 5.8.1.** *If $f''(c) > 0$, the graph of $f(x)$ is concave up at $c$; if $f''(c) < 0$, the graph is concave down at $c$.*

*Example 5.8.1.* Determine the concavity of the graph of $f(x) = x^{4/3}$ at each point.

*Solution.* We find $f'(x) = 4x^{1/3}/3$, $f''(x) = 4x^{-2/3}/9$. Since $x^{-2/3} > 0$ for $x \neq 0$, we have $f''(x) > 0$ for $x \neq 0$. Hence the graph is concave up for $x \neq 0$. What about $x = 0$? The second derivative of $f$ does not exist at $x = 0$, but $f'(0) = 0$, and the graph of $f$ shown in Fig. 5.8.2 indicates that the graph is concave up also at $x = 0$. This is indeed the case and may be shown by Theorem 5.8.2.

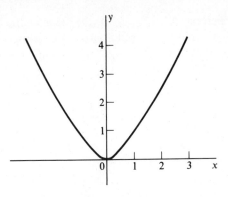

Fig. 5.8.2

The following theorem is analogous to Theorem 5.3.2, and may be proved by use of that theorem.

**THEOREM 5.8.2.** *Let $f'$ be continuous at $x = c$. If $f''(x) > 0$ for $x$ in the intervals $(a, c)$ and $(c, b)$ for some $a$ and $b$, then the graph of $f$ is concave up at $c$; if $f''(x) < 0$ in $(a, c)$ and $(c, b)$ for some $a$ and $b$, then the graph of $f$ is concave down at $c$.*

Theorem 5.8.2 applies directly to the function in Example 5.8.1 at $x = 0$, even with no second derivative there, and tells us that the graph of $x^{4/3}$ is concave up at $x = 0$. In general, Theorem 5.8.2 may be used even when $f''$ does not exist at $x = c$. It may also apply where $f''(c) = 0$, as in the case of the function $f(x) = x^4$ at $x = 0$.

A point where the concavity of a curve changes is of particular significance.

**DEFINITION 5.8.2.** *If $P$ is a point on a graph such that the graph has opposite types of concavity on both sides of $P$, then $P$ is called a **point of inflection** of the graph. We assume that the graph possesses a unique tangent at $P$.*

According to this definition the point $P$ in Fig. 5.8.1 is a point of inflection of the graph shown there.

Definitions 5.8.1 and 5.8.2, and Theorem 5.8.1 lead to the following criterion.

**THEOREM 5.8.3** **Test for a point of inflection.** *Let $c$ be a number such that the graph of $f(x)$ has a unique tangent for $x = c$. If $a$ and $b$ exist such that either*

$$f''(x) < 0, \quad a < x < c \quad \text{and} \quad f''(x) > 0, \quad c < x < b,$$

*or*

$$f''(x) > 0, \quad a < x < c \quad \text{and} \quad f''(x) < 0, \quad c < x < b,$$

*then there is a point of inflection at $c$.*

**Example 5.8.2.** Determine the concavity of the graph of
$$f(x) = 5x^4 - x^5$$
at $x = -1, 0, 1, 2, 3, 4$.

*Solution.* We obtain

(5.8.1) $\quad f'(x) = 20x^3 - 5x^4, \quad f''(x) = 60x^2 - 20x^3 = 20x^2(3 - x)$.

We then find:

$f''(-1) > 0$, the graph is concave up at $-1$;
$f''(0) = 0$, no decision yet at 0;
$f''(1) > 0$, the graph is concave up at 1;
$f''(2) > 0$, the graph is concave up at 2;
$f''(3) = 0$, no decision yet at 3;
$f''(4) < 0$, the graph is concave down at 4.

We now examine $x = 0$ further, and we see from (5.8.1) that $f''(x) > 0$ both for $x < 0$ and for $0 < x < 3$. It follows from Theorem 5.8.2 that the graph is concave up at $x = 0$. To examine $x = 3$ further we determine that

$$f''(x) > 0, \quad \text{if} \quad 0 < x < 3 \quad \text{and} \quad f''(x) < 0, \quad \text{if} \quad x > 3.$$

By Theorem 5.8.3, there is a point of inflection at $x = 3$.

From Definitions 5.8.1 and 5.8.2 and the meaning of extremum, it follows that if $f'$ exists at a point of inflection, then $f'$ (and hence the slope) has an extremum there. Theorem 5.8.3 is essentially the first derivative test for an extremum, applied to the function $f'$. If we apply the second derivative test for an extremum, again to the function $f'$, we obtain another test for a point of inflection, in the following form.

**THEOREM 5.8.4.** *Let $f''(c) = 0$. If $f'''(c)$ exists and $f'''(c) \neq 0$, there is a point of inflection at c.*

This test is sometimes easier to apply. Thus in Example 5.8.2 we found $f''(3) = 0$. Then we obtain

$$f'''(x) = 120x - 60x^2, \quad f'''(3) = -180 \neq 0.$$

Hence, by Theorem 5.8.4, there is a point of inflection at $x = 3$. We also found $f''(0) = 0$; however, $f'''(0) = 0$, and Theorem 5.8.4 does not apply. We saw that the graph in Example 5.8.2 was concave up at $x = 0$, and there is, in fact, a relative minimum there. In this connection you may wish to study Section 5.9.

It is now clear that if $f''$ exists, any change in concavity must occur where $f''(x) = 0$, that is, at zeros of $f''$. On the other hand, a change in concavity may also occur where $f''$ fails to exist. Thus in order to determine intervals of concavity we examine the intervals determined by the zeros of $f''$ and those values for which $f''$ does not exist.

*Example 5.8.3.* Determine the intervals of concavity for the graph of the function
$$f(x) = 4x^{1/3} - x^{4/3}.$$

*Solution.* We find
$$f'(x) = \frac{4}{3}x^{-2/3} - \frac{4}{3}x^{1/3},$$

$$f''(x) = -\frac{8}{9}x^{-5/3} - \frac{4}{9}x^{-2/3} = -\frac{4}{9}\frac{2+x}{x^{5/3}}.$$

Since $f''(-2) = 0$ and $f''$ is undefined at $x = 0$, the possible intervals of concavity are
$$x < -2, \quad -2 < x < 0, \quad 0 < x.$$

We find:

for $x < -2, f''(x) < 0$, the graph is concave down;
for $-2 < x < 0, f''(x) > 0$, the graph is concave up;
for $0 < x, f''(x) < 0$, the graph is concave down.

At $x = -2$ there is a change in concavity and hence a point of inflection. At $x = 0$ there is also a change in concavity, and despite the fact that $f'$ and $f''$ do not exist there, there is a point of inflection since the function is continuous. To illustrate these results, the graph of $f(x)$ is presented in Fig. 5.8.3. Note the vertical tangent at $x = 0$.

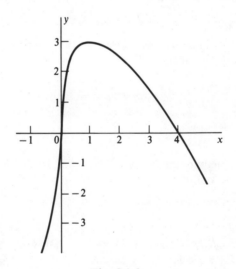

Fig. 5.8.3

**EXERCISE GROUP 5.6**

In each of Exercises 1-14 determine any extrema of the function, examine the equation for concavity and points of inflection of its graph, and sketch the graph.

1. $y = 12x - x^3$
2. $y = x^4 - 4x^2 + 2$
3. $y = x^3 + 2x^2 - 4x$
4. $y = x^4 - 4x + 3$
5. $y = 6x + 2x^3 + 3$
6. $y = x^5 - 2x^4$
7. $y = x^{2/3}$
8. $y = x^{4/3}$
9. $y = x^4 - 4x^2$
10. $y = x(x-2)^2$
11. $y = (x+2)^2(x-1)^3$
12. $y = (x^2+1)^2$
13. $y = (x^2-1)^2$
14. $y = x - x^{2/3}$

15. Prove that if $f''(x_0) > 0$, the graph of $y = f(x)$ lies above its tangent line near $x = x_0$, $x \neq x_0$, while if $f''(x_0) < 0$, the graph lies below. [Suggestion: Examine the function $G(x) = f(x) - f(x_0) - f'(x_0)(x - x_0)$ for a minimum or maximum at $x_0$.]

## *5.9 EXTENDED CRITERION FOR AN EXTREMUM OR POINT OF INFLECTION

In previous sections we have seen that if $f'(c) = 0$ and $f''(c) = 0$, the second derivative test for an extremum fails. Similarly, if $f''(c) = 0$ and $f'''(c) = 0$, the third derivative test for a point of inflection fails. However, if higher derivatives exist, we may evaluate them at $c$ until one of them is different from zero; the situation that exists depends on whether the order of the first nonzero derivative at $c$ is odd or even. This will now be stated explicitly.

**THEOREM 5.9.1.** *Let*

$$f'(c) = f''(c) = \cdots = f^{(n-1)}(c) = 0, \quad f^{(n)}(c) \neq 0, \quad n \geq 2.$$

*If $n$ is odd, the graph of $f$ has a point of inflection at $c$ (even if the condition $f'(c) = 0$ is dropped). If $n$ is even, the function $f$ has a relative maximum at $c$, if $f^{(n)}(c) < 0$, and a relative minimum, if $f^{(n)}(c) > 0$.*

We shall illustrate this before giving a proof.

*Example 5.9.1.* Find the extrema of the function $f(x) = x^4$.

*Solution.* We find $f'(x) = 4x^3$, and $x = 0$ is the only critical value. To test this value we obtain $f''(x) = 12x^2$, $f''(0) = 0$, and the second derivative test fails. We now continue. We have $f'''(x) = 24x$, $f'''(0) = 0$, $f^{(4)}(x) = 24$, $f^{(4)}(0) = 24 > 0$. By Theorem 5.9.1, since $n = 4$ and is even, there is a relative minimum at $x = 0$.

*Example 5.9.2.* Find the extrema of the function $f(x) = x^6 - x^5$.

*Solution.* We find $f'(x) = 6x^5 - 5x^4 = x^4(6x - 5)$. The critical values are $x = 0$, $x = 5/6$. Substituting $x = 5/6$ in $f''(x) = 30x^4 - 20x^3$, we find $f''(5/6) > 0$, and there is a relative minimum at $x = 5/6$ by the second derivative test.

However, $f''(0) = 0$, and the second derivative test fails at $x = 0$. Continuing, we would find $f'''(0) = 0$, $f^{(4)}(0) = 0$, $f^{(5)}(0) = -120$. By Theorem 5.9.1 there is a point of inflection at $x = 0$ (with a horizontal tangent). Hence the only extremum occurs at $x = 5/6$.

Before proving Theorem 5.9.1 we present a lemma, which is simply a preliminary result.

**LEMMA.** *If $f^{(k-1)}(c) = 0$, and $f^{(k)}(x)$ has a relative maximum (or minimum) value of zero at c with $k \geq 2$ (that is, $f^{(k)}(c) = 0$), then $f^{(k-2)}(x)$ likewise has a relative maximum (or minimum) at c.*

*Proof.* We prove the lemma for the case of a maximum; the proof for the case of a minimum is similar. You may refer to Fig. 5.9.1 as you follow the steps. If $f^{(k)}(c) = 0$ is a relative maximum for $f^{(k)}(x)$, then $f^{(k)}(x) < 0$, except at $c$, in some open interval $(a, b)$ containing $c$. It follows that $f^{(k-1)}(x)$ is decreasing in this interval. But $f^{(k-1)}(c) = 0$, by hypothesis. Hence

$$f^{(k-1)}(x) \begin{cases} >0, & a < x < c, \\ <0, & c < x < b. \end{cases}$$

Then

$$f^{(k-2)}(x) \text{ is} \begin{cases} \text{increasing for } a < x < c, \\ \text{decreasing for } c < x < b, \end{cases}$$

and $f^{(k-2)}(x)$ has a relative maximum at $c$. The proof of the lemma is complete.

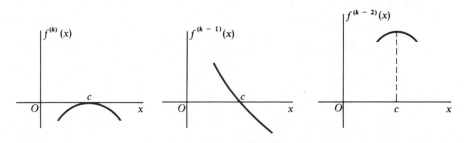

Fig. 5.9.1

*Proof of Theorem 5.9.1.* **Case 1:** *n odd.* By the second derivative test for an extremum, applied for $f^{(n-2)}(x)$, the function $f^{(n-2)}(x)$ has an extremum at $c$. It is sufficient to consider the case of a maximum. The lemma tells us that $f^{(n-4)}(x)$ has a maximum at $c$, then that $f^{(n-6)}(x)$ has a maximum at $c$, and so on. Since $n$ is odd, the reasoning may be continued until we have shown that $f'(x)$ has a maximum at $c$, which is then a point of inflection for the graph of $f$, and the theorem is proved for this case. Note that this does not require that $f'(c) = 0$.

**Case 2:** *n even.* The same reasoning applies, with the added comment that if $f^{(n)}(c) < 0$, then $f^{(n-2)}(x)$ actually has a maximum at $c$. Then it follows that $f^{(n-4)}(x)$ has a maximum at $c$, and so on. Since $n$ is even, the reasoning is continued until we have shown that $f(x)$ has a maximum at $c$, and the theorem is proved. In Case 2 we used the hypothesis that $f'(c) = 0$.

## ★ EXERCISE GROUP 5.7

★ 1. Let $y = (x-r)^n f(x)$, where $f(r) \neq 0$, and $n$ is a positive integer. If $n > 1$, prove that the graph of the equation is tangent to the $x$ axis at $x = r$, with an extremal there, if $n$ is even, and a point of inflection, if $n$ is odd.

In Exercises 2–11 use the result of Exercise 1 to make an approximate sketch of the equation, emphasizing the appearance near the $x$ intercepts. (Concavity and points of inflection need not be determined.)

2. $y = x^2(1-x)$
3. $y = x^3(x-2)^2$
4. $y = (x+1)^2(2x-1)^3$
5. $y = x^5(x+1)^2$
6. $y = x^2(x+1)(x-2)$
7. $y = x^2(x-2)^2(x-3)$
8. $y = x^3(x-3)(x+5)$
9. $y = x^3(x-3)^2(x+5)$
10. $y = x^3(x-3)^2(x+5)^2$
11. $y = x^3(x-3)^2(x+5)^3$

Make an approximate sketch of each of the following. A slight modification of the result of Exercise 1 is applicable.

12. $y = x(x-4)^3 + 2$
13. $y + 3 = x^2(x-3)^3$
14. $y = x^2(x-1) - 3$
15. $y = 4 - x^2(x+2)^2$

## 5.10 RATE OF CHANGE

In Section 4.2 the derivative of a function $f$ at a value $x$ was defined as

$$(5.10.1) \qquad f'(x) = \frac{df}{dx} = \lim_{h \to 0} \frac{f(x+h) - f(x)}{h}.$$

The difference $f(x+h) - f(x)$ is the change in value of the function $f$ as $x$ changes by an amount $h$. The ratio $[f(x+h) - f(x)]/x$ may be considered as the change in $f$ relative to the change in $x$ in the interval $[x, x+h]$. As $h \to 0$, the relative rate of change applies to a smaller and smaller change in $x$. It is then reasonable to define the *rate of change of $f$ at $x$*, sometimes called the *instantaneous rate of change*, by the limit that appears in (5.10.1).

**DEFINITION 5.10.1.** *The **rate of change** of a function $f$ with respect to $x$ is given by the derivative of $f$, that is, by $f'(x)$ or $df/dx$.*

*Example 5.10.1.* The volume of a cube is $V = x^3$, where $x$ is the length of a side. We then have $dV/dx = 3x^2$ as the rate of change of $V$ with respect to $x$. In particular, if $x$ is measured in feet, the rate of change of volume with respect to $x$ is $12 \text{ ft}^3/\text{ft} = 12 \text{ ft}^2$ when $x = 2$.

The area of one of the six faces of the cube is $A = x^2$. We may then write $x = A^{1/2}$, and hence

$$V = (A^{1/2})^3 = A^{3/2},$$

by which the volume has been expressed as a function of $A$. Then

$$\frac{dV}{dA} = \frac{3}{2} A^{1/2}$$

is the rate of change of the volume with respect to $A$. Since $\frac{3}{2} A^{1/2} = \frac{3}{2} x$, we see that $dV/dA \neq dV/dx$.

In general, *a rate of change depends on the variable used*, and is different for different variables.

The second derivative of $f$ with respect to $x$ was defined as

$$\frac{d^2 f}{dx^2} = \frac{d}{dx} \left( \frac{df}{dx} \right);$$

in other words, the second derivative is itself a derivative, but of the first derivative. The second derivative is accordingly the rate of change, with respect to $x$, of the first derivative. Similar statements apply to derivatives of higher order.

**Example 5.10.2.** A point moves along the curve $y = x^2$ in the first quadrant. Find the rate of change of the area of the rectangle indicated in Fig. 5.10.1, (a) with respect to $x$, (b) with respect to $y$.

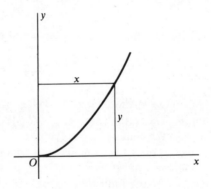

Fig. 5.10.1

*Solution.* The area of the rectangle is given by

$$A = xy.$$

For part (a) we may write, since $y = x^2$,

$$A = x^3.$$

Hence the rate of change of the area with respect to $x$ is $dA/dx = 3x^2$.

For part (b), we express the area as $A = y^{3/2}$, from which we find the rate of change of area with respect to $y$ as $dA/dy = \frac{3}{2} y^{1/2}$.

To compare the results we note that

$$\frac{dA}{dx} = 2x \frac{dA}{dy};$$

this corresponds to the result of using the chain rule, by which

$$\frac{dA}{dx} = \frac{dA}{dy} \cdot \frac{dy}{dx} = 2x \frac{dA}{dy}.$$

## 5.11  LINEAR MOTION

The concept of rate of change has a direct application to the motion of a particle along a straight line. On a coordinate line let the directed distance $s$ of a particle from the origin be given at time $t$ by the function $s = s(t)$, called a *position function*. As $t$ varies, the particle moves along the line. We define the velocity of the particle.

**DEFINITION 5.11.1.** *The* **velocity** *of a particle which moves along a straight line is the rate of change with respect to time of a position function of the particle.*

It follows from this definition, applied to the function $s(t)$, that the velocity $v$ at any time $t$ is given by

(5.11.1) $$v = v(t) = s'(t) = \frac{ds}{dt}.$$

Thus velocity of a particle on a straight line is a particular kind of rate of change, of a function representing a distance relative to the time. Let us think of a coordinate line with $s$ positive to the right. If $v(t) > 0$, it follows from (5.11.1) and Theorem 5.3.1 that $s(t)$ is increasing; hence the particle is moving to the right. If $v(t) < 0$, the particle is moving to the left.

*Example 5.11.1.* Let $s(t) = 2t^3 - t^2$ be the position function of a particle on a straight line. Describe the motion of the particle in the interval $0 \leq t \leq 4$.

*Solution.* We have

$$v(t) = s'(t) = 6t^2 - 2t = 2t(3t - 1).$$

Thus $v(t) < 0$ for $0 < t < 1/3$, while $v(t) > 0$ for $1/3 < t \leq 4$. Accordingly, the particle is at $s = 0$ when $t = 0$, moves a distance $s(1/3) = -1/27$, that is, to the left a distance $1/27$, then moves to the right, and is at the position $s(4) = 112$ at $t = 4$.

In the study of motion the *time-rate of change of velocity* also plays an important role and is called the *acceleration* of the particle. Since the rate of change of velocity is the rate of change of $v(t)$, we may state the following.

*The acceleration a(t) of a particle moving in a straight line is given by*

(5.11.2) $$a(t) = v'(t) = s''(t).$$

It follows that the velocity is increasing if $a(t) > 0$, and the velocity is decreasing if $a(t) < 0$.

**Example 5.11.2.** Let $s(t) = 40t - 16t^2$. Then

$$v(t) = s'(t) = 40 - 32t, \qquad a(t) = s''(t) = -32.$$

Since $a(t) < 0$, the velocity is always decreasing. The particle moves to the right when $v(t) > 0$, that is, for $t < 40/32 = 1.25$, and moves to the left for $t > 1.25$. The motion can be shown schematically in the diagram in Fig. 5.11.1. We see that $s(t)$ has a maximum when $t = 1.25$. This follows also from the criteria for extrema; we have

$$s'(1.25) = 0, \qquad s''(1.25) < 0.$$

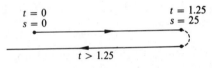

Fig. 5.11.1

**EXERCISE GROUP 5.8**

In Exercises 1–6 describe the motion of a particle with given position function, in the indicated interval for $t$.

1. $s = 60t - 16t^2$, $[0, 3]$
2. $s = 3t^3 + t + 4$, $[0, 2]$
3. $s = t^3 - 3t^2 + 3t + 4$, $[0, 2]$
4. $s = t^2(t - 2)^2 + 2$, $[-1, 4]$
5. $s = 3t + \dfrac{16}{t^3}$, $[1, 4]$
6. $s = \sqrt{4 + t^2}$, $[-2, 2]$

7. The position function of a moving particle is $s = t^2 - 3t^3$. (a) When is the rate of change of $s$ a maximum? (b) Find the minimum rate of change of $s$ in the interval $[0, 1]$.

8. Find the minimum velocity (rate of change of $s$) for a moving particle with position function $s = (t^3 + t - 8)/t$, in the interval $(0, \infty)$.

9. If an object is thrown vertically up with an initial speed $v_0$ ft/sec, its position function is $s = v_0 t - \tfrac{1}{2}gt^2$, where $g$ is a constant (approximately 32 in terms of feet and seconds). Describe the motion of the object.

10. Find the rate of change of the volume $V$ of a sphere (a) with respect to the radius $r$, (b) with respect to the surface area $S$.

11. Find the rate of change of the volume of a cube of side $x$ with respect to the surface area of the cube.

12. A point moves along the curve $xy = 1$ in the first quadrant. Find the rate of change, with respect to $x$, of the distance of the point from the origin when $x = 2$.

13. From a point $P$ on the curve $y = x^2$ in the first quadrant lines are drawn parallel to the coordinate axes, forming a rectangle with these axes. Find the point for which the rates of change of the area of the rectangle with respect to $x$ and with respect to $y$ are equal.

* 14. Show that if the rate of change with respect to $x$ of the distance of a point on a curve from the origin is $c$ ($\neq 0$) times the rate of change with respect to $y$ at a point, then we must have $dy/dx = c$ at this point, provided the rates of change are not zero.

* 15. A point moves along the curve $x^2 + 2y = 4$. (a) Find the points on the curve for which the rate of change with respect to $x$ of the distance from the origin is twice the rate of change of that distance with respect to $y$. (b) Do the same as in (a), but with the two rates of change equal.

## 5.12 RELATED RATES

Of particular interest in connection with rate of change is *time-rate* of change, that is, the rate of change with respect to a time variable of a quantity which is a function of time. The velocity of a particle which moves in a straight line (Section 5.11) is a special case of a time-rate of change. But the idea applies also in other cases.

If $q$ is a function of time $t$, the *time-rate of change* of $q$ is $dq/dt$, *the derivative of $q$ with respect to $t$*. In many applications a variable quantity is defined as a function of $t$ by way of an intermediate variable. The time-rates of change of the two quantities are themselves related, as in the following examples.

*Example 5.12.1.* A ship sails due east at 25 mph. At what rate is the distance of the ship from a point 25 mi due north of the starting point increasing after 3 hr?

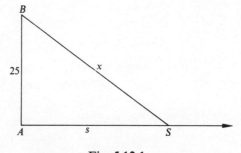

Fig. 5.12.1

## 144 APPLICATIONS OF DERIVATIVES

*Solution.* In Fig. 5.12.1, $A$ is the starting point, $B$ is the stated point due north of $A$, and $S$ is the position of the ship at time $t$. If $x$ denotes the distance of $S$ from $B$, we seek $dx/dt$ at $t = 3$. Let $s$ be the distance of $S$ from $A$. Then

$$x = \sqrt{s^2 + 625}, \quad \frac{dx}{dt} = \frac{s}{\sqrt{s^2 + 625}} \frac{ds}{dt}.$$

Since $ds/dt = 25$ we have, at any time,

$$\frac{dx}{dt} = \frac{25s}{\sqrt{s^2 + 625}}.$$

When $t = 3$, $s = 75$; hence

$$\left.\frac{dx}{dt}\right|_{t=3} = \frac{25 \cdot 75}{\sqrt{5625 + 625}} = \frac{25 \cdot 75}{25\sqrt{10}} = \frac{75}{\sqrt{10}} = \frac{15\sqrt{10}}{2} = 23.7 \text{ (mph)}.$$

We could have expressed $x$ directly in terms of $t$ by substituting $s = 25t$ in the expression for $x$.

**Example 5.12.2.** A point moves along the upper half of the curve $y^2 = x^3$ in such a way that the abscissa is increasing 2 units per second. At what rate is the distance of the point from the origin increasing when $x = 1$?

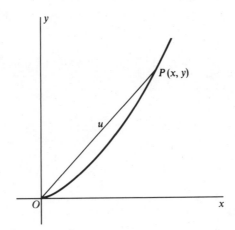

Fig. 5.12.2

*Solution.* Let $|OP| = u$ (Fig. 5.12.2). Then

$$u = \sqrt{x^2 + y^2}.$$

It is clear that both $x$ and $y$ depend on $t$; hence $u$ may be considered as a function of $t$. Differentiating $u$ with respect to $t$, we obtain

(5.12.1) $$\frac{du}{dt} = \frac{x\dfrac{dx}{dt} + y\dfrac{dy}{dt}}{\sqrt{x^2 + y^2}}.$$

From the equation $y^2 = x^3$ we get, on differentiating implicitly with respect to $t$,

$$2y\frac{dy}{dt} = 3x^2 \frac{dx}{dt}.$$

When $x = 1$ we find $y = 1$; these values, together with $dx/dt = 2$, give $(dy/dt)|_{x=1} = 3$. Substituting these values in (5.12.1) we obtain

$$\frac{du}{dt} = \frac{1 \cdot 2 + 1 \cdot 3}{\sqrt{1+1}} = \frac{5}{\sqrt{2}} = 3.5 \text{ (units/sec)}.$$

An alternative method is to use $x = 2t$, to write

$$u = \sqrt{x^2 + y^2} = \sqrt{x^2 + x^3} = \sqrt{4t^2 + 8t^3},$$

and then to find $du/dt$ directly and evaluate it at $t = 1/2$.

We emphasize an important idea that was used in the above examples, but which is often overlooked: *particular values of the variables involved are not used in any way until after the necessary differentiations have been performed.*

**Example 5.12.3.** A rectangle has a diagonal of length 13 cm. If one side is increasing at a rate of 2 cm/sec, find the rate of change of the area when that side is 5 cm.

**Solution.** Let $x$ be the length of the indicated side, so that $dx/dt = 2$. If $y$ is the length of an adjacent side, the area is

$$A = xy,$$

and we obtain

$$\frac{dA}{dt} = x\frac{dy}{dt} + y\frac{dx}{dt}.$$

From the rectangle we obtain $x^2 + y^2 = 169$, which yields, on differentiation.

$$2x\frac{dx}{dt} + 2y\frac{dy}{dt} = 0.$$

Now when $x = 5$ we find $y = 12$; then, with $dx/dt = 2$, we have $dy/dt = -5/6$. Hence

$$\frac{dA}{dt} = 5\left(-\frac{5}{6}\right) + 12(2) = \frac{119}{6} \text{ (cm}^2\text{/sec)}.$$

## EXERCISE GROUP 5.9

1. A plane flying horizontally in a straight line at an altitude of 1000 ft passes point $B$ directly over point $A$ on the ground. If its speed is 120 mph, find the rate in ft/sec at which its distance from $A$ is increasing when its distance from $B$ is 2000 ft.

2. The area of an equilateral triangle is increasing at a rate of 6 in$^2$/sec. Find the rate of increase of the length of a side when the altitude of the triangle is 1 in.

3. A trough 8 ft long has a vertical cross section in the form of an isosceles triangle with side 3 ft and altitude 2 ft. Water is pouring in at the rate of 6 ft$^3$/sec. How fast is the water level rising when the water is 1 ft deep?

4. Do Exercise 3 if the cross section is an isosceles trapezoid of side 3 ft, altitude 2 ft, and smaller base 2 ft.

5. A trough of length 10 ft has as its cross section an isosceles trapezoid with lower base 5 ft, upper base 11 ft, and altitude 4 ft. Water pours into the trough so that the depth increases at 2 ft/min. At what rate is the volume increasing when the depth is 2 ft?

6. Sand pouring out of a hopper at a rate of 20 ft$^3$/min forms a conical mound whose vertex angle is 90°. At what rate is the altitude of the mound increasing when the altitude is 4 ft?

7. A conical mound whose vertex angle is 60° is being formed as sand is poured onto it. If the altitude is increasing at a rate of 1 ft/min, at what rate is the volume increasing when the volume is 81 ft$^3$?

8. Ships A and B start at the same time at the same point with A traveling north at 20 mph, and B traveling east at 24 mph. How fast are they separating at the end of 2 hr?

9. Solve Exercise 8 if $A$ starts 2 hours earlier than $B$.

10. A point moves along the line $2x + 3y = 4$ so that $dx/dt = 3$. At what rate is its distance from the $x$ axis changing when $x = 3/2$?

11. A point moves along the curve $y = x^2$ in such a way that $dx/dt = 2$. How fast is the point receding from the origin when $x = 3$?

12. A point moves along the curve $8(y + 2) = x^2$ with its distance from the $x$ axis increasing at a rate of 2 units per minute. How fast is $x$ changing at the point (8, 6)?

13. A point moves along the line $x - y - 1 = 0$ with its abscissa increasing at a rate of 2 units/sec. (a) At what rate is the distance from the origin changing when $x = 2$? (b) At what point is the rate of change of the distance from the origin equal to 1?

14. One leg of a right triangle is increasing at a rate of 2 cm/sec, while the other leg is decreasing at the rate of 1 cm/sec. Find the relation between the lengths of the legs when the area is neither increasing nor decreasing.

★ 15. A fly is walking up a vertical edge of a room at 2 ft/sec. At what rate is its distance from the far corner of the floor increasing when it is 6 ft high, if the floor is 8 ft × 10 ft?

16. A 12-ft ladder rests against a vertical wall. If the bottom of the ladder is pulled away horizontally at 2 ft/sec, how fast is the top of the ladder moving when the bottom is 6 ft from the wall?

17. A vertical wall makes an angle of 120° with the ground. A 12-ft ladder rests against the wall. If the bottom of the ladder is slipping down the slope at 2 ft/sec, at what rate is the top of the ladder moving down when the top of the ladder strikes the ground?

* 18. A ladder of length 10 ft rests on the horizontal ground and extends over a 4-ft wall. If the lower end is pulled along the ground at a rate of 5 ft/sec, how fast is the upper end dropping when it is 8 ft above the ground?

* 19. A helicopter is flying due east at an altitude of $h$ ft, at a speed of $a$ ft/sec. An auto on the ground is traveling north at a speed of $b$ ft/sec. How fast are the helicopter and auto separating $t$ sec after a time when the one is directly above the other?

# 6 Curve Sketching

## 6.1 INTRODUCTION

The most fundamental procedure in sketching the graph of a function or an equation is that of plotting points. In many cases this procedure becomes very laborious; in other cases it does not yield all the information that may be needed. We wish to consider other methods that can be used to obtain the essential features of a graph with a minimum of plotting of points. Some of these are results of Chapter 5, such as criteria for determining where a function is increasing or decreasing, and criteria for maxima and minima, concavity, and points of inflection. Other methods are those available from the study of analytic geometry.

The graph of a polynomial function or equation can often be obtained reasonably well by plotting a sufficient number of points. A knowledge of any extrema or points of inflection may then be used to make the graph more precise.

*Example 6.1.1.* Draw the graph of the equation

$$y = x^3 - 3x^2 - 9x + 11.$$

*Solution.* We obtain the following table of values:

| $x$ | $-3$ | $-2$ | $-1$ | $0$ | $1$ | $2$ | $3$ | $4$ | $5$ |
|---|---|---|---|---|---|---|---|---|---|
| $y$ | $-16$ | $9$ | $16$ | $11$ | $0$ | $-11$ | $-16$ | $-9$ | $16$ |

The corresponding points are now plotted, and a curve is drawn through them (Fig. 6.1.1). The resulting curve is fairly typical of the graph of a third-degree polynomial. It is clear from the graph that both a relative maximum and a relative minimum exist. They appear to be at the points $(-1, 16)$ and $(3, -16)$, respectively. Let us find out.

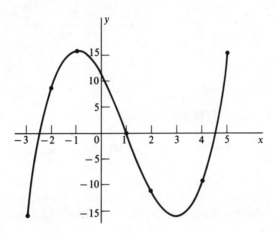

Fig. 6.1.1

We find
$$y' = 3x^2 - 6x - 9 = 3(x - 3)(x + 1).$$
The critical values are $-1$ and $3$. We now obtain
$$y'' = 6x - 6 = 6(x - 1).$$
Hence $y''(-1) < 0$, $y''(3) > 0$, so that there is, in fact, a relative maximum at $(-1, 16)$ and a relative minimum at $(3, -16)$. Furthermore, $y''(1) = 0$, $y'''(1) \neq 0$, and we determine that there is a point of inflection at $(1, 0)$.

The use of calculus thus permits a greater precision than would otherwise be possible in drawing the graph.

**EXERCISE GROUP 6.1**

In each of Exercises 1–10 draw the graph of the given equation.

1. $y = 2x^2 - 3x + 1$
2. $y = 3x^2 - 2x$
3. $y = x^3 - 6x + 5$
4. $y = 2x^3 - x^2 - 6x + 3$
5. $y = x^4 - 14x^2 + 1$
6. $y = 1 + 2x^3 - x^4$
7. $x = y^3 + y + 1$
8. $x = y^2 - 5y + 6$
9. $x = 2y - y^3$
10. $x = y^3 - 6y + 5$

Draw the graph of each of the following equations by using extrema, points of inflection and concavity, and plotting of points as needed. Be sure to include nonintegral values of $x$, particularly near any number for which the function is undefined. (Special methods for such equations will be developed in later sections.)

11. $y = \dfrac{x}{x^2 + 1}$

12. $y = \dfrac{1}{x^2 + 1}$

13. $y = \dfrac{x}{x + 1}$

14. $y = \dfrac{x + 1}{x}$

15. $y = \dfrac{1}{x^2 - 1}$

16. $y = \dfrac{x}{x^2 - 4}$

## 6.2 ADDITIONAL CRITERIA

In this section we present additional criteria which may be employed in sketching the graphs of certain equations or functions.

*Intercepts.* The knowledge of specific points on a graph is very helpful. (The method of plotting points relies on this knowledge exclusively.) In particular, the points where a curve intersects a coordinate axis may be especially useful.

Any point $(a, 0)$ or $(0, b)$ on a curve is called an *intercept* of the curve. The number $a$ is then called an $x$ *intercept*, and $b$ is called a $y$ *intercept*.

Thus the word "intercept," used alone, refers to a point of a curve which lies on a coordinate axis; while "$x$ intercept" is used for the abscissa of such a point on the $x$ axis, and "$y$ intercept" is used for the ordinate of such a point on the $y$ axis. The $x$ intercepts of the graph are found by setting $y = 0$ and solving the resulting equation; the $y$ intercepts are found similarly by setting $x = 0$.

*Example 6.2.1.* The intercepts of the graph of the equation

$$(x^2 + 1)y = x^2 - 1$$

are $(1, 0), (-1, 0)$, and $(0, -1)$. The $x$ intercepts are $-1$ and $1$, and the $y$ intercept is $-1$.

*Symmetry.* If $(m, n)$ is a point in the plane, the point $P_1(-m, n)$ is called its *reflection* or *image* in the $y$ axis; the point $P_2(m, -n)$ is its reflection in the $x$ axis; and $P_3(-m, -n)$ is its reflection in the origin (Fig. 6.2.1). If each point of a graph is reflected in a straight line, the resulting graph is called the *reflection* or *image* of the original curve in the line. It follows from these statements that if $x$ is replaced by $-x$ in an equation, the resulting equation has as its graph the reflection of the original graph in the $y$ axis.

Similarly, replacing $y$ by $-y$ yields an equation whose graph is the reflection of the graph of the original equation in the $x$ axis.

Finally, if at the same time $x$ is replaced by $-x$ and $y$ is replaced by $-y$, the graph of the new equation is called the reflection of the original graph in the origin.

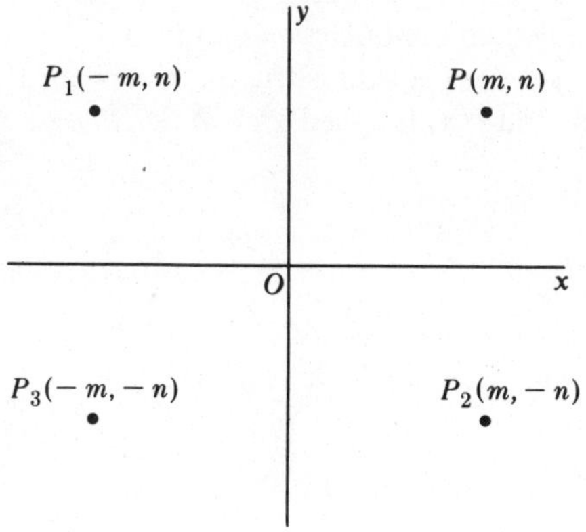

## 6.2 ADDITIONAL CRITERIA

$P_1(-m, n)$      $P(m, n)$

$P_3(-m, -n)$      $P_2(m, -n)$

Fig. 6.2.1

***Example 6.2.2.*** The reflection of the straight line $y = 2x + 3$ in the $y$ axis is the line

$$y = 2(-x) + 3 \quad \text{or} \quad y = -2x + 3.$$

The reflection of the first line in the $x$ axis is the line

$$-y = 2x + 3 \quad \text{or} \quad y = -2x - 3.$$

Finally, the reflection of $y = 2x + 3$ in the origin is the line

$$-y = 2(-x) + 3 \quad \text{or} \quad y = 2x - 3.$$

The four lines are shown in Fig. 6.2.2 on the following page.

It may happen that the reflection of a curve in a line (or point) is the same curve as the original one; in such an event the curve is said to be *symmetric* in that line (or point), or with respect to that line (or point).

We call two equations *equivalent* if one of them can be obtained from the other by a combination of the following operations:

(a) adding the same constant to both members,
(b) multiplying both members by the same nonzero constant.

A test for symmetry is then that the change in variable which produces the reflection yields an equivalent equation. Accordingly, we have the following tests for symmetry.

   I. *If replacing $y$ by $-y$ yields an equivalent equation, the curve is symmetric in the $x$ axis.*

  II. *If replacing $x$ by $-x$ yields an equivalent equation, the curve is symmetric in the $y$ axis.*

 III. *If replacing $x$ by $-x$ and $y$ by $-y$ yields an equivalent equation, the curve is symmetric in the origin.*

If the graph of $y = f(x)$ is symmetric in the $y$ axis, it follows that $f(x) = f(-x)$, and $f(x)$ is called an *even function* of $x$.

If the graph of $y = f(x)$ is symmetric in the origin, then $-y = f(-x)$, so that $f(-x) = -f(x)$, and $f(x)$ is called an *odd function* of $x$.

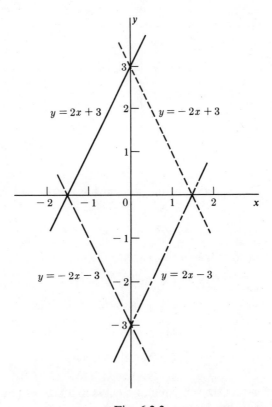

Fig. 6.2.2

**Example 6.2.3.** If $y = (x^2 - 1)/(x^2 + 1)$, the criterion above shows that the graph of the equation is symmetric in the $y$ axis. Furthermore, the function $f(x) = (x^2 - 1)/(x^2 + 1)$ is an even function of $x$, since $f(x) = f(-x)$.

**Example 6.2.4.** Let $f(x) = 2x^2 - x$ be defined only for $x \geq 0$. Extend the definition for $x < 0$ so that the new function is (a) odd, (b) even, for all real values of $x$.

*Solution.* (a) Let the new function be $F(x)$. For $x \geq 0$ we wish to have

(6.2.1) $$F(x) = f(x) = 2x^2 - x, \quad x \geq 0.$$

If $x < 0$, let $x = -u$, where $u > 0$. Since $F$ is to be odd, $F(-u) = -F(u)$. But

for $F(u)$ with $u > 0$ we may use (6.2.1). Then
$$F(u) = f(u) = 2u^2 - u, \quad u > 0,$$
and
$$F(-u) = -F(u) = -(2u^2 - u).$$
Since $u = -x$, we get
(6.2.2) $$F(x) = -2x^2 - x, \quad x < 0.$$
Thus (6.2.1) and (6.2.2) define the desired odd function.

(b) Let the even function be $G(x)$. We wish to have
(6.2.3) $$G(x) = f(x) = 2x^2 - x, \quad x \geq 0.$$
For $x < 0$, let $x = -u$, where $u > 0$. Since $G$ is to be even, $G(-u) = G(u)$. But for $G(u)$ with $u > 0$ we may use (6.2.3). Then
$$G(-u) = G(u) = f(u) = 2u^2 - u, \quad u > 0,$$
and since $-u = x$, we get
(6.2.4) $$G(x) = 2x^2 + x, \quad x < 0.$$
Then (6.2.3) and (6.2.4) define the desired even function.

Figure 6.2.3 shows the results graphically. For $x \geq 0$, the graphs of $f(x)$, $F(x)$, and $G(x)$ are identical. For $x < 0$, the function $f(x)$ is not defined, the graph of $F(x)$ is shown as a broken curve, and the graph of $G(x)$ is shown as a dotted curve.

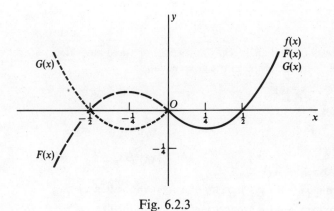

Fig. 6.2.3

When a curve is known to possess any of the types of symmetry, the graph of part of the curve may be suitably reproduced to yield more of the curve.

*Extent.* When two variables are related by an equation, it may happen that one of the variables is undefined for specific values of the other variable, or for intervals of values of that variable. Any information about such *excluded intervals* is of value in sketching the graph of the equation. Such information can often be obtained by direct algebraic analysis.

**Example 6.2.5.** The function $y = (x^2 - 1)/(x^2 + 1)$ is clearly defined for all $x$. On the other hand we may solve the equation for $x^2$ as

(6.2.5) $$x^2 = \frac{1+y}{1-y};$$

hence, for $x$ to be real, the right member of (6.2.5) cannot be negative. By solving the inequality

$$\frac{1+y}{1-y} < 0,$$

we find that the intervals $y < -1$ and $y > 1$ are excluded. In addition, $y = 1$ is excluded since, from (6.2.5), $x$ is undefined for $y = 1$.

### EXERCISE GROUP 6.2

In each of Exercises 1–8 a function is defined with the domain $x \geq 0$. Extend the definition of the function to the interval $x < 0$, (a) so that the new function is even for all real $x$, (b) so that the new function is odd for all real $x$.

1. $2x$
2. $x^2$
3. $x^2 - x$
4. $1 - 3x, x \neq 0; \quad 0, x = 0$
5. $x^3 + x^2$
6. $4x^2 - x + 1, x \neq 0; \quad 0, x = 0$
7. $\dfrac{1}{1+x^2}, x \neq 0; \quad 0, x = 0$
8. $\dfrac{x}{1+x^2}$

In Exercises 9–16 find analytically any values of $x$ and any values of $y$ for which there are no points on the graph of the given equation.

9. $y = 2x^2 - 3x + 1$
10. $y^2 + 4y = x$
11. $4x^2 + y^2 = 4$
12. $4x^2 - y^2 = 4$
13. $x^2 + xy + y^2 = 2$
14. $xy^2 + 4y - x = 0$
15. $y = \dfrac{1}{1+x^2}$
16. $y^2 = \dfrac{1}{1-x}$

17. Determine a test for symmetry of a curve in the line $x = h$.
18. Determine a test for symmetry of a curve in the line $y = k$.
19. Determine a test for symmetry of a curve in the point $(h, k)$.
20. Determine a test for symmetry of a curve in the line $y = x$.
21. Determine a test for symmetry of a curve in the line $x + y = 0$.

22. Show that, if a curve has any two of the three kinds of symmetry discussed in Section 6.2, it must also have the third kind.

23. (a) Find the point of inflection for the curve
$$y = 2x^3 + 6x^2 - 4x.$$
(b) Prove that the point found in (a) is a point of symmetry of the curve.

24. Show that the graph of the equation $y = ax^3 + bx^2 + cx + d$, $a \neq 0$, is symmetric in its point of inflection.

25. Prove (a) that the sum of two even functions is even,
(b) that the sum of two odd functions is odd.

26. Prove (a) that the product of two even functions or of two odd functions is even.
(b) that the product of an odd function and an even function is odd.

* 27. Prove that any function $f(x)$, defined for $-a \leq x \leq a$, may be expressed as the sum of an even function and an odd function. [Suggestion: Consider the functions
$$F(x) = f(x) + f(-x), \qquad G(x) = f(x) - f(-x).]$$

* 28. (a) If $f'(-x)$ is the value of the derivative $f'$ at $-x$, show that
$$f'(-x) = \lim_{h \to 0} \frac{f(-x+h) - f(-x)}{h}.$$
(b) If $f$ is odd and differentiable, prove that $f'$ is even.
(c) If $f$ is even and differentiable, prove that $f'$ is odd.

## 6.3  BEHAVIOR OF FUNCTIONS FOR LARGE $x$

The graph of a polynomial was considered in Section 6.1. An unresolved question (for which we intuitively know the answer) is the behavior of the graph as $x$ becomes large positively or negatively. Again, it is fairly easy to draw the graph of the function $y = (x^2 - 1)/(x^2 + 1)$, but a knowledge of what happens as $x$ gets large is needed for completeness. We are led to a discussion of limits of functions as $x$ becomes positively infinite or negatively infinite. The definitions follow the pattern set in Chapter 3, with necessary modifications.

**DEFINITION 6.3.1.**  1. $\lim_{x \to \infty} f(x) = A$ *if, for each number $\epsilon > 0$, a positive number N exists such that*
$$|f(x) - A| < \epsilon \quad \text{whenever} \quad x > N.$$

2. $\lim_{x \to -\infty} f(x) = A$ *if, for each number $\epsilon > 0$, a positive number N exists such that*
$$|f(x) - A| < \epsilon \quad \text{whenever} \quad x < -N.$$

Definition 6.3.1 applies to the existence of a *finite limit* as $|x|$ becomes infinite; a distinction is made, as $|x|$ becomes large, between the cases where $x$ is positive and $x$ is negative. We also define what is meant by an *infinite limit*, and make a distinction, as the function becomes large, between the cases where the function is positive and where it is negative.

**DEFINITION 6.3.2.** 1. $\lim_{x \to \infty} f(x) = \infty$ *if, for each number* $B > 0$, *a positive number* $N$ *exists such that*

$$f(x) > B \quad \text{whenever} \quad x > N.$$

2. $\lim_{x \to \infty} f(x) = -\infty$ *if, for each number* $B > 0$, *a positive number* $N$ *exists such that*

$$f(x) < -B \quad \text{whenever} \quad x > N.$$

Similar definitions apply to the limits

$$\lim_{x \to -\infty} f(x) = \infty \quad \text{and} \quad \lim_{x \to -\infty} f(x) = -\infty;$$

the statement, "whenever $x > N$," is to be replaced by, "whenever $x < -N$."

The use of the letter $\epsilon$, as in Definition 6.3.1, is customary when we are interested in small values of the variable. The number $N$ is usually large when $\epsilon$ is small.

With the use of the following theorem we can evaluate the limits of many functions as $x$ becomes infinite.

**THEOREM 6.3.1.** *If $n$ is a positive integer, then*

$$\lim_{x \to \infty} x^n = \infty;$$

$$\lim_{x \to -\infty} x^n = \begin{cases} \infty & \text{if } n \text{ is even,} \\ -\infty & \text{if } n \text{ is odd;} \end{cases}$$

$$\lim_{x \to \infty} x^{-n} = 0;$$

$$\lim_{x \to -\infty} x^{-n} = 0.$$

*Proof.* To prove the first part, let $B > 0$ be given. Choose $N = B^{1/n}$. Then if $x > N$, we have $x^n > N^n = B$, and Definition 6.3.2 is satisfied.

To prove the third part, let $\epsilon > 0$ be given. Choose $N = 1/\epsilon^{1/n}$. Then, if $x > N$, we have $0 < x^{-n} < N^{-n} = \epsilon$, and Definition 6.3.1 is satisfied.

The other parts of the theorem may be proved similarly.

It is not our intention to give a complete account of limit theorems as applied to infinite limits. Some of these will be stated as used. However, it should be noted that in general more care is required in anticipating results than was perhaps the case with finite limits. The following result is an illustration of this.

**THEOREM 6.3.2.** *If* $\lim_{x \to \infty} f(x) = \infty$ *and* $\lim_{x \to \infty} g(x) = A \neq 0$, *then*

(6.3.1) $$\lim_{x \to \infty} f(x)g(x) = \begin{cases} \infty & \text{if } A > 0, \\ -\infty & \text{if } A < 0. \end{cases}$$

*On the other hand, nothing specific can be stated about the limit of the product of $f(x)$ and $g(x)$ if $A = 0$.*

The proof of the first part is asked in Exercise 19. Regarding the second part you may refer to Exercise 18 and Section 15.2.

We are now in a position to extend the discussion in Example 6.1.1. We may write

$$y = x^3 - 3x^2 - 9x + 11 = x^3 \left(1 - \frac{3}{x} - \frac{9}{x^2} + \frac{11}{x^3}\right).$$

Using Theorem 6.3.1 (and Theorem 3.5.1, on the limit of a sum, which is still valid in this case), we find that

$$\lim_{x \to \infty} \left(1 - \frac{3}{x} - \frac{9}{x^2} + \frac{11}{x^3}\right) = 1;$$

hence, by Theorem 6.3.2, $\lim_{x \to \infty} y = \infty$. Similarly, $\lim_{x \to -\infty} y = -\infty$. This information tells us the behavior of the graph in Fig. 6.1.1 for large values of $|x|$.

**Example 6.3.1.** Find: $\lim_{x \to \infty} \dfrac{x^2 - 1}{x^2 + 1}$.

*Solution.* We have

$$\lim_{x \to \infty} \frac{x^2 - 1}{x^2 + 1} = \lim_{x \to \infty} \frac{x^2 \left(1 - \frac{1}{x^2}\right)}{x^2 \left(1 + \frac{1}{x^2}\right)} = \lim_{x \to \infty} \frac{1 - \frac{1}{x^2}}{1 + \frac{1}{x^2}} = 1.$$

Here we have used the fact that

$$\lim_{x \to \infty} \left(1 - \frac{1}{x^2}\right) = 1, \quad \lim_{x \to \infty} \left(1 + \frac{1}{x^2}\right) = 1,$$

and that the limit of the quotient of $1 - 1/x^2$ and $1 + 1/x^2$ is equal to the quotient of the limits (the limit of the denominator being different from zero).

**Example 6.3.2.** Find: $\lim_{x \to \infty} \dfrac{x^2 - x}{x^3 + 1}$.

*Solution.* We have

$$\lim_{x \to \infty} \frac{x^2 - x}{x^3 + 1} = \lim_{x \to \infty} \frac{\frac{1}{x} - \frac{1}{x^2}}{1 + \frac{1}{x^3}} = \frac{0 - 0}{1 + 0} = 0.$$

**Example 6.3.3.** Find: $\lim_{x \to -\infty} \dfrac{x^4 - 2}{x^2 + 1}$.

**Solution.** We have

$$\frac{x^4 - 2}{x^2 + 1} = x^2 \cdot \frac{1 - 2/x^4}{1 + 1/x^2}.$$

Since

$$\lim_{x \to -\infty} x^2 = \infty \quad \text{and} \quad \lim_{x \to -\infty} \frac{1 - 2/x^4}{1 + 1/x^2} = 1,$$

it follows from Theorem 6.3.2, which is also valid for the case $x \to -\infty$, that

$$\lim_{x \to -\infty} \frac{x^4 - 2}{x^2 + 1} = \infty.$$

As we have seen from the above examples, a method for finding the limit of a rational function $P(x)/Q(x)$ as $x \to \infty$ or $x \to -\infty$, where $P(x)$ is a polynomial of degree $m$ and $Q(x)$ is a polynomial of degree $n$, is as follows:

If $m \leq n$ first divide numerator and denominator by $x^n$.

If $m > n$ first extract a factor $x^{m-n}$ from the numerator, and then divide the resulting numerator and denominator by $x^n$.

In both cases appropriate limit theorems may then be applied.

### EXERCISE GROUP 6.3

Determine the limit (finite or infinite) in Exercises 1–16.

1. $\lim\limits_{x \to \infty} \dfrac{x+1}{x}$

2. $\lim\limits_{x \to \infty} \dfrac{1}{x^2 + 1}$

3. $\lim\limits_{x \to -\infty} \dfrac{x}{x^2 - 1}$

4. $\lim\limits_{x \to \infty} \dfrac{x}{x^2 - 4}$

5. $\lim\limits_{x \to \infty} \dfrac{3x^2 - x + 4}{2x^2 + 1}$

6. $\lim\limits_{x \to -\infty} \dfrac{x^3 + 3x}{2x^2 - 2x - 1}$

7. $\lim\limits_{x \to \infty} \dfrac{1 - 2x^3}{1 - 2x + x^2}$

8. $\lim\limits_{x \to -\infty} \dfrac{x^2 - 1}{x + 4}$

9. $\lim\limits_{x \to -\infty} \dfrac{1 - 2x^3}{1 - 2x}$

10. $\lim\limits_{|x| \to \infty} \dfrac{1 - 2x^3}{1 - 2x}$

11. $\lim\limits_{|x| \to \infty} \dfrac{(2x + 3)^2(x - 3)^3}{(x + 1)(x + 3)^3(x + 2)}$

12. $\lim\limits_{x \to \infty} \dfrac{(2x - 1)(3x + 4)}{(4x + 9)(7x - 4)}$

13. $\lim\limits_{x \to \infty} \dfrac{\sqrt{x}}{\sqrt{x + 1}}$

14. $\lim\limits_{x \to -\infty} \dfrac{\sqrt{2 - x}}{x - 1}$

15. $\lim\limits_{x \to -\infty} \dfrac{\sqrt{x^2 - 3}}{x - 2}$

16. $\lim\limits_{x \to \infty} \dfrac{\sqrt[3]{2x + 4}}{\sqrt{x - 2}}$

17. Prove: if $\lim\limits_{x \to \infty} f(x) = \infty$, then $\lim\limits_{x \to \infty} [-f(x)] = -\infty$.

18. Illustrate the statement, "If $\lim_{x\to\infty} f(x) = \infty$ and $\lim_{x\to\infty} g(x) = 0$, then nothing specific can be stated about $\lim_{x\to\infty} f(x)g(x)$," by considering the following cases, with $f(x) = x^4$ in each: (a) $g(x) = 1/x^4$, (b) $g(x) = 1/x^5$, (c) $g(x) = 1/x^3$.

19. Prove: if $\lim_{x\to\infty} f(x) = \infty$ and $\lim_{x\to\infty} g(x) = A \neq 0$, then

$$\lim_{x\to\infty} f(x)g(x) = \begin{cases} \infty & \text{if } A > 0, \\ -\infty & \text{if } A < 0. \end{cases}$$

20. A point moves along the curve $xy = 1$ in the first quadrant. Find the limit as $x \to \infty$ of the rate of change with respect to $x$, of the distance of the point from the origin.

## 6.4 HORIZONTAL AND VERTICAL ASYMPTOTES

We have seen that a function may have a limit as $x$ becomes large or $-x$ becomes large. In such a case the following definition applies.

**DEFINITION 6.4.1.** *If* $\lim_{x\to\infty} f(x) = b$, *the line* $y = b$ *is called a* **horizontal asymptote** *of the graph of* $f(x)$; *a similar statement applies if* $\lim_{x\to-\infty} f(x) = b$.

Horizontal asymptotes, if they exist, are an aid in sketching the graph, since the distance between the curve and the line $y = b$ approaches zero as $x$ or $-x$ becomes large. It is implied in Definition 6.4.1 that in some cases a curve may approach a horizontal asymptote for positive $x$ and not for negative $x$, or the reverse. This does indeed occur, as later examples will show. (See Fig. 6.6.1.)

In Example 6.3.1 the behavior of the function $(x^2 - 1)/(x^2 + 1)$ was determined for large $x$. In effect, we thereby established the fact that $y = 1$ is an asymptote of the graph of the function both for $x > 0$ and $x < 0$.

**Example 6.4.1.** Draw the graph of $y = \dfrac{x^2 - 1}{x^2 + 1}$.

*Solution.* Collecting the information about this function from previous discussions, we have the following.

Intercepts: $(1, 0)$, $(-1, 0)$, $(0, -1)$;
Symmetry: In $y$ axis;
Excluded intervals: $y < -1$, $y \geq 1$;
Horizontal asymptote: $y = 1$.

With this information the essential features of the graph may be shown (Fig. 6.4.1). For more precise information we may also determine that there is a relative minimum at $(0, -1)$, and points of inflection at $(\pm\sqrt{3}/3, -1/2)$.

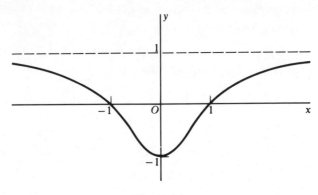

Fig. 6.4.1

We make two additional observations that may be helpful in determining the shape of a graph. It may happen that a graph crosses an asymptote. We may learn whether this occurs by solving simultaneously the equations of the asymptote and the graph. (The graph in Fig. 6.4.1 cannot cross its asymptote, since we know that $y < 1$.)

We illustrate the second observation by referring to the equation in Example 6.4.1. This equation may be written in the form

$$(x^2 + 1)y = x^2 - 1,$$

an equation of degree 3 in $x$ and $y$. Since the substitution $y = mx + b$ gives a third-degree equation in $x$, it follows that *no straight line may intersect the graph of the equation in more than* 3 *points*. This may be verified in Fig. 6.4.1.

In general, *no straight line may intersect the graph of an equation of degree n in more than n points*, since the solution of the equations of the curve and a straight line depends on solving a polynomial equation of degree $n$, and such an equation has at most $n$ real solutions.

*Vertical asymptotes.* We are aware that the function $y = 1/(x - 1)$ is undefined at $x = 1$. In order to study this function near $x = 1$ we find it necessary to study the function both for $x < 1$ and $x > 1$. We see that if $x < 1$ while $x$ is close to 1, $y$ is large in absolute value and negative, while if $x > 1$ and $x$ is close to 1, $y$ is large and positive. This behavior is expressed in the statements

$$\lim_{x \to 1^-} \frac{1}{x - 1} = -\infty, \qquad \lim_{x \to 1^+} \frac{1}{x - 1} = \infty.$$

Such limits are called *one-sided limits*. We formulate these ideas more precisely.

**DEFINITION 6.4.2.** $\lim_{x \to a^-} f(x) = \infty$, *if for any number* $A > 0$ *a number* $\delta > 0$ *exists such that*

$$f(x) > A \quad \text{whenever} \quad -\delta < x - a < 0.$$

*Similarly,* $\lim_{x \to a^+} f(x) = \infty$ *if, for any number* $A > 0$, *a number* $\delta > 0$ *exists such that*

$$f(x) > A \quad \text{whenever} \quad 0 < x - a < \delta.$$

*Similar definitions apply to* $\lim_{x \to a^-} f(x) = -\infty$, *and to* $\lim_{x \to a^+} f(x) = -\infty$, *simply by replacing the inequality* $f(x) > A$ *by* $f(x) < -A$.

With the definitions for the cases $x \to a^-$, $x \to a^+$, and $x \to a$, the following result may be stated.

**THEOREM 6.4.1.** *If both* $\lim_{x \to a^-} f(x) = \infty$ *and* $\lim_{x \to a^+} f(x) = \infty$, *then* $\lim_{x \to a} f(x) = \infty$. *A corresponding statement holds if both one-sided limits are* $-\infty$.

Similar definitions and theorems may be formulated for one-sided finite limits.

We are now in a position to define a vertical asymptote.

**DEFINITION 6.4.3.** *If* $\lim_{x \to a^-} f(x) = \infty$, *or if* $\lim_{x \to a^-} f(x) = -\infty$, *or if a similar statement applies as* $x \to a^+$, *the line* $x = a$ *is called a* **vertical asymptote** *of the graph of* $f(x)$.

The implication of a vertical asymptote is that as $x$ is getting closer to $a$, for $x < a$ or $x > a$ as the case may be, the distance between the graph and the asymptote is approaching zero.

The basic limits in this connection are the following.

(6.4.1) $$\lim_{x \to a^-} \frac{1}{x - a} = -\infty, \quad \lim_{x \to a^+} \frac{1}{x - a} = \infty.$$

To prove the second one, for example, let $A > 0$ be given. We may choose $\delta = 1/A$. Then if $0 < x - a < 1/A$, we have $1/(x - a) > A$, and Definition 6.4.2 is satisfied.

For a rational function, that is, a function of the form

(6.4.2) $$y = \frac{P(x)}{Q(x)},$$

where $P(x)$ and $Q(x)$ are polynomials with no factor in common other than a constant, it can be proved (and we shall accept the fact) that *vertical asymptotes of the graph occur for those values of x for which* $Q(x) = 0$. Furthermore, the graph of (6.4.2) has a horizontal asymptote if the degree of $P(x)$ is less than or equal to the degree of $Q(x)$.

*Example 6.4.2.* Draw the graph of

$$y = \frac{x^2 - x}{(x + 1)(x - 2)^2}.$$

*Solution.* To find where $y = 0$ we solve the equation $x^2 - x = 0$, and we find $x = 0$ or $1$. Thus the points $(0, 0)$ and $(1, 0)$ are the only intercepts (the origin is an intercept on both axes).

There is no symmetry in the $x$ axis, the $y$ axis, or the origin.

Since the denominator of the fraction is a cubic polynomial we may write

$$\lim_{x \to \infty} y = \lim_{x \to \infty} \frac{x^2 - x}{x^3 + ax^2 + bx + c} = \lim_{x \to \infty} \frac{1}{x} \frac{1 - 1/x}{1 + a/x + b/x^2 + c/x^3} = 0.$$

Similarly, $\lim_{x \to -\infty} y = 0$. Hence $y = 0$ is a horizontal asymptote in both directions.

By the above statements concerning rational functions, $x = -1$ and $x = 2$ are vertical asymptotes.

There are no excluded intervals for $x$ (the excluded values $x = -1$ and $x = 2$ yield asymptotes). Finding directly excluded intervals for $y$, if any, is difficult. However, the graph may now be drawn reasonably well with the available information, and appears in Fig. 6.4.2. Any questions about excluded intervals may be examined with reference to the graph, and, in particular, there are no excluded intervals for $y$.

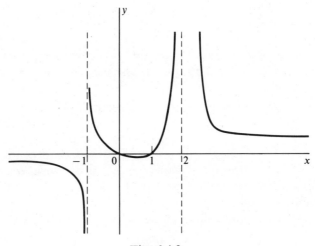

Fig. 6.4.2

It is clear from the graph that there is a relative minimum for some value of $x$ between 0 and 1. Finding this minimum involves solving a third-degree equation (it happens to occur for $x$ slightly less than 0.6). However, a precise knowledge of this minimum does not appreciably affect the drawing of the graph.

It may be pointed out here that if an equation may be solved for $x$ as a rational function of $y$, the previous discussion concerning horizontal and vertical asymptotes still applies, with the roles of $x$ and $y$ interchanged.

## EXERCISE GROUP 6.4

Draw the graph of each of the following equations, including a consideration of intercepts, symmetry, and horizontal and vertical asymptotes.

1. $y = \dfrac{x}{x+1}$
2. $y = \dfrac{x+1}{x}$
3. $y = \dfrac{x^2}{x-3}$
4. $y = \dfrac{x}{x^2-9}$
5. $y = \dfrac{x^2}{x^2-9}$
6. $y = \dfrac{x}{x^2+1}$
7. $y = \dfrac{x^2}{x^2+1}$
8. $y = \dfrac{x^3}{x^2+1}$
9. $y = \dfrac{x^2-1}{x^2-4}$
10. $y = \dfrac{x^2-4}{x^2-1}$
11. $x = \dfrac{4}{y^2-4}$
12. $x = \dfrac{4}{y(y^2-4)}$
13. $x = \dfrac{4}{y^2(y-4)}$
14. $x = \dfrac{y}{y^2-1}$
15. $y = \dfrac{8}{x(x^2-4)}$
16. $y = \dfrac{x^3-8}{2x}$
17. $y = \dfrac{x+1}{x^2+2x}$
18. $y = \dfrac{x-2}{x(x^2-9)}$
19. $y = \dfrac{x^2-4}{2x^2-3x-9}$
20. $y = \dfrac{x^2-16}{2x^2-3x-9}$
21. $y = \dfrac{x^2(x+1)}{x-1}$
22. $y = \dfrac{x^3(2-x)}{(x-1)^2}$
23. $y = \dfrac{x^3}{2-x}$
24. $y = \dfrac{x^2(1+x)}{1-x}$

25. Formulate a definition for (a) $\lim\limits_{x \to a^-} f(x) = A$, (b) $\lim\limits_{x \to a^+} f(x) = A$.
26. Prove the first limit in (6.4.1).
27. Prove Theorem 6.4.1 for the case where the one-sided limits are $\infty$.

## 6.5 VERTICAL TANGENTS

When the derivative of a function $f$ exists at $x = a$, the slope of the graph of $f$ at $x = a$ is the value of the derivative (Section 4.3). We were led to this definition and the definition of tangent line by considering a secant line joining the points $P(a, f(a))$ and $Q(a + h, f(a + h))$ on the graph. We now consider one type of situation that occurs if the derivative does not exist at a value $a$ in the domain of $f$ at which $f$ is continuous. If the fraction

(6.5.1) $$\frac{f(a+h)-f(a)}{h}$$

becomes large in absolute value as $h \to 0$, the inclination of the secant line (Fig. 6.5.1) through the points $P(a, f(a))$ and $Q(a+h, f(a+h))$ approaches $\pi/2$,

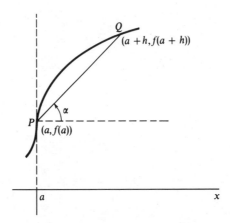

Fig. 6.5.1

and there is a *vertical tangent* at $P$, that is, a tangent line which is parallel to the $y$ axis. We may describe this type of occurrence by writing

$$\lim_{h \to 0} \left| \frac{f(a+h)-f(a)}{h} \right| = \infty,$$

and, perhaps more directly, by considering the reciprocal of (6.5.1). The above discussion leads to the following definition.

**DEFINITION 6.5.1.** *The graph of the function $f$ has a **vertical tangent** at a value $x = a$ at which $f$ is continuous, if*

$$\lim_{h \to 0} \frac{h}{f(a+h)-f(a)} = 0.$$

*Example 6.5.1.* Show that the curve $y = x^{2/3}$ has a vertical tangent at $x = 0$.

*Solution.* We have, with $a = 0$ and $f(x) = x^{2/3}$,

$$\frac{h}{f(a+h)-f(a)} = \frac{h}{h^{2/3} - 0} = h^{1/3}.$$

But $\lim_{h \to 0} h^{1/3} = 0$. Hence, by Definition 6.5.1, there is a vertical tangent at $x = 0$.
See Fig. 6.5.2.

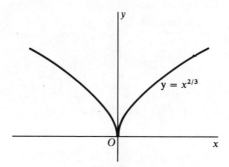

Fig. 6.5.2

In some cases a function may be defined on one side of some value $x = a$, and a vertical tangent may exist on that side. In this event Definition 6.5.1 may be modified by specifying a *one-sided limit* either as $h \to 0^-$ or as $h \to 0^+$.

***Example 6.5.2.*** Show that the curve $y = (1 - x)^{1/2}$ has a vertical tangent at $x = 1$.

*Solution.* The function $g(x) = (1 - x)^{1/2}$ is defined only for $x \le 1$. With $a = 1$ we have

$$\frac{h}{g(a + h) - g(a)} = \frac{h}{(-h)^{1/2} - 0} = (-h)^{1/2}.$$

We must have $h < 0$, and hence the limit is one-sided as $h \to 0^-$, and

$$\lim_{h \to 0^-} (-h)^{1/2} = 0.$$

There is a vertical tangent at $x = 1$ (from the left). See Fig. 6.5.3.

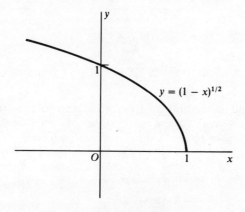

Fig. 6.5.3

In many cases that arise it is unnecessary to apply Definition 6.5.1 to show the existence of a vertical tangent. If the derivative of a function exists in an interval $[a, b]$ except at some value $c$ where $f$ is continuous, and if

$$\lim_{x \to c} |f'(x)| = \infty,$$

then there is a vertical tangent at $x = c$. It is understood that the limit may be one-sided, if $c$ is an endpoint. Thus for the function $f(x) = x^{2/3}$ of Example 6.5.1 we have

$$f'(x) = \frac{2}{3x^{1/3}}, \qquad \lim_{x \to 0} |f'(x)| = \infty.$$

Since $f$ is continuous at 0, there is a vertical tangent at $x = 0$. For the function $g(x) = (1 - x)^{1/2}$ of Example 6.5.2 we have

$$g'(x) = \frac{-1}{2(1 - x)^{1/2}}, \qquad \lim_{x \to 1^-} |g'(x)| = \infty,$$

and there is a vertical tangent at $x = 1$.

Bear in mind the fact that if the graph of a function has a vertical tangent at $x = a$, there may be an extremum or point of inflection at this point, or an endpoint extremum. Indeed, this will generally be the case except for very unusual functions. Figure 6.5.4 shows some typical appearances at a vertical tangent.

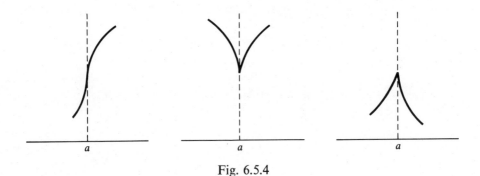

Fig. 6.5.4

## EXERCISE GROUP 6.5

In Exercises 1–10 find any vertical tangents for the graph of the given function. If any such exist, determine whether there is a relative maximum or minimum, endpoint maximum or minimum, or point of inflection at each such point.

1. $(2x - x^2)^{1/2}$
2. $(2x - x^2)^{1/3}$
3. $(2x - x^2)^{-1/2}$
4. $x^{2/3} + x^{1/2}$
5. $(x^2 + 1)^{1/2}$
6. $(x^2 + 1)^{-1/2}$
7. $(x^2 - 1)^{1/2}$
8. $(x^2 - 1)^{-1/2}$
9. $x - \sqrt{x - 1}$
10. $x + \sqrt{x^2 - 1}$

In each of the following find any vertical tangent for the graph of the given equation.

11. $4x^2 - y^2 = 36$
12. $4x^2 + y^2 = 36$
13. $x^2 + xy - y^2 = 1$
14. $x^2 + 4y^2 - 8y = 5$
15. $3x^2 - 2y^2 - 6x - 4y = 11$
16. $x^2 - 4xy + 4y^2 + x = 3$
17. $2x^2 + xy + y^2 - 2y = 4$
18. $x^{1/2} + y^{1/2} = a^{1/2}$
19. $x^{2/3} + y^{2/3} = a^{2/3}$

* 20. $x^3 + y^3 - 3xy = 0$
* 21. $x^2 + y^2 = 3x^2y^2$
* 22. $y^2 = x^2 - x^3$

## 6.6 THE GRAPH OF $y^2 = R(x)$

Many interesting and important curves have equations which may be written in the form

(6.6.1) $$y^2 = R(x) = \frac{P(x)}{Q(x)},$$

where $R(x)$ is a rational function of $x$, and $P(x)$ and $Q(x)$ are polynomials in $x$ with no common factor other than a constant. Several observations may be made at once.

  i. *The graph is symmetric in the x axis.*
  ii. *The graph exists only for values of $x$ for which $R(x) \geq 0$.*
  iii. *If $x = h$ is a vertical asymptote of $y = R(x)$, then it is a vertical asymptote of (6.6.1) provided the graph of (6.6.1) exists near $x = h$.*
  iv. *If $y = k > 0$ is a horizontal asymptote of $y = R(x)$, then $y = \pm\sqrt{k}$ are horizontal asymptotes of (6.6.1). If $y = 0$ is a horizontal asymptote of $y = R(x)$, then $y = 0$ is a horizontal asymptote also of (6.6.1) if $R(x) > 0$ for $x$ large and positive or for $x$ large and negative, or both; the line $y = 0$ may then be an asymptote in only one direction.*

With these observations much of the graph of (6.6.1) can usually be drawn. However, if $R(x) = 0$ for $x = a$, the appearance of the graph of (6.6.1) near $x = a$ requires examination. If $R(a) = 0$, then $P(a) = 0$, and $P(x)$ may be factored in such a way that

$$P(x) = (x - a)^r P_1(x), \qquad P_1(a) \neq 0, \qquad r \text{ a positive integer.}$$

We may then express $R(x)$ as $R(x) = (x - a)^r R_1(x)$, where $R_1(x)$ is a rational function with $R_1(a) \neq 0$; then (6.6.1) becomes

(6.6.2) $$y^2 = (x - a)^r R_1(x), \qquad R_1(a) \neq 0, \qquad r \text{ a positive integer.}$$

If $r$ is even, we must have $R_1(a) > 0$, if there are to be points of the curve near $x = a$. If $r$ is odd, the details may depend on whether $R_1(a)$ is positive or negative, but the conclusions in Theorem 6.6.1 apply in either case.

We shall carry out the discussion for the case $R_1(a) > 0$. By implicit differentiation of (6.6.2) with respect to $x$, we get

$$2yy' = (x - a)^r R_1'(x) + r(x - a)^{r-1} R_1(x),$$

and, substituting $y$ from (6.6.2),

$$y' = \pm \frac{(x - a)^{r/2} R_1'(x) + r(x - a)^{r/2-1} R_1(x)}{2[R_1(x)]^{1/2}}.$$

We know that $R_1(a) \neq 0$. If $r = 1$, we find $\lim_{x \to a^+} y' = \pm \infty$; if $r = 2$ and $R_1(a) > 0$, we get $\lim_{x \to a} y' = \pm [R_1(a)]^{1/2}$; and if $r > 2$, we get $\lim_{x \to a^+} y' = 0$, unless $r$ is even and $R_1(a) < 0$. We state the result in the following form.

**THEOREM 6.6.1.** *Let $R_1(x)$ be a rational function of $x$, and let*

(6.6.3) $\qquad\qquad y^2 = (x - a)^r R_1(x), \qquad R_1(a) \neq 0.$ *r a positive integer.*

*Then, at the intercept $(a, 0)$ on the graph of (6.6.3) one of the following occurs.*

(a) *if $r = 1$, there is a vertical tangent;*

(b) *if $r = 2$ and $R_1(a) > 0$, the curve crosses the x axis with slopes $\pm [R_1(a)]^{1/2}$;*

(c) *if $r > 2$, the graph is tangent to the x axis, unless $r$ is even and $R_1(a) < 0$;*

(d) *if $r$ is even and $R_1(a) < 0$, there is an isolated point at $(a, 0)$; that is, in some interval about $x = a$ there are no other points of the curve.*

**Example 6.6.1.** Draw the graph of $y^2 = \dfrac{x + 4}{x(x + 2)}$.

*Solution.* The graph is symmetric in the $x$ axis, and has $x$ intercept $-4$. By iii above, $x = 0$ and $x = -2$ are vertical asymptotes. By iv above, $y = 0$ is a horizontal asymptote, in the positive $x$ direction.

To see where $R(x) \geq 0$ we may test values of $x$ in the intervals determined by the vertical asymptotes and the $x$ intercepts, that is, in the intervals $(-\infty, -4)$, $(-4, -2)$, $(-2, 0)$, and $(0, \infty)$. If $x = -5$, $R(-5) < 0$ and there is no graph for $x$ in $(-\infty, -4)$. If $x = -3$, $R(-3) > 0$ and the graph exists in $(-4, -2)$. If $x = -1$, $R(-1) < 0$ and there is no graph in $(-2, 0)$. Finally, if $x = 1$, $R(1) > 0$ and the graph exists in the interval $(0, \infty)$.

By Theorem 6.6.1, part (a), there is a vertical tangent at $x = -4$. The graph appears in Fig. 6.6.1.

Entirely similar considerations to those above apply, if an equation can be expressed in the form $x^2$ equal to a rational function of $y$. The following example illustrates this.

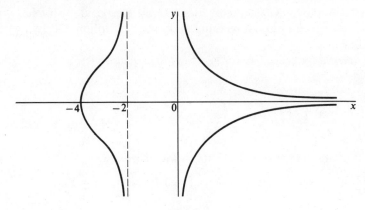

Fig. 6.6.1

***Example 6.6.2.*** Draw the graph of $y^3 + x^2y = 2(3y^2 - x^2)$.

*Solution.* We may write the equation as

$$x^2 = \frac{6y^2 - y^3}{y + 2}.$$

There is symmetry in the $y$ axis, and the $y$ intercepts are 0 and 6. Applying Theorem 6.6.1 with $x$ and $y$ interchanged, we find that the tangents at the origin have slopes $\pm\sqrt{3}/3$ (their slopes with respect to the $y$ axis are $\pm\sqrt{3}$), and guide lines with these slopes may be drawn. Applying Theorem 6.6.1 in the same way, we see that there is a horizontal tangent at $(0, 6)$. There are no vertical asymptotes, but $y = -2$ is a horizontal asymptote. No points of the graph exist for $y \leq -2$ or for $y > 6$. The graph appears in Fig. 6.6.2.

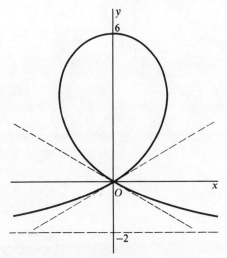

Fig. 6.6.2

Many of the previous observations apply to the graph of an equation $y^2 = R(x)$, where $R$ need not be rational. In the description that follows $R$ may be any function.

An *alternative procedure* for graphing the equation

(6.6.4) $$y^2 = R(x)$$

is to graph the equation

(6.6.5) $$y_s = R(x),$$

and then to graph (6.6.4) with the following observations.

(a) Wherever $y_s < 0$, no point of (6.6.4) exists.

(b) Wherever $y_s > 0$, two points of (6.6.4) exist, with $y = \pm\sqrt{y_s}$.

(c) If $y_s = 0$ at $x = a$, the point $(a, 0)$ may be an isolated point.

(d) We may use all other available information, such as that involving intercepts, asymptotes, and tangents.

**Example 6.6.3.** Draw the graph of

(6.6.6) $$y^2 = x^4(5 + x).$$

*Solution.* The graph of $y_s = x^4(5 + x)$ appears in Fig. 6.6.3 (a). There are $x$ intercepts 0 and $-5$, and $y = 0$ is a tangent at $x = 0$. There are no asymptotes, and there is no symmetry.

Since $y_s < 0$ when $x < -5$, there are no points on the desired graph of (6.6.6) by (a) above for $x < -5$. By Theorem 6.6.1 the graph, in Fig. 6.6.3(b), has a tangent $y = 0$ at the origin, and a vertical tangent at the point $(-5, 0)$.

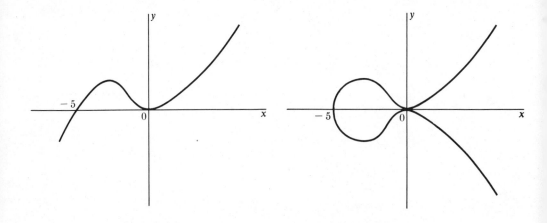

Fig. 6.6.3 (a)  Fig. 6.6.3 (b)

## EXERCISE GROUP 6.6

In Exercises 1–21 draw the graph of the given equation, using any combination of previously discussed methods.

1. $y^2 = 5 + x$
2. $y^2 = 2x^2 - 1$
3. $y^2 = x^2(5 + x)$
4. $y^2 = x^3(5 + x)$
5. $x^4 + x^2y^2 = 4y^2$
6. $(x - 1)^2 y^2 = x^3(2 - x)$
7. $y^2 = \dfrac{x^2 - 1}{(2x + 1)^2}$
8. $y^2 = \dfrac{x}{4 - x}$
9. $y^2 = \dfrac{x}{x^2 - 4}$
10. $y^2 = \dfrac{x^3}{1 - x}$
11. $x^2 = \dfrac{4y - 1}{y - 1}$
12. $x^2 y^2 + 4y^2 = 2x^2$
13. $y^2(2 - x) = x^2(2 + x)$
14. $2xy^2 = x^3 - 8$
15. $(x^2 - 4)y^2 = 8$
16. $(x^2 - 9)y^2 = x^2 - 4$
17. $xy^4 - 4xy^2 - 4 = 0$
18. $y^2 = \dfrac{x(x^2 - 1)}{(x^2 - 4)^2}$
19. $(x^2 - 1)y^2 = x^2$
20. $y^2 = \dfrac{x^2(x + 1)}{x - 1}$
21. $x^2 + y^2 = x^2 y$

\* 22. A tangent to the circle $x^2 + y^2 = 4$ meets the hyperbola $2xy = 1$ in the points $P_1$ and $P_2$. Show that an equation of the locus of the midpoint of the line segment joining $P_1$ and $P_2$ is the equation of Exercise 19.

23. Obtain the graph of the equation $y^2 = 2x + 4$ by making use of the straight line $y = 2x + 4$.

24. Prove part (d) of Theorem 6.6.1.

## \* 6.7 ASYMPTOTIC CURVES

We know that the graph of the equation

$$y = \frac{x - 2}{2x - 1}$$

has a horizontal asymptote $y = 1/2$. If we write the equation of this asymptote as $y_A = 1/2$, we have

$$y - y_A = \frac{x - 2}{2x - 2} - \frac{1}{2} = -\frac{3}{2(2x - 1)}.$$

Hence $\lim(y - y_A) = 0$ as $x \to \infty$ and as $x \to -\infty$, which fact of course underlies the idea of a horizontal asymptote. Moreover, $y - y_A > 0$ when $x$ is large and negative, and $y - y_A < 0$ when $x$ is large and positive. Hence the curve lies above its horizontal asymptote when $x$ is large and negative; it lies below the asymptote when $x$ is large and positive..

Similar methods may often be applied to find an *oblique asymptote*, that is, an asymptote which is neither horizontal nor vertical. Consider the equation

(6.7.1) $$y = \frac{x^2 - 3x}{x - 2}.$$

By ordinary long division we find

(6.7.2) $$y = x - 1 - \frac{2}{x - 2}.$$

If we set

$$y_A = x - 1,$$

then (6.7.2) may be written as

$$y - y_A = -\frac{2}{x - 2}.$$

Then $y - y_A \to 0$ as $x \to \pm\infty$, and therefore the line

(6.7.3) $$y = x - 1$$

is an (oblique) asymptote of the graph of (6.7.1). Moreover, $y - y_A > 0$ if $x$ is large and negative, and $y - y_A < 0$ if $x$ is large and positive. The graph of (6.7.1) and the asymptote (6.7.3) are shown in Fig. 6.7.1.

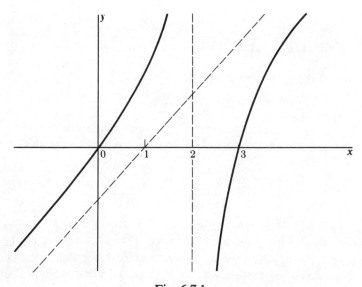

Fig. 6.7.1

We may now use the methods as developed to find *asymptotic curves* other than straight lines. Consider, for example, the equation

(6.7.4) $$y = \frac{x^3 - 3x}{x - 2} = x^2 + 2x + 1 + \frac{2}{x - 2}.$$

If we set

(6.7.5) $$y_A = x^2 + 2x + 1,$$

then

$$y - y_A = \frac{2}{x - 2},$$

and $y - y_A \to 0$ as $x \to \pm \infty$. We then say that the graph of (6.7.4) is *asymptotic* to the graph of (6.7.5). In particular, the graph of (6.7.4) lies above its asymptotic curve (6.7.5) for large positive $x$ and below it for large negative $x$.

The graph of (6.7.4) and its asymptotic curve (6.7.5) are shown in Fig. 6.7.2. The graph of (6.7.4) is the solidly drawn curve. The vertical asymptote $x = 2$ of the graph of (6.7.4) is also shown.

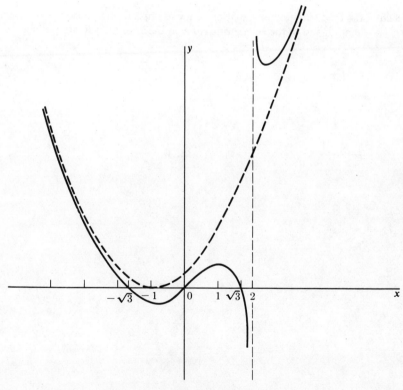

Fig. 6.7.2

## * EXERCISE GROUP 6.7

In each of Exercises 1–12 find an equation of the oblique asymptote; sketch the graph of the given equation and of the asymptote in the same coordinate system.

1. $y = 2x - \dfrac{1}{x}$

2. $y = 2x + 3 - \dfrac{1}{3x}$

3. $y = 2 - x + \dfrac{1}{2-x}$

4. $y = \dfrac{x^2 + 1}{x}$

5. $y = \dfrac{3x^2 + x + 1}{2x}$

6. $y = \dfrac{x^2 + x}{x - 1}$

7. $y = \dfrac{x^2 + 2x + 1}{x - 1}$

8. $y = \dfrac{2x^2}{2x + 1}$

9. $y = x + \dfrac{2}{x^2}$

10. $y = \dfrac{x^3}{x^2 + 1}$

11. $y = \dfrac{2x^3 - x}{2x^2 + 1}$

12. $y = \dfrac{x^4 + 1}{x^3}$

In each of the following find an equation of the asymptotic curve; sketch the graph of the given equation and of the asymptotic curve in the same coordinate system.

13. $y = x^2 - \dfrac{2}{x}$

14. $y = \dfrac{x^3 + 1}{x}$

15. $y = \dfrac{x^3 + x + 1}{x}$

16. $y = \dfrac{x^3 + 1}{x - 1}$

17. $y = \dfrac{x^3}{x + 1}$

18. $y = \dfrac{x^4}{x^2 + 1}$

# 7 Equations of Second Degree

## 7.1 INTRODUCTION

In Chapter 6 we studied and graphed many special cases of polynomial equations in $x$ and $y$. We now study in detail one type of such equations, the *general second-degree or quadratic equation in x and y*, which is of the form

(7.1.1) $$Ax^2 + Bxy + Cy^2 + Dx + Ey + F = 0,$$

where $A$, $B$, $C$, $D$, $E$, and $F$ are constants. This equation is especially interesting, because its graph can be studied completely and has interesting and important locus properties.

We shall first study special cases of (7.1.1) in detail, and then in Chapter 8 we shall relate these cases to the general equation.

Equations of the form (7.1.1) include straight lines (for example, if $A = B = C = 0$. The curves which are not straight lines and whose equations are of the form (7.1.1) are the circle, parabola, ellipse, and hyperbola. These curves are called collectively *conics* or *conic sections*, since they occur as intersections of a right circular cone (Fig. 20.5.5) and a plane.

## 7.2 EQUATION OF A LOCUS

The *graph of an equation* was defined in Section 2.6. The same definition may be interpreted as defining an *equation of a curve*, which we shall make explicit as follows.

An equation in Cartesian coordinates of a curve is an equation in $x$ and $y$ which is satisfied by the coordinates of every point of the curve and only such points.

This definition may be rephrased as follows. *The graph of an equation in $x$ and $y$ is the locus of points $(x, y)$ whose coordinates satisfy the equation.* The implication of these definitions is that a curve and an equation of the curve describe the same set of points, one description being geometric and the other analytic. Here we wish to consider the question of finding an equation of a set of points, or a *locus*, which is described geometrically. A straight line, not parallel to the $y$ axis, may be considered as the locus of points such that the slope of the line joining any one of them and a fixed point is constant. In the next section we shall study the circle, *the locus of points equidistant from a fixed point*.

**Example 7.2.1.** Find an equation of the locus of a point such that the lines joining the point to the points $A(-1, 0)$ and $B(1, 0)$ are perpendicular.

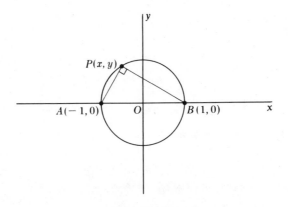

Fig. 7.2.1

*Solution.* Let $P(x, y)$ represent an arbitrary point with the desired property (Fig. 7.2.1). Then $\overrightarrow{PA} \perp \overrightarrow{PB}$. The vectors $\overrightarrow{PA}$ and $\overrightarrow{PB}$ are given by

$$\overrightarrow{PA} = (-1 - x, -y), \qquad \overrightarrow{PB} = (1 - x, -y).$$

We must then have $\overrightarrow{PA} \cdot \overrightarrow{PB} = 0$ (Section 2.2). Hence

$$(-1 - x)(1 - x) + (-y)(-y) = 0,$$

from which we get

(7.2.1) $$x^2 + y^2 - 1 = 0.$$

Moreover, if a point $P(x, y)$ satisfies (7.2.1), the steps may be reversed, and $\overrightarrow{PA} \perp \overrightarrow{PB}$. Hence (7.2.1) is an equation of the locus. *The locus is a circle with AB*

**7.2**                                EQUATION OF A LOCUS

*as diameter*; this follows from plane geometry, or from the next section. We may wish to exclude the points $A$ and $B$ from the locus since, if $P$ is $A$ or $B$, there is no line determined by $P$ and that point.

The general procedure for finding an *equation of a locus* may now be described as follows.

(a) *Let $x$ and $y$ be the coordinates of a representative point on the locus, a point with no special property other than that given in the definition.*
(b) *Use the description of the locus to obtain an equation satisfied by $x$ and $y$, and simplify this equation.*
(c) *Show that if $x$ and $y$ satisfy the equation obtained in* (b) *then the corresponding point has the desired property. If such is the case, the equation obtained is an equation of the locus.*

*Example 7.2.2.*   Find an equation of the locus of a point whose distance from the point (3, 0) is twice its distance from the $y$ axis.

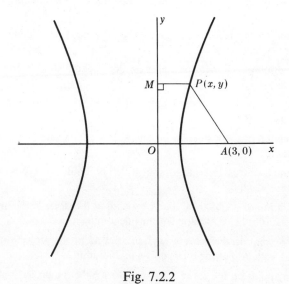

Fig. 7.2.2

*Solution.*   Referring to Fig. 7.2.2, we see that we must have

(7.2.2) $$|PA| = 2|MP|.$$

Now $|PA| = \sqrt{(x-3)^2 + y^2}$, $|MP| = |x|$. Hence (7.2.2) becomes

$$\sqrt{(x-3)^2 + y^2} = 2|x|.$$

On squaring both sides and expanding we have

$$x^2 - 6x + 9 + y^2 = 4x^2,$$

which becomes

(7.2.3) $$3x^2 + 6x - y^2 = 9.$$

Since the steps leading from (7.2.2) to (7.2.3) are reversible, any point $(x, y)$ satisfying (7.2.3) has the desired property, and (7.2.3) is an equation of the locus. This type of curve, a *hyperbola*, is studied in detail in Section 7.9.

**EXERCISE GROUP 7.1**

1. Find an equation of the locus of a point such that the slope of the line joining it to $(2, 0)$ is 2 more than the slope of the line joining it to $(-2, 0)$.

2. Find an equation of the locus of a point such that the slope of the line joining it to $(3, 1)$ is 3 less than the slope of the line joining it to $(-1, 3)$.

3. Find an equation of the locus of a point such that the lines joining it to $(1, 0)$ and $(0, 1)$ are perpendicular.

4. Find an equation of the locus of a point such that the lines joining it to $(2, 2)$ and $(4, 3)$ are perpendicular.

5. Find an equation of the locus of a point which is equidistant from the point $(2, 0)$ and the $y$ axis.

6. Find an equation of the locus of a point which is equidistant from the point $(-2, 0)$ and the $y$ axis.

7. Find an equation of the locus of a point which is equidistant from the line $y = 4/3$ and the point $(0, -4/3)$.

8. Find an equation of the locus of a point which is equidistant from the line $x = 5/2$ and the point $(-5/2, 0)$.

9. Find an equation of the locus of a point such that the slope of the line joining it to the point $(-2, 0)$ is twice the slope of the line joining it to the point $(2, 0)$.

10. Let $P$ be any point on the curve $y = x^2$, and let $M$ be the projection of $P$ on the $x$ axis. Find an equation of the locus of the midpoint of $MP$.

11. Let $P$ be any point on the curve $x^2 = y$, and let $N$ be the projection of $P$ on the $y$ axis. Find an equation of the locus of the midpoint of $NP$.

12. Let $P$ be any point on the curve $y = x^2 - x$. Find an equation of the locus of the midpoint of $OP$.

13. Let $P$ be any point on the curve $x^2 + 2y^2 = 2$, and let $M$ and $N$ be the projections of $P$ on the $x$ and $y$ axes, respectively. Find an equation of the locus of the midpoint of $MN$.

14. Let $y = 2x + 1$ and $y = x^2 - 3$ be the equations of two curves. If a line parallel to the $y$ axis intersects the curves in $P$ and $Q$, find an equation of the locus of the midpoint of $PQ$.

* **15.** A long narrow sheet of paper of width *a* is placed as in Fig. 7.2.3. The sheet is folded along *BC* so that the corner at *A* falls on the opposite edge (*y* axis) at *D*. Find an equation of the locus of *P*.

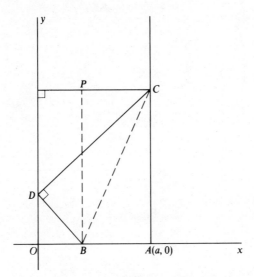

Fig. 7.2.3

## 7.3 THE EQUATION OF A CIRCLE

A circle is defined as the set of points in a plane which are at the same distance, the *radius*, from a fixed point, the *center*. Let $C(h, k)$ be the center, and $a > 0$ the radius (Fig. 7.3.1). Now let $P(x, y)$ be a representative point on the circle. Then we have $|CP| = a$ which gives, with use of the distance formula,

(7.3.1) $$\sqrt{(x - h)^2 + (y - k)^2} = a.$$

Squaring both sides of (7.3.1) we get

(7.3.2) $$(x - h)^2 + (y - k)^2 = a^2.$$

Equation (7.3.2) is the *standard form of the equation of a circle*, and exhibits directly the radius *a* and the coordinates *h* and *k* of the center. (See Section 7.11 for a discussion of standard forms.)

In order to establish (7.3.2) as an equation of the circle it is necessary to show also that if *x* and *y* satisfy (7.3.2), then the distance from $C(h, k)$ to $P(x, y)$ is *a*. This follows immediately by extracting the square root of both sides of (7.3.2), since (7.3.1) must then hold if $a > 0$.

When the origin is the center of a circle, then $h = k = 0$, and (7.3.2) becomes

$$x^2 + y^2 = a^2.$$

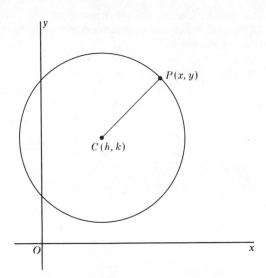

Fig. 7.3.1

Thus the equation $x^2 + y^2 - 1 = 0$, obtained in Example 7.2.1, represents a circle with radius 1 and center at the origin.

In expanded form (7.3.2) becomes

$$x^2 + y^2 - 2hx - 2ky + h^2 + k^2 - a^2 = 0;$$

this may be written as

(7.3.3) $\qquad x^2 + y^2 + Dx + Ey + F = 0,$

where

(7.3.4) $\qquad D = -2h, \quad E = -2k, \quad F = h^2 + k^2 - a^2.$

Equation (7.3.3) is the *general form of the equation of a circle*. It is a *second-degree equation* in $x$ and $y$, with no $xy$ term, and with equal (unity) coefficients of $x^2$ and $y^2$ when these terms are on the same side of the equation.

An equation of a circle in general form may be converted to standard form by a process of *completing the square*. In general, if we wish to complete the square for the sum $u^2 + au$, we add $(a/2)^2 = a^2/4$, and we get

$$u^2 + au + \left(\frac{a}{2}\right)^2 = \left(u + \frac{a}{2}\right)^2.$$

**Example 7.3.1.** Write the following equation in standard form.

$$2x^2 + 2y^2 - 3x + 4y - 7 = 0.$$

*Solution.* We divide both sides by 2 and write the equation as

$$x^2 - \tfrac{3}{2}x + y^2 + 2y = \tfrac{7}{2}.$$

We now add $(3/4)^2$ to the left side to complete the square in the terms involving $x$, and add 1 to complete the square in the terms in $y$, *adding the same numbers to the right member*:

$$[x^2 - \tfrac{3}{2}x + (\tfrac{3}{4})^2] + [y^2 + 2y + 1] = \tfrac{7}{2} + \tfrac{9}{16} + 1.$$

This can now be written in *standard form* as

$$(x - \tfrac{3}{4})^2 + (y + 1)^2 = \tfrac{81}{16}.$$

The graph is a circle with center $(3/4, -1)$, and radius $9/4$.

We may now analyze Equation (7.3.3) completely. By the same process of completing the square as was used in Example 7.3.1, we write (7.3.3) as

$$\left(x + \frac{D}{2}\right)^2 + \left(y + \frac{E}{2}\right)^2 = \frac{D^2 + E^2 - 4F}{4}.$$

Since the left member is a sum of two squares, and therefore positive for a point on a circle, it is necessary for the right member to be positive if (7.3.3) represents a circle. The following is a complete description.

*Equation* (7.3.3) *represents a circle if and only if* $D^2 + E^2 - 4F > 0$; *it represents the single point* $(-D/2, -E/2)$ *if and only if* $D^2 + E^2 - 4F = 0$; *equation* (7.3.3) *has no real graph if and only if* $D^2 + E^2 - 4F < 0$.

If Equation (7.3.3) represents a single point, the graph is sometimes called a *point circle*.

We know from geometry that the tangent line to a circle is perpendicular to the radius drawn to the point of tangency. We shall use this in discussing tangents of a circle.

**Example 7.3.2.** Find an equation of the line which is tangent to the circle

$$x^2 + y^2 + 4x - 6y + 8 = 0$$

at the point $(-3, 5)$.

*Solution.* By completing squares we may write the equation of the circle in standard form as

$$(x + 2)^2 + (y - 3)^2 = 5.$$

The circle, with center at $(-2, 3)$ and radius $\sqrt{5}$ is shown in Fig. 7.3.2. The slope of $CA$ is $(5 - 3)/(-3 + 2) = -2$; hence the slope of the tangent line is $1/2$, and by use of the point-slope form for a straight line (Section 2.8), an equation of the line is

$$y - 5 = \tfrac{1}{2}(x + 3) \quad \text{or} \quad x - 2y + 13 = 0.$$

The slope of the circle at $(-3, 5)$ may also be obtained by implicit differentiation (Section 4.7). We obtain $(y' = dy/dx)$

$$2x + 2yy' + 4 - 6y' = 0.$$

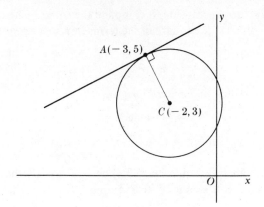

Fig. 7.3.2

Substitution of $x = -3$ and $y = 5$ permits us to solve for $y'$ as $y' = 1/2$; this is the slope of the desired tangent line.

**Example 7.3.3.** Find an equation of each tangent line drawn from the point $(5, 1)$ to the circle $x^2 + y^2 = 13$.

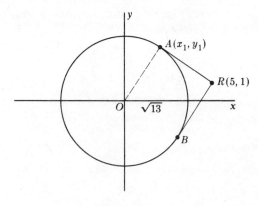

Fig. 7.3.3

*Solution.* Let $(x_1, y_1)$ be a point of tangency for a tangent drawn from $(5, 1)$ to the circle (Fig. 7.3.3). Then since $\overrightarrow{OA} \perp \overrightarrow{RA}$, we have (Section 2.2)

$$x_1(x_1 - 5) + y_1(y_1 - 1) = 0,$$

or

(7.3.5) $$5x_1 + y_1 = x_1^2 + y_1^2.$$

Since $x_1^2 + y_1^2 = 13$, we have, using (7.3.5), the two equations,

(7.3.6) $\qquad 5x_1 + y_1 = 13, \qquad x_1^2 + y_1^2 = 13.$

From the first of these equations we get $y_1 = 13 - 5x_1$, and substituting this into the second equation, we get, after collecting terms and simplifying,

$$x_1^2 - 5x_1 + 6 = 0; \qquad x_1 = 2 \quad \text{or} \quad x_1 = 3.$$

From the first equation of (7.3.6) we then find $y_1 = 3$ if $x_1 = 2$, or $y_1 = -2$ if $x_1 = 3$. Hence the two points of tangency are $A(2, 3)$ and $B(3, -2)$. Equations of the tangents are then obtained as

$$RA: 2x + 3y = 13, \qquad RB: 3x - 2y = 13,$$

by using the method of Example 7.3.2 for each tangent.

### EXERCISE GROUP 7.2

In Exercises 1–8 describe the graph (if it exists) of the equation. In the case of a circle give the center and radius, and draw the graph. In the case of a point circle give the point.

1. $x^2 + y^2 - 10x + 14y + 10 = 0$
2. $x^2 + y^2 + 6x - 2y + 6 = 0$
3. $x^2 + y^2 - 4x + 6y + 13 = 0$
4. $x^2 + y^2 - 4x - 2y + 4 = 0$
5. $18x^2 + 18y^2 - 27x + 24y - 10 = 0$
6. $3x^2 + 3y^2 + 10x - 4y - 7 = 0$
7. $8x^2 + 8y^2 + 12x - 28y = 3$
8. $5x^2 + 5y^2 + 10x - 8y = 1$

In Exercises 9–16 find an equation of the circle satisfying the given conditions.

9. Center $(0, 2)$, radius 3
10. Center $(-4, 2)$, radius 1
11. Center $(-3, 7)$, radius 4
12. Center $(-4, 3)$, radius 5
13. Center at $(2, 3)$, and passing through the origin
14. Center at $(-2, 1)$, and passing through the point $(1, 2)$
15. Having the line segment joining the points $(-1, 1)$ and $(3, 2)$ as a diameter
16. Center at $(-2, 3)$, and tangent to the $y$ axis

In Exercises 17–20 find an equation of the tangent line to the given circle at the given point.

17. $x^2 + y^2 = 5, (-1, 2)$
18. $x^2 + y^2 = 20, (-2, -4)$
19. $x^2 + y^2 - x + y - 26 = 0, (3, 4)$
20. $x^2 + y^2 + 3x + 5y - 12 = 0, (3, -2)$

In Exercises 21–24 find equations of the tangent lines drawn to the given circle from the given point.

21. $x^2 + y^2 = 5, (7, 1)$
22. $x^2 + y^2 = 13, (-1, 5)$
23. $x^2 + y^2 - 3y = 8, (-3, 8)$
24. $x^2 + y^2 + 4y - 6 = 0, (-8, 4)$

25. Show that for any value of $\theta$ the point $(a \cos \theta, a \sin \theta)$ lies on the circle $x^2 + y^2 = a^2$.

26. Show that for any value of $\theta$ the point $(h + a \sin \theta, k + a \cos \theta)$ lies on the circle $(x - h)^2 + (y - k)^2 = a^2$.

27. Find the length of a tangent drawn from the point $(5, -2)$ to the point of tangency on the circle $x^2 + y^2 - 3x + 5y - 2 = 0$.

28. Show that the tangent to the circle $x^2 + y^2 = a^2$ at the point $(x_1, y_1)$ has an equation $x_1 x + y_1 y = a^2$.

29. Show that the tangent to the circle (7.3.2) at the point $(x_1, y_1)$ has an equation $(x_1 - h)(x - h) + (y_1 - k)(y - k) = a^2$.

30. Show that the tangent to the circle (7.3.3) at the point $(x_1, y_1)$ has an equation $x_1 x + y_1 y + (D/2)(x + x_1) + (E/2)(y + y_1) + F = 0$.

31. Let $t$ be the length of the tangent drawn from an exterior point $(x_1, y_1)$ to the point of tangency of the circle (7.3.2). Show that $t^2 = (x_1 - h)^2 + (y_1 - k)^2 - a^2$.

32. If $P(x_1, y_1)$ lies in the interior of the circle (7.3.2), show that $a^2 - (x_1 - h)^2 - (y_1 - k)^2$ is the square of half the length of the chord which is perpendicular to the line drawn from the center to $P$.

33. Show by the methods of calculus that a tangent line to a circle is perpendicular to the radius drawn to the point of tangency.

## *7.4  CIRCLES DETERMINED BY VARIOUS CONDITIONS

In either Equation (7.3.2) or Equation (7.3.3) three essential constants appear, $h, k, a$ in (7.3.2) and $D, E, F$ in (7.3.3). This means that in general a circle may be required to satisfy three conditions, which may be of various types. We shall consider several sets of such conditions. Some introductory problems of this type appeared in Exercise Group 7.2.

**Example 7.4.1.**  Find an equation of a circle passing through the points $(3, -2)$, $(-1, -4)$, and $(2, -5)$.

*Solution.  First method.*  We use the general form (7.3.3) and employ the basic principle that the coordinates of each given point must satisfy the equation

$$x^2 + y^2 + Dx + Ey + F = 0.$$

Substituting the coordinates of each point, in turn, in this equation, we find

$$9 + 4 + 3D - 2E + F = 0, \qquad 1 + 16 - D - 4E + F = 0,$$
$$4 + 25 + 2D - 5E + F = 0.$$

These equations may be rewritten as the system

$$3D - 2E + F = -13,$$
$$-D - 4E + F = -17,$$
$$2D - 5E + F = -29.$$

This system of linear equations in $D, E, F$ may be solved as follows. Subtract the second equation from the first, and divide by 2, to give

$$2D + E = 2.$$

Now subtract the second from the third equation, to give

$$3D - E = 12.$$

The two equations in $D$ and $E$ have the solution $D = -2$, $E = 6$. Then $F$ is found from an equation of the original system as $F = 5$. Substitution in (7.3.3) gives the result

(7.4.1) $$x^2 + y^2 - 2x + 6y + 5 = 0.$$

*Second method.* From plane geometry we know that the center of a circle lies on the perpendicular bisector of each chord. With the points labeled as in Fig. 7.4.1, an equation of the perpendicular bisector of $AB$ is found to be $2x + y + 1 = 0$; while the perpendicular bisector of $BC$ is $x + 3y + 8 = 0$. The solution of the linear system

$$2x + y + 1 = 0, \quad x + 3y + 8 = 0$$

gives the center $(1, -3)$. The radius may be found as the distance from the center to $A$, $B$, or $C$, say $A$, and is $\sqrt{5}$. We now use $h = 1$, $k = -3$, $a = \sqrt{5}$ in (7.3.2), and get

$$(x - 1)^2 + (y + 3)^2 = 5.$$

In expanded form this gives (7.4.1) again.

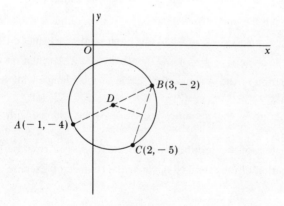

Fig. 7.4.1

*Example 7.4.2.* Find an equation of the circle which passes through the points $(-2, -4)$ and $(4, 6)$, and whose center lies on the line $3x - 2y + 20 = 0$.

*Solution.* With use of Equation (7.3.3), the requirement that the circle pass through the given points yields two equations in $D$, $E$, and $F$,

(7.4.2)   $-2D - 4E + F = -20,$   $4D + 6E + F = -52.$

From equation (7.3.4) the coordinates of the center are $-D/2$ and $-E/2$. Substitution of these values in $3x - 2y = 20$ gives a third equation

(7.4.3)   $3D - 2E = 40.$

The three linear equations in (7.4.2) and (7.4.3) may be solved to give $D = 8$, $E = -8$, $F = -36$, and the desired equation is, from (7.3.3),

$$x^2 + y^2 + 8x - 8y - 36 = 0.$$

* **EXERCISE GROUP 7.3**

In Exercises 1–6 find an equation of the circle which passes through the given points.

1. $(-1, -1), (-2, 1), (2, 3)$
2. $(0, -8), (7, -7), (3, -9)$
3. $(4, 2), (0, 4), (1, 5)$
4. $(5, 8), (3, -2), (6, 7)$
5. $(2, 1), (1, 5/4), (-1/4, 1/2)$
6. $(1/2, -1), (-3/4, 1/4), (2/5, -4/5)$

In Exercises 7–14 find an equation of the circle which satisfies the given conditions.

7. Center at $(2, -3)$, and passing through the point $(1, 4)$.
8. Center at $(-4, 1)$, and passing through the point $(7, 2)$.
9. Center on the lines $x - 2y - 5 = 0$ and $3x + y - 1 = 0$, and passing through the point $(0, 3)$.
10. Center on the lines $x + y = 2$ and $3x + 4y = 4$, and passing through the point $(2, 1)$.
11. Passing through the points $(4, 3)$ and $(2, -1)$, with radius $5/2$.
12. Passing through the points $(-1, 1)$ and $(5, 3)$, with radius $2\sqrt{5}$.
13. Tangent to the circle $x^2 + y^2 = 4$ and the $x$ axis, and with center on the line $x = 4$.
14. Tangent to the circle $x^2 + y^2 = 4$ and the $y$ axis, and with center on the line $y = 3$.
15. Find an equation of the circumscribed circle of the triangle with vertices $(0, 1/2)$, $(2, 9/2)$, and $(-1, 6)$.
16. Find an equation of the circumscribed circle of the triangle with vertices $(2, 1)$, $(1, 0)$, and $(5, -2)$.
* 17. Find an equation of the inscribed circle of the triangle of Exercise 15.
* 18. Find an equation of the inscribed circle of the triangle of Exercise 16.

* **7.5 INTERSECTIONS INVOLVING CIRCLES**

A circle and a straight line in a plane will either have *no point* in common, or *two distinct points* in common, or *exactly one point* in common; in the last case the line is tangent to the circle. These cases may be characterized algebraically.

Equations of the line and circle are

(7.5.1) $\qquad Ax + By + C = 0, \qquad x^2 + y^2 + Dx + Ey + F = 0.$

To find any common solution we may solve the linear equation in (7.5.1) for $y$, if $B \neq 0$, and substitute the resulting expression for $y$ in the second-degree equation in (7.5.1). There results a quadratic equation in $x$. If the roots of this quadratic equation in $x$ are imaginary, the line and the circle do not meet; if the roots are real and distinct, there are two distinct points of intersection; if the roots are equal, there is one point of intersection, and tangency.

**Example 7.5.1.** Find the intersection of the line $x + y - 6 = 0$ and the circle $x^2 + y^2 - 4x - 4 = 0$.

*Solution.* From the linear equation we have $y = 6 - x$; substitution into the quadratic equation then gives

$$x^2 + (6 - x)^2 - 4x - 4 = 0, \quad \text{or} \quad x^2 - 8x + 16 = 0.$$

Since the latter equation can be expressed as $(x - 4)^2 = 0$, there is one root, $x = 4$; then from $y = 6 - x$ we find $y = 2$. Hence the line and circle are *tangent* at the point (4, 2).

**Example 7.5.2.** If the line in Example 7.5.1 were $x + y - 2 = 0$, we would have $y = 2 - x$, and

$$x^2 + (2 - x)^2 - 4x - 4 = 0, \quad \text{or} \quad x^2 - 4x = 0.$$

Then $x = 0$ or 4, and the two points of intersection are (0, 2) and (4, −2).

Let us now consider two circles

(7.5.2) $\qquad \begin{aligned} x^2 + y^2 + D_1 x + E_1 y + F_1 &= 0, \\ x^2 + y^2 + D_2 x + E_2 y + F_2 &= 0. \end{aligned}$

If one equation is subtracted from the other, the square terms cancel and a linear equation in $x$ and $y$ results. The solution of this equation with either equation of (7.5.2) by the method described above for (7.5.1), will yield any points of intersection of the two circles. The circles will have either no point in common, or two distinct points in common, or exactly one point in common which will be a point of tangency of the two circles.

**Example 7.5.3.** Find any intersections of the circles

$$x^2 + y^2 - 3x + 2y - 12 = 0, \qquad x^2 + y^2 + x + 3y - 10 = 0.$$

*Solution.* Subtraction of the first equation from the second one gives

(7.5.3) $\qquad\qquad 4x + y + 2 = 0.$

We solve this for $y = -(4x + 2)$ and substitute in the first given equation:

$$x^2 + (4x + 2)^2 - 3x - 2(4x + 2) - 12 = 0,$$

or

$$17x^2 + 5x - 12 = 0.$$

The latter equation becomes $(x + 1)(17x - 12) = 0$ so that $x = -1$ or $12/17$. Then (7.5.3) gives $y = 2$ or $y = -82/17$. There are two points of intersection, $(-1, 2)$ and $(12/17, -82/17)$.

Let us examine Example 7.5.3 further. Any points which lie on both circles must also lie on the graph of (7.5.3). But (7.5.3) represents a straight line. Hence (7.5.3) must be the equation of the extension of the *common chord* of the two circles.

There is a theorem from geometry that *the lengths of the tangents from any point on the extension of the common chord of two circles to the points of tangency of the circles are equal.* We can prove this theorem and extend the result to *any* two circles, whether they intersect or not.

**THEOREM 7.5.1.** *Let the equation obtained by eliminating the square terms from Equations* (7.5.2) *be*

(7.5.4)  $\quad (D_1 - D_2)x + (E_1 - E_2)y + (F_1 - F_2) = 0.$

*Let P be any point of* (7.5.4) *external to the circles* (7.5.2). *Then the lengths of the tangents drawn from P to the points of tangency of the circles are equal.*

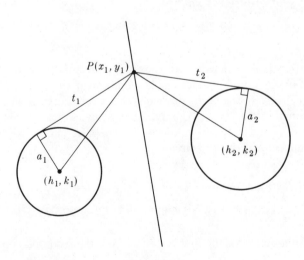

Fig. 7.5.1

*Proof.* Fig. 7.5.1 shows one possible case; the proof, however, applies to all cases. Using the labeling in the figure, we have

$$t_1^2 = (x_1 - h_1)^2 + (y_1 - k_1)^2 - a_1^2$$
$$= x_1^2 - 2h_1 x_1 + h_1^2 + y_1^2 - 2k_1 y_1 + k_1^2 - a_1^2,$$

$$t_2^2 = (x_1 - h_2)^2 + (y_1 - k_2)^2 - a_2^2$$
$$= x_1^2 - 2h_2 x_1 + h_2^2 + y_1^2 - 2k_2 y_1 + k_2^2 - a_2^2.$$

By subtraction we obtain

$$t_1^2 - t_2^2 = 2(h_2 - h_1)x_1 + 2(k_2 - k_1)y_1 + (h_1^2 + k_1^2 - a_1^2) - (h_2^2 + k_2^2 - a_2^2).$$

Now from (7.3.4) we have

$$2h_1 = -D_1, \quad 2h_2 = -D_2, \quad 2k_1 = -E_1, \quad 2k_2 = -E_2,$$
$$h_1^2 + k_1^2 - a_1^2 = F_1, \quad h_2^2 + k_2^2 - a_2^2 = F_2.$$

Using these values we get

$$t_1^2 - t_2^2 = (D_1 - D_2)x_1 + (E_1 - E_2)y_1 + (F_1 - F_2).$$

Since $P(x_1, y_1)$ is on (7.5.4), the last equation gives $t_1^2 - t_2^2 = 0$. Hence $t_1 = t_2$, (since $t_1$ and $t_2$ are positive), and the proof is complete.

The straight line (7.5.4) is called the *radical axis* of the circles (7.5.2). It is characterized by the property described in Theorem 7.5.1.

Equation (7.5.3) is an equation of the radical axis of the two circles given in the statement of Example 7.5.3.

We have seen that the radical axis of two circles which cross is the line of their common chord. The radical axis of two tangent circles is their common tangent. The radical axis of two nonintersecting circles also exists, and does not intersect either circle.

* **EXERCISE GROUP 7.4**

In Exercises 1-8 find any points of intersection of the given curves.

1. $x^2 + y^2 + 4x - 16 = 0$, $2x + 3y - 10 = 0$
2. $9x^2 + 9y^2 + 18x - 9y - 17 = 0$, $4x - y - 2 = 0$
3. $x^2 + y^2 - 2x + 6y + 5 = 0$, $2x + y - 4 = 0$
4. $x^2 + y^2 - 6x - 4y + 13 = 0$, $x + y + 1 = 0$
5. $x^2 + y^2 - 8x + 6y + 20 = 0$, $x - 2y - 5 = 0$
6. $2x^2 + 2y^2 + x - y - 31 = 0$, $3x - 2y - 13 = 0$
7. $3x^2 + 3y^2 - 7y - 23 = 0$, $x - y - 2 = 0$
8. $2x^2 + 2y^2 - 4x - 5y - 14 = 0$, $4x - y + 10 = 0$

In Exercises 9–16 find (a) any points of intersection of the curves, (b) an equation of the radical axis. Then (c) in each exercise draw the circles and their radical axis in the same coordinate system.

9. $x^2 + y^2 + 2x = 0$, $x^2 + y^2 - 2y = 0$
10. $x^2 + y^2 - 2x - 9 = 0$, $x^2 + y^2 - 1 = 0$
11. $x^2 + y^2 - 3x + 7y + 2 = 0$, $x^2 + y^2 - 4x - y - 7 = 0$
12. $x^2 + y^2 + 8x - 6y - 11 = 0$, $x^2 + y^2 - 16x + 4y + 19 = 0$
13. $x^2 + y^2 - 16 = 0$, $x^2 + y^2 - 4x - 2y + 4 = 0$
14. $x^2 + y^2 - 9 = 0$, $x^2 + y^2 - 4x + 3 = 0$
15. $x^2 + y^2 - 6x - 8y = 0$, $x^2 + y^2 - 8x - 4y + 18 = 0$
16. $2x^2 + 2y^2 + 4x - 2y = 1$, $x^2 + y^2 + 3x - y = 4$
17. Prove that the radical axis of two circles is perpendicular to the line joining their centers.
18. Prove that the radical axis of two nonintersecting circles does not intersect either circle.
* 19. Show that each of the circles $x^2 + y^2 = 9$, $x^2 + y^2 - 12x + 27 = 0$, and $x^2 + y^2 - 6x - 8y + 21 = 0$ is tangent to the other two. Do the common tangents meet in a point? If they do, find the point.

## 7.6 THE PARABOLA

Our study of the parabola is based on the following locus definition. Other properties may also be used as the definition. (See Section 8.1.)

**DEFINITION 7.6.1.** *A* **parabola** *is the locus of a point in a plane whose distances from a fixed point, the* **focus,** *and a fixed line, the* **directrix,** *are equal.*

The definition is independent of any coordinate system; once the focus and directrix are specified, a parabola is determined. Most of the properties of a parabola are also independent of any coordinate system. Thus in order to prove many properties of the parabola we may place the curve in a coordinate system in a position that suits our convenience.

We shall in time study the equation of a parabola in its most general form, but we start our discussion with a parabola for which the $x$ axis passes through the focus and is perpendicular to the directrix and the $y$ axis passes through the midpoint of the perpendicular line segment from the focus to the directrix. We designate the coordinates of the focus as $(p/2, 0)$, the directrix as $D: x = -p/2$. We may have either $p > 0$ or $p < 0$; in either case the distance between $D$ and $F$ is $|p|$. Fig. 7.6.1 applies to the case $p > 0$, but the following derivation of the equation is identical in both cases.

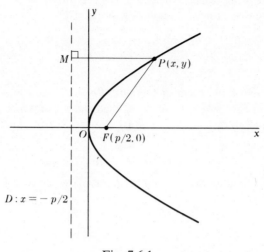

Fig. 7.6.1

Let $P(x, y)$ designate a representative point on the locus. Then, by Definition 7.6.1,

$$|FP| = |MP|,$$

where $MP$ is perpendicular to $D$. By the distance formula this gives

$$\sqrt{(x - p/2)^2 + y^2} = |x + p/2|.$$

By squaring both sides and simplifying, we obtain the equation

(7.6.1) $$y^2 = 2px.$$

Since the steps in this derivation are reversible [Exercise 21], (7.6.1) is established as an equation of the parabola.

From (7.6.1) we see at once that the curve is symmetric in the $x$ axis, passes through the origin, and, if $p > 0$, has no points to the left of the $y$ axis ($x$ cannot be negative).

The line of symmetry of any parabola is called its *axis*; the point of the curve which lies on the axis is the *vertex*. For a parabola with Equation (7.6.1), the $x$ axis is its axis, and the origin is its vertex.

A parabola with directrix perpendicular to the $x$ axis is called a *horizontal* parabola, and may open to the right or left. Accordingly, Equation (7.6.1) is *the standard equation of a horizontal parabola with vertex at the origin*; the parabola opens to the right if $p > 0$ and to the left if $p < 0$.

We now consider a parabola whose directrix is perpendicular to the $y$ axis; this is called a *vertical* parabola. For a vertical parabola with vertex at the origin the roles of $x$ and $y$ in (7.6.1) are interchanged and an equation is

(7.6.2) $$x^2 = 2py.$$

Equation (7.6.2) is *the standard equation of a vertical parabola with vertex at the origin*; it opens upward if $p > 0$ and downward if $p < 0$.

***Example 7.6.1.*** Discuss the equation $y^2 = -6x$.

*Solution.* This is of the form (7.6.1) with $2p = -6$ or $p = -3$. The graph is thus a horizontal parabola with vertex at the origin, and which opens to the left. Its focus is at $(p/2, 0)$ or $(-3/2, 0)$, and its directrix has an equation $x = -p/2$ or $x = 3/2$.

***Example 7.6.2.*** Find an equation of the parabola with focus at $(0, 3)$ and vertex at the origin.

*Solution.* An equation of the specified parabola is of the form (7.6.2) with $p/2 = 3$. Hence $p = 6$, and an equation is

$$x^2 = 12y.$$

**DEFINITION 7.6.2.** *A* **chord** *of a curve is any line segment joining two points of the curve.*

**DEFINITION 7.6.3.** *The* **latus rectum** *of a parabola is the chord through the focus and perpendicular to the axis. (Fig. 7.6.2).*

The latter definition applies to *any* parabola. In Fig. 7.6.2 the latus rectum is shown for a parabola in a particular position.

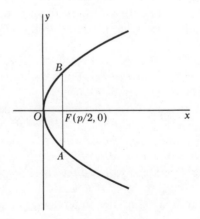

Fig. 7.6.2

*Example 7.6.3.* Find the length of the rectum of a parabola in terms of $p$.

*Solution.* In order to prove a general property of a parabola, we may, as stated earlier, place the parabola in a coordinate system as we choose; let us do so as in Fig. 7.6.2. The latus rectum is then the chord $AB$. Substituting $x = p/2$ in the equation $y^2 = 2px$, we find that the ordinate of $B$ is $p$. Hence $|AB| = 2p$. If $p < 0$, we find $|AB| = -2p$. In general then, the latus rectum is of length $2|p|$.

An equation of a horizontal or vertical parabola with vertex at the origin contains one essential constant. Accordingly, one additional condition may be imposed on such a parabola.

*Example 7.6.4.* Find an equation of the horizontal parabola with vertex at the origin, and passing through the point $(-3, 7)$.

*Solution.* The equation may be taken as $y^2 = 2px$. Then $x = -3$, $y = 7$ must satisfy this equation. Hence $49 = -6p$, $p = -49/6$, and the desired equation may be given as

$$y^2 = -\tfrac{49}{3}x, \quad \text{or} \quad 3y^2 + 49x = 0.$$

### EXERCISE GROUP 7.5

In Exercises 1–10 find an equation of the indicated parabola.

1. Directrix $x = -2$, focus $(2, 0)$
2. Directrix $y = -5$, focus $(0, 5)$
3. Directrix $y = 4$, focus $(0, -4)$
4. Directrix $x = 3$, focus $(-3, 0)$
5. Vertex $(0, 0)$, directrix $y = 4/3$
6. Vertex $(0, 0)$, directrix $x = 3/4$
7. Vertex $(0, 0)$, focus $(-2/3, 0)$
8. Vertex $(0, 0)$, focus $(0, -2/3)$
9. Vertex $(0, 0)$, focus on the $y$ axis, and passing through the point $(8, -4)$
10. Vertex $(0, 0)$, focus on the $x$ axis, and passing through the point $(-7, 7)$

In Exercises 11–16 find the coordinates of the focus and an equation of the directrix for the given parabola, and sketch the graph.

11. $y^2 = 6x$
12. $x^2 - 8y = 0$
13. $3x^2 + 2y = 0$
14. $3y^2 + 4x = 0$
15. $2x^2 + 5y = 0$
16. $2y^2 - 9x = 0$

In Exercises 17–20 find the specified equation.

17. Of the locus of points equidistant from the line $y = 4/3$ and the point $(0, -4/3)$
18. Of the locus of points equidistant from the line $x = 5/2$ and the point $(-5/2, 0)$
19. Of the parabola with $y = 2$ as directrix and $(0, 4)$ as focus
20. Of the parabola with $x = 3$ as directrix and $(-2, 0)$ as focus

21. Carry out the steps leading to equation (7.6.1), and also show that they are reversible.

**194** EQUATIONS OF SECOND DEGREE

**22.** Carry out the derivation of (7.6.2) directly, by using Definition 7.6.1 with the focus $(0, p/2)$ and the directrix $y = -p/2$.

## 7.7 TRANSLATION OF AXES

We now take the first step in the direction of studying a curve in a more general position by use of its equation in a special position.

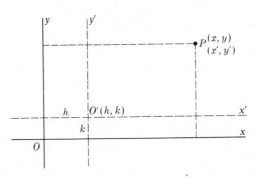

Fig. 7.7.1

In Fig. 7.7.1 two sets of parallel coordinate axes are drawn, the origin $O'$ of the $x'$, $y'$ system having coordinates $h$ and $k$ in the $x$, $y$ system. A point $P$ in the plane has coordinates $(x, y)$ in the $x$, $y$ system, and $(x', y')$ in the $x'$, $y'$ system. It is clear from the figure that *the two pairs of coordinates are related by the equations*

(7.7.1) $\qquad\qquad x = x' + h, \qquad y = y' + k,$

regardless of the position of $O'$ and $P$.

Equations (7.7.1) are called *equations of translation*; they may be used to express $x$ and $y$ in terms of $x'$ and $y'$, or the reverse, for two coordinate systems related as in Fig. 7.7.1. The $x'$, $y'$ system is said to be obtained from the $x$, $y$ system by a *translation of axes* to a position whose origin $O'$ has $x$, $y$ coordinates $h$ and $k$. Similarly, the $x$, $y$ system is obtained from the $x'$, $y'$ system by a translation of axes to a position with origin $(-h, -k)$ in the $x'$, $y'$ system.

Consider now a *horizontal* parabola (Fig. 7.7.2) with vertex at $(h, k)$. If we translate the axes to the new origin $O'(h, k)$, and let $P$ be any point on the curve, an equation of the curve in the $x'$, $y'$ system is, by (7.6.1),

(7.7.2) $\qquad\qquad y'^2 = 2px'.$

But from (7.7.1) we may write

$$x' = x - h, \qquad y' = y - k.$$

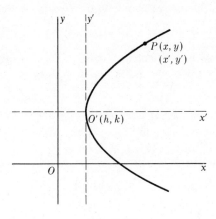

Fig. 7.7.2

Substitution of these into (7.7.2) gives

(7.7.3) $$(y - k)^2 = 2p(x - h).$$

Equation (7.7.3) is the *standard form of the equation of a horizontal parabola with vertex at the point* $(h, k)$. If $p > 0$, the parabola opens to the right; if $p < 0$, it opens to the left. The $x'$, $y'$ axes were introduced as an aid in deriving (7.7.3), and may now be discarded.

In precisely the same manner we obtain

(7.7.4) $$(x - h)^2 = 2p(y - k)$$

as the *standard form of the equation of a vertical parabola with vertex at the point* $(h, k)$. If $p > 0$, the parabola opens upward; if $p < 0$, it opens downward.

Expansion of (7.7.3) or (7.7.4) leads to a second-degree equation in $x$ and $y$ with only one second-degree term actually present; it may be $x^2$ or $y^2$ *but not both*. We may summarize these observations as follows.

An equation of the form

(7.7.5) $$Ax^2 + Cy^2 + Dx + Ey + F = 0$$

*in general represents a horizontal parabola if* $A = 0$ *and* $C \neq 0$, *and a vertical parabola if* $C = 0$ *and* $A \neq 0$.

An equation of the form (7.7.5), with $A = 0$ and $C \neq 0$, or $C = 0$ and $A \neq 0$, may be transformed to standard form by a process of completing the square in the coordinate whose square is present. We shall illustrate this.

**Example 7.7.1.** Discuss and sketch the graph of the equation

$$2x^2 - 8x + 3y + 17 = 0.$$

*Solution.* We write the equation as

$$x^2 - 4x = -\tfrac{3}{2}y - \tfrac{17}{2}.$$

Then 4 is added to both sides to complete the square on the left, and

$$x^2 - 4x + 4 = -\tfrac{3}{2}y - \tfrac{17}{2} + 4 = -\tfrac{3}{2}y - \tfrac{9}{2}.$$

This equation may now be written as

$$(x - 2)^2 = -\tfrac{3}{2}(y + 3).$$

By comparison with (7.7.4) we identify the graph as a vertical parabola opening downward, with *vertex* at $(2, -3)$, and $2p = -3/2$ or $p = -3/4$. The focus is at $(2, -3 + p/2)$ or $(2, -27/8)$. The directrix is the line $y = -3 - p/2$, or

$$D: y = -3 + \frac{3}{8} \quad \text{or} \quad y = \frac{-21}{8}.$$

The graph is shown in Fig. 7.7.3.

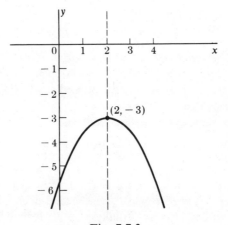

Fig. 7.7.3

Just as a horizontal or vertical parabola with vertex at an arbitrary point was related to a similar parabola with vertex at the origin, we now see that a circle with center at $(h, k)$ may be obtained from a circle with center at the origin. [See Equation (7.3.2).]

Previous statements regarding slopes and tangents of a curve are, of course, applicable to the curves being studied here (see Section 4.3). The necessary differentiation may often be conveniently done implicitly.

*Example 7.7.2.* Find an equation of the tangent to the parabola

$$y^2 + x + 3y - 2 = 0$$

at its intercept on the $x$ axis.

*Solution.* At the intercept on the $x$ axis we have $y = 0$, and we find $x = 2$; the point is $(2, 0)$. By implicit differentiation we obtain

$$2yy' + 1 + 3y' = 0.$$

At the point with $y = 0$ we get $y' = -1/3$, which is the slope of the curve at that point. An equation of the tangent line is then

$$y = -\tfrac{1}{3}(x - 2) \quad \text{or} \quad x + 3y = 2.$$

**EXERCISE GROUP 7.6**

In Exercises 1–6 the $x, y$ axes are translated to $x', y'$ axes with origin $O'(2, -3)$.

1. Find the $x', y'$ coordinates of (a) $(2, -3)$, (b) $(-2, -5)$, (c) $(-1, 4)$.
2. Find the $x', y'$ coordinates of (a) $(7, -4)$, (b) $(-3, 6)$, (c) $(0, -3)$.
3. Find the $x', y'$ equation for $2x - 3y + 4 = 0$.
4. Find the $x', y'$ equation for $-x + 2y = 0$.
5. Find the $x', y'$ equation for $x^2 + y^2 - 2 = 0$.
6. Find the $x', y'$ equation for $x^2 + 2y - 2 = 0$.

In Exercises 7–10 find an equation of the parabola satisfying the given conditions.

7. Vertex $(-1, 3)$, directrix $x = 3$
8. Vertex $(-3, -2)$, focus $(-3, 2)$
9. Focus $(2, -2)$, directrix $x = -2$
10. Focus $(-3, 5)$, directrix $y + 5 = 0$

In each of Exercises 11–20 find the vertex, focus, and directrix of the parabola, and then sketch the graph of the equation.

11. $y^2 - 2x - 1 = 0$
12. $x^2 + 3y + 2 = 0$
13. $2x^2 + 4x - 3y = 0$
14. $y^2 + x + 3y - 2 = 0$
15. $3x^2 - y + 2x = 4$
16. $2y^2 - 4y - 2x = 5$
17. $4y^2 + 6y - 3x = 7$
18. $2x^2 - 5x = 2 + y$
19. $y = 2x^2 + 4x - 3$
20. $x = 3y^2 - 6y + 2$

21. Show that all parabolas of the form $y^2 = 2cx + c^2$ have the same focus.
22. Find an equation of the family of vertical parabolas which pass through the points $(1, 0)$ and $(-1, 0)$.
23. Find an equation of the family of vertical parabolas which pass through the points $(1, 1)$ and $(-1, -1)$.
24. An isosceles triangle has its base on the positive $x$ axis, with one end of the base at the origin. Find an equation of the locus of the third vertex if the perimeter of the triangle is 24. Discuss the equation obtained.

In Exercises 25–28 find an equation of the indicated tangent line.

25. Tangent to $y^2 = x + 1$ at $(3, -2)$

26. Tangent to $x^2 = 4y$ at $(-2, 1)$
27. Tangent to $x^2 + 3y = 4$ at $(4, -4)$
28. Tangent to $2x^2 + 3x + y - 1 = 0$ at $(-2, -1)$
29. Show that the tangent to the parabola $y^2 = 2px$ at the point $(x_1, y_1)$ has an equation $y_1 y = p(x + x_1)$.

## 7.8 THE ELLIPSE

We continue the study of conic sections by turning to the ellipse, which is defined by the following locus property.

**DEFINITION 7.8.1.** *An* **ellipse** *is the locus of a point in a plane the sum of whose distances from two fixed points is constant.*

The distance between the fixed points, which are called *foci*, is taken as $2c$, $c > 0$; the midpoint of the line segment joining the foci is the *center*; and the constant sum is taken as $2a$, $a > 0$.

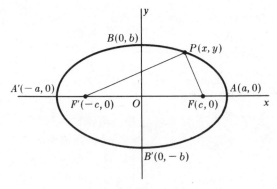

Fig. 7.8.1

To initiate the study of the locus we shall consider the curve in a special position, in which the $x$ axis passes through the foci and the $y$ axis passes through the center, which is now at the origin, (Fig. 7.8.1). The foci are then $F'(-c, 0)$, $F(c, 0)$. Letting $P(x, y)$ denote any point on the locus we have, by Definition (7.8.1),

(7.8.1) $$|F'P| + |FP| = 2a.$$

By the distance formula this becomes

(7.8.2) $$\sqrt{(x + c)^2 + y^2} + \sqrt{(x - c)^2 + y^2} = 2a.$$

Subtracting the second radical from both members, then squaring both sides and simplifying, we get

(7.8.3) $$a\sqrt{(x-c)^2 + y^2} = a^2 - cx;$$

on squaring and collecting terms in this equation, we have

(7.8.4) $$(a^2 - c^2)x^2 + a^2y^2 = a^2(a^2 - c^2).$$

Now from $\triangle PF'F$ it follows that $|F'P| + |FP| > |F'F|$; hence $2a > 2c$ and $a > c$. Therefore, $a^2 - c^2 > 0$ and we may define a *positive quantity b* by the relation

$$a^2 - c^2 = b^2, \quad \text{(we must then have } b < a\text{)}.$$

Substituting this in (7.8.4) we get $b^2x^2 + a^2y^2 = a^2b^2$, or

(7.8.5) $$\frac{x^2}{a^2} + \frac{y^2}{b^2} = 1, \quad a > b.$$

Equation (7.8.5) is called *the standard form of the equation of an ellipse with foci on the x axis and center at the origin.*

We have shown that if $P(x, y)$ satisfies the locus property, then (7.8.5) holds. To complete the proof that (7.8.5) is an equation of the ellipse it is necessary to show that (7.8.1) can be derived from (7.8.5). The steps leading to (7.8.5) are all reversible, but in the step of this reverse process which leads to (7.8.3) (in which square roots of both members are taken), we must admit, tentatively, the equation

(7.8.6) $$a\sqrt{(x-c)^2 + y^2} = \pm(a^2 - cx).$$

Now, from (7.8.5) we get $x^2 \leq a^2$, and $x \leq a$. Then

$$cx \leq ac < a^2.$$

Hence $a^2 - cx > 0$ for any $x$ satisfying (7.8.5), and the minus sign in (7.8.6) cannot apply. Similarly, as the reverse process is continued, in the step which leads to (7.8.2) the equation

(7.8.7) $$|F'P| = \pm(2a - |FP|)$$

must tentatively be admitted. If the minus sign can hold, we should have

$$|FP| - |F'P| = 2a > 2c,$$

and the difference of two sides of a triangle would be greater than the third side, which is impossible. Hence the minus sign in (7.8.7) cannot hold, and the proof is complete.

The graph of (7.8.5) is symmetric in both axes and in the origin, has $x$ intercepts $\pm a$, and $y$ intercepts $\pm b$. At the points of the curve on the $x$ axis there are vertical tangents, and at the points on the $y$ axis there are horizontal tangents (Exercise 31). The graph appears in Fig. 7.8.1.

The points $A'$ and $A$ (Fig. 7.8.1) are the *vertices* of the ellipse. The line segment $A'A$, of length $2a$, is called the *major axis*, and $B'B$, of length $2b$, is the *minor axis* of the ellipse. The points $B'$ and $B$ have no special name, other than the ends of the minor axis.

If $P$ is a point of an ellipse, (Fig. 7.8.1), the vectors $\overrightarrow{F'P}$ and $\overrightarrow{FP}$ are the *focal vectors* of $P$. Their lengths are *focal radii* of $P$. From (7.8.3) we find

(7.8.8) $$|FP| = a - \frac{c}{a}x.$$

Using Definition 7.8.1 we then get

(7.8.9) $$|F'P| = a + \frac{c}{a}x.$$

**Example 7.8.1.** Discuss and graph the equation $4x^2 + 9y^2 = 36$.

*Solution.* Dividing both members of the equation by 36 we obtain the standard form

$$\frac{x^2}{9} + \frac{y^2}{4} = 1.$$

Hence $a = 3$, $b = 2$, $c = \sqrt{a^2 - b^2} = \sqrt{5}$. The foci are $(\pm\sqrt{5}, 0)$, and the vertices are $(\pm 3, 0)$. The graph is shown in Fig. 7.8.2.

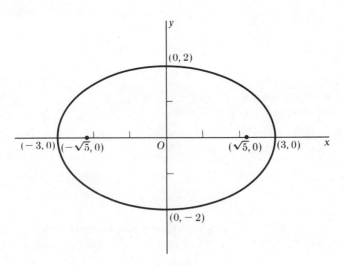

Fig. 7.8.2

If an ellipse with center at the origin has its foci on the $y$ axis, the roles of $x$ and $y$ in (7.8.5) are interchanged. The *standard form* of the equation of such an ellipse is

(7.8.10) $$\frac{x^2}{b^2} + \frac{y^2}{a^2} = 1, \quad a > b,$$

where again $a^2 - c^2 = b^2$. The vertices are now on the $y$ axis.

The major axis of an ellipse is longer than the minor axis, since $2a > 2b$. Thus the graph of (7.8.5) is longer in the $x$ direction and is called a *horizontal ellipse*, while (7.8.10) is a *vertical ellipse*.

Both (7.8.5) and (7.8.10) can be written in the form

(7.8.11) $$Rx^2 + Sy^2 = T,$$

where $R > 0$, $S > 0$, $T > 0$. When (7.8.11) is written in standard form,

$$\frac{x^2}{T/R} + \frac{y^2}{T/S} = 1,$$

we see that *the ellipse is horizontal* if $T/R > T/S$, or $R < S$. The ellipse is vertical if $R > S$. The larger of the numbers $T/R$ and $T/S$ becomes $a^2$, the smaller one becomes $b^2$. *If $R = S$ the ellipse reduces to a circle*, and the foci coalesce in the center of the circle.

**Example 7.8.2.** Discuss the graph of $3x^2 + 2y^2 = 6$.

*Solution.* We may write the equation as

$$\frac{x^2}{2} + \frac{y^2}{3} = 1.$$

The graph is therefore a *vertical ellipse* with center at the origin. We have

$$a = \sqrt{3}, \quad b = \sqrt{2}, \quad c = \sqrt{a^2 - b^2} = 1.$$

Hence the vertices are $(0, \pm\sqrt{3})$, and the foci are $(0, \pm 1)$.

We now consider horizontal and vertical ellipses with center at an arbitrary point. Suppose a horizontal ellipse has its center at a point $C(h, k)$. If we introduce, through $C$, axes parallel to the given ones (Fig. 7.8.3), we have as the $x'$, $y'$ equation of the ellipse

(7.8.12) $$\frac{x'^2}{a^2} + \frac{y'^2}{b^2} = 1, \quad a > b.$$

Using the equations of translation,

$$x' = x - h, \quad y' = y - k,$$

in (7.8.12), we get

(7.8.13) $$\frac{(x-h)^2}{a^2} + \frac{(y-k)^2}{b^2} = 1, \quad a > b, \quad a^2 - b^2 = c^2.$$

*Equation (7.8.13) is the standard equation of a horizontal ellipse with center at $(h, k)$.* The foci are at $(h \pm c, k)$, and the vertices are at $(h \pm a, k)$.

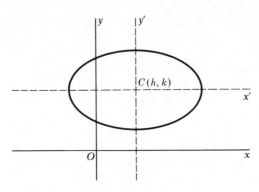

Fig. 7.8.3

In precisely the same way it is shown that *the equation*

(7.8.14) $$\frac{(x-h)^2}{b^2} + \frac{(y-k)^2}{a^2} = 1, \quad a > b, \quad a^2 - b^2 = c^2,$$

*is the standard equation of a vertical ellipse with center at* $(h, k)$. The foci are at $(h, k \pm c)$, and the vertices are at $(h, k \pm a)$.

We can now see how (7.8.13) and (7.8.14) relate to the general second-degree equation (7.1.1). Both of these equations may be expanded and written in the form

(7.8.15) $$Ax^2 + Cy^2 + Dx + Ey + F = 0,$$

in which *A and C have the same sign*, which we may *assume to be plus*. Then, *if* $A < C$ *the ellipse is horizontal*, and *if* $A > C$ *the ellipse is vertical*. (Recall that $A = C$ yields a circle.)

Conversion of (7.8.15) to standard form (7.8.13) or to (7.8.14) again involves *completion of the square*, in the $x$ terms and $y$ terms separately, as in the following example.

***Example 7.8.3.*** Discuss and sketch the graph of

$$3x^2 + 2y^2 + 12x - 4y + 2 = 0.$$

*Solution.* We group the $x$ terms and factor out the coefficient of $x^2$, group the $y$ terms and factor out the coefficient of $y^2$, and write the equation as

$$3(x^2 + 4x) + 2(y^2 - 2y) = -2.$$

We complete the square on $x^2 + 4x$ by adding 4 within the parentheses. This adds $3 \cdot 4 = 12$ to the left side; the same number must be added to the right. Similarly, we add 1 to $y^2 - 2y$ and add $2 \cdot 1 = 2$ to the right. This gives

$$3(x^2 + 4x + 4) + 2(y^2 - 2y + 1) = -2 + 12 + 2 = 12,$$

which we now write as $3(x+2)^2 + 2(y-1)^2 = 12$, and finally as

$$\frac{(x+2)^2}{4} + \frac{(y-1)^2}{6} = 1.$$

The curve is a *vertical ellipse*, with center at $(-2, 1)$. We have $a = \sqrt{6}$, $b = 2$, $c = \sqrt{6-4} = \sqrt{2}$. Hence the foci are at $(-2, 1 \pm \sqrt{2})$, and the vertices are at $(-2, 1 \pm \sqrt{6})$. The graph appears in Fig. 7.8.4.

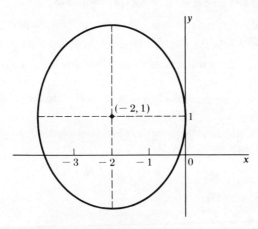

Fig. 7.8.4

## EXERCISE GROUP 7.7

In Exercises 1–10 find an equation of the indicated ellipse.

1. Foci $(\pm 3, 0)$, $a = 5$. Also find the focal radii of a point for which $x = 2$.
2. Foci $(\pm 2, 0)$, vertices $(\pm 3, 0)$. Also find the focal radii of a point for which $x = -2$.
3. Foci $(0, \pm\sqrt{2})$, $b = 4$
4. Foci $(0, \pm\sqrt{3})$, ends of minor axis $(0, \pm\sqrt{5})$
5. Foci $(0, \pm 3)$, and passing through the point $(\sqrt{2}, 2)$
6. Vertices $(\pm 4, 0)$, and passing through the point $(3, -\sqrt{5})$
7. Vertices $(-3, 3)$ and $(7, 3)$, length of minor axis 8
8. Vertices $(-1, 4)$ and $(-1, -8)$, one focus $(-1, -6)$
9. Foci $(0, -3)$ and $(4, -3)$, length of major axis 10
10. Foci $(2 \pm \sqrt{3}, -2)$, one vertex $(5, -2)$

In Exercises 11–26 find the coordinates of the foci and of the vertices, and draw the graph of the equation.

11. $9x^2 + 4y^2 = 36$     12. $x^2 + 4y^2 = 16$

13. $4x^2 + 3y^2 = 12$
14. $2x^2 + 6y^2 = 9$
15. $16x^2 + 25y^2 = 400$
16. $25x^2 + 4y^2 = 100$
17. $2x^2 + 5y^2 = 7$
18. $7x^2 + 3y^2 = 14$
19. $4x^2 + 2y^2 + 8x - 5 = 0$
20. $8x^2 + 4y^2 + 24x - 4y + 1 = 0$
21. $16x^2 + 25y^2 - 32x + 100y - 284 = 0$
22. $4x^2 + 5y^2 - 30y = 0$
23. $2x^2 + y^2 - 16x = 0$
24. $4x^2 + y^2 - 8x - 24y + 48 = 0$
25. $2x^2 + 5y^2 - 16x + 20y + 42 = 0$
26. $3x^2 + 5y^2 - 12x - 10y = 0$

In Exercises 27–30 find an equation of the specified curve.

27. Of the locus of points the sum of whose distances from the points $(0, \pm 2)$ is 6.
28. Of the locus of points the sum of whose distances from the points $(\pm 4, 0)$ is 10.
29. Of the locus of points the sum of whose distances from the points $(-3, 4)$ and $(-1, 4)$ is 6.
30. Of the locus of points the sum of whose distances from the points $(2, 3)$ and $(4, 1)$ is 6.
31. Show that the ellipse with equation (7.8.5) has vertical tangents $x = \pm a$ and horizontal tangents $y = \pm b$.

## 7.9 THE HYPERBOLA

We continue our study of conic sections with the hyperbola, for which the development parallels that for the ellipse. In particular, a hyperbola has a locus definition similar to that of an ellipse.

**DEFINITION 7.9.1.** *A hyperbola is the locus of a point in a plane the difference of whose distances from two fixed points is constant.*

The fixed points are again the *foci*, the distance between them is $2c$, the midpoint of the line segment joining the foci is the *center*, and the constant difference of distances is $2a$.

In contrast with the ellipse we have here $a < c$. We start with a hyperbola for which the $x$ axis passes through the foci, and the $y$ axis passes through the center (Fig. 7.9.1). If $P(x, y)$ is a point on the locus, we have

(7.9.1) $$|F'P| - |FP| = \pm 2a,$$

where the plus sign applies if $x > 0$, and the minus sign applies if $x < 0$. We apply the distance formula in (7.9.1) and get

(7.9.2) $$\sqrt{(x + c)^2 + y^2} - \sqrt{(x - c)^2 + y^2} = \pm 2a.$$

Adding the second radical to both sides, then squaring both sides, and simplifying, we get

(7.9.3) $$\pm a\sqrt{(x - c)^2 + y^2} = cx - a^2,$$

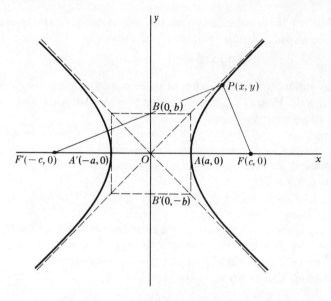

Fig. 7.9.1

with the same agreement about sign as above. On squaring and collecting terms we get precisely the same equation as (7.8.4) for the ellipse,

(7.9.4) $$(a^2 - c^2)x^2 + a^2 y^2 = a^2(a^2 - c^2).$$

However, since $c^2 > a^2$, we now define a *positive* quantity by the relation

$$c^2 - a^2 = b^2,$$

and substituting in (7.9.4) we get $-b^2 x^2 + a^2 y^2 = -a^2 b^2$. Dividing both sides of this equation by $-a^2 b^2$, we get

(7.9.5) $$\frac{x^2}{a^2} - \frac{y^2}{b^2} = 1.$$

Equation (7.9.5) is the *standard form of the equation of a hyperbola with foci on the x axis, and center at the origin.*

What has been shown above is that if $P(x, y)$ satisfies (7.9.1), then $x$ and $y$ satisfy (7.9.5). To establish the equivalence of these equations, it must also be shown that if $x$ and $y$ satisfy (7.9.5), then (7.9.1) holds. We may work backwards from (7.9.5) to (7.9.3), with the plus sign in (7.9.3) applying if $x > 0$ and the minus sign applying if $x < 0$. For example, if $x > 0$ in (7.9.5) then also $x \geq a$. But $c > a$. Hence $cx - a^2 > 0$. Continuing to work backwards from (7.9.3), we get, tentatively,

$$\pm\sqrt{(x + c)^2 + y^2} = \pm 2a + \sqrt{(x - c)^2 + y^2}.$$

If the minus sign on the left can hold, we have

$$-|F'P| = 2a + |FP| \quad \text{or} \quad -|F'P| = -2a + |FP|.$$

The first relation is impossible, since the left side is negative and the right side positive. If the second relation held, we should have

$$|F'P| + |FP| = 2a < 2c,$$

which is also impossible since the sum of two sides of a triangle can not be less than the third side. Hence (7.9.2), and in turn (7.9.1), must hold, and (7.9.1) has been shown to be completely equivalent to (7.9.5).

The graph of (7.9.5) is symmetric in both coordinate axes and in the origin, has $x$ intercepts $\pm a$, but has no $y$ intercepts. There are no points on the graph for $|x| < a$. At the points $(\pm a, 0)$ there are vertical tangents. The graph appears in Fig. 7.9.1.

The points $A'(-a, 0)$ and $A(a, 0)$ are the *vertices* of the hyperbola; the line segment $A'A$ of length $2a$ is the *transverse axis*. The segment $B'B$ joining $B'(0, -b)$ and $B(0, b)$ is the *conjugate axis* even though $B'$ and $B$ are not on the hyperbola,

We state here a distinctive property of the hyperbola which is very useful in drawing the curve, but we shall postpone proofs to the next section.

The hyperbola (7.9.5) *has two asymptotes, which pass through the origin and have equations*

$$(7.9.6) \qquad y = \pm \frac{b}{a} x.$$

If a rectangle is drawn with sides parallel to the coordinate axes and through the ends of the transverse axis and conjugate axis, *the diagonals of the rectangle, extended, are the asymptotes* (see Fig. 7.9.1).

The *focal radii* of the hyperbola (7.9.5) may be found. Using the plus sign in (7.9.3) and then (7.9.1) we find

$$(7.9.7) \qquad \text{if } x > 0, \quad |FP| = \frac{c}{a} x - a, \quad |F'P| = \frac{c}{a} x + a.$$

Similarly, using the minus signs in (7.9.3) and (7.9.1), we have

$$(7.9.8) \qquad \text{if } x < 0, \quad |FP| = a - \frac{c}{a} x, \quad |F'P| = -a - \frac{c}{a} x.$$

**Example 7.9.1.** Discuss and graph the equation $9x^2 - 4y^2 = 36$.

*Solution.* Dividing both members by 36, we obtain

$$\frac{x^2}{4} - \frac{y^2}{9} = 1.$$

This is in the standard form (7.9.5), with $a = 2$, $b = 3$, $c = \sqrt{a^2 + b^2} = \sqrt{13}$. The vertices are $(\pm 2, 0)$, and the foci are $(\pm \sqrt{13}, 0)$. Equations of the asymptotes (7.9.6) are

$$y = \pm \tfrac{3}{2} x.$$

The graph appears in Fig. 7.9.2.

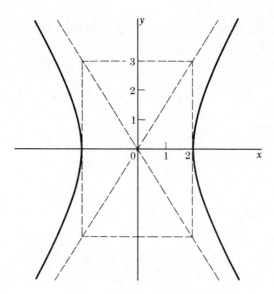

Fig. 7.9.2

If a hyperbola with center at the origin has its foci on the $y$ axis, the roles of $x$ and $y$ in (7.9.5) are interchanged. The *standard form* of the equation of such a hyperbola is

(7.9.9) $$\frac{y^2}{a^2} - \frac{x^2}{b^2} = 1, \qquad c^2 = a^2 + b^2.$$

The vertices are on the $y$ axis, at $(0, \pm a)$, and the foci are $(0, \pm c)$. The asymptotes of the hyperbola (7.9.9) are

$$y = \pm \frac{a}{b} x.$$

Both (7.9.5) and (7.9.9) can be written in the general form of (7.8.11),

(7.9.10) $$Rx^2 + Sy^2 = T.$$

*Any equation of the form* (7.9.10) *with* $T \neq 0$, *represents a hyperbola if R and S have opposite signs.* Since $a$ and $b$ in (7.9.5) and (7.9.9) can be of any relative magnitudes, the criterion in (7.9.10) as to whether the hyperbola is horizontal or vertical is not the relative magnitudes of $R$ and $S$. If (7.9.10) represents a hyperbola, in which case $R$ and $S$ have opposite signs, *the hyperbola is horizontal if R and T have the same sign, and vertical if S and T have the same sign.*

With the use of the same method of translation as in Sections 7.7 and 7.8, it is easy to see that the following equation,

(7.9.11) $$\frac{(x-h)^2}{a^2} - \frac{(y-k)^2}{b^2} = 1, \quad a^2 + b^2 = c^2,$$

is *the standard form of the equation of a horizontal hyperbola with center at* $(h, k)$. The foci are at $(h \pm c, k)$, and the vertices are at $(h \pm a, k)$. The asymptotes have equations

$$y - k = \pm \frac{b}{a}(x - h).$$

Similarly, the equation

(7.9.12) $$\frac{(y-k)^2}{a^2} - \frac{(x-h)^2}{b^2} = 1, \quad a^2 + b^2 = c^2,$$

is *the standard form of the equation of a vertical hyperbola with center at* $(h, k)$. The foci are at $(h, k \pm c)$ and the vertices are at $(h, k \pm a)$. The asymptotes have equations

$$y - k = \pm \frac{a}{b}(x - h).$$

Equations (7.9.11) and (7.9.12) may be expanded and written in the same form (7.8.15) as for the ellipse,

(7.9.13) $$Ax^2 + Cy^2 + Dx + Ey + F = 0,$$

where *A and C now have opposite signs*. We shall leave the determination of whether the hyperbola is horizontal or vertical to an examination of the standard form obtained upon conversion of (7.9.13).

**Example 7.9.2.** Discuss and sketch the graph of

$$3x^2 - 2y^2 + 12x + 4y + 22 = 0.$$

*Solution.* We group the $x$ terms and factor out the coefficient of $x^2$, group the $y$ terms and factor out the coefficient of $y^2$, and write the equation as

$$3(x^2 + 4x) - 2(y^2 - 2y) = -22.$$

We complete the square on $x^2 + 4x$ by adding 4 within the parentheses. This *adds* 12 to the left side, which must also be added to the right. Similarly we add 1 to $y^2 - 2y$; this *subtracts* $2 \cdot 1 = 2$ from the left, which must also be subtracted from the right. This gives

$$3(x^2 + 4x + 4) - 2(y^2 - 2y + 1) = -22 + 12 - 2,$$

which we write as $3(x + 2)^2 - 2(y - 1)^2 = -12$. Dividing both sides by $-12$, we get

$$\frac{(y-1)^2}{6} - \frac{(x+2)^2}{4} = 1.$$

The curve is a *vertical hyperbola* with center at $(-2, 1)$. We have $a = \sqrt{6}$, $b = 2$, $c = \sqrt{6 + 4} = \sqrt{10}$. Hence the foci are at $(-2, 1 \pm \sqrt{10})$, and the vertices are at $(-2, 1 \pm \sqrt{6})$. Equations of the asymptotes are

$$y - 1 = \pm \frac{\sqrt{6}}{2}(x + 2).$$

The graph appears in Fig. 7.9.3.

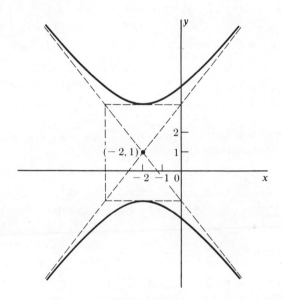

Fig. 7.9.3

## EXERCISE GROUP 7.8

In Exercises 1–14 find an equation of the indicated hyperbola.

1. Foci $(\pm 5, 0)$, $a = 3$. Also find the focal radii of a point for which $x = 6$.
2. Foci $(\pm 3, 0)$, vertices $(\pm 2, 0)$. Also find the focal radii of a point for which $x = -4$.
3. Foci $(0, \pm 4)$, $b = \sqrt{2}$
4. Foci $(0, \pm \sqrt{5})$, ends of conjugate axis $(\pm \sqrt{3}, 0)$
5. Vertices $(0, \pm \sqrt{3})$ and passing through the point $(2/3, 3)$
6. Foci $(\pm 5, 0)$ and passing through the point $(20/3, 4)$
7. Vertices $(\pm 2, 0)$, asymptotes with slopes $\pm 3/2$
8. Vertices $(0, \pm 4)$, asymptotes with slopes $\pm 1$
9. Vertices $(4, 3)$ and $(-2, 3)$, conjugate axis of length 8
10. Foci $(-1, 2)$ and $(-7, 2)$, a vertex at $(-5, 2)$

11. Vertices $(-2, 4)$ and $(10, 4)$, an asymptote with slope $1/2$
12. Foci $(0, -4)$ and $(0, 0)$, and passing through the point $(3, -4)$
13. Vertices $(-1, 1)$ and $(7, 1)$, and passing through the point $(3 + 4\sqrt{2}, 3)$
14. Center $(3, 4)$, a focus at $(8, 4)$, an asymptote $x - 2y + 5 = 0$

In each of Exercises 15–30 find the foci, vertices, and asymptotes, and draw the graph of the equation.

15. $16x^2 - y^2 = 16$
16. $8x^2 - y^2 = 8$
17. $4x^2 - 9y^2 + 36 = 0$
18. $4x^2 - y^2 + 16 = 0$
19. $y^2 - x^2 = 4$
20. $2x^2 - 5y^2 + 4 = 0$
21. $2x^2 - 3y^2 = 6$
22. $4x^2 - 3y^2 + 2 = 0$
23. $x^2 - 2y^2 - 6x - 8y - 3 = 0$
24. $4x^2 - 9y^2 + 8x + 36y + 4 = 0$
25. $4x^2 - y^2 + 8x = 0$
26. $2x^2 - y^2 - 12x - 10y = 0$
27. $2x^2 - 2y^2 - 2x + 2y - 17 = 0$
28. $8x^2 - 12y^2 + 24x + 12y = -27$
29. $4x^2 - 5y^2 + 10y + 15 = 0$
30. $3x^2 - 4y^2 + 18x + 3 = 0$

In Exercises 31–34 find an equation of the specified curve.

31. Of the locus of points the difference of whose distances from the points $(0, \pm 2)$ is 2.
32. Of the locus of points the difference of whose distances from the points $(\pm 4, 0)$ is 6.
33. Of the locus of points the difference of whose distances from the points $(4, -3)$ and $(4, 1)$ is 2.
34. Of the locus of points the difference of whose distances from the points $(2, 3)$ and $(4, 1)$ is 2.
35. Show that all conics of the following form, whether ellipses or hyperbolas, have the same foci.

$$\frac{x^2}{A^2 + C} + \frac{y^2}{B^2 + C} = 1, \qquad A^2 > B^2.$$

36. Show that the hyperbola with equation (7.9.5) has vertical tangents $x = \pm a$.

## *7.10 ASYMPTOTES OF THE HYPERBOLA

Let us consider the horizontal hyperbola with center at the origin,

(7.10.1) $$\frac{x^2}{a^2} - \frac{y^2}{b^2} = 1.$$

If we solve this equation for the function whose graph is the *upper half* of the hyperbola, and then subtract $b|x|/a$, we have

$$y - \frac{b}{a}|x| = \frac{b}{a}(\sqrt{x^2 - a^2} - |x|).$$

Now rationalizing the numerator, and letting $|x| \to \infty$, we get

$$y - \frac{b}{a}|x| = \frac{b}{a} \cdot \frac{-a^2}{\sqrt{x^2 - a^2} + |x|} \to 0.$$

Hence, if we set $y_A = b|x|/a$, we have

$$\lim_{|x| \to \infty} (y - y_A) = 0,$$

and the curve $y = b|x|/a$ is asymptotic to the upper half of the hyperbola. Then, by symmetry in the $x$ axis, the curve $y = -b|x|/a$ is asymptotic to the lower half of the hyperbola, and we have proved the result stated in Section 7.9.

*The lines $y = \pm bx/a$ are asymptotes of the hyperbola* (7.10.1).

The results for a vertical hyperbola with center at the origin, and for horizontal and vertical hyperbolas with center at $(h, k)$, as stated in Section 7.9, follow at once.

An examination of the equations of the asymptotes, as related to the standard forms of the equations of horizontal and vertical hyperbolas, establishes the following.

*For each standard form of the hyperbola considered in* Section 7.9 *equations of the asymptotes may be obtained by replacing the constant term by zero.* For example, for the hyperbola

(7.10.2) $$\frac{x^2}{4} - \frac{y^2}{9} = 1,$$

the asymptotes have an equation

$$\frac{x^2}{4} - \frac{y^2}{9} = 0,$$

since the latter equation is equivalent to $y = \pm 3x/2$. Even when equation (7.10.2) is rewritten as

$$9x^2 - 4y^2 = 36,$$

the asymptotes may be obtained by *replacing* 36 with 0, which gives $9x^2 - 4y^2 = 0$ or $3x = \pm 2y$.

Similarly, if a hyperbola has the equation

$$\frac{(y-3)^2}{2} - \frac{(x+1)^2}{4} = 1,$$

the asymptotes are given immediately by the equation

$$\frac{(y-3)^2}{2} - \frac{(x+1)^2}{4} = 0, \quad \text{or} \quad y - 3 = \pm(x+1)/\sqrt{2}.$$

∗ EXERCISE GROUP 7.9

In Exercises 1–8 find equations of the asymptotes of the given hyperbola in the form of a single second-degree equation.

1. $2x^2 - 3y^2 + 4 = 0$
2. $x^2 - 3(y+1)^2 = 4$
3. $\dfrac{(x+1)^2}{2} - \dfrac{(y-3)^2}{4} = 7$
4. $9(x-2)^2 - 6(y-1)^2 = 19$
5. $2x^2 - 3y^2 + 4x = 0$
6. $4x^2 - 9y^2 - 9y = 0$
7. $3x^2 - 4y^2 + 6x + 8y = 17$
8. $7x^2 - 3y^2 - 7x - 3y + 2 = 0$

9. Find an equation of the horizontal or vertical hyperbola with center at the origin, if the curve passes through the point (1, 2) and $x = 2y$ is an asymptote.
10. Find an equation of the horizontal or vertical hyperbola with center on the $y$ axis, if the curve passes through the point (5/4, 5/2) and $y = 2x + 1$ is an asymptote.
11. Find an equation of the family of horizontal hyperbolas with asymptotes $2y = \pm 5x$.
12. Find an equation of the family of vertical hyperbolas with asymptotes $3y = \pm 4x$.
13. Find an equation of the family of hyperbolas with asymptotes $x + 4y + 11 = 0$ and $x - 4y - 13 = 0$.
14. Find an equation of the family of hyperbolas with asymptotes $3x - 2y + 7 = 0$ and $3x + 2y - 1 = 0$.
15. If $y_A = b|x|/a$ and $(x, y)$ is a point on the hyperbola (7.10.1), show that $y^2 - y_A^2 < 0$. It follows that $y < y_A$ if $y > 0$, and that the asymptotes for the upper half of the hyperbola lie above the curve itself.

## 7.11 A WORD ON STANDARD FORMS

Up to this point we have used standard forms of equations in various connections. The role of any standard form of an equation is to exhibit certain characteristics of the equation, or of the curve it represents, and as such is usually very precise.

The slope-intercept form of the equation of a straight line (Section 2.8) is $y = mx + b$. This is a standard form, in which $m$ is the slope of the line and $b$ is the $y$ intercept. It does not apply to a line parallel to the $y$ axis.

In Example 7.9.1 we called

$$\frac{x^2}{4} - \frac{y^2}{9} = 1$$

the standard form of the equation of the curve in question since it conforms to Equation (7.9.5). On the other hand, the equations

(7.11.1) $\qquad \dfrac{x^2}{4} - \dfrac{y^2}{9} = 2 \quad \text{and} \quad \dfrac{x^2}{4} - \dfrac{2y^2}{9} = 1$

are not in that standard form. The first of these is not in standard form, because the right member is not 1; the standard form is

$$\frac{x^2}{8} - \frac{y^2}{18} = 1,$$

from which we can read $a = \sqrt{8} = 2\sqrt{2}$, $b = \sqrt{18} = 3\sqrt{2}$. The second equation in (7.11.1) is not in standard form, because the numerator of the second term contains a coefficient, 2, which does not appear in (7.9.5). The standard form of this equation is

$$\frac{x^2}{4} - \frac{y^2}{9/2} = 1,$$

from which $a = 2$, $b = \sqrt{9/2} = 3\sqrt{2}/2$.

As another simple illustration, the equation

$$2x^2 + 2y^2 = 7$$

clearly represents a circle with center at the origin, but the standard form of this equation, in conformity with (7.3.2), is

$$x^2 + y^2 = 7/2,$$

from which $r = \sqrt{7/2} = \sqrt{14}/2$.

The use of a standard form in a given context is perhaps never a necessity, but it is often a great aid, and should be made with this distinction in mind. Standard forms can be devised to emphasize almost any feature we choose, but any advantage is minimized when the variety of standard forms is too great. In our work we attempt to label as standard forms those which offer some definite advantage; we then relate other features to a minimum number of such forms.

# 8 Equations of Second Degree, Continued

## 8.1 INTRODUCTION

In the preceding chapter we studied the conic sections with emphasis on the properties of the individual types. The equations were all special cases of (7.1.1), the *general second-degree or quadratic equation in x and y*, with $B = 0$. In the present chapter we learn how to relate the general equation to the ones already studied, and how to identify the graph of a specific equation. This is done by means of a rotation of axes, considered in the next section. In the last three sections some properties which are common to conics are treated.

## 8.2 ROTATION OF AXES

Let the $x$ and $y$ axes in a coordinate plane be rotated about the origin through an angle $\theta$, to yield an $x'$, $y'$ coordinate system (Fig. 8.2.1). Let $\mathbf{i}$ and $\mathbf{j}$ be base vectors in the original system, and let $\mathbf{i}'$ and $\mathbf{j}'$ be base vectors in the new system. A point $P$ in the plane will have coordinates $(x, y)$ in the old system and $(x', y')$ in the new system. If $\mathbf{r}$ is the position vector of $P$, we may write

(8.2.1) $$\mathbf{r} = x'\mathbf{i}' + y'\mathbf{j}' = x\mathbf{i} + y\mathbf{j}.$$

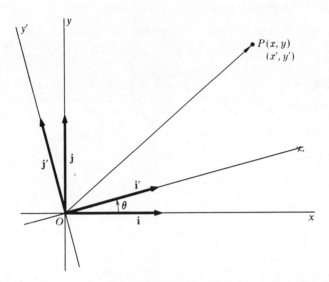

Fig. 8.2.1

Now $\mathbf{i}'$ has direction $\theta$ in the $x, y$ system, and $\mathbf{j}'$ has direction $\theta + \pi/2$. Hence we have as the representation of $\mathbf{i}'$ and $\mathbf{j}'$ in the $x, y$ system

$$\mathbf{i}' = (\cos \theta, \sin \theta),$$
$$\mathbf{j}' = (\cos(\theta + \pi/2), \sin(\theta + \pi/2)) = (-\sin \theta, \cos \theta).$$

Then from (8.2.1) we get

$$\mathbf{r} = x'(\cos \theta, \sin \theta) + y'(-\sin \theta, \cos \theta)$$
$$= (x' \cos \theta - y' \sin \theta, x' \sin \theta + y' \cos \theta).$$

But from (8.2.1) we also have $\mathbf{r} = (x, y)$ in the $x, y$ system. Hence

(8.2.2)
$$x = x' \cos \theta - y' \sin \theta,$$
$$y = x' \sin \theta + y' \cos \theta.$$

Equations (8.2.2) are called the *equations of rotation*. They express the $x, y$ coordinates of a point in terms of the $x', y'$ coordinates and the angle of rotation $\theta$, where $\theta$ is an angle the positive $x'$ axis makes with the positive $x$ axis.

In order to express $x'$ and $y'$ in terms of $x$ and $y$ we have alternative methods.

I. Think of the $x$ and $y$ axes as obtained from the $x'$ and $y'$ axes by a rotation about the origin through an angle $-\theta$. We may then apply (8.2.2) with the primed coordinates interchanged with the unprimed ones, and $\theta$ replaced by $-\theta$. Since $\cos(-\theta) = \cos \theta$ and $\sin(-\theta) = -\sin \theta$, we get

(8.2.3)
$$x' = x \cos \theta + y \sin \theta,$$
$$y' = -x \sin \theta + y \cos \theta.$$

II. Solve (8.2.2) directly for $x'$ and $y'$ (Exercise 25).

III. Express **i** and **j** in the $x'$, $y'$ system, and use the method by which (8.2.2) was derived (Exercise 26).

***Example 8.2.1.*** Find the $x'$, $y'$ coordinates of the point $(9, 5)$ under a rotation of axes through the obtuse angle $\theta$ for which $\tan \theta = -3/4$.

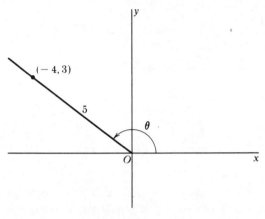

Fig. 8.2.2

*Solution.* Since $\theta$ is obtuse (between $90°$ and $180°$), we find from Fig. 8.2.2, $\sin \theta = 3/5$, $\cos \theta = -4/5$. Substituting these values in (8.2.3), with $x = 9$, $y = 5$, we have

$$x' = 9(-\tfrac{4}{5}) + 5(\tfrac{3}{5}) = -\tfrac{21}{5},$$
$$y' = -9(\tfrac{3}{5}) + 5(-\tfrac{4}{5}) = -\tfrac{47}{5}.$$

The desired coordinates are $(-21/5, -47/4)$.

***Example 8.2.2.*** Transform the equation $3x - 5y = 2$ under the rotation defined in Example 8.2.1.

*Solution.* With $\sin \theta = 3/5$, $\cos \theta = -4/5$, as in Example 8.2.1, we use Equations (8.9.2) for this purpose; that is

$$x = -\tfrac{4}{5}x' - \tfrac{3}{5}y', \qquad y = \tfrac{3}{5}x' - \tfrac{4}{5}y'.$$

Substituting these in the given equation we have

$$3(-\tfrac{4}{5}x' - \tfrac{3}{5}y') - 5(\tfrac{3}{5}x' - \tfrac{4}{5}y') = 2;$$

this reduces to the desired equation

$$27x' - 11y' + 10 = 0.$$

**Example 8.2.3.** Transform the equation $13x^2 - 18xy + 37y^2 = 160$ by rotating the axes through a positive acute angle $\theta$ such that $\tan \theta = 1/3$.

*Solution.* With $\theta$ in the first quadrant we find

$$\cos \theta = 3/\sqrt{10}, \qquad \sin \theta = 1/\sqrt{10}.$$

The equations of rotation (8.2.2) become

$$x = \frac{3x' - y'}{\sqrt{10}}, \qquad y = \frac{x' + 3y'}{\sqrt{10}}.$$

Substituting these into the given equation, we obtain

$$\frac{13(3x' - y')^2}{10} - \frac{18(3x' - y')(x' + 3y')}{10} + \frac{37(x' + 3y')^2}{10} = 160.$$

On expanding and collecting terms this becomes $100x'^2 + 400y'^2 = 1600$, which simplifies to the result,

(8.2.4) $$x'^2 + 4y'^2 = 16.$$

Both sets of axes are shown in Fig. 8.2.3. The graph of (8.2.4) is drawn as a horizontal ellipse in the $x'$, $y'$ system. The same curve is the graph of the given equation in the $x, y$ system, but now with oblique axes (not parallel to the $x$ and $y$ axes).

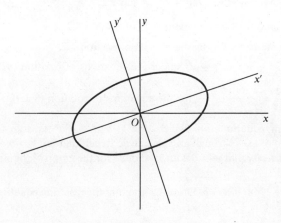

Fig. 8.2.3

## EXERCISE GROUP 8.1

1. Obtain the equations of rotation for rotating the axes through the angle $-30°$.

2. Obtain the equations of rotation for rotating the axes through the acute angle $\theta$ for which $\tan \theta = 5/12$.

In Exercises 3–16 find the point into which the given point is transformed by a rotation of the axes through the indicated angle.

3. $(4, 0)$; $60°$
4. $(-3, 4)$; $90°$
5. $(0, -2)$; $120°$
6. $(-2, 0)$; $-60°$
7. $(6, 2)$; $-225°$
8. $(4, -2)$; $210°$
9. $(3, 0)$; $2\pi/3$
10. $(0, -1)$; $\pi/4$
11. $(1, 1)$; $-\pi/6$
12. $(1, -1)$; $3\pi/4$
13. $(1, 2)$; $45°$
14. $(-1, 2)$; $30°$
15. $(2, 3)$; acute angle $\theta$ such that $\cos \theta = 12/13$
16. $(4, -1)$; obtuse angle $\theta$ such that $\tan \theta = -4/3$

In Exercises 17–24 find an equation into which the given equation is transformed under a rotation of axes through the indicated angle.

17. $3x - 4y = 7$; acute angle $\theta$ such that $\tan \theta = 3/4$
18. $3x - 4y = 7$; obtuse angle $\theta$ such that $\tan \theta = -3/4$
19. $x^2 + 2y^2 = 1$; obtuse angle $\theta$ such that $\sin \theta = 5/13$
20. $2x^2 - y^2 = 1$; acute angle $\theta$ such that $\sin \theta = 5/13$
21. $y^2 - x^2 = a^2$; $45°$
22. $5x^2 + 6xy - 3y^2 = 8$; $\theta = \tan^{-1} 1/3$, $\theta$ acute
23. $5x^2 + 6xy - 3y^2 = 8$; $\theta = \tan^{-1}(-3)$, $\theta$ obtuse
24. $8x^2 + 5xy - 4y^2 = 9$; $\theta = \tan^{-1} 1/5$, $\theta$ acute

25. Obtain Equations (8.2.3) by method II of Section 8.2.
26. Obtain Equations (8.2.3) by method III of Section 8.2.
27. Show that $x^2 + y^2 = x'^2 + y'^2$ under the equations of rotation (8.2.2).
28. Show that $\dfrac{y}{x} = \dfrac{x' \tan \theta + y'}{x' - y' \tan \theta}$ for the equations of rotation (8.2.2).
29. (a) Find an equation into which $b^2x^2 - a^2y^2 = a^2b^2$ is transformed under a rotation of axes through an acute angle $\theta$ such that $\tan \theta = b/a$.
    (b) Show directly that $y' = 0$ is an asymptote for the graph of the equation obtained in (a).
30. Do Exercise 29 (a) if $a = b$. Draw the graph of the resulting equation.

## 8.3 THE GENERAL SECOND-DEGREE EQUATION

The equations studied in Chapter 7 were special cases of the *general second-degree equation*

(8.3.1) $$Ax^2 + Bxy + Cy^2 + Dx + Ey + F = 0$$

for which $B = 0$.

We now consider (8.3.1) where $B \neq 0$. We shall see that a rotation of axes about the origin through a suitable angle $\theta$ transforms (8.3.1) into a similar equation with $B' = 0$.

(8.3.2) $$A'x'^2 + C'y'^2 + D'x' + E'y' + F' = 0.$$

To do this we substitute in (8.3.1) the expressions for $x$ and $y$ from the equations of rotation (8.2.2). Omitting details of the work, we state that upon expanding and collecting terms, we obtain an equation of the form

$$A'x'^2 + B'x'y' + C'y'^2 + D'x' + E'y' + F' = 0.$$

The coefficients are given by

(8.3.3) $$\begin{cases} A' = A\cos^2\theta + B\sin\theta\cos\theta + C\sin^2\theta \\ B' = B(\cos^2\theta - \sin^2\theta) - 2(A - C)\sin\theta\cos\theta \\ C' = A\sin^2\theta - B\sin\theta\cos\theta + C\cos^2\theta \\ D' = D\cos\theta + E\sin\theta \\ E' = -D\sin\theta + E\cos\theta \\ F' = F \end{cases}$$

In order to have $B' = 0$ we must choose $\theta$ such that

$$B(\cos^2\theta - \sin^2\theta) - 2(A - C)\sin\theta\cos\theta = 0.$$

Since $\cos^2\theta - \sin^2\theta = \cos 2\theta$ and $2\sin\theta\cos\theta = \sin 2\theta$, this equation becomes

(8.3.4) $$B\cos 2\theta - (A - C)\sin 2\theta = 0.$$

We may assume $B \neq 0$; otherwise (8.3.1) is already in the desired form. Then (8.3.4) yields the equation

(8.3.5) $$\cot 2\theta = \frac{A - C}{B}, \qquad B \neq 0.$$

*The desired angle of rotation $\theta$ may be found from* (8.3.5). This equation has more than one solution; however, *we may always take $0 < 2\theta < \pi$ so that $0 < \theta < \pi/2$ and $\sin\theta$ and $\cos\theta$ are positive.* We note that *the determination of $\theta$ depends only on the second-degree terms in the equation.* The example indicates how $\sin\theta$ and $\cos\theta$ are found from (8.3.5) (see also Exercise 11).

**Example 8.3.1.** Graph the equation

(8.3.6) $$4x^2 + 24xy + 11y^2 + 20 = 0$$

by use of a rotation of axes which eliminates the $xy$ term.

*Solution.* From (8.3.5) we have

$$\cot 2\theta = \frac{A - C}{B} = -\frac{7}{24}.$$

Since $2\theta$ must be between $\pi/2$ and $\pi$, we see from Fig. 8.3.1 that $\cos 2\theta = -7/25$.

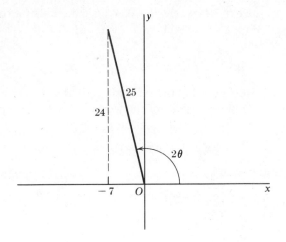

Fig. 8.3.1

Then by the half-angle formulas we have

$$\sin \theta = \sqrt{\frac{1 - \cos 2\theta}{2}} = \sqrt{\frac{1 + 7/25}{2}} = \frac{4}{5}.$$

$$\cos \theta = \sqrt{\frac{1 + \cos 2\theta}{2}} = \sqrt{\frac{1 - 7/25}{2}} = \frac{3}{5}.$$

The equations of rotation (8.2.2) become

$$x = \tfrac{1}{5}(3x' - 4y'), \qquad y = \tfrac{1}{5}(4x' + 3y').$$

Substituting these in (8.3.6) we have

$$\frac{4(3x' - 4y')^2}{25} + \frac{24(3x' - 4y')(4x' + 3y')}{25} + \frac{11(4x' + 3y')^2}{25} + 20 = 0.$$

On expanding and collecting terms this simplifies to

$$4x'^2 - y'^2 + 4 = 0.$$

We may now identify the curve as a vertical hyperbola in the $x'$, $y'$ system with center at the origin. In Fig. 8.3.2 both coordinate systems are drawn, and the hyperbola is sketched on the $x'$, $y'$ axes. The resulting curve is then also the graph of (8.3.6) in the $x$, $y$ system.

*Example 8.3.2.* In Example 8.2.3, the $xy$ term was eliminated by a rotation of axes through a specified angle $\theta$. We may now see how the angle was obtained. For the equation of that example we have by (8.3.5)

$$\cot 2\theta = \frac{13 - 37}{-18} = \frac{4}{3}.$$

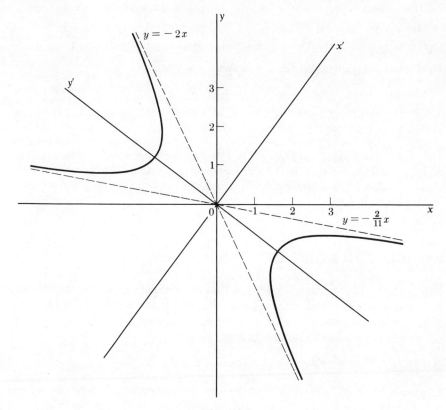

Fig. 8.3.2

Hence $\cos 2\theta = 4/5$. By the half-angle formulas we have

$$\sin \theta = \sqrt{\frac{1 - \cos 2\theta}{2}} = \sqrt{\frac{1 - 4/5}{2}} = \frac{1}{\sqrt{10}},$$

$$\cos \theta = \sqrt{\frac{1 + \cos 2\theta}{2}} = \sqrt{\frac{1 + 4/5}{2}} = \frac{3}{\sqrt{10}},$$

the same values that were used in that example.

### EXERCISE GROUP 8.2

In each of Exercises 1–10 a rotation of axes is made through a positive acute angle to eliminate the $xy$ term. (a) Obtain the new equation. (b) Draw both sets of axes and the graph, except in 9 and 10.

1. $3x^2 + 3xy - y^2 = 5$
2. $7x^2 - 8xy - 8y^2 = 4$
3. $10x^2 - 12xy + 5y^2 = 84$
4. $x^2 + 4xy - 2y^2 = 7$

5. $10x^2 + 40xy + y^2 - 9 = 0$

6. $4x^2 - 4xy + y^2 = 2x$

7. $x^2 - 6xy + 9y^2 + 3x = 3$

8. $16x^2 - 24xy + 9y^2 + 4y = 4$

9. $10x^2 - 24xy + 3y^2 + 2x - 3y + 12 = 0$

10. $120x^2 - 120xy + y^2 + 3y - 3 = 0$

11. Show that Equation (8.3.5) can be solved for $\tan \theta$ to give

$$\tan \theta = \frac{(C - A) \pm \sqrt{(C - A)^2 + B^2}}{B}, \quad B \neq 0.$$

12. (a) Show that the graph of $xy = a$ is a hyperbola.
    (b) Let the lines $L_1 : x = c$, $L_2 : x = 2c$, $L_3 : x = 3c$, ..., where $c > 0$, intersect the hyperbola in (a). Let $M_i$ be the line through the point where $L_{i+1}$ intersects the hyperbola and the point where $L_i$ intersects the $x$ axis. Show that the lines $M_i$, $i = 2, 3, \ldots$, are concurrent, the common point being on the other branch of the hyperbola.

## 8.4 IDENTIFICATION OF CONICS

It was shown in the preceding section that the general second-degree equation in $x$ and $y$ can be transformed into another one with $B' = 0$, if $B$ itself is not zero. It follows that *the graphs of the general second-degree equation are the curves which can occur as the graphs of the equation*

(8.4.1) $$Ax^2 + Cy^2 + Dx + Ey + F = 0.$$

Equations of this form have been treated by completing squares. However, they may be exceptional in the sense of the following illustrations, each of which type may result upon completing squares in (8.4.1).

***Example 8.4.1.*** $2(x + 2)^2 + (y - 3)^2 = -4$ has the form of an equation of an ellipse, but *there is no graph*; that is, the coordinates of no point satisfy this equation.

***Example 8.4.2.*** $4(x + 2)^2 + 7(y - 5)^2 = 0$ has the single point $(-2, 5)$ as the complete graph. This graph is sometimes called a *point ellipse*.

***Example 8.4.3.*** $4(x - 1)^2 - 9(y + 2)^2 = 0$ has the form of an equation of a hyperbola, but the graph consists of the two straight lines $y + 2 = \pm \frac{2}{3}(x - 1)$.

***Example 8.4.4.*** The equation $(y + 2)^2 = 4$ may also result, but it represents two parallel lines $y = -4$ and $y = 0$, not a parabola.

A second-degree equation in $x$ and $y$ for which no graph exists, or for which the graph consists of straight lines or a point, is called *degenerate*.

The type of graph of (8.4.1) is determined by the second-degree terms.

By our previous criteria we may say that if the graph of (8.4.1) is not degenerate, then:

(8.4.2) $\begin{cases} \text{if } AC < 0 \text{ the graph is a hyperbola } (A \text{ and } C \text{ have opposite signs});\\ \text{if } AC = 0 \text{ the graph is a parabola } (A = 0 \text{ or } C = 0);\\ \text{if } AC > 0 \text{ the graph is an ellipse (or circle) } (A \text{ and } C \text{ have the same sign}). \end{cases}$

We shall use this result to determine a similar criterion for the general second-degree equation. From Equations (8.3.3) we obtain the following.

(8.4.3) $$A' + C' = A + C.$$

(8.4.4) $$A' - C' = (A - C)\cos 2\theta + B \sin 2\theta.$$

(8.4.5) $$B' = B \cos 2\theta - (A - C)\sin 2\theta.$$

Squaring both sides of (8.4.4) and (8.4.5) and adding, we have

(8.4.6) $$B'^2 + (A' - C')^2 = B^2 + (A - C)^2.$$

From (8.4.3) we have

(8.4.7) $$(A' + C')^2 = (A + C)^2.$$

On subtracting (8.4.7) from (8.4.6) we find, finally,

(8.4.8) $$B'^2 - 4A'C' = B^2 - 4AC.$$

The quantity $B^2 - 4AC$ is called the *discriminant* of the second-degree equation in $x$ and $y$. By virtue of (8.4.8) we have proved the following statement.

**THEOREM 8.4.1.** *The discriminant of a second-degree equation in $x$ and $y$ remains unchanged in value under any rotation of axes.*

This property is also expressed in the statement: *the discriminant of a second-degree equation in $x$ and $y$ is invariant under a rotation of axes.* In this sense (8.4.3) and (8.4.6) give two additional *invariants*. (See also Exercise 8.)

Note that all statements involving invariants under a rotation of axes are based on the assumption that the constant terms in both equations are equal ($F' = F$). If the intermediate steps in carrying out the rotation have included multiplication of both sides of an equation by a number different from 1, the statements as given no longer hold.

Now suppose that the coefficients of a second-degree equation in $x$ and $y$ are unprimed letters, and a rotation of axes leads to an equation with primed letters such that $B' = 0$. By (8.4.8) we then have

$$B^2 - 4AC = -4A'C'.$$

We may now apply (8.4.2) to the primed coefficients and state the following result.

**THEOREM 8.4.2.** *If a second-degree equation in x and y is not degenerate, then:*

*if $B^2 - 4AC > 0$ the graph is a hyperbola;*

*if $B^2 - 4AC = 0$ the graph is a parabola;*

*if $B^2 - 4AC < 0$ the graph is an ellipse.*

The theorem is valid, of course, also in the case where $B = 0$, in which case it becomes the same as (8.4.2).

The theorem just proved enables us to identify directly the conic section represented by a second-degree equation in $x$ and $y$, provided that the equation is not degenerate.

*Example 8.4.5.* The type of curve represented by the equation in Example 8.2.3 is easily ascertained in advance, if it is assumed that the equation is not degenerate. We have $A = 13$, $B = -18$, $C = 37$. Hence

$$B^2 - 4AC = (-18)^2 - 4(13)(37) = 324 - 1924 = -1600 < 0,$$

and by Theorem 8.4.2 the curve is an ellipse. This is confirmed in the result of that example.

*Example 8.4.6.* Identify the graph of the following equation, assuming that the equation is not degenerate.

$$4x^2 - 4xy + y^2 = 2x.$$

*Solution.* We have $B^2 - 4AC = (-4)^2 - 4(4)(1) = 0$. Hence the graph is a parabola (see also Exercise 7).

There are criteria, in terms of the coefficients of a quadratic equation in $x$ and $y$, to determine whether degeneracy occurs and of what type. However, these are too detailed for our purpose.

We have not explicitly discussed invariants under translations. It is clear that a translation of axes applied to (8.3.1) does not alter the coefficients of the second-degree terms, and therefore does not affect the directions of the axes of the curve relative to the coordinate axes.

**EXERCISE GROUP 8.3**

Each equation in Exercises 1–6 is degenerate. (a) Describe the graph if it exists. (b) If you did not know that the equation is degenerate, how would you identify the graph by use of Theorem 8.4.2?

1. $x^2 = y^2$
2. $y^2 = 4$
3. $x^2 + 2x = 3$
4. $x^2 - 2xy + y^2 - 4 = 0$
5. $x^2 + 2y^2 - 4x + 4y + 6 = 0$
6. $x^2 - 4xy + 4y^2 + 4x - 8y = 0$

7. Show that an equation of a parabola cannot be of the form
$$Ax^2 + Bxy + Cy^2 + F = 0.$$

8. Show that $D^2 + E^2$ is invariant under a rotation of axes, in the sense of Section 8.4.

9. (a) Show that $x^2 + Axy = y$ represents a hyperbola if $A \neq 0$. (b) What is the curve if $A = 0$?

10. Name the graph of $Ax^2 + xy + y^2 = 4$, according to the value of $A$.

11. Name the graph of $Ax^2 + 2xy + Ay^2 = y + 2$, according to the value of $A$, assuming nondegeneracy.

## 8.5 FOCUS-DIRECTRIX PROPERTY OF THE CONICS

The parabola, ellipse, and hyperbola were defined by particular locus properties. It turns out that the ellipse and the hyperbola have a locus property which is similar to that used in the definition of the parabola. This connection can be developed rather quickly with our previous results. From Equation (7.8.8) for a focal radius of a horizontal ellipse with center at the origin we have

(8.5.1) $$|FP| = a - \frac{c}{a}x = \frac{c}{a}\left(\frac{a^2}{c} - x\right).$$

Now, for any point on the ellipse, $x \leq a < a^2/c$, and thus $a^2/c - x > 0$. Hence if $PM$ is perpendicular to the line $x = a^2/c$, we have (Fig. 8.5.1)
$$|PM| = a^2/c - x,$$
and with (8.5.1),

(8.5.2) $$|FP| = \frac{c}{a}|PM|.$$

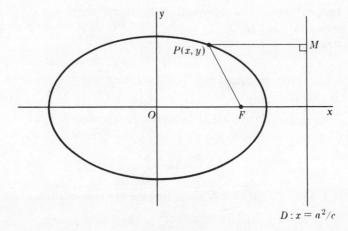

Fig. 8.5.1

If we use Equations (7.9.7) and (7.9.8), which apply to a horizontal hyperbola with center at the origin, we again obtain (8.5.2) (Exercise 24). For both the ellipse and the hyperbola the ratio of $|FP|$ to $|PM|$ is thus the same for each point of the curve, and is equal to $c/a$. The line $x = a^2/c$ is called a *directrix* for either curve.

For a parabola the property $|FP| = |PM|$ was the definition, where $F$ is the focus of the parabola and $PM$ the perpendicular segment from a point $P$ of the curve to the directrix.

Let us now define the *eccentricity e* as follows

$$e = \begin{cases} c/a \text{ for an ellipse or hyperbola;} \\ 1 \text{ for a parabola} \end{cases}$$

We may now write equation (8.5.2) as

$$|FP| = e|PM|,$$

which then holds for all three curves. We combine these results in the following.

**Focus-directrix property of the conics.** *The parabola, ellipse, and hyperbola have the property that for any point on the curve the ratio of the distances from a fixed point and a fixed line is a constant, called the eccentricity e. The fixed point is a focus, and the fixed line is a directrix of the curve. If $e > 1$, the curve is a hyperbola; if $e = 1$, the curve is a parabola; and if $e < 1$, the curve is an ellipse.*

The last statement depends on the fact that

*for a hyperbola $c > a$ and $c/a > 1$,*

*for an ellipse $c < a$ and $c/a < 1$.*

An equation of the directrix in Fig. 8.5.1 is $x = a^2/c = a/e$. This directrix corresponds to the focus $F(c, 0)$. By symmetry there is a second directrix, corresponding to the other focus $F'(-c, 0)$, an equation of the second directrix being $x = -a/e$. The same applies to a hyperbola. We may then state the following property, *which is independent of any coordinate system.*

*The ellipse and the hyperbola have two directrices; these are perpendicular to the line through the foci, and at a distance $a/e$ from the center.*

***Example 8.5.1.*** Find an equation of the hyperbola with foci at $(\pm 2, 0)$ and $x = 1$ as a directrix.

*Solution.* The center of the hyperbola must be at the origin, and $c = 2$. Since $x = 1$ is a directrix, we have, by the above, $a/e = 1$. Therefore, since $e = c/a$,

$$1 = \frac{a}{e} = \frac{a^2}{c} = \frac{a^2}{2}.$$

Hence $a^2 = 2$, $b^2 = c^2 - a^2 = 4 - 2 = 2$, and the desired equation is

$$\frac{x^2}{2} - \frac{y^2}{2} = 1.$$

The eccentricity $e$ of an ellipse or hyperbola is a measure of the flatness of the curve. In the ellipse, if $e$ is close to 1, then $c$ is close to $a$ in value, $b$ is small (near zero), and the ellipse is long and narrow. If $e$ is near zero, then $b$ is close to $a$ in value, and the ellipse is more nearly circular.

Similarly, in the case of a hyperbola ($e > 1$), if $e$ is close to 1, then $b$ is near zero, and the curve is elongated in the direction of the transverse axis. If $e$ is large, then $b$ is large, and the hyperbola is broad.

The parabola may be considered, in a sense which will not be made precise here, as a boundary case between ellipses and hyperbolas. Any variation in the eccentricity from the value of 1 will tip the balance in the direction of one or the other of these curves.

## EXERCISE GROUP 8.4

In Exercises 1–12 find an equation of the curve described.

1. An ellipse with foci at $(\pm 3, 0)$ and $e = 1/2$.
2. An ellipse with vertices at $(\pm 3, 0)$ and $e = 1/2$.
3. An ellipse with vertices at $(2, -1)$ and $(6, -1)$, and $e = 2/3$.
4. An ellipse with foci at $(2, -1)$ and $(2, -5)$, and $e = 1/3$.
5. An ellipse with foci at $(0, \pm 3)$ and directrices $y = \pm 12$.
6. An ellipse with directrices $y = \pm 4$, $e = 3/4$, and center on the $y$ axis.
7. A hyperbola with foci at $(2, -1)$ and $(6, -1)$, and $e = 3/2$.
8. A hyperbola with vertices at $(2, -1)$ and $(2, -5)$, and $e = 2$.
9. A hyperbola with directrices $x = \pm 1/2$ and asymptotes $y = \pm 3x$.
10. A hyperbola with directrices $y = \pm 2$ and asymptotes $y = \pm x$.
11. A hyperbola with directrices $y = 6$ and $y = 2$, $e = 2$, and center on the $y$ axis.
12. A hyperbola with asymptotes $2x - y - 7 = 0$ and $2x + y - 1 = 0$, and directrices $x = 0$ and $x = 4$.
13. Show that for the ellipse $x^2/a^2 + y^2/b^2 = 1$, $a > b$, the lengths of the focal radii for the point $(x, y)$ are $a \pm ex$.
14. Show that for the ellipse $x^2/b^2 + y^2/a^2 = 1$, $a > b$, the lengths of the focal radii for the point $(x, y)$ are $a \pm ey$.
15. Show that for the hyperbola $x^2/a^2 - y^2/b^2 = 1$ the lengths of the focal radii for the point $(x, y)$ are $ex \pm a$ if $x > 0$ and $-ex \pm a$ if $x < 0$.
16. Show that for the hyperbola $x^2/b^2 - y^2/a^2 = -1$ the lengths of the focal radii for the point $(x, y)$ are $ey \pm a$ if $y > 0$ and $-ey \pm a$ if $y < 0$.
17. If $a = b$ in a hyperbola, the hyperbola is called *equilateral*. Find the eccentricity of an equilateral hyperbola.
18. Show that the asymptotes of an equilateral hyperbola (see Exercise 17 for definition) are perpendicular.
19. Find the eccentricity of an ellipse if the lines joining an end of the minor axis to the foci are perpendicular.

20. If $p$ is the distance from a focus to the corresponding directrix in an ellipse or hyperbola, show that $p = b^2/c$.

21. If $p$ is the distance from a focus to the corresponding directrix in an ellipse, show that the length of the major axis is $2ep/(1 - e^2)$.

22. If $p$ is the distance from a focus to the corresponding directrix in a hyperbola, show that the length of the transverse axis is $2ep/(e^2 - 1)$.

23. The asymptotes of a hyperbola have slopes $\pm m$. Express the eccentricity $e$ in terms of $m$ for (a) a horizontal hyperbola, (b) a vertical hyperbola.

24. Prove Equation (8.5.2) for a horizontal hyperbola with center at the origin, considering the cases $x < 0$ and $x > 0$ separately.

## * 8.6  TANGENTS OF THE CONICS

In determining the slope of a conic, implicit differentiation is usually convenient for finding the derivative of the function of $x$ defined by the equation of the curve.

***Example 8.6.1.***  Find an equation of the tangent to the curve (ellipse)
$$2x^2 - 3xy - y^2 - x - 11 = 0$$
at the point $(2, -1)$.

*Solution.*  By implicit differentiation ($y' = dy/dx$) we find
$$4x - 3(xy' + y) - 2yy' - 1 = 0, \qquad y' = \frac{4x - 3y - 1}{3x + 2y}.$$

Hence the slope at $(2, -1)$ is $5/2$, and the tangent line is given by
$$y + 1 = \tfrac{5}{2}(x - 2) \quad \text{or} \quad 5x - 2y - 12 = 0.$$

The following theorem provides a form for an equation of the tangent line to a conic which permits writing the equation almost immediately.

**THEOREM 8.6.1.**  *The tangent line at the point $(x_1, y_1)$ of the (nondegenerate) conic*
$$Ax^2 + Bxy + Cy^2 + Dx + Ey + F = 0$$
*is given by*

(8.6.1) $\quad Ax_1 x + B\dfrac{y_1 x + x_1 y}{2} + Cy_1 y + D\dfrac{x + x_1}{2} + E\dfrac{y + y_1}{2} + F = 0.$

*Proof.*  By implicit differentiation the slope of the curve at the point $(x_1, y_1)$ is found to be
$$m = -\frac{2Ax_1 + By_1 + D}{Bx_1 + 2Cy_1 + E}.$$

If the denominator is not zero, an equation of the tangent line is, by use of the point-slope form,

(8.6.2) $\quad (2Ax_1 + By_1 + D)(x - x_1) + (Bx_1 + 2Cy_1 + E)(y - y_1) = 0.$

Since $x_1$ and $y_1$ satisfy the equation of the conic, we may write

$$2Ax_1^2 + 2Bx_1y_1 + 2Cy_1^2 + 2Dx_1 + 2Ey_1 + 2F = 0,$$

and this equation may be rewritten as

(8.6.3) $\quad (2Ax_1 + By_1 + D)x_1 + (Bx_1 + 2Cy_1 + E)y_1 + Dx_1 + Ey_1 + 2F = 0.$

We now add (8.6.2) and (8.6.3), and get the equation

$$(2Ax_1 + By_1 + D)x + (Bx_1 + 2Cy_1 + E)y + Dx_1 + Ey_1 + 2F = 0,$$

which is equivalent to (8.6.1).

Without amplification we state that (8.6.1) also holds if the denominator in the above expression for the slope $m$ is zero, in which case there is a vertical tangent.

The significance of writing an equation of the tangent line at $(x_1, y_1)$ in the form (8.6.1) is that this equation may be obtained directly from the equation of the conic on replacing $x^2$ by $x_1x$, $xy$ by $(y_1x + x_1y)/2$, $y^2$ by $y_1y$, $x$ by $(x + x_1)/2$, and $y$ by $(y + y_1)/2$. To apply this to the equation in Example 8.6.1, at the point $(2, -1)$, we replace $x^2$ by $2x$, $xy$ by $(-x + 2y)/2$, $y^2$ by $-y$, and $x$ by $(x + 2)/2$; we obtain in this way, as an equation of the tangent line at that point,

$$2(2x) - 3\frac{-x + 2y}{2} - (-y) - \frac{x + 2}{2} - 11 = 0,$$

which reduces to $5x - 2y - 12 = 0$, as in that example. With practice, the use of this method can become very helpful.

## * 8.7  OPTICAL PROPERTIES OF THE CONICS

When a ray of light is reflected from a curve (as though the curve were a mirror) the *incident ray* $RP$ and *reflected ray* $PQ$ (Fig. 8.7.1) make equal angles with the tangent, that is

$$\angle RPA = \angle QPB.$$

The conics have interesting and useful reflective, or *optical properties*, which we shall consider. In establishing such properties we may place the curve in a coordinate system in any convenient position, and the property will apply to any such curve independently of the coordinate system. We start with the optical property of the parabola.

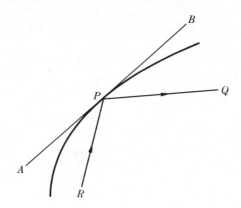

Fig. 8.7.1

**THEOREM 8.7.1.** *Any incident ray through the focus of a parabola is reflected parallel to the axis of the parabola.*

*Proof.* Place the parabola as in Fig. 8.7.2, so that its equation is

$$y^2 = 2px.$$

We must show that $FP$ is reflected parallel to the $x$ axis. Let $PQ$ be parallel to the $x$ axis. Then $\measuredangle FAP = \measuredangle QPB$. If we can show that $\measuredangle FPA = \measuredangle FAP$, it will follow that $PQ$ is the reflected ray, and the property will be established.

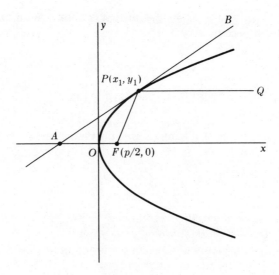

Fig. 8.7.2

We do this by showing that $|FP| = |AF|$. By the defining property of the parabola (Section 7.6) we know that

$$|FP| = x_1 + p/2.$$

To find $|AF|$ we obtain the equation of the tangent from Theorem 8.6.1, which becomes

$$y_1 y = p(x + x_1).$$

Setting $y = 0$, we find the abscissa of $A$ as $x = -x_1$. Hence

$$|AF| = \frac{p}{2} - (-x_1) = \frac{p}{2} + x_1.$$

Therefore, $|FP| = |AF|$ and the proof is complete.

We shall state the optical property of the ellipse as follows.

**THEOREM 8.7.2.** *A ray from either focus of an ellipse is reflected through the other focus.*

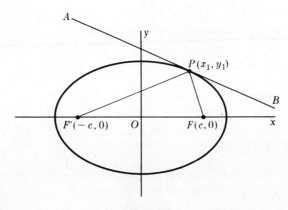

Fig. 8.7.3

*Proof.* Place the ellipse in the horizontal position with center at the origin (Fig. 8.7.3). An equation is then

(8.7.1) $$b^2 x^2 + a^2 y^2 = a^2 b^2.$$

We shall prove that if $APB$ is the tangent line at $P(x_1, y_1)$, then $\angle F'PA = \angle FPB$.

We may express the vectors $\overrightarrow{PF'}$ and $\overrightarrow{PF}$ in component form as

$$\overrightarrow{PF'} = (-c - x_1, -y_1), \qquad \overrightarrow{PF} = (c - x_1, -y_1).$$

If $m$ is the slope of the tangent at $P$, then vectors in the directions of $\overrightarrow{PA}$ and $\overrightarrow{PB}$ are respectively

$$(-1, -m), \quad (1, m).$$

Then (Section 2.2)

$$\cos \measuredangle F'PA = \frac{c_1 + x_1 + my_1}{|\overrightarrow{PF'}|\sqrt{1+m^2}}, \quad \cos \measuredangle FPB = \frac{c - x_1 - my_1}{|\overrightarrow{PF}|\sqrt{1+m^2}}.$$

These cosines, and therefore the angles, are equal, if

$$|\overrightarrow{PF}|(c_1 + x_1 + my_1) = |\overrightarrow{PF'}|(c - x_1 - my_1).$$

This in turn holds if

(8.7.2) $\quad (x_1 + my_1)(|\overrightarrow{PF'}| + |\overrightarrow{PF}|) = c(|\overrightarrow{PF'}| - |\overrightarrow{PF}|).$

Now, $|\overrightarrow{PF'}| = a + ex_1$, $|\overrightarrow{PF}| = a - ex_1$ (Section 7.8). Hence

(8.7.3) $\quad |\overrightarrow{PF'}| + |\overrightarrow{PF}| = 2a, \quad |\overrightarrow{PF'}| - |\overrightarrow{PF}| = 2ex_1.$

The slope of the ellipse at the point $(x_1, y_1)$ is $m = -b^2x_1/(a^2y_1)$. Hence

(8.6.4) $\quad x_1 + my_1 = x_1\left(1 - \frac{b^2}{a^2}\right) = e^2 x_1,$

and (8.7.2) becomes, with use of (8.7.3) and (8.7.4),

$$e^2 x_1 \cdot 2a = c \cdot 2ex_1.$$

This equation clearly holds for any value of $x_1$, since $e = c/a$. The proof is thus complete. (See Exercise 23 for an alternate proof.)

We shall merely state the optical property of the hyperbola; the proof will be called for in the exercises, with comments on the method.

**THEOREM 8.7.3.** *A focal ray $FP$ is reflected from a hyperbola along a line which is the extension of the focal ray $F'P$ through $P$.*

### * EXERCISE GROUP 8.5

In each of Exercises 1–14 find an equation of the indicated tangent line. Use Theorem 8.6.1 for some, and a direct method for some; also do some by both methods.

1. Tangent to $y^2 - 2x^2 + 7 = 0$ at $(2, -1)$
2. Tangent to $x^2 + 3y^2 = 4$ at $(-1, 1)$
3. Tangent to $\dfrac{x^2}{4} + \dfrac{y^2}{9} = 1$ at $(1, 3\sqrt{3}/2)$
4. Tangent to $\dfrac{x^2}{3} - \dfrac{y^2}{2} = 1$ at $(3, -2)$

5. Tangent to $7x^2 - 3y^2 + 5 = 0$ at $(1, 2)$
6. Tangent to $4x^2 + 7y^2 = 8$ at $(1/2, 1)$
7. Tangent to $x^2 + y^2 - 3x + 4y = 25$ at $(-1, 3)$
8. Tangent to $y^2 + 3x - 2y = 6$ at $(2, 2)$
9. Tangent to $3x^2 + 2x - y = 2$ at $(1, 3)$
10. Tangent to $3x^2 + y^2 + 4x - 3y + 1 = 0$ at $(-1, 3)$
11. Tangent to $4x^2 - 3y^2 + 6x = 1$ at $(2, 3)$
12. Tangent to $xy - 3x = 5$ at $(-1, -2)$
13. Tangent to $x^2 + 4xy - 3y^2 = 1$ at $(2, 3)$.
14. Tangent to $10x^2 - 24xy + 3y^2 + 2x - 3y + 12 = 0$ at $(1, 1)$.

15. Find an equation of that tangent line to $y^2 + 3x - 2y = 6$ which has slope $3/4$.
16. Find equations of the tangent lines to $3x^2 + y^2 + 4x - 3y + 1 = 0$ which have slope $2/3$.
17. Find an equation of the family of vertical parabolas which pass through the origin and are tangent to the line $y = x$ there.
18. Show that if two parabolas of the form $y^2 = 2Cx + C^2$ intersect, their tangents at a point of intersection are perpendicular.
19. Prove that the condition that the line $y = mx + \beta$ be tangent to the hyperbola $b^2x^2 - a^2y^2 = a^2b^2$ is $\beta^2 + b^2 = a^2m^2$.
20. Show that if any two conics of the form
$$\frac{x^2}{A^2 + C} + \frac{y^2}{B^2 + C} = 1$$
intersect, their tangents at a point of intersection are perpendicular.
21. Show that if a curve of the form $x^2 - y^2 = k$ intersects a curve of the form $y^2 + 2xy - x^2 = c$, their tangents at a point of intersection make an angle of $45°$.
22. Show that any tangent to the curve $2xy = a^2$ forms with the coordinate axes a triangle whose area is $a^2$.

* 23. Prove Theorem 8.7.2 as follows:
    (a) Find an equation of the line through $P$ in Fig. 8.7.3, perpendicular to the tangent. Let this line intersect the $x$ axis at $Q$.
    (b) Show that the abscissa of $Q$ is $e^2x_1$.
    (c) Then show that $|F'Q| = e(a + ex_1)$, $|QF| = e(a - ex_1)$.
    (d) Use (c) to show that $|F'Q|/|QF| = |F'P|/|FP|$.
    (e) It follows from a theorem of Plane Geometry that the line of (a) bisects $\angle F'PF$.

* 24. Prove Theorem 8.7.3. Suggestion: Start with the equation $b^2x^2 - a^2y^2 = a^2b^2$ and show that $\angle F'PA = \angle FPA$, where $A$ is a point on the tangent. The proof then follows the method of proof of Theorem 8.7.2, with modifications as needed.

# 9 Integration

## 9.1 INTRODUCTION

Much of the preceding work has been concerned with the definition of derivative of a function, and the study of properties and applications of the derivative. In such applications a function was given or obtained in some manner, and the derivative was found, to be used for whatever purpose was appropriate. In other words, the order of events was from a function to its derivative.

In other types of application the order of events needs to be reversed. Because of the properties of derivatives, a physical or geometrical situation may be characterized by some knowledge of the derivative of an unknown function. It becomes necessary then to obtain a function from a knowledge of its derivative. In this way we are led to a consideration of the process of *antidifferentiation*, or, as it is also called, *integration*.

We shall study this process and then relate it to an apparently unrelated topic, the *definite integral*. In the course of this development a fundamental theorem of calculus will emerge. In the next section we consider a needed theorem about derivatives, a theorem which is of great importance in many other connections as well.

## 9.2 THE MEAN-VALUE THEOREM

The main theorem to be presented here is of a type which may be characterized as an *existence theorem*, a theorem which asserts the existence of an entity with a certain property but which usually does not tell us how to find that entity.

One theorem of this type, which you have probably met previously, is the Fundamental Theorem of Algebra concerning the existence of a solution of a polynomial equation. Another such result concerns the existence of a maximum and a minimum of a continuous function in a closed interval, which is cited below. The Mean-Value Theorem which follows has its significant application in the proofs of other theorems, rather than in a direct use.

**THEOREM 9.2.1.  The Mean-Value Theorem.**  *Let $f$ be a function continuous in a closed interval $[a, b]$, and let $f'$ exist in the open interval $(a, b)$. Then there exists a number $c$, $a < c < b$, such that*

(9.2.1) $$f(b) - f(a) = (b - a)f'(c).$$

*Proof.*  Consider the function

$$g(x) = f(x) - f(a) - \frac{f(b) - f(a)}{b - a}(x - a).$$

We compute directly that $g(a) = g(b) = 0$. If $g(x)$ is identically zero in $[a, b]$, then $f(x)$ is a linear function, which may be written as $f(x) = mx + r$, and we find

$$f(b) - f(a) = m(b - a).$$

Since in this case $f'(x) = m$ for all $x$ in $[a, b]$, the result follows with $c$ as any number in $(a, b)$. [Note that the conclusion in the theorem does not state that $c$ is unique.]

Now suppose that $g(x)$ is not identically zero. We know that $g(x)$ is continuous in $[a, b]$, since $f(x)$ is. By Theorem 3.7.1, it follows that $g(x)$ has both a maximum and a minimum in $[a, b]$, at least one of which cannot be zero (why?). If $c$ is a value in $[a, b]$ at which such a nonzero maximum or minimum occurs, then $c$ must be different from $a$ or $b$. We know that $g'$ exists in $(a, b)$, since $f'$ exists there. Hence, by Theorem 5.4.1, $g'(c) = 0$. We then have

$$g'(c) = f'(c) - \frac{f(b) - f(a)}{b - a} = 0, \qquad a < c < b,$$

from which (9.2.1) follows. This completes the proof.

The Mean-Value Theorem is also valid if $b < a$, in which case $b < c < a$.

To illustrate Theorem 9.2.1 geometrically, let the curve in Fig. 9.2.1 represent the graph of $f(x)$. Then

$$\frac{f(b) - f(a)}{b - a}$$

is the slope of chord $PQ$. By the Mean-Value Theorem, there is a point $(c, f(c))$ with $a < c < b$, at which the tangent to the curve is parallel to the chord $PQ$. One such tangent line is drawn in the figure. Actually, a second such tangent line exists in the illustration shown.

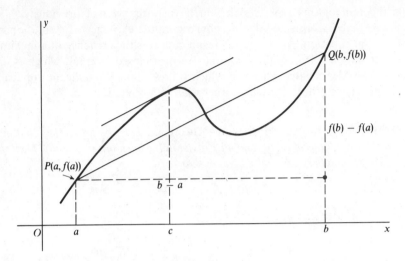

Fig. 9.2.1

It is important to realize what Theorem 9.2.1 states, as well as what it does not state. The theorem states a conclusion if $f$ is continuous in a closed interval and $f'$ exists in the open interval. In a case where one of the hypotheses is not satisfied, the conclusion may fail to hold, but not necessarily so; the theorem simply does not apply in this event. [See Exercises 12 and 13.]

We now give an example to illustrate the meaning of the theorem; however, the Mean-Value Theorem is almost never used in this way. As stated at the beginning of this section, the knowledge of the existence of $c$ with the required property is usually all that is needed. [See Theorem 9.2.2.]

***Example 9.2.1.*** Find a value $c$ as provided by the Mean-Value Theorem, for the function $f(x) = x/(x+1)$ in the interval $[1, 3]$.

*Solution.* We find $f'(x) = (x+1)^{-2}$. Then (9.2.1) gives, since $f(1) = 1/2$, $f(3) = 3/4$,

$$\tfrac{3}{4} - \tfrac{1}{2} = (3-1)(c+1)^{-2}.$$

Then $(c+1)^2 = 8$, and we find $c = -1 \pm 2\sqrt{2}$. Of these values, the value $-1 + 2\sqrt{2}$ is in $(1, 3)$, and is a value as stated in the theorem (there is only one in this case).

A special case of the Mean-Value Theorem is *Rolle's Theorem*, in which there is the additional hypothesis that $f(b) = f(a)$. In this case the conclusion is that there is a number $c$, $a < c < b$, such that $f'(c) = 0$.

The following is an application of the Mean-Value Theorem.

**THEOREM 9.2.2.** *If $f'(x) > 0$ in $(a, b)$, then $f$ is increasing in $(a, b)$ in the sense that if $a < x_1 < x_2 < b$, then $f(x_2) > f(x_1)$.*

*Proof.* The hypotheses of Theorem 9.2.1 are satisfied in the interval $[x_1, x_2]$. Then for some number $c$, $x_1 < c < x_2$,

$$f(x_2) - f(x_1) = (x_2 - x_1)f'(c).$$

But $x_2 - x_1 > 0$ and $f'(c) > 0$. Hence $f(x_2) > f(x_1)$.

In ordinary usage this is often an easy and convenient criterion for an increasing function.

We note that the method of proof just presented also yields a proof of Theorem 5.3.2.

## EXERCISE GROUP 9.1

In Exercises 1–10 determine a value $c$ as provided by the Mean-Value Theorem for the given function in the given interval.

1. $x^2 + 3x$, $[1, 3]$
2. $x^3$, $[-1, 2]$
3. $x^3 + 3$, $[1, 2]$
4. $x^3 + 2x^2 + 1$, $[1, 2]$
5. $x^4$, $[2, 4]$
6. $x^3 - 4x^2 - 3x - 2$, $[0, 3]$
7. $\dfrac{x}{x^2 + 1}$, $[-1, 2]$
8. $\sqrt{2x + 1}$, $[3/2, 4]$
9. $\sqrt{2x^2 + 1}$, $[-2, 0]$
10. $x^{2/3} + x$, $[1, 8]$

11. Find the value provided by the Mean-Value Theorem for the function $ax^2 + bx + c$ in the interval $[x_0, x_1]$.

12. State a reason why the Mean-Value Theorem does not apply to the function $(x - 1)^{2/3}$ in the interval $[0, 2]$. Does the theorem apply in the interval $[0, 1]$? Draw a graph.

13. State a reason why the Mean-Value Theorem does not apply to the function $(x - 1)^{1/3}$ in the interval $[0, 9/8]$. Show, however, that the conclusion of the theorem holds, and find the corresponding value. Draw a graph.

14. Let $f$ and $f'$ be continuous in $[a, b]$, and let $f''$ exist in $(a, b)$. Apply the Mean-Value Theorem to the function

$$F(x) = f(b) - f(x) - (b - x)f'(x) - (b - x)^2 \frac{f(b) - f(a) - (b - a)f'(a)}{(b - a)^2}$$

to show that a number $c$ exists in $(a, b)$ such that

$$f(b) - f(a) = (b - a)f'(a) + (b - a)^2 \frac{f''(c)}{2}.$$

This is an *Extended Form of the Mean-Value Theorem.*

15. Find a value $c$ as provided by the result of Exercise 14 for the function $2x^3 + 3x^2 - 1$ in the interval $[1, 3]$.

16. Show that the result of Exercise 14 is an identity for any quadratic polynomial $f(x) = c_0 + c_1 x + c_2 x^2$ (that is, for any $a$ and $b$, with $c$ not appearing).

* 17. Justify each of the following steps, if $h > 0$:
$$\sqrt{1+h} = 1 + \frac{h}{2\sqrt{c}}, \qquad 1 < c < 1 + h;$$
therefore,
$$\sqrt{1+h} < 1 + \frac{h}{2}.$$

## 9.3 THE INDEFINITE INTEGRAL

We begin this section with a simple example:
$$Dx^3 = 3x^2.$$
This means that $3x^2$ is the derivative of the function $x^3$; starting with $x^3$, we were led to $3x^2$ by some process, that is, differentiation. In the reverse process we wish to start with $3x^2$ and arrive at $x^3$; in effect we arrive at the function $x^3$ by recognizing that $3x^2$ is its derivative. Basically, the reverse process involves a recognition of a function which has the specified derivative, together with the development of methods that will facilitate such recognition. The function $x^3$ is called an *antiderivative* of $3x^2$.

**DEFINITION 9.3.1.** *A function $F(x)$ is an* **antiderivative** *of the function $f(x)$ if $F'(x) = f(x)$.*

We can now see that $3x^2$ has other antiderivatives, such as $x^3 - 7$, $x^3 + 2\pi$, and in fact any function of the form $x^3 + C$, where $C$ is any constant, positive, negative, or zero. In general, if $F(x)$ is an antiderivative of $f(x)$, then $F(x) + C$ is also an antiderivative, for any value of $C$. Are there any antiderivatives which are not of this form? Specifically, is there any antiderivative of $3x^2$ which is not of the form $x^3 + C$? The following result answers such questions in the negative; that is, if two functions have the same derivative, the functions differ only by a constant.

**THEOREM 9.3.1.** *If $F'(x) = G'(x)$ on an interval $(a, b)$, there exists a constant $C$ such that $F(x) - G(x) = C$ in $(a, b)$.*

*Proof.* Let $q(x) = F(x) - G(x)$. Using the hypothesis, we find
(9.3.1) $\qquad q'(x) = F'(x) - G'(x) = 0, \qquad x \text{ in } (a, b).$
Let $a'$ be any number in $(a, b)$. If $\alpha$ is any number such that $a < \alpha < b$, we may apply the Mean-Value Theorem to $q(x)$ in $[a', \alpha]$ and obtain
$$q(\alpha) - q(a') = (\alpha - a')q'(c), \qquad \text{for some } c \text{ in } (a', \alpha).$$
But $q'(c) = 0$ by (9.3.1), and hence $q(\alpha) - q(a') = 0$. Thus, if $\alpha$ is any number in $(a, b)$, $q(\alpha)$ has the same value, $q(a')$, which is the constant $C$ in the statement of the theorem, and the proof is complete.

The import of Theorem 9.3.1 is that as soon as one antiderivative $F(x)$ of a function $f(x)$ is known, *every* antiderivative is of the form $F(x) + C$. The expression $F(x) + C$, which represents *all* antiderivatives of $f(x)$, is called the *indefinite integral* of $f(x)$, and we write

$$\int f(x)\, dx = F(x) + C \quad \text{if} \quad F'(x) = f(x).$$

The symbol $\int$, an elongated $S$, is called an "integral sign." The significance of $dx$ will be explained later (Section 9.5). For the present we may think of it as indicating that the differentiation of $F$ is with respect to $x$. The process of finding an indefinite integral may be called *antidifferentiation* or *integration*. The additive constant which appears in an indefinite integral is called a *constant of integration*. The function $f$ or $f(x)$ is the *integrand*.

Any formula of differentiation may be interpreted as a formula of integration. Conversely, any integration result may be proved by differentiation.

**Example 9.3.1.** (a) $\int 3x^2\, dx = x^3 + C$, since $Dx^3 = 3x^2$

(b) $\int (x-1)^2\, dx = \tfrac{1}{3}(x-1)^3 + C$, since $D[\tfrac{1}{3}(x-1)^3] = (x-1)^2$

(c) $\int \dfrac{1}{(1-x)^2}\, dx = \dfrac{x}{1-x} + C$ (Why?)

The following specific *formulas of integration* are easily verified by appropriate differentiation.

1. $\int dx = x + C.$

2. $\int x^n\, dx = \dfrac{x^{n+1}}{n+1} + C, \quad n+1 \neq 0,\ n$ rational.

In addition, we present two fundamental, general properties of the indefinite integral.

3. $\int kf(x)\, dx = k \int f(x)\, dx, \quad k$ a constant.

4. $\int [f(x) + g(x)]\, dx = \int f(x)\, dx + \int g(x)\, dx.$

Each of these general properties concerns a class or classes of function, and must be interpreted accordingly. Formula 3, for example, means: *If an antiderivative of $f(x)$ is multiplied by $k$, the product is an antiderivative of $kf(x)$*. This suggests the proof.

*Proof* of 3. Let $F'(x) = f(x)$. Then

$$\frac{d}{dx}[kF(x)] = kF'(x) = kf(x),$$

and $kF(x)$ is an antiderivative of $kf(x)$.

Property 4 means that *the sum of an antiderivative of $f(x)$ and an antiderivative of $g(x)$ is an antiderivative of $f(x) + g(x)$.*

*Proof* of 4.   Let $F'(x) = f(x)$, $G'(x) = g(x)$. Then

$$\frac{d}{dx}[F(x) + G(x)] = F'(x) + G'(x) = f(x) + g(x),$$

and 4 is proved.

In practice, Formula 3 means that *a constant factor may be moved inside or outside an integral*; Formula 4 means that *the integral of a sum of functions may be obtained by integrating the individual functions and adding.*

**Example 9.3.2.**   Find $\int (2x^3 + x^{1/2} - 3x^{-2})\, dx$.

*Solution.*   Indicating in detail the steps, for which you may supply the reasons, we have

$$\int (2x^3 + x^{1/2} - 3x^{-2})\, dx = \int 2x^3\, dx + \int x^{1/2}\, dx - \int 3x^{-2}\, dx$$

$$= 2\int x^3\, dx + \int x^{1/2}\, dx - 3\int x^{-2}\, dx$$

$$= 2\frac{x^4}{4} + \frac{x^{3/2}}{3/2} - 3\frac{x^{-1}}{-1} + C$$

$$= \frac{x^4}{2} + \frac{2x^{3/2}}{3} + 3x^{-1} + C.$$

After facility in the process of integration is achieved, the final result in such an integration can be written without the intervening steps.

**Example 9.3.3.**   Evaluate the integral $\int (2v^{1/2} - 1)^2\, dv$.

*Solution.*   The use of $v$ as a variable does not affect any of the preceding discussion. None of the preceding formulas applies directly to the given integral. However, if we expand the integrand, we have

$$\int (2v^{1/2} - 1)^2\, dv = \int (4v - 4v^{1/2} + 1)\, dv$$

$$= 4\int v\, dv - 4\int v^{1/2}\, dv + \int dv$$

$$= 2v^2 - \frac{8v^{3/2}}{3} + v + C.$$

Because of Theorem 9.3.1 the constant of integration may be attached at the very last in the above examples.

We have seen that the constant of integration in an indefinite integral ensures the inclusion of all antiderivatives. However, the statement of a problem may require a particular antiderivative of a function, also called a *particular integral*. A frequently used procedure in such a case is to find the indefinite integral and then to specialize the value of the constant to suit the particular situation.

**Example 9.3.4.** Find an equation of the curve whose slope at any point is equal to the square of the abscissa of the point, and such that the curve passes through the point $(4, -1)$.

*Solution.* If $(x, y)$ is a point on the desired curve $y = y(x)$, the slope is given by $y'(x)$. Hence

$$y'(x) = x^2, \qquad y(4) = -1,$$

and by integration we have

(9.3.2) $$y = y(x) = \frac{x^3}{3} + C.$$

This equation represents different curves with the desired slope property, for different values of $C$. To choose the particular curve passing through the point $(4, -1)$, we select $C$ so that $x = 4$ and $y = -1$ satisfy the equation. We get

$$-1 = \frac{(-4)^3}{3} + C, \qquad C = \frac{61}{3};$$

substitution in (9.3.2) gives the desired equation,

$$y = \frac{x^3}{3} + \frac{61}{3}.$$

**EXERCISE GROUP 9.2**

In Exercises 1–22 evaluate the given integral.

1. $\int (2x^2 - 3x)\, dx$
2. $\int (x^4 + 5x^3 - 2x)\, dx$
3. $\int (x + x^{1/2} + x^{1/3})\, dx$
4. $\int (x^2 - x^{2/3} + 3)\, dx$
5. $\int \sqrt{2t}\, dt$
6. $\int \sqrt[3]{2t}\, dt$
7. $\int \sqrt{au}\, du$
8. $\int \sqrt[4]{2au}\, du$
9. $\int (2x^2 - 1)^2\, dx$
10. $\int (2x - 1)^3\, dx$
11. $\int (x^{1/2} + 1)^2\, dx$
12. $\int (2x^{1/3} - 1)^2\, dx$
13. $\int (t^{1/3} - t^{-1/3})^2\, dt$
14. $\int (t^{1/3} - 1)^3\, dt$
15. $\int (u^{2/3} - u^{-2/3})^2\, du$
16. $\int (u^{1/2} + u + 1)^2\, du$
17. $\int \frac{1+x}{x^{2/3}}\, dx$
18. $\int \frac{(1+x)^2}{x^{2/3}}\, dx$

19. $\displaystyle\int \frac{(1+x)^2}{\sqrt{x}}\,dx$

20. $\displaystyle\int \frac{(1+x)^2}{\sqrt[3]{x}}\,dx$

21. $\displaystyle\int \left(\frac{y+1}{y^{1/3}}\right)^2 dy$

22. $\displaystyle\int \left(\frac{y^{1/2}+1}{y^{1/3}}\right)^2 dy$

In Exercises 23–30 find the function which satisfies the given conditions.

23. $f'(x) = 2x,\ f(0) = 1$
24. $f'(x) = 2x^2,\ f(-1) = 0$
25. $g'(x) = 1 - \sqrt{x},\ g(2) = 2$
26. $f'(x) = (1+x)^2,\ f(-1) = 4$
27. $f'(x) = \dfrac{1-\sqrt{x}}{\sqrt{x}},\ f(4) = 2$
28. $h'(x) = (x^2 - x)^2,\ h(1/2) = -1$
29. $f'(t) = 4t - 3,\ f(3) = 2f(0)$
30. $f'(x) = x^{1/3} + x^{-1/3},\ 4f(1) = f(8)$

31. Find an equation of a curve whose slope at any point is proportional to the abscissa of the point, if the curve passes through the point (1, 1) and has an $x$ intercept 2.

32. Find an equation of a curve whose slope at any point is equal to the abscissa of the point, and such that the tangent line at the point with abscissa 2 passes through the origin.

33. The speed of a particle which moves in a straight line is proportional to $\sqrt{t}$. Find the position function of the particle if the distance is 1 at $t = 1$, and 3 at $t = 4$.

\* 34. A point moves in a plane so that the rate of change with respect to $x$, of its distance from the origin is inversely proportional to that distance. Find an equation of the path if it passes through the points (1, 1) and (2, 3). [*Hint:* You may be able to use the relation $\int yy'\,dx = y^2/2 + C$. Why is this valid?]

## 9.4 USE OF THE CHAIN RULE IN INTEGRATION

In the study of differentiation it was seen that the use of the chain rule (Section 4.5) permitted the extension of the power rule to functions like $(1-x^2)^{17}$ and $(3+x^2)^{1/3}$. We shall examine the significance of this for integration.

We have, for example,

(9.4.1) $$\frac{d}{dx}(3+x^2)^{1/3} = \frac{2x(3+x^2)^{-2/3}}{3}.$$

It follows that

(9.4.2) $$\int x(3+x^2)^{-2/3}\,dx = \frac{3(3+x^2)^{1/3}}{2} + C.$$

But how can we recognize directly that the integral in (9.4.2) can be obtained by the power rule? First of all, we note that a constant factor in an integrand does not affect the type of integration. We then note that in (9.4.1) the factor $x$ on the right arises as a result of differentiating $3 + x^2$. It is the presence of this factor in the integrand of (9.4.2), in addition to the power $(3+x^2)^{-2/3}$, that makes the use of the power rule valid. In general, *the power rule may be applied if the integrand is, apart from a constant factor, the product of a power of a function and the derivative of that function.*

The *power rule of integration*, in its general form, may be written as

(9.4.3) $$\int [f(x)]^n f'(x)\, dx = \frac{[f(x)]^{n+1}}{n+1} + C, \quad n+1 \neq 0;$$

a proof, if $n$ is a rational number, is obtained by differentiating the right member.

The factor $f'(x)$ in the integrand may be called an *enabling factor*. This factor does not appear in the result. Its presence, *in addition to* the power of $f(x)$, enables us to use the power rule.

For the integral in (9.4.2) the presence of the factor $x$ means that we have the enabling factor except for a constant factor. If we introduce a factor 2 into the integrand, and a compensating factor 1/2 before the integrand, the original integral remains unchanged. We write

$$\int x(3 + x^2)^{-2/3}\, dx = \frac{1}{2}\int 2x(3 + x^2)^{-2/3}\, dx.$$

Now (9.4.3) applies with $f(x) = 3 + x^2$, $f'(x) = 2x$, $n = -2/3$, and we get

$$\frac{1}{2}\frac{(3 + x^2)^{1/3}}{1/3} + C = \frac{3(3 + x^2)^{1/3}}{2} + C.$$

**Example 9.4.1.** Evaluate (a) $\int (x - 2)^5\, dx$, (b) $\int (3x - 5)^{4/3}\, dx$.

*Solution.* (a) If $f(x) = x - 2$, then $f'(x) = 1$; the enabling factor is thus present, and we obtain

$$\int (x - 2)^5\, dx = \tfrac{1}{6}(x - 2)^6 + C.$$

The integration can also be performed by first expanding the integrand; this process is lengthy.

(b) If $f(x) = 3x - 5$, then $f'(x) = 3$. We introduce a factor 3 inside and a factor 1/3 outside the integral. We get

$$\int (3x - 5)^{4/3}\, dx = \frac{1}{3}\int (3x - 5)^{4/3}(3\, dx) = \frac{1}{3}\frac{(3x - 5)^{7/3}}{7/3} + C$$

$$= \frac{1}{7}(3x - 5)^{7/3} + C.$$

(Expansion of the integrand is not feasible in this case.)

**Example 9.4.2.** Evaluate $\int x^2(x^3 + a^3)^7\, dx$.

*Solution.* It is understood that $a$ is constant. We may expand the integrand

and apply the basic formulas of integration. However, $3x^2$ is an enabling factor for $x^3 + a^3$. Hence

$$\int x^2(x^3 + a^3)^7 \, dx = \frac{1}{3}\int (x^3 + a^3)^7 (3x^2) \, dx = \frac{1}{3}\frac{(x^3 + a^3)^8}{8} + C$$

$$= \frac{(x^3 + a^3)^8}{24} + C.$$

**Example 9.4.3.** Consider the integral $\int x(x^3 + a^3)^7 \, dx$. The enabling factor for $x^3 + a^3$ is $3x^2$. However, we cannot modify the integrand in such a way as to have the enabling factor by use of a constant factor only. Hence the power rule cannot be used, as applied to $x^3 + a^3$. We may expand the integrand and proceed; this is the most practical method here, but it is lengthy. (We are not interested in the actual result at this time.)

The idea of an enabling factor, while directed here to the power rule, *is fundamental to all integration*. Its use will be amplified as other formulas of integration are developed. In its general form, it is expressed in the following statement, which is proved by differentiation.

**THEOREM 9.4.1.** *If $\int f(x) \, dx = F(x) + C$, then*

(9.4.4) $$\int f[g(x)]g'(x) \, dx = F[g(x)] + C.$$

*Proof.* To prove this we need only show that

$$\frac{d}{dx} F[g(x)] = f[g(x)]g'(x);$$

this is immediately valid by the chain rule of differentiation, since

$$\frac{d}{dx} F[g(x)] = F'[g(x)]g'(x) \quad \text{and} \quad F'(x) = f(x).$$

### EXERCISE GROUP 9.3

In Exercise 1–32 evaluate the integral.

1. $\int \sqrt{2 + x} \, dx$

2. $\int \sqrt{5 - x} \, dx$

3. $\int \sqrt{3 - 2x} \, dx$

4. $\int \sqrt{2x + 3} \, dx$

5. $\int (2x + 3)^{1/2} \, dx$

6. $\int (3 - 2x)^{2/3} \, dx$

7. $\int y(3 - y^2)^4 \, dy$

8. $\int 4x(2x^2 + 7)^{4/9} \, dx$

9. $\int (u-1)(u^2 - 2u - 3)^4 \, du$

10. $\int (u+2)(u^2 + 4u)^{1/7} \, du$

11. $\int x\sqrt{3-2x^2} \, dx$

12. $\int \dfrac{x \, dx}{\sqrt{3-2x^2}}$

13. $\int \dfrac{x^2 \, dx}{\sqrt[3]{2+x^3}}$

14. $\int x^3(x^4-2)^5 \, dx$

15. $\int \dfrac{(y-1) \, dy}{(y^2 - 2y + 2)^2}$

16. $\int \dfrac{(2t-1) \, dt}{\sqrt{5t^2 - 5t}}$

17. $\int \dfrac{t^2 \, dt}{\sqrt{t^4 - t^2}}$

18. $\int \dfrac{(t+2) \, dt}{\sqrt[4]{(t+1)(t+3)}}$

19. $\int \sqrt{x}(1+x^{3/2})^{1/3} \, dx$

20. $\int \dfrac{\sqrt{1-\sqrt{x}}}{\sqrt{x}} \, dx$

21. $\int \sqrt[3]{x}(2x^{4/3} - 3)^{1/2} \, dx$

22. $\int \sqrt{x + 2\sqrt{x}} \left(1 + \dfrac{1}{\sqrt{x}}\right) dx$

23. $\int \dfrac{(x+2) \, dx}{(x^2 + 4x)^{1/2}}$

24. $\int \sqrt{x^4 - 3x^2} \, dx$

25. $\int \dfrac{\sqrt{x^{1/2} - a^{1/2}}}{x^{1/2}} \, dx$

26. $\int \dfrac{\sqrt{x^{1/3} + a^{1/3}}}{x^{2/3}} \, dx$

27. $\int x^{1/3}\sqrt{x^{4/3} + 8} \, dx$

28. $\int \dfrac{\sqrt{x^{2/3} - 23}}{x^{1/3}} \, dx$

29. $\int \dfrac{1}{y^2}\sqrt{1 + \dfrac{1}{y}} \, dy$

30. $\int \dfrac{1}{t^3}\sqrt[3]{2 - \dfrac{3}{t^2}} \, dt$

31. $\int \dfrac{1 + 6x^{2/3}}{x^{2/3}} \sqrt{x^{1/3} + 2x} \, dx$

32. $\int \dfrac{4x+1}{x^{2/3}} \sqrt{x^{4/3} + x^{1/3}} \, dx$

In each of the following determine whether the power rule of integration can be made to apply merely by introducing a constant factor into the integrand. Do not integrate. If the answer is "Yes," give the needed constant factor if it is different from 1.

33. $\int \sqrt{x + a^2} \, dx$

34. $\int \sqrt{y^2 + a^2} \, dy$

35. $\int x\sqrt{x^2 + a^2} \, dx$

36. $\int v^2\sqrt{v^2 - 7} \, dv$

37. $\int x^2\sqrt{x^3 + 1} \, dx$

38. $\int x^{-1/2}(1 + x^{1/2})^{3/2} \, dx$

39. $\int x^2(x+2)^{-1/2} \, dx$

40. $\int \dfrac{x^2 \, dx}{\sqrt{x^2 + 2}}$

41. $\int \dfrac{t^2 \, dt}{\sqrt{t^3 + 2}}$

42. $\int (1 + \sqrt{t})^{3/2} \, dt$

43. $\int [1 + f(x)]^2 \, dx$

44. $\int [1 + f(x)]^2 f'(x) \, dx$

45. $\int \{1 + [f(x)]^2\}^{1/2} f'(x) \, dx$

46. $\int \{1 + [f(x)]^2\}^{1/2} f(x) f'(x) \, dx$

## 9.5 CHANGE OF VARIABLE

Various methods that may be applied to integration are studied later in this book. At this time we present a method or process that is called *substitution* or *change of variable*. The appearance of a differential in the notation for an indefinite integral makes the process of substitution rather automatic. However, the process must be justified. We shall illustrate the method before giving a proof, using for this purpose an integral which can be evaluated more directly by the methods of Section 9.4. As will be seen, however, the use of this method is much more general.

Consider the integral $\int x(3x^2 - 4)^{1/2} \, dx$. We anticipate the use of the power rule, applied to $3x^2 - 4$. We set $u = 3x^2 - 4$. Then, by Section 4.9, we have $du = 6x \, dx$, or $x \, dx = du/6$. Making this substitution (not yet justified), we obtain

$$\int x(3x^2 - 4)^{1/2} \, dx = \int (3x^2 - 4)^{1/2}(x \, dx) = \int u^{1/2}(du/6)$$

$$= \frac{1}{6} \int u^{1/2} \, du = \frac{1}{6} \frac{u^{3/2}}{3/2} + C$$

$$= \frac{u^{3/2}}{9} + C = \frac{(3x^2 - 4)^{3/2}}{9} + C.$$

The last expression, in which $u$ is replaced by its value in terms of $x$, is the desired result.

We could have solved the equation $u = 3x^2 - 4$ for $x$, and proceeded as follows:

$$x = \frac{\sqrt{u+4}}{\sqrt{3}}, \qquad dx = \frac{du}{2\sqrt{3}\sqrt{u+4}},$$

$$\int x(3x^2 - 4)^{1/2} \, dx = \int \frac{\sqrt{u+4}}{\sqrt{3}} \cdot u^{1/2} \cdot \frac{du}{2\sqrt{3}\sqrt{u+4}} = \frac{1}{6} \int u^{1/2} \, du.$$

Clearly the first method is more efficient.

In general, *a substitution involves setting $x = g(u)$ in an integral $\int f(x) \, dx$, replacing $dx$ by $g'(u) \, du$, integrating in terms of $u$, and substituting for $u$ a function $h(x)$ such that $u = h(x)$ is a solution of the equation $x = g(u)$.* This becomes

(9.5.1) $$\int f(x) \, dx = \int f[g(u)]g'(u) \, du \Big|_{u=h(x)}$$

where the notation on the right means that $h(x)$ is substituted for $u$ after the integration on the right is performed. If $F$ is an antiderivative of $f$ so that $F'(x) = f(x)$, (9.5.1) may be expressed in terms of $u$ in the form

$$F[g(u)] + C = \int f[g(u)]g'(u) \, du.$$

But this, with a change of letter, is precisely (9.4.4), which was proved. Hence (9.5.1) is established, and the method of substitution is justified.

## 9.5 CHANGE OF VARIABLE

In Chapter 14 it will be seen that a change of variable is the recommended method for the evaluation of certain integrals. In other cases, where it may not be immediately clear how one should proceed, a change of variable may nevertheless suggest itself. This change may then be carried out, to be followed by a judgment as to its effectiveness.

**Example 9.5.1.** Evaluate $\int \dfrac{x^3 \, dx}{\sqrt{x^2 - 4}}$.

*Solution.* The power rule does not work directly. Let us try the change of variable $x^2 - 4 = u$, $2x \, dx = du$. We get

$$\int \frac{x^3 \, dx}{\sqrt{x^2 - 4}} = \int \frac{x^2(x \, dx)}{\sqrt{x^2 - 4}} = \int \frac{(u + 4) \, du/2}{u^{1/2}} = \frac{1}{2} \int (u^{1/2} + 4u^{-1/2}) \, du$$

$$= \frac{1}{2}\left(\frac{2u^{3/2}}{3} + 8u^{1/2}\right) + C = \frac{1}{3}(x^2 - 4)^{3/2} + 4(x^2 - 4)^{1/2} + C.$$

It worked!

**Example 9.5.2.** Evaluate $\int \dfrac{x^2 \, dx}{\sqrt{x^2 - 4}}$.

*Solution.* Again the power rule does not work directly. Again we try the change of variable $x^2 - 4 = u$, $2x \, dx = du$. We get

$$\int \frac{x^2 \, dx}{\sqrt{x^2 - 4}} = \int \frac{x(x \, dx)}{\sqrt{x^2 - 4}} = \int \frac{\sqrt{u + 4} \, du/2}{\sqrt{u}} = \frac{1}{2} \int \frac{\sqrt{u + 4}}{\sqrt{u}} \, du.$$

We have something that seems no more manageable than the original integral. The change of variable did not work! We avoid the temptation to say that the integration cannot be done. Rather, the method we tried was not effective. The fact is that the integration is possible, but by methods which will be studied later (Section 14.4). We are not in a position to do it now.

**EXERCISE GROUP 9.4**

In each of Exercises 1–10 use a suitable change of variable to perform the integration. (These integrations can also be carried out by the method of Section 9.4.)

1. $\int (4 - 3x)^{1/2} \, dx$

2. $\int (2x - 3)^{1/3} \, dx$

3. $\int y(3 - 2y^2)^9 \, dy$

4. $\int 4x^2(2x^3 - 3)^{4/3} \, dx$

5. $\int (x - 1)\sqrt{x^2 - 2x} \, dx$

6. $\int (u - 2)(u^2 - 4u + 3)^4 \, du$

7. $\int \dfrac{t \, dt}{\sqrt{3 + 2t^2}}$

8. $\int \dfrac{(u - 1) \, du}{\sqrt[3]{u^2 - 2u - 2}}$

9. $\int \dfrac{\sqrt{4 - t^{1/2}}}{t^{1/2}} \, dt$

10. $\int \dfrac{\sqrt{1 - 2u^{1/3}}}{u^{2/3}} \, du$

In each of the following, (a) carry out the given change of variable, obtaining an integral in the new variable; (b) if you can evaluate the resulting integral, complete the process.

11. $\int (x+1)(2x-3)^{1/2}\,dx; \quad u = 2x - 3$

12. $\int (2x-3)(x+1)^{1/2}\,dx; \quad u = x + 1$

13. $\int \dfrac{x^3 - x}{\sqrt{x^2 + 1}}\,dx; \quad u = x^2 + 1$

14. $\int \dfrac{dx}{\sqrt{x^2 - 4}}; \quad u = x^2 - 4$

15. $\int \dfrac{\sqrt{1 - 2u^{1/3}}\,du}{u^{1/3}}; \quad v = 1 - 2u^{1/3}$

16. $\int \dfrac{(x+2)\,dx}{(2x+1)^{5/2}}; \quad z^2 = 2x + 1$

17. $\int \dfrac{x+1}{(2x+1)^{3/2}}\,dx; \quad z^2 = 2x + 1$

18. $\int \dfrac{y^2}{\sqrt{y - 1}}\,dy; \quad u = y - 1$

19. $\int \dfrac{x}{\sqrt{3x + 2}}\,dx; \quad u = 3x + 2$

20. $\int \sqrt{\dfrac{y - 1}{y + 1}}\,dy; \quad u = \dfrac{y - 1}{y + 1}$

## 9.6 INTRODUCTION TO THE DEFINITE INTEGRAL

The definite integral is one of the most fundamental concepts in calculus. The definition involves a rather protracted process, a type of process that will be a new kind of experience in your mathematical development. Patience is required of you, to await the actual definition, to understand its meaning, to see the development of its properties and connection with a derivative, and finally to see its wide use in many types of application.

The definition of definite integral proceeds as follows. Let $f$ be a function defined in an interval $[a, b]$; choose a set of numbers $x_0, x_1, \ldots, x_{n-1}, x_n$, with $x_0 = a$, $x_n = b$, such that

$$a = x_0 < x_1 < \cdots < x_{n-1} < x_n = b;$$

now choose numbers $w_i$ such that

$$x_{i-1} \leq w_i \leq x_i, \quad i = 1, 2, \ldots, n;$$

and form the sum

(9.6.1) $\quad f(w_1)(x_1 - x_0) + f(w_2)(x_2 - x_1) + \cdots + f(w_n)(x_n - x_{n-1}).$

Suppose that a number $A$ exists with the following property: given $\epsilon > 0$, a number $\delta$ exists such that for the sum formed as in (9.6.1) we have

$$|\text{Sum} - A| < \epsilon \quad \text{whenever} \quad \max|x_i - x_{i-1}| < \delta, \quad i = 1, 2, \ldots, n.$$

Then the number $A$ is said to be the *definite integral* of $f$ in the interval $[a, b]$.

We emphasize that this definition is purely analytic, and is independent of any connection with any type of application, geometric or otherwise. Moreover, it is a very formal, abstract type of definition, and seems to have little connection, at the moment, with anything of value. Do not be concerned if the definition does not have a great deal of meaning for you at present, but try to keep it in mind as we proceed.

In the following sections we shall show that it is quite reasonable to want to study such a concept, and shall develop its properties in such a way that they lend themselves readily to the applications to be made. We start by developing some special sums.

## 9.7 SUMMATION

Let $f$ be a function whose domain is the set of integers. The "summation notation" is a convenient way of representing certain sums of values of $f$. For example, we write

$$\sum_{k=1}^{4} f(k) = f(1) + f(2) + f(3) + f(4).$$

This notation implies that the *index* $k$ takes on the integral values between the *lower limit* 1 and the *upper limit* 4, inclusive, that is, the values 1, 2, 3, and 4; the values of $f$ for these values of $k$ are then added. If $f(k) = k^2$, we have

$$\sum_{k=1}^{4} f(k) = \sum_{k=1}^{4} k^2 = 1^2 + 2^2 + 3^2 + 4^2 = 30.$$

In general, we write

$$\sum_{k=1}^{n} f(k) = f(1) + f(2) + \cdots + f(n),$$

for the indicated sum of $n$ values of $f$. The *index* $k$ is called a *dummy index* or *dummy variable*, since the value of the sum is independent of the letter used; thus

$$\sum_{k=1}^{n} f(k) = \sum_{i=1}^{n} f(i) = \sum_{j=1}^{n} f(j).$$

The function $f(k)$ is called the *summand*. The lower limit in a sum may be any integer.

**Example 9.7.1.**  $\displaystyle\sum_{j=-1}^{4} j^3 = (-1)^3 + 0^3 + 1^3 + 2^3 + 3^3 + 4^3$

$$= -1 + 0 + 1 + 8 + 27 + 64 = 99.$$

**Example 9.7.2.** Evaluate $\displaystyle\sum_{k=1}^{n} (2k^2 - 3k + 2)$ directly.

*Solution.* We evaluate the summand $f(k) = 2k^2 - 3k + 2$ for $n = 1, 2, 3,$ and 4, and find

$$f(1) = 1, \quad f(2) = 4, \quad f(3) = 11, \quad f(4) = 22.$$

Hence
$$\sum_{k=1}^{4}(2k^2 - 3k + 2) = 1 + 4 + 11 + 22 = 38.$$

Each of the following properties of sums is easily derived from familiar properties of real numbers, and can be established by expanding the sums on both sides.

1. $\sum_{k=1}^{n} c = nc$, $c$ constant

2. $\sum_{k=1}^{n} [cf(k)] = c \sum_{k=1}^{n} f(k)$

3. $\sum_{k=1}^{n} [f(k) + g(k)] = \sum_{k=1}^{n} f(k) + \sum_{k=1}^{n} g(k)$

4. If $m < n$, then $\sum_{k=1}^{n} f(k) = \sum_{k=1}^{m} f(k) + \sum_{k=m+1}^{n} f(k)$.

The last summation in 4 indicates the sum $f(m+1) + f(m+2) + \cdots + f(n)$.

We present some special sums that are used in what follows.

(9.7.1) $\qquad \sum_{k=1}^{n} k = 1 + 2 + \cdots + n = \dfrac{n(n+1)}{2}.$

(9.7.2) $\qquad \sum_{k=1}^{n} k^2 = 1^2 + 2^2 + \cdots + n^2 = \dfrac{n(n+1)(2n+1)}{6}.$

(9.7.3) $\qquad \sum_{k=1}^{n} k^3 = 1^3 + 2^3 + \cdots + n^3 = \dfrac{n^2(n+1)^2}{4}.$

Formula (9.7.1) expresses the sum of an arithmetic progression and will not be proved here.

*Proof* of (9.7.2). If we write $(k+1)^3 = k^3 + 3k^2 + 3k + 1$, and then sum from 1 to $n$, we obtain

$$3\sum_{k=1}^{n} k^2 = \sum_{k=1}^{n} [(k+1)^3 - k^3] - 3\sum_{k=1}^{n} k - \sum_{k=1}^{n} 1.$$

But

$$\sum_{k=1}^{n} [(k+1)^3 - k^3] = (2^3 - 1^3) + (3^3 - 2^3) + \cdots + [(n+1)^3 - n^3]$$
$$= (n+1)^3 - 1,$$

since the terms in the expansion cancel in pairs, except for the terms $(n+1)^3$ and $1^3$. Such a sum is called a *telescoping sum*. (See Exercises 28 and 29.) Then, using (9.7.1) and Property 1, we have

$$3\sum_{k=1}^{n} k^2 = (n+1)^3 - 1 - \frac{3n(n+1)}{2} - n = (n+1)^3 - \frac{3n^2 + 5n + 2}{2}$$

$$= \frac{2(n+1)^3 - (n+1)(3n+2)}{2} = \frac{(n+1)[2(n+1)^2 - (3n+2)]}{2}$$

$$= \frac{(n+1)(2n^2 + n)}{2} = \frac{n(n+1)(2n+1)}{2},$$

and (9.7.2) follows.

Formula (9.7.3) is proved in a similar manner (Exercise 27).

**Example 9.7.3.** Evaluate $\sum_{k=1}^{4}(2k^2 - 3k + 2)$ by use of the properties of sums.

*Solution.* We have

$$\sum_{k=1}^{4}(2k^2 - 3k + 2) = \sum_{k=1}^{4} 2k^2 - \sum_{k=1}^{4} 3k + \sum_{k=1}^{4} 2 \qquad \text{[by 3.]}$$

$$= 2\sum_{k=1}^{4} k^2 - 3\sum_{k=1}^{4} k + 8 \qquad \text{[by 2. and 1.]}$$

$$= 2 \cdot \frac{4 \cdot 5 \cdot 9}{6} - 3 \cdot \frac{4 \cdot 5}{2} + 8 \qquad \text{[by (9.7.2) and (9.7.1)]}$$

$$= 60 - 30 + 8 = 38.$$

This is the value obtained directly in Example 9.7.2.

At times it is convenient to make a *change of variable* or *change of index* in a sum. Suppose, for example, that we wish to evaluate $\sum_{i=1}^{7}(i-1)^2$. This suggests the use of (9.7.2), but the formula cannot be used directly. We may set $i - 1 = k$. Then as $i$ ranges from 1 to 7, $k$ ranges from 0 to 6, and we have

$$\sum_{i=1}^{7}(i-1)^2 = \sum_{k=0}^{6} k^2 = \sum_{k=1}^{6} k^2 \qquad \text{[since } 0^2 = 0\text{]}$$

$$= \frac{6 \cdot 7 \cdot 13}{6} = 91. \qquad \text{[by (9.7.2)]}$$

More generally, we may write

(9.7.4) $$\sum_{k=m}^{n} f(k) = \sum_{k=m-r}^{n-r} f(k+r).$$

For, by setting $k = i + r$ on the left we obtain

$$\sum_{k=m}^{n} f(k) = \sum_{i=m-r}^{n-r} f(i+r);$$

in the latter sum the dummy index $i$ may be replaced by $k$, and (9.7.4) is obtained. We have the following rule.

We may increase (decrease) the index or variable in the summand of a summation by an integer $r$, if we decrease (increase) both limits by $r$.

**Example 9.7.4.** Evaluate $\sum_{k=3}^{9} 2(k+1)^2$.

*Solution.* Using the rule just stated, we write

$$\sum_{k=3}^{9} 2(k+1)^2 = \sum_{k=4}^{10} 2k^2 = 2\sum_{k=4}^{10} k^2 = 2\left(\sum_{k=1}^{10} k^2 - \sum_{k=1}^{3} k^2\right)$$

$$= 2\left(\frac{10 \cdot 11 \cdot 21}{6} - \frac{3 \cdot 4 \cdot 7}{6}\right) \qquad \text{[by (9.7.2)]}$$

$$= 2 \cdot 371 = 742.$$

### EXERCISE GROUP 9.5

In Exercises 1–12 evaluate the given sum directly.

1. $\sum_{i=1}^{5} (2i+1)^2$

2. $\sum_{i=1}^{7} (i-1)^2$

3. $\sum_{j=1}^{5} (1 - 3j - 2j^2)$

4. $\sum_{k=3}^{6} [(k+2)^2 - k^2]$

5. $\sum_{j=-3}^{3} \frac{1}{2j-1}$

6. $\sum_{k=1}^{4} \frac{1}{k(k+1)}$

7. $\sum_{i=1}^{3} \frac{i}{2i+1}$

8. $\sum_{j=1}^{5} \frac{j+1}{2j-1}$

9. $\sum_{k=1}^{4} 2^k$

10. $\sum_{k=2}^{4} 2^{-k}$

11. $\sum_{k=1}^{7} \sin(k\pi/2)$

12. $\sum_{k=1}^{7} \cos(k\pi/4)$

In Exercises 13–22 evaluate the given sum with the use of appropriate formulas.

13. $\sum_{k=1}^{20} (2k-3)$

14. $\sum_{i=3}^{9} i^2$

15. $\sum_{i=-3}^{5} i^2$

16. $\sum_{i=1}^{7} (2i+1)^2$

17. $\sum_{k=10}^{30} (4-3k)$

18. $\sum_{k=1}^{8} (4k - 3k^2)$

19. $\sum_{k=2}^{7} (k^2 + 2k)$

20. $\sum_{j=1}^{6} j^2(2j-3)$

21. $\sum_{k=3}^{7} k(k^2 + 3k + 3)$

22. $\sum_{k=3}^{8} (k+1)^3$

23. Evaluate (a) $\sum_{k=1}^{n} (2k+1)^2$, (b) $\sum_{k=1}^{n} (2k-1)^2$.

24. Prove that $\sum_{k=1}^{n} [(k+1)^r - k^r] = (n+1)^r - 1$.

25. Evaluate $\sum_{k=1}^{n} r^{k+1} - \sum_{k=1}^{n} r^k$.

26. Prove formula (9.7.1).

27. (a) Prove (9.7.3) by the same type of proof as given for (9.7.2). Start with the expansion of $(k+1)^4$.
    (b) Prove (9.7.3) by mathematical induction.

28. Prove that $\sum_{k=1}^{n} [f(k) - f(k-1)] = f(n) - f(0)$.

29. (a) Verify: $\dfrac{1}{k(k+1)} = \dfrac{1}{k} - \dfrac{1}{k+1}$.

    (b) Use the result in (a) to evaluate the sum $\sum_{k=1}^{n} \dfrac{1}{k(k+1)}$.

30. Prove that $\sum_{k=1}^{n} k^3 = \left( \sum_{k=1}^{n} k \right)^2$ by mathematical induction, without using the formula for $\sum_{k=1}^{n} k^3$. You may use the formula for $\sum_{k=1}^{n} k$.

31. Either derive the formula
$$\sum_{k=1}^{n} k^4 = \frac{n(n+1)(2n+1)(3n^2+3n-1)}{30},$$
or prove it by mathematical induction.

## 9.8 AN AREA PROBLEM

Determination of area in a plane provides a ready example of the kind of analysis that leads to sums of the type introduced in Section 9.6.

In previous work you have found the area of a triangle, the area of a plane polygon, and the area of a circle. Beyond that you probably did not go. We plan to consider more general figures and to define area in a way that is consistent with previous experience. It is not surprising that limiting processes are involved; on close examination it can be seen that such processes are used in deriving the formulas for area even of the simple figures mentioned above. The remainder of this section should be viewed as illustrating a general procedure, as we find the area of a specific figure.

Part of the graph of $y = x^2 + 1$ appears in Fig. 9.8.1. Suppose we wish to find the area of the portion of the plane that is bounded by the curve and the lines $x = 0$, $x = 1$, and $y = 0$. Such a figure is called a *region*, but we omit a formal definition at this time. This is a type of region for which area has not previously been found or even defined. We shall lead up to a definition which is governed, in part, by properties we can reasonably expect area to have. Two rectangles with base on the $x$ axis are shown in Fig. 9.8.1. Since the region includes the smaller rectangle and is included in the larger rectangle, we certainly expect the area $A$ of the region to have a value between the areas of the rectangles; that is, $1 \leq A \leq 2$. We have at once a rough estimate of the area.

We now separate the region into two parts by drawing the line $x = 1/2$ as in Fig. 9.8.2, and do the same kind of thing as above for both parts. If $A_1$ is

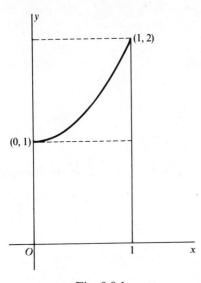
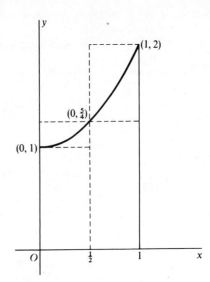

Fig. 9.8.1  Fig. 9.8.2

the area of the part of the region between $x = 0$ and $x = 1/2$, and $A_2$ is the area of the remaining part, we have

$$1 \cdot \frac{1}{2} \leq A_1 \leq \frac{5}{4} \cdot \frac{1}{2}, \quad \frac{5}{4} \cdot \frac{1}{2} \leq A_2 \leq 2 \cdot \frac{1}{2},$$

and hence, by addition, since $A = A_1 + A_2$,

$$1 \cdot \frac{1}{2} + \frac{5}{4} \cdot \frac{1}{2} \leq A \leq \frac{5}{4} \cdot \frac{1}{2} + 2 \cdot \frac{1}{2}.$$

This gives a better estimate for $A$, $9/8 \leq A \leq 13/8$.

We repeat the process again, separating the region into four parts (Fig. 9.8.3). The area of the portion of the region lying above each subinterval of length $1/4$ on the $x$ axis lies between the areas of the two rectangles formed. For the total area $A$ we then have

(9.8.1) $$1 \cdot \frac{1}{4} + \frac{17}{16} \cdot \frac{1}{4} + \frac{5}{4} \cdot \frac{1}{4} + \frac{25}{16} \cdot \frac{1}{4} \leq A \leq \frac{17}{16} \cdot \frac{1}{4} + \frac{5}{4} \cdot \frac{1}{4} + \frac{25}{16} \cdot \frac{1}{4} + 2 \cdot \frac{1}{4},$$

which gives $39/32 \leq A \leq 47/32$. This is the best estimate yet for the area.

We can expect that better and better estimates of the area can be obtained as the process is continued. We now carry out the process in a way that will apply more generally. We subdivide the interval $[0, 1]$ into $n$ equal parts by using the values (see Fig. 9.8.4)

$$0, \frac{1}{n}, \frac{2}{n}, \ldots, \frac{n-1}{n}, \frac{n}{n} = 1.$$

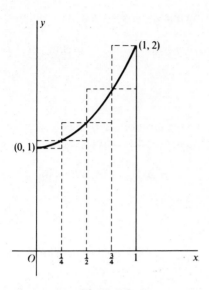

Fig. 9.8.3

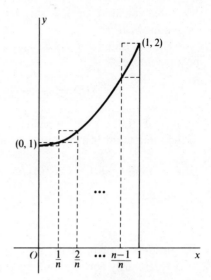

Fig. 9.8.4

The length of each subinterval is $1/n$. The $i$th subinterval is $[(i-1)/n, i/n]$. If $A_i$ is the area of the region above the $i$th subinterval, we have

$$\left[\left(\frac{i-1}{n}\right)^2 + 1\right] \cdot \frac{1}{n} \leq A_i \leq \left[\left(\frac{i}{n}\right)^2 + 1\right] \cdot \frac{1}{n}, \quad i = 1, 2, \ldots, n,$$

and by addition of these inequalities we obtain, since $A = \sum_{i=1}^{n} A_i$,

(9.8.2) $$\sum_{i=1}^{n}\left[\left(\frac{i-1}{n}\right)^2 + 1\right] \cdot \frac{1}{n} \leq A \leq \sum_{i=1}^{n}\left[\left(\frac{i}{n}\right)^2 + 1\right] \cdot \frac{1}{n}.$$

The left side of (9.8.2) becomes

$$\frac{1}{n}\sum_{i=1}^{n}\left[\left(\frac{i-1}{n}\right)^2 + 1\right] = \frac{1}{n^3}\sum_{i=1}^{n}(i-1)^2 + \frac{1}{n}\sum_{i=1}^{n}1 = \frac{1}{n^3}\sum_{i=1}^{n}(i-1)^2 + 1.$$

By a change of index the last sum becomes $\sum_{k=0}^{n-1} k^2$; since the summand is zero for $k = 0$, this sum equals $\sum_{k=1}^{n-1} k^2$, and its value may be obtained from (9.7.2) by replacing $n$ by $n - 1$. The left side of (9.8.2) then becomes

$$\frac{(n-1)n(2n-1)}{6n^3} + 1 = \frac{8n^2 - 3n + 1}{6n^2} = \frac{4}{3} - \frac{1}{2n} + \frac{1}{6n^2}.$$

The right side of (9.8.2) is evaluated in a similar manner and (9.8.2) itself becomes

(9.8.3) $$\frac{4}{3} - \frac{1}{2n} + \frac{1}{6n^2} \leq A \leq \frac{4}{3} + \frac{1}{2n} + \frac{1}{6n^2}.$$

Note that substitution of $n = 1, 2, 4$ in this inequality gives the same estimates as obtained earlier.

The left member of (9.8.3) is a function of $n$, with domain consisting of the positive integers,

$$f(n) = \frac{4}{3} - \frac{1}{2n} + \frac{1}{6n^2}, \qquad n = 1, 2, 3, \ldots.$$

Such a function is called a *sequence*. We define the limit of a sequence as $n \to \infty$ in the same manner as the limit of a function as $x \to \infty$ (Section 6.3).

**DEFINITION 9.8.1.** *Let $f(n)$ be defined for $n = 1, 2, 3, \ldots$ Then $L = \lim_{n \to \infty} f(n)$ if, given $\epsilon > 0$, a positive integer $N$ exists such that*

$$|f(n) - L| < \epsilon \quad \text{whenever} \quad n > N.$$

Properties of limits of sequences are entirely analogous to properties of limits previously considered, and will be assumed. In particular,

$$\lim_{n \to \infty} (1/n) = 0, \qquad \lim_{n \to \infty} (1/n^2) = 0,$$

and for $f(n)$ as above we have $\lim_{n \to \infty} f(n) = 4/3$. Similarly, for the right member of (9.8.3) we have

$$\lim_{n \to \infty} \left( \frac{4}{3} + \frac{1}{2n} + \frac{1}{6n^2} \right) = \frac{4}{3}.$$

Thus from (9.8.3) we see that $A$ lies between two functions which approach the same value as $n \to \infty$, and this can be so only if $A = 4/3$.

We have accomplished the following. Using properties that we require area to possess, we have shown that the area of the region bounded by the curve $y = x^2 + 1$ and the lines $x = 0$, $x = 1$, $y = 0$ can have only the value $4/3$. *This value is then taken as the desired area.*

We emphasize that the sums appearing in (9.8.2) are of the type expressed in (9.6.1). *We chose equally spaced values $x_i = i/n$, and in each interval $[x_{i-1}, x_i]$ we chose a value $w_i$.* For the left sum in (9.8.2), we chose $w_i = x_{i-1}$. Then $f(w_i)$ was found for the function $f(x) = x^2 + 1$, and the sum (9.6.1) expressed the sum of the areas of the smaller rectangles. The right sum in (9.8.2), which expresses the sum of the areas of the larger rectangles, is also a sum of the type (9.6.1), where $w_i = x_i$. For each sum the limit as $n \to \infty$ is the desired area.

It is also true that different choices of $w_i$, and different values of $x_i$, will lead to sums whose limits as $n \to \infty$ are the same value, $4/3$ (see Exercise 12).

**EXERCISE GROUP 9.6**

In each of Exercises 1–4 estimate the area of the region bounded by the given curves, by using a subdivision of the indicated number of equal parts and choosing the smallest abscissa in each subinterval.

1. $y = 1/x$, $x = 1$, $x = 4$, $y = 0$;  $n = 3$
2. $y = (2x - 1)/(2x + 1)$, $x = 1$, $x = 5$, $y = 0$;  $n = 4$
3. $y = x^2$, $x = -2$, $x = 1$, $y = 0$;  $n = 6$
4. $y = \sqrt{x}$, $x = 2$, $y = 0$;  $n = 4$

For each of the functions in Exercises 5–10 find in simplified form the sum of the type (9.6.1), in the interval [0, 1], (a) using $w_i = x_{i-1} = (i-1)/n$, (b) using $w_i = x_i = i/n$. (c) Using the results of (a) and (b), find the area of the region bounded by the graph of the function and the lines $x = 0$, $x = 1$, $y = 0$.

5. $x$
6. $2x + 3$
7. $3x + 2$
8. $x^2$
9. $x^2 + x$
10. $x^3$

11. Verify that (9.8.3), with $n = 4$, is the same as (9.8.1).

12. Express the sum (9.6.1), for the function $x^2 + 1$ in the interval [0, 1], in the same manner as in Section 9.8, using the midpoint of the interval $[(i-1)/n, i/n]$; that is, $w_i = (2i-1)/(2n)$. Simplify the result by use of the formulas of Section 9.7.

## 9.9 THE DEFINITE INTEGRAL

In Section 9.6 we introduced certain sums (9.6.1). In Section 9.8 we showed that in some cases at least the finding of areas could be related to such sums. *However, the use of sums like (9.6.1) is independent of the concept of area.* We wish to define the appropriate ideas in a rather general way; these will lead to a variety of applications, of which finding of areas is one.

Let $f$ be a function defined in the closed interval $[a, b]$. We choose a set of numbers $x_i$, $i = 0, 1, \ldots, n$, called a *subdivision* of $[a, b]$, and denoted by $\Delta$, such that

$$a = x_0 < x_1 < \cdots < x_{n-1} < x_n = b.$$

The $i$th *subinterval* is $[x_{i-1}, x_i]$, with length denoted by $\Delta x_i$,

$$\Delta x_i = x_i - x_{i-1}, \qquad i = 1, 2, \ldots, n.$$

The largest of the numbers $\Delta x_i$ is called the *norm of* $\Delta$, and is denoted by $\|\Delta\|$. Choose an *arbitrary* number $w_i$ in the $i$th subinterval for each $i$,

$$x_{i-1} \leq w_i \leq x_i, \qquad i = 1, 2, \ldots, n,$$

and form the sum

$$(9.9.1) \qquad \sum_{i=1}^{n} f(w_i) \Delta x_i = f(w_1)\Delta x_1 + f(w_2)\Delta x_2 + \cdots + f(w_n)\Delta x_n.$$

Such sums are called *Riemann sums*. The concept we have been leading up to, the *definite integral*, will now be defined.

**DEFINITION 9.9.1.** *Let $f$ be a function defined in an interval $[a, b]$. Let $\Delta$ denote a subdivision of the interval $[a, b]$, with norm $\|\Delta\|$, and let $w_i$ denote any number such that $x_{i-1} \leq w_i \leq x_i$, $i = 1, 2, \ldots, n$. Suppose that a number $A$ exists with the following property: given any number $\epsilon > 0$ a number $\delta$ exists such that*

$$\left| \sum_{i=1}^{n} f(w_i) \Delta x_i - A \right| < \epsilon \quad \text{whenever} \quad \|\Delta\| < \delta.$$

*Then $f$ is said to be* **integrable** *in $[a, b]$; the number $A$ is called the* **definite integral** *of $f$ in $[a, b]$; and we write*

$$A = \int_a^b f(x)\, dx.$$

As was pointed out earlier, this is a rather elaborate definition, and it is not feasible to apply it directly except in the most simple cases. Even for functions where the definite integral is known to exist, the definition does not afford a convenient means of evaluation. Fortunately, there are properties of the definite integral to be developed, which reduce the problem of evaluation to manageable proportions in many instances that are of considerable importance.

Definition 9.9.1 applies to an interval $[a, b]$, where $a < b$. We extend this definition to cover other cases.

**DEFINITION 9.9.2.** (a) $\int_a^a f(x)\, dx = 0.$

(b) *If $f$ is integrable in an interval $[a, b]$, then*

$$\int_b^a f(x)\, dx = -\int_a^b f(x)\, dx.$$

This definition permits us to write, for example,

$$\int_4^2 f(x)\, dx = -\int_2^4 f(x)\, dx.$$

Our work with definite integrals will mainly concern continuous functions, for which the following, straightforward criterion applies.

**THEOREM 9.9.1.** *If $f$ is a continuous function in $[a, b]$, then $f$ is integrable in $[a, b]$.*

The proof of this basic theorem is quite abstract, and is omitted.

Definition 9.9.1 for a definite integral is phrased in terms of *all* Riemann sums of a specified type. In practice, it is often convenient to relate a definite integral to a *sequence* of Riemann sums, for then the relation may be expressed in the form of a limit.

**THEOREM 9.9.2.** *Let $R_k$, $k = 1, 2, \ldots$, be a sequence of Riemann sums for a continuous function $f$ in an interval $[a, b]$, where each $R_k$ has associated with it a specific subdivision $\Delta_k$ and a specific set of numbers $w_i$. If $\lim_{k \to \infty} \|\Delta_k\| = 0$, then*

$$\lim_{k \to 0} R_k = \int_a^b f(x)\, dx.$$

*Proof.* According to Definition 9.8.1 we must show that, given $\epsilon > 0$, there exists an integer $N$ such that

(9.9.2) $$\left| R_k - \int_a^b f(x)\, dx \right| < \epsilon, \quad \text{whenever} \quad k > N.$$

By Theorem 9.9.1 the definite integral of $f$ exists in $[a, b]$. Hence, given $\epsilon > 0$, there exists, by Definition 9.9.1, a number $\delta > 0$ such that

(9.9.3) $$\left| \sum_{i=1}^n f(w_i)\Delta x_i - \int_a^b f(x)\, dx \right| < \epsilon$$

for any subdivision $\Delta$, for which $\|\Delta\| < \delta$, and any choice of $w_i$. For this $\delta$ there exists, since $\lim_{k \to \infty} \|\Delta_k\| = 0$, an integer $N$ such that

$$\|\Delta_k\| < \delta, \quad \text{whenever} \quad k > N.$$

Hence, for any $k > N$, the norm of $\Delta_k$ is less than $\delta$; inequality (9.9.3) applies with the corresponding $R_k$ so that (9.9.2) holds, and the proof is complete.

**Example 9.9.1.** Prove that $\int_a^b dx = b - a$.

*Proof.* The integrand is the function $f(x) = 1$. For any subdivision of the interval we have

$$R_n = \sum_{i=1}^n f(w_i)(x_i - x_{i-1}) = \sum_{i=1}^n (x_i - x_{i-1}) = x_n - x_0 = b - a.$$

Hence $R_n$ is constant for all $n$, and

$$\int_a^b dx = \lim_{n \to \infty} R_n = b - a.$$

**Example 9.9.2.** In Section 9.8 we have, in effect, established the following integral,

$$\int_0^1 (x^2 + 1)\, dx = \frac{4}{3};$$

for the function $x^2 + 1$ is continuous in $[0, 1]$, and hence its integral exists. In (9.8.3) the left member is a sequence of Riemann sums whose norms approach zero as $n \to \infty$, and the limit of these Riemann sums is 4/3.

The following theorem contains some of the important properties of definite integrals, proofs of which may be based on Theorem 9.9.2. We shall prove one of the properties. All of these correspond to properties that hold for sums.

**THEOREM 9.9.3.** 1. *If $f$ is continuous in $[a, b]$, and $k$ is constant, then*

$$\int_a^b kf(x)\,dx = k\int_a^b f(x)\,dx.$$

2. *If $f$ and $g$ are both continuous in $[a, b]$, then*

$$\int_a^b [f(x) + g(x)]\,dx = \int_a^b f(x)\,dx + \int_a^b g(x)\,dx.$$

3. *If $f$ is continuous in an interval containing $a$, $b$, $c$, then*

$$\int_a^b f(x)\,dx = \int_a^c f(x)\,dx + \int_c^b f(x)\,dx.$$

4. *If $f$ and $g$ are continuous and $f(x) \le g(x)$ in $[a, b]$, then*

$$\int_a^b f(x)\,dx \le \int_a^b g(x)\,dx.$$

*Proof* of 3. We shall prove this property for the case $a < c < b$. Other cases can be proved on the basis of this case and Definition 9.9.2 (Exercise 13). We know that each integral exists and that each one is a limit of appropriate Riemann sums (Theorem 9.9.2). Consider a subdivision of $[a, b]$ which includes $c$ (Fig. 9.9.1),

$$a = x_0 < x_1 < \cdots < x_m = c < x_{m+1} < \cdots < x_n = b.$$

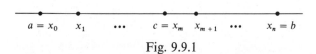

Fig. 9.9.1

Then $x_0, x_1, \ldots, x_m$ form a subdivision of $[a, c]$, and $x_m, x_{m+1}, \ldots, x_n$ form a subdivision of $[c, b]$. Choose a value $w_i$ in each subinterval. Then

(9.9.4) $$\sum_{i=1}^n f(w_i)\Delta x_i = \sum_{i=1}^m f(w_i)\Delta x_i + \sum_{i=m+1}^n f(w_i)\Delta x_i.$$

Repeat this process for a sequence of subdivisions $\Delta_k$ of $[a, b]$, each subdivision including $c$, and chosen in such a way that the norms approach zero. Then as $k \to \infty$ each sequence of Riemann sums in (9.9.4) approaches the corresponding definite integral, and part 3 is established.

*Note.* Theorems 9.9.2 and 9.9.3 can be stated in a more general way so as to apply to integrable functions. We shall need to use them, however, mainly for continuous functions.

With the use of Theorem 9.9.3, part 4 and Example 9.9.1 we are in a position to prove an important result which is particularly useful in proving other results.

**THEOREM 9.9.4. Law of the Mean for Integrals.** *Let $f$ be continuous in $[a, b]$. There then exists a number $w$, $a < w < b$, such that*

$$\int_a^b f(x)\, dx = f(w)(b - a).$$

*Proof.* By Theorem 3.7.1 there exist numbers $s$ and $t$ in $[a, b]$ such that $f(s)$ is a minimum and $f(t)$ is a maximum; that is

$$f(s) \leq f(x) \leq f(t), \qquad x \text{ in } [a, b].$$

Then, by Theorem 9.9.3, part 4,

$$\int_a^b f(s)\, dx \leq \int_a^b f(x)\, dx \leq \int_a^b f(t)\, dx.$$

But $f(s)$ and $f(t)$ do not depend on $x$; hence by part 1 of Theorem 9.9.3, we may write

(9.9.5) $$f(s) \int_a^b dx \leq \int_a^b f(x)\, dx \leq f(t) \int_a^b dx.$$

By Example 9.9.1, $\int_a^b dx = b - a$, and we obtain from (9.9.5),

$$f(s) \leq \frac{1}{b - a} \int_a^b f(x)\, dx \leq f(t).$$

The middle member of this inequality is thus a number between $f(s)$ and $f(t)$. By the Intermediate Value Theorem (Theorem 3.7.2) it follows that a number $w$, $a < w < b$, must exist such that

$$f(w) = \frac{1}{b - a} \int_a^b f(x)\, dx,$$

and the theorem follows.

Note that nothing in the statement of Theorem 9.9.4 precludes the existence of more than one value with the stated property. In general, one should not read into the statement of any theorem more than is actually stated.

*Example 9.9.3.* Find a number $w$ as provided in Theorem 9.9.4 for the definite integral in Example 9.9.2.

*Solution.* In that example we have $f(x) = x^2 + 1$, $a = 0$, $b = 1$. The result of Theorem 9.9.4 becomes, since $b - a = 1$,

$$\tfrac{4}{3} = w^2 + 1.$$

There are two solutions, $\pm\sqrt{3}/3$. But $w$ must be in the interval $[0, 1]$. Thus the desired value is $w = \sqrt{3}/3$.

In a broad sense the application of the definite integral to finding a certain quantity depends on approximating the quantity for a small part of some configuration by an expression of the form $f(w_i)\Delta x_i$; then the quantity for the entire configuration is approximated by a sum of the form $\sum f(w_i)\Delta x_i$, and the quantity itself can be expressed as a definite integral.

**EXERCISE GROUP 9.7**

1. Obtain the result $\int_a^b x\, dx = (b^2 - a^2)/2$, by using Riemann sums.

* 2. Obtain the result $\int_a^b x^2\, dx = (b^3 - a^3)/3$, by using Riemann sums.

In Exercises 3–6 use the result of Exercise 2 to find $w$ as provided in Theorem 9.9.4, the Law of the Mean for Integrals, for $f(x) = x^2$, in the specified interval.

3. $a = -1, b = 1$
4. $a = 0, b = 1$
5. $a = 1, b = 4$
6. $a = -1, b = 2$

7. (a) By using the results in Exercises 1 and 2, find the value of $\int_1^2 (2x - x^2)\, dx$.

    (b) Find $w$ as provided in the Law of the Mean for Integrals for the integral in (a).

8. Given $\int_a^b x^3\, dx = (b^4 - a^4)/4$, find $w$ as provided in the Law of the Mean for Integrals for $f(x) = x^3$, in each of the following cases.

    (a) $a = -1, b = 0$  (b) $a = -1, b = 3$  (c) $a = -2, b = 1$

9. (a) Evaluate $\int_0^2 (2x - x^3)\, dx$.

    (b) Find $w$ as provided in the Law of the Mean for Integrals for the integral in (a).

10. Prove Theorem 9.9.3, Part 1.
11. Prove Theorem 9.9.3, Part 2.
12. Prove Theorem 9.9.3, Part 4.
13. Prove Theorem 9.9.3, Part 3, in each of the following cases.

    (a) $c < a < b$  (b) $a < b < c$  (c) $c < a \leq b$

14. Prove: $\int_a^{x+h} f(u)\, du - \int_a^x f(u)\, du = \int_x^{x+h} f(u)\, du$.

## 9.10 FUNDAMENTAL THEOREM OF INTEGRAL CALCULUS

Recent sections have developed the concept and some properties of the definite integral. These were of a general nature, but were not especially directed to the evaluation of definite integrals. In the present section we will develop a basic result which is of immediate use in the evaluation of definite integrals; at the same time this result is of fundamental significance in that it connects the previously unconnected concepts of the definite integral and the derivative.

If $f$ is continuous in $[a, b]$ we know that the definite integral of $f$ exists in $[a, b]$. The use of a particular letter for the variable is of no real significance. Thus

$$\int_a^b f(x)\,dx, \quad \int_a^b f(t)\,dt, \quad \int_a^b f(u)\,du$$

all have the same value, depending only on the values of the function $f$ in $[a, b]$. In fact, the notation $\int_a^b f$, in which the variable is not indicated, is also valid. In particular, the definite $\int_a^x f(u)\,du$ is a function of the upper limit $x$, provided $f$ is integrable in the interval $[a, x]$, and we may write in this case

$$F(x) = \int_a^x f(u)\,du.$$

This function is treated in the following result.

**THEOREM 9.10.1.** *If $f$ is continuous in $[a, b]$, then the function $F$, defined as $F(x) = \int_a^x f(u)\,du$, is differentiable for any $x$ in $[a, b]$, and*

$$F'(x) = f(x).$$

*In words, $F$ is an antiderivative of $f$.*

*Proof.* We revert to the definition of derivative. If $x$ and $x + h$ are in $[a, b]$ we have (see Exercise 14 of Exercise Group 9.7)

$$F(x + h) - F(x) = \int_a^{x+h} f(u)\,du - \int_a^x f(u)\,du = \int_x^{x+h} f(u)\,du.$$

Hence, by the Law of the Mean for Integrals (Theorem 9.9.4), applied to the integral on the right, we obtain

(9.10.1) $\qquad F(x + h) - F(x) = hf(t),$ for some $t$ between $x$ and $x + h$.

As $h \to 0$ we must have $t \to x$, and hence $f(t) \to f(x)$ by the continuity of $f$; thus

$$F'(x) = \lim_{h \to 0} \frac{F(x + h) - F(x)}{h} = \lim_{t \to x} f(t) = f(x),$$

and the result is established.

**COROLLARY.** *If $f$ is continuous in $[a, b]$, then $\int_a^x f(u)\,du$ is continuous for $x$ in $[a, b]$.*

This follows immediately from Theorem 4.2.1 since the integral is differentiable (Theorem 9.10.1). It also follows from (9.10.1) if we let $h \to 0$.

*Example 9.10.1.* Let us define $F$ as

$$F(x) = \int_1^x du.$$

Then, by Theorem 9.10.1, we have $F'(x) = 1$. This may also be verified directly, for from the result of Example 9.9.1 we have

$$F(x) = \int_1^x du = x - 1.$$

We are now ready for the main result.

**THEOREM 9.10.2.** *The Fundamental Theorem of Integral Calculus. Let $f$ be continuous in $[a, b]$, and let $g$ be any antiderivative of $f$ ($g$ exists by Theorem 9.10.1). Then*

$$\int_a^b f(x) = g(b) - g(a).$$

*Proof.* Let $F(x) = \int_a^x f(u)\,du$. Then, by Theorem 9.10.1, $F'(x) = f(x)$. Thus both $F(x)$ and $g(x)$ are antiderivatives of the same function $f$, and there must exist, by Theorem 9.3.1, a constant $K$ such that

$$F(x) = g(x) + K, \quad x \text{ in } [a, b].$$

In particular, when $x = a$ this equation yields $K = F(a) - g(a)$, and since $F(a) = 0$, we get $K = -g(a)$, $F(x) = g(x) - g(a)$, and $F(b) = g(b) - g(a)$. We then have

$$\int_a^b f(x)\,dx = \int_a^b f(u)\,du = F(b) = g(b) - g(a),$$

and the proof is complete.

The significance of Theorem 9.10.2 is that it provides a link between the concepts of definite integral and antiderivative. If an antiderivative of $f$ can be found, then $\int_a^b f(x)\,dx$ can be found at once. Thus the earlier methods for finding antiderivatives may be utilized, as well as others to be developed later.

A notation that is commonly used for $g(b) - g(a)$ is

$$g(b) - g(a) = g(x)\Big|_a^b.$$

Thus, if $g$ is an antiderivative of $f$, we may write

$$\int_a^b f(x)\,dx = g(x)\Big|_a^b = g(b) - g(a).$$

**Example 9.10.2.** Evaluate $\int_2^4 \sqrt{x-1}\,dx$.

*Solution.* By earlier methods we find $2(x-1)^{3/2}/3$ as an antiderivative of $\sqrt{x-1}$; hence

$$\int_2^4 \sqrt{x-1}\,dx = \frac{2}{3}(x-1)^{3/2}\Big|_2^4 = \frac{2}{3}\cdot 3^{3/2} - \frac{2}{3}\cdot 1^{3/2} = 2\sqrt{3} - \frac{2}{3}.$$

**Example 9.10.3.** Evaluate $\int_2^4 x\sqrt{x^2-2}\,dx$.

**Solution.** We make a change of variable in the indefinite integral, $u = x^2 - 2$, $du = 2x\,dx$. Then

$$\int x\sqrt{x^2-2}\,dx = \frac{1}{2}\int u^{1/2}\,du = \frac{u^{3/2}}{3} + C = \frac{(x^2-2)^{3/2}}{3} + C.$$

Using one antiderivative, with $C = 0$, we have

$$\int_2^4 x\sqrt{x^2-2}\,dx = \frac{(x^2-2)^{3/2}}{3}\bigg|_2^4 = \frac{14^{3/2}}{3} - \frac{2^{3/2}}{3} = \frac{14\sqrt{14}-2\sqrt{2}}{3}.$$

Note the feature of Theorem 9.10.2 that $g$ is *any* antiderivative of $f$. *Different antiderivatives of $f$ lead to the same value of the definite integral.* (See Exercise 26.)

### EXERCISE GROUP 9.8

Evaluate the definite integrals in Exercises 1–20.

1. $\int_1^3 (2x^2 - 3x + 2)\,dx$

2. $\int_1^4 (2x - x^3)\,dx$

3. $\int_0^8 \sqrt{2u}\,du$

4. $\int_{1/3}^3 \sqrt[3]{3u}\,du$

5. $\int_{-1}^2 (y-1)^2\,dy$

6. $\int_{2a}^{4a} \sqrt{x-a}\,dx$

7. $\int_1^4 (x^{1/2} + 1)^2\,dx$

8. $\int_{2a}^{4a} u\sqrt{u^2-a^2}\,du$

9. $\int_1^9 \frac{dx}{\sqrt{x}}$

10. $\int_{-7}^0 \frac{dx}{\sqrt{1-x}}$

11. $\int_1^8 \frac{dx}{x^{2/3}}$

12. $\int_{1/8}^1 \frac{dx}{x^{4/3}}$

13. $\int_0^{a/2} \frac{dx}{\sqrt{(a-x)^3}}$

14. $\int_0^{a/2} \frac{dx}{(a-x)^{1/3}}$

15. $\int_0^{a/3} \frac{t\,dt}{\sqrt{a^2-t^2}}$

16. $\int_0^{a/2} \frac{x\,dx}{(a^2-x^2)^{3/2}}$

17. $\int_3^5 \frac{x\,dx}{\sqrt{x^2-4}}$

18. $\int_{1/4}^{3/4} \frac{2x-1}{(x^2-x)^2}\,dx$

19. $\int_a^b [f(x)]^n f'(x)\,dx$

20. $\int_a^b \sqrt{1+f(x)}f'(x)\,dx$

21. Evaluate $\lim_{t\to 0+} \int_t^1 \frac{dx}{x^{2/3}}$

22. Evaluate $\lim_{t\to a-} \int_0^t \frac{dx}{(a-x)^{1/3}}$

23. Find a value of $w$ as provided in the Law of the Mean for Integrals for the integral in Exercise 11.

24. Find a value of $w$ as provided in the Law of the Mean for Integrals for the integral in Exercise 6.

* 25. Find a value of $w$ as provided in the Law of the Mean for Integrals for the integral in Exercise 17.

26. If $r(x)$ and $s(x)$ are antiderivatives of the same function, show that
$$r(x)\Big|_a^b = s(x)\Big|_a^b.$$

27. Prove that if $F'(x) = f(x)$ and $F(a) = 0$, then $F(x) = \int_a^x f(t)\,dt$.

28. If $F(x) = x \int_0^x f(t)\,dt$, prove that $xF'(x) - F(x) = x^2 f(x)$.

29. If $F(x) = x \int_1^x \dfrac{dt}{t} + 2x$, $x > 0$, prove that $xF'(x) - F(x) = x$.

## 9.11 CHANGE OF VARIABLE IN A DEFINITE INTEGRAL

In Example 9.10.3 we used a change of variable to find an antiderivative of a given function so that Theorem 9.10.2 could be used to evaluate a definite integral. A change of variable may also be used directly in a definite integral, according to the following statement.

**THEOREM 9.11.1.** *Let $g$ be a function such that $g'$ is continuous in $[a, b]$. Let $f$ be continuous in the range of $g$. Then*
$$\int_a^b f[g(u)]g'(u)\,du = \int_{g(a)}^{g(b)} f(x)\,dx.$$

*Proof.* Both integrands are continuous, and both integrals exist. Let $F$ be an antiderivative of $f$; that is, $F'(x) = f(x)$. Then $F[g(u)]$ is an antiderivative of $f[g(u)]g'(u)$, for, using the chain rule, we have
$$D_u F[g(u)] = F'[g(u)]g'(u) = f[g(u)]g'(u).$$
By Theorem 9.10.1 we may write
$$\int_a^b f[g(u)]g'(u)\,du = F[g(u)]\Big|_a^b;$$
$$\int_{g(a)}^{g(b)} f(x)\,dx = F(x)\Big|_{g(a)}^{g(b)}.$$
The expressions on the right are equal, since each is equal to
$$F[g(b)] - F[g(a)],$$
and the proof is complete.

In practice the effect of the theorem is that we may make a change of variable $x = g(u)$ in a definite integral, change the limits of integration accordingly, and evaluate the definite integral in terms of the new variable. We shall carry out this process for Example 9.10.3.

**Example 9.11.1.** Evaluate $\int_2^4 x\sqrt{x^2-2}\,dx$.

*Solution.* We set $u = x^2 - 2$, $du = 2x\,dx$. This is equivalent to setting $x = \sqrt{u+2}$. For $x = 2$, we have $u = 2$, and for $x = 4$, $u = 14$. Then,

$$\int_2^4 x\sqrt{x^2-2}\,dx = \frac{1}{2}\int_2^{14} u^{1/2}\,du = \frac{u^{3/2}}{3}\bigg|_2^{14} = \frac{14^{3/2}}{3} - \frac{2^{3/2}}{3}$$

$$= \frac{14\sqrt{14} - 2\sqrt{2}}{3}.$$

**Example 9.11.2.** Use the change of variable $x^2 + 4 = u$ to evaluate

$$\int_0^{\sqrt{2}} \frac{x^3}{\sqrt{x^2+4}}\,dx.$$

*Solution.* We have $2x\,dx = du$, and $u = 4$ for $x = 0$; while $u = 6$ for $x = \sqrt{2}$. Hence

$$\int_0^{\sqrt{2}} \frac{x^3}{\sqrt{x^2+4}}\,dx = \int_0^{\sqrt{2}} \frac{x^2 \cdot x\,dx}{\sqrt{x^2+4}} = \int_4^6 \frac{(u-4)(du/2)}{\sqrt{u}}$$

$$= \frac{1}{2}\int_4^6 (u^{1/2} - 4u^{-1/2})\,du = \frac{1}{2}\left(\frac{2}{3}u^{3/2} - 8u^{1/2}\right)\bigg|_4^6$$

$$= \frac{1}{2}\left(\frac{2}{3}\cdot 6\sqrt{6} - 8\sqrt{6}\right) - \frac{1}{2}\left(\frac{2}{3}\cdot 8 - 8\cdot 2\right)$$

$$= \frac{16}{3} - 2\sqrt{6}.$$

**EXERCISE GROUP 9.9**

Use a change of variable to evaluate the definite integrals in Exercise 1–12.

1. $\int_0^1 \dfrac{x\,dx}{\sqrt{4-x^2}}$

2. $\int_0^1 t(2-t^2)^{1/2}\,dt$

3. $\int_0^a x^3(a^4 - x^4)^{1/2}\,dx$

4. $\int_0^1 x^2(1+x^3)^3\,dx$

5. $\int_0^1 v^{1/2}(1-v^{3/2})^4\,dv$

6. $\int_2^3 \dfrac{x\,dx}{(x-1)^5}$

7. $\int_2^3 \dfrac{x^3}{(x-1)^5}\,dx$

8. $\int_{1.1}^2 \dfrac{t\,dt}{(t^2-1)^3}$

9. $\int_0^3 \dfrac{x\,dx}{(x+2)^3}$

10. $\int_0^4 (1+x)\sqrt{2x+x^2}\,dx$

11. $\int_2^4 \dfrac{1}{(2x-3)^2}\,dx$

12. $\int_0^1 \dfrac{\sqrt{u}}{\sqrt[3]{1+u^{3/2}}}\,du$

Evaluate each of the following definite integrals by use of the suggested change of variable.

13. $\int_{-3}^{0} \dfrac{x^2}{\sqrt{x+4}}\,dx;\ x+4=u$

14. $\int_{1}^{4} \dfrac{dt}{\sqrt{t}+t\sqrt{t}};\ t=u^2$

15. $\int_{0}^{1} \dfrac{u^3}{\sqrt{4-u^2}}\,du;\ 4-u^2=v$

16. $\int_{0}^{3\sqrt{3}} (1+t^{-2/3})^{1/2}\,dt;\ t=u^{3/2}$

17. $\int_{1}^{2} \left(x+\dfrac{1}{x}\right)^2 \dfrac{x^2-1}{x^2}\,dx;\ u=x+\dfrac{1}{x}$

18. $\int_{0}^{1} \dfrac{x^5}{\sqrt{x^3+3}}\,dx;\ u=x^3+3$

## 9.12 AREA OF A PLANE REGION

We shall evaluate areas in the plane as a first application of the definite integral. Let $f$ be a continuous, nonnegative function in an interval $[a, b]$; that is, $f(x) \geq 0$, $a \leq x \leq b$. We are interested in the area of the region of the plane which is bounded by the curve $y = f(x)$, the $x$ axis, and the lines $x = a$, $x = b$ (Fig. 9.12.1). We subdivide the interval $[a, b]$ as in Section 9.9. In the subinterval $[x_{i-1}, x_i]$ we now choose a number $s_i$ such that $f$ is a minimum, and a number $t_i$ such that $f$ is a maximum. If $A_i$ is the area of the part of the region between $x = x_{i-1}$ and $x = x_i$, then $A_i$ is between the areas of two rectangles of heights $f(s_i)$ and $f(t_i)$, and base $\Delta x_i = x_i - x_{i-1}$, that is,

$$f(s_i)\Delta x_i \leq A_i \leq f(t_i)\Delta x_i.$$

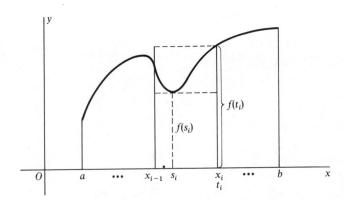

Fig. 9.12.1

Figure 9.12.1 shows a typical subinterval. If we do the same thing for each subinterval, we get

(9.12.1) $$\sum_{i=1}^{n} f(s_i)\Delta x_i \leq A = \sum_{i=1}^{n} A_i \leq \sum_{i=1}^{n} f(t_i)\Delta x_i.$$

Both sums in (9.12.1) are Riemann sums for the function $f$ in the interval $[a, b]$. If we choose a sequence of subdivisions $\Delta_k$ such that $\|\Delta_k\| \to 0$ as $k \to \infty$, each sum in (9.12.1) approaches the same limit, and we are led to the following definition.

**DEFINITION 9.12.1.** *The* **area** *of the region bounded by the graph of a non-negative continuous function $f$, the $x$ axis, and the lines $x = a$ and $x = b$, where $a < b$, is*

$$A = \int_a^b f(x)\, dx$$

*The region described is called the* **region under the curve** $y = f(x)$ *between $x = a$ and $x = b$.*

Let us again find the area treated in Section 9.8. In the present terminology we see that it is the area under the curve $y = x^2 + 1$ between $x = 0$ and $x = 1$. Hence

$$A = \int_0^1 (x^2 + 1)\, dx = \left(\frac{x^3}{3} + x\right)\Big|_0^1 = \frac{4}{3}.$$

The evaluation was done by use of an antiderivative.

**Example 9.12.1.** Find the area of the region bounded by the curve $y^2 = x - 1$ and the line $x = 5$.

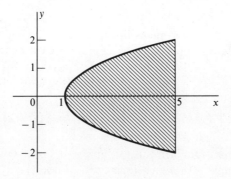

Fig. 9.12.2

*Solution.* From Fig. 9.12.2 we see that the region is symmetric in the $x$ axis, and hence the area is twice the area of that portion of the region above the $x$ axis, that is, twice the area under the curve $y = f(x) = \sqrt{x - 1}$ between $x = 1$ and $x = 5$. Note that $f(1) = 0$, which is permissible; we may have $f(a) = 0$, or $f(b) = 0$, or both. Hence, if $A$ is the desired area,

$$A = 2 \int_1^5 \sqrt{x - 1}\, dx = \frac{4}{3}(x - 1)^{3/2}\Big|_1^5 = \frac{4}{3} \cdot 4^{3/2} = \frac{32}{3}.$$

By analogy with Definition 9.12.1, it is clear that if $g$ is continuous and non-negative in $[c, d]$, then the area of the region bounded by the curve $x = g(y)$, the $y$ axis, and the lines $y = c$ and $y = d$, is defined as

$$A = \int_c^d g(y)\,dy.$$

**Example 9.12.2.** Find the area of the region bounded by the curve $y^2 = x - 1$, the $y$ axis, and the lines $y = -2$ and $y = 2$.

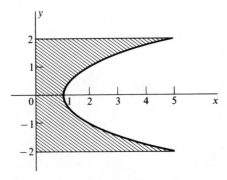

Fig. 9.12.3

*Solution.* The curve is the same as in Fig. 9.12.2. The region is now shown in Fig. 9.12.3. The desired area may be expressed as a definite integral of the function $x = g(y) = y^2 + 1$. We may integrate between $-2$ and $2$, or between $0$ and $2$ and make use of symmetry. We choose the latter. Thus

$$A = 2\int_0^2 (y^2 + 1)\,dy = 2\left(\frac{y^3}{3} + y\right)\bigg|_0^2 = 2\left(\frac{8}{3} + 2\right) = \frac{28}{3}.$$

(The sum of the areas found in this example and Example 9.12.1 should equal the area of a rectangle with dimensions $4 \times 5$; it does!)

The basic formula of Definition 9.12.1 may be extended to a region between two curves. In particular, consider the region (Fig. 9.12.4) bounded by $y = f(x)$, $y = g(x)$, $x = a$, and $x = b$, where $f(x) \geq g(x)$ in $[a, b]$.

The area of the indicated region is given by

(9.12.2) $$\int_a^b [f(x) - g(x)]\,dx.$$

*Proof.* Draw a line $y = r$ below the lowest point of the curve $y = g(x)$ in $[a, b]$; such a number $r$ exists since $g$ is assumed continuous in $[a, b]$ (see Theorem 3.7.1).

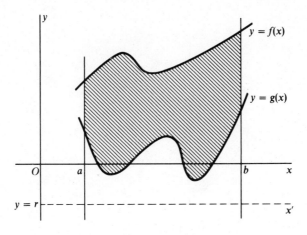

Fig. 9.12.4

Now think of that line and the $y$ axis as defining a coordinate system. Then for any point in the plane the coordinates relative to the $x'$, $y$ axes are $(x, y' - r)$, and equations of the two curves in the new coordinate system are

$$y = f(x) - r \quad \text{and} \quad y = g(x) - r.$$

Still thinking of the $x'$, $y$ axes, we see that the desired area is the area under the upper curve minus the area under the lower curve; hence

$$\text{Area} = \int_a^b [f(x) - r]\,dx - \int_a^b [g(x) - r]\,dx$$

$$= \int_a^b \{[f(x) - r] - [g(x) - r]\}\,dx \quad \text{[by Theorem 9.9.3, Part 3]}$$

$$= \int_a^b [f(x) - g(x)]\,dx.$$

This result is independent of the new coordinate system, which has served its purpose and is now discarded. This completes the proof. (We have assumed that area of a region between two curves behaves as expected.)

It is easy to state an analogous result relative to the $y$ axis, but this is left to the student.

In an area problem it is desirable to draw a figure and then choose one of the available procedures. In most cases there will be a choice of procedure; this choice must be made in the light of the experience gained as you work additional problems.

**Example 9.12.3.** Find the area of the region bounded by the curves $2x^2 = 2y + 1$ and $2y = x + 2$.

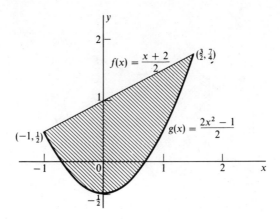

Fig. 9.12.5

*Solution.* The graph appears in Fig. 9.12.5. The described region is between the graphs of the given equations,

$$y = f(x) = \frac{x+2}{2} \quad \text{and} \quad y = g(x) = \frac{2x^2 - 1}{2},$$

a straight line and a parabola. We solve simultaneously the two equations to find the points of intersection.

$$2x^2 - x - 3 = 0, \quad (2x - 3)(x + 1) = 0, \quad x = -1, 3/2;$$

the ordinates of these points are not needed. We also obtain

$$f(x) - g(x) = \frac{x+2}{2} - \frac{2x^2 - 1}{2} = \frac{3}{2} + \frac{x}{2} - x^2.$$

Then the area in question is given by (9.12.2) as

$$A = \int_{-1}^{3/2} [f(x) - g(x)]\, dx = \int_{-1}^{3/2} \left(\frac{3}{2} + \frac{x}{2} - x^2\right) dx$$

$$= \left(\frac{3x}{2} + \frac{x^2}{4} - \frac{x^3}{3}\right)\Big|_{-1}^{3/2} = \frac{27}{16} + \frac{11}{12} = \frac{125}{48}.$$

**Example 9.12.4.** Find the area of the region bounded by the $x$ axis and the curve $y = (x - 1)(x^2 - 9)$, between $x = 1$ and $x = 3$.

*Solution.* The part of the curve for $x$ in $[1, 3]$ is shown in Fig. 9.12.6. Since $y \le 0$ for $x$ in $[1, 3]$, we may use (9.12.2) with

$$f(x) = 0, \quad g(x) = (x - 1)(x^2 - 9).$$

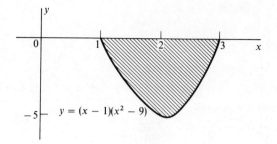

Fig. 9.12.6

The area is given by

$$A = \int_1^3 [-g(x)]\, dx = \int_1^3 (-x^3 + x^2 + 9x - 9)\, dx$$
$$= \left(-\frac{x^4}{4} + \frac{x^3}{3} + \frac{9x^2}{2} - 9x\right)\bigg|_1^3 = \frac{20}{3}.$$

In general, if $g(x) \leq 0$ for $x$ in $[a, b]$, the area of the region bounded by $x = a$, $x = b$, $y = 0$, and $y = g(x)$ is

$$A = -\int_a^b g(x)\, dx.$$

## EXERCISE GROUP 9.10

In Exercises 1–26 find the area of the indicated region.

1. Bounded by $y = 9 - x^2$ and the $x$ axis.
2. Bounded by $y = 2x^2 - 3x - 9$ and the $x$ axis.
3. Bounded by $y = \dfrac{1}{\sqrt{x-1}}$, $y = 0$, $x = 2$, and $x = 5$.
4. Bounded by $y = (x^4 - 4)(x + 1)$ and $y = 0$, above the $x$ axis.
5. Bounded by $y = x^2(3 - x)$ and $y = 0$.
6. Bounded by $y = 1 + x^2(3 - x)$, $x = -1$, $x = 3$, and $y = 0$.
7. Bounded by $y = x^3 - x$ and $y = 0$, above the $x$ axis.
8. Bounded by $y = 5 - x^2$ and $y = 1$.
9. Bounded by $y^2 = 2x + 4$ and $x = 0$.
10. Bounded by $x = 2y - y^2$ and the $y$ axis.
11. Bounded by $x^{1/2} + y^{1/2} = a^{1/2}$ and the coordinate axes.
12. Under the curve $y^3 = x^3(9 - x^2)$, in the interval $[0, 3]$.
13. Bounded by $xy^2 = 1$, $x = 1$, and $x = 9$.
14. Bounded by $y = (x + 1)^2(x - 2)^2$ and $y = 0$.

15. Bounded by $y = x^{3/2}$ and $y = x$.
16. Bounded by $y = 4 - x^2$ and $y = x + 2$.
17. Bounded by $y = -2x^2 + 7x$ and $y = 2x$.
18. Bounded by $x^2 = 4y + 1$ and $2x^2 + 2y - 7 = 0$.
19. Bounded by $x = y^2 - 4y + 3$ and $y = x + 1$.
★ 20. Bounded by $y^2 = x^3$ and $y = 3x - 4$.
★ 21. Bounded by $x^2 y = 2$ and $10x + 2y = 21$.
★ 22. Bounded by $y = x^3 - 5x^2 + 5x + 2$ and $y = -2x + 5$.
★ 23. Formed by the loop of the curve $y^2 = (x - 1)(x - 2)^2$.
★ 24. Formed by the loop of the curve $y^2 = x^4(1 - x^3)$.
★ 25. Formed by the loop of the curve $x^2 = 4y^4 - y^5$.
★ 26. Formed by one loop of the curve $4y^2 = x^2(4 - x^2)$.
27. (a) Find the area of the region bounded by $y = 1/(x + 1)^{3/2}$, $y = 0$, $x = 0$, and $x = t > 0$.
    (b) Find the limit of the area in (a) as $t \to \infty$.
28. A trough of length 10 ft has a cross section in the shape of the parabola $y = x^2$. If water is poured into the trough at a rate of 10 cu ft/min, how fast is the depth increasing when the water is 1/2 ft deep?
29. Prove, without evaluating any integrals, that the region bounded by the curves $y = x^2 + 1$, $y = x^2 - 2$, $x = 0$, and $x = 2$ has the same area as a rectangle of dimensions $2 \times 3$.
30. Prove, without evaluating any integrals, that the region bounded by the curves $y = x^2 - 4x + 3$ and $y = x - 1$ has the same area as the region bounded by the curves $y = -x^2 + 5x - 4$ and $y = 0$.

# 10 Trigonometric Functions and Their Inverses

## 10.1 INTRODUCTION

The specific differentiation formulas which have been previously obtained applied directly to certain *algebraic functions*. The class of algebraic functions contains the functions which are expressible in terms of roots of polynomials, but it contains many functions in addition. By implicit differentiation we were able to find derivatives of functions defined by algebraic equations; such functions are algebraic and cannot, in general, be expressed merely by roots of polynomials. Thus with our previous methods we are in a position to differentiate functions of a broad class.

We turn now to another broad class of very important functions which are not included in those mentioned above. We consider here the trigonometric and inverse trigonometric functions, and in the next chapter the exponential and logarithmic functions, all of which are part of the class of *transcendental functions*.

In general, an *algebraic function* is defined as any function $y = y(x)$ which satisfies an equation

$$p_0(x)y^n + p_1(x)y^{n-1} + \cdots + p_n(x) = 0,$$

where $p_i(x)$, $i = 0, 1, \ldots, n$ are polynomials. A *transcendental function* is defined as any function which is not algebraic.

The derivative of sin $x$ will be obtained by use of the definition of derivative, since sin $x$ cannot be expressed in terms of previously differentiable functions. Some needed properties of trigonometric functions will be established initially. A brief trigonometric review appears in the Appendix.

## 10.2 FUNDAMENTAL TRIGONOMETRIC LIMITS

The derivation of some fundamental trigonometric limits will be based on geometric limits obtained from Fig. 10.2.1, in which $BC$ is an arc of a circle with center at $O$ and with radius 1. Angle $DOC$ is an acute angle whose radian measure is taken as $x > 0$. Then

$$AB < BC < \text{arc } BC.$$

But $AB = \sin x$, and arc $BC = x$ (see Appendix). Hence

(10.2.1) $\qquad 0 < \sin x < x \quad \text{if} \quad 0 < x < \pi/2.$

From (10.2.1) and the fact that $\sin(-x) = -\sin x$, it follows that

(10.2.2) $\qquad x < \sin x < 0 \quad \text{if} \quad -\pi/2 < x < 0.$

Then (10.2.1) and (10.2.2) may be combined to give

(10.2.3) $\qquad |\sin x| \le |x| \quad \text{if} \quad |x| < \pi/2,$

where the equal sign has been included so that the relation is also valid for $x = 0$.

An important consequence of (10.2.3) is that it leads to the continuity properties of the trigonometric functions. We show first that

(10.2.4) $\qquad \lim_{x \to 0} \sin x = 0.$

By Definition 3.4.1 we must show that, given $\epsilon > 0$, there exists $\delta > 0$ such that

$|\sin x - 0| < \epsilon$ when $|x - 0| < \delta$, or $|\sin x| < \epsilon$ when $|x| < \delta$.

Given $\epsilon > 0$, let $\delta = \epsilon$; then if $|x| < \delta$ we have, by (10.2.3),

$$|\sin x| \le |x| < \epsilon \quad \text{when} \quad |x| < \delta = \epsilon,$$

and (10.2.4) is established.

We now write, using a half-angle formula,

$$\cos x = 1 - 2 \sin^2 x/2.$$

Since $\lim_{x \to 0} \sin x = 0$, it follows that $\lim_{x \to 0} \sin x/2 = 0$, $\lim_{x \to 0} \sin^2 x/2 = 0$, and

(10.2.5) $\qquad \lim_{x \to 0} \cos x = 1.$

The general result follows.

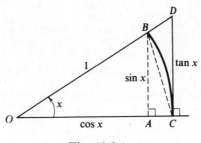

Fig. 10.2.1

**THEOREM 10.2.1.**  1. $\sin x$ and $\cos x$ are continuous for all real $x$.

2. $\tan x$ and $\sec x$ are continuous if $\cos x \neq 0$.

3. $\cot x$ and $\csc x$ are continuous if $\sin x \neq 0$.

*Proof* of 1.  We may show that $\sin x$ is continuous at a value $x$ by showing that

$$\lim_{h \to 0} \sin(x + h) = \sin x.$$

Now, by an addition formula of trigonometry,

$$\sin(x + h) = \sin x \cos h + \cos x \sin h.$$

As $h \to 0$, the value of $x$ does not vary; hence, by properties of limits,

$$\lim_{h \to 0} \sin(x + h) = \sin x \lim_{h \to 0} \cos h + \cos x \lim_{h \to 0} \sin h$$

$$= (\sin x) \cdot 1 + (\cos x) \cdot 0 \qquad \text{by (10.2.5) and (10.2.4)}$$

$$= \sin x,$$

and $\sin x$ is continuous for all real $x$.

Since $\cos x = \sin(\pi/2 - x)$, we see that $\cos x$ is expressed as a continuous function of a continuous function, and is therefore continuous (Section 3.6).

*Proof* of 2 and 3.  We illustrate the proofs of 2 and 3 by writing

$$\tan x = \frac{\sin x}{\cos x}.$$

Since $\sin x$ and $\cos x$ are continuous for all $x$, the quotient is continuous for all $x$ for which the denominator is not zero.

We now return to Fig. 10.2.1. We see from the figure that

$$\text{area}(\triangle OAB) < \text{area}(\text{sector } OBC) < \text{area}(\triangle OCD).$$

Since area (sector $OBC$) $= x/2$, we obtain

$$\tfrac{1}{2} \sin x \cos x < \tfrac{1}{2} x < \tfrac{1}{2} \tan x, \qquad 0 < x < \pi/2,$$

and hence

(10.2.6)
$$\cos x < \frac{x}{\sin x} < \frac{1}{\cos x}.$$

If $-\pi/2 < x < 0$, the same inequality holds, since $\sin(-x) = -\sin x$, $\cos(-x) = \cos x$, and we can say that (10.2.6) holds if $0 < |x| < \pi/2$. Now, $\lim_{x\to 0} \cos x = 1$, and hence $\lim_{x\to 0} 1/\cos x = 1$. Hence $x/\sin x$ lies between two functions each of which approaches 1 as $x \to 0$. The same is true of $\sin x/x$, and we have the *fundamental trigonometric limit*,

(10.2.7)
$$\lim_{x\to 0} \frac{\sin x}{x} = 1.$$

An additional important limit derives from (10.2.7). We write

$$\frac{1 - \cos x}{x^2} = \frac{2\sin^2 x/2}{x^2} = \frac{1}{2}\left(\frac{\sin x/2}{x/2}\right)^2.$$

But by (10.2.7), $\lim_{x\to 0} (\sin x/2)/(x/2) = 1$. Hence

(10.2.8)
$$\lim_{x\to 0} \frac{1 - \cos x}{x^2} = \frac{1}{2}.$$

**Example 10.2.1.** Evaluate $\lim_{x\to 0} \dfrac{1 - \cos x}{x}$.

*Solution.* We write

$$\frac{1 - \cos x}{x} = \frac{1 - \cos x}{x^2} \cdot x.$$

Then, by the product theorem for limits,

$$\lim_{x\to 0} \frac{1-\cos x}{x} = \lim_{x\to 0} \frac{1-\cos x}{x^2} \cdot \lim_{x\to 0} x = \frac{1}{2} \cdot 0 = 0.$$

**Example 10.2.2.** Evaluate $\lim_{x\to 0} \dfrac{\sin 3x}{x}$.

*Solution.* We write

$$\frac{\sin 3x}{x} = 3 \cdot \frac{\sin 3x}{3x};$$

hence

$$\lim_{x\to 0} \frac{\sin 3x}{x} = 3 \lim_{x\to 0} \frac{\sin 3x}{3x} = 3 \cdot 1 = 3.$$

The idea in the above examples was to involve the limits (10.2.7) and (10.2.8), noting at the same time that the limits are valid if the argument in the expression approaches zero as $x \to 0$.

## EXERCISE GROUP 10.1

Use the fundamental trigonometric limits, together with properties of limits and of the trigonometric functions, to evaluate the limits in Exercises 1–18.

1. $\lim\limits_{x \to 0} \dfrac{x}{\sin 2x}$
2. $\lim\limits_{x \to 0} \dfrac{3x}{\sin 4x}$
3. $\lim\limits_{x \to 0} \dfrac{\tan x}{2x}$
4. $\lim\limits_{x \to 0} \dfrac{2x}{\tan 4x}$
5. $\lim\limits_{x \to 0} \dfrac{\tan x}{\sin 2x}$
6. $\lim\limits_{x \to 0} \dfrac{\tan 2x}{\sin x}$
7. $\lim\limits_{x \to 0} \dfrac{\sin^2 x}{\tan 2x}$
8. $\lim\limits_{x \to 0} \dfrac{\sin^2 x/2}{x^2}$
9. $\lim\limits_{x \to 0} \dfrac{\tan x^2}{x}$
10. $\lim\limits_{x \to \infty} \dfrac{\sin 1/x}{2/x}$
11. $\lim\limits_{\theta \to 2} \dfrac{2 - \theta}{\sin \pi \theta}$
12. $\lim\limits_{\theta \to 0} \dfrac{\theta - \sin \theta}{\tan \theta}$
13. $\lim\limits_{\theta \to 0} \dfrac{1 - \cos \theta}{\tan \theta}$
14. $\lim\limits_{x \to \pi/2} \dfrac{\cos x}{\pi/2 - x}$
15. $\lim\limits_{x \to \pi} \dfrac{1 + \cos x}{(x - \pi)^2}$
16. $\lim\limits_{x \to 1} \dfrac{x - 1}{\sin(x - 1)}$
17. $\lim\limits_{x \to 0} \dfrac{1 - \cos^3 x}{x \sin x}$
18. $\lim\limits_{x \to \pi/2} \left(\dfrac{\pi}{2} - x\right) \sec x$

19. Devise a derivation of (10.2.7) by comparing the lengths of $AB$, arc $BC$, and $CD$ in Fig. 10.2.1.

20. Expand $\cos(x + h)$ and show that $\cos x$ is continuous for all $x$ by a method similar to that used for $\sin x$.

21. Use Fig. 10.2.1 to show that $|1 - \cos x| < |x|$ for $|x| < \pi/2$, and hence show directly that $\lim\limits_{x \to 0} \cos x = 1$.

## 10.3 DERIVATIVES OF TRIGONOMETRIC FUNCTIONS

To differentiate $\sin x$ we revert to the definition of derivative (Section 4.2), and write

$$\frac{\sin(x + h) - \sin x}{h} = \frac{\sin x \cos h + \cos x \sin h - \sin x}{h}$$

$$= \sin x \, \frac{\cos h - 1}{h} + \cos x \, \frac{\sin h}{h}.$$

By Example 10.2.1 and (10.2.7), we have

$$\lim_{h \to 0} \frac{\cos h - 1}{h} = 0, \qquad \lim_{h \to 0} \frac{\sin h}{h} = 1.$$

Applying these limits to the equation above, we obtain

$$D \sin x = \lim_{h \to 0} \frac{\sin(x + h) - \sin x}{h} = \cos x.$$

Now that the derivative of $\sin x$ has been obtained, the derivatives of the other trigonometric functions may be obtained without direct use of the definition. Thus

$$\cos x = \sin(\pi/2 - x), \qquad D \cos x = \cos(\pi/2 - x)D(\pi/2 - x)$$
$$= -\cos(\pi/2 - x) = -\sin x;$$

$$\tan x = \frac{\sin x}{\cos x}, \qquad D \tan x = \frac{\cos x D \sin x - \sin x D \cos x}{\cos^2 x}$$
$$= \frac{\cos^2 x + \sin^2 x}{\cos^2 x} = \frac{1}{\cos^2 x} = \sec^2 x.$$

The remaining formulas are obtained in a similar way (Exercises 39–41):

$$D \cot x = -\csc^2 x, \qquad D \sec x = \sec x \tan x, \qquad D \csc x = -\csc x \cot x.$$

The basic formulas for the derivatives of the trigonometric functions may be used in conjunction with the power rule and chain rule of differentiation to permit the differentiation of a wide variety of functions. We list these differentiation formulas for reference and also include the corresponding formulas in terms of differentials; in each case $u$ is a differentiable function of $x$

$$D_x \sin u = \cos u D_x u \qquad\qquad d \sin u = \cos u\, du$$
$$D_x \cos u = -\sin u D_x u \qquad\qquad d \cos u = -\sin u\, du$$
$$D_x \tan u = \sec^2 u D_x u \qquad\qquad d \tan u = \sec^2 u\, du$$
$$D_x \cot u = -\csc^2 u D_x u \qquad\qquad d \cot u = -\csc^2 u\, du$$
$$D_x \sec u = \sec u \tan u D_x u \qquad\qquad d \sec u = \sec u \tan u\, du$$
$$D_x \csc u = -\csc u \cot u D_x u \qquad\qquad d \csc u = -\csc u \cot u\, du$$

**Example 10.3.1.** Find $D \sin^2 x/2$.

*Solution.* The function $\sin^2 x/2$ is a power (of $\sin x/2$), and we start with the power rule of differentiation. We get

$$D \sin^2 x/2 = 2 \sin x/2 \; D \sin x/2 = 2 \sin x/2 \cos x/2 \; D(x/2)$$
$$= \sin x/2 \cos x/2.$$

**Example 10.3.2.** Find $dy/dx$ if $y = x \tan^2 3x$.

*Solution.* We have, by the product formula,

$$\frac{dy}{dx} = x \frac{d}{dx}(\tan^2 3x) + \tan^2 3x \frac{dx}{dx}$$
$$= x \cdot 2 \tan 3x \cdot \sec^2 3x \cdot 3 + \tan^2 3x$$
$$= 6x \tan 3x \sec^2 3x + \tan^2 3x.$$

***Example 10.3.3.*** Find $y'$ and $y''$ if $y = \sin^3 2x$.

*Solution.* Differentiating as a power, we obtain

$$y' = 3 \sin^2 2x D \sin 2x = 3 \sin^2 2x \cdot 2 \cos 2x$$
$$= 6 \sin^2 2x \cos 2x;$$
$$y'' = 6[\sin^2 2x(-2 \sin 2x) + \cos 2x \cdot 2 \sin 2x \cdot 2 \cos 2x]$$
$$= -12 \sin^3 2x + 24 \sin 2x \cos^2 2x.$$

***Example 10.3.4.*** Show that the function $y = \sin 3x + 2 \cos 3x$ satisfies the relation $y'' + 9y = 0$.

*Solution.* The given relation is an example of a *differential equation.* that is, an equation involving a function and its derivatives. All that is required in this example is to find $y''$, and to substitute in the differential equation. We find

$$y' = (\cos 3x) \cdot 3 + 2(-\sin 3x) \cdot 3$$
$$= 3 \cos 3x - 6 \sin 3x;$$
$$y'' = -9 \sin 3x - 18 \cos 3x.$$

Then $y'' + 9y = (-9 \sin 3x - 18 \cos 3x) + 9(\sin 3x + 2 \cos 3x) = 0.$

## EXERCISE GROUP 10.2

Find $dy/dx$ in each of Exercises 1–26

1. $y = \sin 2x$
2. $y = \sin x/2$
3. $y = x + \sin 3x$
4. $y = 2x \sin 2x + \cos 2x$
5. $y = \sqrt{\sin 2x}$
6. $y = \tan \sqrt{x}$
7. $y = \sqrt{\tan x}$
8. $y = \cot x + \csc x$
9. $y = \csc^2 3x$
10. $y = \cos(3\pi/2 - x)$
11. $y = \sec 3x$
12. $y = x \sec x$
13. $y = (x^2 - 2) \cos x$
14. $y = \sin 2x \cos 2x$
15. $y = \tan x - \sec x$
16. $y = \tan^2 3x$
17. $y = \sin^2 x/2$
18. $y = \cos^3 x/3$
19. $y = \dfrac{2}{1 + \cos x}$
20. $y = \dfrac{\sin x}{1 - \cos x}$
21. $y = \dfrac{\tan x}{1 + \sec x}$
22. $y = \dfrac{\cos x}{1 - \sin x}$
23. $y = \dfrac{\tan x}{1 + \tan x}$
24. $y = 2x \cos x + (x^2 - 2) \sin x$
25. $y = x^2 \sin x + 2x \cos x + \sin x$
26. $y = (3x^2 - 6) \sin x - (x^3 - 6x) \cos x$

In each of Exercises 27–32 show that the given function satisfies the accompanying differential equation, that is, an equation which involves a function and its derivatives.

27. $y = 2 \cos 3x - 3 \sin 3x$; $y'' + 9y = 0$
28. $y = 2 \cos 2x - 4 \sin 2x$; $y'' - y' + 5y = 10 \cos 2x$
29. $y = x \sin x$; $x^2 y'' - 2xy' + (x^2 + 2)y = 0$
30. $y = \tan(x + 2)$; $y'' - 2yy' = 0$
31. $y = 2 \tan(2x + 1) - 4x + 1$; $y' = (y + 4x - 1)^2$
32. $y = \sin x - 3 \sin 2x + 5 \cos 2x$; $y'' + 4y = 3 \sin x$

In Exercises 33–38 find the relative extrema of the function in the interval $[0, 2\pi]$.

33. $\sin x + \cos x$
34. $\sin x - \cos x$
35. $2 \sin x - 3 \cos x$
36. $a \sin x + b \cos x$
37. $2 \sin x + \sin 2x$
38. $2 \cos x + \cos 2x$

39. Derive: $D \cot x = - \csc^2 x$
40. Derive: $D \sec x = \sec x \tan x$
41. Derive: $D \csc x = - \csc x \cot x$
42. Assuming the formula for the derivative of $\sin x$, obtain the formula for the derivative of $\cos x$ by using the relation $\cos x = 1 - 2 \sin^2 x/2$.
43. Use the definition of derivative to find the derivative of $\cos x$ directly.
* 44. We may use a trigonometric identity to write

$$\sin(x + h) - \sin x = 2 \cos\left(x + \frac{h}{2}\right) \sin \frac{h}{2}.$$

Use this relation to derive the formula for the derivative of $\sin x$. Note particularly that this derivation uses the continuity of $\cos x$; whereas the one at the beginning of Section 10.3 does not.

## 10.4 INTEGRALS OF TRIGONOMETRIC FUNCTIONS

Any differentiation formula may be rewritten in the form of an integration formula. Not all such integration formulas are useful. However, the following, obtained directly from the differentiation formulas of Section 10.3, are useful and important.

$$\int \sin x \, dx = - \cos x + C, \qquad \int \cos x \, dx = \sin x + C$$

$$\int \sec^2 x \, dx = \tan x + C, \qquad \int \csc^2 x \, dx = - \cot x + C$$

$$\int \sec x \tan x \, dx = \sec x + C, \qquad \int \csc x \cot x \, dx = - \csc x + C$$

Notice that, because of their source, these formulas have the basic trigonometric functions as integrals. Only two of them have basic trigonometric functions as integrands. The other basic trigonometric functions cannot be integrated yet.

**Example 10.4.1.** Evaluate $\int \sin(2x - 1) \, dx$.

*Solution.* We recognize that $dx$ is, apart from a constant factor, the differential of $2x - 1$. Accordingly, if we let $2x - 1 = u$, then $2 \, dx = du$, and

$$\int \sin(2x - 1) \, dx = \int \sin u \cdot \tfrac{1}{2} \, du = \tfrac{1}{2} \int \sin u \, du = -\tfrac{1}{2} \cos u + C$$

$$= -\tfrac{1}{2} \cos(2x - 1) + C.$$

**Example 10.4.2.** Evaluate $\int \sqrt{\tan x} \, \sec^2 x \, dx$.

*Solution.* We see that the integrand contains a power of $\tan x$; furthermore, $\sec^2 x$ is the derivative of $\tan x$ and is hence an enabling factor (Section 9.4). The power rule (9.4.3) may be used and we obtain

$$\int \sqrt{\tan x} \, \sec^2 x \, dx = \frac{\tan^{3/2} x}{3/2} + C$$

$$= \tfrac{2}{3} \tan^{3/2} x + C.$$

**Example 10.4.3.** Evaluate $\int \sec^2 x \tan x \, dx$.

*Solution.* One method is to set $u = \tan x$, $du = \sec^2 x \, dx$. Then

$$\int \sec^2 x \tan x \, dx = \int u \, du = \tfrac{1}{2} u^2 + C = \tfrac{1}{2} \tan^2 x + C.$$

An alternate method is to set $v = \sec x$, $dv = \sec x \tan x \, dx$. Then

$$\int \sec^2 x \tan x \, dx = \int \sec x \cdot \sec x \tan x \, dx = \int v \, dv$$

$$= \tfrac{1}{2} v^2 + C = \tfrac{1}{2} \sec^2 x + C.$$

Each of these forms could have been obtained directly with the use of an enabling factor.

The different forms of the result in Example 10.4.3 are due to the different methods used. This situation arises frequently, and it often becomes important to compare the results. The use of the constant in the indefinite integral is not meant to imply that the two expressions are equal for the same value of $C$. What is meant is simply that $\tfrac{1}{2} \tan^2 x$ and $\tfrac{1}{2} \sec^2 x$ are both antiderivatives of the integrand; this means that they differ by a constant. Now

$$\tfrac{1}{2} \tan^2 x = \tfrac{1}{2}(\sec^2 x - 1) = \tfrac{1}{2} \sec^2 x - \tfrac{1}{2}.$$

Hence $\tfrac{1}{2} \tan^2 x$ and $\tfrac{1}{2} \sec^2 x$ differ by $-\tfrac{1}{2}$, and the two results are equivalent; that is, both are correct.

As another example of a similar situation we observe the two forms of the following integral:

$$\int \sin x \cos x \, dx = \begin{cases} \frac{1}{2} \sin^2 x + C, \\ -\frac{1}{2} \cos^2 x + C. \end{cases}$$

**Example 10.4.4.** Evaluate $\int \dfrac{\sin x}{\sqrt{1-\cos x}} \, dx$.

**Solution.** We let $u = 1 - \cos x$, $du = \sin x \, dx$, and obtain

$$\int \frac{\sin x}{\sqrt{1 - \cos x}} \, dx = \int \frac{du}{\sqrt{u}} = 2u^{1/2} + C$$

$$= 2(1 - \cos x)^{1/2} + C.$$

**EXERCISE GROUP 10.3**

Evaluate the integrals in Exercises 1–36.

1. $\int \sin 3x \, dx$

2. $\int \sin (\pi/4 - x) \, dx$

3. $\int \cos (1 - 2x) \, dx$

4. $\int \sec 2x \tan 2x \, dx$

5. $\int \sec^2(3x - 1) \, dx$

6. $\int x \cos x^2 \, dx$

7. $\int \sin 3x \cos 3x \, dx$

8. $\int \dfrac{\sin x}{\cos^3 x} \, dx$

9. $\int \dfrac{\cos x}{\sin^4 x} \, dx$

10. $\int \tan^2 x \, dx$

11. $\int \dfrac{\sin x}{\sqrt{\cos x}} \, dx$

12. $\int \dfrac{dx}{\sec 2x}$

13. $\int \dfrac{\sin \sqrt{x}}{\sqrt{x}} \, dx$

14. $\int \dfrac{1}{\sqrt{x}} \sec \sqrt{x} \tan \sqrt{x} \, dx$

15. $\int \sec x \tan x \sqrt{\sec x} \, dx$

16. $\int \sec^2 x \cdot \sec x \tan x \, dx$

17. $\int \sec^2 2x \sqrt{\tan 2x} \, dx$

18. $\int \csc^2 3x \cot^{2/3} 3x \, dx$

19. $\int \csc^2 3x \cot^2 3x \, dx$

20. $\int \sin 2x \cos^2 2x \, dx$

21. $\int_0^{\pi/6} \sin 2x \sqrt{\cos 2x} \, dx$

22. $\int_0^{\pi/4} \tan^{1/3} x \sec^2 x \, dx$

23. $\int_0^{\pi/3} \sec^4 x \tan x \, dx$

24. $\int_0^2 x \sin (x^2 - 2) \, dx$

25. $\int \tan^2 x \sec^2 x \, dx$

26. $\int \tan x \sec^3 x \, dx$

27. $\int \sin^3 x/2 \cos x/2 \, dx$

28. $\int \dfrac{\sec x \tan x}{\sqrt{\sec x - 1}} \, dx$

29. $\int \dfrac{\csc^2 x \, dx}{\sqrt{1 + \cot x}}$

30. $\int \dfrac{\sin 2x \, dx}{\sqrt{2 - \cos^2 x}}$

31. $\int \dfrac{\sin x \cos x \, dx}{\sqrt{1 + \sin^2 x}}$

32. $\int \sin^{1/2} 3x \cos 3x \, dx$

33. $\int \dfrac{\sin 2x}{(2 - \sin^2 x)^3} \, dx$

34. $\int \sec^{3/2} x \tan x \, dx$

35. $\int \dfrac{\sec^2 x \, dx}{\sqrt{2 - \tan x}}$

36. $\int \dfrac{\sin 2x \, dx}{\sqrt{3 - \sin^2 x}}$

37. Find the area under one arch of the sine curve $y = \sin x$.
38. Find the area under one arch of the curve $y = \cos(3x - \pi/2)$.
39. Find the area under one arch of the curve $y = 2 \sin x - \cos x$.
40. Find the area between the curves $y = x$ and $y = \sin x$ in the interval $[0, \pi]$.

## 10.5  INVERSE FUNCTIONS

In Section 2.6 a *relation* was defined as a set of number pairs $R: \{(a, b)\}$. If we reverse the numbers of each pair, the new set of pairs also defines a relation, called the *inverse relation* of $R$, and designated as $R^{-1}$. (The superscript is *not* an exponent.) The graph of $R^{-1}$ is symmetric to the graph of $R$ in the line $y = x$.

In Section 3.1 a *function* was defined as a relation such that no two distinct pairs of the function have the same first element. The inverse of a function is therefore a relation such that no two distinct pairs of the inverse relation have the same second element. Thus (1, 3) and (2, 3) cannot both belong to the inverse of a function, but (3, 1) and (3, 2) can both belong to the inverse of a function. For example, for the function

$$f: \{(1, 3), (2, 3), (3, 2), (4, -1)\},$$

the inverse is

$$f^{-1}: \{(3, 1), (3, 2), (2, 3), (-1, 4)\}.$$

Thus, we see that *the inverse of a function need not itself be a function*.

**DEFINITION 10.5.1.** *If the inverse $f^{-1}$ of a function $f$ is itself a function, then $f^{-1}$ is called an* **inverse function** *of $f$*.

We have made a distinction between an *inverse of a function* and an *inverse function*. This rather fussy distinction will cause no confusion in what follows.

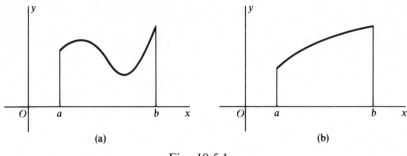

Fig. 10.5.1

It is clear that the inverse of the function shown in Fig. 10.5.1(a) is not a function but that the inverse of the function shown in Fig. 10.5.1(b) is a function. What distinguishes the function shown in Fig. 10.5.1(b) is that it is increasing in [a, b]. In fact, the proof of the following result is not difficult and is left to the exercises.

**THEOREM 10.5.1.** *If the function $f$ is increasing in $[a, b]$, or decreasing in $[a, b]$, then $f^{-1}$ is an inverse function.*

***Example 10.5.1.*** Consider the function $f(x) = x^2$, $x$ in $[1, 2]$. To find the inverse function we may set $y = x^2$ and solve for $x$ as $x = y^{1/2}$. The solution $x = -y^{1/2}$ is not valid, since $x > 0$ for $x$ in $[1, 2]$. Thus the function $g(y) = y^{1/2}$, $y$ in $[1, 4]$, is the inverse of $f$. It does not matter whether we write $g(y) = y^{1/2}$ or $g(x) = x^{1/2}$ or $g(u) = u^{1/2}$, provided $y$, $x$, or $u$ is in $[1, 4]$; the same inverse function is defined in each case.

The inverse of the function $f(x) = x^2$, $x$ in $[-1, 2]$, is not an inverse function, since for some values in the range of $f$, two values of $x$ exist in the domain of $f$. (See Fig. 10.5.2.)

For the function $f$ in Example 10.5.1 and its inverse function $g$, we have

$$f[g(x)] = [g(x)]^2 = (x^{1/2})^2 = x; \qquad g[f(x)] = (x^2)^{1/2} = x.$$

In general it is true for a function $f$ and its inverse function $f^{-1}$ that

(10.5.1) $$f[f^{-1}(x)] = x; \qquad f^{-1}[f(x)] = x.$$

In effect, $f$ and $f^{-1}$ undo one another; applied in succession, in either order, they lead to $x$. This implies that $f$ is the inverse of $f^{-1}$; that is,

(10.5.2) $$[f^{-1}(x)]^{-1} = f(x).$$

It appears that the inverse of the function in Fig. 10.5.1(b) is continuous and differentiable. The continuity of $f^{-1}$ will be assumed if $f$ is increasing (decreasing) and continuous in an interval. We present a theorem on differentiability.

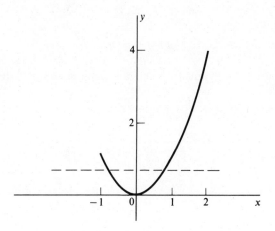

Fig. 10.5.2

**THEOREM 10.5.2.** *Let $f$ be a function such that either $f'(x) > 0$, $x$ in $[a, b]$, or $f'(x) < 0$, $x$ in $[a, b]$. Then $f^{-1}$ is an inverse function in the range of $f$, $f^{-1}$ is differentiable, and*

(10.5.3) $$\frac{d}{dx} f^{-1}(x) = \frac{1}{f'[f^{-1}(x)]}, \qquad x \text{ in the range of } f.$$

*Proof.* We assume $f'(x) > 0$, $x$ in $[a, b]$. (The proof is similar if $f'(x) < 0$.) Then $f$ is increasing in $[a, b]$, and $f^{-1}$ is an inverse function, also increasing, in the interval $[c, d]$, where $c = f(a)$, $d = f(b)$. Let $y = f(x)$, $x$ in $[a, b]$. Then

$$\frac{f^{-1}(y) - f^{-1}(y_0)}{y - y_0} = \frac{x - x_0}{f(x) - f(x_0)} = 1 \bigg/ \left[\frac{f(x) - f(x_0)}{x - x_0}\right].$$

Let $y \to y_0$; then $x \to x_0$ by the continuity of $f^{-1}$, and the quantity in brackets on the right approaches $1/f'(x_0)$. Thus, using the definition of derivative, we have

$$\frac{d}{dy} f^{-1}(y) \bigg|_{y=y_0} = \lim_{y \to y_0} \frac{f^{-1}(y) - f^{-1}(y_0)}{y - y_0} = \frac{1}{f'(x_0)} = \frac{1}{f'[f^{-1}(y_0)]}.$$

This holds for any $y_0$ in $(c, d)$. Since this statement is merely (10.5.3) with a change of letter, the proof is complete.

*Example 10.5.2.* Find the derivative of the inverse of $f(x) = x^2$, $x$ in $(1, 2)$.

*Solution.* Let the inverse function be $g(x)$. Then by Theorem 10.5.2 we have, since $f'(x) = 2x$,

(10.5.4) $$g'(x) = \frac{1}{2g(x)}, \qquad x \text{ in } (1, 4).$$

Knowing that $g(x) = x^{1/2}$, we may write, from (10.5.4),

$$g'(x) = \frac{1}{2x^{1/2}}.$$

This is what would be obtained by differentiating $g(x) = x^{1/2}$ directly.

In practice, if $y = f(x)$, $x = g(y)$, we may write

$$g'(y) = \frac{dx}{dy} = \frac{1}{\frac{dy}{dx}} = \frac{1}{f'(x)} = \frac{1}{f'[g(y)]},$$

where $dx$ and $dy$ are differentials. Moreover, the method of implicit differentiation (Section 4.7) may be used. Thus if $y = x^2$ we may differentiate both sides with respect to $y$ and get

$$1 = 2x\frac{dx}{dy}, \quad \text{and} \quad \frac{dx}{dy} = \frac{1}{2x} = \frac{1}{2y^{1/2}}.$$

## 10.6 INVERSE TRIGONOMETRIC FUNCTIONS

The basic procedures for handling derivatives of inverse functions have been developed in Section 10.5. What is needed here is a definition of the inverse trigonometric functions so that the available procedures apply.

The graph of the function $y = \sin x$ is assumed to be familiar to you. Interchanging $x$ and $y$, we get $x = \sin y$, whose graph is the reflection of the graph of $y = \sin x$ in the line $y = x$. We now write the solution of the equation $x = \sin y$ as

$$y = \arcsin x,$$

an equation which is completely equivalent to $x = \sin y$ in that the number pairs $(x, y)$ which satisfy each equation are the same. However, this set does not constitute a function, since (infinitely) many values of $y$ correspond to each value of $x$. For example, if $x = 1/2$, $y$ may have any value $\pi/6 + 2n\pi$ or $5\pi/6 + 2n\pi$, for any integral value of $n$. On the other hand, if we restrict the values of $y$ to the interval $[-\pi/2, \pi/2]$, then for any value of $x$ with $-1 \leq x \leq 1$, one and only one value of $y$ in $[-\pi/2, \pi/2]$ applies. We shall designate this choice of $y$ as the *inverse function* of $\sin x$, or as the *inverse sine*, and write

$$y = \sin^{-1} x, \quad -\pi/2 \leq y \leq \pi/2.$$

With this restriction $\sin^{-1} x$ is a function of $x$, with domain $[-1, 1]$ and range $[-\pi/2, \pi/2]$. We then know that $\sin^{-1} x$ is an increasing function and that

$$\sin^{-1}(\sin x) = x, \quad -\pi/2 \leq x \leq \pi/2; \quad \sin(\sin^{-1} x) = x, \quad -1 \leq x \leq 1.$$

As thus defined, $\sin^{-1} x$ is an inverse function of $\sin x$. Note that $\sin^{-1} x$ is a function; whereas $\arcsin x$ is not.

Inverses may be defined in a similar way for the other trigonometric functions. We indicate the range for each inverse trigonometric function in the following.

$$-\pi/2 \leq \sin^{-1} x \leq \pi/2 \qquad 0 \leq \cos^{-1} x \leq \pi$$
$$-\pi/2 < \tan^{-1} x < \pi/2 \qquad 0 < \cot^{-1} x < \pi$$
$$0 \leq \sec^{-1} x \leq \pi \qquad -\pi/2 \leq \csc^{-1} x \leq \pi/2.$$

Of course, $\sec^{-1} x$ cannot equal $\pi/2$, and $\csc^{-1} x$ cannot equal zero. The graphs of these functions appear in Fig. 10.6.1.

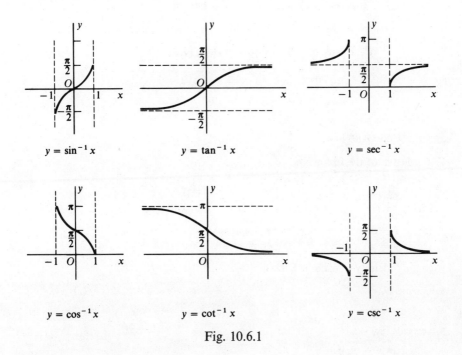

Fig. 10.6.1

**Example 10.6.1.** Prove: $\tan (\tan^{-1} x) = x$.

*Solution.* From the above we know that

$$y = \tan^{-1} x \Rightarrow \tan y = x.$$

Substitution of $y$ from the first equation into the second one gives

$$\tan (\tan^{-1} x) = x,$$

This is the desired result.

It is important to understand the implications of the above definitions, as illustrated by the following statements in one case.

$$y = \tan^{-1} x \Rightarrow \tan y = x;$$
$$\tan y = x \Rightarrow y = \tan^{-1} x \quad \text{only if } -\pi/2 < y < \pi/2.$$

It should be noted that it is not valid to state that arctan(tan $x$) is equal to $x$. For a given value of $x$, tan $x$ has a specified value, say $u$, but arctan $u$ has infinitely many values, only one of which is $x$.

**Example 10.6.2.** Find (a) sin (cos$^{-1}$ $x$), (b) sin (2 sin$^{-1}$ $x$).

*Solution.* (a) Let $u = \cos^{-1} x$, cos $u = x$. Then $0 \le u \le \pi$; hence sin $u \ge 0$, and

$$\sin (\cos^{-1} x) = \sin u = \sqrt{1 - \cos^2 u} = \sqrt{1 - x^2}.$$

(b) Let $u = \sin^{-1} x$, sin $u = x$. Then $-\pi/2 \le u \le \pi/2$, and cos $u \ge 0$. Hence

$$\sin (2 \sin^{-1} x) = \sin 2u = 2 \sin u \cos u$$
$$= 2 \sin u \sqrt{1 - \sin^2 u} = 2x\sqrt{1 - x^2}.$$

### EXERCISE GROUP 10.4

1. Find each of the following.
   (a) $\sin^{-1}(-1/2)$
   (b) $\cos^{-1}(-1/2)$
   (c) $\pi/2 - \tan^{-1} \sqrt{3}$
   (d) $\cot^{-1}(-\sqrt{3}/3)$
   (e) $\sin^{-1} 1/2 + \cos^{-1} 1/2$
   (f) $\tan^{-1} \sqrt{3} + \cot^{-1} \sqrt{3}$

2. (a) If $y = \cos x$ and $\pi/2 < x < \pi$, express $x$ in terms of $y$.
   (b) If $y = \sin x$ and $\pi/2 < x < \pi$, express $x$ in terms of $y$.

Prove the indentity in each of Exercises 3–16.

3. $\cos(\cos^{-1} x) = x$
4. $\cot(\cot^{-1} x) = x$
5. $\sec(\sec^{-1} x) = x$
6. $\sin^{-1} x + \cos^{-1} x = \pi/2$
7. $\tan^{-1} x + \cot^{-1} x = \pi/2$
8. $\sec^{-1} x + \csc^{-1} x = \pi/2$
9. $\sin^{-1}(\sin x) = x$, if $-\pi/2 \le x \le \pi/2$
10. $\cos^{-1}(\cos x) = x$, if $0 \le x \le \pi$
11. $\tan^{-1}(\tan x) = x$, if $-\pi/2 < x < \pi/2$
12. $\cot^{-1}(\cot x) = x$, if $0 < x < \pi$
13. $\sec^{-1}(\sec x) = x$, if $0 \le x \le \pi$
14. $\csc^{-1}(\csc x) = x$, if $-\pi/2 \le x \le \pi/2$
15. $\cos^{-1}(\cos x) = -x$, if $-\pi/2 \le x \le 0$
16. $\sin^{-1}(\sin x) = \pi - x$, if $\pi/2 \le x \le \pi$

Simplify the given function in each of Exercises 17–24. Each one can be expressed algebraically in terms of $x$.

17. $\cos(\sin^{-1} x)$
18. $\cos(2 \sin^{-1} x)$
19. $\tan(2 \sin^{-1} x)$
20. $\sin(\tan^{-1} x)$
21. $\sin(2 \tan^{-1} x)$
22. $\cos(2 \cos^{-1} x)$
23. $\cos(2 \tan^{-1} x)$
24. $\tan(2 \cos^{-1} x)$

* 25. Prove by trigonometric methods that

$$2 \sin^{-1} \sqrt{x} - \sin^{-1}(2x-1) = \pi/2, \text{ if } 0 \leq x \leq 1.$$

* 26. Prove by trigonometric methods that

$$\cot^{-1} \frac{2x}{1-x^2} = 2 \cot^{-1} x + \frac{\pi}{2}.$$

27. Prove Theorem 10.5.1.
28. Prove formula (10.5.2).
29. Use Theorem 10.5.2 to show that $x^{1/n}$ is differentiable for $x > 0$, if $n$ is a positive integer.

## 10.7 DERIVATIVES OF INVERSE TRIGONOMETRIC FUNCTIONS

As defined in the previous section, $\sin^{-1} x$ is an increasing function of $x$ for $-1 \leq x \leq 1$. By Section 10.5 we know that the derivative of $\sin^{-1} x$ then exists, if the derivative of the sine function does not vanish for the argument $\sin^{-1} x$; that is, if

$$\cos(\sin^{-1} x) \neq 0, \quad \text{or} \quad x \neq \pm 1.$$

To find the derivative we use (10.5.3) with $f(x) = \sin x$, $f^{-1}(x) = \sin^{-1} x$. Then $f'(x) = \cos x$ and we get

(10.7.1) $$\frac{d}{dx} \sin^{-1} x = \frac{1}{\cos(\sin^{-1} x)}.$$

Now let $\sin^{-1} x = u$, $\sin u = x$. Then $-\pi/2 \leq u \leq \pi/2$, and we have

$$\cos u = \sqrt{1 - \sin^2 u} = \sqrt{1 - x^2}$$

so that

$$\cos(\sin^{-1} x) = \sqrt{1 - x^2}.$$

With this relation, (10.7.1) gives the result

(10.7.2) $$D \sin^{-1} x = \frac{1}{\sqrt{1-x^2}}, \quad |x| < 1.$$

Each of the other inverse trigonometric functions either increases everywhere in its domain or decreases everywhere in its domain, and each is, accordingly,

differentiable in an appropriate domain. Let us find $D\tan^{-1}x$ as follows. We have
$$y = \tan^{-1} x \Rightarrow \tan y = x.$$
By implicit differentiation in the latter equation, we find
$$\sec^2 y \frac{dy}{dx} = 1, \qquad D\tan^{-1} x = \frac{dy}{dx} = \cos^2 y.$$
Now, $\sec^2 y = 1 + \tan^2 y = 1 + x^2$. Hence $\cos^2 y = 1/(1+x^2)$, and we obtain
$$D\tan^{-1} x = \frac{1}{1+x^2}, \qquad \text{all real } x.$$

In a similar manner we may use implicit differentiation in the equation $\sec y = x$, and get
$$\sec y \tan y \frac{dy}{dx} = 1, \qquad D\sec^{-1} x = \frac{dy}{dx} = \frac{1}{\sec y \tan y}.$$
But $\tan y = \pm\sqrt{x^2 - 1}$ (+ if $x > 0$, $-$ if $x < 0$). Hence we may write
$$D\sec^{-1} x = \frac{1}{|x|\sqrt{x^2 - 1}}, \qquad |x| > 1.$$

The derivatives of the other inverse trigonometric functions may be obtained in a similar manner. The six formulas are as follows.

$$D\sin^{-1} x = \frac{1}{\sqrt{1-x^2}} \qquad D\cos^{-1} x = -\frac{1}{\sqrt{1-x^2}}, \quad |x| < 1$$

$$D\tan^{-1} x = \frac{1}{1+x^2} \qquad D\cot^{-1} x = -\frac{1}{1+x^2}, \quad \text{all real } x$$

$$D\sec^{-1} x = \frac{1}{|x|\sqrt{x^2-1}} \qquad D\csc^{-1} x = -\frac{1}{|x|\sqrt{x^2-1}}, \quad |x| > 1.$$

It is clear that (i) each of these formulas may be used in conjunction with the chain rule, and (ii) for each of these differentiation formulas there is a corresponding differential formula.

**Example 10.7.1.** Find $dy/dx$ if $y = x\tan^{-1} x$.

*Solution.* The product formula of differentiation is used; thus
$$\frac{dy}{dx} = x\frac{d}{dx}(\tan^{-1} x) + \tan^{-1} x = \frac{x}{1+x^2} + \tan^{-1} x.$$

**Example 10.7.2.** Find the derivative of $\cos^{-1} x^2$.

*Solution.* We have

$$D \cos^{-1} x^2 = -\frac{1}{\sqrt{1-(x^2)^2}} Dx^2 = -\frac{2x}{\sqrt{1-x^4}}.$$

## EXERCISE GROUP 10.5

Find the derivative of the given function in Exercises 1–32.

1. $\cos^{-1} 2x$
2. $\sin^{-1}(x-1)$
3. $\sin^{-1} x^2$
4. $\cos^{-1} 1/x$
5. $x \sin^{-1} x$
6. $x^2 \sin^{-1} x$
7. $\cot^{-1} 1/x$
8. $\sec^{-1} 2x$
9. $(\cos^{-1} x)/x$
10. $\tan^{-1} \sqrt{x}$
11. $\cos^{-1} 1/\sqrt{x}$
12. $\sec^{-1} \sqrt{x}$
13. $\sin^{-1} \sqrt{x}$
14. $\tan^{-1}(\sin x)$
15. $\sin^{-1}(\sin 2x)$
16. $\tan^{-1} \sqrt{x-1}$
17. $\cos^{-1} \sqrt{\dfrac{2a-y}{2a}}$
18. $\cos^{-1} \dfrac{2-x}{2}$
19. $\tan^{-1} \dfrac{x-c}{1+cx}$
20. $\tan^{-1} x + \tan^{-1} 1/x$
21. $\cos^{-1} \sqrt{1-x^2}$
22. $\tan^{-1} \sqrt{1-x^2}$
23. $x \sin^{-1} x + \sqrt{1-x^2}$
24. $\cos^{-1}(1-x) - \sqrt{2x-x^2}$
25. $\cos^{-1}(1-x) + (x-1)\sqrt{2x-x^2}$
26. $\sqrt{x^2-9} - 3\cos^{-1} 3/x$
27. $4 \sin^{-1} x/2 - x\sqrt{4-x^2}$
28. $\dfrac{\sqrt{1-x^2}}{x} - \sin^{-1} x$
29. $\sin^{-1} x - \sqrt{1-x^2}$
30. $x \cos^{-1} x/a - \sqrt{a^2-x^2}$
31. $x\sqrt{a^2-x^2} + a^2 \cos^{-1} x/a$
* 32. $\sin^{-1}(x\sqrt{1-c^2} + c\sqrt{1-x^2})$

33. Derive the formula for the derivative of $\cot^{-1} x$.
34. Derive the formula for the derivative of $\sec^{-1} x$.
35. Derive the formula for the derivative of $\csc^{-1} x$.
36. Obtain the formula for the derivative of $\sin^{-1} x$ by starting with the relation

$$\cos(\sin^{-1} x) = \sqrt{1-x^2}.$$

37. By using $f'(x)$, prove that $f(x)$ is constant, if

$$f(x) = 2 \sin^{-1} \sqrt{x} - \sin^{-1}(2x-1), \quad 0 \le x \le 1.$$

(See also Exercise 25 of Exercise Group 10.4.)

**38.** By using $f'(x)$, prove that $f(x)$ is constant, if

$$f(x) = \cot^{-1}\frac{2x}{1-x^2} - 2\cot^{-1} x.$$

(See also Exercise 26 of Exercise Group 10.4.)

*  **39.** Show that $b/(1+b^2) < \tan^{-1} b < b$ if $b > 0$. [*Hint*: Use the Mean-Value Theorem with $a = 0$.]

## 10.8 INTEGRALS LEADING TO INVERSE TRIGONOMETRIC FUNCTIONS

The derivatives of inverse trigonometric functions lead to three basic integration formulas.

(10.8.1) $$\int \frac{dx}{\sqrt{1-x^2}} = \sin^{-1} x + C = -\cos^{-1} x + C'.$$

(10.8.2) $$\int \frac{dx}{1+x^2} = \tan^{-1} x + C = -\cot^{-1} x + C'.$$

(10.8.3) $$\int \frac{dx}{|x|\sqrt{x^2-1}} = \sec^{-1} x + C = -\csc^{-1} x + C'.$$

The two forms of each indefinite integral are equivalent in the sense described in Section 10.4. A different constant is used in each form for the sake of accuracy. In most cases the first form is used.

The above formulas are our first exposure to integrals of algebraic functions which are not themselves algebraic, and are noteworthy for this characteristic. Other integrals exhibiting the same feature will occur later.

**Example 10.8.1.** Evaluate $\int \frac{dx}{\sqrt{1-4x^2}}$.

*Solution.* This integral suggests formula (10.8.1). Let $u = 2x$, $du = 2\,dx$; then

$$\int \frac{dx}{\sqrt{1-4x^2}} = \frac{1}{2}\int \frac{du}{\sqrt{1-u^2}} = \frac{1}{2}(\sin^{-1} u + C') = \frac{1}{2}\sin^{-1} 2x + C,$$

where $C = C'/2$. Note the use of constants of integration. The first such constant was written as $C'$, and then $C'/2$ was called $C$. Usually the constant of integration is desired as a single letter. In practice the constants of integration are not included in the intermediate steps, and then a constant of integration is tacked on after an antiderivative of the original function is obtained.

**Example 10.8.2.** Evaluate $\int \frac{dx}{1+(2x-1)^2}$.

*Solution.* The form of the integrand suggests the use of (10.8.2) with the substitution $u = 2x - 1$, $dx = du/2$. We get

$$\int \frac{dx}{1 + (2x - 1)^2} = \frac{1}{2} \int \frac{du}{1 + u^2} = \frac{1}{2} \tan^{-1} u + C$$

$$= \frac{1}{2} \tan^{-1}(2x - 1) + C.$$

**Example 10.8.3.** Evaluate $\int \dfrac{dx}{a^2 + x^2}$.

*Solution.* Let $x = au$, $dx = a\, du$. Then

$$\int \frac{dx}{a^2 + x^2} = \frac{1}{a} \int \frac{du}{1 + u^2} = \frac{1}{a} \tan^{-1} u + C$$

$$= \frac{1}{a} \tan^{-1} \frac{x}{a} + C.$$

**Example 10.8.4.** Evaluate $\int \dfrac{dx}{\sqrt{x}(x + 1)}$.

*Solution.* This integral does not appear to fit any known form. However, let us use the substitution $x = u^2$, $dx = 2u\, du$. Then

$$\int \frac{dx}{\sqrt{x}(x + 1)} = \int \frac{2u\, du}{u(u^2 + 1)} = 2 \int \frac{du}{u^2 + 1}$$

$$= 2 \tan^{-1} u + C$$

$$= 2 \tan^{-1} \sqrt{x} + C.$$

The integration formulas at the beginning of this section should be studied not only from the point of view of the type of function to which they apply, but also of certain types to which they do not apply. In particular, none of the above formulas applies to any of the integrals

$$\int \frac{dx}{\sqrt{x^2 \pm 1}}, \quad \int \frac{dx}{1 - x^2}, \quad \int \frac{dx}{x\sqrt{x^2 + 1}},$$

which, at first glance, have some similarity to the integrals appearing in the formulas.

**EXERCISE GROUP 10.6**

Evaluate each of the following integrals.

1. $\int \dfrac{dx}{\sqrt{4 - x^2}}$

2. $\int \dfrac{du}{4u^2 + 1}$

3. $\displaystyle\int \frac{du}{9u^2 + 4}$

4. $\displaystyle\int \frac{3u\,du}{1 + 4u^4}$

5. $\displaystyle\int \frac{dx}{\sqrt{1 - 4x^2}}$

6. $\displaystyle\int \frac{\cot^{-1} x}{1 + x^2}\,dx$

7. $\displaystyle\int \frac{\tan^{-1} x}{1 + x^2}\,dx$

8. $\displaystyle\int \frac{\sin^{-1} x}{\sqrt{1 - x^2}}\,dx$

9. $\displaystyle\int \frac{\cos^{-1} x}{\sqrt{1 - x^2}}\,dx$

10. $\displaystyle\int \frac{dx}{\sqrt{1 - (x - 1)^2}}$

11. $\displaystyle\int \frac{du}{\sqrt{9 - 4u^2}}$

12. $\displaystyle\int \frac{u\,du}{\sqrt{1 - u^4}}$

13. $\displaystyle\int \frac{x\,dx}{1 + x^4}$

14. $\displaystyle\int \frac{x + 1}{\sqrt{1 - x^2}}\,dx$

15. $\displaystyle\int \frac{\cos x\,dx}{1 + \sin^2 x}$

16. $\displaystyle\int \frac{dx}{x\sqrt{4x^2 - 1}}$

17. $\displaystyle\int \frac{\cos x\,dx}{\sqrt{1 + \cos^2 x}}$

18. $\displaystyle\int_{-2}^{-1} \frac{dx}{x\sqrt{x^2 - 1}}$

19. $\displaystyle\int_{0}^{\pi/6} \frac{\sec^2 x\,dx}{\sqrt{1 - \tan^2 x}}$

20. $\displaystyle\int \frac{dx}{x\sqrt{x^4 - 1}}$

21. $\displaystyle\int \frac{dx}{\sqrt{x}\sqrt{1 - x}}$

22. $\displaystyle\int \frac{dx}{x\sqrt{x - 1}}$

23. $\displaystyle\int \sqrt{\frac{a + x}{a - x}}\,dx$

24. $\displaystyle\int \frac{\sin x\,dx}{\sqrt{4 - \cos^2 x}}$

25. $\displaystyle\int \frac{dx}{x^{2/3}(1 + x^{2/3})}$

* 26. $\displaystyle\int \frac{d\theta}{\sin^2 \theta + a^2 \cos^2 \theta}$

# 11 Logarithmic and Exponential Functions

## 11.1 INTRODUCTION

In more elementary courses the exponential function is developed first and then the logarithm function. To do this with some degree of rigor requires the introduction of more sophisticated ideas and methods than one is usually ready for at that time. Even at the level indicated by our current work, such development requires considerable effort. We shall use a different approach, with the goal of arriving at the same concepts. What we are aiming at is a definition of the exponential function $a^x$, $a > 0$, $x$ real, and then the definition of a logarithm function as the inverse of an exponential function. With available methods, however, it turns out that we can do this very effectively by reversing the process, defining the logarithm function first, following with the exponential function, and then showing that, as defined, these functions have the properties expected of them.

As we shall see in the next section, the basis of the method is to use a definite integral to define a function, a method which is novel at this stage but very effective.

## 11.2 THE LOGARITHM FUNCTION

From Section 9.9 we know that if a function is continuous in a closed interval, the definite integral of that function exists in the interval. Thus, if $f$ is continuous in an interval $[a, b]$, and $c$ is in $[a, b]$, we may define a function $F$ as

$$F(x) = \int_c^x f(u)\, du,$$

for $x$ in $[a, b]$. Then not only does $F$ exist in $[a, b]$, but it is differentiable, and by Theorem 9.10.1 we have $F'(x) = f(x)$.

In this manner we *define* a function $L(x)$ as

(11.2.1) $$L(x) = \int_1^x \frac{du}{u}, \qquad x > 0.$$

This function will turn out to be the logarithm function, but we do not yet make this claim.

It is clear that $L(x)$ as thus defined is a function with domain $(0, \infty)$. Other properties follow.

(i) $L(1) = 0$.

(ii) $L(x)$ is differentiable (hence continuous), and $L'(x) = \dfrac{1}{x}$.

(iii) $L(x)$ is increasing in $(0, \infty)$.

The proof of (i) follows at once by substituting $x = 1$ in (11.2.1). Part (ii) is an immediate consequence of Theorem 9.10.1. To prove (iii) we note that $L'(x) = 1/x > 0$ for each value of $x$ in the domain of $L$, and the increasing nature of $L$ is a consequence of Section 5.3.

We now include some additional properties that are suggestive of the logarithm function.

(iv) $L(x) + L(y) = L(xy)$, $x > 0$, $y > 0$.
(iv') $L(x) - L(y) = L(x/y)$, $x > 0$, $y > 0$.
(v) $L(x^k) = kL(x)$, $x > 0$, $k$ rational.
(vi) $\lim\limits_{x \to \infty} L(x) = \infty$, $\lim\limits_{x \to 0^+} L(x) = -\infty$.

To prove (iv) we express $L(x)$ and $L(y)$ by means of (11.2.1). Thus

$$L(x) + L(y) = \int_1^x \frac{du}{u} + \int_1^y \frac{du}{u}.$$

In the second integral we let $u = v/x$, $du = dv/x$. Then

$$\int_1^y \frac{du}{u} = \int_x^{xy} \frac{dv/x}{v/x} = \int_x^{xy} \frac{dv}{v} = \int_x^{xy} \frac{du}{u}.$$

Hence

$$L(x) + L(y) = \int_1^x \frac{du}{u} + \int_x^{xy} \frac{du}{u} = \int_1^{xy} \frac{du}{u} = L(xy),$$

and (iv) is proved.

(iv') is proved by writing $x = y(x/y)$, and then applying (iv) as follows.

$$L(x) = L\left(y \cdot \frac{x}{y}\right) = L(y) + L\left(\frac{x}{y}\right).$$

To prove (v) we let $y = x$ in (iv) and get $L(x^2) = 2L(x)$. Then we find

$$L(x^3) = L(x^2 \cdot x) = L(x^2) + L(x) = 2L(x) + L(x) = 3L(x).$$

In general, if $n$ is a positive integer, we get

(11.2.2) $$L(x^n) = nL(x);$$

the rigorous proof is carried out by mathematical induction. We also show that Equation (11.2.2) holds if $n$ is a negative integer by letting $n = -m$, $m > 0$. Then

$$L(x^n) = L(x^{-m}) = L(1/x^m) = -L(x^m)$$
$$= -mL(x) = nL(x);$$

we used (iv') and (i) in getting $L(1/x^m) = -L(x^m)$, and (11.2.2) in getting $L(x^m) = mL(x)$. We now show that (11.2.2) holds if $n = 1/m$, where $m$ is a positive integer, by writing

$$L(x) = L[(x^{1/m})^m] = mL(x^{1/m}), \quad \text{hence} \quad L(x^{1/m}) = \frac{1}{m}L(x).$$

Now let $n$ be a rational number, $n = p/q$, where $p$ and $q$ are integers with $q > 0$. We have

$$L(x^{p/q}) = L[(x^{1/q})^p] = pL(x^{1/q}) = p \cdot \frac{1}{q} L(x) = \frac{p}{q} L(x).$$

In this manner (v) has been proved if $k$ is any rational number, positive, negative, or zero.

We may now prove (vi). We have, by (11.2.2), if $n$ is a positive integer,

$$L(2^n) = nL(2).$$

But $L(2) > 0$, and hence $\lim_{n \to \infty} L(2^n) = \infty$. Thus the first limit in (vi) holds if $x = 2^n$. Since $L(x)$ is an increasing function of $x$, it follows that $\lim_{x \to \infty} L(x) = \infty$. The second part of (vi) is proved in a similar manner by considering $L(2^{-n})$, $n$ a positive integer.

Up to this point it has been shown that the function $L(x)$ has all the properties expected of a logarithm, except for its relation to exponents. This relation must await the following development.

## 11.3 THE EXPONENTIAL FUNCTION

The function $L(x)$ defined in Section 11.2 is continuous, differentiable, and increasing in the domain $(0, \infty)$. The range of $L(x)$ has been seen to be $(-\infty, \infty)$. From the discussion of Section 10.5 it follows that $L(x)$ has an inverse function which we call $E(x)$. Then $E(x)$ is continuous, differentiable, and increasing, with domain $(-\infty, \infty)$ and range $(0, \infty)$.

**DEFINITION 11.3.1.** *$E(x)$ is the inverse function of $L(x)$.*

It follows from properties of inverse functions that

(11.3.1) $\quad E(L(x)) = x, \quad 0 < x < \infty; \quad L(E(x)) = x, \quad -\infty < x < \infty;$

all properties of $E(x)$ are a consequence of this inverse relation. We list some of these properties.

(a) $E(0) = 1$.
(b) $E(x)$ is differentiable (hence continuous), and $E'(x) = E(x)$.
(c) $E(x)$ is increasing in $(-\infty, \infty)$.
(d) $E(x + y) = E(x)E(y)$.
(e) $\lim_{x \to \infty} E(x) = \infty, \quad \lim_{x \to -\infty} E(x) = 0$.
(f) If $e = E(1)$, then $E(x) = [E(1)]^x = e^x$, $x$ rational.

Property (a) follows from (11.3.1), since $E(L(1)) = 1$, and $L(1) = 0$.

To prove (b) we note that the differentiability of $E(x)$ was stated above. To find $E'(x)$, we differentiate, by the chain rule, the relation

$$L(E(x)) = x,$$

and get

$$L'(E(x)) \cdot E'(x) = 1.$$

But $L'(E(x)) = 1/E(x)$, by property (ii) of $L(x)$. Hence $E'(x) = E(x)$.

(c) follows from the inverse relation of $E(x)$ and $L(x)$ and the increasing nature of $L(x)$.

To prove (d), let $u = E(x)$, $v = E(y)$; then $x = L(u)$, $y = L(v)$, and we have, using the relation $L(u) + L(v) = L(uv)$,

$$E(x + y) = E[L(u) + L(v)] = E[L(uv)] = uv = E(x)E(y).$$

Property (e) follows from the relation between $E(x)$ and $L(x)$, and property (vi) for $L(x)$.

We turn to (f). If we let $y = x$ in (d) we get $E(2x) = [E(x)]^2$. Then, letting $y = 2x$, we get $E(3x) = [E(x)]^3$, and by induction

(11.3.2) $\qquad E(nx) = [E(x)]^n, \qquad n$ a positive integer.

With $x = 1$ in (11.3.2) we have $E(n) = [E(1)]^n$, which establishes (f) when $x$ is a positive integer.

Now let $x = 1/n$ in (11.3.2); then

$$E(1) = [E(1/n)]^n \quad \text{and} \quad E(1/n) = [E(1)]^{1/n}.$$

With $x = 1/m$ in (11.3.2), we get

$$E(n/m) = [E(1/m)]^n = \{[E(1)]^{1/m}\}^n = [E(1)]^{n/m},$$

and (f) has been proved when $x$ is any positive rational number. Now let $y = -x$ in (d). Then

$$E(x)E(-x) = E(0) = 1 \quad \text{and} \quad E(-x) = 1/E(x);$$

from this it follows that (f) holds also for $x$ any negative rational number, and property (f) is established.

Property (f) states the equality of $E(x)$ and $e^x$ for any *rational* number $x$, where $e = E(1)$. But $E(x)$ is defined for any *real* number $x$. We may then use the same relation to *define* $e^x$ if $x$ is any real number.

**DEFINITION 11.3.2** *For any real number $x$, we define $e^x$ as*

$$e^x = E(x).$$

Property (v) of Section 11.2 yields the relation, in terms of the function $E$,

$$x^k = E(kL(x)), \quad k \text{ rational}.$$

We use this relation to define $x^r$ if $r$ is any real number, and combine the resulting relation with Definition 11.3.2.

**DEFINITION 11.3.3.** *If $x > 0$ and $r$ is any real number, we define*

(11.3.3) $$x^r = e^{rL(x)}.$$

Property (v) of Section 11.2 may now be extended so as to apply to any real exponent in the following result.

**THEOREM 11.3.1.** *If $x > 0$ and $r$ is real, then $L(x^r) = rL(x)$.*

*Proof.* The desired relation is simply an equivalent form of equation (11.3.3).

Now we can extend the power rule of differentiation, which we previously proved for rational exponents, to any real exponent.

**THEOREM 11.3.2.** *If $r$ is any real number, then*

$$\frac{d}{dx} x^r = rx^{r-1}.$$

*Proof.* By Definition 11.3.3, $x^r = e^{rL(x)}$. We know that

$$\frac{d}{dx} e^u = \frac{d}{dx} E(u) = E(u) \frac{du}{dx} = e^u \frac{du}{dx},$$

by using Property (b). Hence

$$\frac{d}{dx} x^r = \frac{d}{dx} [e^{rL(x)}] = e^{rL(x)} \frac{d}{dx} [rL(x)] = e^{rL(x)} \cdot \frac{r}{x} = x^r \cdot \frac{r}{x} = rx^{r-1}.$$

(Familiar laws of exponents hold. See Exercise 7.) This completes the proof.

We have now established that $E(x)$ is an exponential function, $e^x$, for any real $x$, where $e = E(1)$ is the number such that

$$\int_1^e \frac{du}{u} = 1.$$

It then follows that $L(e) = 1$, and that $L(x)$ is a logarithm function in the sense that, for any number $x$, $L(x)$ is the exponent of the power of $e$ that equals $x$; that is,

(11.3.4) $$x = e^{L(x)}.$$

In particular, $L(x)$ is the logarithm of $x$ to the base $e$,

$$L(x) = \log_e x.$$

To make the development complete it remains to show that, as defined, $e$ is the same number that is used as the base of the natural or Naperian logarithms.

We know from the properties of $L(x)$ that $L'(x) = 1/x$, and that $L'(1) = 1$. Using the definition of derivative, we express the latter statement as

(11.3.5) $$\lim_{h \to 0} \frac{L(1+h) - L(1)}{h} = 1.$$

Now $L(1) = 0$, and $(1/h)L(1+h) = L[(1+h)^{1/h}]$ by Property (v). Hence (11.3.5) becomes

$$\lim_{h \to 0} L[(1+h)^{1/h}] = 1.$$

Since the inverse of $L(x)$ is continuous, and $L(e) = 1$, it follows that

(11.3.6) $$\lim_{h \to 0} (1+h)^{1/h} = e.$$

This limit is one of the basic forms in which the base of Naperian logarithms is expressed, and we have now shown that the number $E(1)$, which we have denoted by $e$, is in fact the number $e$ used for Naperian logarithms. This is what we set out to do.

We are accordingly justified in the use of the letter $e$ as it is standardly used in mathematics, and we may henceforth write $E(x)$ as $e^x$, and $L(x)$ as $\ln x$, the logarithm of $x$ to the base $e$, or *natural logarithm* of $x$.

Let us extend the definitions of exponential function and logarithmic function. If $a > 0$, we *define $a^x$* as

(11.3.7) $$a^x = e^{x \ln a}.$$

Then $a^x$ is an exponential function, with similar properties to $e^x$. Further, if $x = a^u$, we *define u as the logarithm of x to the base a*, and write

(11.3.8) $$u = \log_a x.$$

## EXERCISE GROUP 11.1

1. From (11.3.6) we know that, if $n$ is a positive integer,
$$e = \lim_{n \to \infty} \left(1 + \frac{1}{n}\right)^n$$
The function $(1 + 1/n)^n$ thus approximates $e$. The larger $n$ is, the better the approximation is. Evaluate this function for $n = 3, 5, 7$, to three decimal places.

2. It will be shown in Chapter 23 that $e$ can be approximated as closely as we wish by the sum
$$2 + \frac{1}{2!} + \frac{1}{3!} + \cdots + \frac{1}{n!},$$
where $n$ is a positive integer. Evaluate this sum to three decimal places for $n = 3, 5, 7$.

3. Determine the values of $x$ for which each equation is valid:
   (a) $\ln(4x^2 - 9) = \ln(2x + 3) + \ln(2x - 3)$
   (b) $\ln(x^3 - 2x^2) = 2 \ln x + \ln(x - 2)$
   (c) $\ln(2 - x) = \ln(6 - x - x^2) - \ln(x + 3)$
   (d) $\ln x = \ln(x^3 + x^2 - 2x) - \ln(x - 1) - \ln(x + 2)$

4. Show that $f''(x) = 2xf'(x)$ for the function $f$ defined as $f(x) = \int_0^x e^{u^2} du$.

5. If $f(x) = x \int_0^x e^{t^2} dt$, prove that $xf'(x) - f(x) = x^2 e^{x^2}$.

6. In each of the following a function of $x$ is defined by means of a definite integral. Find an explicit expression for the function which does not involve integrals.
   (a) $\int_0^x \frac{du}{\sqrt{u+1}}$
   (b) $\int_0^{x^2-1} \frac{du}{\sqrt{u+1}}$
   (c) $\int_x^{x^2-1} \frac{du}{\sqrt{u+1}}$
   (d) $\int_0^x e^u \, du$
   (e) $\int_1^x \frac{\ln u}{u} du$
   (f) $\int_0^{x^2} \sin u \cos u \, du$

7. Prove the following properties of real exponents, $x > 0$:
   (a) $x^a x^b = x^{a+b}$   (b) $x^a/x^b = x^{a-b}$   (c) $(x^a)^b = x^{ab}$

8. Give an alternate proof that the function $L(x)$ is increasing by using (11.2.1) to determine $L(b) - L(a)$, and showing that this difference is positive if $a < b$.

9. Show that if $a > e$, then $\log_a x > \ln x$ if $0 < x < 1$, and $\log_a x < \ln x$ if $x > 1$.

10. Give an alternate proof of property (f) as follows. Show that $L\{[E(1)]^x\} = x$. Then deduce $E(x) = [E(1)]^x$.

11. Use the binomial expansion to show, if $n$ is a positive integer,
$$\left(1 + \frac{1}{n}\right)^n = 1 + 1 + \frac{1 - \frac{1}{n}}{2!} + \frac{\left(1 - \frac{1}{n}\right)\left(1 - \frac{2}{n}\right)}{3!} + \cdots$$
$$+ \frac{\left(1 - \frac{1}{n}\right)\left(1 - \frac{2}{n}\right) \cdots \left(1 - \frac{n-1}{n}\right)}{n!}.$$

12. Derive (11.3.6) in the following manner. Apply the Law of the Mean for Integrals to the integral

$$\int_1^{1+h} \frac{du}{u} = L(1+h)$$

in order to derive the result $(1+h)^{1/h} = e^{1/\xi}$, $1 < \xi < 1+h$. Use this to derive (11.3.6).

13. Find $F'(x)$ if $F(x)$ is defined as

$$F(x) = \int_0^{x^2} f(u)\,du,$$

where $f$ is continuous in a suitable domain. [Suggestion: Find $g'(x)$ for $g(x) = \int_0^x f(u)\,du$, and use the chain rule.]

14. Define $F(x) = \int_{r(x)}^{s(x)} f(u)\,du$, where $f$, $r$, and $s$ have suitable continuity and differentiability properties.
    (a) If $g(x) = \int_a^x f(u)\,du$, express $F$ in terms of $g$.
    (b) Find $F'(x)$.

* 15. Apply Theorem 9.9.3 to show that

$$e^{1/(1+h)} < (1+h)^{1/h} < e, \quad \text{if} \quad h > 0,$$

and

$$e < (1+h)^{1/h} < e^{1/(1+h)}, \quad \text{if} \quad -1 < h < 0.$$

16. Prove: $\log_a x = \log_b x / \log_b a$.

## 11.4 GRAPHS OF LOGARITHMIC AND EXPONENTIAL FUNCTIONS

The domain of $\ln x$ is the interval $(0, \infty)$. From the representation

$$\ln x = \int_1^x \frac{du}{u},$$

we see that $\ln x < 0$ if $0 < x < 1$, and $\ln x > 0$ if $x > 1$. Since $\lim_{x \to 0^+} \ln x = -\infty$, the negative $y$ axis is a vertical asymptote of the graph of $\ln x$. Furthermore,

$$D \ln x = \frac{1}{x}, \qquad D^2 \ln x = -\frac{1}{x^2}.$$

Thus, $D \ln x$ is never zero, and there are no critical points or extrema for the function. Also, $D^2 \ln x < 0$ for $x > 0$, and the graph of $\ln x$ is concave down everywhere in the domain. Finally, $\lim_{x \to \infty} \ln x = \infty$. With this information the graph of $\ln x$ may be drawn, and appears in Fig. 11.4.1.

The graph of the function $e^x$ is symmetric to that of $\ln x$ in the line $y = x$. The negative $x$ axis is a horizontal asymptote; the curve is concave up for all $x$; and $\lim_{x \to \infty} e^x = \infty$. The graph of $e^x$ also appears in Fig. 11.4.1.

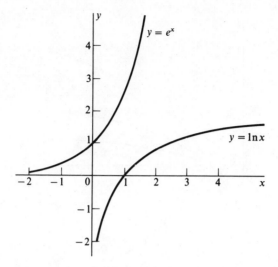

Fig. 11.4.1

We now relate the graphs of other logarithmic and exponential functions to those in Fig. 11.4.1. If $a > e$, we have

$$\log_a x \begin{cases} > \ln x & \text{if } 0 < x < 1 \\ < \ln x & \text{if } x > 1. \end{cases}$$

If $1 < a < e$, then

$$\log_a x \begin{cases} < \ln x & \text{if } 0 < x < 1 \\ > \ln x & \text{if } x > 1. \end{cases}$$

If $0 < a < 1$, let $b = 1/a$. Then $b > 1$ and

$$\log_a x = -\log_{1/a} x = -\log_b x.$$

From these relations the general appearance of the graph of $\log_a x$ may be determined in each case. Fig. 11.4.2 (page 306) shows the general appearance of the graph for each type described above, in relation to the graph of $\ln x$.

In a similar manner Fig. 11.4.3 (page 306) shows the graph of $a^x$ for different values of $a > 0$, in relation to the graph of $e^x$.

It is usually advantageous to try to relate the graph of a function involving logarithms, or of a function involving exponentials, to the graph of $\ln x$ or of $e^x$.

*Example 11.4.1.* Draw the graph of $\ln x^2$.

*Solution.* We have

$$\ln x^2 = \begin{cases} 2 \ln x & \text{if } x > 0 \\ 2 \ln (-x) & \text{if } x < 0. \end{cases}$$

The graph may now be drawn, and appears in Fig. 11.4.4 (page 307).

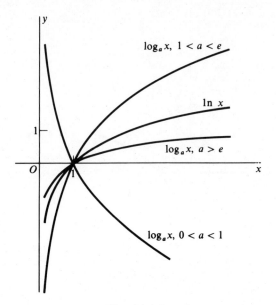

Fig. 11.4.2

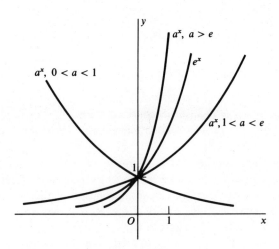

Fig. 11.4.3

**Example 11.4.2.** Draw the graph of $e^{2x+1}$.

*Solution.* We have
$$e^{2x+1} = e \cdot e^{2x} = e \cdot (e^2)^x.$$
The graph of $(e^2)^x$ is of the type $a^x$, where $a = e^2 > e$; then $e(e^2)^x$ will have a similar graph, but with each ordinate multiplied by $e$ (Fig. 11.4.5).

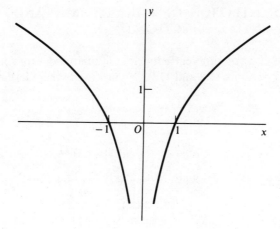

Fig. 11.4.4

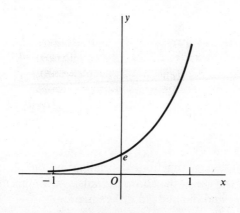

Fig. 11.4.5

**EXERCISE GROUP 11.2**

Draw the graph of each of the following functions. In each case try to relate the type of graph to that of $\ln x$ or of $e^x$. The essential appearance of each graph may be found without recourse to a table of values.

1. $\ln(-x)$
2. $\ln(2x)$
3. $\ln(3-x)$
4. $\ln(x^2 - x)$
5. $\ln|x|$
6. $\log_{1/2}(2x - 1)$
7. $3^{-x}$
8. $3^{1-x}$
9. $(1/2)^{1-x}$

## 11.5 DIFFERENTIATION OF LOGARITHMIC AND EXPONENTIAL FUNCTIONS

The basic formulas for differentiating logarithmic and exponential functions were derived in Sections 11.2 and 11.3. We now write them in the notation of $e$ and $\ln$ as

(11.5.1) $$D \ln x = \frac{1}{x}, \qquad d \ln u = \frac{du}{u},$$

(11.5.2) $$De^x = e^x, \qquad de^u = e^u \, du.$$

Let us extend these formulas. With $a > 0$, the function $a^x$ was defined as

$$a^x = e^{x \ln a}.$$

Then, using (11.5.2), we have

(11.5.3) $$Da^x = e^{x \ln a} \ln a = a^x \ln a.$$

In Section 11.3 we defined $u = \log_a x$ as equivalent to $x = a^u$. To find $D_x u$ we differentiate the relation $x = a^u$ with respect to $x$, making use of (11.5.3), and find

$$1 = a^u \ln a \, D_x u, \qquad D_x u = \frac{1}{a^u \ln a} = \frac{1}{x \ln a}.$$

We have thus obtained the following generalizations of (11.5.1) and (11.5.2):

(11.5.4) $$D \log_a x = \frac{1}{x \ln a}, \qquad d \log_a u = \frac{du}{u \ln a}.$$

(11.5.5) $$Da^x = a^x \ln a, \qquad da^u = a^u \ln a \, du.$$

The differentiation of functions depending on logarithmic or exponential functions involves the use of the above formulas, in conjunction with previously developed formulas and methods of differentiation.

**Example 11.5.1** Find $D(x \ln x)$.

*Solution.* Using the product formula, we get

$$D(x \ln x) = x \, D \ln x + \ln x \, Dx = x \cdot \frac{1}{x} + \ln x$$
$$= 1 + \ln x.$$

**Example 11.5.2.** Find $\dfrac{d}{dx}(xe^{2x})$.

*Solution.* We get

$$\frac{d}{dx}(xe^{2x}) = x \frac{d}{dx} e^{2x} + e^{2x} = 2xe^{2x} + e^{2x} = e^{2x}(2x + 1).$$

**Example 11.5.3.** Find $\dfrac{d}{dx} \ln \sqrt{1-x^2}$.

*Solution.* It is convenient to use a property of logarithms:
$$\ln \sqrt{1-x^2} = \tfrac{1}{2} \ln(1-x^2).$$
Then
$$\frac{d}{dx} \ln \sqrt{1-x^2} = \frac{1}{2} \frac{d}{dx} \ln(1-x^2) = \frac{1}{2} \cdot \frac{-2x}{1-x^2} = -\frac{x}{1-x^2}.$$

The method of *logarithmic differentiation* is often conveniently used to find the derivative of a function involving products and quotients of powers of functions. This method, which provides for the introduction of logarithms and the use of their properties, is illustrated in the following example.

**Example 11.5.4.** Find $dy/dx$ if $y = \dfrac{\sqrt{x^2-1}}{\sqrt[3]{x+2}}$.

*Solution.* We equate logarithms of both members,
$$\ln y = \tfrac{1}{2} \ln(x^2-1) - \tfrac{1}{3} \ln(x+2),$$
and differentiate implicitly. This gives
$$\frac{1}{y} \frac{dy}{dx} = \frac{1}{2} \frac{2x}{x^2-1} - \frac{1}{3} \frac{1}{x+2}.$$
Now multiplying both sides by $y$, we have
$$\frac{dy}{dx} = y \left[ \frac{x}{x^2-1} - \frac{1}{3(x+2)} \right] = \frac{\sqrt{x^2-1}}{\sqrt[3]{x+2}} \left[ \frac{x}{x^2-1} - \frac{1}{3(x+2)} \right].$$

We consider, finally, a function of the form
$$y = u^v,$$
where both $u$ and $v$ are differentiable functions of $x$. We may find $y'$ either from the form $y = e^{v \ln u}$ or by logarithmic differentiation. Using the latter method, we have
$$\ln y = v \ln u, \qquad \frac{y'}{y} = v \frac{u'}{u} + v' \ln u,$$
where primes denote derivatives with respect to $x$; hence

(11.5.6) $\qquad y' = u^v \left( \dfrac{vu'}{u} + v' \ln u \right) = v u^{v-1} u' + (u^v \ln u) v'.$

This result is of interest in that the first term on the right is what we get by using the power rule to differentiate $u^v$ (as though $v$ were constant); while the second term would result from using the formula for the derivative of the exponential function (as though $u$ were constant). In practice, one generally uses one of the methods described above, rather than the formula.

**Example 11.5.5.** Differentiate $x^{x^2+1}$.

*Solution.* Set $y = x^{x^2+1}$. Then $\ln y = (x^2 + 1)\ln x$, and we get

$$\frac{y'}{y} = (x^2 + 1)\frac{1}{x} + (\ln x)(2x);$$

we then find

$$y' = x^{x^2+1}\left(\frac{x^2+1}{x} + 2x \ln x\right) = x^{x^2}(x^2 + 1 + 2x^2 \ln x).$$

### EXERCISE GROUP 11.3

Find the derivative of the given function in Exercises 1–30.

1. $xe^{2x}$
2. $e^x(x-1)$
3. $xe^{x^2}$
4. $x \ln 2x$
5. $(\ln x)^2$
6. $x(\ln x - 1)$
7. $\ln \sin x$
8. $\ln \cos x$
9. $e^x \sin x$
10. $e^{2x} \ln x$
11. $e^{2 \sin x}$
12. $\ln \tan x$
13. $e^{2x}(2 \sin 3x - 3 \cos 3x)$
14. $e^{3x}(3 \cos 2x + 2 \sin 2x)$
15. $e^x \ln(x-1)$
16. $e^{-x}(\sin x - \cos x)$
17. $\ln \dfrac{e^x}{1+e^x}$
18. $\tan^{-1} e^x$
19. $\sin^{-1} e^{2x}$
20. $\ln(x + \sqrt{x^2 - a^2})$
21. $\ln \dfrac{a + \sqrt{x^2 + a^2}}{x}$
22. $\ln(\sec x + \tan x)$
23. $\ln \sqrt{\dfrac{1+x^2}{1-x^2}}$
24. $\ln \sqrt{\dfrac{1-\cos x}{1+\cos x}}$
25. $x \cot x - \ln \sin x$
26. $\tan^{-1} \tfrac{1}{2}(e^x - e^{-x})$
27. $x \tan^{-1} x - \ln \sqrt{1+x^2}$
28. $\ln \cos^{-1} x$
29. $\ln \tan\left(\dfrac{x}{2} + \dfrac{\pi}{4}\right)$
30. $\sec x \tan x + \ln(\sec x + \tan x)$

In Exercises 31–35 show that the given function satisfies the differential equation which follows it.

31. $f(x) = xe^{2x}$; $f'' - 4f' + 4f = 0$
32. $f(x) = (6x - 5)e^x$; $f'' + 3f' + 2f = 36xe^x$
33. $f(x) = x^{1/2} \ln x$; $4x^2 f'' + f = 0$

34. $f(x) = e^{3x} \cos 2x$; $f'' - 6f' + 13f = 0$
35. $f(x) = \sin(2 \ln x)$; $x^2 f'' + xf' + 4f = 0$
36. Evaluate $x^2 f'' - 2xf' + 2f$ if $f(x) = 3 \ln x + 9/2$.
37. Show that a function $y$ defined by $(y+2)[\ln(y+2) - 1] = x - 1$ satisfies the equation
$$\frac{dy}{dx} = \frac{y+2}{x+y+1}.$$
38. Show that each of the functions
$$y = e^x \cos x, \quad y = e^x \sin x, \quad y = e^{-x} \cos x, \quad y = e^{-x} \sin x$$
satisfies the equation $y^{(4)} + 4y = 0$.
39. If $y = x \int_0^x e^{t^2} \, dt$, show that $xy' - y = x^2 e^{x^2}$.
40. Derive the formula in (11.5.6) by differentiating $y = e^{v \ln u}$ directly.

In Exercises 41–49 find the derivative of the given function. Logarithmic differentiation is suitable.

41. $\sqrt[3]{\dfrac{1+x^2}{1-x^2}}$
42. $\dfrac{\sqrt[3]{x^2-1}}{\sqrt{x^2+1}}$
43. $\left(1 + \dfrac{1}{x}\right)^x$
44. $x^x$
45. $(\ln x)^x$
46. $x^{\ln x}$
47. $(\sin x)^x$
48. $x^{\sin x}$
49. $x^{e^x}$

In Exercises 50–56 draw the graph of the given function, making use of extrema, points of inflection, etc.

50. $x - \ln x$
51. $\ln \sin x$
52. $x \ln x$
53. $x^2 \ln x$
54. $x^2 e^x$
55. $(2x^2 - 3x)e^x$

* 56. Show that the rectangle of maximum area which can be inscribed between the curve $y = e^{-x^2}$ and the $x$ axis has two vertices at the points of inflection of the curve.

## 11.6 INTEGRALS LEADING TO LOGARITHMIC AND EXPONENTIAL FUNCTIONS

The differentiation formula for $\ln u$ applies when $u > 0$. If $u < 0$ we obtain the same result for $\ln(-u)$; thus, $-u > 0$ and we have

$$\frac{d}{du} \ln(-u) = \frac{-1}{-u} = \frac{1}{u}.$$

Hence

$$\int \frac{du}{u} = \begin{cases} \ln u & \text{if } u > 0, \\ \ln(-u) & \text{if } u < 0, \end{cases}$$

and we have the following formula which is valid for both $u > 0$ and $u < 0$:

(11.6.1) $$\int \frac{du}{u} = \ln|u| + C.$$

The differentiation formulas for $e^u$ and $a^u$ yield the following integration formulas directly.

(11.6.2.) $$\int e^u \, du = e^u + C.$$

(11.6.3) $$\int a^u \, du = \frac{a^u}{\ln a} + C, \quad a > 0.$$

Formula (11.6.1) fills the gap left in the power formula of integration (9.4.3), in which the value $n = -1$ was excluded.

**Example 11.6.1.** Evaluate $\int \frac{dx}{2x+1}$.

*Solution.* Set $u = 2x + 1$, $du = 2 \, dx$. Then

$$\int \frac{dx}{2x+1} = \frac{1}{2} \int \frac{du}{u} = \frac{1}{2} \ln|u| + C$$

$$= \frac{1}{2} \ln|2x+1| + C.$$

**Example 11.6.2.** Evaluate $\int \cot x \, dx$.

*Solution.* We have

$$\int \cot x \, dx = \int \frac{\cos x}{\sin x} \, dx.$$

Now let $u = \sin x$, $du = \cos x \, dx$, and we find

$$\int \cot x \, dx = \int \frac{du}{u} = \ln|u| + C$$

$$= \ln|\sin x| + C.$$

This result furnishes the integral of one of the trigonometric functions that we could not find in Section 10.4.

**Example 11.6.3.** Evaluate $\int \frac{\sin x \, dx}{1 - \cos x}$.

*Solution.* Here we let $u = 1 - \cos x$, $du = \sin x \, dx$, and we find

$$\int \frac{\sin x \, dx}{1 - \cos x} = \int \frac{du}{u} = \ln|u| + C$$

$$= \ln(1 - \cos x) + C.$$

We omit the absolute value signs since $1 - \cos x$ is never negative.

The clue to the integration in each of the above examples was the fact that the numerator of the fraction was (possibly apart from a constant factor) the derivative of the denominator.

**Example 11.6.4.** Evaluate $\int \dfrac{e^x \, dx}{e^x + 1}$.

**Solution.** By letting $u = e^x + 1$ we obtain easily,
$$\int \frac{e^x \, dx}{e^x + 1} = \ln(e^x + 1) + C.$$

**Example 11.6.5.** Evaluate $\int \dfrac{dx}{e^x + e^{-x}}$.

**Solution.** We multiply numerator and denominator by $e^x$,
$$\int \frac{dx}{e^x + e^{-x}} = \int \frac{e^x \, dx}{e^{2x} + 1},$$
and let $e^x = u$, $e^x \, dx = du$. Then we get
$$\int \frac{e^x \, dx}{e^{2x} + 1} = \int \frac{du}{u^2 + 1} = \tan^{-1} u + C$$
$$= \tan^{-1} e^x + C,$$
which is the desired result.

## EXERCISE GROUP 11.4

Evaluate the given integral in Exercises 1–34.

1. $\int \dfrac{du}{2u - 3}$

2. $\int \dfrac{du}{au + b}$

3. $\int \dfrac{(x - 2) \, dx}{x^2 - 4x - 5}$

4. $\int \dfrac{e^{ax} \, dx}{e^{ax} + b}$

5. $\int \dfrac{x^2 \, dx}{x^3 - 2}$

6. $\int 3^x \, dx$

7. $\int_0^1 3^{x^2} x \, dx$

8. $\int 2^{\tan x} \sec^2 x \, dx$

9. $\int \dfrac{e^{\sqrt{x}}}{\sqrt{x}} \, dx$

10. $\int \dfrac{dx}{\sqrt{x} e^{\sqrt{x}}}$

11. $\int \dfrac{\sec^2 x \, dx}{1 + \tan x}$

12. $\int \dfrac{\sec^2 x \, dx}{2 - \tan x}$

13. $\int \tan x \, dx$

14. $\int \tan x \ln \cos x \, dx$

15. $\int \cot x \ln \sin x \, dx$

16. $\int \dfrac{dx}{x \ln x}$

17. $\int \dfrac{1 + \ln x}{x} \, dx$

18. $\int \dfrac{\ln x}{x} \, dx$

19. $\int \dfrac{(\ln x)^n}{x} \, dx$

20. $\int \dfrac{\csc x \cot x}{1 + \cot x} \, dx$

21. $\int \dfrac{e^t \, dt}{\sqrt{e^t - 1}}$

22. $\int \dfrac{\sqrt{u} \, du}{u \sqrt{u} - 2}$

23. $\int \dfrac{\sqrt{x} \, dx}{a^{3/2} - x^{3/2}}$

24. $\int \dfrac{dx}{x^{2/3}(a^{1/3} - x^{1/3})}$

25. $\int_{1}^{\sqrt{3}} \frac{\cot^{-1} x}{1+x^2} dx$

26. $\int \frac{dx}{\sqrt{1-x^2}\sin^{-1} x}$

27. $\int \frac{dx}{(1+x^2)\sqrt{\tan^{-1} x}}$

28. $\int \frac{dx}{(1+x^2)\tan^{-1} x}$

29. $\int \frac{\sin x\, dx}{\sqrt{3-\cos^2 x}}$

★ 30. $\int \frac{e^x\, dx}{\sqrt{e^{-2x}-e^{2x}}}$

31. $\int \frac{dx}{x(1+\ln x)}$

32. $\int_{0}^{(\ln 3)/4} \frac{e^{2x}\, dx}{1+e^{4x}}$

33. $\int_{1}^{x} 2^u\, du$

34. $\int_{0}^{1} e^x \sqrt{1+e^x}\, dx$

35. Find the area under the curve $y = (\ln x)/x$ in the interval $[1, e]$.

36. Find the area under the curve $2xy - 3y - 1 = 0$ in the interval $[2, 4]$.

## 11.7 THE HYPERBOLIC FUNCTIONS

The *hyperbolic functions* are defined directly in terms of $e^x$. It turns out that the properties of these functions are similar to those of the trigonometric functions, and in fact the names and abbreviations used are similar. For example, the *hyperbolic sine* of $x$ is abbreviated as sinh $x$; similar abbreviations are used for the other hyperbolic functions.

*The hyperbolic functions are defined as follows.*

$$\sinh x = \frac{e^x - e^{-x}}{2} = \frac{e^{2x} - 1}{2e^x}, \qquad -\infty < x < \infty$$

$$\cosh x = \frac{e^x + e^{-x}}{2} = \frac{e^{2x} + 1}{2e^x}, \qquad -\infty < x < \infty$$

$$\tanh x = \frac{\sinh x}{\cosh x} = \frac{e^x - e^{-x}}{e^x + e^{-x}} = \frac{e^{2x} - 1}{e^{2x} + 1}, \qquad -\infty < x < \infty$$

$$\coth x = \frac{\cosh x}{\sinh x} = \frac{e^x + e^{-x}}{e^x - e^{-x}} = \frac{e^{2x} + 1}{e^{2x} - 1}, \qquad x \neq 0$$

$$\operatorname{sech} x = \frac{1}{\cosh x} = \frac{2}{e^x + e^{-x}} = \frac{2e^x}{e^{2x} + 1}, \qquad -\infty < x < \infty$$

$$\operatorname{csch} x = \frac{1}{\sinh x} = \frac{2}{e^x - e^{-x}} = \frac{2e^x}{e^{2x} - 1}, \qquad x \neq 0$$

The fundamental identities connecting the hyperbolic functions differ from those for the trigonometric functions only in an occasional sign. We list some of them. The proofs are a direct application of the definitions and are left to the exercises.

(11.7.1) $\cosh^2 x - \sinh^2 x = 1$, $\tanh^2 x + \operatorname{sech}^2 x = 1$, $\coth^2 x - \operatorname{csch}^2 x = 1$

(11.7.2) $\sinh(x \pm y) = \sinh x \cosh y \pm \cosh x \sinh y$

(11.7.3)  $\cosh(x \pm y) = \cosh x \cosh y \pm \sinh x \sinh y$

(11.7.4)  $\sinh 2x = 2 \sinh x \cosh x$

(11.7.5)  $\cosh 2x = \cosh^2 x + \sinh^2 x = 2 \cosh^2 x - 1 = 1 + 2 \sinh^2 x$

The graphs of the hyperbolic functions may be obtained from their properties as derived from their definitions. We shall sketch the graphs of $y = \sinh x$ and $y = \cosh x$.

*The graph of $y = \sinh x$.*

We have $\sinh 0 = 0$. The function $\sinh x$ is an odd function of $x$ (Exercise 6), and the graph of this function is symmetric in the origin. We may construct a table of values, or use addition of ordinates. The graph is shown in Fig. 11.7.1.

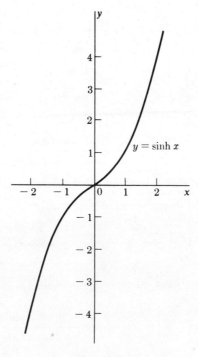

Fig. 11.7.1

*The graph of $y = \cosh x$.*

We have $\cosh 0 = 1$. The function $\cosh x$ is an even function of $x$ (Exercise 7), and the graph of this function is symmetric in the $y$ axis. Addition of ordinates again may be used. The graph appears in Fig. 11.7.2.

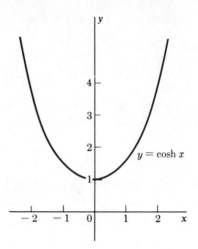

Fig. 11.7.2

The limits of $e^x$ as $x \to \infty$ and $x \to -\infty$ may be used to establish some of the asymptotic properties of the hyperbolic functions. These properties will be pursued in the exercises.

The formulas for the derivatives of the hyperbolic functions closely resemble those for the trigonometric functions and are:

$$D \sinh x = \cosh x \qquad D \cosh x = \sinh x$$
$$D \tanh x = \text{sech}^2 x \qquad D \coth x = -\text{csch}^2 x$$
$$D \text{sech } x = -\text{sech } x \tanh x \qquad D \text{csch } x = -\text{csch } x \coth x.$$

We shall derive some of them. We have

$$D \sinh x = D \frac{e^x - e^{-x}}{2} = \frac{e^x + e^{-x}}{2} = \cosh x;$$

$$D \cosh x = D \frac{e^x + e^{-x}}{2} = \frac{e^x - e^{-x}}{2} = \sinh x;$$

$$D \tanh x = D \frac{\sinh x}{\cosh x} = \frac{\cosh x \, D \sinh x - \sinh x \, D \cosh x}{\cosh^2 x}$$

$$= \frac{\cosh^2 x - \sinh^2 x}{\cosh^2 x} = \frac{1}{\cosh^2 x} = \text{sech}^2 x;$$

$$D \text{sech } x = D \frac{1}{\cosh x} = -\frac{1}{\cosh^2 x} D \cosh x = -\frac{\sinh x}{\cosh^2 x}$$

$$= -\frac{1}{\cosh x} \frac{\sinh x}{\cosh x} = -\text{sech } x \tanh x.$$

The other two formulas may be derived in the same way as those for tanh $x$ and sech $x$.

**Example 11.7.1.** Differentiate $\ln(\cosh ax)$.

Solution. We have

$$D \ln(\cosh ax) = \frac{D \cosh ax}{\cosh ax} = \frac{a \sinh ax}{\cosh ax} = a \tanh ax.$$

**Example 11.7.2.** Show that the graph of $y = \tanh x$ has a point of inflection at $x = 0$.

Solution. We have $y' = \text{sech}^2 x$, and

$$y'' = 2 \,\text{sech}\, x \, D \,\text{sech}\, x = -2 \,\text{sech}^2 x \tanh x.$$

Since $\tanh 0 = 0$ it follows that $y''(0) = 0$. Further,

$$\tanh x < 0 \quad \text{if} \quad x < 0 \quad \text{and} \quad \tanh x > 0 \quad \text{if} \quad x > 0.$$

Hence $y'' > 0$ if $x < 0$ and $y'' < 0$ if $x > 0$; there must then be a point of inflection at $x = 0$.


### EXERCISE GROUP 11.5

In Exercises 1–4 draw the graph of the specified function, giving particular attention to any horizontal or vertical asymptotes.

1. $y = \tanh x$
2. $y = \coth x$
3. $y = \text{sech}\, x$
4. $y = \text{csch}\, x$
5. Find the values of the hyperbolic functions at $x = 0$.
6. Prove that sinh $x$, tanh $x$, coth $x$, and csch $x$ are odd functions of $x$.
7. Prove that cosh $x$ and sech $x$ are even functions of $x$.
8. Prove formula (11.7.1).
9. Prove formula (11.7.2).
10. Prove formula (11.7.3).
11. Prove formula (11.7.4).
12. Prove formula (11.7.5).
13. Prove: $\tanh \dfrac{x}{2} = \dfrac{\sinh x}{\cosh x - 1}$.

\* 14. Prove that $y = e^x/2$ is an asymptotic curve to $y = \sinh x$ for $x > 0$, and $y = -e^{-x}/2$ is an asymptotic curve for $x < 0$.

\* 15. Prove that $y = e^x/2$ is an asymptotic curve to $y = \cosh x$ for $x > 0$, and $y = e^{-x}/2$ is an asymptotic curve for $x < 0$.

16. Prove that the graph of $y = e^x/2$ lies between the graphs of $y = \sinh x$ and $y = \cosh x$ for $x > 0$.

Find the derivative of the given function in Exercises 17–32.

17. $\sinh(3x - 2)$
18. $\cosh x^2$
19. $\cosh \sqrt{x}$
20. $\tanh(3x^2 - 1)$
21. $x \sinh x - \cosh x$
22. $x \cosh x - \sinh x$
23. $\sinh e^x$
24. $e^x \sinh x$
25. $\operatorname{sech}^3 ax$
26. $\sinh(\sin x)$
27. $(\sinh x)^x$
28. $\ln \operatorname{sech} x$
29. $\sin^{-1}(\cosh x)$
30. $\sin^{-1}(\tanh ax)$
31. $\ln|\sinh ax|$
32. $\ln\left|\tanh \dfrac{ax}{2}\right|$

33. Define $\theta$ by $\tan \theta = \sinh x$. Find $d\theta/dx$.

34. Show that if $y = a \sinh kx + b \cosh kx$, where $a$, $b$, and $k$ are constants, then $y'' = k^2 y$.

35. Show that each of the functions

$y = \cos x \cosh x$, $\quad y = \cos x \sinh x$, $\quad y = \sin x \cosh x$, $\quad y = \sin x \sinh x$

satisfies the equation $y^{(4)} + 4y = 0$.

36. Prove the identity $2 \tan^{-1} e^x = \tan^{-1}(\sinh x) + \pi/2$.

## 11.8 INTEGRALS OF HYPERBOLIC FUNCTIONS

Each of the differentiation formulas of Section 11.7 leads directly to an integration formula. The formulas, and the methods of working with them, are very similar to the corresponding formulas and methods for trigonometric functions. We may write the basic formulas immediately as:

$$\int \sinh x \, dx = \cosh x + C \qquad \int \cosh x \, dx = \sinh x + C$$

$$\int \operatorname{sech}^2 x \, dx = \tanh x + C \qquad \int \operatorname{csch}^2 x \, dx = -\coth x + C$$

$$\int \operatorname{sech} x \tanh x \, dx = -\operatorname{sech} x + C \qquad \int \operatorname{csch} x \coth x \, dx = -\operatorname{csch} x + C.$$

**Example 11.8.1.** Evaluate $\int \cosh (2 - 3x) \, dx$.

*Solution.* Let $2 - 3x = u$, $dx = -du/3$. Then

$$\int \cosh(2 - 3x) \, dx = -\frac{1}{3} \int \cosh u \, du = -\frac{1}{3} \sinh u + C$$

$$= -\frac{1}{3} \sinh(2 - 3x) + C.$$

**Example 11.8.2.** Evaluate $\displaystyle\int \frac{\cosh 2x}{\sinh^3 2x} \, dx$.

*Solution.* One method is to write

$$\int \frac{\cosh 2x}{\sinh^3 2x}\, dx = \int \frac{1}{\sinh^2 2x} \frac{\cosh 2x}{\sinh 2x}\, dx = \int \text{csch}^2\, 2x \coth 2x\, dx.$$

Then substitute $2x = u$, $dx = du/2$, and get

$$\int \text{csch}^2\, 2x \coth 2x\, dx = \frac{1}{2}\int \text{csch}^2\, u \coth u\, du = -\frac{1}{2}\int \text{csch}\, u\, d(\text{csch}\, u)$$

$$= -\frac{1}{2}\frac{\text{csch}^2 u}{2} + C = -\frac{1}{4}\text{csch}^2\, 2x + C.$$

A second method uses the substitution $u = \sinh 2x$, $du = 2\cosh 2x\, dx$. Then

$$\int \frac{\cosh 2x}{\sinh^3 2x}\, dx = \frac{1}{2}\int \frac{du}{u^3} = \frac{1}{2}\frac{u^{-2}}{-2} + C = -\frac{1}{4}\frac{1}{\sinh^2 2x} + C = -\frac{1}{4}\text{csch}^2\, 2x + C.$$

**EXERCISE GROUP 11.6**

Evaluate the integrals in Exercises 1–10.

1. $\int \sinh \frac{3x}{2}\, dx$
2. $\int \cosh(3x - 1)\, dx$
3. $\int \frac{1}{\sqrt{x}} \cosh \sqrt{x}\, dx$
4. $\int \text{sech}^2\, 3x\, dx$
5. $\int \tanh^2 x\, dx$
6. $\int \text{sech}^2\, 2x \tanh 2x\, dx$
7. $\int \sinh^2 x \cosh^3 x\, dx$
8. $\int \sinh^3 x \cosh^2 x\, dx$
9. $\int \tanh x\, dx$
10. $\int \coth 2x\, dx$

11. Find the area under the curve $y = \sinh x$ from $x = 0$ to $x = 1$.
12. Find the area under the curve $y = \cosh x$ from $x = 0$ to $x = 1$.
13. Draw the graph of the equation $x = \tanh y$, and find the area of the region bounded by the curve, the $y$ axis, and the line $y = 2$.

## 11.9 THE INVERSE HYPERBOLIC FUNCTIONS

Inverse functions may be defined directly for $\sinh x$, $\tanh x$, $\coth x$, and $\text{csch}\, x$. A restriction is required for the inverse functions of $\cosh x$ and $\text{sech}\, x$. Two of the inverses will be discussed; the others will be left to the exercises. Interestingly, it turns out that the inverse hyperbolic functions are directly expressible in terms of logarithmic functions.

It is clear from Fig. 11.7.1 that only one value of $x$ exists for each value of

sinh $x$. Thus the inverse hyperbolic sine, $\sinh^{-1} x$, exists for all real $x$. Let $y = \sinh^{-1} x$. Then
$$\sinh y = x.$$
This equation may be written, with use of the definition of sinh, as
$$\frac{e^{2y} - 1}{2e^y} = x \quad \text{or} \quad e^{2y} - 2xe^y - 1 = 0.$$
The latter equation is quadratic in $e^y$, and is solved for $e^y$ as
$$e^y = x \pm \sqrt{x^2 + 1}.$$
The minus sign cannot apply, since this would imply $e^y < 0$, which cannot hold. Hence the plus sign applies and we have

(11.9.1) $\qquad y = \sinh^{-1} x = \ln(x + \sqrt{x^2 + 1}), \qquad -\infty < x < \infty.$

We see from Fig. 11.7.2 that two values of $x$ exist for any value of $\cosh x$ (if $\cosh x > 1$). We *define the nonnegative value as the inverse hyperbolic cosine*. We write $y = \cosh^{-1} x$, $y \geq 0$. Then
$$\cosh y = x, \qquad x \geq 1.$$
This yields, with the definition of cosh,
$$\frac{e^{2y} + 1}{2e^y} = x \quad \text{or} \quad e^{2y} - 2xe^y + 1 = 0.$$
The solution of the latter equation for $e^y$ is

(11.9.2) $\qquad\qquad\qquad e^y = x \pm \sqrt{x^2 - 1}.$

Since $y \geq 0$, the larger value on the right of (11.9.2) is desired, the plus sign thus applies, and we have

(11.9.3) $\qquad y = \cosh^{-1} x = \ln(x + \sqrt{x^2 - 1}), x \geq 1.$

The graph of $\sinh^{-1} x$ is the reflection of the curve $y = \sinh x$ in the line $y = x$. The graph of $\cosh^{-1} x$ is the reflection in the line $y = x$ of the portion of the graph in Fig. 11.7.2 for which $x \geq 0$.

No restriction on the value of the inverse function is needed for the remaining inverse hyperbolic functions, except in the case of $\text{sech}^{-1} x$. For this function we *define*
$$y = \text{sech}^{-1} x, \qquad 0 < x \leq 1, \qquad y \geq 0.$$

### EXERCISE GROUP 11.7

Draw the graph of each function in Exercises 1–6.

1. $y = \sinh^{-1} x$
2. $y = \cosh^{-1} x$
3. $y = \tanh^{-1} x$
4. $y = \coth^{-1} x$
5. $y = \text{sech}^{-1} x$
6. $y = \text{csch}^{-1} x$

7. Show that the minus sign in (11.9.2) leads to a value of $y$ which is the negative of the value of $y$ in (11.9.3).

8. Show that $\tanh^{-1} x = \dfrac{1}{2} \ln \dfrac{1+x}{1-x}$, $|x| < 1$.

9. Show that $\coth^{-1} x = \dfrac{1}{2} \ln \dfrac{x+1}{x-1}$, $|x| > 1$.

10. Show that $\operatorname{sech}^{-1} x = \ln \dfrac{1 + \sqrt{1-x^2}}{x}$, $0 < x \le 1$.

11. Show that $\operatorname{csch}^{-1} x = \ln \left( \dfrac{1}{x} + \sqrt{\dfrac{1}{x^2} + 1} \right)$, $x \ne 0$. This function may also be written as $\ln \dfrac{1 \pm \sqrt{x^2 + 1}}{x}$. When does each sign apply?

12. Show that $\sinh^{-1} x = \cosh^{-1} \sqrt{x^2 + 1}$.

13. Show that $\cosh^{-1} x = \sinh^{-1} \sqrt{x^2 - 1}$.

14. Let $x = \ln(\sec \theta + \tan \theta)$. Prove:
    (a) $\tan \theta = \sinh x$    (b) $\sec \theta = \cosh x$
    (c) $\sin \theta = \tanh x$    (d) $\tan \theta/2 = \tanh x/2$

*Note:* The given equation defines $\theta$ as a function of $x$, which is written as $\theta = \operatorname{gd} x$, the *gudermannian* of $x$.

★ 15. Exercise 14 may be used to evaluate $\int \sec \theta \, d\theta$ as follows. Make the change of variable $\sec \theta = \cosh x$, and justify each of the following steps. We have $d\theta = \operatorname{sech} x \, dx$; then
$$\int \sec \theta \, d\theta = \int dx = x + c = \ln(\sec \theta + \tan \theta) + C, \qquad -\pi/2 < \theta < \pi/2.$$

★ 16. Use the method of Exercise 15 to show that
$$\int \sec^3 \theta \, d\theta = \tfrac{1}{2}[\sec \theta \tan \theta + \ln(\sec \theta + \tan \theta)] + C.$$

## 11.10 DERIVATIVES OF INVERSE HYPERBOLIC FUNCTIONS

The derivatives of the inverse hyperbolic functions are the following.

$$D \sinh^{-1} x = \dfrac{1}{\sqrt{x^2 + 1}}, \qquad -\infty < x < \infty.$$

$$D \cosh^{-1} x = \dfrac{1}{\sqrt{x^2 - 1}}, \qquad x > 1.$$

$$D \tanh^{-1} x = \dfrac{1}{1 - x^2}, \qquad |x| < 1.$$

$$D \coth^{-1} x = \dfrac{1}{1 - x^2}, \qquad |x| > 1.$$

$$D \operatorname{sech}^{-1} x = -\frac{1}{x\sqrt{1-x^2}}, \quad 0 < x < 1.$$

$$D \operatorname{csch}^{-1} x = -\frac{1}{x\sqrt{x^2+1}}, \quad x \neq 0.$$

Derivations of these formulas may be based on the appropriate inverse relation in each case. For example, to obtain $D \sinh^{-1} x$ let us write

$$y = \sinh^{-1} x, \quad \sinh y = x.$$

Differentiating both sides with respect to $x$, we have

(11.10.1) $$\cosh y \frac{dy}{dx} = 1, \quad \frac{dy}{dx} = \frac{1}{\cosh y}.$$

Now

$$\cosh y = \sqrt{\sinh^2 y + 1} = \sqrt{x^2 + 1}.$$

Then from (11.10.1) we get $dy/dx = D \sinh^{-1} x = 1/\sqrt{x^2 + 1}$.

To find $D \operatorname{sech}^{-1} x$ we can use the same method. However, let us use the relation

$$\operatorname{sech}(\operatorname{sech}^{-1} x) = x.$$

Differentiation gives

$$-\operatorname{sech}(\operatorname{sech}^{-1} x) \cdot \tanh(\operatorname{sech}^{-1} x) \cdot D \operatorname{sech}^{-1} x = 1.$$

But, using the fact that $\tanh^2 y = 1 - \operatorname{sech}^2 y$, we have

$$\operatorname{sech}(\operatorname{sech}^{-1} x) = x, \quad \tanh(\operatorname{sech}^{-1} x) = \sqrt{1 - x^2}.$$

Hence

$$D \operatorname{sech}^{-1} x = -\frac{1}{x\sqrt{1-x^2}}.$$

The method for differentiating the other inverse hyperbolic functions should now be clear.

Clearly, the differentiation formulas for the inverse hyperbolic functions lead immediately to corresponding integration formulas. However, we shall postpone the treatment of these until Chapter 14.

**EXERCISE GROUP 11.8**

Find the derivative of the given function in Exercises 1–12.

1. $\sinh^{-1} 2x$
2. $\cosh^{-1} 3x$
3. $\tanh^{-1} \sqrt{x}$
4. $\coth^{-1}(x - 1)$
5. $\operatorname{sech}^{-1}(2x - 1)$
6. $\operatorname{csch}^{-1}(x + 2)$
7. $e^{\sinh^{-1} x}$
8. $e^{2\cosh^{-1} x}$
9. $x \sinh^{-1} x$
10. $\coth^{-1}(\operatorname{sech} x)$
11. $\tanh^{-1}(\cosh x)$
12. $x e^{\cosh^{-1} 2x}$

13. Derive the logarithmic expression for each of the following by the method used in the text.

    (a) $\cosh^{-1} x$,  (b) $\tanh^{-1} x$,  (c) $\coth^{-1} x$,  (d) $\operatorname{csch}^{-1} x$

14. Obtain the derivative of each inverse hyperbolic function by differentiating the logarithmic function which expresses it.

15. As a variation of the method used in the text, start with the relation
$$\cosh(\sinh^{-1} x) = \sqrt{1 + x^2},$$
and find $D \sinh^{-1} x$.

16. If $\theta$ is defined by $x = \ln|\sec \theta + \tan \theta|$, show that $d\theta/dx = \operatorname{sech} x$. (See also Exercise 14 in Exercise Group 11.7.)

# 12 Geometric Applications of the Definite Integral

## 12.1 INTRODUCTION

The definite integral was defined and developed in Chapter 9. An application of this concept was then made in finding the area of a region in a plane. After developing a formula for the area under a curve as a definite integral, we saw that areas of fairly general regions could be evaluated.

In this chapter and the next one we will consider a variety of additional applications which can be treated by the definite integral, applications which can be effectively handled in no other way. The underlying aspect of the entire treatment is that some quantity under consideration can be approximated for a small portion of a given configuration, and that then the evaluation of the quantity for the complete configuration can be made by proceeding to an appropriate definite integral. For the area of a region under a curve this takes the following form. The area of a thin strip of the region, bounded by a curve $y = f(x) \geq 0$ and the lines $x = x_{i-1}$, $x = x_i$, and $y = 0$ is approximately $f(x)\Delta x_i$, for any value of $x$ in the interval $[x_{i-1}, x_i]$, and then the desired area is $\int_a^b f(x)\,dx$.

In each case to be considered, a justification in the form of either a proof or a definition is required for proceeding to the definite integral. We shall elaborate upon the justification more in some applications than in others.

## 12.2 VOLUME OF A SOLID WITH KNOWN CROSS SECTION

We consider a solid, that is, a region in space, such that the intersection with a plane perpendicular to some fixed line is a plane region with known area. Specifically, we show in Fig. 12.2.1 a solid such that a plane perpendicular to the $x$ axis intersects the solid in a region whose area is a function $A(x)$ of $x$, for $x$ in $[a, b]$. The interval $[a, b]$ is then subdivided in the usual way, and planes perpendicular to the $x$ axis are drawn at the points of subdivision. In this way the solid is divided into slabs, one of which is shown in the figure, and the volume of the solid is the sum of the volumes of the slabs. Let $A(u_i)$ and $A(v_i)$, for some $u_i$ and $v_i$ in $[x_{i-1}, x_i]$, be the smallest and largest cross-sectional area for $x$ in $[x_{i-1}, x_i]$.

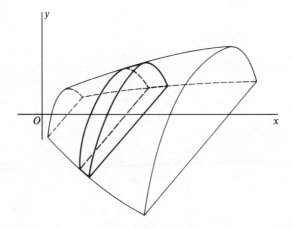

Fig. 12.2.1

If the volume of the slab is denoted by $\Delta V_i$, we clearly have, with $\Delta x_i = x_i - x_{i-1}$,

$$A(u_i)\Delta x_i \leq \Delta V_i \leq A(V_i)\,\Delta x_i, \qquad i = 1, 2, \ldots, n,$$

and hence, for the total volume $V$,

$$\sum_{i=1}^{n} A(u_i)\,\Delta x_i \leq \sum_{i=1}^{n} \Delta V_i = V \leq \sum_{i=1}^{n} A(v_i)\,\Delta x_i.$$

The sums on the left and right are Riemann sums for the function $A(x)$ in $[a, b]$. Thus, if $A(x)$ is continuous in $[a, b]$, both sums approach the same definite integral as the norms of the subdivisions approach zero, and we obtain

(12.2.1) $$V = \int_a^b A(x)\,dx.$$

Similar formulas may easily be written if the sections of known area are perpendicular to a different line, in particular the $y$ axis.

In the derivation of (12.2.1) we may think of the volume $\Delta V_i$ of a slab as given approximately by $A(x)\Delta x_i$ for any $x$ in $[x_{i-1}, x_i]$. It is convenient to say that the *element of volume* is $A(x)\,dx$; that is,

$$dV = A(x)\,dx.$$

This will be expanded upon later.

**Example 12.2.1.** Find the volume of a pyramid whose base has area $K$, and whose altitude is $h$.

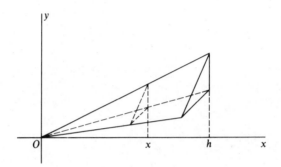

Fig. 12.2.2

*Solution.* In Fig. 12.2.2 we show a pyramid with vertex at $O$, and base perpendicular to the $x$ axis. A triangular pyramid in a special position is shown for simplicity in drawing, but the derivation is valid for any pyramid. The cross-section corresponding to any value of $x$ between $0$ and $h$ is a plane figure similar to the base, and its area is clearly a function of $x$. If $A(x)$ denotes the area of the cross-section, we have by properties of similar figures, (areas of corresponding regions are proportional to the squares of corresponding lengths),

$$\frac{A(x)}{K} = \frac{x^2}{h^2}, \qquad A(x) = \frac{Kx^2}{h^2}.$$

Hence, by (12.2.1), the volume $V$ of the pyramid is

$$V = \int_0^h A(x)\,dx = \frac{K}{h^2}\int_0^h x^2\,dx = \frac{K}{h^2}\frac{x^3}{3}\bigg|_0^h = \frac{Kh}{3}.$$

Thus *the volume is one-third the product of the area of the base and the altitude.*

*Note:* The derivation in Example 12.2.1 and hence the result are valid also for a cone with altitude $h$ and area of base $K$, whatever type of region the base may be.

## EXERCISE GROUP 12.1

In Exercises 1–5 a solid is described as follows. Each cross section perpendicular to the $x$ axis is a square with center on the $x$ axis and one pair of opposite sides perpendicular to the $xy$ plane. The midpoint of the upper side of the square lies on the given curve. Find the volume of the solid.

1. $y^2 - x^2 = 1$, between $x = 1$ and $x = 2$
2. $x^2 + y^2 = 4$, for the entire curve
3. $y = \sin x$, one arch of the curve
4. $y = 4 - x^2$, between $x = 0$ and $x = 2$
5. $y^2 = x^2(2 - x)$, for the upper half of the loop of the curve

In Exercises 6–10 cross sections of a solid by planes perpendicular to the $xy$ plane are circular regions, with the extremities of a diameter on the given curves. Find the volume of the solid for $x$ in the specified interval.

6. $y = 2x$, $y = 0$; $0 \leq x \leq 3$
7. $y = (\sin x)^{1/2}$, $y = 0$; $0 \leq x \leq \pi$
8. $y = x$, $y = 3x$; $0 \leq x \leq 3$
9. $y = \sqrt{x}$, $y = 2\sqrt{x}$; $0 \leq x \leq 2$
10. $y = \sqrt{x}$, $y = x$; $0 \leq x \leq 1$
11. Find the volume common to two right circular cylinders of radius $a$, whose axes intersect at right angles.

## 12.3 VOLUME OF A SOLID OF REVOLUTION

An important application of the result expressed in Equation (12.2.1) is finding the *volume of a solid of revolution*. Let $f(x)$ be nonnegative and continuous in the interval $[a, b]$. If the region under the curve $y = f(x)$ in $[a, b]$ is revolved about the $x$ axis, a *solid of revolution* is generated. A cross-section by a plane perpendicular to the $x$ axis is a circle. Thus in Fig. 12.3.1, if $|OM| = x$, the cross-section is the circle with radius $|MP| = y = f(x)$, and the area of this circle is

$$A(x) = \pi y^2 = \pi [f(x)]^2.$$

By (12.2.1) the *volume of the solid of revolution* is

(12.3.1) $$V = \pi \int_a^b y^2 \, dx = \pi \int_a^b [f(x)]^2 \, dx.$$

The volume of a slab of the solid is approximated by the disc (cylindrical in shape) of radius $f(x)$ and altitude $dx$, as shown in Fig. 12.3.1, and we may write, for the element of volume,

(12.3.2) $$dV = \pi y^2 \, dx = \pi [f(x)]^2 \, dx.$$

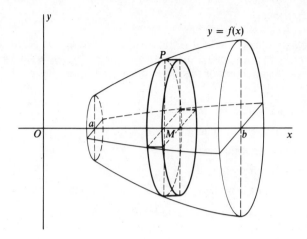

Fig. 12.3.1

***Example 12.3.1.*** The region bounded by the circle $x^2 + y^2 = a^2$ and to the right of the line $x = b$, $-a < b < a$, is revolved about the $x$ axis. Find the volume of the solid generated.

*Solution.* The solid is a solid of revolution (Fig. 12.3.2). The desired volume is then

$$V = \pi \int_b^a y^2 \, dx = \pi \int_b^a (a^2 - x^2) \, dx$$

$$= \pi \left( a^2 x - \frac{x^3}{3} \right) \bigg|_b^a = \frac{(2a^3 - 3a^2 b + b^3)\pi}{3}.$$

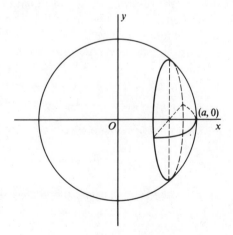

Fig. 12.3.2

*Comment* on Example 12.3.1. The number $b$ may be negative. If $b = -a$ in particular, the solid is a sphere of radius $a$, and the resulting volume is $V = 4\pi a^3/3$.

It follows from the above development that if the region which is bounded by the curve $x = g(y)$, where $g(y) \geq 0$ in $[c, d]$, and the lines $y = c$, $y = d$, and $x = 0$, is revolved about the $y$ axis, the volume of the solid thus generated is

$$(12.3.3) \qquad V = \pi \int_c^d x^2 \, dy = \pi \int_c^d [g(y)]^2 \, dy.$$

*Example 12.3.2.* Find the volume of the solid generated by revolving about the $x$ axis the region bounded by $y = x$ and $x = y^2$.

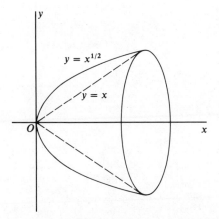

Fig. 12.3.3

*Solution.* Referring to Fig. 12.3.3, we see that the desired volume $V$ is the difference of two volumes (the smaller solid is a cone) generated by the regions under the curve $y = x^{1/2}$ and the curve $y = x$ between $x = 0$ and $x = 1$. Hence

$$V = \pi \int_0^1 x \, dx - \pi \int_0^1 x^2 \, dx = \frac{\pi x^2}{2}\bigg|_0^1 - \frac{\pi x^3}{3}\bigg|_0^1 = \frac{\pi}{2} - \frac{\pi}{3} = \frac{\pi}{6}.$$

Due to the type of figure used as an element of volume in the above method for finding the volume of a solid of revolution, the method is often referred to as the *method of cylindrical discs.*

Consider now a region bounded by the curves $y_1 = g(x)$ and $y_2 = f(x)$, and the lines $x = a$ and $x = b$, such that

$$0 \le g(x) \le f(x), \qquad a \le x \le b.$$

We wish to find the volume of the solid generated as this region is revolved about the $x$ axis. The rectangle of width $\Delta x$ shown in Fig. 12.3.4 generates a cylinder with a hole removed at the center, that is, a *hollow disc*. Fig. 12.3.5 shows the

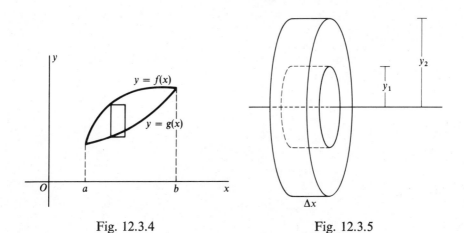

Fig. 12.3.4          Fig. 12.3.5

appearance of a typical hollow disc, with inner radius $y_1$, outer radius $y_2$, and thickness $\Delta x$. The volume of this hollow disc may be used as an element of volume for the solid generated, and we have

(12.3.4) $\qquad dV = \pi y_2^2 \, dx - \pi y_1^2 \, dx = \pi(y_2^2 - y_1^2) \, dx.$

The volume of the solid generated is then given by

(12.3.5) $\qquad V = \pi \int_a^b (y_2^2 - y_1^2) \, dx = \pi \int_a^b \{[f(x)]^2 - [g(x)]^2\} \, dx.$

The use of (12.3.4) and (12.3.5) suggests the name *hollow-disc method*.

The hollow-disc method may be used to find the volume in Example 12.3.2; the required functions are $y_1 = x$, $y_2 = x^{1/2}$, and we find

$$V = \pi \int_0^1 (x - x^2) \, dx = \pi \left( \frac{x^2}{2} - \frac{x^3}{3} \right) \bigg|_0^1 = \frac{\pi}{6}.$$

There is a decided similarity in the two methods.

If a solid is generated by revolving about the $y$ axis the region bounded by the curves $x_1 = r(y)$, $x_2 = s(y)$, and the lines $y = c$ and $y = d$, where

$$0 \le r(y) \le s(y), \qquad c \le y \le d,$$

the hollow-disc method may be used and yields the formulas

(12.3.6) $$V = \pi \int_c^d (x_2^2 - x_1^2)\, dy, \qquad dV = \pi(x_2^2 - x_1^2)\, dy.$$

**Example 12.3.3.** Find the volume generated by revolving the region defined in Example 12.3.2 about the $y$ axis.

*Solution.* The hollow-disc method and (12.3.6) are applicable, with $x_1 = y^2$, $x_2 = y$, and the volume is given by

$$V = \pi \int_0^1 (x_2^2 - x_1^2)\, dy = \pi \int_0^1 (y^2 - y^4)\, dy = 2\pi/15.$$

### EXERCISE GROUP 12.2

In Exercises 1–8 the region bounded by the given curves is revolved about the $x$ axis. Find the volume of the solid generated, using a disc or hollow-disc method.

1. $y = x + 1$, $x = 0$, $x = 2$, $y = 0$
2. $y^2 = x + 1$, $x = 2$
3. $y = 2 - x^2$, $y = 0$
4. $y = \sin x$, $y = 0$, $0 \le x \le \pi$
5. $y = \sec x$, $x = 0$, $y = 0$, $x = \pi/4$
6. $y = e^x$, $x = 0$, $y = 0$, $x = 2$
7. $y = \dfrac{1}{x+1}$, $x = 0$, $y = 0$, $x = 2$
8. $y^2 = 4x$, $y = x$

In Exercises 9–16 the region bounded by the given curves is revolved about the $y$ axis. Find the volume of the solid generated, using a disc or hollow-disc method.

9. $y = 2 - x^2$, $y = 0$
10. $x^2 = y^3$, $x = 2$, $y = 0$
11. $y^2 = 4x$, $y = x$
12. $x^2 = \sin y$, $x = 0$, $0 \le y \le \pi$
13. $x = \sinh y$, $x = 0$, $y = 1$
14. $y = \ln x$, $x = 0$, $y = 0$, $y = 2$
15. $y = \dfrac{1}{x+1}$, $x = 0$, $y = 0$, $x = 2$
16. $y = \dfrac{1}{4 - x^2}$, $y = 2$

17. Find the volume of the solid generated by revolving about the $x$ axis the region bounded by the curves $y = x^2$ and $x = y^2$.

18. Find the volume of the solid generated by revolving the loop of the curve $y^2 = x(2 - x)$ about the $x$ axis.

19. Find the volume of the solid generated by revolving the loop of the curve $y^2 = x^2(8 + x^3)$ about the $x$ axis.

\* 20. The formula for the volume of a frustum of a cone is

$$V = \frac{\pi}{3} [r_1^2 + r_1 r_2 + r_2^2] h,$$

where $r_1$ and $r_2$ are the radii of the bases and $h$ is the altitude. Show that, in generating a solid of revolution about the $x$ axis, by revolving a trapezoid instead of a rectangle to get an element of volume, we get

$$\Delta V_i = \frac{\pi}{3} [y_{i-1}^2 + y_{i-1} y_i + y_i{}^2] \Delta x_i.$$

Then obtain the same formula for the volume of the solid as in (12.3.1). [Assume that $\sum y_{i-1} y_i \Delta x_i$ acts as a Riemann sum. See Theorem 12.6.1.]

## 12.4 VOLUME OF A SOLID OF REVOLUTION BY CYLINDRICAL SHELLS

We present an alternate method for finding the volume of a solid of revolution. Consider the region in the $xy$ plane under the curve $y = f(x)$ in the interval $[a, b]$, where $0 < a < b$. We assume that $f$ is continuous and nonnegative in this interval. If this region is revolved about the $y$ axis, a solid of revolution is generated (Fig. 12.4.1). We now subdivide the interval $[a, b]$, and consider the subinterval $[x_{i-1}, x_i]$. At the midpoint, with coordinate $u_i = (x_{i-1} + x_i)/2$, we draw the vertical line of length $f(u_i)$ to intersect the curve. The rectangle above the interval $[x_{i-1}, x_i]$, with altitude $f(u_i)$, generates a *cylindrical shell*, the region between two cylinders with the same axis and altitude. We assume that the volume of this shell approximates the volume generated by the region under the arc $PQ$, and we obtain, for the approximation $\Delta V_i$,

$$\Delta V_i = \pi(x_i{}^2 - x_{i-1}^2) f(u_i) = \pi(x_i + x_{i-1})(x_i - x_{i-1}) f(u_i)$$
$$= \pi \cdot 2 u_i f(u_i) \Delta x_i, \quad \text{where} \quad \Delta x_i = x_i - x_{i-1}.$$

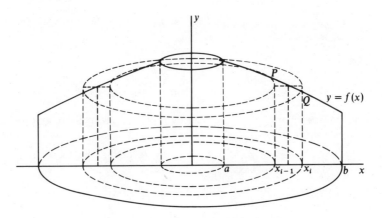

Fig. 12.4.1

If we define the volume $V$ as $V = \lim_{n \to \infty} \sum_{i=1}^{n} \Delta V_i$, where $(\max \Delta x_i) \to 0$ as $n \to \infty$, we have

(12.4.1) $$V = 2\pi \lim_{n \to \infty} \sum_{i=1}^{n} u_i f(u_i) \Delta x_i.$$

The sum in (12.4.1) is a Riemann sum for the integral of the continuous function $xf(x)$ from $a$ to $b$; hence the limit exists, and we have

(12.4.2) $$V = 2\pi \int_a^b xf(x)\,dx = 2\pi \int_a^b xy\,dx.$$

*This result expresses the volume of the solid obtained by revolving the region under the curve $y = f(x)$, from $x = a$ to $x = b$, about the $y$ axis.*

For the solid described the *element of volume*, that is, the approximate volume of the cylindrical shell, may be written as

(12.4.3) $$dV = 2\pi x f(x)\,dx = 2\pi xy\,dx.$$

It should be observed that the derivation of (12.4.2) remains valid even when $f$ is not increasing in the entire interval $[a, b]$, or decreasing in the entire interval.

**Example 12.4.1.** Find the volume of the solid obtained by revolving about the $y$ axis the region bounded by the curve $y = 2x - x^2$ and the $x$ axis.

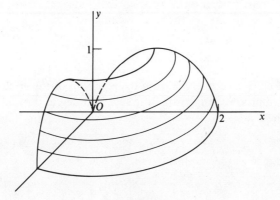

Fig. 12.4.2

*Solution.* One-fourth of the solid is shown in Fig. 12.4.2. The method of cylindrical shells applies, and from (12.4.2) the volume $V$ is given by

$$V = 2\pi \int_0^2 x(2x - x^2)\,dx = 8\pi/3.$$

*Note.* The volume in Example 12.4.1 can also be found by the hollow-disc method, but the use of cylindrical shells is simpler in this case. In other cases the reverse may be true.

**Example 12.4.2.** Find the volume of the solid generated by revolving about the $y$ axis the interior of the circle $(x - b)^2 + y^2 = a^2$, $0 < a < b$. (The surface of the doughnut-shaped solid formed in this way is called a *torus*.)

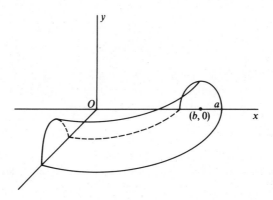

Fig. 12.4.3

*Solution.* We use the method of cylindrical shells to set up an integral for the volume of the upper half of the solid. Fig. 12.4.3 shows one-fourth of the upper half of the solid. If $V$ denotes the total volume, we get

$$V/2 = 2\pi \int_{b-a}^{b+a} xy \, dx = 2\pi \int_{b-a}^{b+a} x\sqrt{a^2 - (x-b)^2} \, dx,$$

where $y$ was found in terms of $x$ from the equation of the circle.

A standard procedure in evaluating such an integral is to substitute $x - b = u$, $dx = du$. Then

(12.4.4)   $$\int_{b-a}^{b+a} x\sqrt{a^2 - (x-b)^2} \, dx = \int_{-a}^{a} (u+b)\sqrt{a^2 - u^2} \, du$$

$$= \int_{-a}^{a} u\sqrt{a^2 - u^2} \, du + b \int_{-a}^{a} \sqrt{a^2 - u^2} \, du.$$

For the first integral in the right member of (12.4.4) we have

$$\int_{-a}^{a} u\sqrt{a^2 - u^2} \, du = -\frac{1}{3}(a^2 - u^2)^{3/2} \Big|_{-a}^{a} = 0.$$

The second integral in the right member of (12.4.4) yields the area of a semicircle of radius $a$, and is $\pi a^2/2$. Hence

$$\frac{V}{2} = 2\pi \frac{\pi a^2 b}{2}, \quad \text{and} \quad V = 2\pi^2 a^2 b.$$

We have now considered the hollow-disc method and the cylindrical shell method for finding volumes of solids of revolution. In both methods the element of volume is the same type of solid, a hollow cylinder. What distinguishes one method from the other? In the hollow-disc method the altitude is $\Delta x$ or $\Delta y$ or something similar, and approaches zero, while the thickness, the difference between the inner and outer radii, does not approach zero; in the cylindrical shell method the thickness of the shell approaches zero, while the altitude does not. In many cases an expression for the volume may be set up by either method. However, one method may have the advantage of leading to an integral that is simpler to evaluate.

In general, suppose that a solid is generated by revolving a plane region about some axis in the plane, and that the plane region is subdivided into rectangles with dimensions $h_i$ and $du_i$, where $h$ is some function of $u$. Then the hollow-disc method comes into play if the axis of revolution is parallel to the side of length $du_i$; and the cylindrical shell method applies if the axis of revolution is parallel to the side of length $h_i$.

**EXERCISE GROUP 12.3**

In Exercises 1–4 find the volume of the solid of revolution described, using the method of cylindrical shells.

1. About the $y$ axis, generated by the region bounded by the curves $y = x$, $y = 2x$, $x = 2$.

2. About the $y$ axis, generated by the region bounded by $y = x^2$, $y^2 = x$.

3. About the $x$ axis, generated by the region in the first quadrant bounded by the curves $y = x^2$, $y = 2x^2$, $y = 8$.

4. About the $x$ axis, generated by the region in the first quadrant bounded by the curves $y = x$, $y = x^3$.

5. Find the volume of a sphere of radius $a$ by the use of cylindrical shells.

6. Find the volume of the solid generated by revolving about the $y$ axis the smaller region bounded by the curves $x^2 + y^2 = a^2$, $x = b$, $0 < b < a$. (You might try doing this one by two methods.)

7. The region bounded by the curves $y = x^2$ and $y^2 = x$ is revolved about the line $x = 1$, generating a solid. Find the volume of the solid, using the methods of the text to develop an appropriate formula.

8. Work Exercise 7 if the axis of revolution is $x = c > 1$.

* 9. Find the volume described in Example 12.4.2 by the hollow-disc method, as follows. Show that you may take

$$dV = 4\pi b \sqrt{a^2 - y^2}\, dy.$$

The integral can then be evaluated by using the result (which arises from the area of a circle)

$$\int_0^a \sqrt{a^2 - y^2}\, dy = \pi a^2/4.$$

* 10. Find the volume of the solid generated by revolving about the line $y = x$ the region bounded by that line and the curve $y = x^2$.

* 11. Try some of the exercises of Exercise Group 12.2 by the method of cylindrical shells.

## 12.5 ARC LENGTH

Length of a curve will be defined in a manner that recalls the definition of the length of a circle in terms of lengths of inscribed (regular) polygons.

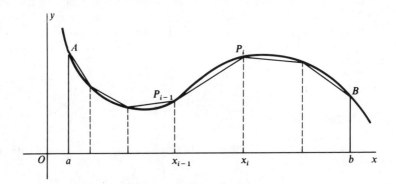

Fig. 12.5.1

It is adequate for the present to think of a curve as the graph of a continuous function $f$ (Fig. 12.5.1). We subdivide the interval $[a, b]$ in the usual way, and consider the corresponding points on the curve, $P_i(x_i, f(x_i))$. The broken line joining successive points $P_i$ is called an *inscribed polygon*. The length of the chord $P_{i-1}P_i$ is

(12.5.1) $\qquad |P_{i-1}P_i| = \sqrt{(x_i - x_{i-1})^2 + [f(x_i) - f(x_{i-1})]^2}.$

and the length of the inscribed polygon is

(12.5.2) $\qquad \sum_{i=1}^{n} |P_{i-1}P_i| = \sum_{i=1}^{n} \sqrt{(x_i - x_{i-1})^2 + [f(x_i) - f(x_{i-1})]^2}.$

This length approximates the length of the arc $AB$. As the subdivision becomes finer, the approximation improves, and the following definition is reasonable.

**DEFINITION 12.5.1.** *If the sum in (12.5.2) approaches the same limit $L$ for every sequence of subdivisions whose norms approach zero, then $L$ is defined as the **length** of the arc $AB$ in Fig. 12.5.1.*

This definition is essentially the same type as used for a definite integral (Section 9.9). In fact, we show that $L$ exists by showing that it is a definite integral. We apply the Mean-Value Theorem (Section 9.2) to $f$ in the interval $[x_{i-1}, x_i]$, assuming that $f'$ exists; then for some $w_i$,

$$f(x_i) - f(x_{i-1}) = f'(w_i)(x_i - x_{i-1}), \qquad x_{i-1} < w_i < x_i.$$

Hence, from (12.5.1),

$$|P_{i-1}P_i| = \sqrt{1 + [f'(w_i)]^2}\,(x_i - x_{i-1}) = \sqrt{1 + [f'(w_i)]^2}\,\Delta x_i,$$

and from (12.5.2),

(12.5.3) $$\sum_{i=1}^n |P_{i-1}P_i| = \sum_{i=1}^n \sqrt{1 + [f'(w_i)]^2}\,\Delta x_i.$$

This is a Riemann sum for the function $\sqrt{1 + [f'(x)]^2}$ in the interval $[a, b]$. If $f'$ is continuous, so is the function $\sqrt{1 + [f'(x)]^2}$, and we know that the definite integral of the latter function exists. By the relation of the Riemann sums (12.5.3) to the definite integral, we can then assert that the number $L$ in Definition 12.5.1 exists, and is a definite integral, and we have proved the following.

**THEOREM 12.5.1.** *If $f'$ is continuous in the interval $[a, b]$, the length $L$ of the arc joining the points $(a, f(a))$ and $(b, f(b))$ along the curve $y = f(x)$ is*

(12.5.4) $$L = \int_a^b \sqrt{1 + [f'(x)]^2}\,dx.$$

It is fairly straightforward to set up the integral expression for arc length for a curve given in the form $y = f(x)$. However, the actual integration may present a problem. Many simple curves lead to integrals which cannot be evaluated in elementary form; others lead to integrals for which methods of evaluation have not yet been presented. We are thus restricted in the examples and exercises available to us at this time.

***Example 12.5.1.*** Find the length of the curve $y = x^{3/2}$ between the points $(1, 1)$ and $(4, 8)$.

*Solution.* Let $f(x) = x^{3/2}$. Then $f'(x) = 3x^{1/2}/2$, and the integral for the length of arc is

$$L = \int_1^4 \sqrt{1 + \frac{9x}{4}}\,dx = \frac{1}{2}\int_1^4 \sqrt{4 + 9x}\,dx$$

$$= \frac{1}{2} \cdot \frac{1}{9} \frac{(4+9x)^{3/2}}{3/2}\bigg|_1^4 = \frac{1}{27}(4+9x)^{3/2}\bigg|_1^4$$

$$= \frac{1}{27}(40^{3/2} - 13^{3/2}) = \frac{80\sqrt{10} - 13\sqrt{13}}{27}.$$

We shall return to such problems after additional methods of integration have been presented.

If the equation of a curve between two points can be expressed in the form

$x = g(y)$, where $g'$ is continuous in the closed interval $[c, d]$, then by analogy with (12.5.4) the length of arc between the points $(g(c), c)$ and $(g(d), d)$ is given by

(12.5.5) $$L = \int_c^d \sqrt{1 + [g'(y)]^2}\, dy.$$

We now return to Theorem 12.5.1 and consider the length of arc on a curve $y = f(x)$ between a fixed point $(a, f(a))$ and a variable point $(x, f(x))$. For each point of the curve for which $x \geq a$ the length is determined by $x$, and is therefore a function of $x$ which we now denote by $s(x)$, and we may write

(12.5.6) $$s(x) = \int_a^x \sqrt{1 + [f'(u)]^2}\, du.$$

Note that the variable $x$ appears only in the upper limit of the integral; any letter may be used for the variable in the integrand.

Equation (12.5.6) defines a function, *the arc length function*, by means of a definite integral. The properties of the function are prescribed by the integral; in particular, *if $f'$ is continuous then $s'$ exists and is continuous*, and, by Theorem 9.10.1, we have

(12.5.7) $$s'(x) = \sqrt{1 + [f'(x)]^2}.$$

Since $s'(x) > 0$ for $a \leq x \leq b$, it follows that *s is an increasing function* of $x$. Moreover, by Section 10.5 we know that the *inverse function exists*; hence we may write $x = x(s)$, and this function is differentiable in the interval $[s(a), s(b)]$.

The arc length function was defined in (12.5.6.) in such a way that $s(a) = 0$. It could have been defined so that it was zero for some other specified value. Any two such functions satisfy (12.5.7) and differ by a constant.

We may write $s'(x) = ds/dx$. Hence from (12.5.7), we have, for the *element of arc length ds*,

(12.5.8) $$ds = \sqrt{1 + [f'(x)]^2}\, dx = \sqrt{1 + y'^2}\, dx.$$

**EXERCISE GROUP 12.4**

1. Find the length of the curve $y = (x - 1)^{3/2}$ from $x = 1$ to $x = 5$.
2. Find the length of the curve $(y - 1)^3 = x^2$ from $x = 1$ to $x = 8$.
3. Find the length of the curve $y^3 = x^2$ between the points $(1, 1)$ and $(8, 4)$.
4. Find the length of the curve $y = \dfrac{x^2}{16} - 2 \ln x$ from $x = 4$ to $x = 8$.
5. Find the length of the curve $y = \cosh x$ from $x = 0$ to $x = 1$.
6. Find the length of the curve $y = \ln\left(\coth \dfrac{x}{2}\right)$ between $x = 1$ and $x = 2$.
7. Find the length of the curve $y = x^4/4 + 1/(8x^2)$ from $x = 1$ to $x = 2$.
8. Find the length of the curve $y = x^5/5 + 1/(12x^3)$ from $x = 1/2$ to $x = 1$.
9. Find the length of the curve $x = y^3/3 + 1/(4y)$ from $y = 1$ to $y = 2$.

10. Find the length of the curve $y = (2/3)x^{3/2} - (1/2)x^{1/2}$ from $x = 4$ to $x = 9$.
11. Find the length of the entire curve $x^{2/3} + y^{2/3} = a^{2/3}$.
12. Find the length of the curve $y = \ln \cos x$ from $x = 0$ to $x = \pi/6$.
* 13. Find the length of the curve $y = \ln \dfrac{e^x - 1}{e^x + 1}$ from $x = 1$ to $x = 3$.
14. If $s(x)$ is an arc length function for the curve $y = x^2$, show that
$$\frac{ds}{dx}\frac{d^2s}{dx^2} = 4x.$$
15. For the curve $y = f(x)$ show that $dy/ds = f'(x)/\sqrt{1 + [f'(x)]^2}$.
16. If $x = g(y)$, express $ds/dx$ in terms of the function $g$.
* 17. Water is entering a spherical container of radius 5 in. through a hole at the top, at the rate of 3 in.³/sec. Let $s$ be the distance from the top of the sphere to the water surface along a great circle (a circle through the center). Find the rate of change of $s$ when the water level is 3 in. above the center.
* 18. A hemispherical dome of radius 200 ft rests on the ground. Outside the dome a ball is thrown vertically up with such velocity that it reaches precisely the height of the dome before falling. If $y$ is the height of the ball, you may use $y = -16t^2 + 200$, with $t = 0$ at the top of the dome. The sun's rays are horizontal, and the sun and the path of the ball are in the same plane as the center line of the dome.
    (a) Find the speed of the shadow ($ds/dt$ along the circle in the plane described) of the ball at any time $t$ during the motion.
    (b) Find the speed of the shadow when the ball reaches its maximum height.

## 12.6 AREA OF A SURFACE OF REVOLUTION

In Section 12.3 we developed a formula for the volume of a solid of revolution. We consider here the *area* of a *surface of revolution*. We mean by such a surface the set of points generated by the curve $y = f(x)$, between $x = a$ and $x = b$, as it is revolved about a line in its plane. In particular, we consider the surface generated by revolving this curve about the $x$ axis. Equations of such surfaces will be considered in Section 20.4, but they are not needed for the present purpose.

Fig. 12.6.3 shows a surface generated by revolving the curve $y = f(x)$, between $x = a$ and $x = b$, about the $x$ axis. We subdivide the interval $[a, b]$, and consider the polygon formed by joining the corresponding points $P_i$ on the curve (as in the discussion of arc length). The chord $P_{i-1}P_i$ generates a *frustum* of a cone as the surface is formed, and we consider the area of the curved part of the frustum as an approximation to the area of the part of the surface generated by the arc $P_{i-1}P_i$.

Now, the area of the curved part of a right circular cone is $\pi rs$, where $r$ is the radius of the base, and $s$ is the *slant height* (Fig. 12.6.1(a)). We may see this by cutting the cone along an *element* (one of the line segments of length $s$) and unwrapping the cone so that it lies in a plane; in this way we form a sector of a circle of radius $s$ and arc $2\pi r$ (Fig. 12.6.1(b)). The central angle of the sector is

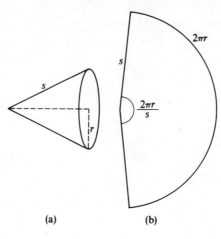

(a)  (b)

Fig. 12.6.1

$2\pi r/s$, and the area is $\frac{1}{2}(2\pi r/s)s^2 = \pi rs$. Using this result, we may find the area $A$ of the frustum in Fig. 12.6.2 to be

(12.6.1) $$A = \pi(r_1 + r_2)s,$$

where $s$ is now the slant height of the frustum, and $r_1$ and $r_2$ are the radii of the bases (see Exercise 9).

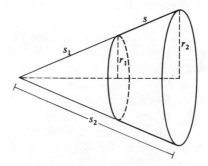

Fig. 12.6.2

Turning now to Fig. 12.6.3 and using (12.6.1), we find the area of the frustum in the figure to be

(12.6.2) $$\pi(y_{i-1} + y_i)|P_{i-1}P_i| = \pi(y_{i-1} + y_i)\sqrt{(x_i - x_{i-1})^2 + (y_i - y_{i-1})^2}.$$

The sum of this expression from $i = 1$ to $i = n$ is then an approximation to the area of the surface, and the limit of the sum as the norm of the subdivision of

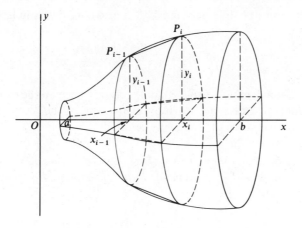

Fig. 12.6.3

$[a, b]$ approaches zero, if the limit exists, is the area of the surface. We wish to show that this limit can be represented by a definite integral.

We first apply the Mean-Value Theorem (Section 9.2) to $y = f(x)$ in the interval $[x_{i-1}, x_i]$. On the assumption that $f'$ exists in $(a, b)$, there is a number $w_i$ such that

$$y_{i-1} - y_i = f(x_i) - f(x_{i-1}) = f'(w_i)(x_i - x_{i-1}), \qquad x_{i-1} < w_i < x_i.$$

Then (12.6.2) becomes, with $\Delta x_i = x_i - x_{i-1}$,

(12.6.3) $\pi(y_{i-1} + y_i)\sqrt{1 + [f'(w_i)]^2}\,(x_i - x_{i-1})$
$$= \pi y_{i-1}\sqrt{1 + [f'(w_i)]^2}\,\Delta x_i + \pi y_i\sqrt{1 + [f'(w_i)]^2}\,\Delta x_i.$$

The sum of each term on the right is almost a Riemann sum. For example, the sum from $i = 1$ to $i = n$ of the first term may be written

(12.6.4) $$\sum_{i=1}^{n} \pi f(x_{i-1})\sqrt{1 + [f'(w_i)]^2}\,\Delta x_i.$$

This is *not* a Riemann sum, because $x_{i-1}$ and $w_i$, although both are in the interval $[x_{i-1}, x_i]$, are not necessarily the same. This is the first time this situation has been encountered. The resulting difficulty is resolved by the following theorem, which is stated without proof.

**THEOREM 12.6.1** *Let $f$ and $g$ be continuous in the interval $[a, b]$. Let both $v_i$ and $w_i$ be in the ith subinterval of a subdivision of $[a, b]$. Then if the norm of the subdivision approaches zero,*

$$\lim_{n \to \infty} \sum_{i=1}^{n} f(w_i)g(v_i)\,\Delta x_i = \int_a^b f(x)g(x)\,dx.$$

This is precisely what is needed for the limit of (12.6.4); it does not matter that $x_{i-1}$ and $w_i$ may be different; and if $f'$ is also continuous in $[a, b]$, the limit of this sum is

$$\pi \int_a^b f(x)\sqrt{1 + [f'(x)]^2}\, dx.$$

The limit of the sum from $i = 1$ to $i = n$ of the second term on the right in (12.6.3) is the same integral, and hence the limit of the sum of the terms in (12.6.2) is known. We have obtained the following result.

**THEOREM 12.6.2.** *Let the curve $y = f(x)$ between $x = a$ and $x = b$ be revolved about the x axis. If $f'$ is continuous in the interval $[a, b]$, the area $S$ of the surface generated is*

(12.6.5) $$S = 2\pi \int_a^b f(x)\sqrt{1 + [f'(x)]^2}\, dx.$$

Formula (12.6.5) is also conveniently written as

$$S = 2\pi \int_a^b y\, ds,$$

if it is clearly understood that $y$ and $ds$ refer to the generating curve. We may also write

(12.6.6) $$dS = 2\pi y\, ds = 2\pi f(x)\sqrt{1 + [f'(x)]^2}\, dx$$

for the *element of surface area dS* for a surface of revolution about the $x$ axis.

***Example 12.6.1.*** Find the area of the surface generated by revolving about the $x$ axis the upper half of the parabola $y^2 = 4x$ between $x = 0$ and $x = 1$.

*Solution.* We have $y = 2x^{1/2}$, $y' = x^{-1/2}$. Then the element of arc length is

$$ds = \sqrt{1 + y'^2}\, dx = \sqrt{1 + x^{-1}}\, dx,$$

and the element of surface area is

$$dS = 2\pi y\, ds = 4\pi x^{1/2}\sqrt{1 + x^{-1}}\, dx = 4\pi\sqrt{x + 1}\, dx.$$

Hence the surface area is

$$S = 4\pi \int_0^1 \sqrt{x + 1}\, dx = \frac{8\pi}{3}(x + 1)^{3/2}\Big|_0^1 = \frac{8\pi}{3}(2\sqrt{2} - 1).$$

The surface appears in Fig. 12.6.4.

The element of surface area $dS = 2\pi y\, ds$, for a surface of revolution about the $x$ axis, is not restricted to the form given in (12.6.6). We may sometimes express the element of arc length $ds$ in terms of $dy$. Thus in Example 12.6.1 we also have $x = y^2/4$, $dx/dy = y/2$, and

$$ds = \sqrt{1 + (y/2)^2}\, dy, \qquad dS = 2\pi y\sqrt{1 + (y/2)^2}\, dy = \pi y\sqrt{4 + y^2}\, dy.$$

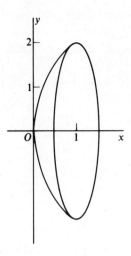

Fig. 12.6.4

Then, since $y = 0$ when $x = 0$, and $y = 2$ when $x = 1$,

$$S = \pi \int_0^2 y\sqrt{4 + y^2}\, dy = \frac{\pi}{3}(4 + y^2)^{3/2}\Big|_0^2 = \frac{\pi}{3}(8^{3/2} - 4^{3/2}) = \frac{8\pi}{3}(2\sqrt{2} - 1).$$

If a curve $x = g(y)$ is revolved about the $y$ axis, the element of area for the surface generated is

$$dS = 2\pi x\, ds = 2\pi g(y)\sqrt{1 + [g'(y)]^2}\, dy,$$

from which the corresponding integral follows. It may also be possible in this case to express $ds$ in terms of $dx$ instead of $dy$.

## EXERCISE GROUP 12.5

1. The arc of the parabola $y^2 = x - 1$ from $(1, 0)$ to $(3, \sqrt{2})$ is revolved about the $x$ axis. Find the area of the surface generated.

2. Find the area of a sphere by revolving the upper half of the circle $x^2 + y^2 = a^2$ about the $x$ axis.

3. Find the area of the surface obtained by revolving about the $y$ axis the arc of the parabola $x^2 = 4ay$ from the origin to the point $(2a, a)$.

4. Find the area of the surface generated by revolving the upper half of the curve $x^{2/3} + y^{2/3} = a^{2/3}$ about the $x$ axis.

5. Find the area of the surface generated by revolving about the $x$ axis the part of the curve $6xy = x^4 + 3$ from $x = 1$ to $x = 2$.

6. Find the area of the surface generated when the curve $y = 2\ln x - x^2/16$ between $x = 4$ and $x = 8$ is revolved about the $y$ axis.

7. The curve $y = e^x + e^{-x}/4$, between $x = 0$ and $x = 1$, is revolved about the $x$ axis. Find the area of the surface generated.

* 8. Find the area of the surface generated by revolving the upper half of the loop of the curve $3y^2 = x(1-x)^2$ about the $x$ axis.

9. Derive formula (12.6.1) for the area of a frustum of a cone. (*Hint*: Use similar triangles in Fig. 12.6.2.)

* 10. Find the area of the surface obtained by revolving the circle $(x-b)^2 + y^2 = a^2$, $0 < a < b$, about the $y$ axis.

# 13 Physical Applications of the Definite Integral

## 13.1 INTRODUCTION

In this chapter we consider various applications which are of a physical rather than geometric nature and for which the evaluation can be made by the use of definite integrals. We repeat here the statement of Section 12.1: the underlying aspect of the entire treatment is that some quantity under consideration can be approximated for a small portion of a given configuration, and that then the evaluation of the quantity for the complete configuration can be made by proceeding to an appropriate definite integral. In fact, the quantities to be studied here can sometimes be defined initially for individual particles, and this will be done.

Some of the concepts to be presented here will be considered in a more general context in later chapters; accordingly, we shall make an effort to make the presentation applicable to the later study, at least insofar as feasible.

## 13.2 WORK

As used in physics and mathematics *work* is a precisely defined concept which will be developed in this section in certain cases.

Work is performed when a force acts and motion results. Force is a vector, which involves both a magnitude and a direction. We consider at this time only forces for which the direction does not change, and motion which is along a straight line. If a force is constant and acts through a distance $d$ in the direction of the force, *the work $W$ done is defined as the product of the magnitude $F$ of the force and the distance*:

$$W = F \cdot d.$$

The unit for measuring work is *lb ft* or *dyne cm* or something similar, that is, a force unit times a distance unit depending on the system of units used. For example, if a weight of 5 lb is raised a distance of 7 ft, the work done is $5 \cdot 7$ lb ft or 35 lb ft. If a force of magnitude $F$ acts at an angle $\theta$ to the direction of motion, the component of the force in the direction of motion is $F \cos \theta$ (Fig. 13.2.1) and the work done is $W = F \cos \theta \cdot d$.

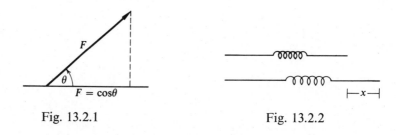

Fig. 13.2.1        Fig. 13.2.2

The interesting problems involving work occur in situations where the force is variable. A typical such case arises with an elastic spring. Fig. 13.2.2 shows a spring in its natural (unstretched) position, and the same spring stretched by an amount $x$. The applicable law of physics (Hooke's Law) states that *the force required to stretch or compress a spring by an amount $x$ from its natural length is proportional to $x$,*

$$F(x) = kx, \quad k \text{ constant.}$$

As the spring stretches (or compresses), the applied force must increase; we are interested in the amount of work done as the spring is stretched from $x = a$ to $x = b$. The method, as usual, is to consider first the work done in stretching the spring during a small part of the motion (Fig. 13.2.3). Suppose the spring is stretched from $x_{i-1}$ to $x_i$ in Fig. 13.2.3. At $x_{i-1}$ the force is $F(x_{i-1})$, and at $x_i$ it is $F(x_i)$. The work $\Delta W_i$ done in moving from $x_{i-1}$ to $x_i$ lies between $F(x_{i-1}) \Delta x_i$ and $F(x_i) \Delta x_i$,

$$F(x_{i-1}) \Delta x_i \leq \Delta W_i \leq F(x_i) \Delta x_i.$$

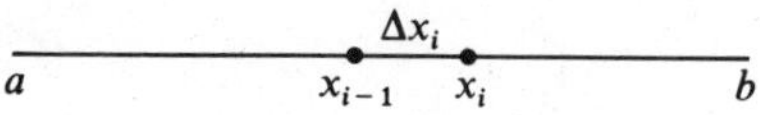

Fig. 13.2.3

Summing from $i = 1$ to $i = n$, to include each subinterval, we have for the total amount of work $W$,

(13.2.1) $$\sum_{i=1}^{n} F(x_{i-1})\,\Delta x_i \le W = \sum_{i=1}^{n} \Delta W_i \le \sum_{i=1}^{n} F(x_i)\,\Delta x_i.$$

Each sum in (13.2.1) is a Riemann sum for the same definite integral; as $n \to \infty$ both sums approach this integral, and we have

(13.2.2) $$W = \int_a^b F(x)\,dx.$$

***Example 13.2.1.*** A force of 1000 lb is required to stretch a spring 2 in. from its natural length. Find the work done in stretching the spring an additional 3 in.

*Solution.* The force is $F = kx$, where $x$ measures the amount of stretching. We have $x = 2$ when $F = 1000$. Hence $1000 = k \cdot 2$, $k = 500$, and $F = 500x$. The spring is stretched from an initial position with $x = 2$ to its final position with $x = 5$, and we have

$$W = \int_2^5 F(x)\,dx = \int_2^5 500x\,dx = 250x^2 \Big|_2^5 = 5250 (\text{in. lb}).$$

Another application of work concerns the emptying of a tank which contains some substance; we will explain that by means of an example.

***Example 13.2.2.*** The tank in Fig. 13.2.4 has the shape of a paraboloid of revolution with dimensions as shown, and is filled with a liquid of density $w$ to a depth of 2 ft. Find the work done in pumping the water (a) over the edge of the tank, and (b) to a height 3 ft above the top of the tank.

*Solution.* We use as an element of mass a horizontal slab, determine the approximate amount of work required to raise the slab to the desired level, and integrate to find the total amount of work.

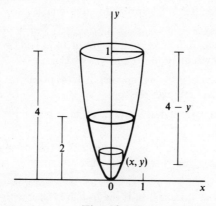

Fig. 13.2.4

The volume of a slab is $dV = \pi x^2 \, dy$ (by Section 12.3), and the mass is $dm = w \, dV = \pi w x^2 \, dy$. The variables $x$ and $y$ are related by the equation of the parabola in the $xy$ plane, which is found to be $y = 4x^2$. The only difference in the two parts of the problem is the approximate distance through which each slab is raised.

(a) In this case each part of the slab is lifted approximately the amount $4 - y$. The approximate amount of work required to lift the slab through this distance is

$$dW = (4 - y) \, dm = \pi w (4 - y) x^2 \, dy,$$

where $x$ and $y$ are related by the equation $y = 4x^2$. Hence

$$W = \pi w \int_0^2 (4 - y) x^2 \, dy = \pi w \int_0^2 (4 - y)(y/4) \, dy$$

$$= \frac{\pi w}{4} \int_0^2 (4y - y^2) \, dy = \frac{\pi w}{4} \left( 2y^2 - \frac{y^3}{3} \right) \bigg|_0^2 = \frac{4 \pi w}{3}.$$

(b) The distinction here is that each slab is lifted an amount $7 - y$ instead of $4 - y$. Thus

$$W = \pi w \int_0^2 (7 - y) x^2 \, dy = \pi w \int_0^2 (7 - y)(y/4) \, dy$$

$$= \frac{\pi w}{4} \int_0^2 (7y - y^2) \, dy = \frac{\pi w}{4} \left( \frac{7 y^2}{2} - \frac{y^3}{3} \right) \bigg|_0^2 = \frac{17 \pi w}{6}.$$

**EXERCISE GROUP 13.1**

1. A spring of natural length 6 in. is stretched $\frac{1}{4}$ in. by a force of 500 lb. Find the work done in stretching the spring from a length of 7 in. to a length of 10 in.

2. A force of 10 lb stretches a spring 2 in. Find the work done in compressing the spring 2 in. from its natural length.

3. It requires 200 in. lb of work to stretch a spring 10 in. from its natural length. Find the work required to stretch the spring an additional 2 in.

4. Find the natural length $L$ of a spring, if 64 in. lb of work are required to stretch the spring from a length of 10 in. to a length of 14 in., and 90 in. lb are needed to stretch it from a length of 10 in. to a length of 15 in.

5. A square tank with base 4 ft × 4 ft and depth 5 ft is filled with water. Find the work done in emptying the tank over the top.

6. Find the work done in emptying a cylindrical tank over the top if the radius is 2 ft and the height is 5 ft.

7. A conical tank with radius 2 ft and altitude 6 ft is half-filled with water. Find the work done in emptying the tank over the top.

8. A trough is 10 ft long and has, as vertical cross-sections, isosceles trapezoids with bases 2 ft and 6 ft and altitude 2 ft. If the trough is filled with water, find the work done in emptying the trough through a discharge pipe 2 ft above the top.

9. The cross-section of a trough of length 10 ft is an isosceles triangle of width 4 ft and depth 4 ft. Find the work done in emptying the trough if the water is 2 ft deep and the discharge is 2 ft above the top of the trough.

10. The force of attraction between two particles of masses $m_1$ and $m_2$ acts along the line joining them, and its magnitude is given by $F = km_1m_2/r^2$, where $k$ is a constant and $r$ is the distance between the particles. Let each particle have unit mass, and let one of them be at $(0, 0)$.
    (a) Find the work done if the other particle moves from $(a, 0)$ to $(b, 0)$ along the $x$ axis, $0 < a < b$.
    (b) Find the work done if the second particle moves from $(a, 0)$ to $(a, b)$ along the line $x = a$, $a > 0$, $b > 0$.
*   (c) Find the work done if the second particle moves from $(1, 0)$ to $(3, 4)$ along the line $y = 2x - 2$.

## 13.3 FLUID PRESSURE

If a column of liquid of depth $h$ rests on a horizontal plane of area 1 (Fig. 13.3.1) the total weight of the column is $wh$, where $w$ is the density of the liquid. This weight is defined as the *pressure* at depth $h$ due to the liquid. In general, *the pressure at a given depth is the weight per unit area*, that is, the total weight of liquid resting on a horizontal plate divided by the area of the plate. If $p$ is the pressure at a given depth, the force acting on an area $A$ at that depth is $F = pA = wkA.$

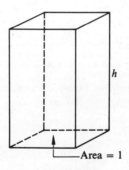

Fig. 13.3.1

It is a fact of physics that at a given depth below the surface of a liquid *the pressure acts equally in all directions*, not only vertically. Now suppose that a vertical surface is suspended in a liquid. The horizontal force due to pressure on one side of the surface varies with the depth, and the total horizontal force on the surface can be calculated.

The procedure for determining the total force on a submerged surface follows a now familiar pattern. The surface is subdivided into horizontal strips;

an approximation to the force acting on such a strip is found; and the total force is then determined by a limiting process which leads to a definite integral.

**Example 13.3.1.** A vertical gate is in the form of an isosceles trapezoid, 4 ft wide at the top, 3 ft wide at the bottom, and 3 ft deep. Find the force acting on one side of the gate if the top of the gate is 2 ft below the surface of the liquid.

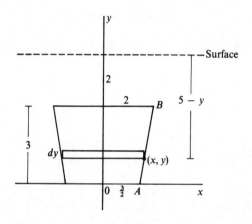

Fig. 13.3.2

*Solution.* We place the trapezoid in a coordinate system as in Fig. 13.3.2. A typical horizontal strip is of width $2x$ and height $dy$, where $(x, y)$ is on the line $AB$ whose equation we find to be

$$AB: y = 6x - 9.$$

The approximate depth of the strip below the surface is $5 - y$. We then have, for the approximate force $dF$ acting on the strip,

$$dF = p\, dA = w(5 - y) \cdot 2x\, dy,$$

and the total force $F$ is

$$F = \int_0^3 w(5 - y) \cdot 2x\, dy = \frac{2w}{6} \int_0^3 (5 - y)(9 + y)\, dy$$

$$= \frac{w}{3} \int_0^3 (45 - 4y - y^2)\, dy$$

$$= 36w.$$

**Example 13.3.2.** A vertical gate is bounded by the parabola $y = 4x^2$ and a horizontal line 4 ft above the vertex of the parabola. Find the force due to pressure on the gate, if the top of the gate is in the surface of the liquid.

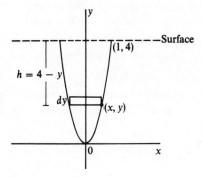

Fig. 13.3.3

*Solution.* The gate is shown in Fig. 13.3.3, with the surface of the liquid at $y = 4$. A representative horizontal strip of the gate is also shown. The area of the strip is

$$dA = 2x \, dy,$$

where $(x, y)$ is a point on the parabola in the first quadrant. Each point in the strip is at approximately the same depth $4 - y$, and the horizontal force on the strip is approximately

$$dF = p \, dA = wh \, dA = w(4 - y) \cdot 2x \, dy,$$

where $p$ is the pressure at depth $h = 4 - y$, and $w$ is the density of the liquid. Since $x = \sqrt{y}/2$ we find

$$F = 2w \int_0^4 (4 - y)(\sqrt{y}/2) \, dy = w \int_0^4 (4y^{1/2} - y^{3/2}) \, dy$$

$$= w \left( \frac{8}{3} y^{3/2} - \frac{2}{5} y^{5/2} \right) \Big|_0^4 = w \left( \frac{64}{3} - \frac{64}{5} \right)$$

$$= \frac{128w}{15}.$$

## EXERCISE GROUP 13.2

1. Find the force on the gate described in Example 13.3.1, if the top of the trapezoid is in the surface of the liquid.

2. Find the force on a vertical rectangle of width $a$ and depth $b$, if the upper edge of the rectangle is in the surface of the water.

3. The cross-section of a trough is an isosceles triangle of width 4 ft and depth 4 ft. Find the force due to water pressure on one end, if the trough is full.

4. A cylindrical tank of diameter 8 ft lies horizontally and is half-filled with water. Find the force due to water pressure on one end of the tank.

5. Find the force against an elliptical dam, if the major axis, of length 15 ft, is at the surface of the water and the semiminor axis has length 5 ft.

6. The end of a horizontal tank is a vertical parabola 4 ft deep and 8 ft across the top. Find the force against this end, if there is water in the tank to a depth of 2 ft.

7. By what amount must the rectangle of Exercise 2 be lowered to double the force?

8. A square of side $a$ is submerged in water with one vertex in the surface, the diagonal through this vertex being vertical. Find the force on one side of the square.

9. Find the horizontal force on a gate such as in Example 13.3.1 if the top of the trapezoid is in the surface of the liquid, and the entire trapezoid is tilted at an angle of 30° with the vertical.

10. At what angle with the vertical should the rectangle of Exercise 2 be tilted to reduce the force by half?

## 13.4 MOMENTS IN A PLANE. CENTER OF MASS

Consider a system of particles or point-masses in a plane, by which we mean an idealized situation whereby a mass can be assigned to a point. The *moment* of a single particle of mass $m_i$, $i = 1, 2, \ldots, n$, about a line $L$ (Fig. 13.4.1) is defined as the product $r_i m_i$, where $r_i$ is the directed distance of $m_i$ from $L$. The moment of the system of $n$ particles about $L$ is then *defined* as

$$(13.4.1) \qquad M_L = \sum_{i=1}^{n} r_i m_i,$$

it being understood that the distances $r_i$ are measured so that they are positive for particles on one side of the line and negative for particles on the other side of $L$. (The line $L$ separates the plane into two distinct parts.) The moment $M_L$ may be positive, negative, or zero. For a fixed system of particles in a plane, moving $L$ parallel to itself tends to decrease or increase the value of the moment according as the motion is toward the positive side or negative side, respectively. It is clear that *for some line parallel to $L$ the moment must be zero.*

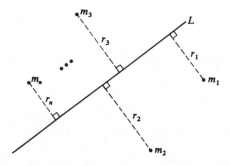

Fig. 13.4.1

## 13.4 MOMENTS IN A PLANE 353

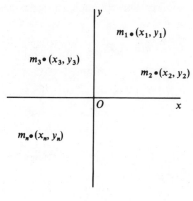

Fig. 13.4.2

We now place a coordinate system in the plane of the particles (Fig. 13.4.2), and consider moments about lines parallel to the coordinate axes. There is one line of each type such that the moment about this line is zero. Let these lines be $x = \bar{x}$, $y = \bar{y}$. We have $M_{x=\bar{x}} = 0$, and by (13.4.1) this becomes, if horizontal distances to the right and vertical distances upward are taken positive,

$$M_{x=\bar{x}} = \sum_{i=1}^{n}(x_i - \bar{x})m_i = \sum_{i=1}^{n} x_i m_i - \bar{x}\sum_{i=1}^{n} m_i = 0.$$

Hence

$$\bar{x} = \frac{\sum_{i=1}^{n} x_i m_i}{\sum_{i=1}^{n} m_i} = \frac{M_y}{M},$$

where $M_y = \sum_{i=1}^{n} x_i m_i$ is the moment of the system about the $y$ axis and $M = \sum_{i=1}^{n} m_i$ is the total mass of the system. We find $\bar{y}$ similarly, and have

(13.4.2) $$\bar{x} = \frac{M_y}{M} = \frac{\sum_{i=1}^{n} x_i m_i}{\sum_{i=1}^{n} m_i}, \qquad \bar{y} = \frac{M_x}{M} = \frac{\sum_{i=1}^{n} y_i m_i}{\sum_{i=1}^{n} m_i}.$$

The point $P(\bar{x}, \bar{y})$ is called the *center of mass* of the system of particles, and is the intersection of lines parallel to the coordinate axes such that the moment of the system about each of these lines is zero.

A line through the center of mass that is parallel to a coordinate axis should be no more significant than any other line through this point. The following theorem contains an accurate statement of the relevant property.

**THEOREM 13.4.1.** *The moment of a system of plane particles is zero about* **any** *line in the plane that passes through the center of mass.*

*Proof.* The equation of any line through the center of mass $(\bar{x}, \bar{y})$ may be written in the form

$$L: a(x - \bar{x}) + b(y - \bar{y}) = 0.$$

By Exercise 34 of Exercise Group 2.6 and Exercise 17 of Exercise Group 2.7, the directed distance $r_i$ of the point $(x_i, y_i)$ from $L$ may be taken as

$$r_i = \frac{a(x_i - \bar{x}) + b(y_i - \bar{y})}{\sqrt{a^2 + b^2}}.$$

Then, for the moment $M_L$ of the system of particles about $L$, we have

$$M = \sum_{i=1}^{n} r_i m_i = \frac{1}{\sqrt{a^2 + b^2}} \sum_{i=1}^{n} [a(x_i - \bar{x}) + b(y_i - \bar{y})] m_i$$

$$= \frac{1}{\sqrt{a^2 + b^2}} \left[ a \sum_{i=1}^{n} (x_i - \bar{x}) m_i + b \sum_{i=1}^{n} (y_i - \bar{y}) m_i \right].$$

Now,

$$\sum_{i=1}^{n} (x_i - \bar{x}) m_i = M_{x=\bar{x}} = 0, \qquad \sum_{i=1}^{n} (y_i - \bar{y}) m_i = M_{y=\bar{y}} = 0.$$

Hence $M_L = 0$, and the proof is complete.

The following interpretation of center of mass of a system of particles is both useful and characteristic. *For purposes involving moments, a system of particles may be replaced by a single particle at the center of mass with a mass equal to the total mass of the system.* The following example should be studied carefully.

**Example 13.4.1.** Find the center of mass of the following system of particles; in each case the mass is given first, and then the position.

$$2, (4, 3); \quad 4, (-1, 2); \quad 7, (-2, -3); \quad 3, (2, -2); \quad 6, (5, 2).$$

*Solution.* *First method.* We use (13.4.2) and find

$$M = 2 + 4 + 7 + 3 + 6 = 22,$$
$$M_y = 4 \cdot 2 + (-1) \cdot 4 + (-2) \cdot 7 + 2 \cdot 3 + 5 \cdot 6 = 26,$$
$$M_x = 3 \cdot 2 + 2 \cdot 4 + (-3) \cdot 7 + (-2) \cdot 3 + 2 \cdot 6 = -1.$$

Hence the coordinates of the center of mass are

$$\bar{x} = \frac{M_y}{M} = \frac{26}{22} = \frac{13}{11}, \qquad \bar{y} = \frac{M_x}{M} = \frac{-1}{22}.$$

*Second method.* For the second method we shall illustrate the principle stated immediately preceding this example. We shall consider the first three particles as a system $A$, and the other two particles as a system $B$ (this choice is arbitrary). Applying (13.4.2), we find for the system $A$,

$$M = 2 + 4 + 7 = 13, \quad M_y = 4 \cdot 2 + (-1) \cdot 4 + (-2) \cdot 7 = -10,$$
$$M_x = 3 \cdot 2 + 2 \cdot 4 + (-3) \cdot 7 = -7,$$

and hence

$$\bar{x}_A = -\frac{10}{13}, \quad \bar{y}_A = -\frac{7}{13}.$$

In a similar way we find for the system $B$, $M = 9$, $M_y = 36$, $M_x = 6$; hence

$$\bar{x}_B = 4, \quad \bar{y}_B = \tfrac{2}{3}.$$

We now replace system $A$ by a particle of mass 13 at the point $(-10/13, -7/13)$ and system $B$ by a particle of mass 9 at the point $(4, 2/3)$, and find the center of mass of this *system of two particles*, by (13.4.2),

$$\bar{x} = \frac{-10/13 \cdot 13 + 4 \cdot 9}{13 + 9} = \frac{26}{22} = \frac{13}{11}, \quad \bar{y} = \frac{-7/13 \cdot 13 + 2/3 \cdot 9}{13 + 9} = -\frac{1}{22};$$

this is the same point as obtained by the first method.

### EXERCISE GROUP 13.3

In Exercises 1–4 find the center of mass of the system of particles. The mass of each particle is written first, then the position.

1. $7/3, (4, -3); \quad 1/3, (2, 4)$
2. $0.3, (-1, 4); \quad 1.7, (-2, -4)$
3. $2, (1, 3); \quad 4, (-1, 4); \quad 8, (0, -2)$
4. $4, (7, 2); \quad 1, (-4, -2); \quad 2, (3, 5)$

In each of Exercises 5–10 two systems of particles are designated as $A$ and $B$. For each particle the mass is written first and then its position. (a) Find the center of mass of $A$ and of $B$. (b) Use the result of (a) to find the center of mass of the combined system.

5. $\begin{cases} A: 2, (4, 3); \quad 1, (2, 1) \\ B: 7, (-1, 2); \quad 4, (4, 0) \end{cases}$

6. $\begin{cases} A: 3, (-1, -2); \quad 9, (2, -2) \\ B: 5, (0, -2); \quad 4, (5, 0) \end{cases}$

7. $\begin{cases} A: 4, (1, 1); \quad 5, (5, 0) \\ B: 1, (-4, 0); \quad 6, (0, 5) \end{cases}$

8. $\begin{cases} A: 3, (0, 0); \quad 2, (-1, 3) \\ B: 1, (-2, -2); \quad 5, (-2, 4) \end{cases}$

9. $\begin{cases} A: 1, (2, 4); \quad 3, (-1, 2); \quad 7, (0, 4); \quad 4, (-5, 0) \\ B: 1.5, (4, 4); \quad 0.5, (2, -2); \quad 2, (7, -3) \end{cases}$

10. $\begin{cases} A: 0.2, (4, -3); \quad 1.2, (2, -3); \quad 2, (-2, 0) \\ B: 5, (0, -3); \quad 4, (-3, 0); \quad 2, (2, 2); \quad 1, (-3, -3) \end{cases}$

In Exercises 11 and 12 find (a) the center of mass of each system of particles, (b) the center of mass of the combined system.

11. $\begin{cases} A: 4, (2, 3); \quad 2, (-1, 5) \\ B: 2.5, (6, 8); \quad 5, (2, -2) \\ C: 1, (0, 2); \quad 2, (3, 0) \end{cases}$

12. $\begin{cases} A: 3, (-2, -3); \quad 4, (-1, 0); \quad 2, (7, 1) \\ B: 1, (0, 0); \quad 6, (-3, 0); \quad 3, (4, 2) \\ C: 2, (0, 4); \quad 4, (0, 0); \quad 2, (-2, -1) \end{cases}$

## 13.5 CENTER OF MASS OF A LAMINA

We wish to extend the ideas of the preceding section to finding the center of mass of a flat plate, or *lamina*. We consider an idealized situation in which the plate has no thickness, but has mass which we take to be uniformly distributed; that is, the density does not vary from point to point. To define density at a point $P$ (Fig. 13.5.1) we divide the mass of a small plate containing $P$ by the area of the plate, and define the density $\rho$ as the limit of this ratio as the plate shrinks down to the point,

$$\rho = \lim_{\Delta A \to 0} \frac{\Delta m}{\Delta A}.$$

Fig. 13.5.1

If the mass is a function of the area we have, accordingly,

$$\rho = \frac{dm}{dA},$$

the derivative of the mass with respect to the area. *At this time we are restricting the discussion to the case where $\rho$ is constant.*

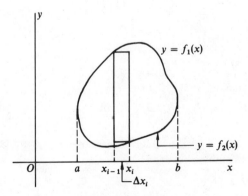

Fig. 13.5.2

Consider the lamina in Fig. 13.5.2, and subdivide it into vertical strips, one of which is shown in the figure. We assume that the lamina is such that the upper and lower boundaries may be represented by functions $f_1(x)$ and $f_2(x)$, respectively. Then, if $dA$ is the element of area, $dA$ is represented approximately as

$$dA = [f_1(x) - f_2(x)]\, dx.$$

Further, if $dm$ is the element of mass for the strip, then

$$dm = \rho\, dA = \rho[f_1(x) - f_2(x)]\, dx,$$

and the *mass of the lamina* is

(13.5.1) $$m = \int_a^b \rho[f_1(x) - f_2(x)]\, dx.$$

The moment of the strip about the $y$ axis is approximately $x\, dm$, for any $x$ such that $x_{i-1} \leq x \leq x_i$, and the moment of the lamina about the $y$ axis is approximately $\sum x_i\, dm_i$, summed over the strips. It is then reasonable to *define the moment of the lamina* about the $y$ axis as

(13.5.2) $$M_y = \int_a^b x\, dm = \int_a^b x\rho[f_1(x) - f_2(x)]\, dx.$$

A modification is needed to get $M_x$, if we wish to use a vertical strip as the element of area. The center of mass of the strip in Fig. 13.5.2 is, approximately, at the point $[x_i, \tfrac{1}{2}(f_1(x) + f_2(x))]$. Hence the moment of the strip about the $x$ axis is, approximately,

$$\frac{1}{2}[f_1(x) + f_2(x)]\, dm = \frac{1}{2}[f_1(x) + f_2(x)] \cdot \rho[f_1(x) - f_2(x)]\, dm$$

$$= \frac{\rho}{2}[(f_1(x))^2 - (f_2(x))^2]\, dx;$$

and

(13.5.3) $$M_x = \frac{1}{2}\int_a^b [f_1(x) + f_2(x)]\, dm = \frac{1}{2}\int_a^b \rho[(f_1(x))^2 - (f_2(x))^2]\, dx.$$

We now *define the center of mass of the lamina* as the point $(\bar{x}, \bar{y})$, where

(13.5.4) $$\bar{x} = \frac{M_y}{m}, \qquad \bar{y} = \frac{M_x}{m}.$$

Let us examine the reason why the expressions for $M_y$ and $M_x$ appear so different. The element of area in Fig. 13.5.2 is such that $x$ is approximately the same for each point in the strip, and hence the moment of the strip about the $y$ axis is approximately $x\, dm$. However, it is clear that $y$ varies between its value at the bottom of the strip and its value at the top, and no single number can approximate all of these, even when $\Delta x_i$ is very small. We get around this difficulty by replacing the strip by a single particle, of the same mass, at its center of mass.

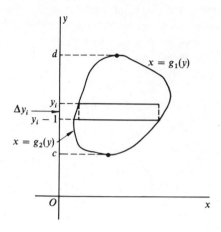

Fig. 13.5.3

The roles of $x$ and $y$ are reversed if we consider the lamina as made up of horizontal strips (Fig. 13.5.3). We now think of the lamina as having a boundary on the right, $x = g_1(y)$, and a boundary on the left, $x = g_2(y)$. Then

$$dA = [g_1(y) - g_2(y)]\, dy, \qquad dm = \rho\, dA = \rho[g_1(y) - g_2(y)]\, dy$$

and by the same reasoning as above we obtain

(13.5.5) $$m = \int_c^d \rho[g_1(y) - g_2(y)]\, dy,$$

(13.5.6) $$M_y = \frac{1}{2}\int_c^d [g_1(y) + g_2(y)]\, dm = \frac{1}{2}\int_c^d \rho[g_1{}^2(y) - g_2{}^2(y)]\, dy,$$

(13.5.7) $$M_x = \int_c^d y\, dm = \int_c^d y\rho[g_1(y) - g_2(y)]\, dy.$$

The center of mass is still given by (13.5.4), where we have a choice of (13.5.1) or (13.5.5) to represent $m$, a choice of (13.5.2) or (13.5.6) for $M_y$, and a choice of (13.5.3) or (13.5.7) for $M_x$.

Since we are taking the density $\rho$ as constant, the center of mass is independent of $\rho$, and in this case is also called the *centroid* of the corresponding region. In particular, the above formulas may be used with $\rho = 1$ to find a centroid.

The following *properties* are capable of proof, if $\rho$ is constant.

1. *The moment of a region about an axis of symmetry is zero.*
2. *If a region has an axis of symmetry, the centroid lies on this axis.*

**Example 13.5.1.** Find the centroid of a semicircular region.

*Solution.* We take the semicircular region as in Fig. 13.5.4. By symmetry we

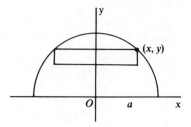

Fig. 13.5.4

have $\bar{x} = 0$. From the equation of the circle $x^2 + y^2 = a^2$ we have, for the right boundary and the left boundary of the circle, respectively,

$$x = g_1(y) = \sqrt{a^2 - y^2}, \qquad x = g_2(y) = -\sqrt{a^2 - y^2}.$$

Then, from (13.5.7)

$$M_x = \int_0^a y \cdot 2\sqrt{a^2 - y^2}\, dy = -\frac{2}{3}(a^2 - y^2)^{3/2} \Big|_0^a = \frac{2a^3}{3}.$$

Since we are taking $\rho = 1$, we have $m =$ area of the region $= \pi a^2/2$. Hence

$$\bar{y} = \frac{2a^3/3}{\pi a^2/2} = \frac{4a}{3\pi}.$$

The centroid of the semicircular region of Fig. 13.5.4 is $(0, 4a/3\pi)$.

**Example 13.5.2.** Find the centroid of the region in the first quadrant bounded by the circle $x^2 + y^2 = a^2$ and the coordinate axes.

*Solution.* We may refer to the region in the first quadrant in Fig. 13.5.4. The moment about the $x$ axis is one-half the moment of the region in Example 13.5.1, and is $a^3/3$. Hence

$$\bar{y} = \frac{a^3/3}{\pi a^2/4} = \frac{4a}{3\pi},$$

and by symmetry, $\bar{x} = \bar{y}$. The centroid is at $(4a/(3\pi), 4a/(3\pi))$.

Note that $\bar{y}$ is the same in Examples 13.5.1 and 13.5.2. Can you formulate a general statement that applies to such situations?

**Example 13.5.3.** Find the centroid of the region bounded by the curves $y^2 = x$ and $y = x$.

*Solution.* To find the area it is convenient to use a horizontal strip as the element (Fig. 13.5.5). For the straight line we have $x = g_1(y) = y$, and for the parabola $x = g_2(y) = y^2$. Hence

$$dA = (y - y^2)\, dy,$$

and
$$A = \int_0^1 (y - y^2)\, dy = \left(\frac{y^2}{2} - \frac{y^3}{3}\right)\bigg|_0^1 = \frac{1}{6}.$$

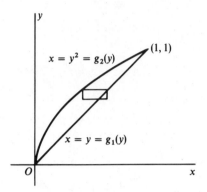

Fig. 13.5.5

We also use the horizontal strip to find $M_x$ and $M_y$. By (13.5.6) and (13.5.7),

$$M_y = \frac{1}{2}\int_0^1 (y^2 - y^4)\, dy = \frac{1}{2}\left(\frac{y^3}{3} - \frac{y^5}{5}\right)\bigg|_0^1 = \frac{1}{15},$$

$$M_x = \int_0^1 y(y - y^2)\, dy = \int_0^1 (y^2 - y^3)\, dy = \frac{1}{12}.$$

Hence

$$\bar{x} = \frac{M_y}{A} = \frac{1/15}{1/6} = \frac{2}{5}, \qquad \bar{y} = \frac{M_x}{A} = \frac{1/12}{1/6} = \frac{1}{2}.$$

### EXERCISE GROUP 13.4

1. Find the centroid of the region bounded by the curves $x = 0, y = 0, x = 1, y = x + 1$.
2. Find the centroid of the region bounded by the curves $x = 0$, $x = 1$, $y = 1 - x$, $y = 2 + 2x$.
3. Find the centroid of the region in the first quadrant bounded by the curves $x = 0$, $y = 0, y = 1 - x^2$.
4. Find the centroid of the region in the first quadrant bounded by the curves $y^2 = x^3$, $x = 0$, and $y = 1$.
5. Find the centroid of the region in the first quadrant bounded by the curves $x^2 = 2y$, $x = 0$, and $x + y = 4$.
6. Find the centroid of the region bounded by the curves $y^2 = 5 - x$ and $y = x + 1$.
7. Find the centroid of the region bounded by the curves $y = x^2$ and $y = 3 - 2x^2$.
8. Find the centroid of the region in the first quadrant bounded by the curves $y = x^3$ and $y = 4x$.

9. Show that the centroid of a triangular region is at a distance from any side equal to one-third the altitude to that side. [*Hint*: Take the vertices at $(0, 0)$, $(a, 0)$, $(b, c)$ and show that $\bar{y} = c/3$.]

* 10. Find the centroid of the region bounded by the curves $y = x^3$ and $y = 3x + 2$.

* 11. If a plane region of area $A$ is submerged vertically in a liquid, prove that the force acting on one side is $wh A$, where $\bar{h}$ is the depth of the centroid of the region below the surface of the liquid.

* 12. A rectangular gate of width $a$ and depth $b$ is submerged vertically in water, the upper edge lying in the surface. If lines are drawn from the midpoint of the lower edge to the ends of the upper edge, show that the forces on the three triangles formed are equal. [*Suggestion*: Use Exercise 11 and the result for the centroid of a triangular region (Exercise 9).]

## * 13.6  CENTER OF MASS OF A COMPOSITE BODY

The methods of the preceding sections may be extended to find the center of mass or centroid of a *composite figure*, that is, a figure which is composed of two or more parts which do not overlap except possibly on boundaries. The regions in Fig. 13.6.1 may be thought of as constituting a single object, whose center of

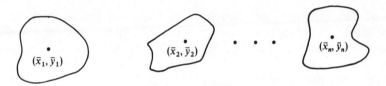

Fig. 13.6.1

mass is desired. In accordance with the principle stated in Section 13.4 whereby, for purposes involving moments, a body may be replaced by a single particle with the same mass at the center of mass of the body, we state the following, for which a formal proof can be given:

*To find the center of mass of a composite body, replace each part by a single particle of equal mass at its center of mass, and find the center of mass of the resulting system of particles.*

This principle also applies to centroids and is particularly useful when the parts of a figure are simple geometric configurations whose centroids and areas are known, such as circles, rectangles, and triangles. (See Exercise 9 of Exercise Group 13.4 for the centroid of a triangular region.)

*Example 13.6.1.* Find the centroid of the figure shown in Fig. 13.6.2, a semi-circular region mounted on a rectangular region.

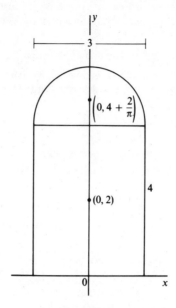

Fig. 13.6.2

*Solution.* By symmetry the centroid is on the center line of the figure. For convenience we have included a coordinate system. Then $\bar{x} = 0$, and $\bar{y}$ is needed. The centroid of the semicircle (see Example 13.5.1) is at a distance $2/\pi$ above the diameter shown, and the centroid of the rectangle is at its center. We replace each part by a particle at its centroid with mass equal to its area, yielding the particles

$$m_1 = 9\pi/8, \text{ at } (0, 4 + 2/\pi); \qquad m_2 = 12, \text{ at } (0, 2).$$

Hence

$$\bar{y} = \frac{\dfrac{9\pi}{8}\left(4 + \dfrac{2}{\pi}\right) + 12 \cdot 2}{\dfrac{9\pi}{8} + 12} = \frac{12\pi + 70}{3\pi + 32} \, (= 2.60 \text{ approx}).$$

*The principle stated above may also be used for figures with parts deleted. The part removed need only be treated as having a negative mass.* This is easily justified upon noting that in computing the moment of the figure about any axis the moment of the deleted part is subtracted; while, for the mass, the mass of the deleted part is subtracted.

**Example 13.6.2.** Find the centroid of the region between the two circles shown in Fig. 13.6.3.

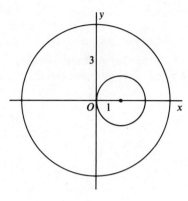

Fig. 13.6.3

*Solution.* Replacing each circular region by an equivalent particle, we have a system of two particles:

$$m_1 = 9\pi, \text{ at } (0, 0); \qquad m_2 = -\pi, \text{ at } (1, 0).$$

Clearly $\bar{y} = 0$. Then

$$\bar{x} = \frac{9\pi \cdot 0 - \pi \cdot 1}{9\pi - \pi} = -\frac{1}{8}.$$

### ★ EXERCISE GROUP 13.5

In the following exercises you may use any symmetries and any previous result.

1. Find the centroid of a region consisting of a vertical square of side 4 mounted by an isosceles triangle of altitude 3 (and base 4).

2. An isosceles trapezoid has bases 4 and 8 and altitude 2. Find its centroid by considering the trapezoid as made up of a rectangle and two right triangles.

3. Consider the region between two concentric circles of radii 4 and 8. Find the centroid of one of the halves formed by a line through the common center.

4. Find the centroid of the region between a circle of radius 4 with center at the origin, and a circle of radius 1 with center at (3, 0).

5. Find the centroid of a system consisting of a circular region of radius 4 with center at $(-4, 0)$ and a circular region of radius 2 with center at $(4, 0)$.

6. Find the centroid of the region between a semicircle of radius 4 and the circumscribed rectangle.

7. Find the centroid of the region bounded by the upper half of the circle $x^2 + y^2 = 4$, the upper half of the circle $(x - 1)^2 + y^2 = 1$, and the lower half of the circle $(x + 1)^2 + y^2 = 1$.

8. Find the centroid of the region in the first quadrant bounded by the curves $x^2 + y^2 = 4$, $y = 2x + 2$, $x = 3$, and $y = 0$.

9. Consider the region between the square $|x| + |y| = 1$ and the square formed by the lines $x = \pm 2$, $y = \pm 2$. (a) Find the centroid of the upper half of this region. (b) Find the centroid of the part of the region which lies in the first quadrant.

10. Find the centroid of a body consisting of a square region of side 2, on three sides of which semicircular regions are constructed outside the square.

## 13.7 CENTROIDS OF OTHER FIGURES

Let $r$ represent the distance from a point of some plane figure to a line $L$ in the plane. The figure shown in Fig. 13.7.1 is an arc, but the discussion applies also to a region. If $\rho$ is the density, which may be mass per unit length for an arc or mass per unit area for a region, it is reasonable to expect from our earlier work that the distance of the center of mass from $L$ is given by

(13.7.1) $$\bar{r} = \frac{\int r \, dm}{\int dm}.$$

We have $dm = \rho \, dA$ for a region or $dm = \rho \, ds$ for an arc.

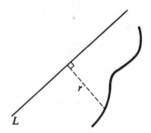

Fig. 13.7.1

The generic formula (13.7.1) may also be used for a space figure if $r$ represents the distance of a point in the figure from a plane. We then have $dm = \rho \, dV$ for a solid region, $dm = \rho \, dS$ for a surface, and again $dm = \rho \, ds$ for an arc.

It is essential in each case that for the moment in the numerator of (13.7.1) the element of mass be chosen so that $r$ is approximately the same for each point in the element. This requirement permits the limiting process which leads to a definite integral.

As in the earlier sections, if the density is constant we refer to the center of mass as the *centroid*, and we may use the value $\rho = 1$.

*Centroid of an arc*

In the case of an arc in the plane we deduce from (13.7.1) that the centroid $(\bar{x}, \bar{y})$ is given by

$$\bar{x} = \frac{\int x \, ds}{\int ds}, \quad \bar{y} = \frac{\int y \, ds}{\int ds}.$$

**Example 13.7.1.** Find the centroid of a semicircular arc of radius $a$.

*Solution.* We place the arc in the position shown in Fig. 13.7.2. By symmetry we have $\bar{x} = 0$. The equation of the arc is $x^2 + y^2 = a^2$, $y \geq 0$. We find

$$\frac{dy}{dx} = -\frac{x}{y}, \quad ds = \sqrt{1 + \frac{x^2}{y^2}}\, dx = \sqrt{\frac{y^2 + x^2}{y^2}}\, dx = \frac{a\, dx}{y}.$$

Hence

$$\int y\, ds = \int_{-a}^{a} y(a/y)\, dx = \int_{-a}^{a} a\, dx = 2a^2.$$

We know that $\int ds$ is the length of arc and is $\pi a$. Thus

$$\bar{y} = \frac{2a^2}{\pi a} = \frac{2a}{\pi}.$$

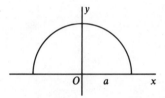

Fig. 13.7.2

*Centroid of a solid of revolution*

The centroid of a solid of revolution lies on the axis of the solid, and it is necessary only to find its position on the axis.

**Example 13.7.2.** Find the centroid of the solid generated when the curve $y = x^2$, $0 \leq x \leq 2$ is revolved about the $x$ axis.

*Solution.* If we use as an element of volume the disc shown in Fig. 13.7.3 we have $dV = \pi y^2\, dx$. The use of this element is permissible since the distance to the plane perpendicular to the $x$ axis at the origin is approximately the same for each point in the disc ($dx$ is small), and may be taken as the value of $x$ for a point on the curve. We may use (13.7.1), with $\rho = 1$, in the form

$$\bar{x} = \frac{\int x\, dm}{\int dm} = \frac{\int x\, dV}{\int dV} = \frac{\pi \int xy^2\, dx}{\pi \int y^2\, dx} = \frac{\pi \int_0^2 x^5\, dx}{\pi \int_0^2 x^4\, dx}.$$

We find

$$\int_0^2 x^5\, dx = 32/3, \quad \int_0^2 x^4\, dx = 32/5.$$

Hence $\bar{x} = 5/3$, and we have the position of the centroid on the axis of the solid.

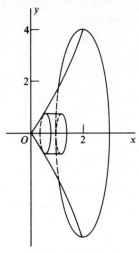

Fig. 13.7.3

*Centroid of a surface of revolution*

The centroid of a surface of revolution lies on the axis of symmetry. If the axis of symmetry is the $x$ axis, we have

$$\bar{x} = \frac{\int x \, dS}{\int dS},$$

where the use of the usual frustrum of a cone (see Section 12.6) is valid for the element of surface area. We then have $dS = 2\pi y \, ds$, and

(13.7.2) $$\bar{x} = \frac{2\pi \int xy \, ds}{2\pi \int y \, ds}.$$

*Example 13.7.3.* Find the centroid of the surface of a right circular cone of altitude $h$.

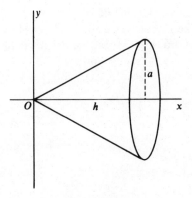

Fig. 13.7.4

*Solution.* We place the cone as in Fig. 13.7.4, and use (13.7.2) with $y = ax/h$. Then

$$ds = \frac{\sqrt{h^2 + a^2}}{h} dx,$$

and (13.7.2) reduces to

$$\bar{x} = \frac{\int_0^h x^2\, dx}{\int_0^h x\, dx} = \frac{h^3/3}{h^2/2} = \frac{2h}{3}.$$

This result does not depend on the radius of the base.

## EXERCISE GROUP 13.6

1. Find the coordinates of the centroid of the solid generated by revolving about the $x$ axis the region bounded by the curves $y^2 = 2x$ and $x = 2$.

2. Find the centroid of the solid generated by revolving about the $x$ axis the region bounded by $x + y = 2$, $x = 1$, $y = 0$.

3. Find the centroid of the solid generated by revolving about the $y$ axis the region bounded by $x + y = 2$, $x = 1$, $y = 0$.

4. Find the centroid of the surface generated by revolving about the $x$ axis the part of the line $x + y = 2$ from $x = 0$ to $x = 2$.

5. Find the centroid of the part of the curve $x^{2/3} + y^{2/3} = a^{2/3}$ in the first quadrant.

6. Find the centroid of a thin uniform wire in the shape of a triangle with vertices (0, 0), (1, 0), and (3, 4).

7. Adapt formula (13.7.1) to apply in the case of variable density, and find the center of gravity of a straight wire of length $L$ if the density is proportional to the distance from one end.

8. Find the centroid of the body consisting of the sides of the quadrilateral $ABCD$, with vertices $A(2, 0)$, $B(0, \sqrt{3})$, $C(5, 0)$, $D(0, -\sqrt{3})$.

9. Find the center of gravity of a thin uniform wire formed by the semicircle $x^2 + y^2 = a^2$, $y \geq 0$, and the line segment $y = 0$ between $x = a$ and $x = -a$.

10. Find the centroid of the curve consisting of the semicircle $x^2 + y^2 = 4$, $x \leq 0$, and the line $y = 2$ from $x = 0$ to $x = 10$.

*11. *Theorems of Pappus* (a) If the curve $y = f(x)$, $f(x) \geq 0$, $a \leq x \leq b$, is revolved about the $x$ axis, the area of the surface generated is $2\pi \int_a^b f(x)\, ds$. Express $\int_a^b f(x)\, ds$ in terms of $\bar{y}_L$, the ordinate of the centroid of the arc, and prove the following Theorem of Pappus.

*If a plane curve is revolved about an axis in the plane which does not cut the curve, the area of the surface generated is the product of the length of the arc and the circumference of the circle described by the centroid of the arc.*

(b) By considering the volume generated by revolving a plane region about an axis in the plane, prove the following Theorem of Pappus.

*If a plane region is revolved about an axis in the plane which does not cut through the region, the volume of the solid generated is the product of the area of the region and the circumference of the circle described by the centroid of the region.*

Use a Theorem of Pappus to find each of the following.

* 12. The centroid of a semicircular region.
* 13. The centroid of a semicircular arc.
* 14. The volume of the solid obtained by revolving the region bounded by $x^2 + (y-b)^2 = a^2$, $0 < a < b$, about the $x$ axis.
* 15. The area of the surface generated by revolving the curve of Exercise 14 about the $x$ axis.

## 13.8 MOMENT OF INERTIA

In Section 13.4 the moment of a system of particles about a line was defined. Such a moment is also called a *first moment*. At this time we define *moment of inertia*, also called *second moment*. Figure 13.4.1 may again be used for this purpose.

Consider a system of particles in a plane, of respective masses $m_i$, $i = 1, 2, \ldots, n$, and distances $r_i$, $i = 1, 2, \ldots, n$, from a line $L$ in the plane. The *moment of inertia* of the system of particles about the line $L$ is *defined* as

$$(13.8.1) \qquad I_L = \sum_{i=1}^{n} r_i^2 m_i.$$

Note that the distance $r_i$ is squared, so that it need not be a directed distance.

Suppose now that we have a flat plate or *lamina*. We restrict the discussion to the case of constant density $\rho$. Let us subdivide a region as shown in Fig. 13.5.3 into horizontal strips. If $dm_i$ is the element of mass for the strip, the moment of inertia of the strip about the $x$ axis is approximately $y_i^2 \, dm_i$, and it is reasonable to *define* the moment of inertia $I_x$ about the $x$ axis as

$$(13.8.2) \qquad I_x = \int_a^b y^2 \, dm = \rho \int_a^b y^2 [g_1(y) - g_2(y)] \, dy.$$

In a similar manner, by using vertical strips (Fig. 13.5.2) we obtain, for the moment of inertia about the $y$ axis,

$$(13.8.3) \qquad I_y = \rho \int_c^d x^2 [f_1(x) - f_2(x)] \, dx.$$

For convenience in the examples and in the exercises we shall now assume $\rho = 1$.

***Example 13.8.1.*** Find the moment of inertia of a rectangular region of sides $a$ and $b$ about a side of length $a$.

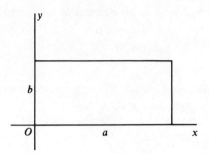

Fig. 13.8.1

*Solution.* With the rectangle placed as in Fig. 13.8.1, we seek $I_x$. We then have $g_1(y) = a$, $g_2(y) = 0$, and by (13.8.2), with $\rho = 1$,

$$I_x = \int_0^b y^2 \cdot a \, dy = ab^3/3.$$

By analogy we may immediately write

$$I_y = a^3 b/3.$$

**Example 13.8.2.** Find $I_x$ and $I_y$ for the region in the first quadrant bounded by $y^2 = 4x$, $y = 0$, $x = 2$.

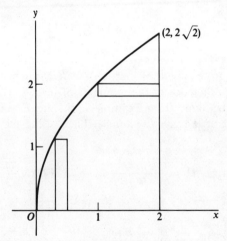

Fig. 13.8.2

*Solution.* The region is shown in Fig. 13.8.2. To find $I_x$ we use the horizontal strip as element of area and find, by (13.8.2),

$$I_x = \int_0^{2\sqrt{2}} y^2 \left(2 - \frac{y^2}{4}\right) dy = \left(\frac{2y^3}{3} - \frac{y^5}{20}\right)\bigg|_0^{2\sqrt{2}} = \frac{64\sqrt{2}}{15}.$$

For $I_y$ we use the vertical strip as an element of area, and with (13.8.3) we find

$$I_y = \int_0^2 x^2 \cdot 2\sqrt{x}\, dx = 2\int_0^2 x^{5/2}\, dx = \frac{4x^{7/2}}{7}\bigg|_0^2 = \frac{32\sqrt{2}}{7}.$$

Moment of inertia about an axis has an additive property, which is that *the moment of inertia of a composite body about an axis is equal to the sum of the moments of inertia of the parts about the same axis*. This property may also be used as in the following example.

***Example 13.8.3.*** Find the moment of inertia of a rectangular region of sides $a$ and $b$ about a line parallel to a side of length $a$, and not intersecting the rectangle.

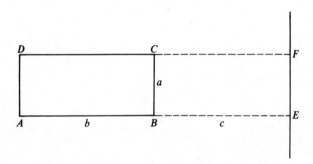

Fig. 13.8.3

*Solution.* We seek the moment of inertia of *ABCD* about $L$ (see Fig. 13.8.3). This is equal to the difference of the moments of inertia of *AEFD* and *BEFC*, each of which may be found from Example 13.8.1. Let $c$ be the distance from $L$ to the nearest point of the given rectangle. Then, using the result of Example 13.8.1, we have

$$I_L(AEFD) = \frac{a(b+c)^3}{3}, \qquad I_L(BEFC) = \frac{ac^3}{3},$$

and hence

$$I_L(ABCD) = \frac{a(b+c)^3 - ac^3}{3} = \frac{a(b^3 + 3b^2c + 3bc^2)}{3}.$$

If the total mass $m$ of a body were concentrated at a point, the moment of inertia about a line $L$ would be $mr^2$, where $r$ is the distance from the point to $L$. For the moment of inertia of the particle to be the same as that of the body about $L$, we must have

$$mr_L^2 = I_L \quad \text{or} \quad r_L = \sqrt{I_L/m}.$$

The distance $r_L$ is called the *radius of gyration* with respect to $L$. The *radius of gyration* of a body with respect to a line $L$ is *defined as the distance from $L$ at which the mass of the body may be concentrated so that the particle has the same moment of inertia as the body*. If $\rho = 1$ the mass of the object is equal to the area of the region.

In Example 13.8.1 we may take $m$ as $ab$. Then the radii of gyration about the $x$ and $y$ axes are

$$r_x = \sqrt{\frac{ab^3}{3ab}} = \frac{b}{\sqrt{3}}, \qquad r_y = \sqrt{\frac{a^3b}{3ab}} = \frac{a}{\sqrt{3}}.$$

In Example 13.8.2 we find

$$m = \text{Area} = \int_0^2 2\sqrt{x}\, dx = 8\sqrt{2}/3.$$

Hence

$$r_x = \sqrt{(64\sqrt{2}/15)/(8\sqrt{2}/3)} = 2\sqrt{2}/\sqrt{5},$$

$$r_y = \sqrt{(32\sqrt{2}/7)/(8\sqrt{2}/3)} = 2\sqrt{3}/\sqrt{7}.$$

*Optional discussion.* Horizontal strips were used as elements of area in deriving (13.8.2), while vertical strips were used in deriving (13.8.3). We may in fact use horizontal or vertical strips to find either $I_x$ or $I_y$, but modifications in method are needed. Let us find $I_x$ in Example 13.8.2 with the use of vertical strips. We find the approximate moment of inertia $dI_x$ of the vertical strip in Fig. 13.8.2 from the result of Example 13.8.1. Thus

$$dI_x = \frac{y^3\, dx}{3} = \frac{8x^{3/2}}{3}\, dx,$$

and

$$I_x = \int dI_x = \int_0^2 8x^{3/2}/3\, dx = 64\sqrt{2}/15,$$

which is the same result as previously obtained.

To find $I_y$ in Example 13.8.2 by use of horizontal strips (such a strip is shown in Fig. 13.8.2), we use the method employed in Example 13.8.3 to find the moment of inertia $dI_y$ of a typical horizontal strip. Thus

$$dI_y = \left[\frac{2^3}{3} - \frac{1}{3}\left(\frac{y^2}{4}\right)^3\right] dy = \frac{1}{192}(512 - y^6)\, dy,$$

and

$$I_y = \int dI_y = \frac{1}{192}\int_0^{2\sqrt{2}} (512 - y^6)\, dy;$$

this evaluates to $32\sqrt{2}/7$, the same as $I_y$ in Example 13.8.2.

## EXERCISE GROUP 13.7

1. Find (a) the moment of inertia, (b) the radius of gyration, of a triangular region of base $b$ and altitude $h$ about its base.
2. Find the moment of inertia of a square region of side $a$ about a diagonal.
3. Find $I_x$ and $I_y$ for the region bounded by the curves $y = x^2$, $y = 0$, and $x = 2$.
4. Find (a) $I_x$, (b) $I_y$, (c) the moment of inertia about the line $x = 1$, for the region bounded by $y = 2x^2$ and $y = 2x$.
5. Assume that the moment of inertia of a circular region of radius $a$ about a diameter is $\pi a^4/4$.
   (a) Find the moment of inertia about the common diameter of a region bounded by two internally tangent circles of radii $a$ and $b$ with $a > b$.
   (b) The same as (a) if the circles are externally tangent.

   Consider a thin wire in the shape of a curve. If the density (mass per unit length) is 1, the moments of inertia about the $x$ axis and $y$ axis, respectively, are seen to be

   $$I_x = \int y^2 \, ds, \qquad I_y = \int x^2 \, ds$$

   with suitable limits of integration, where $ds$ is the element of arc length.

6. Find (a) the moment of inertia, (b) the radius of gyration, of a line of length $L$ about an axis perpendicular to the line at its center.
7. Work Exercise 6 if the axis is perpendicular to the line at one end.
8. Find $I_x$ for the part of the line $x + y = 2$ from $x = 0$ to $x = 2$.
9. The *polar moment of inertia* $I_0$ is the moment of inertia of a plane body about an axis perpendicular to the plane at the origin. It will be shown (Section 22.8) that $I_0 = I_x + I_y$.

   Using this result and the moment of inertia of a circular region about a diameter given in Exercise 5, show that the moment of inertia of a circular region about a line through its center and perpendicular to its plane is $\pi a^4/2$.

* 10. Using Exercise 9, show that the moment of inertia with respect to the $x$ axis of a solid of revolution obtained by revolving the region bounded by $y = f(x) \geq 0$, $x = a$, $x = b$ about the $x$ axis is given by

   $$I_x = \frac{\pi}{2} \int_a^b y^4 \, dx.$$

11. The region bounded by $y = x^{1/2}$, $y = 0$, $x = 2$ is revolved about the $x$ axis. Use Exercise 10 to find $I_x$ for the solid generated.
12. Use Exercise 10 to find the moment of inertia of a solid sphere about a diameter.

# 14 Methods of Integration

## 14.1 FORMAL INTEGRATION

Formal integration has thus far consisted primarily of the development of specific formulas of integration, together with a few basic methods. We present a list of the specific formulas already obtained, some with small modifications, and add to it a few formulas which should also be considered as part of the basic list.

Most of the formulas are immediately familiar. The usual derivations of 8 and 9 are rather artificial and are not presented; a valid proof is obtained by differentiating the right member of each. (You may wish to look at Exercise 24 of Exercise Group 14.6, where a derivation is suggested which, although not elegant, is not artificial.) Formula 17 may be derived as follows. Let $u = av$, $du = a\, dv$. Then

$$\int \frac{du}{\sqrt{a^2 - u^2}} = \int \frac{a\, dv}{a\sqrt{1 - v^2}} = \int \frac{dv}{\sqrt{1 - v^2}} = \sin^{-1} v + C$$
$$= \sin^{-1} u/a + C.$$

Formulas 18 and 19 may be derived similarly. These forms are more convenient in practice than the original ones. Formulas 20 and 21, while not easily derived, occur in practice and are included in the list; the first of these will be obtained in Example 14.8.5. (See also Exercise 25 of Exercise Group 14.6.)

## Formulas of Integration

1. $\int a\, du = au + C.$

2. $\int u^n\, du = \dfrac{u^{n+1}}{n+1} + C,\ n \neq -1.$

3. $\int [r(u) + s(u)]\, du = \int r(u)\, du + \int s(u)\, du.$

4. $\int \sin u\, du = -\cos u + C.$

5. $\int \cos u\, du = \sin u + C.$

6. $\int \tan u\, du = \ln|\sec u| + C = -\ln|\cos u| + C.$

7. $\int \cot u\, du = -\ln|\csc u| + C = \ln|\sin u| + C.$

8. $\int \sec u\, du = \ln|\sec u + \tan u| + C.$

9. $\int \csc u\, du = \ln|\csc u - \cot u| + C.$

10. $\int \sec^2 u\, du = \tan u + C.$

11. $\int \csc^2 u\, du = -\cot u + C.$

12. $\int \sec u \tan u\, du = \sec u + C.$

13. $\int \csc u \cot u\, du = -\csc u + C.$

14. $\int \dfrac{du}{u} = \ln|u| + C.$

15. $\int e^u\, du = e^u + C.$

16. $\int a^u\, du = \dfrac{a^u}{\ln a} + C,\ a > 0,\ a \neq 1.$

17. $\int \dfrac{du}{\sqrt{a^2 - u^2}} = \sin^{-1}\dfrac{u}{a} + C = -\cos^{-1}\dfrac{u}{a} + C,\ a > 0.$

18. $\int \dfrac{du}{a^2 + u^2} = \dfrac{1}{a}\tan^{-1}\dfrac{u}{a} + C = -\dfrac{1}{a}\cot^{-1}\dfrac{u}{a} + C,\ a > 0.$

19. $\int \dfrac{du}{u\sqrt{u^2 - a^2}} = \dfrac{1}{a}\sec^{-1}\dfrac{u}{a} + C = -\dfrac{1}{a}\csc^{-1}\dfrac{u}{a} + C,\ a > 0.$

20. $\int \sec^3 u\, du = \tfrac{1}{2}(\sec u \tan u + \ln|\sec u + \tan u|) + C.$

21. $\int \csc^3 u\, du = \tfrac{1}{2}(-\csc u \cot u + \ln|\csc u - \cot u|) + C.$

## 14.1 EXERCISES

The only significant method of integration previously used was that of substitution, or change of variable (Sections 9.5 and 9.11). Other steps usually consisted in changing the form of a given integrand by expanding, by long division of polynomials, by addition or breaking up of fractions, by the use of trigonometric identities, and so on. These methods remain as part of the technique of integration, but in the present chapter we shall enlarge the scope of the technique by presenting new methods for specific types of integral that include many of those frequently met.

The first set of exercises is designed as a review of those aspects of integration already studied. Some of these exercises perhaps go beyond what you have worked out earlier, but all of them can be done without the use of any new methods.

### EXERCISE GROUP 14.1

Perform the integrations in Exercises 1–32

1. $\int (2x-1)^3 \, dx$
2. $\int \sqrt{1-2x} \, dx$
3. $\int \dfrac{2x^2 - x + 3}{\sqrt{x}} \, dx$
4. $\int \dfrac{x \, dx}{\sqrt{1-4x^2}}$
5. $\int \dfrac{dx}{\sqrt{1+4x}}$
6. $\int \dfrac{2x^2 \, dx}{\sqrt{1-x^3}}$
7. $\int \dfrac{(2x-1) \, dx}{x^2 - x + 1}$
8. $\int \sin 3x \cos 3x \, dx$
9. $\int x \sin x^2 \, dx$
10. $\int \tan^2 x \sec^2 x \, dx$
11. $\int \dfrac{dx}{\sqrt{4-x^2}}$
12. $\int \dfrac{dx}{4+x^2}$
13. $\int \dfrac{\sin^{-1} x \, dx}{\sqrt{1-x^2}}$
14. $\int \dfrac{\tan^{-1} 2x \, dx}{1+4x^2}$
15. $\int \dfrac{\sin x}{\cos^2 x} \, dx$
16. $\int x^2 e^{x^3} \, dx$
17. $\int 5^x \, dx$
18. $\int x^2 \, 5^{x^3} \, dx$
19. $\int \dfrac{\ln x \, dx}{x}$
20. $\int \dfrac{e^x \, dx}{1-e^x}$
21. $\int \dfrac{(1 + \ln x) \, dx}{x}$
22. $\int \dfrac{dx}{5 + 3x^2}$
23. $\int \dfrac{x \, dx}{1+x}$
24. $\int \dfrac{\sec^2 2x \, dx}{1 + \tan 2x}$
25. $\int \dfrac{\sin x \, dx}{2 - \cos x}$
26. $\int \dfrac{4 - 3x}{4 + 3x} \, dx$
27. $\int \dfrac{2^x}{3 - 2^x} \, dx$
28. $\int_1^4 \dfrac{1 - 3x}{x} \, dx$
29. $\int_{\pi/4}^{\pi/3} \tan x \sec x \, dx$
30. $\int_0^{3/4} x\sqrt{1 + x^2} \, dx$
31. $\int_{\pi/4}^{\pi/3} \sin^2 3x \cos 3x \, dx$
32. $\int_{\pi/4}^{\pi/3} \sin 3x \cos^2 3x \, dx$

33. Prove formula 8 by differentiation.
34. Prove formula 20 by differentiation.
35. Find the length of the curve $y = \ln \sec x$ from $x = 0$ to $x = \pi/3$.

## 14.2 POWERS AND PRODUCTS OF SINE AND COSINE

Methods for evaluating integrals of the form

(14.2.1) $$\int \sin^m x \cos^n x \, dx$$

depend on the exponents $m$ and $n$.

(a) *m or n (or both) odd*

Suppose $m$ is an odd integer. We may then separate a factor $\sin x$ and use the squares identity $\sin^2 u + \cos^2 u = 1$ on the remaining even power of $\sin x$. A similar procedure applies when $n$ is odd.

**Example 14.2.1.** Evaluate $\int \cos^3 2x \, dx$.

*Solution.* We have

$$\int \cos^3 2x \, dx = \int \cos^2 2x \cdot \cos 2x \, dx = \int (1 - \sin^2 2x) \cos 2x \, dx.$$

Substituting $u = \sin 2x$, $du = 2 \cos 2x \, dx$, we get

$$\int (1 - \sin^2 2x) \cos 2x \, dx = \frac{1}{2} \int (1 - u^2) \, du = \frac{1}{2} \left( u - \frac{u^3}{3} \right) + C.$$

Hence

$$\int \cos^3 2x \, dx = \frac{\sin 2x}{2} - \frac{\sin^3 2x}{6} + C.$$

**Example 14.2.2.** Evaluate $\int \sin^5 x \cos^2 x \, dx$.

*Solution.* We have

$$\int \sin^5 x \cos^2 x \, dx = \int \sin^4 x \cos^2 x (\sin x \, dx)$$

$$= \int (\sin^2 x)^2 \cos^2 x (\sin x \, dx)$$

$$= \int (1 - \cos^2 x)^2 \cos^2 x (\sin x \, dx).$$

We now substitute $u = \cos x$, $du = -\sin x \, dx$ and we get, continuing the above,

$$\int (1 - u^2)^2 u^2 (-du) = -\int (u^2 - 2u^4 + u^6) \, du$$

$$= -\frac{u^3}{3} + \frac{2u^5}{5} - \frac{u^7}{7} + C$$

$$= -\frac{\cos^3 x}{3} + \frac{2 \cos^5 x}{5} - \frac{\cos^7 x}{7} + C.$$

If both $m$ and $n$ are odd, the procedure may be used with either exponent. It should be noted that if one of the exponents is a positive odd integer, the other exponent may be any number at all, negative, nonintegral, irrational, or zero.

The method just presented may also be attempted in cases where the odd exponent is negative. See Exercise 25 of Exercise Group 14.6 for $\int \cos^{-3} x \, dx$.

**(b)** *Both m and n nonnegative, even integers*

In this case the integral (14.2.1) may be evaluated by using half-angle formulas and double-angle formulas in the forms

$$\sin^2 u = \frac{1 - \cos 2u}{2}, \qquad \cos^2 u = \frac{1 + \cos 2u}{2}, \qquad \sin u \cos u = \frac{\sin 2u}{2}.$$

With use of these identities the integral (14.2.1) is expressed in terms of similar integrals with smaller values of $m$ and $n$, and with multiples of $x$ in place of $x$. The process is continued until $\int \sin^2 kx \, dx$ or $\int \cos^2 kx \, dx$ is involved. Then for the second of these, for example, we have

$$\int \cos^2 kx \, dx = \frac{1}{2} \int (1 + \cos 2kx) \, dx = \frac{1}{2} \left( x + \frac{\sin 2kx}{2k} \right) + C.$$

**Example 14.2.3.** Evaluate $\int \sin^6 x \, dx$.

*Solution.* We have

$$\int \sin^6 x \, dx = \int [\sin^2 x]^3 \, dx = \frac{1}{8} \int (1 - \cos 2x)^3 \, dx$$

$$= \frac{1}{8} \int (1 - 3 \cos 2x + 3 \cos^2 2x - \cos^3 2x) \, dx.$$

The first and second terms in the integrand integrate directly. For the third term we have (omitting a constant of integration),

$$\int \cos^2 2x \, dx = \frac{1}{2} \int (1 + \cos 4x) \, dx = \frac{x}{2} + \frac{\sin 4x}{8}.$$

For the fourth term we have, from Example 14.2.1,

$$\int \cos^3 2x \, dx = \frac{\sin 2x}{2} - \frac{\sin^3 2x}{6}.$$

Hence

$$\int \sin^6 x \, dx = \frac{1}{8} \left[ x - \frac{3 \sin 2x}{2} + \frac{3x}{2} + \frac{3 \sin 4x}{8} - \frac{\sin 2x}{2} + \frac{\sin^3 2x}{6} \right] + C$$

$$= \frac{1}{192} [60x - 48 \sin 2x + 4 \sin^3 2x + 9 \sin 4x] + C.$$

**Example 14.2.4.** Evaluate $\int \sin^2 x \cos^4 x \, dx$.

*Solution.* We have

$$\int \sin^2 x \cos^4 x \, dx = \int (\sin x \cos x)^2 \cos^2 x \, dx = \int \frac{\sin^2 2x}{4} \cdot \frac{1 + \cos 2x}{2} \, dx$$

$$= \frac{1}{8} \left[ \int \sin^2 2x \, dx + \int \sin^2 2x \cos 2x \, dx \right]$$

$$= \frac{1}{8} \left[ \left( \frac{x}{2} - \frac{\sin 4x}{8} \right) + \frac{\sin^3 2x}{6} \right] + C$$

$$= \frac{1}{192} [12x + 4 \sin^3 2x - 3 \sin 4x] + C.$$

The following integrals of products of sines and cosines arise in some important applications:

$$\int \sin mx \cos nx \, dx, \quad \int \sin mx \sin nx \, dx, \quad \int \cos mx \cos nx \, dx.$$

A standard procedure for evaluating such integrals is to make direct use of the following trigonometric identities.

(14.2.2) $\quad \sin A \cos B = \frac{1}{2}[\sin(A + B) + \sin(A - B)],$

(14.2.3) $\quad \sin A \sin B = \frac{1}{2}[\cos(A - B) - \cos(A + B)],$

(14.2.4) $\quad \cos A \cos B = \frac{1}{2}[\cos(A + B) + \cos(A - B)].$

***Example 14.2.5.*** Evaluate $\int \sin 4x \sin 3x \, dx.$

*Solution.* We apply (14.2.3) to the integrand and find

$$\int \sin 4x \sin 3x \, dx = \frac{1}{2} \int [\cos(4x - 3x) - \cos(4x + 3x)] \, dx$$

$$= \frac{1}{2} \int (\cos x - \cos 7x) \, dx$$

$$= \frac{\sin x}{2} - \frac{\sin 7x}{14} + C.$$

**EXERCISE GROUP 14.2**

Evaluate each of the following integrals.

1. $\int (1 + \sqrt{\cos x}) \sin x \, dx$

2. $\int \sin^4(2x + 3) \cos(2x + 3) \, dx$

3. $\int \sin^2 x \cos^5 x \, dx$

4. $\int \sin^5 2x \, dx$

5. $\int \cos^2 2x \, dx$

6. $\int \sin^{1/3} x \cos^3 x \, dx$

7. $\int \sin^7 x \, dx$

8. $\int \cos^4 2x \, dx$

9. $\int \frac{\sin^3 x \, dx}{1 + \cos x}$

10. $\displaystyle\int \frac{\cos^3 x \, dx}{1 + \sin x}$

11. $\displaystyle\int \sin^2 x \cos^2 x \, dx$

12. $\displaystyle\int \sin^4 x \cos^2 x \, dx$

13. $\displaystyle\int \cos^6 3x \, dx$

14. $\displaystyle\int_0^{\pi/2} \sin^2 x \cos^3 x \, dx$

15. $\displaystyle\int \frac{\cos^3 x}{\sin^2 x} \, dx$

16. $\displaystyle\int \sin^3 x \cos^{-1/2} x \, dx$

17. $\displaystyle\int \tan^3 x \cos x \, dx$

18. $\displaystyle\int \sin 3x \cos 4x \, dx$

19. $\displaystyle\int \cos 2x \cos x \, dx$

20. $\displaystyle\int \cos \frac{5x}{2} \cos \frac{3x}{2} \, dx$

21. $\displaystyle\int \sin 2x \cos \frac{5x}{2} \, dx$

22. $\displaystyle\int_0^1 \sin 2\pi x \cos 2\pi x \, dx$

23. $\displaystyle\int_0^3 \cos \frac{2\pi x}{3} \cos \frac{5\pi x}{3} \, dx$

24. $\displaystyle\int_0^{2\pi} \sin 4x \cos 6x \, dx$

25. $\displaystyle\int_0^{2\pi} \sin mx \sin nx \, dx$, $m, n$ nonnegative integers

26. $\displaystyle\int_0^{2\pi} \cos mx \cos nx \, dx$, $m, n$ nonnegative integers

## 14.3 POWERS OF TANGENT AND SECANT

Integrals of the forms

(14.3.1) $\qquad \displaystyle\int \tan^m x \sec^n x \, dx \quad \text{and} \quad \int \cot^m x \csc^n x \, dx$

may be evaluated by appropriate use of the squares identities

(14.3.2) $\qquad \sec^2 x = 1 + \tan^2 x, \quad \csc^2 x = 1 + \cot^2 x$

in those cases where either $m$ is odd or $n$ is even.

(a) *m odd.* In this case we split off a factor $\sec x \tan x$ or $\csc x \cot x$ and express the remaining part of the integrand in terms of $\sec x$ or $\csc x$ by use of (14.3.2).

***Example 14.3.1.*** Evaluate $\int \tan^3 x \sec^3 x \, dx$.

*Solution.* We have

$$\int \tan^3 x \sec^3 x \, dx = \int \tan^2 x \sec^2 x \, (\sec x \tan x \, dx)$$

$$= \int (\sec^2 x - 1) \sec^2 x \, (\sec x \tan x \, dx).$$

Making the substitution $u = \sec x$, $du = \sec x \tan x \, dx$, we continue the above as

$$\int (u^2 - 1) u^2 \, du = \int (u^4 - u^2) \, du = \frac{u^5}{5} - \frac{u^3}{3} + C$$

$$= \frac{\sec^5 x}{5} - \frac{\sec^3 x}{3} + C.$$

When $m$ is odd, $n$ need not be an integer. Also, see Exercise 18 for an illustration where $n = 0$.

(b) *n even*. In this case we may split off a factor $\sec^2 x$ or $\csc^2 x$ and express the remaining part in terms of $\tan x$ or $\cot x$.

***Example 14.3.2.*** Evaluate $\int \tan^3 x \sec^4 x\, dx$.

*Solution.* The method for odd $m$ applies, but we shall illustrate the procedure for even $n$. We have

$$\int \tan^3 x \sec^4 x\, dx = \int \tan^3 x \sec^2 x\, (\sec^2 x\, dx)$$

$$= \int \tan^3 x\, (1 + \tan^2 x) \sec^2 x\, dx.$$

With the substitution $u = \tan x$, $du = \sec^2 x\, dx$, we continue with

$$\int u^3(1 + u^2)\, du = \int (u^3 + u^5)\, du = \frac{u^4}{4} + \frac{u^6}{6} + C$$

$$= \frac{\tan^4 x}{4} + \frac{\tan^6 x}{6} + C.$$

Although the method just described applies also when $n = 0$, the method used in the following example may be simpler.

***Example 14.3.3.*** Evaluate $\int \cot^4 x\, dx$.

*Solution.* We write

$$\int \cot^4 x\, dx = \int (\csc^2 x - 1) \cot^2 x\, dx = \int \csc^2 x \cot^2 x\, dx - \int \cot^2 x\, dx$$

$$= -\frac{\cot^3 x}{3} - \int (\csc^2 x - 1)\, dx$$

$$= -\frac{\cot^3 x}{3} + \cot x + x + C.$$

In the case where $n$ is even, $m$ need not be an integer.

The above cases do not include all possibilities, even for positive integers $m$ and $n$. Thus $\int \tan^2 x \sec x\, dx$ and $\int \sec^3 x\, dx$ are not covered by either case. In fact, these integrals are closely related by virtue of the following.

$$\int \tan^2 x \sec x\, dx = \int (\sec^2 x - 1) \sec x\, dx$$

$$= \int \sec^3 x\, dx - \int \sec x\, dx$$

$$= \int \sec^3 x\, dx - \ln|\sec x + \tan x|.$$

The integral $\int \sec^3 x \, dx$ is treated in Section 14.8, and its value is included in the formulas of Section 14.1. It turns out that integrals of the forms (14.3.1), where $m$ is even and $n$ is odd, in general require a different treatment, such as the use of the method of Section 14.8.

**EXERCISE GROUP 14.3**

Evaluate the integrals in Exercises 1–18

1. $\int \sec(2x+4) \tan(2x+4) \, dx$
2. $\int \tan^4(x-1) \sec^2(x-1) \, dx$
3. $\int \cot(5x-8) \, dx$
4. $\int \tan^2(2x+3) \, dx$
5. $\int (\tan x + \cot x)^2 \, dx$
6. $\int \tan^3 3x \sec^2 3x \, dx$
7. $\int \cot x \csc^3 x \, dx$
8. $\int \tan^3 x \sec x \, dx$
9. $\int \cot^3 x \csc^3 x \, dx$
10. $\int \tan^2 x \sec^4 x \, dx$
11. $\int \tan^4 x \sec^4 x \, dx$
12. $\int \tan x \sqrt{\sec x} \, dx$
13. $\int \tan^3 x \sqrt{\sec x} \, dx$
14. $\int \sqrt{\tan x} \sec^4 x \, dx$
15. $\int \cot^{3/2} x \csc^4 x \, dx$
16. $\int \csc^5 x \cot x \, dx$
17. $\int \dfrac{\sin^3 x}{\cos^6 x} \, dx$
18. $\int \dfrac{\cos^{1/2} x}{\sin^{5/2} x} \, dx$

19. Evaluate $\int \cot^4 x \, dx$ by expressing it as $\int (\cot^4 x / \csc^2 x) \csc^2 x \, dx$ and using the substitution $u = \cot x$. Compare the result with Example 14.3.3.
20. Evaluate $\int \tan^3 x \, dx$ by expressing it as $\int (\tan^2 x / \sec x) \sec x \tan x \, dx$ and using the substitution $u = \sec x$.
21. Evaluate the integral in Example 14.3.2 by using the method for odd $m$.

## 14.4 TRIGONOMETRIC SUBSTITUTION

An integral of a function containing one of the expressions

$$\sqrt{a^2 - u^2}, \quad \sqrt{u^2 - a^2}, \quad \sqrt{u^2 + a^2}$$

can often be handled successfully by a trigonometric substitution which serves to rationalize the expression. For example, if $u = a \sin \theta$, then

$$\sqrt{a^2 - u^2} = \sqrt{a^2(1 - \sin^2 \theta)} = \sqrt{a^2 \cos^2 \theta} = a \cos \theta.$$

Similarly,

$$\text{if} \quad u = a \sec \theta, \quad \text{then} \quad \sqrt{u^2 - a^2} = a \tan \theta,$$
$$\text{if} \quad u = a \tan \theta, \quad \text{then} \quad \sqrt{u^2 + a^2} = a \sec \theta.$$

It is assumed in this discussion that $u > 0$, $a > 0$, and $0 \leq \theta \leq \pi/2$, in which case the above relations are valid.

The present method thus involves the substitution

$$u = a \sin \theta \quad \text{if} \quad \sqrt{a^2 - u^2} \quad \text{is present,}$$
$$u = a \sec \theta \quad \text{if} \quad \sqrt{u^2 - a^2} \quad \text{is present,}$$
$$u = a \tan \theta \quad \text{if} \quad \sqrt{u^2 + a^2} \quad \text{is present.}$$

In each case there results a trigonometric integral, to which previous methods may be applied. Once the integration has been performed in terms of $\theta$, it is possible to express the result in terms of $u$ by the use of trigonometric identities. However, the same purpose may be accomplished by considering a suitable right triangle with acute angle $\theta$; this will be illustrated in the examples.

**Example 14.4.1.** Evaluate $\int \dfrac{x^2 \, dx}{\sqrt{4 - x^2}}$.

*Solution.* We use the substitution

$$x = 2 \sin \theta, \qquad dx = 2 \cos \theta \, d\theta.$$

Then

$$\int \frac{x^2 \, dx}{\sqrt{4 - x^2}} = \int \frac{4 \sin^2 \theta \, (2 \cos \theta \, d\theta)}{2 \cos \theta} = 4 \int \sin^2 \theta \, d\theta$$

$$= 2 \int (1 - \cos 2\theta) \, d\theta = 2\left(\theta - \frac{\sin 2\theta}{2}\right) + C$$

$$= 2(\theta - \sin \theta \cos \theta) + C.$$

To obtain the last expression we used the identity $\sin 2\theta = 2 \sin \theta \cos \theta$. We now refer to Fig. 14.4.1, with the acute angle $\theta$ such that $x = 2 \sin \theta$. Then we note that

$$\sin \theta = x/2, \qquad \cos \theta = \sqrt{4 - x^2}/2,$$

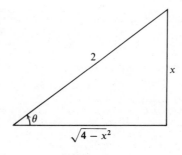

Fig. 14.4.1

and the above integral becomes

$$\int \frac{x^2\,dx}{\sqrt{4-x^2}} = 2\left(\sin^{-1}\frac{x}{2} - \frac{1}{4}x\sqrt{4-x^2}\right) + C.$$

**Example 14.4.2.** Evaluate $\int \dfrac{dx}{\sqrt{4x^2-9}}$.

**Solution.** Here we use the substitution

$$2x = 3\sec\theta, \qquad 2\,dx = 3\sec\theta\tan\theta\,d\theta.$$

Then

$$\int \frac{dx}{\sqrt{4x^2-9}} = \int \frac{(3/2)\sec\theta\tan\theta\,d\theta}{3\tan\theta} = \frac{1}{2}\int \sec\theta\,d\theta$$

$$= \frac{1}{2}\ln|\sec\theta + \tan\theta| + C'$$

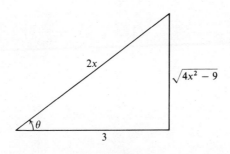

Fig. 14.4.2

From Fig. 14.4.2 the last expression gives

$$\int \frac{dx}{\sqrt{4x^2-9}} = \frac{1}{2}\ln\left|\frac{2x}{3} + \frac{\sqrt{4x^2-9}}{3}\right| + C' = \frac{1}{2}\ln\left|2x + \sqrt{4x^2-9}\right| + C.$$

The last form of the result was obtained from the relation

$$\frac{1}{2}\ln\left|\frac{2x+\sqrt{4x^2-9}}{3}\right| = \frac{1}{2}\ln\left|2x+\sqrt{4x^2-9}\right| - \frac{1}{2}\ln 3;$$

then $(1/2)\ln 3$, being a constant, was incorporated with $C'$ to give $C$:

$$C' - \tfrac{1}{2}\ln 3 = C.$$

In fact, either form of the answer in Example 14.4.2 is correct; the difference between them emphasizes the fact that the form of the result in an indefinite integral may vary and still be correct.

Although the above trigonometric substitutions were introduced primarily to eliminate certain radicals, they may be effective where an integrand contains a quadratic expression not in a radical.

**Example 14.4.3.** Evaluate $\int \dfrac{dx}{(a^2 + x^2)^2}$

*Solution.* We try the following substitution, suggested by the presence of $a^2 + x^2$.

$$x = a \tan \theta, \qquad dx = a \sec^2 \theta \, d\theta.$$

Then, since $a^2 + x^2 = a^2(1 + \tan^2 \theta) = a^2 \sec^2 \theta$, we have

$$\int \frac{dx}{(a^2 + x^2)^2} = \int \frac{a \sec^2 \theta \, d\theta}{a^4 \sec^4 \theta} = \frac{1}{a^3} \int \cos^2 \theta \, d\theta$$

$$= \frac{1}{2a^3} \int (1 + \cos 2\theta) \, d\theta = \frac{1}{2a^3} (\theta + \sin \theta \cos \theta) + C.$$

With $\sin \theta$ and $\cos \theta$ obtained from Fig. 14.4.3, the result is

$$\int \frac{dx}{(a^2 + x^2)^2} = \frac{1}{2a^3} \left( \tan^{-1} \frac{x}{a} + \frac{ax}{a^2 + x^2} \right) + C.$$

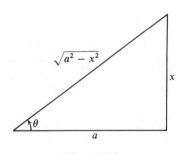

Fig. 14.4.3

**EXERCISE GROUP 14.4**

Evaluate the integrals in Exercise Group 1–25. If $\int \sec^3 \theta \, d\theta$ appears, use formula 20. If $\int \tan^2 \theta \sec \theta \, d\theta$ appears, use the discussion at the end of Section 14.3.

1. $\int \dfrac{dx}{(4x^2 - 1)^{3/2}}$

2. $\int \dfrac{dx}{a^2 - x^2}$

3. $\int \sqrt{a^2 - x^2} \, dx$

4. $\int \dfrac{dx}{x^3 \sqrt{x^2 - 1}}$

5. $\int \dfrac{\sqrt{x^2 - 9}}{x} \, dx$

6. $\int \dfrac{x^3}{\sqrt{4 - x^2}} \, dx$

7. $\int x^2 \sqrt{a^2 - x^2}\, dx$

8. $\int y^3 \sqrt{9 - y^2}\, dy$

9. $\int_0^2 x^2 \sqrt{4 - x^2}\, dx$

10. $\int_3^5 \sqrt{4 - (x-3)^2}\, dx$

11. $\int \dfrac{x^2\, dx}{(x^2 + 1)^2}$

12. $\int \dfrac{\sqrt{(3x-1)^2 - 4}}{3x - 1}\, dx$

13. $\int_0^3 x^3 \sqrt{9 - x^2}\, dx$

14. $\int \dfrac{\sqrt{4 - v^2}}{v^2}\, dv$

15. $\int \dfrac{\sqrt{16 - x^2}}{x}\, dx$

16. $\int \dfrac{dx}{x^2(x^2 + 9)^{3/2}}$

17. $\int \dfrac{u^3\, du}{(a^2 + u^2)^{3/2}}$

18. $\int \sqrt{4x^2 + 9}\, dx$

19. $\int \dfrac{dx}{(4 - x^2)^2}$

20. $\int \dfrac{x^3\, dx}{\sqrt{x^2 + 4}}$

21. $\int x^3 \sqrt{1 + x^2}\, dx$

22. $\int \dfrac{e^x\, dx}{\sqrt{e^{2x} + e^{-2x}}}$

23. $\int \dfrac{e^x\, dx}{\sqrt{e^{2x} - e^{-2x}}}$

24. $\int \sqrt{e^{2x} - 1}\, dx$

25. $\int \dfrac{\sqrt{x}\, dx}{\sqrt{1 - x}}$ by substituting $x = \sin^2 \theta$

26. Find the area of the surface obtained by revolving one arch of the curve $y = \sin x$ about the $x$ axis.

27. Find the area of the surface obtained by revolving the curve $y = \ln x$ from $x = 1$ to $x = 4$ about the $y$ axis.

28. Find the area of the surface obtained by revolving the ellipse $x^2 + 4y^2 = 4$ about the $x$ axis.

29. Show that the area of the surface obtained by revolving the ellipse $b^2 x^2 + a^2 y^2 = a^2 b^2$, $a > b$, about the $x$ axis is $2\pi \left( b^2 + \dfrac{ab}{e} \sin^{-1} e \right)$, where $e$ is the eccentricity of the ellipse.

30. The water level just reaches the top of a vertical circular gate 8 ft in diameter. Find the force of the water on the gate.

31. Find the moment of inertia of the arc of a circle of radius $a$ about a diameter.

32. Find the moment of inertia of a circular region of radius $a$ about a diameter.

33. Find the moment of inertia about the major axis of the region interior to the ellipse $b^2 x^2 + a^2 y^2 = a^2 b^2$, $a > b$.

## 14.5 INTEGRALS CONTAINING COMPLETE QUADRATIC FUNCTIONS

A complete quadratic function,

$$ax^2 + bx + c, \quad \text{with} \quad a \neq 0,\, b \neq 0,$$

may be expressed as a sum or difference of squares by a process of completing squares. Previous methods may then be applied.

**Example 14.5.1.** Evaluate $\int \dfrac{dx}{\sqrt{3 + 2x - x^2}}$.

*Solution.* We write the quadratic as $3 - (x^2 - 2x)$. We now add 1 within the parentheses, to give a complete square, and add 1 outside to compensate, and we get

$$3 + 2x - x^2 = 3 + 1 - (x^2 - 2x + 1) = 4 - (x - 1)^2.$$

We now set $x - 1 = u$, $dx = du$, and we have

$$\int \frac{dx}{\sqrt{3 + 2x - x^2}} = \int \frac{du}{\sqrt{4 - u^2}} = \sin^{-1}\frac{u}{2} + C = \sin^{-1}\frac{x-1}{2} + C.$$

**Example 14.5.2.** Evaluate $\int \dfrac{x\,dx}{\sqrt{3x^2 + 2x - 1}}$.

*Solution.* By completing squares,

$$3x^2 + 2x - 1 = 3(x^2 + \tfrac{2}{3}x + \tfrac{1}{9}) - 1 - \tfrac{1}{3} = 3(x + \tfrac{1}{3})^2 - \tfrac{4}{3},$$

and setting $x + 1/3 = u$, $dx = du$, we get

$$\int \frac{x\,dx}{\sqrt{3x^2 + 2x - 1}} = \int \frac{(u - 1/3)\,du}{\sqrt{3u^2 - 4/3}} = \frac{1}{3\sqrt{3}}\int \frac{(3u - 1)\,du}{\sqrt{u^2 - 4/9}}.$$

We now substitute $u = (2/3)\sec\theta$, $du = (2/3)\sec\theta\tan\theta\,d\theta$, and the above continues as

$$\frac{1}{3\sqrt{3}}\int \frac{(2\sec\theta - 1)(2/3)\sec\theta\tan\theta\,d\theta}{(2/3)\tan\theta} = \frac{1}{3\sqrt{3}}\int (2\sec^2\theta - \sec\theta)\,d\theta$$

$$= \frac{1}{3\sqrt{3}}(2\tan\theta - \ln|\sec\theta + \tan\theta|) + C'$$

$$= \frac{1}{3\sqrt{3}}\left(\sqrt{9u^2 - 4} - \ln\left|3u + \sqrt{9u^2 - 4}\right|\right) + C$$

[From Fig. 14.5.1].

We now replace $u$ by $x + 1/3$, and obtain the result

$$\int \frac{x\,dx}{\sqrt{3x^2 + 2x - 1}} = \frac{1}{3\sqrt{3}}\left(\sqrt{9x^2 + 6x - 3} - \ln\left|3x + 1 + \sqrt{9x^2 + 6x - 3}\right|\right) + C.$$

The method may also be applied when a complete quadratic function appears without a radical sign.

**Example 14.5.3.** Evaluate $\int \dfrac{x\,dx}{x^2 - 4x + 8}$.

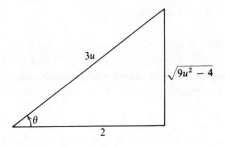

Fig. 14.5.1

*Solution.* By completing squares and then making the substitution $x - 2 = u$, $dx = du$, we obtain

$$\int \frac{x\,dx}{x^2 - 4x + 8} = \int \frac{x\,dx}{(x-2)^2 + 4} = \int \frac{(u+2)\,du}{u^2 + 4}$$

$$= \int \frac{u\,du}{u^2 + 4} + 2\int \frac{du}{u^2 + 4}$$

$$= \frac{1}{2}\ln(u^2 + 4) + \tan^{-1}\frac{u}{2} + C$$

$$= \frac{1}{2}\ln(x^2 - 4x + 8) + \tan^{-1}\frac{x-2}{2} + C.$$

## EXERCISE GROUP 14.5

Evaluate each of the following integrals.

1. $\int \dfrac{dx}{\sqrt{5 - 4x - x^2}}$

2. $\int \dfrac{dx}{x^2 - 6x + 13}$

3. $\int \dfrac{x\,dx}{x^2 - 6x + 13}$

4. $\int \dfrac{x^2\,dx}{x^2 - 6x + 13}$

5. $\int \dfrac{(x - 2)\,dx}{\sqrt{5x - 6 - x^2}}$

6. $\int \dfrac{(3x + 2)\,dx}{\sqrt{5 - 4x - x^2}}$

7. $\int \dfrac{(x - 3)\,dx}{\sqrt{7 - 12x - 2x^2}}$

8. $\int \dfrac{x\,dx}{\sqrt{2x - x^2}}$

9. $\int \dfrac{x\,dx}{\sqrt{x^2 - 2x}}$

10. $\int_{-1/2}^{1/4} \dfrac{x\,dx}{\sqrt{2 - x - x^2}}$

11. $\int \dfrac{(3x + 2)\,dx}{\sqrt{4x^2 - 4x - 3}}$

12. $\int \dfrac{x\,dx}{\sqrt{8x^2 + 6x + 1}}$

13. $\int_{4}^{7} \dfrac{dx}{\sqrt{x^2 - 8x + 25}}$

14. $\int \dfrac{(2x + 1)\,dx}{\sqrt{x^2 - 8x + 25}}$

## 14.6 PARTIAL FRACTIONS. Case 1

The method of partial fractions is employed in the integration of a rational function, that is, a function of the form $p(x)/q(x)$, where $p(x)$ and $q(x)$ are polynomials in $x$. By this method a suitable rational function can be expressed as a sum of simpler rational functions, to each of which previous methods of integration can be applied.

First of all, *we assume that the degree of $p(x)$ is less than that of $q(x)$*; if such is not the case, ordinary long division is applied initially. For example, we obtain the following by long division.

$$(14.6.1) \qquad \frac{x^4 + x^3 - 4x^2 - 2x - 1}{x^3 - x^2 - 2x} = x + 2 + \frac{2x - 1}{x^3 - x^2 - 2x}.$$

The degree of the numerator on the left is greater than the degree of the denominator. On the right, the degree of the numerator is less than that of the denominator.

*We also assume that $p(x)$ and $q(x)$ have no factor in common, other than a constant.*

Any polynomial with real coefficients can be expressed as a product of linear and quadratic factors all of which also have real coefficients. In what follows we describe how a rational function $p(x)/q(x)$, under the above assumptions, can be decomposed into a sum of *partial fractions*, of types determined by the factors in $q(x)$. The applicable theorems will be stated without proof.

*Case 1. Partial fractions corresponding to a factor $(x - r)^m$*

Here $m$ is a positive integer such that $(x - r)^m$ is a factor of $q(x)$, while $(x - r)^{m+1}$ is not.

**THEOREM 14.6.1.** *Corresponding to a factor $(x - r)^m$ of $q(x)$, the partial fraction decomposition of $p(x)/q(x)$ contains the sum*

$$(14.6.2) \qquad \frac{a_0}{(x - r)^m} + \frac{a_1}{(x - r)^{m-1}} + \cdots + \frac{a_{m-1}}{x - r},$$

*where $a_0, a_1, \ldots, a_{m-1}$ are constants.*

Take particular note of the fact that each numerator in (14.6.2) is a constant. The determination of these constants is taken up in the examples. If a factor $x - r$ in $q(x)$ is unrepeated, then $m = 1$ and the form of (14.6.2) is simply $A/(x - r)$ for some constant $A$.

**Example 14.6.1.** Evaluate $\int \dfrac{x^4 + x^3 - 4x^2 - 2x - 1}{x^3 - x^2 - 2x}\, dx$.

*Solution.* The first step is to express the fraction as in (14.6.1). The denominator of the fraction in the right member of (14.6.1) factors as

$$x^3 - x^2 - 2x = x(x + 1)(x - 2).$$

Since the denominator contains only distinct linear factors, we have, by Theorem 14.6.1,

(14.6.3) $$\frac{2x-1}{x(x+1)(x-2)} = \frac{A}{x} + \frac{B}{x+1} + \frac{C}{x-2},$$

where the constants $A$, $B$, and $C$ are to be determined. Equation (14.6.3) is an identity so that, if we combine the fractions on the right into one with denominator $x(x+1)(x-2)$, the resulting numerator must be identical with the one on the left. We obtain in this manner the identity

$$2x - 1 = A(x+1)(x-2) + Bx(x-2) + Cx(x+1).$$

Since this equation holds for all values of $x$, we may set $x = 0, -1, 2$ in turn, and we find

$$A = 1/2, \quad B = -1, \quad C = 1/2.$$

With these values (14.6.3) furnishes the partial fraction decomposition of the left member, and we may write

$$\int \frac{2x-1}{x(x+1)(x-2)} dx = \frac{1}{2}\int \frac{dx}{x} - \int \frac{dx}{x+1} + \frac{1}{2}\int \frac{dx}{x-2}$$

$$= \frac{1}{2}\ln|x| - \ln|x+1| + \frac{1}{2}\ln|x-2| + C$$

$$= \frac{1}{2}\ln\left|\frac{x(x-2)}{(x+1)^2}\right| + C.$$

From (14.6.1) we see that $\int (x+2)\, dx$ must be added. We conclude that

$$\int \frac{x^4 + x^3 - 4x^2 - 2x - 1}{x^3 - x^2 - 2x} dx = \frac{x^2}{2} + 2x + \frac{1}{2}\ln\left|\frac{x(x-2)}{(x+1)^2}\right| + C.$$

The following example illustrates the method as applied in the case of a repeated linear factor.

**Example 14.6.2.** Evaluate $\int \frac{3x^2 + 8x - 1}{(x+1)^3(x-2)} dx$.

**Solution.** By Theorem 14.6.1 we write, with $A$, $B$, $C$, and $D$ constant,

(14.6.4) $$\frac{3x^2 + 8x - 1}{(x+1)^3(x-2)} = \frac{A}{(x+1)^3} + \frac{B}{(x+1)^2} + \frac{C}{x+1} + \frac{D}{x-2};$$

this yields the identity (on adding the fractions on the right)

(14.6.5) $$3x^2 + 8x - 1 = A(x-2) + B(x+1)(x-2)$$
$$+ C(x+1)^2(x-2) + D(x+1)^3.$$

Setting $x = -1$ we find $A = 2$, and with $x = 2$ we find $D = 1$. There is no further *convenient* value of $x$ to use, that is, a value of $x$ which yields one of the constants directly. However, since (14.6.5) is an identity, we may also equate coefficients of like powers of $x$ on both sides of (14.6.5). Doing this for $x^3$ and $x^0$ (the constant term), we obtain

$$0 = C + D, \quad -1 = -2A - 2B - 2C + D;$$

with the known values of $A$ and $D$ these equations give $C = -1$, $B = 0$. The values $A = 2$, $B = 0$, $C = -1$, $D = 1$ having been found, the explicit expansion (14.6.1) is known, and we have

$$\int \frac{3x^2 + 8x - 1}{(x+1)^3(x-2)} dx = 2 \int \frac{dx}{(x+1)^3} - \int \frac{dx}{x+1} + \int \frac{dx}{x-2}$$

$$= -\frac{1}{(x+1)^2} - \ln|x+1| + \ln|x-2| + C$$

$$= -\frac{1}{(x+1)^2} + \ln\left|\frac{x-2}{x+1}\right| + C.$$

## EXERCISE GROUP 14.6

Evaluate the integrals in Exercises 1–20.

1. $\int \dfrac{7x-2}{x^2 - x - 2} dx$

2. $\int \dfrac{3x-8}{(2x-3)(x+2)} dx$

3. $\int \dfrac{x^2+1}{x^3 - x} dx$

4. $\int \dfrac{x^2-7}{x^2 - x - 2} dx$

5. $\int \dfrac{dx}{x^3 + x^2 - 2x}$

6. $\int \dfrac{-x^2 + 4x - 1}{x^3 - 6x^2 + 11x - 6} dx$

7. $\int_0^2 \dfrac{x+7}{x^2 - x - 6} dx$

8. $\int_0^1 \dfrac{x+2}{x^2 + 2x - 8} dx$

9. $\int \dfrac{dx}{x^3 - x^2}$

10. $\int \dfrac{x+2}{x^3(x-1)} dx$

11. $\int \dfrac{dx}{(x^2 - 4)^2}$

12. $\int \dfrac{19x + 32}{(x^2 - 2x - 8)(x+2)} dx$

13. $\int \dfrac{2x^3 + 3x^2 + 2x + 2}{x^3(x+1)} dx$

14. $\int \dfrac{2x-1}{(x^2 - x - 2)^2} dx$

15. $\int \dfrac{x^4 - 4x^3 + 3x^2 + x + 4}{x(x-2)^2} dx$

16. $\int_{-1}^0 \dfrac{x^2 + 2}{(x-1)^2} dx$

17. $\int_0^{1/2} \dfrac{3x^2 + 2x + 1}{x^3 - 2x^2 - x + 2} dx$

18. $\int \dfrac{\sqrt{x}}{x - 4} dx$

19. $\int \dfrac{\sqrt{x}}{(x-1)^2} dx$

*20. $\int \dfrac{\sqrt{x^2 + 9}}{x} dx$

21. Prove: $\displaystyle\sum_{k=1}^{n} \dfrac{1}{k(k+1)} = \dfrac{n}{n+1}$

**22.** Prove: $\sum_{k=1}^{n} \dfrac{2k+1}{k^2(k+1)^2} = \dfrac{n^2+2n}{(n+1)^2}$   **23.** Prove: $\sum_{k=1}^{n} \dfrac{1+k-k^2}{k^2(k+1)^2} = \dfrac{n}{(n+1)^2}$

*  **24.** Evaluate $\int \sec x \, dx$ by writing

$$\int \sec x \, dx = \int \dfrac{dx}{\cos x} = \int \dfrac{\cos x}{1 - \sin^2 x} \, dx$$

and then substituting $u = \sin x$. Obtain the result in the form of formula 8 of Section 14.1.

*  **25.** Evaluate $\int \sec^3 x \, dx$ by writing

$$\int \sec^3 x \, dx = \int \dfrac{dx}{\cos^3 x} = \int \dfrac{\cos x}{(1 - \sin^2 x)^2} \, dx$$

and then substituting $u = \sin x$. Compare the result with formula 20 of Section 14.1.

## 14.7  PARTIAL FRACTIONS.  *Case 2*

The remaining consideration in partial fraction decompositions concerns the appearance of quadratic factors in the denominator of a rational function.

*Case 2.*  *Partial fractions corresponding to a factor $(ax^2 + bx + c)^m$.*

Here $m$ is a positive integer such that $(ax^2 + bx + c)^m$ is a factor of the denominator $q(x)$, while $(ax^2 + bx + c)^{m+1}$ is not.

**THEOREM 14.7.1.**  *Corresponding to a factor $(ax^2 + bx + c)^m$ of $q(x)$, the partial fraction decomposition of $p(x)/q(x)$ contains the sum*

$$(14.7.1) \quad \dfrac{a_0 x + b_0}{(ax^2 + bx + c)^m} + \dfrac{a_1 x + b_1}{(ax^2 + bx + c)^{m-1}} + \cdots + \dfrac{a_{m-1} x + b_{m-1}}{ax^2 + bx + c},$$

*where $a_i$ and $b_i$ are constants.*

The numerator of each fraction in (14.7.1) is a linear function of $x$, with coefficients to be determined. The method of such determination is discussed in the examples.

**Example 14.7.1.**  Evaluate $\displaystyle\int \dfrac{7x - 3}{x^4 + 2x^3 + 3x^2} \, dx$.

*Solution.*  The denominator factors completely as $x^2(x^2 + 2x + 3)$. Hence we may write

$$(14.7.2) \quad \dfrac{7x - 3}{x^4 + 2x^3 + 3x^2} = \dfrac{Ax + B}{x^2 + 2x + 3} + \dfrac{C}{x^2} + \dfrac{D}{x},$$

where the first term on the right is due to Theorem 14.7.1 with $m = 1$, while the remaining two terms are due to Theorem 14.6.1 with $m = 2$. By multiplying both sides of (14.7.2) by $x^2(x^2 + 2x + 3)$, we obtain the identity

$$7x - 3 = (Ax + B)x^2 + C(x^2 + 2x + 3) + Dx(x^2 + 2x + 3);$$

this equation may be rewritten as

$$7x - 3 = (A + D)x^3 + (B + C + 2D)x^2 + (2C + 3D)x + 3C.$$

Comparing coefficients of similar powers of $x$ in both members, we have

$$A + D = 0, \quad B + C + 2D = 0, \quad 2C + 3D = 7, \quad 3C = -3;$$

these equations are now solved to give $A = -3$, $B = -5$, $C = -1$, $D = 3$. Substituting these values in (14.7.2) and integrating, we have

(14.7.3) $$\int \frac{7x - 3}{x^4 + 2x^3 + 3x^2} dx = -\int \frac{3x + 5}{x^2 + 2x + 3} dx - \int \frac{dx}{x^2} + 3 \int \frac{dx}{x}.$$

To determine the first integral on the right we write (see Section 14.5)

$$x^2 + 2x + 3 = (x + 1)^2 + 2$$

and substitute $u = x + 1$, $du = dx$. Then, omitting a constant of integration, we have

$$\int \frac{3x + 5}{x^2 + 2x + 3} dx = \int \frac{3u + 2}{u^2 + 2} du = 3 \int \frac{u\, du}{u^2 + 2} + 2 \int \frac{du}{u^2 + 2}$$

$$= \frac{3}{2} \ln(u^2 + 2) + \sqrt{2} \tan^{-1} \frac{u}{\sqrt{2}}$$

$$= \frac{3}{2} \ln(x^2 + 2x + 3) + \sqrt{2} \tan^{-1} \frac{x + 1}{\sqrt{2}}.$$

We use this result in (14.7.3), perform the remaining integrations, and obtain the result,

$$\int \frac{7x - 3}{x^4 + 2x^3 + 3x^2} dx = -\frac{3}{2} \ln(x^2 + 2x + 3) - \sqrt{2} \tan^{-1} \frac{x + 1}{\sqrt{2}} + \frac{1}{x} + 3 \ln|x| + C$$

$$= 3 \ln \frac{|x|}{\sqrt{x^2 + 2x + 3}} - \sqrt{2} \tan^{-1} \frac{x + 1}{\sqrt{2}} + \frac{1}{x} + C.$$

**Example 14.7.2.** Evaluate $\int \dfrac{x\, dx}{(x - 1)(x^2 + 1)^2}$

*Solution.* By Theorems 14.6.1 and 14.7.1 we may write

$$\frac{x}{(x - 1)(x^2 + 1)^2} = \frac{Ax + B}{(x^2 + 1)^2} + \frac{Cx + D}{x^2 + 1} + \frac{E}{x - 1}.$$

We then get

(14.7.4) $$x = (Ax + B)(x - 1) + (Cx + D)(x - 1)(x^2 + 1) + E(x^2 + 1)^2.$$

We shall find $A$, $B$, $C$, $D$, and $E$ by a combination of methods. Using the only convenient value, $x = 1$, we find $E = 1/4$. Comparing the coefficients of $x^4$ and $x^0$ in (14.7.4), we have

(14.7.5) $\qquad 0 = C + E, \qquad 0 = -B - D + E.$

Since (14.7.4) holds for all values of $x$ we may use $x = -1$ and $x = 2$, for example, and we get

(14.7.6) $\qquad -1 = 2A - 2B + 4C - 4D + 4E,$

(14.7.7) $\qquad 2 = 2A + B + 10C + 5D + 25E.$

From (14.7.5), (14.7.6), and (14.7.7), with $E = 1/4$, we find

$$A = -1/2, \quad B = 1/2, \quad C = -1/4, \quad D = -1/4, \quad E = 1/4.$$

Then

$$\int \frac{x}{(x-1)(x^2+1)^2} dx = -\frac{1}{2}\int \frac{x\,dx}{(x^2+1)^2} + \frac{1}{2}\int \frac{dx}{(x^2+1)^2} - \frac{1}{4}\int \frac{x\,dx}{x^2+1}$$

$$-\frac{1}{4}\int \frac{dx}{x^2+1} + \frac{1}{4}\int \frac{dx}{x-1}.$$

The first term on the right integrates by the power rule, using $u = x^2 + 1$, $du = 2x\,dx$. To evaluate the second integral on the right we substitute

$$x = \tan\theta, \qquad dx = \sec^2\theta\, d\theta.$$

Then

$$\int \frac{dx}{(x^2+1)^2} = \int \frac{\sec^2\theta\, d\theta}{\sec^4\theta} = \int \cos^2\theta\, d\theta = \frac{1}{2}\int (1 + \cos 2\theta)\, d\theta$$

$$= \frac{1}{2}(\theta + \sin\theta\cos\theta) = \frac{1}{2}\left(\tan^{-1} x + \frac{x}{x^2+1}\right).$$

Therefore,

$$\int \frac{x\,dx}{(x-1)(x^2+1)^2} = \frac{1}{4(x^2+1)} + \frac{1}{4}\tan^{-1} x + \frac{x}{4(x^2+1)} - \frac{1}{8}\ln(x^2+1)$$

$$-\frac{1}{4}\tan^{-1} x + \frac{1}{4}\ln|x-1| + C$$

$$= \frac{x+1}{4(x^2+1)} + \frac{1}{8}\ln\frac{(x-1)^2}{x^2+1} + C.$$

It may be helpful at this time to summarize the discussion of the decomposition of a rational function $p(x)/q(x)$. If the degree of $p(x)$ is less than that of $q(x)$, and if $p(x)$ and $q(x)$ have no common factor other than a constant, then $p(x)/q(x)$ may be expressed as a sum of fractions according to Theorems 14.6.1 and 14.7.1. The determination of the constants in the numerators of these fractions is made by adding the fractions and setting the resulting numerator identically equal to $p(x)$. The latter may be achieved by a combination of the following methods:

(a) the use of *convenient* values—these are the numbers $r$ corresponding to the linear factors $x - r$ of $q(x)$;

(b) the use of arbitrary values of $x$;

(c) equating coefficients of like powers of $x$.

In this manner as many equations are obtained as there are constants to be found. The integrations of the resulting fractions can then be attempted by the methods already presented and to be presented.

**EXERCISE GROUP 14.7**

Evaluate each of the following integrals.

1. $\int \dfrac{3\,dx}{x^3 - 1}$

2. $\int \dfrac{3x\,dx}{x^3 + 1}$

3. $\int \dfrac{3x\,dx}{(x^3 - 1)(x - 1)}$

4. $\int \dfrac{2x^3 + 2x + 1}{x^3 + x}\,dx$

5. $\int \dfrac{1 - x}{(x + 1)(x^2 + 1)}\,dx$

6. $\int \dfrac{x^2 + 2x + 2}{(x^2 - 2)(x^2 + x - 1)}\,dx$

7. $\int \dfrac{x^2 + 1}{x^3 + x^2 + x}\,dx$

8. $\int \dfrac{x\,dx}{x^4 + 3x^2 + 2}$

9. $\int \dfrac{x^3 + 3}{(x + 1)^2(x^2 - 2)}\,dx$

10. $\int \dfrac{x^3 + x^2 + 5x + 3}{x^4 + 2x^3 + 3x^2}\,dx$

11. $\int \dfrac{x^2}{(x^2 + 4)^2}\,dx$

12. $\int \dfrac{1 - x^2}{(x^2 + 1)^2}\,dx$

13. $\int \dfrac{x^3}{(x^2 + 4)^2}\,dx$

14. $\int \dfrac{x\,dx}{(x - 1)(x^2 + 1)^2}$

15. $\int \dfrac{2x^2 + 1}{(x^2 - x + 1)^2}\,dx$

16. $\int \dfrac{x^4 + 4x^3 + 4x + 4}{x(x^2 + 2x + 2)^2}\,dx$

17. $\int \dfrac{x^2\,dx}{(x^2 + 2)^3}$

18. $\int \dfrac{x^5 + x^4 + 2x^3 + 2x^2 + 2x + 1}{(x^2 + 1)^4}\,dx$

## 14.8 INTEGRATION BY PARTS

There is no formula for the direct integration of a product of functions. Perhaps the closest to such a formula is the one for *integration by parts*, which is derived from the formula for differentiating a product.

In terms of differentials the product formula for differentiation is

(14.8.1) $\qquad d(uv) = u\,dv + v\,du \quad \text{or} \quad u\,dv = d(uv) - v\,du,$

where $u$ and $v$ are functions of some variable. Since $\int d(uv)$ has $uv$ as one value, integration of the second equation in (14.8.1) yields the relation

(14.8.2) $$\int u\, dv = uv - \int v\, du.$$

Equation (14.8.2) is known as the *formula for integration by parts*. An equation such as (14.8.2) should be interpreted to mean that the arbitrary constants may be chosen so that the equation holds.

*Integration by parts* constitutes one of the most powerful tools in the study of integration. It is often a method of last resort, to be attempted when other methods offer little hope of success. However, certain types of integral lend themselves particularly well to evaluation by integration by parts, and the recognition of these types is very helpful. Some of the ways in which the formula may be used will be considered here and in following sections.

The most direct application of (14.8.2) to an integral $\int f(x)\, dx$ involves writing $f(x)\, dx$ as a product of two factors $u$ and $dv$, such that $dv$ integrates immediately to give $v$, while $v\, du$ is also integrable. This criterion can often suggest the choice of $u$ and $dv$.

***Example 14.8.1.*** Evaluate $\int x \sin 2x\, dx$.

*Solution.* We choose $u = x$, $dv = \sin 2x\, dx$. Then

$$du = dx, \qquad v = -\frac{\cos 2x}{2},$$

and with (14.8.2) we have

$$\int x \sin 2x\, dx = -\frac{x \cos 2x}{2} + \frac{1}{2} \int \cos 2x\, dx = -\frac{x \cos 2x}{2} + \frac{\sin 2x}{4} + C.$$

Example 14.8.1 indicates one typical manner in which integration by parts is used. An alternative procedure might attempt to use $u = \sin 2x$, $dv = x\, dx$, but then $v = x^2/2$, $du = 2 \cos 2x\, dx$, and the form of $\int v\, du$, while similar to the original integral, is more difficult in that the exponent of $x$ has been increased.

***Example 14.8.2.*** Evaluate $\int x^3 \ln x\, dx$.

*Solution.* We use

$$u = \ln x, \qquad dv = x^3\, dx; \qquad du = dx/x, \qquad v = x^4/4.$$

Then (14.8.2) gives

$$\int x^3 \ln x\, dx = \frac{1}{4} x^4 \ln x - \frac{1}{4} \int x^3\, dx = \frac{x^4 \ln x}{4} - \frac{x^4}{16} + C.$$

Note that the intermediate steps in any integration by parts can be directed at finding a particular integral. The constant of integration may then be added at the very end.

The next example illustrates a second way in which integration by parts is used, whereby the formula is applied more than once in order to effect the integration.

**Example 14.8.3.** Evaluate $\int x^2 \sin 2x \, dx$.

*Solution.* We choose $u = x^2$, $dv = \sin 2x \, dx$. Then

$$du = 2x \, dx, \qquad v = -\frac{\cos 2x}{2},$$

and (14.8.2) gives

(14.8.3) $$\int x^2 \sin 2x \, dx = -\frac{x^2 \cos 2x}{2} + \int x \cos 2x \, dx.$$

The integral on the right is similar to that in Example 14.8.1, and we proceed similarly. We take $U = x$, $dV = \cos 2x \, dx$. Then

$$dU = dx, \qquad V = \frac{\sin 2x}{2},$$

and we get, as a particular integral,

(14.8.4) $$\int x \cos 2x \, dx = \frac{x \sin 2x}{2} - \frac{1}{2} \int \sin 2x \, dx$$

$$= \frac{x \sin 2x}{2} + \frac{\cos 2x}{4}.$$

Combining (14.8.3) and (14.8.4) we get (adding $C$ at the end),

$$\int x^2 \sin 2x \, dx = -\frac{x^2 \cos 2x}{2} + \frac{x \sin 2x}{2} + \frac{\cos 2x}{4} + C.$$

A third way in which integration by parts may be used is illustrated in the following example.

**Example 14.8.4.** Evaluate $\int e^{2x} \cos 3x \, dx$.

*Solution.* A good choice for $u$ and $dv$ is not immediately apparent. Let us choose $u = e^{2x}$, $dv = \cos 3x \, dx$. Then

$$du = 2e^{2x} \, dx, \qquad v = \tfrac{1}{3} \sin 3x,$$

and we get, by (14.8.2),

(14.8.5) $$\int e^{2x} \cos 3x \, dx = \frac{e^{2x} \sin 3x}{3} - \frac{2}{3} \int e^{2x} \sin 3x \, dx.$$

We proceed similarly with the integral on the right. Thus

$$U = e^{2x}, \quad dV = \sin 3x \, dx; \qquad dU = 2e^{2x} \, dx, \; V = -\tfrac{1}{3} \cos 3x,$$

and we obtain

(14.8.6) $$\int e^{2x} \sin 3x \, dx = -\frac{e^{2x} \cos 3x}{3} + \frac{2}{3} \int e^{2x} \cos 3x \, dx.$$

The integral on the right of (14.8.6) is similar to the original integral. Substituting (14.8.6) into (14.8.5), we have

$$\int e^{2x} \cos 3x \, dx = \frac{e^{2x} \sin 3x}{3} + \frac{2e^{2x} \cos 3x}{9} - \frac{4}{9} \int e^{2x} \cos 3x \, dx.$$

This is an equation in $\int e^{2x} \cos 3x \, dx$ which can be solved for a particular integral; we get

$$\frac{13}{9} \int e^{2x} \cos 3x \, dx = \frac{e^{2x} \sin 3x}{3} + \frac{2e^{2x} \cos 3x}{9}.$$

Multiplying through by 9/13 (and then adding $C$), we have

$$\int e^{2x} \cos 3x \, dx = \frac{e^{2x}(3 \sin 3x + 2 \cos 3x)}{13} + C.$$

*Note:* We may also choose $u = \cos 3x$, $dv = e^{2x} \, dx$, if the corresponding choice is made for the second integration by parts.

In brief, the method illustrated in Example 14.8.4 revolves around finding an equation in the desired integral, from which the integral can be found. This was achieved by using integration by parts a second time. In the next example the overall method is similar, but a second integration by parts is not needed.

***Example 14.8.5.*** Evaluate $\int \sec^3 x \, dx$.

*Solution.* A seemingly artificial choice is made for $u$ and $dv$. We write $\sec^3 x \, dx = \sec x \cdot \sec^2 x \, dx$ and choose

$$u = \sec x, \quad dv = \sec^2 x \, dx.$$

Then $du = \sec x \tan x \, dx$, $v = \tan x$, and thus (14.8.2) gives

(14.8.7) $$\int \sec^3 x \, dx = \sec x \tan x - \int \sec x \tan^2 x \, dx.$$

We set $\tan^2 x = \sec^2 x - 1$ in the integral on the right and obtain

(14.8.8) $$\int \sec x \tan^2 x \, dx = \int (\sec^3 x - \sec x) \, dx$$

$$= \int \sec^3 x \, dx - \ln|\sec x + \tan x|.$$

Combining (14.8.7) and (14.8.8), we get

$$\int \sec^3 x \, dx = \sec x \tan x - \int \sec^3 x \, dx + \ln|\sec x + \tan x|.$$

We solve this equation for $\int \sec^3 x \, dx$ (and add $C$), and we have

(14.8.9) $\qquad \int \sec^3 x \, dx = \tfrac{1}{2}(\sec x \tan x + \ln|\sec x + \tan x|) + C.$

It is very helpful to be aware of the existence of the formula in Equation (14.8.9), since $\int \sec^3 x \, dx$ arises in connection with trigonometric substitutions. The same is true of the formula

(14.8.10) $\qquad \int \sec x \tan^2 x \, dx = \tfrac{1}{2}(\sec x \tan x - \ln|\sec x + \tan x|) + C,$

obtained by combining (14.8.7) and (14.8.9). Both (14.8.9) and (14.8.10) are cases of $\int \tan^m x \sec^n x \, dx$ which were not treated in Section 14.3. Formula (14.8.9) was included in the list of formulas in Section 14.1.

Integration by parts is a powerful method of integration; it may be used for integrations that typically lend themselves to this method, and it may be attempted when other methods do not appear promising.

It should also be mentioned that the formula for integration by parts is often of theoretical value in expressing an integral in a different form, without actually completing the integration. Some of the exercises will touch upon this. In addition, many *reduction formulas*, taken up in the next section, are based on integration by parts.

**EXERCISE GROUP 14.8**

Evaluate the integrals in Exercises 1–32.

1. $\int x \cos 3x \, dx$
2. $\int x^2 \cos x \, dx$
3. $\int xe^{2x} \, dx$
4. $\int_{-2}^{2} x^2 e^x \, dx$
5. $\int x^3 e^x \, dx$
6. $\int \sqrt{x} \sin \sqrt{x} \, dx$
7. $\int \sin^{-1} \sqrt{v} \, dv$
8. $\int \dfrac{x \, dx}{\cos^2 x}$
9. $\int_{0}^{1} \sqrt{x} e^{\sqrt{x}} \, dx$
10. $\int x^n \ln x \, dx$
11. $\int (\ln x)^2 \, dx$
12. $\int x(\ln x)^2 \, dx$
13. $\int \ln(4 + x^2) \, dx$
14. $\int e^{2x} \sin x \, dx$
15. $\int e^{-x} \cos 3x \, dx$
16. $\int \sin^{-1} x \, dx$
17. $\int \cot^{-1} x \, dx$
18. $\int_{0}^{1} \tan^{-1} x \, dx$
19. $\int \sec^{-1} x \, dx$
20. $\int x \sin^{-1} x \, dx$

21. $\int x \tan^{-1} x \, dx$

22. $\int x \cot^{-1} x \, dx$

23. $\int \cos x \cos 2x \, dx$

24. $\int \sin x \sin 4x \, dx$

25. $\int \sin x \cos 3x \, dx$

* 26. $\int e^{x^2}(2x^2 + 1) \, dx$

* 27. $\int \left(\frac{1}{x^2} - \frac{6}{x^4}\right) e^x \, dx$

* 28. $\int \frac{5x^3 + 2}{\sqrt{1 + x^3}} \, dx$

29. $\int \sec^5 x \, dx$

30. $\int \sec^3 x \tan^2 x \, dx$

31. $\int \sec x \tan^4 x \, dx$

32. $\int \sin(\ln x) \, dx$

33. Referring to Example 14.8.4 see what happens if in the integral on the right of (14.8.5) we make the "wrong" choice: $U = \sin 3x$, $dV = e^{2x} \, dx$.

34. Show that $\int \frac{\sin x}{x} \, dx = -\frac{\cos x}{x} - \int \frac{\cos x}{x^2} \, dx$.

35. Show that $\int \frac{\cos x}{\ln x} \, dx = \frac{\sin x}{\ln x} + \int \frac{\sin x}{x(\ln x)^2} \, dx$.

36. If $u$ and $v$ are functions of $x$ show that
$$\int_a^b (uv'' - u''v) \, dx = (uv' - u'v)\Big|_a^b.$$

37. Find the area of the region bounded by the curves $y = \ln x$, $y = 0$, $x = e$.

38. Find the area under the curve $y = x \ln x$ in the interval $[1, e]$.

39. Find the area under the curve $y = x \sin x$ in the interval $[0, \pi]$.

40. Find the centroid of the region under the curve $y = \sin x$ in the interval $[0, \pi]$.

41. The region in the first quadrant bounded by $y^2 = xe^x$, $y = 0$, $x = 1$ is revolved about the $x$ axis. Find the volume of the solid generated.

42. The region under the curve $y = e^x$ in the interval $[0, 1]$ is revolved about the $y$ axis. Find the volume of the solid generated.

43. Find the length of the arc of the parabola $y = 4x^2$ from $x = 0$ to $x = 2$.

* 44. Let $f(x)$ be increasing in the interval $[a, b]$, $a \geq 0$, with $f(a) = c$, $f(b) = d$. The region bounded by $y = f(x)$, $x = b$, $y = c$ is revolved about the $y$ axis, generating a solid.
   (a) By using the hollow-disc method and the cylindrical shell method (Sections 12.3, 12.4) show that the volume of the solid may be expressed in the two forms
   $$V = \pi \int_c^d (b^2 - x^2) \, dy = 2\pi \int_a^b x(y - c) \, dx, \quad \text{where} \quad y = f(x).$$
   (b) Show that the first form reduces directly to the second form, by letting $dy = f'(x) \, dx$ and applying the method of integration by parts.
   [*Suggestion*: Take $f(x) - c$ as an antiderivative of $f'(x)$.]

## * 14.9 REDUCTION FORMULAS

The method that was used in Example 14.8.5 can also be used to obtain a formula for $\int \sec^n x \, dx$ in terms of a similar integral. We choose
$$u = \sec^{n-2} x, \quad dv = \sec^2 x \, dx.$$

Then
$$du = (n-2)\sec^{n-2} x \tan x\, dx, \qquad v = \tan x,$$

and, using the formula for integration by parts, we obtain

$$\int \sec^n x\, dx = \sec^{n-2} x \tan x - (n-2) \int \sec^{n-2} x \tan^2 x\, dx$$

$$= \sec^{n-2} x \tan x - (n-2) \int \sec^{n-2} x (\sec^2 x - 1)\, dx$$

$$= \sec^{n-2} x \tan x - (n-2) \int \sec^n x\, dx + (n-2) \int \sec^{n-2} x\, dx.$$

We now solve the resulting equation for $\int \sec^n x\, dx$, and have

(14.9.1) $$\int \sec^n x\, dx = \frac{\sec^{n-2} x \tan x}{n-1} + \frac{n-2}{n-1} \int \sec^{n-2} x\, dx.$$

Equation (14.9.1) is an example of a *reduction formula of integration*. Such a formula expresses an integral in terms of a similar integral which is often simpler in the sense that some index has been reduced. Note that (14.9.1) can be used to evaluate $\int \sec^3 x\, dx$, with $n = 3$, and can also be used in evaluating $\int \sec^5 x\, dx$, with $n = 5$. By repeated use of a reduction formula we can hopefully arrive at an integral with an index that then permits a direct evaluation.

We give another illustration, by deriving a reduction formula for the integral $\int \sin^m x \cos^n x\, dx$. We choose

$$u = \cos^{n-1} x, \qquad dv = \sin^m x \cos x\, dx.$$

Then we find

$$du = -(n-1) \sin x \cos^{n-2} x\, dx, \qquad v = \frac{\sin^{m+1} x}{m+1}.$$

Integration by parts now gives

$$\int \sin^m x \cos^n x\, dx = \frac{\sin^{m+1} x \cos^{n-1} x}{m+1} + \frac{n-1}{m+1} \int \sin^{m+2} x \cos^{n-2} x\, dx$$

$$= \frac{\sin^{m+1} x \cos^{n-1} x}{m+1} + \frac{n-1}{m+1} \int (1 - \cos^2 x) \sin^m x \cos^{n-2} x\, dx$$

$$= \frac{\sin^{m+1} x \cos^{n-1} x}{m+1} + \frac{n-1}{m+1} \int (\sin^m x \cos^{n-2} x - \sin^m x \cos^n x)\, dx.$$

Solving the resulting equation for the desired integral, we obtain the reduction formula

$$\int \sin^m x \cos^n x\, dx = \frac{\sin^{m+1} x \cos^{n-1} x}{m+n} + \frac{n-1}{m+n} \int \sin^m x \cos^{n-2} x\, dx.$$

Another reduction formula for the same integral appears in Exercise 1; in that formula $m$ is reduced by 2.

\* EXERCISE GROUP 14.9

Derive the reduction formulas in Exercises 1–9.

1. $\int \sin^m x \cos^n x \, dx = -\dfrac{\sin^{m-1} x \cos^{n+1} x}{m+n} + \dfrac{m-1}{m+n} \int \sin^{m-2} x \cos^n x \, dx$

2. $\int x^m \sin x \, dx = -x^m \cos x + m \int x^{m-1} \cos x \, dx$

3. $\int \dfrac{\sin x}{x^m} \, dx = -\dfrac{\sin x}{(m-1)x^{m-1}} + \dfrac{1}{m-1} \int \dfrac{\cos x}{x^{m-1}} \, dx$

4. $\int x^m e^{ax} \, dx = \dfrac{x^m e^{ax}}{a} - \dfrac{m}{a} \int x^{m-1} e^{ax} \, dx$

5. $\int \tan^n x \, dx = \dfrac{\tan^{n-1} x}{n-1} - \int \tan^{n-2} x \, dx$

6. $\int \dfrac{\sin^n x}{\cos^m x} \, dx = \dfrac{1}{m-1} \left[ \dfrac{\sin^{n-1} x}{\cos^{m-1} x} - (n-1) \int \dfrac{\sin^{n-2} x}{\cos^{m-2} x} \, dx \right]$

7. $\int x^m (\ln x)^n \, dx = \dfrac{x^{m+1}(\ln x)^n}{m+1} - \dfrac{n}{m+1} \int x^m (\ln x)^{n-1} \, dx$

8. $\int \dfrac{x^n \, dx}{\sqrt{a+bx}} = \dfrac{2x^n \sqrt{a+bx}}{(2n+1)b} - \dfrac{2na}{(2n+1)b} \int \dfrac{x^{n-1} \, dx}{\sqrt{a+bx}}$

9. $\int x^n \sqrt{2ax - x^2} \, dx = -\dfrac{x^{n-1}\sqrt{(2ax-x^2)^3}}{n+2} + \dfrac{(2n+1)a}{n+2} \int x^{n-1} \sqrt{2ax - x^2} \, dx$

10. Evaluate $\int \sec^3 x \, dx$ by the use of (14.9.1).

11. Evaluate $\int \sec^5 x \, dx$ by the use of (14.9.1).

## 14.10 RATIONAL FUNCTIONS OF $\sin \theta$ AND $\cos \theta$

A rational function of $\sin \theta$ and $\cos \theta$ is a quotient of two polynomials in $\sin \theta$ and $\cos \theta$. Many of the trigonometric integrals previously considered have involved such functions, and special methods were developed for those integrals. Some have been basic formulas, such as those for $\int \cos \theta \, d\theta$ and for $\int \cot \theta \, d\theta = \int (\cos \theta / \sin \theta) \, d\theta$. However, there are many such integrals to which earlier methods do not apply directly, for example,

$$\int \dfrac{d\theta}{2 + \cos \theta}.$$

We present a method involving a substitution which can be used effectively for many such integrals.

We use the substitution

(14.10.1) $\qquad z = \tan \dfrac{\theta}{2}, \quad -\pi < \theta < \pi.$

Then, using the identities

$$\sin\theta = 2\sin\frac{\theta}{2}\cos\frac{\theta}{2}, \qquad \cos\theta = \cos^2\frac{\theta}{2} - \sin^2\frac{\theta}{2},$$

and referring to Fig. 14.10.1 for $\sin\theta/2$ and $\cos\theta/2$, we find

(14.10.2) $$\sin\theta = \frac{2z}{1+z^2}, \qquad \cos\theta = \frac{1-z^2}{1+z^2}.$$

Finally, since $\theta = 2\tan^{-1} z$, we obtain

(14.10.3) $$d\theta = \frac{2\,dz}{1+z^2}.$$

By virtue of (14.10.2) and (14.10.3), an integral of a rational function of $\sin\theta$ and $\cos\theta$ necessarily reduces to an integral of a rational function of $z$, and previous methods apply, the method of partial fractions in particular.

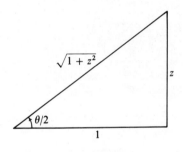

Fig. 14.10.1

**Example 14.10.1.** Evaluate $\displaystyle\int \frac{d\theta}{2+\cos\theta}$.

*Solution.* We make the substitution $z = \tan\theta/2$. With (14.10.2) and (14.10.3) we find

$$\int \frac{d\theta}{2+\cos\theta} = \int \frac{\dfrac{2\,dz}{1+z^2}}{2+\dfrac{1-z^2}{1+z^2}} = 2\int \frac{dz}{3+z^2} = \frac{2}{\sqrt{3}}\tan^{-1}\frac{z}{\sqrt{3}} + C$$

$$= \frac{2}{\sqrt{3}}\tan^{-1}\left(\frac{\tan\theta/2}{\sqrt{3}}\right) + C.$$

**Example 14.10.2.** Evaluate $\displaystyle\int \frac{d\theta}{2\sin\theta - \cos\theta}$.

*Solution.* The substitution $z = \tan \theta/2$, with (14.10.2) and (14.10.3), leads to the equation

(14.10.4) $$\int \frac{d\theta}{2\sin\theta - \cos\theta} = 2\int \frac{dz}{z^2 + 4z - 1}.$$

We may factor $z^2 + 4z - 1$, using radicals, as

$$z^2 + 4z - 1 = (z + 2 - \sqrt{5})(z + 2 + \sqrt{5}),$$

and use partial fractions to obtain

$$\frac{1}{z^2 + 4z - 1} = \frac{1}{2\sqrt{5}}\left(\frac{1}{z+2-\sqrt{5}} - \frac{1}{z+2+\sqrt{5}}\right).$$

Then with (14.10.4)

$$\int \frac{d\theta}{2\sin\theta - \cos\theta} = \frac{1}{\sqrt{5}} \int \left(\frac{1}{z+2-\sqrt{5}} - \frac{1}{z+2+\sqrt{5}}\right) dz$$

$$= \frac{1}{\sqrt{5}} \ln\left|\frac{z+2-\sqrt{5}}{z+2+\sqrt{5}}\right| + C$$

$$= \frac{1}{\sqrt{5}} \ln\left|\frac{\tan\theta/2 + 2 - \sqrt{5}}{\tan\theta/2 + 2 + \sqrt{5}}\right| + C.$$

The integral in the right member of (14.10.4) can also be evaluated by the substitution $z + 2 = \sqrt{5}\sec\alpha$.

As a general rule the substitution $z = \tan\theta/2$ should be used only when other methods do not apply. It would be sheer folly to use this substitution for $\int \cos\theta\, d\theta$ or $\int \cot\theta\, d\theta$, for example, except to prove the point or for the practice (see Exercises 11 and 12).

**EXERCISE GROUP 14.10**

Use the substitution $z = \tan x/2$ to perform the integrations in Exercises 1–10.

1. $\int \dfrac{\cos x\, dx}{\cos x + 1}$

2. $\int \dfrac{2\, dx}{1 + \tan x}$

3. $\int \dfrac{dx}{\cos x - \sin x}$

4. $\int \dfrac{dx}{\sin x + \tan x}$

5. $\int \dfrac{dx}{2 - \sin x}$

6. $\int \dfrac{dx}{2 + \sin x}$

7. $\int \dfrac{dx}{1 + \sin x - \cos x}$

8. $\int \dfrac{dx}{1 - \sin x + \cos x}$

9. $\int \dfrac{dx}{4 + 5\cos x}$

10. $\int \dfrac{dx}{5 + 4\sin x}$

* 11. We know that $\int \cos x \, dx = \sin x + C$. Obtain this result by using the substitution $z = \tan x/2$.
* 12. Obtain the result $\int \cot x \, dx = \ln|\sin x| + C$ by using the substitution $z = \tan x/2$.
* 13. The substitution $w = \cot x/2$ may also be used to integrate a rational function of $\sin x$ and $\cos x$.
    (a) Find $\sin x$, $\cos x$, $dx$.
    (b) Use this substitution to obtain the result $\int \cos x \, dx = \sin x + C$.
* 14. Use the substitution $z = \tan x/2$ to obtain the formula

$$\int \sec x \, dx = \ln|\sec x + \tan x| + C.$$

* 15. Use the substitution $z = \tan x/2$ to obtain the formula

$$\int \sec^3 x \, dx = \tfrac{1}{2}[\sec x \tan x + \ln|\sec x + \tan x|] + C.$$

This is a roundabout method, but it works eventually.

### EXERCISE GROUP 14.11  Supplementary integration exercises

We have developed in this chapter various techniques that may be employed in the evaluation of integrals. The various sets of exercises were developed to illustrate the use of these methods. In many cases the evaluation of an integral may be made in a variety of ways. The present set of exercises further illustrates the use of the methods, but in addition the choice of method must be made carefully.

The form of the integral will often suggest at once the method to be used. When this is not the case, an initial step of some kind must be made; the result of the initial step may then suggest subsequent steps.

Evaluate the integrals in Exercises 1–36.

1. $\int \dfrac{dx}{1 - \sqrt{x}}$

2. $\int \dfrac{dx}{\sqrt{x(x-1)}}$

3. $\int \dfrac{x^3 - 4x + 2}{2x^3 - x^2} dx$

4. $\int \dfrac{dx}{x\sqrt{1 + \ln x}}$

5. $\int \dfrac{x \, dx}{\sqrt{4 - x^4}}$

6. $\int \sqrt{\dfrac{H}{y} - 1} \, dy$

7. $\int \sqrt{\dfrac{y}{2a - y}} \, dy$

8. $\int \dfrac{dx}{(x+1)\sqrt{x+2}}$

9. $\int \dfrac{\sqrt{u^2 + 4}}{u} du$

10. $\int \dfrac{dx}{x^4 \sqrt{x^2 + 9}}$

11. $\int \dfrac{x^2 \, dx}{(x^2 + 1)^{3/2}}$

12. $\int \dfrac{(4 + x^2)^{3/2}}{x^2} dx$

13. $\int \dfrac{dt}{t^3 \sqrt{t^2 + 4}}$

14. $\int \dfrac{x^2 \, dx}{(9 - x^2)^2}$

15. $\int \dfrac{dx}{(x^2 - 2x + 2)^2}$

16. $\int_0^1 \sqrt{x^3 - x^4} \, dx$

17. $\int \dfrac{d\theta}{\sin \theta \cos^2 \theta}$

18. $\int \dfrac{\cos x \, dx}{2 - \cos^2 x}$

19. $\int \dfrac{(1 + \sin x) \, dx}{\sin x (1 + \cos x)}$

20. $\int \dfrac{\ln x}{(1 + \ln x)^2} dx$

21. $\int \dfrac{dx}{x(x^2 - 1)^{3/2}}$

22. $\int \left(\dfrac{2}{x} - \dfrac{1}{x^2}\right) e^{2x}\, dx$

23. $\int x^2 \sin^{-1} x\, dx$

24. $\int x^2 \tan^{-1} x\, dx$

25. $\int e^{\sin^{-1} x}\, dx$

26. $\int (1 - 2x^2) e^{-x^2}\, dx$

27. $\int \dfrac{dx}{x(1 - x^{1/3})}$

28. $\int \dfrac{\sqrt{1 + x^{2/3}}}{x^{1/3}}\, dx$

29. $\int \sqrt{1 - x^{2/3}}\, dx$

30. $\int \dfrac{dx}{x^{2/3}(1 - x^{2/3})^{1/2}}$

31. $\int \dfrac{e^x\, dx}{1 - e^{2x}}$

32. $\int \dfrac{e^{2x}\, dx}{e^{4x} - 1}$

33. $\int \dfrac{dx}{e^x - e^{-x}}$

34. $\int \sqrt{1 + e^{2x}}\, dx$

35. $\int e^{2x + e^x}\, dx$

36. $\int x \cosh^{-1} x\, dx$

★ 37. If $p(x)$ is a polynomial of degree $n$ and $a \neq 0$, prove that
$$\int p(x) e^{-ax}\, dx = -\left[\dfrac{p(x)}{a} + \dfrac{p'(x)}{a^2} + \cdots + \dfrac{p^{(n)}(x)}{a^{n+1}}\right] e^{-ax} + C.$$

38. Find the length of the curve $y = \ln x$ from $x = 1$ to $x = 2\sqrt{2}$.

39. Find the length of the curve $y = 2\ln(4 - x^2)$ from $x = 0$ to $x = 1$.

40. Find the area of the surface generated by revolving about the $x$ axis the arc of the parabola $2y = x^2$ in the first quadrant between $x = 0$ and $x = 1$.

41. Find the area of the surface generated by revolving about the $x$ axis the part of the curve $y = \cosh x$ from $x = 0$ to $x = 2$.

42. Find the surface area of a torus, generated by revolving the circle $(x - b)^2 + y^2 = a^2$, $a < b$, about the $y$ axis.

43. Find the moment of inertia with respect to the $y$ axis of the arc of the parabola $y = 4x^2$ in the first quadrant between $x = 0$ and $x = 2$.

★ 44. The right branch of the hyperbola $y^2 = x^2 - 1$ from the vertex to the point $(3, 2\sqrt{2})$ is revolved about the $y$ axis. Find the area of the surface generated.

# 15 Indeterminate Forms. Improper Integrals

## 15.1 INTRODUCTION

Limits of functions have been encountered in various connections in our work; in fact, the study of limits is fundamental to the subject of calculus. As the study of analysis, of which calculus is a part, is continued, additional facets of the limit concept are pursued. This chapter deals with two additional topics—*indeterminate forms* and *improper integrals*.

## 15.2 INDETERMINATE FORMS

At a value $x = a$, where a function $u(x)$ is continuous, the limit of $u(x)$ can be obtained by direct substitution, by virtue of the definition of continuity at $x = a$,

$$u(a) = \lim_{x \to a} u(x).$$

On the other hand, consider the derivative of a function $u(x)$, defined by

$$u'(x) = \lim_{h \to 0} \frac{u(x+h) - u(x)}{h}.$$

The function on the right, $[u(x + h) - u(x)]/h$, is a function of $h$, since $x$ is held fixed. At $h = 0$ this function is undefined, and additional steps are needed to obtain the limit. For example, for the functions $x^3$ and $\sqrt{x}$ we require, respectively,

$$\lim_{h \to 0} \frac{(x + h)^3 - x^3}{h} \quad \text{and} \quad \lim_{h \to 0} \frac{\sqrt{x + h} - \sqrt{x}}{h}.$$

Each fraction is of the form $f(h)/g(h)$ where, as $h \to 0$, both $f(h) \to 0$ and $g(h) \to 0$. We were able to find the limits in these cases by suitable algebraic manipulation.

Another example is the fundamental trigonometric limit,

$$\lim_{x \to 0} \frac{\sin x}{x}.$$

Here $\sin x \to 0$ and, of course, $x \to 0$, as $x \to 0$. We obtained the limit in this case (Section 10.2) by resorting to a kind of geometric argument.

The above limits are all of the form

(15.2.1) $$\lim_{x \to a} \frac{f(x)}{g(x)}, \quad \text{where} \quad \lim_{x \to a} f(x) = \lim_{x \to a} g(x) = 0.$$

This limit is one type of *indeterminate form*. This designation describes a type of limit, soon to be examined more closely, characterized by the feature that the form of the function involved does not permit any conclusion as to whether a limit exists or not, or as to the value of the limit if a limit does exist.

The indeterminate form in (15.2.1) is usually designated by the symbol $0/0$; *this symbol denotes precisely what is stated in* (15.2.1), *and nothing more*. We are not attempting a definition of any type of division by zero.

The indeterminate forms to be studied are designated by the symbols

$$\frac{0}{0}, \frac{\infty}{\infty}, 0 \cdot \infty, \infty - \infty, 0^0, 1^\infty, \infty^0.$$

We have seen that $0/0$ describes (15.2.1). As another example, the symbol $1^\infty$ describes the following.

$$\lim_{x \to a} [f(x)]^{g(x)}, \quad \text{where} \quad \lim_{x \to a} f(x) = 1, \lim_{x \to a} g(x) = \infty.$$

We shall study the various types of indeterminate form, including the cases where $x \to \infty$.

The following set of exercises includes indeterminate forms which can be evaluated by previous methods, as well as some limits which are not indeterminate.

**EXERCISE GROUP 15.1**

Evaluate each of the following limits. Earlier methods are sufficient for each of them.

1. $\lim\limits_{x \to 2} \dfrac{x - 2}{x^2 - 4}$

2. $\lim\limits_{x \to 1} \dfrac{x^2 - 1}{x^3 - 1}$

3. $\lim\limits_{x \to 4} \dfrac{\sqrt{x}-2}{x-4}$

4. $\lim\limits_{x \to 2} \dfrac{x^3-8}{x^4-16}$

5. $\lim\limits_{x \to 1} \dfrac{x^2-2x-3}{2x^2-3x-5}$

6. $\lim\limits_{x \to 3} \dfrac{3-x}{\sqrt{x+1}-2}$

7. $\lim\limits_{x \to 0} \left(\dfrac{1}{x} - \dfrac{2}{x^2}\right)$

8. $\lim\limits_{x \to \infty} \dfrac{x^2+1}{x+1}$

9. $\lim\limits_{x \to -\infty} \dfrac{(x-2)^2}{x^3-3x}$

10. $\lim\limits_{x \to -\infty} \dfrac{x+8}{\sqrt[3]{x+2}}$

11. $\lim\limits_{x \to \infty} (\sqrt{x^2+1} - x)$

12. $\lim\limits_{x \to -\infty} \dfrac{2x^3+3x-2}{5x^3+4}$

13. $\lim\limits_{x \to -\infty} \dfrac{2x^3-3x^2+1}{x^2-2}$

14. $\lim\limits_{|x| \to \infty} \dfrac{2x^2-3x}{x^2-2}$

15. $\lim\limits_{|x| \to \infty} \dfrac{2x^4+3}{3x^2-x-1}$

16. $\lim\limits_{x \to 0} \dfrac{x}{\sin 3x}$

17. $\lim\limits_{x \to 0} \dfrac{2x}{\tan 3x}$

18. $\lim\limits_{x \to \pi/2} \dfrac{\sin x}{\tan x}$

19. $\lim\limits_{x \to 0} \sin 3x \cot 2x$

20. $\lim\limits_{x \to 0} \dfrac{\tan x - \sin x}{\sin 2x}$

21. $\lim\limits_{x \to \pi/2} (\tan x - \sec x)$

22. $\lim\limits_{x \to \pi/2} \dfrac{1-\sin x}{x-\pi/2}$

* 23. $\lim\limits_{x \to 0} (1-3x)^{1/x}$

* 24. $\lim\limits_{x \to \infty} \left(1 + \dfrac{1}{x^2}\right)^x$

## 15.3  L'HOSPITAL'S RULE

In the preceding set of exercises we considered examples of the indeterminate forms 0/0 and ∞/∞ which could be evaluated by algebraic methods. When such methods are not possible or feasible, L'Hospital's Rule, presented below, leads to a method which has wide application. The proof of this rule makes use of the following result.

**THE GENERALIZED MEAN-VALUE THEOREM.** *Let $f(x)$ and $g(x)$ be continuous in the interval $[a, b]$, and let $f'(x)$ and $g'(x)$ exist in $(a, b)$, with $g'(x) \neq 0$ in $(a, b)$. Then there exists a number $c$ in $(a, b)$ such that*

(15.3.1) $$\dfrac{f(b)-f(a)}{g(b)-g(a)} = \dfrac{f'(c)}{g'(c)}.$$

*Proof.*  Consider the function

$$R(x) = g(x)[f(b)-f(a)] - f(x)[g(b)-g(a)].$$

The hypotheses of the Mean-Value Theorem (Section 9.2) hold for $R(x)$; that is, $R(x)$ is continuous in $[a, b]$, and $R'(x)$ exists in $(a, b)$. Hence a number $c$ in $(a, b)$ exists such that $R(b) - R(a) = (b - a)R'(c)$. Now, it is easily verified that $R(a) = R(b)$. Hence $R'(c) = 0$, and we obtain

$$g'(c)[f(b) - f(a)] - f'(c)[g(b) - g(a)] = 0.$$

Since $g'(x) \neq 0$ in $(a, b)$, we must have $g(b) - g(a) \neq 0$, again by the Mean-Value Theorem, and (15.3.1) follows.

The significant feature of (15.3.1) is that both numerator and denominator of the right member are evaluated at the same number. A geometric interpretation of the result will be presented in Section 16.2. We present one form of L'Hospital's Rule.

**L'HOSPITAL'S RULE.** *Let $f'(x)$ and $g'(x)$ exist at each point of an interval $(a, b)$, with $g'(x) \neq 0$ in $(a, b)$. If $f(x) \to 0$ and $g(x) \to 0$ as $x \to a^+$, and if*

$$\lim_{x \to a^+} \frac{f'(x)}{g'(x)} = L,$$

*then*

$$\lim_{x \to a^+} \frac{f(x)}{g(x)} = L.$$

*This means that if the first limit exists, then the second limit exists and the two limits are equal.*

*Proof.* Since $f(x) \to 0$ and $g(x) \to 0$ as $x \to a^+$ we may extend these functions and define $f(a^+) = 0$, $g(a^+) = 0$. The extended functions are now continuous in $[a, x]$ if $a < x < b$, and we may apply the Generalized Mean-Value Theorem to give

(15.3.2) $\qquad \dfrac{f(x)}{g(x)} = \dfrac{f'(c)}{g'(c)}, \qquad$ for some $c$ in $(a, x)$.

As $x \to a^+$, we also have $c \to a^+$. Hence, if the limit of $f'(c)/g'(c)$ exists, the limit of $f(x)/g(x)$ exists, and the two limits are equal. This completes the proof.

We have proved one form of L'Hospital's Rule. The presentation, with proofs, of the various forms would be lengthy and somewhat difficult. We shall be content here with a list of comments which help explain the rule and its applications.

1. The rule as stated applies as $x \to a^+$. A similar statement applies as $x \to a^-$. In practice, if the same limit is obtained in both cases, we may evaluate a limit as $x \to a$ in a single application.
2. L'Hospital's Rule applies also if the limit of $f'(x)/g'(x)$ is $\infty$ or $-\infty$.

3. L'Hospital's Rule applies also if $x \to \infty$ or if $x \to -\infty$.
4. Comments 1, 2, and 3 apply to the indeterminate form $\infty/\infty$, that is, if $\lim f(x) = \infty$ and $\lim g(x) = \infty$ (or if either limit is $-\infty$).
5. When L'Hospital's Rule is applied, $\lim f'(x)/g'(x)$ may itself be indeterminate. The Rule, if applicable, may then be applied to $f'(x)/g'(x)$.

**Example 15.3.1.** Evaluate $\lim\limits_{x \to 0} \dfrac{e^x - e^{-x}}{\sin 3x}$.

*Solution.* If we set $f(x) = e^x - e^{-x}$, $g(x) = \sin 3x$, we have $f(0) = g(0) = 0$. This limit is of the form 0/0; L'Hospital's Rule applies; and we get

$$\lim_{x \to 0} \frac{e^x - e^{-x}}{\sin 3x} = \lim_{x \to 0} \frac{e^x + e^{-x}}{3 \cos 3x} = \frac{2}{3}.$$

We note that $f'(x)/g'(x)$ is not indeterminate, and its limit can be obtained by direct substitution.

**Example 15.3.2.** Evaluate $\lim\limits_{x \to 0} \dfrac{\tan x - x}{1 - \cos x}$.

*Solution.* This is of the form 0/0. An application of L'Hospital's Rule leads to another indeterminate form of the type 0/0, and the Rule may be applied again. We get

$$\lim_{x \to 0} \frac{\tan x - x}{1 - \cos x} = \lim_{x \to 0} \frac{\sec^2 x - 1}{\sin x} = \lim_{x \to 0} \frac{2 \sec^2 x \tan x}{\cos x} = \frac{0}{1} = 0.$$

Before applying L'Hospital's Rule another time it should be ascertained that an indeterminate form is present, to which the rule applies. Furthermore, before applying L'Hospital's Rule at any step, it may be useful to try a direct method which does not use the Rule. For example, the second limit in Example 15.3.2 may also be evaluated as follows:

$$\lim_{x \to 0} \frac{\sec^2 x - 1}{\sin x} = \lim_{x \to 0} \frac{\tan^2 x}{\sin x} = \lim_{x \to 0} \frac{\sin x}{\cos^2 x} = \frac{0}{1} = 0,$$

which is the same result.

**Example 15.3.3.** Evaluate $\lim\limits_{x \to 0^+} \dfrac{\cot x}{\ln x}$.

*Solution.* If $f(x) = \cot x$ and $g(x) = \ln x$,

$$\lim_{x \to 0^+} f(x) = \infty, \quad \lim_{x \to 0^+} g(x) = -\infty.$$

We apply L'Hospital's Rule and continue as follows.

$$\lim_{x\to 0^+} \frac{\cot x}{\ln x} = \lim_{x\to 0^+} \frac{-\csc^2 x}{1/x} = \lim_{x\to 0^+} \frac{-x}{\sin^2 x} = \lim_{x\to 0^+} \frac{-1}{2\sin x \cos x} = -\infty.$$

The second form was changed to $-x/\sin^2 x$ as a simpler form to which to apply L'Hospital's Rule.

The forms $0 \cdot \infty$ and $\infty - \infty$ are handled by conversion to one of the forms $0/0$ or $\infty/\infty$, as illustrated in what follows.

**Example 15.3.4.** Evaluate $\lim_{x\to 0^+} (x^2 \ln x)$.

*Solution.* We have the form $0 \cdot (-\infty)$, and we may proceed as follows.

$$\lim_{x\to 0^+} (x^2 \ln x) = \lim_{x\to 0^+} \frac{\ln x}{x^{-2}} = \lim_{x\to 0^+} \frac{x^{-1}}{-2x^{-3}} = \lim_{x\to 0^+} \left(-\frac{x^2}{2}\right) = 0.$$

We changed $x^2 \ln x$ to the form $(\ln x)/x^{-2}$, which is of the form $-\infty/\infty$, and L'Hospital's Rule was applicable.

**Example 15.3.5.** Evaluate $\lim_{x\to 1} \left(\frac{1}{\ln x} - \frac{1}{x-1}\right)$.

*Solution.* As $x \to 1^+$ we have the form $\infty - \infty$, and as $x \to 1^-$ we have $-\infty + \infty$. We may treat both cases together and get

$$\lim_{x\to 1}\left(\frac{1}{\ln x} - \frac{1}{x-1}\right) = \lim_{x\to 1} \frac{x - 1 - \ln x}{(x-1)\ln x} = \lim_{x\to 1} \frac{1 - \frac{1}{x}}{\frac{x-1}{x} + \ln x}$$

$$= \lim_{x\to 1} \frac{x-1}{x - 1 + x\ln x} = \lim_{x\to 1} \frac{1}{2 + \ln x} = \frac{1}{2}.$$

The first step was to combine the fractions, leading to the form $0/0$; then L'Hospital's Rule was applied, an algebraic simplification was made, L'Hospital's Rule was applied again, and the last limit was evaluated directly.

Another method for evaluating certain indeterminate forms, by the use of infinite series, will be presented in Section 23.18.

**EXERCISE GROUP 15.2**

Evaluate the limits in Exercises 1–41.

1. $\lim_{x\to 0} \dfrac{\sin 2x}{\tan 5x}$

2. $\lim_{x\to \infty} \dfrac{x}{e^x}$

3. $\lim_{x \to \pi/4} \dfrac{\sqrt{2} - 2\cos x}{x - \pi/4}$

4. $\lim_{x \to 0} \tan 3x \cot 2x$

5. $\lim_{x \to 0} \dfrac{1 - \cos 2x}{x^2}$

6. $\lim_{x \to 0} \dfrac{1 + (x-1)e^x}{x^2}$

7. $\lim_{x \to 1^-} \dfrac{x^2 - x}{x - 1 - \ln x}$

8. $\lim_{x \to 0} \dfrac{x \sec^2 x - \tan x}{x^2}$

9. $\lim_{x \to 0} \dfrac{x \sin x + 2\cos x - 2}{x^3}$

10. $\lim_{x \to 0} \dfrac{x \cos x - \sin x}{x^3}$

11. $\lim_{x \to 0} \dfrac{x^2 \sin x + 2x \cos x - 2 \sin x}{x^3}$

12. $\lim_{x \to 0} \dfrac{x^3}{2x^2 \sec^2 x \tan x - 2x \sec^2 x + 2 \tan x}$

13. $\lim_{x \to -\infty} xe^x$

14. $\lim_{x \to 0^+} \dfrac{x \ln x}{1 - e^x}$

15. $\lim_{x \to 0^+} \dfrac{x \ln x}{\sin x}$

16. $\lim_{x \to 0} \dfrac{x^2 \sin x}{x - \sin^{-1} x}$

17. $\lim_{x \to \infty} x \ln \dfrac{x - 1}{x + 1}$

18. $\lim_{x \to \infty} x[\ln(1 + x) - \ln x]$

19. $\lim_{x \to 0^-} \dfrac{\ln(\sin x/x)}{\sin x - x}$

20. $\lim_{x \to 1} \dfrac{x^x - x}{x - 1 - \ln x}$

21. $\lim_{x \to \pi/2} (\sec x - \tan x)$

22. $\lim_{x \to 1^-} \dfrac{\sin(1/3 \cos^{-1} x)}{\sqrt{1 - x^2}}$

23. $\lim_{x \to \pi/2} \dfrac{\sin(\cos x)}{\tan x - \sec x}$

24. $\lim_{x \to 0} (\cot x - \csc x)$

25. $\lim_{x \to 0} \dfrac{\tan x - \sin x}{x^2 \ln(1 + x)}$

26. $\lim_{x \to 0} \dfrac{x^2 - 2 + 2\cos x}{x^4}$

27. $\lim_{x \to 0} \left( \csc^2 x - \dfrac{1}{x^2} \right)$

28. $\lim_{x \to 0} \left( \dfrac{1}{1 - e^x} + \dfrac{1}{x} \right)$

29. $\lim_{x \to \infty} (\sqrt{x^2 + x} - x)$

30. $\lim_{x \to \infty} x(\sqrt{x^2 + x} - x - \tfrac{1}{2})$

31. $\lim_{x \to 0} \left( \csc x - \dfrac{1}{x} \right)$

32. $\lim_{x \to 0} \left[ \dfrac{1}{x} - \dfrac{\ln(1 + x)}{x^2} \right]$

33. $\lim_{x \to \infty} \dfrac{e^x}{\cosh x}$

34. $\lim_{x \to 0} \dfrac{\tanh x - x}{1 - e^x}$

35. $\lim_{x \to 0} \dfrac{\sinh x}{\ln(1 + x)}$

36. $\lim_{x \to \infty} \dfrac{\ln \cosh x}{x}$

37. $\lim_{x \to 0} \dfrac{\sinh x - x}{(e^x - 1)^3}$

38. $\lim_{x \to 0} \dfrac{1 - \cos x \cosh x}{x^4}$

* 39. $\lim_{x \to \infty} \dfrac{2e^x(2x - 1)}{x^{1/2}(x + e^{2x})}$

* 40. $\lim_{x \to \infty} (x - \ln \cosh x)$

41. $\lim_{h \to 0} \dfrac{f(x + h) - 2f(x) + f(x - h)}{h^2}$, assuming needed differentiability

42. Prove that $\lim_{x \to \infty} x^n/e^x = 0$ if $n$ is a positive integer.

43. Prove that $\lim_{x \to 0^+} x^n \ln x = 0$ if $n$ is a positive integer.

44. Prove that $\lim_{x \to x_0} \dfrac{f(x) - f(x_0) - f'(x_0)(x - x_0)}{(x - x_0)^2} = \dfrac{1}{2} f''(x_0)$ if $f''(x)$ exists in an interval containing $x_0$.

45. Prove: $\lim_{x \to x_0} \dfrac{f(x) - \sum_{k=0}^{n} \dfrac{1}{k!} f^{(k)}(x_0)(x - x_0)^k}{(x - x_0)^{n+1}} = \dfrac{1}{(n+1)!} f^{(n+1)}(x_0)$ if $f^{(n+1)}(x)$ exists and is continuous in an open interval containing $x_0$.

46. Draw the graph of the function $x \ln x$.
47. Draw the graph of the function $x^2 \ln x$.
48. Draw the graph of the function $xe^x$.

## 15.4 INDETERMINATE FORMS OF EXPONENTIAL TYPE

The indeterminate forms considered here are of the types $0^0$, $1^\infty$, and $\infty^0$. They represent limits of functions $F(x)$, where

$$F(x) = f(x)^{g(x)},$$

and $f$ and $g$ have various limits. Since we may write

$$F(x) = e^{\ln F(x)},$$

and since $e^u$ is a continuous function of $u$, it is sufficient to evaluate $\lim \ln F(x)$. Now,

$$\ln F(x) = g(x) \ln f(x),$$

so that $\lim \ln F(x)$, if $\lim F(x)$ is of one of the forms $0^0$, $1^\infty$, and $\infty^0$, will be of one of the forms considered in the preceding sections, for which the methods considered there are available.

***Example 15.4.1.*** Evaluate $\lim_{x \to 0^+} (-\ln x)^x$.

*Solution.* The limit is of the form $\infty^0$. Setting $F(x) = (-\ln x)^x$, we obtain $\ln F(x) = x \ln(-\ln x)$, whose limit is of the form $0 \cdot \infty$. We write

$$\ln F(x) = \frac{\ln(-\ln x)}{x^{-1}},$$

whose limit is of the form $\infty/\infty$ as $x \to 0^+$. Applying L'Hospital's Rule, we have

$$\lim_{x \to 0^+} \ln F(x) = \lim_{x \to 0^+} \frac{\dfrac{1}{\ln x} \cdot x^{-1}}{-x^{-2}} = \lim_{x \to 0^+} \left(-\frac{x}{\ln x}\right) = 0,$$

the last limit of which is no longer indeterminate. Then

$$\lim_{x \to 0^+} (-\ln x)^x = \lim_{x \to 0^+} F(x) = e^{\lim_{x \to 0^+} \ln F(x)} = e^0 = 1.$$

**Example 15.4.2.** Evaluate $\lim_{x \to \infty} \left(1 + \dfrac{3}{x}\right)^x$.

**Solution.** This is of the form $1^\infty$. We write

$$\ln\left(1 + \frac{3}{x}\right)^x = x \ln\left(1 + \frac{3}{x}\right) = \frac{\ln(1 + 3x^{-1})}{x^{-1}};$$

this is of the form 0/0 as $x \to \infty$ and we apply L'Hospital's Rule:

$$\lim_{x \to \infty} \frac{\ln(1 + 3x^{-1})}{x^{-1}} = \lim_{x \to \infty} \frac{\dfrac{-3x^{-2}}{1 + 3x^{-1}}}{-x^{-2}} = \lim_{x \to \infty} \frac{3}{1 + 3x^{-1}} = 3.$$

Hence

$$\lim_{x \to \infty} \left(1 + \frac{3}{x}\right)^x = e^3.$$

The limit in the preceding example is related to the limit often used in the definition of $e$, the base of the natural logarithm system. This definition may be given in either of the forms

$$e = \lim_{x \to \infty} \left(1 + \frac{1}{x}\right)^x = \lim_{v \to 0} (1 + v)^{1/v},$$

the two forms being equivalent. This limit was obtained in Section 11.3 for the number $e$ as defined there. Other examples of related limits will appear in the exercises.

**EXERCISE GROUP 15.3**

Evaluate each of the following limits.

1. $\lim_{x \to 0} (1 + 2x)^{1/x}$
2. $\lim_{x \to 0^-} (1 - x)^{1/x}$
3. $\lim_{x \to 0} (1 - x + x^2)^{1/x}$
4. $\lim_{x \to -\infty} \left(1 + \dfrac{1}{x}\right)^x$
5. $\lim_{x \to 1} (x - 1)^{x^2 - 1}$
6. $\lim_{x \to 0^+} x^x$
7. $\lim_{x \to 0^+} (\tan x)^x$
8. $\lim_{x \to 0} (\sin x)^{\tan x}$
9. $\lim_{x \to \pi/2} (\sin x)^{\tan x}$
10. $\lim_{x \to 1} x^{1/(1-x)}$
11. $\lim_{x \to 0} (x + e^x)^{1/x}$
12. $\lim_{x \to \pi/4} (\tan x)^{1/(x - \pi/4)}$
13. $\lim_{x \to 0} (1 - \cos x)^{\sin x}$
14. $\lim_{x \to 0} \left(\dfrac{2^x + 3^x}{2}\right)^{2/x}$

## 15.5 IMPROPER INTEGRALS

The definite integral $\int_a^b f(x)\,dx$ is known to exist (Section 9.9) if $f(x)$ is continuous in the closed interval $[a, b]$. It is a fact merely stated here that a function can be *integrable* in an interval $[a, b]$, according to the definition in Section 9.9, even with many discontinuities in that interval. However, if a function $f$ is unbounded in the interval, then $f$ cannot be integrable there; moreover, if one or both of the limits of integration are infinite, the definition of the definite integral does not apply. If any of these conditions pertains, the integral is called *improper*. In some cases, it is possible to assign to an improper integral a value which is consistent with previous definitions, in which case the integral will still be called a definite integral. The procedure to be followed will be to isolate the troublesome values, in a sense, and to examine the behavior of the integral near them. We consider separate cases.

*Case 1.* $f(x)$ *is unbounded near an endpoint of* $[a, b]$

If $f(x)$ is continuous for $a \leq x < b$ and unbounded near $b$, we examine the integral $\int_a^c f(x)\,dx$, $a < c < b$. If $\lim_{c \to b^-} \int_a^c f(x)\,dx$ exists, we say that $\int_a^b f(x)\,dx$ exists, or *converges*, and we *define*

$$\int_a^b f(x)\,dx = \lim_{c \to b^-} \int_a^c f(x)\,dx.$$

If the limit on the right does not exist, the improper integral *diverges*.

**Example 15.5.1.** We consider the integral $\int_0^1 (1-x)^{-1/2}\,dx$. The integrand is continuous in the interval of integration except at $x = 1$, at which it becomes infinite. We then consider the integral from 0 to $c$, $0 < c < 1$. This integral exists and we have

$$\int_0^c (1-x)^{-1/2}\,dx = -2(1-x)^{1/2}\Big|_0^c = -2(1-c)^{1/2} + 2.$$

Since $\lim_{c \to 1^-} [-2(1-c)^{1/2} + 2] = 2$, the integral converges, and we have

$$\int_0^1 (1-x)^{-1/2}\,dx = 2.$$

Note that we could have considered $\lim_{\epsilon \to 0^+} \int_0^{1-\epsilon} (1-x)^{-1/2}\,dx$.

If $f(x)$ is continuous for $a < x \leq b$ and unbounded near the lower limit $a$, we examine the integral $\int_c^b f(x)\,dx$, $a < c < b$. If its limit exists as $c \to a^+$, we say that $\int_a^b f(x)\,dx$ exists, or *converges*, and we define

$$\int_a^b f(x)\,dx = \lim_{c \to a^+} \int_c^b f(x)\,dx.$$

*Case 2.* $f(x)$ *is unbounded near an interior point of* $[a, b]$

The ideas will be presented by consideration of the integral

$$\int_0^3 (x - 2)^{-1/3}\, dx.$$

The integrand is discontinuous at $x = 2$, near which the function is unbounded. The value 2 must be approached from both sides independently. We say that the improper integral *converges*, and we *define* its value as

$$\int_0^3 (x - 2)^{-1/3}\, dx = \lim_{\alpha \to 2^-} \int_0^\alpha (x - 2)^{-1/3}\, dx + \lim_{\beta \to 2^+} \int_\beta^3 (x - 2)^{-1/3}\, dx,$$

if *both* limits on the right exist. If even one of the limits fails to exist, the improper integral diverges. We find

$$\lim_{\alpha \to 2^-} \int_0^\alpha (x - 2)^{-1/3}\, dx = \lim_{\alpha \to 2^-} \tfrac{3}{2}(x - 2)^{2/3} \Big|_0^\alpha = \lim_{\alpha \to 2^-} \tfrac{3}{2}[(\alpha - 2)^{2/3} - 2^{2/3}]$$

$$= -\tfrac{3}{2}(2^{2/3}).$$

Similarly, we find

$$\lim_{\beta \to 2^+} \int_\beta^3 (x - 2)^{-1/3}\, dx = \lim_{\beta \to 2^+} \tfrac{3}{2}[1 - (\beta - 2)^{2/3}] = \tfrac{3}{2}.$$

Hence the improper integral converges, and

$$\int_0^3 (x - 2)^{-1/3}\, dx = -\tfrac{3}{2}(2^{2/3}) + \tfrac{3}{2} = \tfrac{3}{2}(1 - 2^{2/3}).$$

**Example 15.5.2.** To examine the improper integral $\int_{-1}^1 x^{-1}\, dx$ we note that $x^{-1}$ is unbounded near zero, and we consider the sum

$$\lim_{\epsilon \to 0^+} \int_{-1}^{-\epsilon} x^{-1}\, dx + \lim_{\eta \to 0^+} \int_\eta^1 x^{-1}\, dx.$$

We have

$$\lim_{\epsilon \to 0^+} \int_{-1}^{-\epsilon} x^{-1}\, dx = \lim_{\epsilon \to 0^+} \ln|x| \Big|_{-1}^{-\epsilon} = \lim_{\epsilon \to 0^+} \ln \epsilon = -\infty.$$

*The integral diverges*; it is unnecessary to examine the second limit above.

If we had not separated the consideration of the two limits, we might have written

$$\lim_{\epsilon \to 0^+} \left[ \int_{-1}^{-\epsilon} x^{-1}\, dx + \int_\epsilon^1 x^{-1}\, dx \right] = \lim_{\epsilon \to 0^+} \left[ \ln|x| \Big|_{-1}^{-\epsilon} + \ln|x| \Big|_\epsilon^1 \right]$$

$$= \lim_{\epsilon \to 0} [\ln \epsilon - \ln \epsilon] = 0,$$

which tells us nothing about the convergence of the integral.

## 15.5 IMPROPER INTEGRALS

*Case 3.* *a or b (or both) infinite*

If $b = \infty$, we have an integral of the form $\int_a^\infty f(x)\,dx$, where $f(x)$ is continuous for $a \leq x$. The improper integral *converges*, and we *define*

$$\int_a^\infty f(x)\,dx = \lim_{t \to \infty} \int_a^t f(x)\,dx,$$

if the limit on the right exists. In the case where $a = -\infty$, we say that $\int_{-\infty}^b f(x)\,dx$ *converges*, and we *define*

$$\int_{-\infty}^b f(x)\,dx = \lim_{t \to -\infty} \int_t^b f(x)\,dx,$$

if the limit on the right exists.

**Example 15.5.3.** To study $\int_0^\infty \dfrac{dx}{4 + x^2}$, we proceed as follows.

$$\lim_{t \to \infty} \int_0^t \frac{dx}{4+x^2} = \lim_{t \to \infty} \frac{1}{2} \tan^{-1} \frac{x}{2} \bigg|_0^t = \lim_{t \to \infty} \frac{1}{2} \tan^{-1} \frac{t}{2} = \frac{1}{2}\left(\frac{\pi}{2}\right) = \frac{\pi}{4}.$$

Hence the integral converges and

$$\int_0^\infty \frac{dx}{4+x^2} = \frac{\pi}{4}.$$

If $f(x)$ is continuous for $-\infty < x < \infty$, we choose some number $a$. We then say that $\int_{-\infty}^\infty f(x)\,dx$ *converges*, and we *define*

$$\int_{-\infty}^\infty f(x)\,dx = \lim_{s \to -\infty} \int_s^a f(x)\,dx + \lim_{t \to \infty} \int_a^t f(x)\,dx,$$

*if both limits on the right exist.* The definition is independent of the choice of $a$. It is important to consider both limits separately.

**Example 15.5.4.** Examine $\int_{-\infty}^\infty \dfrac{dx}{4+x^2}$ for convergence.

*Solution.* Let us choose $a = 0$. We have

$$\lim_{s \to -\infty} \int_s^0 \frac{dx}{4+x^2} = \lim_{s \to -\infty} \frac{1}{2} \tan^{-1} \frac{x}{2} \bigg|_s^0 = -\frac{1}{2} \lim_{s \to -\infty} \tan^{-1} \frac{s}{2} = -\frac{1}{2}\left(-\frac{\pi}{2}\right) = \frac{\pi}{4}.$$

Together with Example 15.5.3 this gives

$$\int_{-\infty}^\infty \frac{dx}{4+x^2} = \frac{\pi}{4} + \frac{\pi}{4} = \frac{\pi}{2}.$$

The following example illustrates the use of L'Hospital's Rule in studying certain definite integrals.

**Example 15.5.5.** Examine $\int_0^2 x \ln x \, dx$ for convergence or divergence.

*Solution.* The integrand is undefined, hence discontinuous, at $x = 0$. We then proceed as follows, performing the integration by parts.

$$(15.5.1) \qquad \int_c^2 x \ln x \, dx = \left(\frac{x^2 \ln x}{2} - \frac{x^2}{4}\right)\bigg|_c^2 = 2 \ln 2 - 1 - \left(\frac{c^2 \ln c}{2} - \frac{c^2}{4}\right).$$

We note that $c^2/4 \to 0$ as $c \to 0^+$. However, $c^2 \ln c$ is indeterminate as $c \to 0^+$. We have, using L'Hospital's Rule,

$$\lim_{c \to 0^+} (c^2 \ln c) = \lim_{c \to 0^+} \frac{\ln c}{c^{-2}} = \lim_{c \to 0^+} \frac{c^{-1}}{-2c^{-3}} = \lim_{c \to 0^+} \frac{-c^2}{2} = 0.$$

Hence the limit of the right member of (15.5.1) in $2 \ln 2 - 1$, and we have

$$\int_0^2 x \ln x \, dx = \lim_{c \to 0^+} \left[2 \ln 2 - 1 - \left(\frac{c^2 \ln c}{2} - \frac{c^2}{4}\right)\right] = 2 \ln 2 - 1.$$

Since $\lim_{x \to 0^+} (x \ln x) = 0$, the integrand may be defined as 0 at $x = 0$, and is then continuous in $[0, 1]$. The definite integral thus exists, and the procedure followed is a method for evaluating the integral.

More than one of the above cases may apply in the same improper integral. It then becomes necessary to isolate the troublesome values and to consider various limits. For example, consider the integral $\int_0^\infty f(x) \, dx$, where $f(x)$ is discontinuous at $x = 0$ and $x = 2$. Referring to Fig. 15.5.1, we select a specific number between 0 and 2, perhaps 1, and a specific number greater than 2, say 3. We consider each of the limits

$$\lim_{\alpha \to 0^+} \int_\alpha^1 f(x) \, dx, \quad \lim_{\beta \to 2^-} \int_1^\beta f(x) \, dx, \quad \lim_{\gamma \to 2^+} \int_\gamma^3 f(x) \, dx, \quad \lim_{t \to \infty} \int_3^t f(x) \, dx.$$

Fig. 15.5.1

If all of these limits exist, the improper integral converges and its value is the sum of the limits. If even one of the limits fails to exist, the original integral diverges, and it is unnecessary to evaluate the other limits. The choice of a specific number between 0 and 2, and a specific number greater than 2 is necessary so that the interior discontinuity may be examined from both sides, while the endpoint discontinuity and the infinite upper limit are also examined separately. The choice of the numbers 1 and 3 was otherwise arbitrary.

The use of the concept of an improper integral permits us to extend the area concept to certain regions to which it was not previously applicable. The graph of the function

$$f(x) = xe^{-x}, \quad x \geq 0,$$

is shown in Fig. 15.5.2. There is a maximum at $x = 1$, a point of inflection at $x = 2$, and a horizontal asymptote, $y = 0$. Can we find the area of the region bounded by the curve and the $x$ axis? The area under the curve between 0 and $t$ is given by

$$A(t) = \int_0^t f(x)\, dx.$$

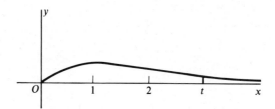

Fig. 15.5.2

As $t$ increases, $A(t)$ is clearly an increasing function. If $A(t)$ approaches a limit as $t \to \infty$, it is reasonable to *define* the area $A$ as this limit, and we have

(15.5.2) $$A = \lim_{t \to \infty} A(t) = \int_0^\infty f(x)\, dx.$$

Thus the area is the value of the improper integral in (15.5.2).

For the function $f(x) = xe^{-x}$ we have (using integration by parts),

$$\int_0^t xe^{-x}\, dx = -e^{-x}(x + 1)\Big|_0^t = 1 - e^{-t}(t + 1).$$

Using L'Hospital's Rule we find that $e^{-t}(t + 1) \to 0$ as $t \to \infty$. Hence, by (15.5.2),

$$A = \lim_{t \to \infty} \int_0^t xe^{-x}\, dx = \lim_{t \to \infty} [1 - e^{-t}(t + 1)] = 1.$$

## EXERCISE GROUP 15.4

In each of Exercises 1–28 evaluate the improper integral if it is convergent.

1. $\int_0^1 \dfrac{dx}{\sqrt{1 - x}}$

2. $\int_0^{\pi/2} \cot \theta\, d\theta$

3. $\int_0^{\pi/2} \dfrac{\sin x}{1 - \cos x}\, dx$

4. $\int_0^1 \dfrac{dx}{1 - x^2}$

5. $\int_0^{1/2} \dfrac{x^2\, dx}{\sqrt{1 - 4x^2}}$

6. $\int_{-1}^0 \dfrac{dx}{x^2 - 2x - 3}$

7. $\int_0^1 \dfrac{dx}{\sqrt{1 - x^2}}$

8. $\int_{-a}^a \dfrac{x\, dx}{\sqrt{a^2 - x^2}}$

9. $\int_0^1 \dfrac{\ln x}{x}\, dx$

10. $\int_0^1 \dfrac{1}{x \ln x}\, dx$

11. $\int_0^{1/2} \dfrac{1}{x(\ln x)^2}\, dx$

12. $\int_{-1}^1 \dfrac{dx}{x^2}$

13. $\int_{-1}^{1} \dfrac{dx}{x^{2/3}}$

14. $\int_{-\infty}^{0} e^{x}\, dx$

15. $\int_{0}^{\infty} \dfrac{dx}{x^{2}+a^{2}}$

16. $\int_{-\infty}^{\infty} \dfrac{dx}{x^{2}+a^{2}}$

17. $\int_{0}^{\infty} \dfrac{dx}{\sqrt{x^{2}+a^{2}}}$

18. $\int_{-\infty}^{\infty} \dfrac{dx}{x^{2}-1}$

19. $\int_{1}^{\infty} \dfrac{dx}{x^{2}-1}$

20. $\int_{3}^{\infty} \dfrac{dx}{\sqrt{x^{2}-9}}$

21. $\int_{0}^{\infty} \dfrac{dx}{\sqrt{x}(1+x)}$

22. $\int_{0}^{\infty} \cos x\, dx$

23. $\int_{1}^{\infty} \dfrac{dx}{x^{2}+x^{4}}$

24. $\int_{-\infty}^{\infty} \dfrac{dx}{e^{x}+e^{-x}}$

*25. $\int_{0}^{\pi/4} \dfrac{dx}{\cos x - \sin x}$

*26. $\int_{0}^{\infty} (1 - \tanh x)\, dx$

27. $\int_{1}^{\infty} \dfrac{dx}{2x(2x-1)}$

28. $\int_{0}^{\infty} x^{2} e^{-x}\, dx$

In Exercises 29–34 determine whether an area can be assigned to the region defined, and if so, find the area.

29. The region under the curve $y = e^{-x}$ to the right of the $y$ axis.
30. The region between the curve $y = \ln x$, the $x$ axis, and the negative $y$ axis.
31. The region under the curve $x^{2}y = 4$ to the right of $x = 1$.
32. The region bounded by the witch $y = 8a^{3}/(x^{2}+4a^{2})$ and its asymptote.
33. The region under the curve $(x-1)^{2}y^{3} = 1$ between $x = 0$ and $x = 2$.
* 34. The region bounded by the cissoid $y^{2} = x^{3}/(2a-x)$ and its asymptote.

35. Evaluate the improper integral, if convergent, for (a) $k > 1$, (b) $k = 1$, (c) $0 < k < 1$:

$$\int_{a}^{b} \dfrac{dx}{(x-a)^{k}}, \quad a < b.$$

36. Prove that $\int_{3}^{\infty} \dfrac{(\ln x)^{k}}{x}\, dx$ converges if and only if $k < -1$.

* 37. The force of attraction between two unit masses acts along the line joining them and has magnitude $F = k/r^{2}$, where $k$ is a constant and $r$ is the distance between them. Let one of the particles be at $(0, 0)$. Find the work done in bringing the second particle from "a point at infinity" to the point $(a, 0)$, $a > 0$, along the $x$ axis.

* 38. Find the centroid of the surface obtained by revolving the part of the curve $x^{2/3} + y^{2/3} = a^{2/3}$ in the first quadrant about the $x$ axis.

# 16 Parametric Equations. Vector Functions

## 16.1 EQUATIONS IN PARAMETRIC FORM

We have previously defined a curve as a set of points whose coordinates satisfy an equation in two variables. We consider here a representation of a curve as a set of points whose coordinates are expressed in terms of an auxiliary variable, called a *parameter*. For example, if a point moves along the parabola $y = x^2$ in such a way that $x = 2t$, where $t$ measures the time, then $y = 4t^2$. The equations

$$x = 2t, \qquad y = 4t^2$$

are called *parametric equations* of the parabola in terms of the *parameter t*. In some cases a parameter may represent time or another physical variable, or it may represent a geometric variable; however, it may also be a variable selected for convenience, with no special significance.

**DEFINITION 16.1.1.** *If the functions*

(16.1.1) $$x = x(t), \qquad y = y(t)$$

*have the same domain, the set of points $(x, y)$ for $t$ in this domain is called the graph of* (16.1.1), *which in turn is called a* **parametric representation**, *or* **parametric equations**, *of the graph.*

The direct method of sketching a curve represented by (16.1.1) is to plot the points $(x, y)$ determined from (16.1.1) by assigning values to $t$, and then to join these points in a suitable manner.

**Example 16.1.1.** Sketch the curve $x = 2t$, $y = t^2$.

*Solution.* The following table is easily obtained.

| $t$ | $-2$ | $-1$ | $0$ | $1$ | $2$ |
|---|---|---|---|---|---|
| $x$ | $-4$ | $-2$ | $0$ | $2$ | $4$ |
| $y$ | $4$ | $1$ | $0$ | $1$ | $4$ |

The points $(-4, 4)$, $(-2, 1)$, ... are then plotted and connected to give the curve in Fig. 16.1.1. The figure also indicates how the curve is traced as $t$ increases.

The following examples illustrate the use of more closely spaced values of the parameter than the integral values that were adequate in Example 16.1.1.

**Example 16.1.2.** Sketch the curve defined by

$$x = \frac{1}{t-1}, \qquad y = 2t + 1.$$

*Solution.* We note that $x$ is undefined for $t = 1$. We include in the table several values of $t$ near, and on either side of, $t = 1$.

| $t$ | $-2$ | $-1$ | $0$ | $\frac{1}{2}$ | $\frac{3}{4}$ | $\frac{7}{8}$ | $\frac{9}{8}$ | $\frac{5}{4}$ | $2$ | $3$ |
|---|---|---|---|---|---|---|---|---|---|---|
| $x$ | $-\frac{1}{3}$ | $-\frac{1}{2}$ | $-1$ | $-2$ | $-4$ | $-8$ | $8$ | $4$ | $1$ | $\frac{1}{2}$ |
| $y$ | $-3$ | $-1$ | $1$ | $2$ | $\frac{5}{2}$ | $\frac{11}{4}$ | $\frac{13}{4}$ | $\frac{7}{2}$ | $5$ | $7$ |

The graph in Fig. 16.1.2 is, in fact, a hyperbola. The figure indicates how each branch is traced as $t$ increases.

**Example 16.1.3.** Graph the curve defined parametrically by

$$x = t^3, \qquad y = t^2.$$

*Solution.* We obtain the following table.

| $t$ | $-\frac{5}{4}$ | $-1$ | $-\frac{3}{4}$ | $-\frac{1}{2}$ | $0$ | $\frac{1}{2}$ | $\frac{3}{4}$ | $1$ | $\frac{5}{4}$ |
|---|---|---|---|---|---|---|---|---|---|
| $x$ | $-\frac{125}{64}$ | $-1$ | $-\frac{27}{64}$ | $-\frac{1}{8}$ | $0$ | $\frac{1}{8}$ | $\frac{27}{64}$ | $1$ | $\frac{125}{64}$ |
| $y$ | $\frac{25}{16}$ | $1$ | $\frac{9}{16}$ | $\frac{1}{4}$ | $0$ | $\frac{1}{4}$ | $\frac{9}{16}$ | $1$ | $\frac{25}{16}$ |

Several values of $t$ near 0 are included, to determine the corresponding part of the graph. The points $(x, y)$ are then plotted and joined, giving the curve in Fig. 16.1.3.

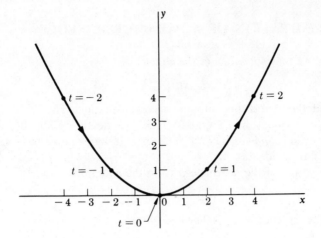

Fig. 16.1.1

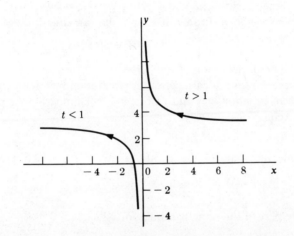

Fig. 16.1.2.

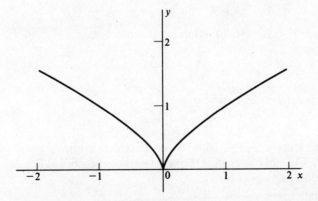

Fig. 16.1.3

## 16.2 ELIMINATION OF A PARAMETER. SLOPE

In general we may solve one of the equations

(16.2.1) $$x = x(u), \quad y = y(u)$$

for $u$, and then substitute into the second equation. The resulting Cartesian equation in $x$ and $y$ has a graph which includes the graph of (16.2.1). This process is called *elimination of the parameter*; it may be used to sketch the graph of (16.2.1), or to identify it.

Elimination of the parameter $t$ from the equations

$$x = 2t, \quad y = 4t^2$$

leads to the equation $y = x^2$. The equations

$$x = u, \quad y = u^2$$

lead to the same equation in $x$ and $y$. As we see, *a curve has more than one parametric representation*. In fact, if a curve has an equation $y = f(x)$, then

$$x = t, \quad y = f(t)$$

is a parametric representation of the curve.

As suggested above, a parametric representation may represent only part of the graph of an equation in $x$ and $y$ obtained by eliminating the parameter. For example, the equations

$$x = \cos \theta, \quad y = \cos^2 \theta$$

represent only that part of the parabola $y = x^2$ for which $|x| \leq 1$, and the equations

$$x = \sqrt{s}, \quad y = s$$

represent the part of the same parabola for which $x \geq 0$.

**Example 16.2.1.** Identify the curve given by

$$x = 2 - 3u^2, \quad y = -1 + 2u^2.$$

*Solution.* We solve each equation for $u^2$, and obtain

$$u^2 = \frac{2-x}{3} = \frac{y+1}{2}.$$

From the second equality we find

$$2x + 3y = 1,$$

which represents a straight line. Since $u^2 \geq 0$, we must have $2 - x \geq 0$ (and $y + 1 \geq 0$). The actual graph is then that part of the straight line $2x + 3y = 1$ for which $x \leq 2$ (and $y \geq -1$).

**Example 16.2.2.** Identify the curve $x = 2 \cos t$, $y = 3 \sin t$.

*Solution.* We obtain $\cos t = x/2$, $\sin t = y/3$. Then from the identity $\cos^2 t + \sin^2 t = 1$ we obtain

$$\frac{x^2}{4} + \frac{y^2}{9} = 1.$$

The curve is a vertical ellipse with center at the origin.

If the equation $x = x(u)$ is solved for $u = u(x)$, and $u$ is substituted in $y = y(u)$, we have

(16.2.2) $\qquad y = y(u(x)) = f(x) = f(x(u)).$

The equation $y = f(x)$ is an ordinary Cartesian equation of a curve which contains the graph of (16.2.1), and the slope of this curve is given by $f'(x)$. Using the chain rule we find, from (16.2.2),

$$\frac{dy}{du} = f'(x(u)) \cdot x'(u), \quad \text{or} \quad \frac{dy}{du} = \frac{df}{dx}\frac{dx}{du}.$$

If $dx/du \neq 0$ we obtain

(16.2.3) $\qquad \dfrac{dy}{dx} = \dfrac{df}{dx} = \dfrac{dy/du}{dx/du} = \dfrac{y'(u)}{x'(u)};$

this relation enables us to find the *slope of the curve* (16.2.1) in terms of the derivatives of $x(u)$ and $y(u)$.

*Example 16.2.3.* Find an equation of the tangent line to the curve of Example 16.2.2 at the point where $t = \pi/3$.

*Solution.* With $t = \pi/3$, the point is $(1, 3\sqrt{3}/2)$. With (16.2.3) we have, with $t$ in place of $u$,

$$\frac{dy}{dx} = \frac{dy/dt}{dx/dt} = \frac{3\cos t}{-2\sin t} = -\frac{3}{2}\cot t.$$

Hence the slope at the given point is $-\sqrt{3}/2$. An equation of the tangent line is

$$y - \frac{3\sqrt{3}}{2} = -\frac{\sqrt{3}}{2}(x - 1) \quad \text{or} \quad \sqrt{3}x + 2y = 4\sqrt{3}.$$

Formula (16.2.3) has a bearing on a geometric interpretation of the Generalized Mean-Value Theorem of Section 15.3. Let $x = f(t)$, $y = g(t)$ be a parametric representation of a curve between $P(f(a), g(a))$ and $Q(f(b), g(b))$ [Fig. 16.2.1]. By that theorem we have

(16.2.4) $\qquad \dfrac{g(b) - g(a)}{f(b) - f(a)} = \dfrac{g'(t')}{f'(t')},$ for some $t'$ between $a$ and $b$.

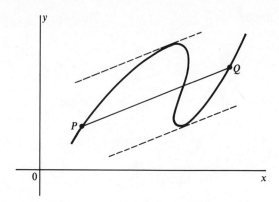

Fig. 16.2.1

The left member of (16.2.4) is the slope of the chord $PQ$. The right member is, by (16.2.3), the slope of the curve at the point for which $t = t'$. Equation (16.2.4) then implies that *the slope of the curve at some point between P and Q is equal to the slope of the chord.* (There are two such points in the figure.)

**EXERCISE GROUP 16.1**

In each of Exercises 1–12, (a) sketch the curve represented by the given parametric equations, (b) eliminate the parameter.

1. $x = 2t + 1, y = 1 - t$
2. $x = t^2 - 1, y = 2t$
3. $x = 4t, y = 6t - t^2$
4. $x = 2t, y = \dfrac{2}{t}$
5. $x = 1 - \dfrac{1}{t}, y = t + \dfrac{1}{t}$
6. $x = \tfrac{1}{4} t^3, y = \tfrac{1}{2} t^2$
7. $x = t^2 - t, y = t^2$
8. $x = t^2 - 2t, y = t^2 - t$
9. $x = \dfrac{1}{(t-1)^2}, y = 2t + 1$
10. $x = \dfrac{1}{t^2 - 1}, y = 2t + 1$
11. $x = t^2 - t, y = t^3$
12. $x = t^2 - t, y = t^3 - 3t$

13. Eliminate $t$ and identify the graph: $x = \dfrac{1 - t^2}{1 + t^2}, y = \dfrac{2t}{1 + t^2}$.
14. Eliminate $t$ and identify the graph: $x = 2 \cos^2 t, y = 3 \sin^2 t$.
15. Eliminate $t$ and identify the graph: $x = a + m \tan kt, y = b + n \sec kt$.
16. Eliminate $\theta$ and identify the graph: $x = a + b \sin \theta, y = c + d \sin \theta$.
17. Eliminate $\theta$: $x = a \cos^3 \theta, y = a \sin^3 \theta$.
18. Eliminate $t$: $x = 2 \cos^4 t, y = 2 \sin^4 t$.
19. Eliminate $\theta$: $x = a\theta - a \sin \theta, y = a - a \cos \theta$.
20. Eliminate $\theta$: $x = \cos 2\theta, y = \cos \theta$.

**21.** Eliminate $t$: $x = 2 \sin^2 t \cos t$, $y = 2 \sin t \cos^2 t$.

**∗ 22.** Eliminate $\theta$: $x = 2 \cos \theta - \cos 2\theta$, $y = 2 \sin \theta - \sin 2\theta$.

In Exercises 23–28 find an equation of the tangent line to the specified curve at the indicated point.

**23.** The curve of Exercise 5 at $t = 1$  
**24.** The curve of Exercise 6 at $t = 2$  
**25.** The curve of Exercise 13 at $t = 1$  
**26.** The curve of Exercise 14 at $t = \pi/6$  
**27.** The curve of Exercise 19 at $\theta = \pi/6$  
**28.** The curve of Exercise 22 at $\theta = \pi/2$  

If $x = x(t)$, $y = y(t)$ are parametric equations of a curve, a horizontal tangent occurs if $y'(t_0) = 0$ and $x'(t_0) \neq 0$; a vertical tangent occurs if $x'(t_0) = 0$ and $y'(t_0) \neq 0$. In Exercises 29–32, find (a) an equation of any horizontal tangent, (b) an equation of any vertical tangent.

**29.** The curve of Exercise 3  
**30.** The curve of Exercise 7  
**31.** The curve of Exercise 8  
**32.** The curve of Exercise 10  

**∗ 33.** For the curve of Exercise 12,
   (a) find an equation of the tangent line at the origin;
   (b) find any vertical tangent lines;
   (c) find two values of $t$ for the point $(2, 2)$. There is a *double point* at $(2, 2)$, the situation at the origin of Fig. 6.6.2.
   (d) find equations of the two tangent lines at $(2, 2)$.

**C∗ 34.** Given the tractrix: $x = a(t - \tanh t)$, $y = a \operatorname{sech} t$:
   (a) show that the length of any tangent line from the point of tangency to the $x$ axis is constant;
   (b) eliminate the parameter $t$;
   (c) sketch the curve.

**∗ 35.** Lines $L_1$ and $L_2$ have $x$ intercepts $a$ and $-a$, respectively. The $y$ intercept of $L_1$ is $a$ units more than that of $L_2$.
   (a) Show that a parametric representation of the locus of the intersection of $L_1$ and $L_2$ in terms of $t$, the $y$ intercept of $L_2$, is
   $$x = \frac{a^2}{2t + a}, \quad y = \frac{2t(t + a)}{2t + a}.$$
   (b) Show that the elimination of the parameter in (a) yields the equation $2xy + x^2 = a^2$.
   (c) We know from Section 8.4 that the graph of the equation in (b) is a hyperbola. Sketch the graph by the methods of Chapter 6, noting the vertical asymptote and the oblique asymptote.

**∗ 36.** Given $x = f(t)$, $y = g(t)$ with $f(t_0) = a$, $f(t_1) = b$, and $f'(t) \neq 0$ in $(t_0, t_1)$. Express $y$ as $y = g[f^{-1}(x)] = F(x)$ and apply the Mean-Value Theorem to $F$ in $[a, b]$. Prove that there is a number $c$ in $(t_0, t_1)$ such that
$$g(t_1) - g(t_0) = \frac{g'(c)}{f'(c)} [f(t_1) - f(t_0)].$$

Compare with the Generalized Mean-Value Theorem in Section 15.3.

## 16.3 THE USE OF A PARAMETER

The study of equations in $x$ and $y$ is often facilitated by the introduction of suitable parameters. An equation of the straight line through the point $(x_1, y_1)$ and parallel to the vector $(a, b)$ is

$$\frac{x - x_1}{a} = \frac{y - y_1}{b}. \tag{16.3.1}$$

If we represent the equal fractions in (16.3.1) by $t$, we have

$$\frac{x - x_1}{a} = t, \qquad \frac{y - y_1}{b} = t,$$

or

$$x = x_1 + at, \qquad y = y_1 + bt. \tag{16.3.2}$$

Equations (16.3.2) are *parametric equations* of the straight line through the point $(x_1, y_1)$ and parallel to the vector $(a, b)$.

*Example 16.3.1.* Find the intersection of the lines

$$x = -3 + 5t, \quad y = 1 - 2t \quad \text{and} \quad x = -4 + 3u, \quad y = -3 + u.$$

*Solution.* Note the different parameters for the two lines. In fact, if the same letter is given as a parameter for both lines, the letter should be changed for one of the lines, since the values of the parameters for the common point need not be the same.

There are four equations in $x$, $y$, $t$, and $u$. We desire the values for $x$ and $y$ of the solution. Eliminating $x$ from the first and third equations, and $y$ from the second and fourth equations, we have

$$-3 + 5t = -4 + 3u \quad \text{and} \quad 1 - 2t = -3 + u,$$

or

$$5t - 3u = -1, \qquad 2t + u = 4.$$

We solve these as $t = 1$, $u = 2$. With $t = 1$ in the equations of the first line we find the point of intersection $x = 2$, $y = -1$. The value $u = 2$ in the equations of the second line yields the same coordinates.

*Example 16.3.2.* If we let $x = 3 \cos \theta$ in the equation of the circle $x^2 + y^2 = 9$, we get $y = \pm 3 \sin \theta$. We choose the plus sign; parametric equations of the circle are then

$$x = 3 \cos \theta, \qquad y = 3 \sin \theta.$$

(The choice of the minus sign leads to another parametric representation.) If $P$ is the point $(x, y)$ on the circle, it is seen that $\theta$ can be taken as the angle which $\overrightarrow{OP}$ makes with the positive $x$ axis. As $\theta$ increases from 0 to $2\pi$, the point $P$ traces the circle in the counterclockwise direction, starting at the point $(3, 0)$.

### 16.3 USE OF A PARAMETER

It is often advantageous to have a curve traced continuously as the parameter varies, as for example, in finding the length of a curve (Section 16.4).

The introduction of a parameter $t$ by the relation $y = tx$ may simplify drawing a graph which passes through the origin (see Exercises 5–9). We present an illustration which really does not require this treatment.

**Example 16.3.3.** Introduce a parameter $t$ such that $y = tx$ in the equation of the circle

(16.3.3) $$x^2 + y^2 - 3y = 0.$$

*Solution.* With $y = tx$ we get

$$x^2 + t^2x^2 - 3tx = 0, \quad \text{or} \quad x[(1 + t^2)x - 3t] = 0.$$

If $x \neq 0$, we get $x$ from the second equation and $y$ from $y = tx$, and we have

(16.3.4) $$x = \frac{3t}{1 + t^2}, \quad y = \frac{3t^2}{1 + t^2}.$$

If $x = 0$, Equation (16.3.3) yields $y = 0$ or $y = 3$. The point $(0, 0)$ is on (16.3.4) for $t = 0$. The point $(0, 3)$ is actually not on (16.3.4) for any value of $t$; this point is approached as $t \to \pm\infty$. Hence (16.3.4) represents parametric equations of (16.3.3) except for the point $(0, 3)$. The graph and the relation of the values of $t$ to the points on the circle are shown in Fig. 16.3.1.

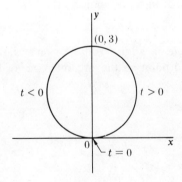

Fig. 16.3.1

The use of a parameter can often facilitate finding an equation of a locus.

**Example 16.3.4.** In Fig. 16.3.2, $|AB| = a$, and $|AP|:|PB| = 2:1$. Find an equation of the locus of $P$.

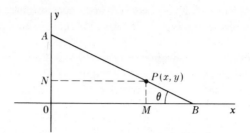

Fig. 16.3.2

*Solution.* We have $|PB| = a/3$, $|AP| = 2a/3$. Let angle $OBP = \theta$. Then

$$\overline{MP} = y = \frac{a}{3} \sin \theta, \qquad \overline{NP} = x = \frac{2a}{3} \cos \theta.$$

Hence parametric equations of the locus are

$$x = \frac{2a}{3} \cos \theta, \qquad y = \frac{a}{3} \sin \theta.$$

We may eliminate $\theta$ and obtain the equation

$$9x^2 + 36y^2 = 4a^2,$$

which represents a horizontal ellipse with center at the origin and vertices at $(\pm 2a/3, 0)$.

The derivation is valid for $P$ in all quadrants. The reader may draw the figure for $P$ in other quadrants, and indicate the angle $\theta$ in each case.

**Example 16.3.5.** A circle of radius $a$ rolls along a straight line without slipping. Find the locus of a fixed point on the circumference of the circle. (The locus is called a *cycloid*.)

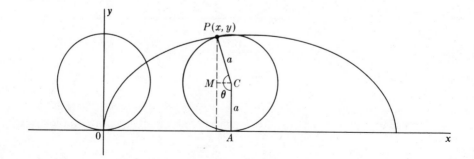

Fig. 16.3.3

## 16.3 USE OF A PARAMETER

*Solution.* Let the circle start with the center on the $y$ axis, as in Fig. 16.3.3, with the fixed point $P$ at the origin, and let the circle roll along the $x$ axis. Let $\theta$ be the angle through which the radius $CP$ has turned. From the circle in Fig. 16.3.3, we see that

$$x = \overline{OA} + \overline{CM} = a\theta - a\cos\left(\theta - \frac{\pi}{2}\right) = a\theta - a\sin\theta,$$

$$y = \overline{AC} + \overline{MP} = a + a\sin\left(\theta - \frac{\pi}{2}\right) = a - a\cos\theta.$$

The key is the fact that the length of $OA$ is equal to the length of arc $AP$ in the circle, and the length of arc $AP$ is $a\theta$. Thus *parametric equations of the cycloid* are

(16.3.5) $\qquad x = a(\theta - \sin\theta), \qquad y = a(1 - \cos\theta).$

The intermediate steps are different for different positions of $P$, but the resulting equations are always the same. For example, if $0 < \theta < \pi/2$, we have directly

$$x = a\theta - a\sin\theta, \qquad y = a - a\cos\theta.$$

(See Exercise 18.)

Determination of the area under a curve involves no basically new methods when parametric equations are used. The area under a curve $y = f(x)$ between $a$ and $b$ is given by (Section 9.12)

(16.3.6) $\qquad\qquad A = \int_a^b f(x)\, dx.$

If the curve is also given by $x = x(t)$, $y = y(t)$, then $x = x(t)$ may be considered as a change of variable in the integral in (16.3.6). We have

$$y = f(x) = f(x(t)) = y(t), \qquad dx = x'(t)\, dt,$$

and (16.3.6) becomes

$$A = \int_{t_0}^{t_1} y(t) x'(t)\, dt, \quad \text{where } x(t_0) = a,\ x(t_1) = b.$$

**Example 16.3.6.** Find the area of the region enclosed by the ellipse

$$x = a\cos t, \qquad y = b\sin t.$$

*Solution.* The ellipse is symmetric in both axes. The total area is twice the area between the $x$ axis and the upper half of the curve. Hence

$$\frac{A}{2} = \int_\pi^0 (b\sin t)(-a\sin t\, dt) = ab\int_0^\pi \sin^2 t\, dt$$

$$= \frac{ab}{2}\int_0^\pi (1 - \cos 2t)\, dt = \frac{\pi ab}{2}.$$

The area enclosed by the ellipse is then $\pi ab$.

**EXERCISE GROUP 16.2**

In Exercises 1–4 find the intersection of the straight lines, by use of the parametric equations.

1. $x = 2 - 3t$, $y = 1 + 2t$ and $x = 2u$, $y = -2 - u$
2. $x = -1 - 4t$, $y = 3 - 3t$ and $x = 2 - u$, $y = 4 + u$
3. $x = 7 - 2t$, $y = 3 - 4t$ and $x = -5 + 3t$, $y = 2 + 2t$
4. $x = 5 - u$, $y = 4 + 3u$ and $x = -2 - u$, $y = -3 - u$

In Exercises 5–9 obtain parametric equations for the given curve by setting $y = tx$. Draw the graph either from the given equation, or from the parametric equations. Show the values of $t$ on the various parts of the curve.

5. $x^2 + y^2 + 2y = 0$
6. $x^2 + y^2 + 2x = 0$
7. $x^2 - y^2 + 3x + 3y = 0$
8. $x^3 + xy^2 - 2ay^2 = 0$
9. $x^3 + y^3 - 3axy = 0$ [Note that there is a point on the graph which corresponds to two distinct values of $t$. This is an advantage, since the point is reached in two different ways.]
10. How is the circle in Example 16.3.2 traced as $\theta$ increases, if the parametric equations are taken as $x = 3 \cos \theta$, $y = -3 \sin \theta$?
11. Let $x = 3 \sin \theta$ in the circle of Example 16.3.2. Find corresponding parametric equations. How is the circle traced as $\theta$ increases?
12. (a) Find parametric equations for the circle of Example 16.3.2 by setting $x = 3 \cos 2\theta$.
    (b) What interval for $\theta$ will trace the circle once?
    (c) Determine what happens as $\theta$ increases from 0 to $2\pi$.
13. Find parametric equations of the ellipse $x^2/4 + y^2/9 = 1$ by setting $y = 3 \cos \theta$.
14. Find parametric equations of the hyperbola $x^2/4 - y^2/9 + 1 = 0$ by setting $x = 2 \tan \theta$.
15. A variable line through the origin intersects the line $y = a$ in a point $Q$. Find an equation of the locus of a point $P$ on $OQ$ if the ordinate of $P$ equals the abscissa of $Q$. [Suggestion: Let $t$ be the abscissa of $Q$.]
16. A triangle has two vertices at $(-a, 0)$ and $(a, 0)$. A variable third vertex is on the line $y = k$. Find an equation of the locus of the intersection of the altitudes of the triangle. [Suggestion: Let $t$ be the abscissa of the intersection.]
17. A variable line through the origin intersects the line $y = a$ in the point $Q$. Find an equation of the locus of a point $P$ on the variable line such $|PQ| = b$. $P$ may be above or below the line $y = a$. Take the angle which $OP$ makes with the positive $x$ axis as the parameter $\theta$.
18. What are the intermediate expressions for $x$ and $y$ in the derivation of the parametric equations of the cycloid if $\pi < \theta < 3\pi/2$?
19. Show that parametric equations of a locus as in Example 16.3.5, if the fixed point starts on the $y$ axis but at a distance $b$ below the center, are

    (A) $\qquad x = a\theta - b \sin \theta, \qquad y = a - b \cos \theta.$

20. Draw the graph of (A) if $b < a$ (e.g., $a = 2$, $b = 1$).

21. Draw the graph of (A) if $b > a$ (e.g., $a = 1$, $b = 2$).
22. Eliminate the parameter $\theta$ in the equations of Exercise 19.
23. (a) A ladder of length 10 ft rests on the ground and extends over an 8-ft wall. Find an equation of the locus of the free end of the ladder. Place the origin at the bottom of the wall, and let $\theta$ be the angle the ladder makes with the wall.
    (b) Eliminate $\theta$ from the answer to (a).
    (c) Draw the complete graph of the equations in (a) or (b), and indicate which part applies to the physical problem.
24. Find an equation of the locus of midpoints of a family of parallel chords of slope $m$ of the parabola $y^2 = 2px$. [Suggestion: Write an equation of the family of parallel chords as $y = mx + t$, $m$ fixed.]
25. Find an equation of the locus of midpoints of a family of parallel chords of slope $m$ ($m$ fixed) of the ellipse $x^2/a^2 + y^2/b^2 = 1$.
26. Find an equation of the locus of midpoints of a family of parallel chords of slope $m$ ($m$ fixed) of the hyperbola $x^2/a^2 - y^2/b^2 = 1$.
27. Find the area under one arch of the cycloid (16.3.5).
28. Find the area of the region enclosed by the ellipse

$$x = a \sin \theta, \qquad y = b \cos \theta, \qquad a > 0, \qquad b > 0.$$

29. Find the area of the region bounded by the curve $x = \sin \theta$, $y = \cos 2\theta$, and the $x$ axis.

\* 30. (a) Sketch the curve (a *cissoid*) given by

$$x = 2a \sin^2 \theta, \qquad y = 2a \sin^2 \theta \tan \theta.$$

   (b) Find the area under the upper part of the curve between the points for which $\theta = 0$ and $\theta = \pi/3$.
   (c) Note that the line $x = 2a$ is an asymptote of the curve. Find the area of the region between the curve and its asymptote.

## 16.4 ARC LENGTH. AREA OF A SURFACE OF REVOLUTION

The definition of arc length for a curve given parametrically is similar to that for a curve given in rectangular coordinates. Let $x = x(t)$ and $y = y(t), a \le t \le b$, be a parametric representation of a curve joining the points $A(x(a), y(a))$ and $B(x(b), y(b))$. [Fig. 16.4.1]. We assume that $x(t)$ and $y(t)$ have continuous derivatives in $[a, b]$. Let $t_0, t_1, \ldots, t_n$, with

$$t_0 < t_1 < \cdots < t_n, \qquad \Delta t_i = t_i - t_{i-1},$$

be a subdivision $\Delta$ of the interval $[a, b]$. If successive points of the curve corresponding to the values $t_i$ are joined by straight line segments, an *inscribed polygon* $\Pi$ is formed, whose length $L(\Pi)$ is

(16.4.1) $$L(\Pi) = \sum_{i=1}^{n} \sqrt{[x(t_i) - x(t_{i-1})]^2 + [y(t_i) - y(t_{i-1})]^2}.$$

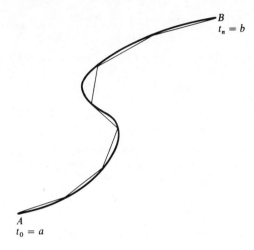

Fig. 16.4.1

Suppose that there is a number $L$ with the following property: given any $\epsilon > 0$, there exists a number $\delta$ such that

$$|L(\Pi) - L| < \epsilon \quad \text{whenever} \quad \|\Delta\| < \delta.$$

Then $L$ is *defined* as the length of the curve from $A$ to $B$, and the curve is said to be *rectifiable*.

We show how the existence of $L$ may be proved. Since $x(t)$ and $y(t)$ are assumed differentiable, the Mean-Value Theorem may be applied, and there exist numbers $\tau_i'$ and $\tau_i''$ in $(t_{i-1}, t_i)$ such that

$$x(t_i) - x(t_{i-1}) = x'(\tau_i')\Delta t_i, \qquad y(t_i) - y(t_{i-1}) = y'(\tau_i'')\Delta t_i.$$

Then (16.4.1) becomes

$$L(\Pi) = \sum_{i=1}^{n} \sqrt{x'^2(\tau_i') + y'^2(\tau_i'')} \, \Delta t_i.$$

This sum would be a Riemann sum if $\tau_i'$ and $\tau_i''$ were the same. Nevertheless, if $x'(t)$ and $y'(t)$ are continuous, it can be shown that a unique number $L$, as specified above, exists. We can then state the following result.

**THEOREM 16.4.1.** *If $x(t)$ and $y(t)$ have continuous derivatives in $[a, b]$, the curve $x = x(t)$, $y = y(t)$ between $(x(a), y(a))$ and $(x(b), y(b))$ has length $L$ given by*

(16.4.2) $$L = \int_a^b \sqrt{x'^2(t) + y'^2(t)} \, dt.$$

The arc length formula (Section 12.5) for a curve in the form $y = f(x)$ is a special case of (16.4.2) for the representation $x = t$, $y = f(t)$.

## 16.4 AREA OF SURFACE OF REVOLUTION

***Example 16.4.1.*** Find the length of the curve

$$x = t^2, \quad y = t^3$$

between the points (0, 0) and (1/4, 1/8).

*Solution.* We have $x'(t) = 2t$, $y'(t) = 3t^2$. The given points occur for $t = 0$ and $t = 1/2$. Hence, by (16.4.2),

$$L = \int_0^{1/2} \sqrt{4t^2 + 9t^4}\, dt = \int_0^{1/2} t\sqrt{4 + 9t^2}\, dt$$

$$= \frac{1}{27}(4 + 9t^2)^{3/2}\bigg|_0^{1/2} = \frac{61}{216}.$$

Now let the function $s(t)$ represent the length of arc of the curve $x = x(t)$, $y = y(t)$ from a fixed point with $t = a$ to the point for variable $t$. By Theorem 16.4.1. we have (with $u$ as the variable of integration)

$$s(t) = \int_a^t \sqrt{x'^2(u) + y'^2(u)}\, du.$$

Applying Theorem 9.10.1, on the derivative of a definite integral, we find

$$s'(t) = \sqrt{x'^2(t) + y'^2(t)} = \sqrt{\left(\frac{dx}{dt}\right)^2 + \left(\frac{dy}{dt}\right)^2},$$

which, since $ds = s'(t)\, dt$, leads to the relation for $ds$,

(16.4.3) $$ds = \sqrt{x'^2(t) + y'^2(t)}\, dt = \sqrt{dx^2 + dy^2}.$$

Thus the relation for $ds$ in terms of $dx$ and $dy$ is the same for a parametric representation as for a Cartesian representation of a curve.

With (16.4.3) we may apply the basic methods of Section 12.6 for finding the area of a surface of revolution to curves which are defined parametrically.

***Example 16.4.2.*** Find the area of the surface generated by revolving the curve

$$x = 4\sqrt{t}, \quad y = t - \ln t,$$

between $t = 1$ and $t = 4$, about the $x$ axis.

*Solution.* We find

$$x'(t) = \frac{2}{\sqrt{t}}, \quad y'(t) = 1 - \frac{1}{t}.$$

Then from (16.4.3),

$$\frac{ds}{dt} = \sqrt{\frac{4}{t} + \left(1 - \frac{1}{t}\right)^2} = 1 + \frac{1}{t}.$$

For the area $S$ of the surface we have (Section 12.6)

$$S = 2\pi \int y\, ds = 2\pi \int_1^4 (t - \ln t)\left(1 + \frac{1}{t}\right) dt$$

$$= 2\pi \int_1^4 \left(t + 1 - \ln t - \frac{\ln t}{t}\right) dt$$

$$= 2\pi \left[\frac{t^2}{2} + 2t - t \ln t - \frac{1}{2}(\ln t)^2\right]\Big|_1^4$$

$$= \pi[27 - 16 \ln 2 - 4(\ln 2)^2].$$

The area of the surface generated by revolving the curve about the $y$ axis can be found similarly and more easily.

### EXERCISE GROUP 16.3

1. Find the length of the circle $x = a \cos t$, $y = a \sin t$.
2. (a) Find the length of the curve $x = 4\sqrt{t}$, $y = t - \ln t$ between $t = 1$ and $t = 4$.
   (b) Find the area of the surface generated when the curve of (a) is revolved about the $y$ axis.
3. Find the length of the curve $x = t$, $y = t^2$ for $0 \leq t \leq 1$.
4. Find the length of the curve $x = \cos t + t \sin t$, $y = \sin t - t \cos t$ from $t = 0$ to $t = \pi$.
5. (a) Find the length of one arch of the cycloid

   $$x = a(\theta - \sin \theta), \qquad y = a(1 - \cos \theta).$$

   (b) Find the area of the surface generated by revolving one arch of the cycloid about the $x$ axis.
   (c) Find the area of the surface generated by revolving about the $y$ axis an arch starting at the origin.
6. (a) Find the length of the curve $x = \cos^2 \theta$, $y = \sin \theta$ in the first quadrant.
   (b) Find the area of the surface generated when the curve in (a) is revolved about the $x$ axis.
7. (a) Find the length of the *hypocycloid* $x = a \cos^3 u$, $y = a \sin^3 u$.
   (b) Find the area of the surface generated by revolving the part of the hypocycloid in the first quadrant about the $x$ axis.
8. Find the length of the curve $x = e^{-2t} \cos t$, $y = e^{-2t} \sin t$ from $t = 0$ to $t = \pi$.
9. Find the area of the surface generated by revolving about the $y$ axis the part of the ellipse $x = a \cos \theta$, $y = b \sin \theta$, $a > b$, to the right of the $y$ axis.
10. (a) Find the length of the curve

    $$x = a(2 \cos \theta - \cos 2\theta), \qquad y = a(2 \sin \theta - \sin 2\theta).$$

    (b) Find the area of the surface generated by revolving the curve of (a) about the $x$ axis.

**11.** Find, for the curve given by

$$x = e^t \sin t, \quad y = e^t \cos t, \quad 0 \le t \le \pi/2,$$

(a) the length of the curve,
(b) the area of the surface generated by revolving the curve about the $x$ axis,
(c) the area of the surface generated by revolving the curve about the $y$ axis,
(d) a Cartesian equation of the curve.

## 16.5 VECTOR FUNCTIONS

If $f(t)$ and $g(t)$ are functions of $t$ defined in some domain, the vector representation $(f(t), g(t))$ defines a (two-dimensional) *vector function* $\mathbf{F}(t)$. Vector functions may be used in various connections. The functions $f(t)$ and $g(t)$ are the *components* of $\mathbf{F}(t)$. In terms of the unit vectors $\mathbf{i}$ and $\mathbf{j}$ we have

$$\mathbf{F}(t) = (f(t), g(t)) = f(t)\mathbf{i} + g(t)\mathbf{j}.$$

In particular, let $\mathbf{R} = x\mathbf{i} + y\mathbf{j}$ be the *position vector* of the point $P(x, y)$. The equation

(16.5.1) $$\mathbf{R} = f(t)\mathbf{i} + g(t)\mathbf{j}$$

specifies the position of $P$ in terms of $t$ and is equivalent to the pair of equations

(16.5.2) $$x = f(t), \quad y = g(t).$$

Thus the vector equation (16.5.1) is equivalent to the parametric equations (16.5.2) and, as such, defines a curve in the plane.

The parameter in (16.5.1) is in general an arbitrary auxiliary variable. However, if $t$ represents time, then (16.5.1) is a vector representation of the path of a moving point or particle in the plane, and (16.5.2) is an equivalent (Cartesian) parametric representation.

***Example 16.5.1.*** Sketch the path of a moving particle whose position vector is given by

$$\mathbf{R}(t) = (2t)\mathbf{i} + (4t - t^2)\mathbf{j}.$$

*Solution.* Parametric equations of this path are

$$x = 2t, \quad y = 4t - t^2.$$

With $t$ eliminated we find

$$y = 2x - x^2/4,$$

so that the curve is the parabola shown in Fig. 16.5.1. At $t = 0$ the point is at the origin. As $t$ increases, the point moves in the direction of the arrow, reaching its highest position when $t = 2$, and returning to the original height $y = 0$ at $t = 4$ (at which time $x = 8$).

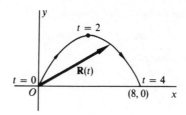

Fig. 16.5.1

The definitions for limit, continuity, and derivative as they relate to vector functions will be formulated in terms of vector operations. They are very similar to the corresponding definitions for real-valued functions, and lead to expected results in terms of the components of vector function In the notation that follows, when vertical bars are used for a vector, the magnitude of the vector is meant.

**DEFINITION 16.5.1.** *The function* $\mathbf{R}(t)$ *has a limit* $\mathbf{A}$ *as* $t \to t_0$, *and we write*

$$\lim_{t \to t_0} \mathbf{R}(t) = \mathbf{A},$$

*if, given $\epsilon > 0$ there exists $\delta$ such that*

$$|\mathbf{R}(t) - \mathbf{A}| < \epsilon \quad \text{whenever} \quad 0 < |t - t_0| < \delta.$$

The definition of continuity follows immediately.

**DEFINITION 16.5.2.** *The vector function* $\mathbf{R}(t)$ *is continuous at $t_0$ if*

$$\lim_{t \to t_0} \mathbf{R}(t) = \mathbf{R}(t_0).$$

We now define the derivative of a vector function.

**DEFINITION 16.5.3.** *The vector function* $\mathbf{R}(t)$ *possesses a derivative* $\mathbf{R}'(t)$, *and we write*

(16.5.3) $$\mathbf{R}'(t) = \lim_{h \to 0} \frac{\mathbf{R}(t+h) - \mathbf{R}(t)}{h},$$

*if the limit in (16.5.3) exists.*

With these definitions the following results express certain relations between a vector function and its components.

**THEOREM 16.5.1.** *Let* $\mathbf{R}(t) = f(t)\mathbf{i} + g(t)\mathbf{j}$. *Then the following hold.*

(a) $\lim_{t \to t_0} \mathbf{R}(t) = \mathbf{A} = a\mathbf{i} + b\mathbf{j}$ *if and only if both*

$$\lim_{t \to t_0} f(t) = a \quad \text{and} \quad \lim_{t \to t_0} g(t) = b.$$

(b) $\mathbf{R}(t)$ is continuous at $t_0$ if and only if both $f(t)$ and $g(t)$ are continuous at $t_0$.

(c) $\mathbf{R}'(t)$ exists if and only if both $f'(t)$ and $g'(t)$ exist, in which case
$$\mathbf{R}'(t) = f'(t)\mathbf{i} + g'(t)\mathbf{j}.$$

*Proof* of (a). We have

(16.5.4) $$|\mathbf{R}(t) - \mathbf{A}| = \{[f(t) - a]^2 + [g(t) - b]^2\}^{1/2}.$$

If $|\mathbf{R}(t) - \mathbf{A}| < \epsilon$ when $0 < |t - t_0| < \delta$, we must have, from (16.5.4), both
$$|f(t) - a| < \epsilon \quad \text{and} \quad |g(t) - b| < \epsilon.$$

On the other hand, if both
$$|f(t) - a| < \epsilon/2 \quad \text{and} \quad |g(t) - b| < \epsilon/2,$$

when $0 < |t - t_0| < \delta$, then from (16.5.4)
$$|\mathbf{R}(t) - \mathbf{A}| < (\epsilon^2/4 + \epsilon^2/4)^{1/2} = \epsilon/\sqrt{2} < \epsilon.$$

From these statements the conclusion of part (a) follows. The proofs of (b) and (c) are left to the exercises.

The significance of the properties expressed in Theorem 16.5.1 is that such questions are handled by working directly with the components of the vector function as though (16.5.1) were an ordinary algebraic expression.

The representation of $\mathbf{R}'(t)$ in Theorem 16.5.1 (c), together with (16.2.3), establishes the following result. *If $\mathbf{R}'(t) \neq 0$, the vector $\mathbf{R}'(t)$ is parallel to the tangent to the curve given by $\mathbf{R}(t)$ at the point corresponding to the value $t$.*

Higher derivatives of vector functions may be treated in a similar fashion.

*Example 16.5.2.* If $\mathbf{R}(t) = \cos t\,\mathbf{i} + \sin t\,\mathbf{j}$, then
$$\mathbf{R}'(t) = -\sin t\,\mathbf{i} + \cos t\,\mathbf{j}, \quad \mathbf{R}''(t) = -\cos t\,\mathbf{i} - \sin t\,\mathbf{j}.$$

It is then clear that we may conclude that
$$\mathbf{R}''(t) + \mathbf{R}(t) = \mathbf{0}.$$

Further properties of vector differentiation are expressed in the following statement.

**THEOREM 16.5.2.** (a) $\dfrac{d}{dt}[a\mathbf{R}(t)] = a\mathbf{R}'(t)$, $a$ constant

(b) $\dfrac{d}{dt}[\mathbf{R}(t) + \mathbf{U}(t)] = \mathbf{R}'(t) + \mathbf{U}'(t)$

(c) $\dfrac{d}{dt}[a(t)\mathbf{R}(t)] = a(t)\mathbf{R}'(t) + a'(t)\mathbf{R}(t)$

(d) $\dfrac{d}{dt}[\mathbf{R}(t) \cdot \mathbf{U}(t)] = \mathbf{R}(t) \cdot \mathbf{U}'(t) + \mathbf{R}'(t) \cdot \mathbf{U}(t)$

*Proof* of (d). Let $\mathbf{R}(t) = f(t)\mathbf{i} + g(t)\mathbf{j}$, $\mathbf{U}(t) = u(t)\mathbf{i} + v(t)\mathbf{j}$. Then
$$\mathbf{R}'(t) = f'(t)\mathbf{i} + g'(t)\mathbf{j}, \qquad \mathbf{U}'(t) = u'(t)\mathbf{i} + v'(t)\mathbf{j}.$$
We have
$$\mathbf{R}(t) \cdot \mathbf{U}(t) = f(t)u(t) + g(t)v(t).$$
Hence
$$\begin{aligned}[] [\mathbf{R}(t) \cdot \mathbf{U}(t)]' &= f(t)u'(t) + f'(t)u(t) + g(t)v'(t) + g'(t)v(t) \\ &= [f(t)u'(t) + g(t)v'(t)] + [f'(t)u(t) + g'(t)v(t)] \\ &= \mathbf{R}(t) \cdot \mathbf{U}'(t) + \mathbf{R}'(t) \cdot \mathbf{U}(t). \end{aligned}$$

The proofs of (a), (b), and (c) may also be effected by working with the appropriate components.

**Example 16.5.3.** Find $[\mathbf{R}(t) \cdot \mathbf{U}(t)]'$, if
$$\mathbf{R}(t) = \cos t\, \mathbf{i} + \sin t\, \mathbf{j}, \qquad \mathbf{U}(t) = 2\mathbf{i} + t^2 \mathbf{j}.$$
*Solution.* We find
$$\mathbf{R}'(t) = -\sin t\, \mathbf{i} + \cos t\, \mathbf{j}, \qquad \mathbf{U}'(t) = 2t\, \mathbf{j}.$$
Hence
$$\begin{aligned}[] [\mathbf{R}(t) \cdot \mathbf{U}(t)]' &= (\cos t\, \mathbf{i} + \sin t\, \mathbf{j}) \cdot (2t\, \mathbf{j}) + (-\sin t\, \mathbf{i} + \cos t\, \mathbf{j}) \cdot (2\mathbf{i} + t^2 \mathbf{j}) \\ &= 2t \sin t - 2 \sin t + t^2 \cos t \\ &= 2(t-1)\sin t + t^2 \cos t. \end{aligned}$$
Of course, we may also obtain directly,
$$\mathbf{R}(t) \cdot \mathbf{U}(t) = 2 \cos t + t^2 \sin t,$$
and differentiate the ordinary function on the right.

A useful consequence of the derivative of a scalar product is expressed in the following.

**THEOREM 16.5.3.** *If $\mathbf{U}(t)$ is a unit vector function, then $\mathbf{U}(t)$ and $\mathbf{U}'(t)$ are orthogonal, or $\mathbf{U}'(t) = \mathbf{0}$.*

*Proof.* Since $\mathbf{U}(t)$ is a unit vector function we have
$$|\mathbf{U}(t)|^2 = \mathbf{U}(t) \cdot \mathbf{U}(t) = 1.$$
Hence, by Theorem 16.5.2, part (d),
$$0 = [\mathbf{U}(t) \cdot \mathbf{U}(t)]' = \mathbf{U}(t) \cdot \mathbf{U}'(t) + \mathbf{U}'(t) \cdot \mathbf{U}(t) = 2\mathbf{U}(t) \cdot \mathbf{U}'(t).$$
Thus $\mathbf{U}(t) \cdot \mathbf{U}'(t) = 0$, and either $\mathbf{U}(t)$ and $\mathbf{U}'(t)$ are orthogonal, or $\mathbf{U}'(t) = \mathbf{0}$.

The same conclusion follows if $\mathbf{U}(t)$ is a vector function of *any* constant magnitude.

**EXERCISE GROUP 16.4**

In Exercises 1–4 evaluate the indicated limit.

1. $\lim_{t \to 2} [(t^2 - 2t)\mathbf{i} + (t^3 - t^2)\mathbf{j}]$

2. $\lim_{t \to \pi/3} [(1 - \cos t)\mathbf{i} + \sin t \, \mathbf{j}]$

3. $\lim_{t \to \infty} \left[ \frac{t^2 - 1}{2t^2 - 1} \mathbf{i} + \left( \tan \frac{1}{t} \right) \mathbf{j} \right]$

4. $\lim_{t \to 0} \left[ (t^2 - 1)\mathbf{i} + \frac{\sin 2t}{t} \mathbf{j} \right]$

For each vector function in Exercises 5–8 find (a) $\mathbf{R}'(t)$, (b) $\mathbf{R}''(t)$, (c) $\frac{d}{dt}[|\mathbf{R}(t)|^2]$.

5. $\mathbf{R}(t) = (t^2 - 1)\mathbf{i} + (t^2 + 1)\mathbf{j}$
6. $\mathbf{R}(t) = t^2\mathbf{i} + t^3\mathbf{j}$
7. $\mathbf{R}(t) = 2\cos t \, \mathbf{i} + 3 \sin t \, \mathbf{j}$
8. $\mathbf{R}(t) = (t - \sin t)\mathbf{i} + (1 - \cos t)\mathbf{j}$
9. If $\mathbf{R}(t) = (t + a \cos 2t)\mathbf{i} + (b \sin 2t)\mathbf{j}$, $a$ and $b$ constant, show that

$$\mathbf{R}''(t) + 4\mathbf{R}(t) = 4t\mathbf{i}.$$

10. Let $\mathbf{U}(t)$ be a normalization of the vector function $(t^2 - 1)\mathbf{i} + (t + 1)\mathbf{j}$. Verify directly that $\mathbf{U}(t)$ and $\mathbf{U}'(t)$ are orthogonal.
11. If $\mathbf{U}(t) = \cos t \, \mathbf{i} + \sin t \, \mathbf{j}$, verify that $\mathbf{U}(t)$ is a unit vector function and that $\mathbf{U}(t)$ and $\mathbf{U}'(t)$ are orthogonal.
12. Let $\mathbf{R}(t) = (a \cos t)\mathbf{i} + (b \sin t)\mathbf{j}$, $a \neq b$. Find the points on the corresponding curve for which $\mathbf{R}(t)$ and $\mathbf{R}'(t)$ are orthogonal.
13. Prove Theorem 16.5.1, part (b).
14. Prove Theorem 16.5.1, part (c).

## 16.6 VELOCITY AND ACCELERATION

When a particle moves along a straight line, with its position on the line at any time $t$ given by a position function $r(t)$, the velocity at time $t$ is defined as the time rate of change of $r(t)$,

$$v(t) = r'(t) = \lim_{h \to 0} \frac{r(t + h) - r(t)}{h}.$$

This formulation does not emphasize the fact that *velocity is a vector*.

For a particle moving along a curve in a plane, with *position vector* $\mathbf{R}(t) = x(t)\mathbf{i} + y(t)\mathbf{j}$, where $t$ represents time, the velocity is *defined* similarly, as the time rate of change of $\mathbf{R}(t)$, and is the *vector function*

$$\mathbf{v}(t) = \mathbf{R}'(t) = \lim_{h \to 0} \frac{\mathbf{R}(t + h) - \mathbf{R}(t)}{h}.$$

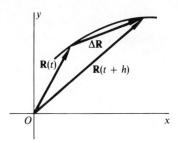

Fig. 16.6.1

The *displacement vector* $\Delta \mathbf{R} = \mathbf{R}(t + h) - \mathbf{R}(t)$, shown in Fig. 16.6.1, denotes the difference in position of the particle at times $t$ and $t + h$. We have

(16.6.1) $\qquad \mathbf{v}(t) = \mathbf{R}'(t) = x'(t)\mathbf{i} + y'(t)\mathbf{j}$.

Hence $\mathbf{v}(t)$ makes an angle $\alpha$ with the positive $x$ axis such that

$$\tan \alpha = y'(t)/x'(t).$$

But the slope of the curve $x = x(t)$, $y = y(t)$ is also $y'(t)/x'(t)$, by (16.2.3). We have thus proved the following result.

**THEOREM 16.6.1.** *At any point on the path of a moving particle the velocity vector is in the direction of the tangent to the curve.*

The *speed* $v(t)$ of the particle is *defined* as the magnitude $|\mathbf{v}(t)|$ of the velocity. Thus, if $\mathbf{T}(t)$ is a *unit vector* parallel to the tangent to the path of motion, we have

(16.6.2) $\qquad \mathbf{v}(t) = |\mathbf{v}(t)|\mathbf{T}(t) = v(t)\mathbf{T}(t)$.

Let $s(t)$ be the length of arc on a curve from some fixed point $P_0$ to the point corresponding to $t$. Then by (16.4.3) we have

(16.6.3) $\qquad s'(t) = \dfrac{ds}{dt} = \sqrt{x'^2(t) + y'^2(t)}$.

But with $\mathbf{v}(t)$ as in (16.6.1) we see that the right side of (16.6.3) is also the speed $v = |\mathbf{v}(t)|$. Hence in general

(16.6.4) $\qquad v = |\mathbf{v}(t)| = \dfrac{ds}{dt} = s'(t)$.

Accordingly, *the speed of a moving particle is equal to the rate of change of arc length with respect to time*, consistent with the idea of speed as a rate of change of distance.

The *acceleration* $\mathbf{a}(t)$ of a moving particle is *defined* as the time rate of change of velocity, is also a vector, and we have

(16.6.5) $$\mathbf{a}(t) = \mathbf{v}'(t) = \mathbf{R}''(t) = x''(t)\mathbf{i} + y''(t)\mathbf{j}.$$

*Example 16.6.1.* Let the position vector of a moving particle be

$$\mathbf{R}(t) = \cos t\, \mathbf{i} + \sin t\, \mathbf{j}.$$

Then

$$\mathbf{v}(t) = \mathbf{R}'(t) = -\sin t\, \mathbf{i} + \cos t\, \mathbf{j}, \qquad \mathbf{a}(t) = \mathbf{R}''(t) = -\cos t\, \mathbf{i} - \sin t\, \mathbf{j}.$$

The path is the unit circle of Fig. 16.6.2, in the counterclockwise direction. Both **v** and **a** are unit vectors at each point, with **v** tangent to the circle, and **a** directed toward the center, perpendicular to **v**. The speed along the curve is constant and equal to 1; however, the velocity varies from point to point since the direction is changing.

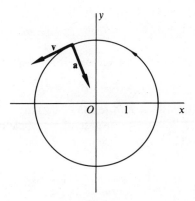

Fig. 16.6.2

The representation of acceleration in (16.6.5) determines the acceleration of a moving particle completely. This representation is in terms of the components of the acceleration in two fixed perpendicular directions, parallel to the $x$ and $y$ axes.

*We may also represent the acceleration of a moving particle in terms of its components in two variable perpendicular directions.* This representation is important in the study of mechanics. It may be developed as follows.

By combining (16.6.2) and (16.6.4) we have

$$\mathbf{v}(t) = \frac{ds}{dt}\mathbf{T}(t),$$

where $\mathbf{T}(t)$ is a unit vector tangent to the path. Differentiating with respect to $t$, we get

(16.6.6) $$\mathbf{a}(t) = \frac{d^2s}{dt^2}\mathbf{T}(t) + \frac{ds}{dt}\frac{d\mathbf{T}}{dt};$$

then, since $d\mathbf{T}/dt = (d\mathbf{T}/ds)(ds/dt)$, Equation (16.6.6) becomes

(16.4.7) $$\mathbf{a}(t) = \frac{d^2s}{dt^2}\mathbf{T}(t) + \left(\frac{ds}{dt}\right)^2 \frac{d\mathbf{T}}{ds}.$$

We know from Theorem 16.5.3 that $d\mathbf{T}/ds$ is orthogonal to $\mathbf{T}$. If we denote by $\mathbf{N}$ the unit vector in the direction of $d\mathbf{T}/ds$, we have, for some constant $k$,

(16.6.8) $$\frac{d\mathbf{T}}{ds} = k\mathbf{N}, \qquad k = \left|\frac{d\mathbf{T}}{ds}\right|,$$

and (16.6.7) becomes

(16.6.9) $$\mathbf{a}(t) = \frac{d^2s}{dt^2}\mathbf{T}(t) + k\left(\frac{ds}{dt}\right)^2 \mathbf{N}(t).$$

We have thereby expressed $\mathbf{a}(t)$ in terms of its *tangential component*, $a_T$, and its *normal component*, $a_N$, as

(16.6.10) $$\mathbf{a} = a_T\mathbf{T} + a_N\mathbf{N}, \qquad a_T = \frac{d^2s}{dt^2}, \qquad a_N = k\left(\frac{ds}{dt}\right)^2.$$

Since $\mathbf{T}$ and $\mathbf{N}$ are orthogonal unit vectors, it follows that

(16.6.11) $$|\mathbf{a}|^2 = a_T^2 + a_N^2.$$

The representation (16.6.10) of **a** in terms of its tangential and normal components turns out to be valid also in the case of a particle moving along a curve in space (Section 20.10). The significance of $k$ will be shown in the next section.

We shall illustrate one characteristic application of (16.6.10) and (16.6.11).

**Example 16.6.2.** Find the tangential and normal components, $a_T$ and $a_N$, of the acceleration for the motion described by the equation

$$\mathbf{R} = (t^3 - t^2)\mathbf{i} + t^2\mathbf{j},$$

at $t = 1$.

*Solution.* We find

$$\mathbf{v} = \frac{d\mathbf{R}}{dt} = (3t^2 - 2t)\mathbf{i} + 2t\mathbf{j}, \qquad \mathbf{a} = \frac{d\mathbf{v}}{dt} = (6t - 2)\mathbf{i} + 2\mathbf{j}.$$

From **v** we find the speed as

$$|\mathbf{v}| = \frac{ds}{dt} = \sqrt{(3t^2 - 2t)^2 + 4t^2} = \sqrt{9t^4 - 12t^3 + 8t^2};$$

and, since $a_T = d^2s/dt^2$, we obtain

$$a_T = \frac{18t^3 - 18t^2 + 8t}{\sqrt{9t^4 - 12t^3 + 8t^2}}.$$

Since the necessary differentiations have been performed, we may substitute the particular value $t = 1$, and find

$$\mathbf{a} = 4\mathbf{i} + 2\mathbf{j}, \quad |\mathbf{a}|^2 = 20, \quad a_T = 8/\sqrt{5}.$$

We now use (16.6.11) to find $a_N$:

$$a_N^2 = |\mathbf{a}|^2 - a_T^2 = 20 - 64/5 = 36/5.$$

Thus at $t = 1$ we have obtained

$$a_T = 8/\sqrt{5}, \quad a_N = 6/\sqrt{5}.$$

The method used in Example 16.6.2 may be summarized as follows. To find $a_T$ obtain $ds/dt$, and then $d^2s/dt^2$, which is $a_T$. To find $a_N$ find $|\mathbf{a}|$ directly, and then $a_N$ from (16.6.11).

We restate some of the preceding as follows. If $\mathbf{R} = \mathbf{R}(t)$ defines a curve in terms of *any* parameter $t$, a *unit tangent vector* $\mathbf{T}(t)$ and a *unit normal vector* $\mathbf{N}(t)$ are

$$\mathbf{T}(t) = \frac{d\mathbf{R}/dt}{|d\mathbf{R}/dt|}, \quad \mathbf{N}(t) = \frac{d\mathbf{T}/dt}{|d\mathbf{T}/dt|}.$$

If $\mathbf{a}(t) = d^2\mathbf{R}/dt^2$ then (16.6.10) and (16.6.11) hold.

We now illustrate an alternative method for the type of question asked in Example 16.6.2.

***Example 16.6.3.*** Let us redo Example 16.6.2 in the following manner. We find $\mathbf{T}(t)$ from that example.

$$\mathbf{T}(t) = \frac{(3t^2 - 2t)\mathbf{i} + 2t\mathbf{j}}{\sqrt{9t^4 - 12t^3 + 8t^2}}, \quad \mathbf{T}(1) = \frac{\mathbf{i} + 2\mathbf{j}}{\sqrt{5}}.$$

Since $\mathbf{T}(1) \cdot \mathbf{N}(1) = 0$, we must have $\mathbf{N}(1) = \pm(2\mathbf{i} - \mathbf{j})/\sqrt{5}$. We determine the correct sign by noting from (16.6.10) that $\mathbf{a} \cdot \mathbf{N} = k(ds/dt)^2 \geq 0$, since $k \geq 0$. With $\mathbf{N}(1)$ as above we find $\mathbf{a}(1) \cdot \mathbf{N}(1) = \pm 6/\sqrt{5}$. Thus we must choose the plus sign for $\mathbf{N}(1)$, and we have

$$\mathbf{N}(1) = \frac{2\mathbf{i} - \mathbf{j}}{\sqrt{5}}.$$

From (16.6.10) we obtain $a_T = \mathbf{a} \cdot \mathbf{T}$, $a_N = \mathbf{a} \cdot \mathbf{N}$, from which we find

$$a_T = 8/\sqrt{5}, \quad a_N = 6/\sqrt{5}.$$

**EXERCISE GROUP 16.5**

For the motion described by the vector function in Exercises 1–10, (a) find $\mathbf{v}$, $v$, $\mathbf{a}$, $a_T$, and $a_N$ at the given point; (b) use formula (16.6.10) for $a_N$ to find $k$ at the given point.

1. $\mathbf{R}(t) = t^3\mathbf{i} - t^2\mathbf{j}$; $t = 2$

2. $\mathbf{R}(t) = \dfrac{t}{t+1}\mathbf{i} + \left(1 - \dfrac{1}{t}\right)\mathbf{j}$; $t = 1$

3. $\mathbf{R}(t) = (4\sqrt{t})\mathbf{i} + (t - \ln t)\mathbf{j}$; $t = 4$

4. $\mathbf{R}(t) = (a \cos t)\mathbf{i} + (b \sin t)\mathbf{j}$; $t = 0$

5. $\mathbf{R}(t) = (\cos 2t)\mathbf{i} + (\sin 2t)\mathbf{j}$; at any point

6. $\mathbf{R}(t) = (2\cos t - \cos 2t)\mathbf{i} + (2\sin t - \sin 2t)\mathbf{j}$; $t = \pi/2$

7. $\mathbf{R}(t) = (e^t \sin t)\mathbf{i} + (e^t \cos t)\mathbf{j}$; $t = \pi/2$

8. $\mathbf{R}(t) = (\cosh t)\mathbf{i} + (\sinh t)\mathbf{j}$; $t = 0$

9. $y = x^2$, $x = 2t$; $t = 1$

10. $x - y^2 = 4$, $y = t^2$; $t = 1$

11. For the cycloid $\mathbf{R}(t) = a[(t - \sin t)\mathbf{i} + (1 - \cos t)\mathbf{j}]$, $0 \le t \le 2\pi$,
    (a) show that a unit tangent vector is $\mathbf{T}(t) = (\sin t/2)\mathbf{i} + (\cos t/2)\mathbf{j}$,
    (b) find the unit normal vector $\mathbf{N}(t)$.

12. A particle moves along the curve $8(y + 2) = x^2$ with its distance from the origin decreasing at the rate of 2 units/min. Find the velocity when the particle is at the point $(-8, 6)$.

13. If $\mathbf{R} = \mathbf{R}(t)$ is a vector representation of a curve in a plane:
    (a) show that the unit tangent vector is $\mathbf{T}(t) = \mathbf{R}'(t)/(ds/dt)$;
    (b) show that the length of the curve between the points with $t = t_0$ and $t = t_1$ is
    $$s = \int_{t_0}^{t_1} |\mathbf{R}'(t)|\, dt.$$

14. Prove that $a_T = (\mathbf{R}' \cdot \mathbf{R}'')/|\mathbf{R}'|$.

\* 15. For the curve $x = x(t)$, $y = y(t)$ prove that the $k$ in (16.6.9) is given by
$$k = \dfrac{(x''^2 + y''^2 - s''^2)^{1/2}}{x'^2 + y'^2},$$
where primes denote derivatives with respect to $t$, and $s = s(t)$ denotes an arc length function for the curve.

16. Prove (16.6.11).

## 16.7 CURVATURE

The intuitive concept of *curvature* of a curve can be made precise. The curvature of a straight line should be zero at any point, the curvature of a circle should be the same at each point of the circle, and the curvature of a curve at any point should be independent of the choice of coordinate axes.

In Fig. 16.7.1 a curve is shown with a tangent to the curve at $P$, the inclination of the tangent being $\alpha$. As $P$ moves along the curve, the rate of change of $\alpha$ appears to reflect an "amount of turning" of the curve. More specifically, the rate of change of $\alpha$ per unit length of arc seems to be inherent in the curve itself.

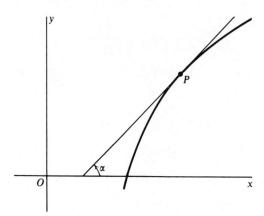

Fig. 16.7.1

**DEFINITION 16.7.1.** *If $\alpha$ is the inclination of the tangent to a curve at a point $P$, the* **curvature** *at $P$ is*

$$k = \left|\frac{d\alpha}{ds}\right|,$$

*where $s$ is an arc length function for the curve.*

We now assert that *the value of $k$ in (16.6.9), which originates from (16.6.8), is in fact the curvature.* For, the unit tangent vector $\mathbf{T}$ may be written as

$$\mathbf{T} = (\cos \alpha)\mathbf{i} + (\sin \alpha)\mathbf{j}.$$

Hence

(16.7.1) $$\frac{d\mathbf{T}}{ds} = [(-\sin \alpha)\mathbf{i} + (\cos \alpha)\mathbf{j}]\frac{d\alpha}{ds}.$$

Now, the $k$ of (16.6.8) was defined as $|d\mathbf{T}/ds|$, and from (16.7.1) we find $|d\mathbf{T}/ds| = |d\alpha/ds|$. Hence the $k$ of (16.6.9) is the same as the $k$ of Definition 16.7.1, and the assertion is proved.

Accordingly, we may use the method of Example 16.6.2 as one way of finding curvature. In that example we found $a_N = 6/\sqrt{5}$. We also find $[ds/dt]_{t=1} = \sqrt{5}$. Hence, substituting $t = 1$ in the relation

$$a_N = k(ds/dt)^2,$$

we have

$$6/\sqrt{5} = k(\sqrt{5})^2, \quad \text{and} \quad k = 6/(5\sqrt{5}) = 6\sqrt{5}/25.$$

Explicit formulas may also be obtained for curvature. Let $x = x(t)$, $y = y(t)$ be a parametric representation of a curve, with $t$ any parameter, not necessarily time. We shall use a subscript to denote the variable of differentiation. If $\alpha$ is the inclination, we have

$$\alpha = \tan^{-1} y_t'/x_t'$$

and hence

$$\frac{d\alpha}{dt} = \frac{x_t' y_t'' - x_t'' y_t'}{x_t'^2 + y_t'^2}.$$

Now, $ds/dt = (x_t'^2 + y_t'^2)^{1/2}$. Hence, and with the relation

$$\frac{d\alpha}{ds} = \frac{d\alpha/dt}{ds/dt},$$

we find

(16.7.2) $$k = \left|\frac{d\alpha}{ds}\right| = \frac{|x_t' y_t'' - x_t'' y_t'|}{(x_t'^2 + y_t'^2)^{3/2}}.$$

**Example 16.7.1.** Show that the curvature of the circle

$$x = a \cos \theta, \qquad y = a \sin \theta$$

is constant.

*Solution.* We obtain

$$x_\theta' = -a \sin \theta, \qquad y_\theta' = a \cos \theta, \qquad x_\theta'' = -a \cos \theta, \qquad y_\theta'' = -a \sin \theta.$$

Hence

$$x_\theta'^2 + y_\theta'^2 = a^2, \qquad x_\theta' y_\theta'' - x_\theta'' y_\theta' = a^2,$$

and from (16.2.7) we have at each point of the circle,

$$k = a^2/a^3 = 1/a.$$

Thus the constant curvature of the circle is equal to the reciprocal of the radius.

Formula (16.7.2) may be used to obtain formulas for curvature if a curve is given in the form $y = f(x)$ or $x = g(y)$. A curve $y = f(x)$ may be represented parametrically as

$$x = x, \qquad y = f(x),$$

with $x$ being considered also as the parameter. Then

$$x_x' = 1, \qquad y_x' = f_x', \qquad x_x'' = 0, \qquad y_x'' = f_x'',$$

and from (16.7.2) we obtain

(16.7.3) $$k = \frac{|f_x''|}{(1 + f_x'^2)^{3/2}}.$$

In a similar manner, if the curve is $x = g(y)$, we find

(16.7.4) $$k = \frac{|g_y''|}{(1 + g_y'^2)^{3/2}}.$$

**Example 16.7.2.** Find the curvature of $y^2 = 2x - 3$ at the point $(6, 3)$.

*Solution.* To use (16.7.4) we solve the given equation for $x$,

$$x = \tfrac{1}{2}(y^2 + 3) = g(y).$$

Then $g_y' = y$, $g_y'' = 1$, and (16.7.4) gives, for $y = 3$,

$$k = \frac{1}{(1+3^2)^{3/2}} = \frac{1}{10\sqrt{10}}.$$

## EXERCISE GROUP 16.6

Find the curvature of the curves in Exercises 1–16, as specified.

1. $x = t^3$, $y = t^2$ at $t = 1$
2. $x = 4\sqrt{t}$, $y = t - \ln t$ at $t = 1$
3. $x = a \cos t$, $y = b \sin t$ at $t = \pi/2$
4. $x = e^t \sin t$, $y = e^t \cos t$ at $t = 0$
5. $x = a \tan t$, $y = a \sec t$, any $t$
6. $x = a \cos^3 t$, $y = a \sin^3 t$, any $t$
7. $\mathbf{R}(t) = \left(1 - \frac{1}{t}\right)\mathbf{i} + \frac{t}{t-1}\mathbf{j}$ at $t = 2$
8. $\mathbf{R}(t) = (t - \sin t)\mathbf{i} + (1 - \cos t)\mathbf{j}$ at $t = \pi$
9. $2y^3 = 3x^2$ at $(12, 6)$
10. $4x^2 - y^2 = 16$ at $(2, 0)$
11. $xy = 8$ at $(-2, -4)$
12. $x^{1/2} + y^{1/2} = a^{1/2}$ at $(0, a)$
13. $x^{2/3} + y^{2/3} = a^{2/3}$ at $(\sqrt{2}a/4, \sqrt{2}a/4)$
14. $y \ln x$, $x = > 0$
15. $y = \ln \cos x$, $0 < x < \pi/2$
16. $y = a \cosh x/a$, any $x$

17. Show that the curvature of the cycloid $x = t - \sin t$, $y = 1 - \cos t$ is $|\csc t/2|/4$.
18. Find the maximum curvature for the curve $y = x^3$.
19. Find the maximum curvature for the curve $y = x^2$.
20. Use the value of $k$ in Exercise 15 of Exercise Group 16.5. Find $s''(t)$ by differentiating $s'(t) = (x'^2 + y'^2)^{1/2}$ and obtain the formula (16.7.2) for $k$.
21. If a particle moves along a curve $x = x(t)$, $y = y(t)$ with constant speed $v(t) = c$, show that the magnitude of the acceleration is

$$|\mathbf{a}(t)| = \frac{|x'y'' - x''y'|}{c}.$$

22. Derive (16.7.4).

## 16.8 ARC LENGTH AS PARAMETER

An arc length function for a curve may serve as a parameter for a parametric representation of the curve. In some cases this representation may be found explicitly, but even if this is not the case, such a representation is valid.

**Example 16.8.1.** Find a representation of the curve $x = t$, $y = t^{3/2}$ in terms of the arc length function measured from $t = 0$.

**Solution.** We have $x_t' = 1$, $y_t' = (3/2)t^{1/2}$. Hence

$$s = \int_0^t \sqrt{1 + \tfrac{9}{4}u}\, du = \tfrac{8}{27}[(1 + \tfrac{9}{4}t)^{3/2} - 1].$$

We now solve for $t$ in terms of $s$,

$$t = \tfrac{1}{9}[(27s + 8)^{2/3} - 4],$$

and substitution into the given equations yields the result

$$x = \tfrac{1}{9}[(27s + 8)^{2/3} - 4], \qquad y = \tfrac{1}{27}[(27s + 8)^{2/3} - 4]^{3/2}.$$

This representation is not as compact as the original one, but it is expressed explicitly in terms of the arc length as parameter.

We present some implications of the use of arc length $s$ as a parameter. A curve is now represented parametrically as

(16.8.1) $\qquad\qquad x = x(s), \qquad y = y(s).$

From the equation $ds^2 = dx^2 + dy^2$ we have at once

(16.8.2) $\qquad\qquad x_s'^2 + y_s'^2 = 1.$

In vector form the curve (16.8.1) is

$$\mathbf{R} = \mathbf{R}(s) = x(s)\mathbf{i} + y(s)\mathbf{j},$$

and

$$\mathbf{R}'(s) = x_s'\mathbf{i} + y_s'\mathbf{j}.$$

Now, using (16.8.2) we find

$$|\mathbf{R}'(s)| = (x_s'^2 + y_s'^2)^{1/2} = 1,$$

and $\mathbf{R}'(s)$ *is directly a unit vector.*

For the above example we have

$$\mathbf{R}(s) = \tfrac{1}{9}[(27s + 8)^{2/3} - 4]\mathbf{i} + \tfrac{1}{27}[(27s + 8)^{2/3} - 4]^{3/2}\mathbf{j},$$

and we find

$$\mathbf{R}'(s) = \frac{2\mathbf{i} + [(27s + 8)^{2/3} - 4]^{1/2}\mathbf{j}}{(27s + 8)^{1/3}}.$$

It may be verified directly that $|\mathbf{R}'(s)| = 1$.


### EXERCISE GROUP 16.7

1. Find parametric equations of the circle $x = a\cos\theta$, $y = a\sin\theta$ in terms of arc length as parameter, measured counterclockwise from $\theta = 0$.

2. Find parametric equations of the hypocycloid $x = a\cos^3 u$, $y = a\sin^3 u$, $0 \leq u \leq \pi/2$, in terms of arc length as parameter, measured from $u = 0$.

3. If $s(t)$ is the arc length function for the curve

$$x = 4\sqrt{t}, \qquad y = t - \ln t,$$

measured from $t = 1$, show that $8[s(t) + y] = x^2 - 8$.

4. If $s(x)$ is an arc length function for the curve $y = (2/3)x^{3/2} - (1/2)x^{1/2}$, measured from $x = 0$, show that $s(x) - y = x^{1/2}$.

5. If arc length $s$ is the parameter for a curve, prove that $x_s' x_s'' + y_s' y_s'' = 0$.

6. If arc length $s$ is the parameter for a curve, show that the curvature is given by $k = |x_s' y_s'' - x_s'' y_s'|$.

7. (a) If arc length $s$ is the parameter, and $\alpha$ is the inclination, show that

$$\frac{dx}{ds} = \cos \alpha, \qquad \frac{dy}{ds} = \sin \alpha.$$

(b) Use the relations in (a) to show that the curvature is given by

$$k = \left| \frac{x_s''}{y_s'} \right| = \left| \frac{y_s''}{x_s'} \right|.$$

(c) Also use the relations in (a) to show that the curvature is given by

$$k = (x_s''^2 + y_s''^2)^{1/2}.$$

8. If $\mathbf{R} = \mathbf{R}(s)$ is a vector representation of a plane curve with $s$ as arc length parameter,
   (a) show that the unit tangent vector is $\mathbf{T}(s) = \mathbf{R}'(s)$.
   (b) Is $d\mathbf{T}/ds$ a unit vector?

*9. If $d\alpha/ds = c > 0$, constant, where $\alpha$ is the inclination, show that the curve is a circle as follows. From $d\alpha/ds = c$ obtain $\alpha = cs + d$. Then from part (a) of Exercise 7 find $x$ and $y$ in terms of the parameter $s$, and then show that the curve is a circle of radius $1/c$.

# 17  Polar Coordinates

## 17.1  THE POLAR COORDINATE SYSTEM

We introduce here a new manner of representing points in a plane, a *polar coordinate system*. Consider a half-line terminating at a point $O$ (Fig. 17.1.1). The point $O$ is called the *pole*, or again, the *origin*, and the half-line is the *polar axis*. If $P$ is any point in the plane other than the pole, let $r = |OP|$, and let $\theta$ be an angle that $\overrightarrow{OP}$ makes with the polar axis. Then $r$ and $\theta$ are *polar coordinates* of $P$.

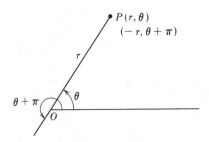

Fig. 17.1.1

In Fig. 17.1.1 the extension of $OP$ through $O$ makes an angle of $\theta + \pi$ with the polar axis. Then $-r$ and $\theta + \pi$ are also taken as polar coordinates of

$P$, and the point $P$ may be designated as $(r, \theta)$ or $(-r, \theta + \pi)$. It is seen that $r$ and $\theta$ may be any numbers, positive, negative, or zero. The polar coordinates of the pole are $(0, \alpha)$, where $\alpha$ represents *any* angle. Fig. 17.1.2 shows two points, with two polar coordinate representations for each. In most connections we express $\theta$ in radian measure, but we use degree measure occasionally.

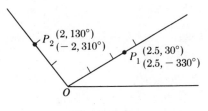

Fig. 17.1.2

If $(r, \theta)$ are polar coordinates of a point $P$ other than $O$, all possible polar coordinate representations of $P$ are given by

(17.1.1) $\quad (r, \theta + 2n\pi) \quad$ or $\quad (-r, \theta + (2n+1)\pi), \qquad n = 0, \pm 1, \pm 2, \ldots .$

Thus *to any point in the plane there correspond infinitely many polar representations*, but *to any pair of polar coordinates there corresponds one and only one point*. This differs from the situation in rectangular coordinates, and often requires special care.

The language of rectangular coordinates may also be used in connection with polar coordinates; thus in a discussion of polar coordinates we may refer to the $x$ and $y$ axes, and label them accordingly. The relevant fact is that we are talking about points in a plane, and we may impose on the plane a rectangular coordinate system, a polar coordinate system, or both.

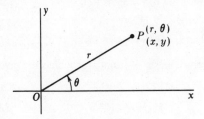

Fig. 17.1.3

In Fig. 17.1.3 both systems are used. If $r > 0$, we have from the properties of the sine and cosine functions,

(17.1.2) $\qquad\qquad x = r \cos \theta, \qquad y = r \sin \theta.$

These relations also hold for any polar representation (17.1.1) of $P$, and also for the pole. If we solve (17.1.2) for $r$ and $\theta$, we get

(17.1.3) $\qquad r = \pm\sqrt{x^2 + y^2}, \qquad \theta = \arctan y/x,$

where the value of arctan $y/x$ must be consistent with the sign of $r$.

Equations (17.1.2) and (17.1.3) are equations of transformation connecting rectangular and polar coordinates.

**Example 17.1.1.** If $P$ has polar coordinates $(-3, 5\pi/6)$, then by (17.1.2) the rectangular coordinates of $P$ are

$$x = -3 \cos 5\pi/6 = 3\sqrt{3}/2, \qquad y = -3 \sin 5\pi/6 = -3/2.$$

**Example 17.1.2.** If $P$ has rectangular coordinates $(-3, 4)$, then for polar coordinates of $P$ we must have, by (17.1.3),

$$r = \pm\sqrt{3^2 + 4^2} = \pm 5, \qquad \theta = \arctan(-4/3).$$

Two possible polar representations in degree measure are $(5, 127°)$ and $(-5, -53°)$; for example, if $r = -5$ then from (17.1.2), since $x < 0$ and $y > 0$, we find $\cos \theta > 0$, $\sin \theta < 0$, so that $\theta$ is in the fourth quadrant.

In terms of the function $\tan^{-1}$ the above polar representations may be precisely written as $(5, \pi - \tan^{-1} 4/3)$ and $(-5, -\tan^{-1} 4/3)$.

**EXERCISE GROUP 17.1**

In Exercises 1–12 plot the point given in polar coordinates, and obtain the rectangular coordinates for each.

1. $(4, 30°)$
2. $(3, 150°)$
3. $(-2, 60°)$
4. $(-4, 210°)$
5. $(7, 23°)$
6. $(-2, 76°)$
7. $(-3, 140°)$
8. $(5, 250°)$
9. $(8, \pi/2)$
10. $(-5, \pi/4)$
11. $(-7, 7\pi/6)$
12. $(9, -\pi/4)$

In Exercises 13–24 describe the set of points for which the given condition holds.

13. $r = 2$
14. $r = -4$
15. $\theta = 30°$
16. $\theta = 210°$
17. $\theta = 30°, r \geq 0$
18. $\theta = 210°, r \leq 0$
19. $r = 3, 0 \leq \theta \leq 90°$
20. $r = 3, \sin \theta > 0$
21. $0 \leq r \leq 2$
22. $1 \leq r \leq 3$
23. $0° \leq \theta \leq 45°$
24. $20° \leq \theta \leq 110°$

25. Show that Equations (17.1.2) hold for the points $(r, \theta + 2n\pi)$, $n$ an integer.
26. Show that Equations (17.1.2) hold for the points $(-r, \theta + (2n+1)\pi)$, $n$ an integer.
27. Derive the following *distance formula*: The distance between the points $P_1(r_1, \theta_1)$ and $P_2(r_2, \theta_2)$ is $|P_1P_2| = \sqrt{r_1^2 + r_2^2 - 2r_1r_2 \cos(\theta_2 - \theta_1)}$.

## 17.2 EQUATIONS OF CURVES IN POLAR FORM

An equation in polar coordinates $r$ and $\theta$ defines a curve in the plane in the following sense: *the set of points $(r, \theta)$ whose coordinates satisfy the equation is called the locus or graph of the equation.* One way to obtain the graph of an equation is to construct a table of values of $r$ and $\theta$ which satisfy the equation, plot the corresponding points, and join them.

*Example 17.2.1.* Draw the graph of the equation

$$r = 2(1 - \cos \theta).$$

*Solution.* We construct the following table, using values of $\theta$ such that $0 \le \theta \le \pi$ for which a table of trigonometric values is not required.

| $\theta$ | 0 | $\pi/6$ | $\pi/4$ | $\pi/3$ | $\pi/2$ | $2\pi/3$ | $3\pi/4$ | $5\pi/6$ | $\pi$ |
|---|---|---|---|---|---|---|---|---|---|
| $r$ | 0 | .27 | .60 | 1 | 2 | 3 | 3.41 | 3.73 | 4 |

We note that for $\pi < \theta \le 2\pi$ the values of $r$ are the reverse of those in the table. From the periodicity of $\cos \theta$ we have

$$\cos(\theta + 2\pi) = \cos \theta.$$

Hence values of $\theta$ outside the interval $[0, 2\pi]$ yield no new points. Fig. 17.2.1 shows the result of plotting the points and joining them. The curve is called a *cardioid*.

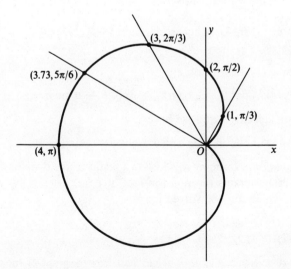

Fig. 17.2.1

A curve may have more than one polar equation. For example, the graphs of $r = 3$ and $r = -3$ are the same circle, with center at the origin and radius 3. We emphasize the fact that these equations of the same curve have no common solution.

The graph of a polar equation may sometimes be found by transforming the equation into rectangular form by use of (17.1.2) and (17.1.3), and then drawing the graph of the resulting equation.

***Example 17.2.2.*** Draw the graph of $r = \cos \theta$.

***Solution.*** We multiply through by $r$; this introduces no new point into the graph, since $r = 0$ corresponds to the pole, and the pole is on the graph of the given equation. We get $r^2 = r \cos \theta$. By (17.1.2) this becomes

$$x^2 + y^2 = x,$$

which we identify as a circle of radius 1/2, with center at (1/2, 0). We dispense with the actual drawing of the graph.

***Example 17.2.3.*** Transform the equation $r = 2/(2 + \cos \theta)$ into rectangular coordinates.

***Solution.*** From the given equation we get $2r + r \cos \theta = 2$; with (17.1.2) and (17.1.3) this becomes

$$\pm 2\sqrt{x^2 + y^2} + x = 2.$$

Subtracting $x$ from both sides, then squaring both sides and simplifying, we get

$$3x^2 + 4y^2 + 4x - 4 = 0,$$

whose graph is a horizontal ellipse.

***Example 17.2.4.*** Transform the equation $x^2 - 3y^2 + x + 2y = 0$ into one in polar coordinates.

***Solution.*** Using (17.1.2) we obtain

$$r^2 \cos^2 \theta - 3r^2 \sin^2 \theta + r \cos \theta + 2r \sin \theta = 0.$$

We divide both sides by $r$ and we have

(17.2.1) $\qquad r \cos^2 \theta - 3r \sin^2 \theta + \cos \theta + 2 \sin \theta = 0,$

the desired equation. No point was lost as a result of the division by $r$, since the graph of (17.2.1) contains the pole; in fact, if $r = 0$ in (17.2.1), we have $\cos \theta + 2 \sin \theta = 0$, which may be solved for $\theta$.

### EXERCISE GROUP 17.2

In Exercises 1–10 construct a table of values, and draw the graph of the given equation.

**1.** $r = 10 \sin \theta$ **2.** $r + 4 \cos \theta = 0$

3. $r(2\sin\theta + 3\cos\theta) = 4$

4. $r = \dfrac{8}{2\cos\theta + \sin\theta}$

5. $r = \dfrac{2}{\sin\theta + \cos\theta}$

6. $r = 2(1 + \sin\theta)$

7. $r = 3(1 + \cos\theta)$

8. $r = 2 - \cos\theta$

9. $r = 1 - 2\cos\theta$

10. $r = \dfrac{6}{2 - \cos\theta}$

In Exercises 11–22 transform the equation to one in rectangular coordinates.

11. $\theta = \pi/3$
12. $r = 4$
13. $r(3\cos\theta - 2\sin\theta) = 1$
14. $r = 10\cos\theta$
15. $r = 1 + \cos\theta$
16. $r = 2 + \cos\theta$
17. $r(1 + \cos\theta) = 8$
18. $r^2 + 3r\cos\theta - r\sin\theta = 18$
19. $r = 2\sin\theta + 3\cos\theta$
20. $r = \dfrac{5}{1 - \sin\theta}$
21. $r = \dfrac{4}{1 + 2\cos\theta}$
22. $r = 2\sin 2\theta$

In Exercises 23–30 transform the equation to one in polar coordinates.

23. $2x - 5y + 4 = 0$
24. $4x + 7y = 2$
25. $x^2 + y^2 - 2x + 4y = 0$
26. $x^2 + y^2 + 3x - 5y + 4 = 0$
27. $y^2 = 8x + 16$
28. $x^2 - 3y^2 - 4y - 1 = 0$
29. $x^2 = (x^2 + y^2)[2(x^2 + y^2) - 1]^2$
30. $(x^2 + y^2)^3 = 4x^2 y^2$

## 17.3 GRAPHING POLAR EQUATIONS

The graphing of a number of specific equations in polar coordinates will be left to the exercises. An often fruitful procedure is to start with a fixed value of $\theta$, usually $\theta = 0$, and then to study the variation in $r$ as $\theta$ increases. The points of intersection of a curve with the $x$ and $y$ axes, that is, for $\theta = 0, \pi/2, \pi, \ldots$, are often useful. *Particular attention is paid to values of $\theta$ for which $r = 0$ since, if $\theta_0$ is such a value, the line $\theta = \theta_0$ is tangent to the curve at the origin.* To see this we refer to Fig. 17.3.1. The slope of the curve at $O$, if it exists, is given by

$$m(O) = \lim_{P \to O} m(OP) = \lim_{\theta \to \theta_0} \tan\theta = \tan\theta_0,$$

provided only that $\theta \to \theta_0$ as $r \to 0$.

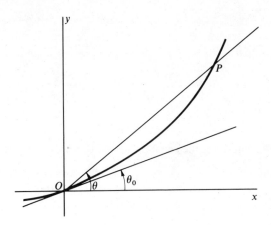

Fig. 17.3.1

The discussion of *symmetry* is governed by the realization that more than one test exists for the same type of symmetry. Thus, *if the replacement of $\theta$ by $-\theta$ (see Fig. 17.3.2) yields an equivalent equation, the graph is symmetric in the $x$ axis.* According to this test, the graph of

(17.3.1) $$r^2 = \cos \theta$$

is symmetric in the $x$ axis. The test fails however for the equation

(17.3.2) $$r = \sin 2\theta,$$

whose graph is nevertheless symmetric in the $x$ axis. Another test for such symmetry is the following (see again Fig. 17.3.2): *if the replacement of $r$ by $-r$ and $\theta$ by $\pi - \theta$ yields an equivalent equation, the graph is symmetric in the $x$ axis.* This test does apply to (17.3.2). Similarly, more than one test applies to symmetry in the $y$ axis and to symmetry in the origin (see Exercises 25 and 26, page 462).

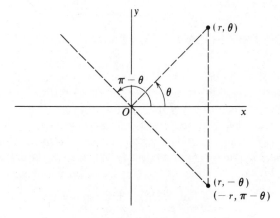

Fig. 17.3.2

Questions of *extent* are illustrated by the following. If we have the equation

$$r^2 = 1 - 2 \sin \theta,$$

then $\sin \theta \leq 1/2$ since $r^2 \geq 0$; we must also have $r^2 \leq 3$. In the equation (17.3.1) it is clear that we must have $\cos \theta \geq 0$; hence $\theta$ may not be an angle in the second or third quadrants. (However, points of the graph exist in the second and third quadrants. Why?)

The above ideas are used in the following examples.

***Example 17.3.1.*** Sketch the graph of $r = 1 - 2 \cos \theta$.

*Solution.* We have $r = -1$ when $\theta = 0$; as $\theta$ increases, $r$ is negative and increasing; at $\theta = \pi/3$ we have $r = 0$, and the line $\theta = \pi/3$ is tangent to the curve at the origin. As $\theta$ increases further, $r$ is positive and increasing, with $r = 1$ at $\theta = \pi/2$ and $r = 3$ at $\theta = \pi$. The curve is symmetric in the $x$ axis (by which test?), and we obtain the graph in Fig. 17.3.3, a *limaçon with loop* (Exercise 10 has a limaçon without loop).

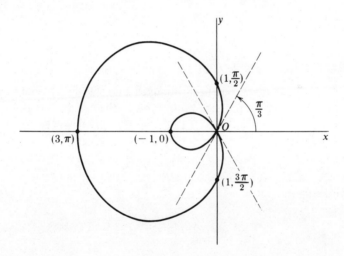

Fig. 17.3.3

***Example 17.3.2.*** Sketch the graph of $r^2 = 4 \sin 2\theta$.

*Solution.* When $\theta = 0$ we have $r = 0$, so that the line $\theta = 0$ is tangent to the curve at the origin. As $\theta$ increases, $r^2$ increases to a maximum of 4 at $\pi/2$ and decreases to 0 again at $\theta = \pi/2$. Thus the line $\theta = \pi/2$ is tangent to the curve at the origin, and since $r$ may be negative, we obtain loops in the first and third quadrants (see Fig. 17.3.4). No points of the graph exist for $\pi/2 < \theta < \pi$, since then $\pi < 2\theta < 2\pi$ and $\sin 2\theta < 0$. We also determine that the graph is symmetric in the origin (by which test?). The curve is called a *lemniscate*.

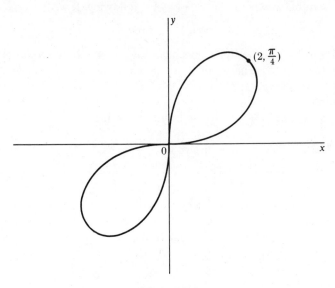

Fig. 17.3.4

## *17.4  ROTATION OF A CURVE

In Fig. 17.4.1 the curve $C$ has been obtained by rotating $C_1$ about the origin through an angle $\alpha$; thus, for each point $P$ on $C$, there is a point $P_1$ on $C_1$ such that $\measuredangle P_1 OP = \alpha$ and $|OP_1| = |OP|$. If $r = f(\theta)$ is an equation of $C_1$, we wish to find an equation of $C$. A line $\overline{Ox'}$ is drawn through $O$ making an angle $\alpha$ with $\overline{Ox}$. Then $C$ is related to $\overline{Ox'}$ as polar axis as $C_1$ is related to $\overline{Ox}$; if $\theta'$ is the angle $\overline{OP}$ makes with $\overline{Ox'}$, an equation of $C$ relative to $\overline{Ox'}$ is $r = f(\theta')$. We see from the figure that

(17.4.1) $$\theta' = \theta - \alpha.$$

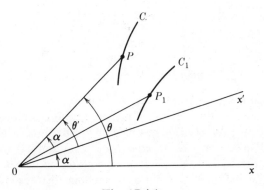

Fig. 17.4.1

Thus an equation of $C$ relative to $\overline{Ox}$ is $r = f(\theta - \alpha)$, and the following result has been proved.

**THEOREM 17.4.1.** *If a curve with polar equation $r = f(\theta)$ is rotated about the origin through an angle $\alpha$, the new curve will have an equation*

$$r = f(\theta - \alpha).$$

*Example 17.4.1.* The graph of the equation $r = 2(1 - \cos \theta)$, a cardioid, appears in Fig. 17.2.1. By Theorem 17.4.1 the graph of the equation

$$r = 2[1 - \cos(\theta - \pi/2)] = 2(1 - \sin \theta)$$

is a similar curve rotated about the origin through the angle $\pi/2$.

Similarly, rotations of the original curve through angles of $\pi$ and $3\pi/2$ give, respectively, the equations

$$r = 2[1 - \cos(\theta - \pi)] = 2(1 + \cos \theta)$$

and

$$r = 2[1 - \cos(\theta - 3\pi/2)] = 2(1 + \sin \theta).$$

Thus equations of the form

$$r = 2\left(1 \pm \begin{Bmatrix} \cos \theta \\ \sin \theta \end{Bmatrix}\right)$$

represent the same curve (*cardioid*) rotated about the origin through multiples of $\pi/2$.

In general, when the graph of an equation of the form $r = f(\cos \theta)$ is rotated through angles of $\pi/2$, $\pi$, and $3\pi/2$ about the origin, equations of the curve in the new positions are, respectively,

$$r = f(\sin \theta), \qquad r = f(-\cos \theta), \quad \text{and} \quad r = f(-\sin \theta).$$

Corresponding statements apply to curves whose equations are of the form $r = f(\sin \theta)$.

**EXERCISE GROUP 17.3**

In Exercises 1–18 sketch the graph of the given equation, with attention to symmetry, and tangents at the origin when applicable.

1. $r = 3 \sin 2\theta$
2. $r = 2 \cos 3\theta$
3. $r = -2 \cos 2\theta$
4. $r = -4 \sin 5\theta$
5. $r = 2\theta$
6. $r = -3\theta$
7. $r = 2^\theta$
8. $r = 2^{-\theta}$
9. $r = 2 + 4 \cos \theta$
10. $r = 4 - 2 \sin \theta$
11. $r^2 = 8 \cos \theta$
12. $r^2 = 8 \sin 2\theta$
13. $r = 1 + \tan \theta$
14. $r = 1 - \sec \theta$
15. $r = \sin 2\theta/3$
16. $r = \sin 3\theta/2$
17. $r = \cos 2\theta/3$
18. $r = \sin \theta/2$

In Exercises 19–24 graph the two equations on the same polar coordinate system.

**19.** $r = 1 + \cos\theta$, $r = 1 - \sin\theta$

**20.** $r = 8\sin\theta$, $r = -8\cos\theta$

**21.** $r = \dfrac{2}{2 + \cos\theta}$, $r = \dfrac{2}{2 - \cos\theta}$

**22.** $r = \dfrac{4}{1 - \sin\theta}$, $r = \dfrac{4}{1 + \cos\theta}$

**23.** $r = 2\sin\theta$, $r = 2\cos(\theta + \pi/4)$

**24.** $r = 2 - \sin\theta$, $r = 2 - \sin(\theta + \pi/3)$

**25.** Show that the graph of a polar equation is symmetric in the $y$ axis if (a) replacing $\theta$ by $\pi - \theta$ yields an equivalent equation, or if (b) replacing $\theta$ by $-\theta$ and $r$ by $-r$ yields an equivalent equation.

**26.** Show that the graph of a polar equation is symmetric in the origin or pole if (a) replacing $r$ by $-r$ yields an equivalent equation, or if (b) replacing $\theta$ by $\pi + \theta$ yields an equivalent equation.

**27.** The line $x = p$, $p > 0$, becomes $r\cos\theta = p$ in polar form. Rotate the line about the origin through an angle $\omega$, and then show that a Cartesian form of the resulting equation is $x\cos\omega + y\sin\omega = p$. This is called the *normal form of the equation of a straight line*.

**\* 28.** (a) Prove that $|x| > a/2$, $|y| > a/2$ for any point on the graph of the polar equation $r = a\csc 2\theta$.

(b) Sketch the graph of the equation $r = a\csc 2\theta$.

## 17.5 CONICS IN POLAR FORM

We shall develop equations of parabolas, ellipses, and hyperbolas in the restricted case where a focus is at the pole. Let $x = -p$ be the corresponding directrix (Fig. 17.5.1). If $e$ is the eccentricity and $P(r, \theta)$ a point on the curve, the focus-directrix property of conics (Section 8.5) gives

$$|OP| = e|MP|,$$

or

$$r = e(|MN| + |NP|) = e(p + r\cos\theta).$$

We now solve for $r$, and obtain

(17.5.1) $$r = \dfrac{ep}{1 - e\cos\theta}.$$

Equation (17.5.1) is a *standard form of the equation of a conic with focus at the pole and directrix $x = -p$*. As in Section 17.4, equation (17.5.1), with $\cos\theta$ replaced by $\sin\theta$, $-\cos\theta$, $-\sin\theta$, or in fact by $\cos(\theta - \alpha)$ for any value of $\alpha$, represents a conic with focus at the pole in a position rotated about the pole

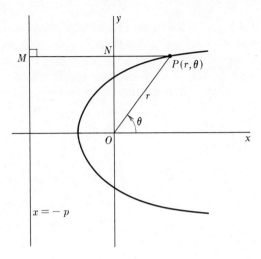

Fig. 17.5.1

from that in Fig. 17.5.1. If $e \neq 1$, the ellipse or hyperbola represented by (17.5.1) has a second focus and corresponding directrix symmetric to the previous one with respect to the center.

**Example 17.5.1.** Draw the graph of the equation

$$r = \frac{4}{2 + \cos \theta}.$$

**Solution.** From the form

$$r = \frac{2}{1 + \tfrac{1}{2} \cos \theta},$$

we see that $e = 1/2 < 1$, and the graph is an ellipse with focus at the origin. A reasonable sketch may now be made by finding those points on the curve which lie on the axes; these points appear in Fig. 17.5.2 in rectangular coordinates.

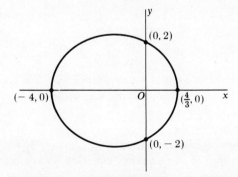

Fig. 17.5.2

Since $ep = 2$ we find $p = 4$. The directrix corresponding to the focus at the origin is the line $x = 4$ (since the ellipse is in a position rotated through an angle $\pi$ from the position in Fig. 17.5.1). The center is halfway between the vertices, at the point $(-4/3, 0)$ in rectangular coordinates. Then $a = 8/3$ (half the major axis), $c = 4/3$ (the distance between the center and $O$), and $b = \sqrt{a^2 - c^2} = 4\sqrt{3}/3$.

**Example 17.5.2.** Sketch the curve

(17.5.2) $$r = \frac{8}{1 - 2\sin\theta}.$$

*Solution.* This is a hyperbola with $e = 2$. The intercepts are shown in Fig. 17.5.3 (in rectangular form). The intercept $(0, -8/3)$ must be the vertex on the upper branch; while $(0, -8)$ is the vertex on the lower branch. We find $a = 8/3$, $c = 16/3$. With $ep = 8$ and $e = 2$ we find $p = 4$; hence the directrix corresponding to the focus at the origin is the line $y = -4$.

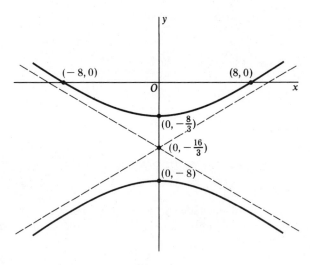

Fig. 17.5.3

We may also find the asymptotes from (17.5.2). Since $r$ is undefined for $\sin\theta = 1/2$, the asymptotes are in the directions $\pi/6$ and $5\pi/6$, and pass through the center $(0, -16/3)$.

From Exercise 26 of Exercise Group 17.3 a test for symmetry of a curve in the pole is the occurrence of an equivalent equation if $r$ is replaced by $-r$. However, *if the replacement of $r$ by $-r$ leads to a different curve, the new curve is symmetrical to the original one in the pole.* Applying this to (17.5.1) we see that the equation

$$r = -\frac{ep}{1 - e \cos \theta}$$

also represents a conic with a focus at the pole; more generally, the same is true for the equation

$$r = -\frac{ep}{1 - e \cos(\theta - \alpha)}.$$

For example, the graph of the equation $r = -4/(2 + \cos \theta)$ is symmetric to the graph of Example 17.5.1 in the pole.

Equations of other curves which are described geometrically may often be found rather directly in polar coordinates. Several such equations appear in the exercises. We shall obtain an equation of an ellipse with focus at the pole, this time using the definition of Section 7.8 rather than the focus-directrix property.

**Example 17.5.3.** Find a polar equation of a curve if the sum of the distances from any point on it to the origin and the point $F(2c, \pi)$ is $2a$ $(a > c)$.

*Solution.* Fig. 17.5.4 shows a typical point $P(r, \theta)$ on the curve. We then have

(17.5.3) $$|OP| + |FP| = 2a.$$

Now, $|OP| = r$, and $|FP|$ is found by the distance formula (Exercise 27 of Exercise Group 17.1). Then (17.5.3) becomes

$$r + \sqrt{r^2 + (2c)^2 - 2r(2c)\cos(\theta - \pi)} = 2a.$$

Subtracting $r$, using $\cos(\theta - \pi) = -\cos \theta$, squaring both sides, and solving for $r$, we get

$$r = \frac{a^2 - c^2}{a + c \cos \theta} = \frac{b^2/a}{1 + e \cos \theta}, \quad b^2 = a^2 - c^2.$$

This is one of the forms obtained earlier. Indeed, the relation $b^2/a = ep$ holds (see Exercise 13).

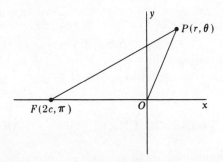

Fig. 17.5.4

**EXERCISE GROUP 17.4**

In Exercises 1–10 draw the graph of the given equation, with particular attention to the intercepts. In the case of an ellipse or hyperbola give the rectangular coordinates of the center, and for a parabola the rectangular coordinates of the vertex. In the case of a hyperbola find also the slopes of the asymptotes.

1. $r = \dfrac{2}{1 - \cos\theta}$

2. $r = -\dfrac{2}{1 + \sin\theta}$

3. $r = \dfrac{8}{3 + 4\cos\theta}$

4. $r = \dfrac{4}{4 - 3\cos\theta}$

5. $r = -\dfrac{4}{1 - 2\sin\theta}$

6. $r = \dfrac{6}{2 - \sin\theta}$

7. $r(2 + \cos\theta) + 4 = 0$

8. $r(2 - 2\sin\theta) = 5$

* 9. $r = \dfrac{3}{1 + \cos(\theta - \pi/3)}$

* 10. $r = \dfrac{5}{2 + \sin(\theta + \pi/3)}$

11. Derive directly the equation $r = ep/(1 + e\cos\theta)$ for a conic with focus at the pole and directrix $x = p$.

12. Derive directly the equation $r = ep/(1 - e\sin\theta)$ for a conic with focus at the pole and directrix $y = -p$.

13. Show that for an ellipse $p = b^2/c$.

14. Show that for a hyperbola $p = b^2/c$.

*Each of the following exercises is to be carried out in a polar coordinate system.*

15. Find an equation of the locus of a point $P$ such that $OP$ is perpendicular to the line joining $P$ to the point $(2, \pi/2)$.

16. A chord $OP$ of the circle $r = a\sin\theta$ is extended a fixed amount $b$. Find an equation of the locus of the point reached. Sketch this locus, and the circle, if $b = a$.

17. Find an equation of the locus of a point $P$ whose distance from the line $x = -2$ is twice its distance from the point $(1, 0)$ in rectangular coordinates. Identify the curve.

18. A tangent is drawn to the circle $r = 2a\cos\theta$ at the point $r = 2a$, $\theta = 0$. A line through the origin intersects the circle at $A$ and the tangent at $B$. Find an equation of the locus of the point $P$ on this line such that $|OP| = |AB|$.

19. Find an equation of the locus described in Exercise 23 of Exercise Group 16.2. Place the origin at the top of the wall.

20. Find an equation of the locus described in Exercise 17 of Exercise Group 16.2.

## 17.6 INTERSECTIONS OF CURVES IN POLAR COORDINATES

Since a point in a plane does not have a unique polar representation, the points of intersection of the graphs of two polar equations may not be obtainable as simultaneous solutions of the given equations. It was pointed out in Section 17.2 that $r = 3$ and $r = -3$ have no common solution, although they represent the same circle. Any solution of a pair of polar equations represents a point of intersection of the corresponding curves, but a point of intersection may have different polar coordinates on the two curves.

***Example 17.6.1.*** Find any intersections of the circles

$$r = 2 \cos \theta \quad \text{and} \quad r = 2\sqrt{3} \sin \theta.$$

*Solution.* For a solution of the equations we must have

$$2 \cos \theta = 2\sqrt{3} \sin \theta, \quad \text{hence} \quad \tan \theta = 1/\sqrt{3}.$$

The solution $\theta = \pi/6$ gives $r = \sqrt{3}$, and $(\sqrt{3}, \pi/6)$ is a point of intersection. Any other solution of $\tan \theta = 1/\sqrt{3}$ yields the same point, but not necessarily the same polar coordinates, for example, $(-\sqrt{3}, 7\pi/6)$ which is also point $R$ in Fig. 17.6.1. The method used does not yield the pole, which is clearly also a point of intersection. Of the possible polar representations of the pole, $(0, \pi/2)$ satisfies the first equation and $(0, 0)$ satisfies the second one.

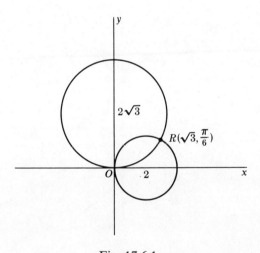

Fig. 17.6.1

Due to the unique properties of the pole, *the pole should be checked separately as a possible point of intersection of two curves.*

The key to the situation is that *one of the forms of one equation will have a common solution with the other equation for a point of intersection.* By Section 17.1, if $r = f(\theta)$ is one of the given equations, we may consider in its place any of the following equations of the same curve

$$r = f(\theta + 2n\pi) \quad \text{or} \quad -r = f(\theta + (2n + 1)\pi), \qquad n = 0, \pm 1, \pm 2, \ldots.$$

*Example 17.6.2.* Find the intersections of the parabolas

$$r = -\frac{1}{1+\cos\theta} \quad \text{and} \quad r = \frac{1}{1+\cos\theta}.$$

*Solution.* We are tempted to set $r = -r$, and get $r = 0$. However, *there is no point on either curve with $r = 0$.*

If $r$ is replaced by $-r$ and $\theta$ by $\theta + \pi$ in the first equation, we get another equation of the same curve,

$$r = \frac{1}{1-\cos\theta}.$$

Solving this equation with the second given one, we find

$$1 + \cos\theta = 1 - \cos\theta, \quad \cos\theta = 0, \quad \theta = \pi/2, 3\pi/2.$$

We have $r = 1$ for each of these values, and the points $(1, \pi/2)$ and $(1, 3\pi/2)$ are obtained.

The curves, horizontal parabolas with focus at the pole, are shown in Fig. 17.6.2. It is clear that there are only two points of intersection.

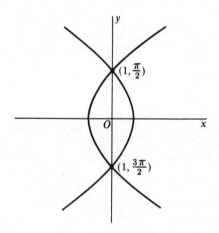

Fig. 17.6.2

**EXERCISE GROUP 17.5**

Find the points of intersection of the graphs of the given equations.

1. $r = 1 + \cos\theta$, $r = 1 - \cos\theta$
2. $r = 2 - 3\cos\theta$, $r = 2\cos\theta$
3. $r = \sin\theta$, $r = \sin 2\theta$
4. $r = \cos\theta$, $r = \sin 2\theta$
5. $r(1 + \cos\theta) = 2$, $r = \cos\theta$
6. $r(1 + \sin\theta) = 9$, $r = 4(1 + \sin\theta)$
7. $r = 1 + \cos\theta$, $r = 3\cos\theta$
8. $r = 1 + \cos\theta$, $r = 3\sin\theta$

9. $r = 2$, $r\cos\theta + \sqrt{3}\,r\sin\theta = 4$
10. $r = 1 + \cos\theta/2$, $r = -1 + \cos\theta$
11. $r(2 - \sin\theta) = 1$, $r(3 + \sin\theta) = 1$
12. $r(1 + \cos\theta) = 1$, $r(1 + \cos\theta) + 2 = 0$
13. $r = a(1 - \sin\theta)$, $r = a(\sin\theta + \cos\theta)$

## 17.7 DIRECTION OF A CURVE IN POLAR COORDINATES

The direction of a curve at a point is determined by the direction of the tangent to the curve at that point. In rectangular coordinates the direction of a curve is most conveniently given by the slope, that is, the tangent of the inclination. For a curve in polar coordinates, however, it turns out that a different angle leads to a simpler formula.

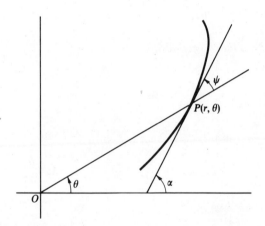

Fig. 17.7.1

Fig. 17.7.1 shows a curve, the radial angle $\theta$ for a point $P$ on the curve, the inclination $\alpha$ of the tangent to the curve at $P$, and the angle $\psi$ between the extended radial line to $P$ and the tangent line. We shall obtain a formula for $\tan\psi$. For the curve in the figure we have $\psi = \alpha - \theta$. The relation for other points or other curves may vary by a multiple of $\pi$, but in any case we have

(17.7.1) $$\tan\psi = \tan(\alpha - \theta) = \frac{\tan\alpha - \tan\theta}{1 + \tan\alpha\tan\theta}.$$

Let the curve be given by $r = f(\theta)$. By virtue of (17.1.2) and (17.1.3) the curve also has a representation in rectangular coordinates. In fact, *with $r = f(\theta)$ we may consider the equations*

(17.7.2) $$x = r\cos\theta, \qquad y = r\sin\theta,$$

*as a parametric representation of the curve in rectangular coordinates.* We then have $\tan\alpha = dy/dx$, $\tan\theta = y/x$, and from (17.7.1)

(17.7.3) $$\tan\psi = \frac{\dfrac{dy}{dx} - \dfrac{y}{x}}{1 + \dfrac{dy}{dx}\dfrac{y}{x}} = \frac{x\,dy - y\,dx}{x\,dx + y\,dy}.$$

From (17.7.2) we obtain

(17.7.4) $\quad dx = -r\sin\theta\,d\theta + \cos\theta\,dr, \qquad dy = r\cos\theta\,d\theta + \sin\theta\,dr.$

From (17.7.2) and (17.7.4) it turns out that

$$x\,dy - y\,dx = r^2\,d\theta, \qquad x\,dx + y\,dy = r\,dr,$$

and hence from (17.7.3) that

(17.7.5) $$\tan\psi = \frac{r\,d\theta}{dr} = \frac{r}{dr/d\theta} = \frac{f(\theta)}{f'(\theta)}.$$

Equation (17.7.5) *is the desired formula for* $\tan\psi$.

**Example 17.7.1.** Given the cardioid $r = 2(1 - \cos\theta)$, (a) find $\tan\psi$, (b) find $\psi$ when $\theta = 2\pi/3$.

*Solution.* (a) We find $dr/d\theta = 2\sin\theta$. Hence

$$\tan\psi = \frac{r}{dr/d\theta} = \frac{1 - \cos\theta}{\sin\theta} = \tan\frac{\theta}{2}.$$

(b) For $\theta = 2\pi/3$ we have

$$\tan\psi = \tan\pi/3 = \sqrt{3}, \qquad \psi = \pi/3.$$

The reader should examine Fig. 17.2.1 to verify that this value of $\psi$ appears reasonable.

We now illustrate the use of (17.7.5) in answering other types of question concerning the direction of a curve.

**Example 17.7.2.** Find the points on the lemniscate $r^2 = \cos 2\theta$ at which the tangent is horizontal.

*Solution.* For a horizontal tangent we have $\alpha = 0$. Hence

$$\tan\psi = \tan(-\theta) = -\tan\theta.$$

By implicit differentiation in the given equation we find

$$2r\frac{dr}{d\theta} = -2\sin 2\theta, \qquad \frac{dr}{d\theta} = -\frac{\sin 2\theta}{r}.$$

Hence
$$\tan \psi = \frac{r}{dr/d\theta} = -\frac{r^2}{\sin 2\theta} = -\frac{\cos 2\theta}{\sin 2\theta} = -\cot 2\theta.$$

Since $\tan \psi = -\tan \theta$, we obtain the equation
$$-\tan \theta = -\cot 2\theta = -\frac{1 - \tan^2 \theta}{2 \tan \theta}.$$

The resulting equation in $\tan \theta$ yields the values $\tan \theta = \pm 1/\sqrt{3}$, and $\theta = \pi/6$, $5\pi/6$, $7\pi/6$, $11\pi/6$. For each value of $\theta$ the value of $r$ is $\sqrt{2}/2$. There are thus four points at which the tangent is horizontal, as may be confirmed by examining a graph; the graph is the curve of Fig. 17.3.4 rotated $\pi/4$ in the clockwise direction (page 460).

Formula (17.7.5) is sometimes conveniently remembered by referring to the pseudo-right triangle $PQR$ in Fig. 17.7.2, with sides $r\, d\theta$ and $dr$, and angle $\psi$ as shown.

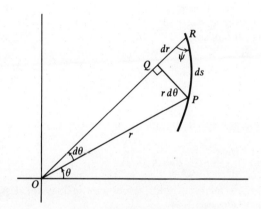

Fig. 17.7.2

## EXERCISE GROUP 17.6

In Exercises 1–8 find an angle $\psi$ for each value of $\theta$ specified. A graph can be helpful for each curve.

1. $r = 2 - \cos \theta$; $\theta = 0, \pi/2$
2. $r = 1 - 2 \cos \theta$; $\theta = \pi/3, \pi/2$
3. $r = 2/(1 + \sin \theta)$; $\theta = 0, \pi/6$
4. $r = 2/(2 - \cos \theta)$; $\theta = \pi/2, \pi$
5. $r = a \cos 2\theta$; $\theta = \pi/6, \pi/4$
6. $r^2 = a^2 \sin 2\theta$; $\theta = \pi/4, \pi/3$
7. $r = a \sin 3\theta$; $\theta = \pi/6, \pi/4$
8. $r\theta = a$; any $\theta > 0$

9. Prove that $\tan \psi = 1/b$ for the curve $r = ae^{b\theta}$.
10. Prove that $\tan \psi = \tan \theta/2$ for the cardioid $r = a(1 - \cos \theta)$.
11. Let a line through the origin intersect the cardioid $r = a(1 - \cos \theta)$ in points $A$ and $B$ other than the origin. Prove that the tangents to the curve at $A$ and $B$ are perpendicular.
12. Deduce from the formula for $\tan \psi$ that if $r = 0$ for $\theta = \theta_0$, then the line $\theta = \theta_0$ is tangent to the curve at the origin.
13. Show that the slope $m$ of a curve with polar equation $r = f(\theta)$ is

$$m = \frac{\sin \theta \frac{dr}{d\theta} + r \cos \theta}{\cos \theta \frac{dr}{d\theta} - r \sin \theta}.$$

For each of the curves in Exercises 14–17 find the slope at each indicated point.

14. $r = 1 + \cos \theta$; $\theta = \pi/3, \pi/2$
15. $r = a \sin 2\theta$; $\theta = 0, \pi/4$
16. $r^2 = a^2 \sin 2\theta$; $\theta = 0, \pi/4$
17. $r = 2/(1 + \cos \theta)$; $\theta = \pi/6, \pi/2$

18. Prove that $\tan \psi = \cot \theta$ at a point of a curve at which there is a vertical tangent.

Use Exercise 18 and the relation in Example 17.7.2, in conjunction with a graph, to find for each of the curves in Exercises 19–24: (a) the points of horizontal tangency, (b) the points of vertical tangency.

19. $r = 1 + \cos \theta$
20. $r = 2(\sin \theta + \cos \theta)$
21. $r^2 = 2 \sin 2\theta$
22. $r^2 = 2 \cos 2\theta$
23. $r = e^\theta$, $0 \leq \theta \leq 2\pi$
* 24. $r = 2 - \sin \theta$

25. Use Exercise 13 to show that conditions for a horizontal and vertical tangent are, respectively,

$$\sin \theta \frac{dr}{d\theta} + r \cos \theta = 0, \qquad \cos \theta \frac{dr}{d\theta} - r \sin \theta = 0,$$

where, if one of these relations holds, the other one does not. Use these relations to work some of Exercises 19–24.

* 26. Prove that the value of $\tan \psi$ is unchanged by a rotation of the curve about the origin.

## 17.8 AREA OF A REGION IN POLAR COORDINATES

In rectangular coordinates the basic formula for area was obtained for a region under a curve, that is, a region bounded by the $x$ axis, the lines $x = a$ and $x = b$, and the curve $y = f(x)$ with $f(x) \geq 0$ for $a \leq x \leq b$. The areas of other types of region could then be found with the use of this formula.

The basic formula for area in polar coordinates applies to a region bounded by a curve $r = f(\theta)$ and two radial lines $\theta = \alpha$, $\theta = \beta$, as in Fig. 17.8.1. We assume

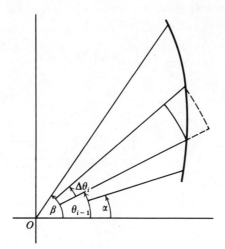

Fig. 17.8.1

that $f$ is continuous in $[\alpha, \beta]$. The interval between $\alpha$ and $\beta$ is subdivided, leading to sector-shaped subregions, one of which is shown in the figure. If $r_1$ is the smallest $r$, and $r_2$ is the largest $r$ for $\theta$ in the interval $[\theta_{i-1}, \theta_{i-1} + \Delta\theta_i]$, then for the area $\Delta A_i$ of the sector we have

$$\tfrac{1}{2}r_1^2 \,\Delta\theta_i < \Delta A_i < \tfrac{1}{2}r_2^2 \,\Delta\theta_i.$$

Hence for some $u_i$ between $r_1$ and $r_2$ we have

(17.8.1) $$\Delta A_i = \tfrac{1}{2}u_i^2 \,\Delta\theta_i.$$

If $A$ is the total area, then

$$A = \sum_{i=1}^n \Delta A_i = \frac{1}{2} \sum_{i=1}^n u_i^2 \Delta\theta_i = \frac{1}{2} \lim_{n \to \infty} \sum_{i=1}^n \mu_i^2 \,\Delta\theta_i,$$

since the $u_i$ were always chosen so that (17.8.1) held. The right member is a Riemann sum for the function $r^2 = f^2(\theta)$, and it follows that

(17.8.2) $$A = \frac{1}{2} \int_\alpha^\beta r^2 \,d\theta = \frac{1}{2} \int_\alpha^\beta f^2(\theta) \,d\theta.$$

This formula also applies if $r = 0$ for one or both of the values $\theta = \alpha$, $\theta = \beta$.

*Example 17.8.1.* Find the area of one loop of the lemniscate $r^2 = 4 \sin 2\theta$.

*Solution.* The graph appears in Fig. 17.3.4. We consider the loop in the first quadrant, for which the bounding radial lines have equations $\theta = 0$, $\theta = \pi/2$. Then

$$A = \frac{1}{2} \int_0^{\pi/2} r^2 \,d\theta = \frac{1}{2} \int_0^{\pi/2} 4 \sin 2\theta \,d\theta = -\cos 2\theta \Big|_0^{\pi/2} = 2.$$

We can just as well make use of the symmetry of the region in the line $\theta = \pi/4$. Then one-half of the area is given by

$$\frac{1}{2} A = \frac{1}{2} \int_0^{\pi/4} 4 \sin 2\theta \, d\theta = 1,$$

and again $A = 2$.

**Example 17.8.2.** Find the area of the region in the first quadrant which lies inside the curve $r = a \sin 2\theta$ and outside the circle $r = a/2$.

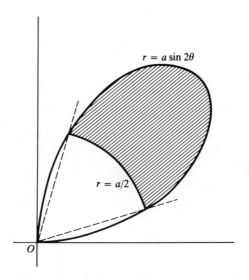

Fig. 17.8.2

*Solution.* The region is shown in Fig. 17.8.2. The intersections of the curves occur for $\theta = \pi/12$ and $\theta = 5\pi/12$. The area is the difference of two areas, each of which can be found by the basic formula (17.8.2). Then

$$A = \frac{1}{2} \int_{\pi/12}^{5\pi/12} a^2 \sin^2 2\theta \, d\theta - \frac{1}{2} \int_{\pi/12}^{5\pi/12} a^2/4 \, d\theta$$

$$= \frac{a^2}{4} \int_{\pi/12}^{5\pi/12} (1 - \cos 4\theta) \, d\theta - \pi a^2/24$$

$$= \frac{a^2}{4} \left( \theta - \frac{\sin 4\theta}{4} \right) \Big|_{\pi/12}^{5\pi/12} - \pi a^2/24$$

$$= \frac{a^2}{4} \left( \frac{\pi}{3} + \frac{\sqrt{3}}{4} \right) - \frac{\pi a^2}{24} = \frac{a^2}{48} (2\pi + 3\sqrt{3}).$$

Example 17.8.2 relates to a region bounded by the curves (Fig. 17.8.3)
$$r_1 = f(\theta), \quad r_2 = g(\theta), \quad \theta = \alpha, \quad \theta = \beta,$$
where $g(\theta) \leq f(\theta)$ for $\alpha \leq \theta \leq \beta$. In such case, in general, the area is the difference of two areas,
$$A = \frac{1}{2}\int_\alpha^\beta r_1^2 \, d\theta - \frac{1}{2}\int_\alpha^\beta r_2^2 \, d\theta.$$
The difference may be combined into one integral and we have, for the area of this type of region,
$$A = \frac{1}{2}\int_\alpha^\beta (r_1^2 - r_2^2) \, d\theta = \frac{1}{2}\int_\alpha^\beta [f^2(\theta) - g^2(\theta)] \, d\theta.$$

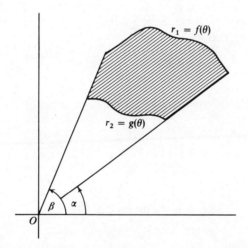

Fig. 17.8.3

## EXERCISE GROUP 17.7

Find the area of the region described in each of the following. Draw a figure in each case.

1. The region enclosed by one loop of the lemniscate $r^2 = a^2 \cos 2\theta$.
2. The region enclosed by one loop of the curve $r = 4 \sin 4\theta$.
3. The region enclosed by one loop of the curve $r = a \sin 3\theta$.
4. The region enclosed by the curve $r = a \cos 2\theta$.
5. The region enclosed by the curve $r^2 = \sin 4\theta$.
6. The region enclosed by the cardioid $r = a(1 + \cos \theta)$.
7. The larger region bounded by the line $r = \sec \theta$ and the circle $r = 4 \cos \theta$.

8. The region bounded by the parabola $r(1 - \cos \theta) = 2$ and $x = 0$.
9. The region enclosed by the curve $r = a \sin^2 \theta$.
10. The region bounded by the curve $r = 2 + \cos 2\theta$.
11. The region interior to both circles $r = a$ and $r = 2a \sin \theta$.
12. The region inside the cardioid $r = 2a(1 + \sin \theta)$ and above the parabola $r(1 + \sin \theta) = 2a$.
* 13. The region enclosed by the ellipse $r(5 - 4 \cos \theta) = 6$.
* 14. The region inside the smaller loop of the limaçon $r = 1 + 2 \sin \theta$.

## 17.9 ARC LENGTH; AREA OF A SURFACE OF REVOLUTION

Arc length for a curve given in polar coordinates may be found from the result for a curve in parametric form. Suppose the curve is given in the form $r = f(\theta)$. We then have the equations

(17.9.1.) $$\begin{cases} x = r \cos \theta = f(\theta) \cos \theta, \\ y = r \sin \theta = f(\theta) \sin \theta, \end{cases}$$

as a pair of parametric equations of the curve in terms of the parameter $\theta$, and we may use the formula

(17.9.2) $$s = \int_\alpha^\beta \left[ \left( \frac{dx}{d\theta} \right)^2 + \left( \frac{dy}{d\theta} \right)^2 \right]^{1/2} d\theta.$$

From (17.9.1) we find

$$dx/d\theta = -f(\theta) \sin \theta + f'(\theta) \cos \theta,$$
$$dy/d\theta = f(\theta) \cos \theta + f'(\theta) \sin \theta,$$

and hence

$$(dx/d\theta)^2 + (dy/d\theta)^2 = f^2(\theta) + f'^2(\theta).$$

Substituting in (17.9.2) we obtain the formula

(17.9.3) $$s = \int_\alpha^\beta [f^2(\theta) + f'^2(\theta)]^{1/2} d\theta$$

for the length of arc on the curve $r = f(\theta)$, between two points corresponding to the values $\theta = \alpha$ and $\theta = \beta$.

The evaluation of the integral in (17.9.3) may present difficulties because of the square root. In some cases where the evaluation is possible, the key is in being able to express the radicand as a perfect square. The following example shows how this is done in a particular instance.

*Example 17.9.1.* Find the length of the cardioid $r = 2(1 - \cos \theta)$.

*Solution.* The curve is shown in Fig. 17.2.1 (page 455). We have
$$f(\theta) = 2(1 - \cos\theta), \qquad f'(\theta) = 2\sin\theta.$$
Hence
$$\begin{aligned}f^2(\theta) + f'^2(\theta) &= 4(1 - \cos\theta)^2 + 4\sin^2\theta \\ &= 4(1 - 2\cos\theta + \cos^2\theta + \sin^2\theta) \\ &= 8(1 - \cos\theta) \\ &= 16\sin^2\theta/2.\end{aligned}$$

Making use of the symmetry of the curve in the $x$ axis, we have, from (17.9.3),
$$s = 2\int_0^\pi 4\sin\theta/2 \, d\theta = -16\cos\theta/2 \Big|_0^\pi = 16.$$

If (17.9.3) is applied with upper limit $\theta$,
$$s(\theta) = \int_\alpha^\theta [f^2(u) + f'^2(u)]^{1/2} \, du,$$
we find for the derivative of $s(\theta)$,

(17.9.4) $$ds/d\theta = [f^2(\theta) + f'^2(\theta)]^{1/2},$$

from which we obtain

(17.9.5) $$ds^2 = f^2(\theta) \, d\theta^2 + f'^2(\theta) \, d\theta^2.$$

We also have $dr = f'(\theta) \, d\theta$. Hence (17.9.5) leads to the equation

(17.9.6) $$ds^2 = r^2 \, d\theta^2 + dr^2$$

for the *element of arc in polar coordinates*. This formula may be remembered by referring to the pseudo-right triangle in Fig. 17.7.2.

The relation (17.9.6) may be used in other integrals in which $ds$ appears. The following example illustrates its use in finding the area of a surface of revolution.

*Example 17.9.2.* Find the area of the surface generated when the upper half of the cardioid $r = 2(1 - \cos\theta)$ is revolved about the $x$ axis.

*Solution.* The formula for area of a surface of revolution may be used,
$$S = 2\pi \int y \, ds,$$
with $y$ and $ds$ interpreted for polar coordinates. For the given cardioid we have
$$y = r\sin\theta = 2(1 - \cos\theta)\sin\theta.$$
The element of arc was found in Example 17.9.1. Hence the area $S$ is as follows:

$$S = 2\pi \int_0^\pi 2(1 - \cos\theta)\sin\theta \cdot 4\sin\theta/2 \, d\theta$$

$$= 16\pi \int_0^\pi 2\sin^2\theta/2 \cdot 2\sin\theta/2 \cos\theta/2 \cdot \sin\theta/2 \, d\theta$$

$$= 64\pi \int_0^\pi \sin^4\theta/2 \cos\theta/2 \, d\theta = \frac{128\pi}{5} \sin^5\theta/2 \Big|_0^\pi = \frac{128\pi}{5}.$$

It is convenient at this time to find the *curvature* for the curve $r = f(\theta)$. Referring to Fig. 17.7.1, we see that $\alpha = \psi + \theta$, and hence

$$\frac{d\alpha}{d\theta} = \frac{d\psi}{d\theta} + 1.$$

Now $\psi = \tan^{-1}[r/(dr/d\theta)]$. Hence

$$\frac{d\psi}{d\theta} = \frac{r_\theta'^2 - rr_\theta''}{r^2 + r_\theta'^2}, \qquad \frac{d\alpha}{d\theta} = \frac{r^2 + 2r_\theta'^2 - rr_\theta''}{r^2 + r_\theta'^2}.$$

We now use the relation

$$\frac{d\alpha}{ds} = \frac{d\alpha/d\theta}{ds/d\theta},$$

and (17.9.4) for $ds/d\theta$, to obtain for the *curvature k* of a curve in polar form, the result

(17.9.7) $$k = \left|\frac{d\alpha}{ds}\right| = \frac{|r^2 + 2r_\theta'^2 - rr_\theta''|}{(r^2 + r_\theta'^2)^{3/2}}.$$

### EXERCISE GROUP 17.8

In Exercises 1–5 find the length of the curve as indicated.

1. $r = 2\cos\theta$
2. $r = 4\sin\theta$
3. $r = 3\sec\theta$, $0 \le \theta \le \pi/3$
4. $r = e^\theta$, $0 \le \theta \le \pi$
5. $r = 1/(\sin\theta + \cos\theta)$, (a) for $0 \le \theta \le \pi/6$, (b) for $0 \le \theta \le \pi/4$

In Exercises 6–12 find the area of the surface of revolution, generated as indicated.

6. By revolving the circle $r = 2\sin\theta$ about the $x$ axis
7. By revolving the upper half of $r = 2\sin\theta$ about the $x$ axis
8. By revolving $r = 6\sec\theta$, $0 \le \theta \le \pi/4$, about the $y$ axis
9. By revolving $r = e^\theta$, $0 \le \theta \le \pi$, about the $x$ axis
10. By revolving $r(\sin\theta + \cos\theta) = 1$, $0 \le \theta \le \pi/2$, about the $x$ axis
11. By revolving $r = a(1 + \cos\theta)$ about the $x$ axis
12. By revolving $r(1 + \cos\theta) = 1$, $0 \le \theta \le \pi/2$, about the $x$ axis

Find the curvature of each of the following curves.

13. $r = a(1 - \cos\theta)$
14. $r = 2a\cos^2 \theta/2$
15. $r^2 = a^2 \sin 2\theta$
16. $r = \dfrac{4a}{1 - \sin\theta}$

## * 17.10 VELOCITY AND ACCELERATION IN POLAR COORDINATES

Suppose that the path of a particle, moving in a plane, is given in polar form,

$$r = f(\theta), \qquad \theta = \theta(t), \qquad (t \text{ denotes time}).$$

Then the equations

(17.10.1) $\qquad x = f(\theta)\cos\theta, \qquad y = f(\theta)\sin\theta, \qquad \theta = \theta(t),$

represent the path parametrically, and the methods of Section 16.6 may be used to study the velocity and acceleration.

We may also study the motion in terms of a different pair of unit, orthogonal vectors, a unit vector $\mathbf{u}_r$ in the radial direction, that of $\overrightarrow{OP}$ (Fig. 17.10.1), and a unit vector $\mathbf{u}_\theta$ perpendicular to $\mathbf{u}_r$. The vector $\mathbf{u}_r$ makes an angle $\theta$, and $\mathbf{u}_\theta$ is taken as making an angle $\theta + \pi/2$, with the polar axis. Hence we have

$$\mathbf{u}_r = \cos\theta\,\mathbf{i} + \sin\theta\,\mathbf{j}, \qquad \mathbf{u}_\theta = -\sin\theta\,\mathbf{i} + \cos\theta\,\mathbf{j}.$$

By direct differentiation we obtain

(17.10.2) $\qquad d\mathbf{u}_r/d\theta = \mathbf{u}_\theta, \qquad d\mathbf{u}_\theta/d\theta = -\mathbf{u}_r.$

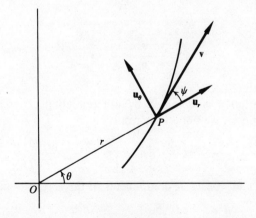

Fig. 17.10.1

Any vector in the plane may be expressed in terms of its components in the directions of $\mathbf{u}_r$ and $\mathbf{u}_\theta$. In particular, the velocity and acceleration may be written as

$$\mathbf{v} = v_r \mathbf{u}_r + v_\theta \mathbf{u}_\theta, \qquad \mathbf{a} = a_r \mathbf{u}_r + a_\theta \mathbf{u}_\theta.$$

We obtain these representations directly by differentiating the position vector $\mathbf{R}$, which we express as

$$\mathbf{R} = r\mathbf{u}_r.$$

Thus

$$\mathbf{v} = d\mathbf{R}/dt = r(d\mathbf{u}_r/dt) + (dr/dt)\mathbf{u}_r.$$

But $d\mathbf{u}_r/dt = (d\mathbf{u}_r/d\theta)(d\theta/dt) = \mathbf{u}_\theta (d\theta/dt)$, by (17.10.2), and we have

(17.10.3) $$\mathbf{v} = (dr/dt)\mathbf{u}_r + (r\, d\theta/dt)\mathbf{u}_\theta.$$

This representation shows that the *radial* and *transverse* components $v_r$ and $v_\theta$ of the velocity are, respectively,

$$v_r = \frac{dr}{dt} \quad \text{and} \quad v_\theta = r\frac{d\theta}{dt}.$$

Since $ds/dt = |\mathbf{v}|$, we obtain from (17.10.3) the relation

$$|\mathbf{v}|^2 = \left(\frac{ds}{dt}\right)^2 = \left(\frac{dr}{dt}\right)^2 + \left(r\frac{d\theta}{dt}\right)^2,$$

which is in accordance with previous results for arc length in polar coordinates. Further, (see Fig. 17.10.1) $\mathbf{v}$ is tangent to the curve at $P$; it makes an angle $\psi$ with $\mathbf{u}_r$, and an angle $\pi/2 - \psi$ with $\mathbf{u}_\theta$. Hence

$$\cos(\pi/2 - \psi) = \sin\psi = \frac{\mathbf{v}\cdot\mathbf{u}_\theta}{|\mathbf{v}|}, \qquad \cos\psi = \frac{\mathbf{v}\cdot\mathbf{u}_r}{|\mathbf{v}|},$$

and

$$\tan\psi = \frac{\mathbf{v}\cdot\mathbf{u}_\theta}{\mathbf{v}\cdot\mathbf{u}_r} = \frac{v_\theta}{v_r} = \frac{r\, d\theta/dt}{dr/dt},$$

in accordance with (17.7.5).

The acceleration $\mathbf{a}$ is obtained by differentiating (17.10.3). Leaving the details to the exercises, we obtain the result

(17.10.4) $$\mathbf{a} = \left[\frac{d^2r}{dt^2} - r\left(\frac{d\theta}{dt}\right)^2\right]\mathbf{u}_r + \left[r\frac{d^2\theta}{dt^2} + 2\frac{dr}{dt}\frac{d\theta}{dt}\right]\mathbf{u}_\theta,$$

so that the *radial* and *transverse* components of the acceleration are

$$a_r = \frac{d^2r}{dt^2} - r\left(\frac{d\theta}{dt}\right)^2, \qquad a_\theta = r\frac{d^2\theta}{dt^2} + 2\frac{dr}{dt}\frac{d\theta}{dt}.$$

*Example 17.10.1.* Express the velocity and acceleration for the motion represented by

$$r = \sin^2 \theta, \qquad d\theta/dt = 2$$

in terms of their radial and transverse components.

*Solution.* We may use (17.10.3) and (17.10.4) directly, but we shall use the method by which these formulas were derived, utilizing (17.10.2) as needed. We start with the relation

$$\mathbf{R} = r\mathbf{u}_r = \sin^2 \theta \, \mathbf{u}_r.$$

We then differentiate, to get

$$\mathbf{v} = \frac{d\mathbf{R}}{dt} = \sin^2 \theta \frac{d\mathbf{u}_r}{d\theta} \frac{d\theta}{dt} + 2 \sin \theta \cos \theta \frac{d\theta}{dt} \mathbf{u}_r$$

$$= 2 \sin 2\theta \, \mathbf{u}_r + 2 \sin^2 \theta \, \mathbf{u}_\theta.$$

Differentiating again, we obtain

$$\mathbf{a} = 2\left( \sin 2\theta \frac{d\mathbf{u}_r}{d\theta} \frac{d\theta}{dt} + 2 \cos 2\theta \frac{d\theta}{dt} \mathbf{u}_r \right)$$

$$+ 2\left( \sin^2 \theta \frac{d\mathbf{u}_\theta}{d\theta} \frac{d\theta}{dt} + 2 \sin \theta \cos \theta \frac{d\theta}{dt} \mathbf{u}_\theta \right)$$

$$= 2(2 \sin 2\theta \, \mathbf{u}_\theta + 4 \cos 2\theta \, \mathbf{u}_r) + 2(-2 \sin^2 \theta \, \mathbf{u}_r + 2 \sin 2\theta \, \mathbf{u}_\theta)$$

$$= 4(2 \cos 2\theta - \sin^2 \theta)\mathbf{u}_r + 8 \sin 2\theta \, \mathbf{u}_\theta.$$

The reader may verify that (17.10.3) and (17.10.4) yield the same results.

**EXERCISE GROUP 17.9**

In Exercises 1–3 find **v** and **a** in terms of $\mathbf{u}_r$ and $\mathbf{u}_\theta$ for the motion described.

1. $r = 1 - \cos \theta$, $d\theta/dt = 2$, for $\theta = \pi/2$
2. $r = \sin 3\theta$, $\theta = t$, at $t = \pi/2$
3. $r = \dfrac{2}{2 + \sin \theta}$, $d\theta/dt = 3$, for $\theta = 0$

4. (a) Find $a_r$ and $a_\theta$, if a particle moves along the circle $r = a$.
   (b) What may be said additionally, if the angular speed $d\theta/dt$ is constant?
5. For a particle moving along the cardioid $r = 1 + \sin \theta$, it is known that $d\theta/dt = 2$ and $d^2\theta/dt^2 = -2$ at the point for which $\theta = \pi/3$. Find $a_r$ and $a_\theta$ at that point.
6. Find **v** and **a** for the motion given by $r = e^{2\theta}$, $\theta = t$.
7. If $r^2 \, d\theta/dt$ is constant and $r \neq 0$, prove that $a_\theta = 0$.
8. Derive (17.10.4) by differentiating (17.10.3).

* **9.** The following is an alternate derivation of (17.10.3) and (17.10.4). Using the equations $x = r\cos\theta$, $y = r\sin\theta$, where $r = f(\theta)$, $\theta = \theta(t)$,

(a) show that

$$v_x = \cos\theta \frac{dr}{dt} - r\sin\theta \frac{d\theta}{dt}, \quad v_y = \sin\theta \frac{dr}{dt} + r\cos\theta \frac{d\theta}{dt}.$$

Then find $v_r$ and $v_\theta$ as components of **v** in the directions of $\mathbf{u}_r$ and $\mathbf{u}_\theta$.

(b) Show that

$$a_x = \cos\theta \frac{d^2r}{dt^2} - r\sin\theta \frac{d^2\theta}{dt^2} - 2\sin\theta \frac{dr}{dt}\frac{d\theta}{dt} - r\cos\theta \left(\frac{d\theta}{dt}\right)^2,$$

$$a_y = \sin\theta \frac{d^2r}{dt^2} + r\cos\theta \frac{d^2\theta}{dt^2} + 2\cos\theta \frac{dr}{dt}\frac{d\theta}{dt} - r\sin\theta \left(\frac{d\theta}{dt}\right)^2.$$

Then find $a_r$ and $a_\theta$ as components of **a** in the directions of $\mathbf{u}_r$ and $\mathbf{u}_\theta$.

# 18  Vectors in Three Dimensions

## 18.1   THE RECTANGULAR COORDINATE SYSTEM

A rectangular coordinate system is formed by three mutually perpendicular planes; each plane is perpendicular to the other two. These *coordinate planes* meet in three mutually perpendicular straight lines, the *coordinate axes*. The intersection of coordinate axes is the *origin*. We shall use a right-handed system, as in Fig. 18.1.1. The name signifies that if the fingers of the right hand, other than the thumb, point in the direction of rotation from the positive $x$ axis to the positive $y$ axis, the thumb points in the direction of the positive $z$ axis. In the Fig. 18.1.1 the positive direction of the $x$ axis is toward the viewer; the positive direction of the $y$ axis is to the right; and the positive direction of the $z$ axis is up. The plane containing the $x$ axis and the $y$ axis is the $xy$ plane; the $xz$ and $yz$ planes are defined similarly. A point in space is specified by three *coordinates* $(x, y, z)$, which are the perpendicular distances from the $yz$ plane, $xz$ plane, and $xy$ plane, respectively. The $x$ coordinate of a point is positive if the perpendicular line segment from the $yz$ plane to the point is in the same sense as the positive $x$ direction. We define the signs of the $y$ and $z$ coordinates similarly.

Fig. 18.1.1 shows two particular points, with an indication of how they are reached from the origin. Care is required in plotting points, since the use of a two-dimensional drawing for a three-dimensional configuration means that the placing of a point in the drawing does not determine its position in space, without some additional indication.

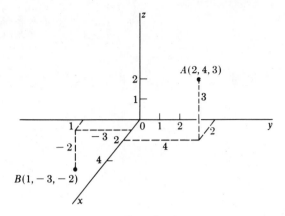

Fig. 18.1.1

The three coordinate planes separate the points in space which are not in these planes into eight *octants*, each of which is characterized by the signs of the coordinates of a point in it. For example, point $A$ (2, 4, 3) in Fig. 18.1.1 is in the first octant, where the coordinates may be described as $(+, +, +)$. Octants other than the first are usually not numbered. Point $B\,(1, -3, -2)$ is in the octant described as $(+, -, -)$, the octant which is to the left of, and below, the first octant.

## 18.2  THREE-DIMENSIONAL VECTORS

The algebraic development of vectors in space follows so closely that of vectors in a plane, that it may be abbreviated. It is suggested that Sections 1.7 and 1.8 be reviewed here. The proofs of most of the theorems parallel those of the earlier sections and thus will be left for the exercises.

**DEFINITION 18.2.1.** *A* **three-dimensional vector a** *is an ordered triple of real numbers* $\mathbf{a} = (a_1, a_2, a_3)$ *such that*:

1. *If* $\mathbf{a} = (a_1, a_2, a_3)$ *and* $\mathbf{b} = (b_1, b_2, b_3)$, *then*

   $$\mathbf{a} = \mathbf{b} \Leftrightarrow a_1 = b_1,\ a_2 = b_2,\ \text{and}\ a_3 = b_3.$$

2. *The* **sum** *of the vectors* **a** *and* **b** *is*

   $$\mathbf{a} + \mathbf{b} = (a_1 + b_1,\ a_2 + b_2,\ a_3 + b_3).$$

3. *If c is any real number, then*

   $$c\mathbf{a} = (ca_1, ca_2, ca_3).$$

**DEFINITION 18.2.2.**   1.  *The **negative** of the vector* $\mathbf{a} = (a_1, a_2, a_3)$ *is*
$$-\mathbf{a} = (-1)\mathbf{a} = (-a_1, -a_2, -a_3).$$
2. *The **difference** of* $\mathbf{a} = (a_1, a_2, a_3)$ *and* $\mathbf{b} = (b_1, b_2, b_3)$ *is*
$$\mathbf{a} - \mathbf{b} = \mathbf{a} + (-\mathbf{b}) = (a_1 - b_1, a_2 - b_2, a_3 - b_3).$$
3. *The **magnitude** of* $\mathbf{a}$ *is*
$$|\mathbf{a}| = \sqrt{a_1^2 + a_2^2 + a_3^2}$$
4. $\mathbf{a}$ *is a **unit vector** if* $|\mathbf{a}| = 1$.
5. *The **zero vector** is*
$$\mathbf{0} = (0, 0, 0).$$

**THEOREM 18.2.1.**   (a)  $\mathbf{a} + \mathbf{b} = \mathbf{b} + \mathbf{a}$,
(b)  $(\mathbf{a} + \mathbf{b}) + \mathbf{c} = \mathbf{a} + (\mathbf{b} + \mathbf{c})$,
(c)  $c(\mathbf{a} + \mathbf{b}) = c\mathbf{a} + c\mathbf{b}$,
(d)  $\mathbf{a} = \mathbf{0} \Leftrightarrow |\mathbf{a}| = 0$.

**DEFINITION 18.2.3.**   *To **normalize** a vector* $\mathbf{a}$ *means to find a vector* $\mathbf{u} = c\mathbf{a}$, *where c is a real number, such that* $|\mathbf{u}| = 1$.

It follows, as in Exercise 23 of Exercise Group 1.4, that if $\mathbf{a} \neq \mathbf{0}$, a normalization of $\mathbf{a}$ is $\mathbf{a}/|\mathbf{a}|$ or $-\mathbf{a}/|\mathbf{a}|$.

We define three *base vectors* $\mathbf{i}$, $\mathbf{j}$, and $\mathbf{k}$ as follows:
$$\mathbf{i} = (1, 0, 0), \quad \mathbf{j} = (0, 1, 0), \quad \mathbf{k} = (0, 0, 1).$$
Then $\mathbf{i}$, $\mathbf{j}$, and $\mathbf{k}$ are unit vectors.

**THEOREM 18.2.2.**   *Any vector* $\mathbf{a} = (a_1, a_2, a_3)$ *may be expressed in the form* $\mathbf{a} = a_1\mathbf{i} + a_2\mathbf{j} + a_3\mathbf{k}$.

Two-dimensional vectors may be considered as special cases of three-dimensional vectors in the following sense: if the two-dimensional vector $(a_1, a_2)$ is made to correspond to the three-dimensional vector $(a_1, a_2, 0)$ then all properties of, and operations with, the two-dimensional vectors correspond exactly to the same properties and operations for those particular three-dimensional vectors.

## 18.3   GEOMETRICAL INTERPRETATION OF VECTORS; DISTANCE FORMULA

The geometrical interpretation of vectors in three dimensions carries over from Section 1.9 with very little change. There are now three components, which are the components in the x, y, and z directions, respectively. In Fig. 18.3.1 let the directed line segment $\overline{P_1P_2}$ represent the vector $\mathbf{a} = (a_1, a_2, a_3)$. Then

$$a_1 = \overline{P_1M}, \qquad a_2 = \overline{P_1N}, \qquad a_3 = \overline{P_1Q}.$$

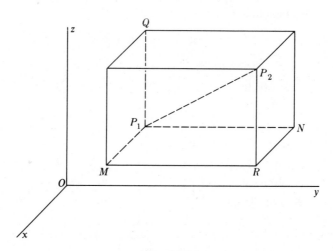

Fig. 18.3.1

Let $d$ denote the length of the line segment $P_1P_2$. Then by the Theorem of Pythagoras we have

$$d^2 = |P_1R|^2 + |RP_2|^2 = (|P_1M|^2 + |MR|^2) + |RP_2|^2.$$

Since $\overline{P_1M} = a_1$, $\overline{MR} = \overline{P_1N} = a_2$, $\overline{RP_2} = \overline{P_1Q} = a_3$, we get

(18.3.1) $$d^2 = a_1^2 + a_2^2 + a_3^2.$$

A comparison with Definition 18.2.2 shows that *the magnitude of the vector* $\mathbf{a}$ *is the length of any directed line segment which represents* $\mathbf{a}$.

If $P_1$ and $P_2$ have the coordinates $(x_1, y_1, z_1)$ and $(x_2, y_2, z_2)$, respectively, it is clear that

$$a_1 = x_2 - x_1, \qquad a_2 = y_2 - y_1, \qquad a_3 = z_2 - z_1.$$

The vector from $P_1$ to $P_2$ may then be written as

(18.3.2) $$\overrightarrow{P_1P_2} = (x_2 - x_1, y_2 - y_1, z_2 - z_1).$$

Combining the above results, we have

(18.3.3) $$d = |P_1P_2| = \sqrt{(x_2 - x_1)^2 + (y_2 - y_1)^2 + (z_2 - z_1)^2}$$

as the *formula for the distance between the points* $P_1(x_1, y_1, z_1)$ *and* $P_2(x_2, y_2, z_2)$.

***Example 18.3.1.*** Show that the points $A(-4, 3, 2)$, $B(0, 1, 4)$, and $C(-6, 4, 1)$ lie on a straight line.

*Solution.* It is sufficient to show that the length of one of the segments formed by the points is equal to the sum of the other two (why?). By the distance formula (18.3.3) we find

$$|AB| = \sqrt{(0+4)^2 + (1-3)^2 + (4-2)^2} = \sqrt{16+4+4} = 2\sqrt{6},$$
$$|BC| = \sqrt{36+9+9} = 3\sqrt{6}, \qquad |AC| = \sqrt{4+1+1} = \sqrt{6}.$$

Hence $|BC| = |AB| + |AC|$. Thus the points lie on a straight line, with $A$ located between $B$ and $C$.

**Example 18.3.2.** Show that the points $A(2, 4, -1)$, $B(3, 2, 4)$, and $C(5, 13, 8)$ are vertices of a right triangle.

*Solution.* It is sufficient to show that the Pythagorean relation holds for the triangle. Using (18.3.3), we find

$$|AB|^2 = (3-2)^2 + (2-4)^2 + (4+1)^2 = 1 + 4 + 25 = 30,$$
$$|AC|^2 = 9 + 81 + 81 = 171, \qquad |BC|^2 = 4 + 121 + 16 = 141.$$

Hence $|AC|^2 = |AB|^2 + |BC|^2$. Thus the triangle is a right triangle, with the right angle at $B$.

The *triangle law* and *parallelogram law for addition of vectors* (Section 1.9) apply as well to three-dimensional vectors.

It is also clear that *two nonzero vectors are parallel if and only if one of them is a multiple of the other.*

The distance formula leads immediately to an equation of the *sphere*, which is the locus of points $(x, y, z)$ at a fixed distance $a$ from a fixed point $(x_0, y_0, z_0)$. Thus the *standard equation of the sphere* is

(18.3.4) $$(x - x_0)^2 + (y - y_0)^2 + (z - z_0)^2 = a^2.$$

In expanded form the equation becomes

$$x^2 + y^2 + z^2 + Ax + By + Cz + D = 0;$$

this is a second-degree equation in $x$, $y$, and $z$ with no mixed product terms.

**Example 18.3.3.** Find the center and radius of the sphere

$$x^2 + y^2 + z^2 - 8x + 3z - 2 = 0.$$

*Solution.* Conversion to standard form is by completing squares in each variable. We get

$$(x^2 - 8x + 16) + y^2 + (z^2 + 3z + 9/4) = 2 + 16 + 9/4,$$

or

$$(x - 4)^2 + y^2 + (z + 3/2)^2 = 81/4.$$

Hence the center is $(4, 0, -3/2)$, and the radius is $9/2$.

## EXERCISE GROUP 18.1

The questions in Exercises 1–8 apply to the vectors

$$\mathbf{a} = 2\mathbf{i} - 5\mathbf{j} + \mathbf{k}, \qquad \mathbf{b} = -3\mathbf{i} + 3\mathbf{j} - 2\mathbf{k}, \qquad \mathbf{c} = 5\mathbf{i} + 3\mathbf{j}.$$

1. Find (a) $\mathbf{a} - \mathbf{b} + 2\mathbf{c}$, (b) $|\mathbf{a} - \mathbf{b} + 2\mathbf{c}|$.
2. Find (a) $2\mathbf{a} + 3\mathbf{b} - \mathbf{c}$, (b) $|2\mathbf{a} + 3\mathbf{b} - \mathbf{c}|$.
3. Normalize (a) $\mathbf{a}$, (b) $\mathbf{b}$, (c) $\mathbf{a} + \mathbf{b}$.
4. Normalize (a) $\mathbf{a} + \mathbf{c}$, (b) $\mathbf{a} - \mathbf{c}$, (c) $\mathbf{a} - \mathbf{b} + 2\mathbf{c}$.
5. Find $\mathbf{u}$ if $\mathbf{u} + \mathbf{a} = \mathbf{b} - \mathbf{c}$.
6. Find $\mathbf{u}$ if $3\mathbf{u} + 2\mathbf{a} = \mathbf{b}$.
7. Find $\mathbf{u}$ if $\mathbf{a} + 2\mathbf{u} = 3\mathbf{b}$.
8. Find $\mathbf{u}$ if $\mathbf{a} + 2\mathbf{b} + 3\mathbf{c} + 4\mathbf{u} = \mathbf{0}$.

9. Show that the points $(4, 2, 1)$, $(1, -4, 10)$, and $(5, 4, -2)$ lie on a straight line.
10. Show that the points $(2, 0, -1)$, $(3/2, -3/2, 0)$, and $(-1/2, -15/2, 4)$ lie on a straight line.
11. Show that the points $(4, -2, 1)$, $(8, 8, 9)$, and $(1, -8, 10)$ are vertices of a right triangle.
12. Show that the points $(\sqrt{2}, 1, 1)$, $(1 + \sqrt{2}, 4, -1)$, and $(-3 + \sqrt{2}, 0, -2)$ are vertices of a right triangle.
13. Show that the points $(3, 1, -1)$, $(5, -2, 1)$, and $(-1, 1, 0)$ are vertices of an isosceles triangle.
14. Show that the points $(4, 1, 4)$, $(2, -2, -3)$, and $(1, 4, 2)$ are vertices of an isosceles triangle.
15. Show that the points $A(-2, 4, 3)$, $B(4, 2, -1)$, $C(0, 3, -2)$, and $D(-6, 5, 2)$ are vertices of a parallelogram.
16. Show that the points $A(-1, 3, 2)$, $B(3, -3, 4)$, $C(5, 0, 3)$, and $D(1, 6, 1)$ are vertices of a parallelogram.
17. Verify that $\overrightarrow{AB} + \overrightarrow{BC} + \overrightarrow{CA} = \mathbf{0}$ for the points $A(3, 4, -1)$, $B(2, -7, 4)$, and $C(-4, 3, -4)$.
18. Verify that $\overrightarrow{AB} + \overrightarrow{BC} + \overrightarrow{CD} + \overrightarrow{DA} = \mathbf{0}$ for the points $A(1, 0, 4)$, $B(2, -3, 0)$, $C(4, -7, 2)$, and $D(0, -3, -2)$.
19. Find the vector sum $\overrightarrow{QA} + \overrightarrow{QB} + \overrightarrow{QC} + \overrightarrow{QD}$ for the points $A(1, 3, 4)$, $B(5, -7, 8)$, $C(5, 9, 15)$, $D(3, 17, -3)$, and $Q(-3, 7, -3)$.
20. (a) Show geometrically that $|\mathbf{a} + \mathbf{b}| \leq |\mathbf{a}| + |\mathbf{b}|$.
    (b) Write out the equivalent inequality in terms of the components of $\mathbf{a}$ and $\mathbf{b}$, for $\mathbf{a} = (a_1, a_2, a_3)$ and $\mathbf{b} = (b_1, b_2, b_3)$.
21. Show that $|c\mathbf{a}| = |c|\,|\mathbf{a}|$, where $c$ is a real number.
22. Prove that $|\mathbf{a}| = 0 \Leftrightarrow \mathbf{a} = \mathbf{0}$.
23. Prove Theorem 18.2.1, part (b).

24. Prove Theorem 18.2.1, part (c).
25. Prove Theorem 18.2.2.
26. Find the center and radius of the sphere $x^2 + y^2 + z^2 - 2x - 8y + 4z + 5 = 0$.
27. Find the center and radius of the sphere $8x^2 + 8y^2 + 8z^2 + 8x - 12y + 4z = 1$.
28. What is the locus of the equation $36x^2 + 36y^2 + 36z^2 - 48x - 36y + 25 = 0$?

## * 18.4 THE MIDPOINT FORMULA; POINT OF DIVISION FORMULAS

The derivation of the midpoint formula is entirely analogous to that in Section 2.1, and leads to the following result:

*The midpoint of the line segment joining $P_1(x_1, y_1, z_1)$ and $P_2(x_2, y_2, z_2)$ has the coordinates*

(18.4.1) $\qquad [\tfrac{1}{2}(x_1 + x_2), \tfrac{1}{2}(y_1 + y_2), \tfrac{1}{2}(z_1 + z_2)].$

***Example 18.4.1.*** In the triangle with vertices $A(2, 4, -1)$, $B(3, 2, 4)$, and $C(5, 13, 8)$ show that the midpoint of $AC$ is equidistant from all three vertices.

***Solution.*** By (18.4.1) the midpoint $M$ of $AC$ has coordinates

$$[\tfrac{1}{2}(2 + 5), \tfrac{1}{2}(4 + 13), \tfrac{1}{2}(-1 + 8)].$$

Hence the midpoint is $M(7/2, 17/2, 7/2)$. It is sufficient to show that $|AM|^2 = |BM|^2$ (why?). By the distance formula (18.3.2) we find

$$|AM|^2 = \left(\frac{7}{2} - 2\right)^2 + \left(\frac{17}{2} - 4\right)^2 + \left(\frac{7}{2} + 1\right)^2 = \frac{9}{4} + \frac{81}{4} + \frac{81}{4} = \frac{171}{4},$$

$$|BM|^2 = \left(\frac{7}{2} - 3\right)^2 + \left(\frac{17}{2} - 2\right)^2 + \left(\frac{7}{2} - 4\right)^2 = \frac{1}{4} + \frac{169}{4} + \frac{1}{4} = \frac{171}{4}.$$

Thus $|AM|^2 = |BM|^2$, and the proof is complete.

A point $P$ is said to divide a directed line segment $\overline{P_1P_2}$ in the ratio $r_1 : r_2$ if

(18.4.2) $\qquad r_2 \overrightarrow{P_1P} = r_1 \overrightarrow{PP_2}.$

This equation implies that $P$ is on the straight line containing $P_1$ and $P_2$. The position of $P$ on the line depends on the ratio $r_1 : r_2$, and will be considered in the exercises.

Let $O$ be any point, called a *reference point*. From Fig. 18.4.1 we have

$$\overrightarrow{P_1P} = \overrightarrow{OP} - \overrightarrow{OP_1}, \qquad \overrightarrow{PP_2} = \overrightarrow{OP_2} - \overrightarrow{OP}.$$

Substituting these in (18.4.2) we get

$$r_2(\overrightarrow{OP} - \overrightarrow{OP_1}) = r_1(\overrightarrow{OP_2} - \overrightarrow{OP}),$$

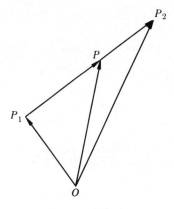

Fig. 18.4.1

which may be solved for $\overrightarrow{OP}$ to give

(18.4.3) $$\overrightarrow{OP} = \frac{r_2 \overrightarrow{OP_1} + r_1 \overrightarrow{OP_2}}{r_1 + r_2}.$$

Equation (18.4.3) is called the *point of division formula*.

If the points are $P_1(x_1, y_1, z_1)$ and $P_2(x_2, y_2, z_2)$, the coordinates of the point $P$ which divides $\overline{P_1 P_2}$ in the ratio $r_1 : r_2$ are

(18.4.4) $$x = \frac{r_2 x_1 + r_1 x_2}{r_1 + r_2}, \qquad y = \frac{r_2 y_1 + r_1 y_2}{r_1 + r_2}, \qquad z = \frac{r_2 z_1 + r_1 z_2}{r_1 + r_2}.$$

(See Exercise 10.)

The midpoint formula (18.4.1) is the special case of (18.4.4) where $r_1 = r_2$.

**Example 18.4.2.** Find the points which trisect the line segment joining the points $A(2, 3, 4)$ and $B(-1, 4, 2)$.

*Solution.* The points of trisection are obtained from (18.4.4) with $r_1 : r_2 = 1 : 2$ and $r_1 : r_2 = 2 : 1$. For the first ratio we may take $r_1 = 1$ and $r_2 = 2$. We get

$$x = \frac{2 \cdot 2 + 1 \cdot (-1)}{1 + 2} = 1, \qquad y = \frac{2 \cdot 3 + 1 \cdot 4}{1 + 2} = \frac{10}{3}, \qquad z = \frac{2 \cdot 4 + 1 \cdot 2}{1 + 2} = \frac{10}{3}.$$

The second ratio, similarly, gives the point $(0, 11/3, 8/3)$.

### ∗ EXERCISE GROUP 18.2

1. Find the lengths of the medians of the triangle with vertices $(2, 3, -1)$, $(0, 2, 1)$, and $(4, 0, 2)$.

2. Given the points $A(1, 2, 2)$, $B(4, 7, 0)$, $C(5, 1, 4)$, and $D(1, 0, -2)$, let $P$, $Q$, $R$, and $S$ be the midpoints of $AB$, $BC$, $CD$, and $DA$, respectively. Show that $\overrightarrow{PQ} = \overrightarrow{SR}$.

3. Let $A$, $B$, and $C$ be any three points in space. If $M$ and $N$ are the midpoints of $AC$ and $BC$, respectively, show that $\overrightarrow{MN} = \frac{1}{2}\overrightarrow{AB}$.

4. Prove the property stated in Exercise 2 if $A$, $B$, $C$, and $D$ are any four points in space.

5. Given the points $P_1(2, -3, 2)$, $P_2(3, -2, 3)$, find the point which divides $\overrightarrow{P_1 P_2}$ in the ratio (a) $1:2$, (b) $2:-1$, (c) $-1:3$, (d) $4:0$.

6. Given the points $P_1(2, 4, 2)$, $P_2(-2, 5, -1)$, find the point $P$ such that $\overrightarrow{P_1 P} = t\overrightarrow{P_1 P_2}$ if (a) $t=1$, (b) $t=0$, (c) $t=-1/3$, (d) $t=-3$.

7. Find the points which trisect the line segment joining the points $(-1, 3, 4)$ and $(2, -1, 2)$.

8. Find the points which divide the line segment joining the points $(4, 7, -3)$ and $(14, -8, 2)$ into fourths.

9. Given the points $A(4, 2, 1)$, $B(2, -3, 0)$, $C(1, -1, -2)$, and $D(3, 4, -1)$, (a) find the points which trisect $BD$; (b) if $M$ is the midpoint of $BC$ and $N$ the midpoint of $CD$, show that one point found in (a) lies on $AM$ and the other point lies on $AN$.

10. (a) If $\overrightarrow{P_1 P} = t\overrightarrow{PP_2}$ show that formula (18.4.4) becomes

$$x = \frac{x_1 + tx_2}{1+t}, \quad y = \frac{y_1 + ty_2}{1+t}, \quad z = \frac{z_1 + tz_2}{1+t}, \quad t \neq -1.$$

(b) Describe the location of $P$ if (i) $t=1$, (ii) $t=0$.
(c) Describe the location of $P$ if $t<0$.
(d) Describe the location of $P$, if $t>0$.
(e) Which point of the line through $P_1$ and $P_2$ is not given by the above formula for any value of $t$?

11. Given any three points $A$, $B$, and $C$, let $P$ be the midpoint of $AB$, let $Q$ be the midpoint of $BC$, and let $M$ be the midpoint of $PQ$. Show that

$$\overrightarrow{MA} + 2\overrightarrow{MB} + \overrightarrow{MC} = 0.$$

12. Carry out the derivation of the midpoint formula (18.4.1).

13. Carry out the derivation of the point of division formula (18.4.3).

## 18.5 SCALAR PRODUCT; DIRECTIONS IN SPACE

The *dot product* or *scalar product* $\mathbf{a} \cdot \mathbf{b}$ of two vectors $\mathbf{a}$ and $\mathbf{b}$ is defined in a manner similar to that of the definition for two-dimensional vectors.

**DEFINITION 18.5.1.** *If* $\mathbf{a} = (a_1, a_2, a_3)$ *and* $\mathbf{b} = (b_1, b_2, b_3)$, *then*

(18.5.1) $\qquad \mathbf{a} \cdot \mathbf{b} = a_1 b_1 + a_2 b_2 + a_3 b_3.$

Scalar product is *commutative, distributive with respect to addition*, and yields a real number.

If **a** = **b** we have (see Section 2.2),

(18.5.2) $$\mathbf{a} \cdot \mathbf{a} = \mathbf{a}^2 = |\mathbf{a}|^2 = a_1^2 + a_2^2 + a_3^2.$$

For the base vectors **i**, **j**, and **k** (Section 18.2) we have

(18.5.3)
$$\mathbf{i} \cdot \mathbf{i} = \mathbf{j} \cdot \mathbf{j} = \mathbf{k} \cdot \mathbf{k} = 1,$$
$$\mathbf{i} \cdot \mathbf{j} = \mathbf{j} \cdot \mathbf{k} = \mathbf{k} \cdot \mathbf{i} = 0.$$

If **a** and **b** are written in terms of the base vectors **i**, **j**, and **k**, the scalar product may be obtained by expanding

$$\mathbf{a} \cdot \mathbf{b} = (a_1 \mathbf{i} + a_2 \mathbf{j} + a_3 \mathbf{k}) \cdot (b_1 \mathbf{i} + b_2 \mathbf{j} + b_3 \mathbf{k})$$

according to the ordinary rules of multiplication, in conjunction with (18.5.3).

*Example 18.5.1.*

$$(2\mathbf{i} + \mathbf{j} - 2\mathbf{k}) \cdot (3\mathbf{i} - 4\mathbf{j} - \mathbf{k})$$
$$6\mathbf{i}^2 + 3\mathbf{j} \cdot \mathbf{i} - 6\mathbf{k} \cdot \mathbf{i} - 8\mathbf{i} \cdot \mathbf{j} - 4\mathbf{j}^2 + 8\mathbf{k} \cdot \mathbf{j} - 2\mathbf{i} \cdot \mathbf{k} - \mathbf{j} \cdot \mathbf{k} + 2\mathbf{k}^2$$
$$= 6 - 4 + 2 = 4.$$

Note that all terms in the expansion are zero, except those in $\mathbf{i}^2$, $\mathbf{j}^2$, and $\mathbf{k}^2$.

A vector may be represented geometrically by any one of a family of parallel directed line segments, drawn in the same sense. If two vectors are given, consider representative directed line segments of each, drawn with the same initial point. *The angle $\theta$ between the vectors is defined as the smallest nonnegative angle whose initial and terminal lines are along these directed line segments.*

Because the position of a vector in space is unspecified, we may think of the angle between two vectors as the angle between the directions determined by the vectors. The same applies to the angle between two directed line segments. It is proper to speak of the angle between directed line segments even when they do not intersect. It may sometimes help to think of each directed line segment as moved parallel to itself (each one retaining its sense) until they assume the same initial point.

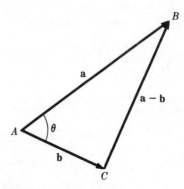

Fig. 18.5.1

Let $\mathbf{a} = \overrightarrow{AB}$ and $\mathbf{b} = \overrightarrow{AC}$ be two vectors, drawn with the same initial point $A$ (Fig. 18.5.1). Then $\overrightarrow{CB} = \mathbf{a} - \mathbf{b}$. If $\theta$ is the angle between $\mathbf{a}$ and $\mathbf{b}$, the Law of Cosines applied to $\triangle ABC$ gives

(18.5.4) $$|\mathbf{a} - \mathbf{b}|^2 = \mathbf{a}^2 + \mathbf{b}^2 - 2|\mathbf{a}||\mathbf{b}|\cos\theta.$$

We also have $|\mathbf{a} - \mathbf{b}|^2 = (\mathbf{a} - \mathbf{b}) \cdot (\mathbf{a} - \mathbf{b}) = \mathbf{a}^2 - 2\mathbf{a} \cdot \mathbf{b} + \mathbf{b}^2$.

Substituting this in (18.5.4) and simplifying, we are led to the equation

(18.5.5) $$\mathbf{a} \cdot \mathbf{b} = |\mathbf{a}||\mathbf{b}|\cos\theta.$$

*Equation (18.5.5) may be considered as a geometric interpretation of the scalar product.* We may also write

(18.5.6) $$\cos\theta = \frac{\mathbf{a} \cdot \mathbf{b}}{|\mathbf{a}||\mathbf{b}|}.$$

*Equation (18.5.6) is a formula for the angle between the nonzero vectors $\mathbf{a}$ and $\mathbf{b}$.* If $\mathbf{a} = (a_1, a_2, a_3)$ and $\mathbf{b} = (b_1, b_2, b_3)$, equation (18.5.6) yields

(18.5.7) $$\cos\theta = \frac{a_1 b_1 + a_2 b_2 + a_3 b_3}{\sqrt{a_1^2 + a_2^2 + a_3^2}\sqrt{b_1^2 + b_2^2 + b_3^2}},$$

*the formula for the angle between two vectors in terms of their components.*

**Example 18.5.1.** Given the points

$$A(-1, 3, 4), \quad B(2, 0, -1), \quad C(4, 1, 3), \quad D(1, -3, 2),$$

find the angle between $\overrightarrow{AB}$ and $\overrightarrow{CD}$.

**Solution.** We have, by (18.3.2),

$$\overrightarrow{AB} = (3, -3, -5), \quad \overrightarrow{CD} = (-3, -4, -1).$$

Hence we have, by (18.5.7), if $\theta$ is the angle between the vectors,

$$\cos\theta = \frac{3(-3) + (-3)(-4) + (-5)(-1)}{\sqrt{9+9+25}\sqrt{9+16+1}} = \frac{8}{\sqrt{43}\sqrt{26}}.$$

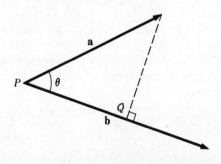

Fig. 18.5.2

Let **a** and **b** (Fig. 18.5.2) represent two vectors drawn with the same initial point, with $\mathbf{b} \neq \mathbf{0}$. If $Q$ is the foot of the perpendicular from the terminal point of **a** to **b**, the directed length $\overline{PQ}$ is called the *projection of* **a** *on* **b**. The sign of $\overline{PQ}$ is taken as plus if $\overline{PQ}$ has the same sense as **b**.

If $\theta$ is the angle between **a** and **b**, we have

(18.5.8) $$\overline{PQ} = |\mathbf{a}|\cos\theta.$$

We know from (18.5.5) that $\mathbf{a} \cdot \mathbf{b} = |\mathbf{a}||\mathbf{b}|\cos\theta$. Hence we get

$$|\mathbf{a}|\cos\theta = \frac{\mathbf{a}\cdot\mathbf{b}}{|\mathbf{b}|}.$$

Substituting this value in (18.5.8) we obtain the relation

(18.5.9) $$\overline{PQ} = \frac{\mathbf{a}\cdot\mathbf{b}}{|\mathbf{b}|}$$

for the *projection of the vector* **a** *on the vector* **b**.

If **a** and **b** are arbitrary vectors, representations of the vectors may be chosen with the same initial point, and the projection of **a** on **b** is still given by (18.5.9).

**Example 18.5.2.** Find the projection of $\overrightarrow{AB}$ on $\overrightarrow{CD}$ for the points

$$A(4, 2, 2), \quad B(-3, 1, 3), \quad C(1, -2, -1), \quad D(3, -3, 2).$$

*Solution.* We find $\overrightarrow{AB} = (-7, -1, 1)$, $\overrightarrow{CD} = (2, -1, 3)$. Hence the projection of $\overrightarrow{AB}$ on $\overrightarrow{CD}$ is, by (18.5.9),

$$\frac{\overrightarrow{AB}\cdot\overrightarrow{CD}}{|\overrightarrow{CD}|} = \frac{-14+1+3}{\sqrt{14}} = -\frac{10}{\sqrt{14}}.$$

Because the value is negative, the projection has the opposite sense to that of $\overrightarrow{CD}$.

## 18.6 DIRECTION ANGLES

Through the initial point of a vector **a** draw vectors **i**, **j**, and **k** (Fig. 18.6.1). The angles that **a** makes with **i**, **j**, and **k**, respectively, are called the *direction angles* $\alpha$, $\beta$, and $\gamma$, of **a**, and their cosines are called the *direction cosines* of **a**. Using $\mathbf{i} = (1, 0, 0)$, and the corresponding representations for **j** and **k**, we obtain from Equation (18.5.6) the formulas

(18.6.1) $$\cos\alpha = \frac{a_1}{|\mathbf{a}|}, \quad \cos\beta = \frac{a_2}{|\mathbf{a}|}, \quad \cos\gamma = \frac{a_3}{|\mathbf{a}|}.$$

We now square each equation in (18.6.1) and add:

$$\cos^2 \alpha + \cos^2 \beta + \cos^2 \gamma = \frac{a_1{}^2 + a_2{}^2 + a_3{}^2}{|\mathbf{a}|^2}.$$

But $a_1{}^2 + a_2{}^2 + a_3{}^2 = |\mathbf{a}|^2$, by Definition 18.2.2. Hence we derive the basic relation

(18.6.2) $$\cos^2 \alpha + \cos^2 \beta + \cos^2 \gamma = 1.$$

Thus the direction cosines of a vector are not independent, but are related by (18.6.2). The relations in (18.6.1) may also be seen geometrically, by projecting the vector $\mathbf{a}$ on the vectors $\mathbf{i}$, $\mathbf{j}$, and $\mathbf{k}$ in Fig. 18.6.1.

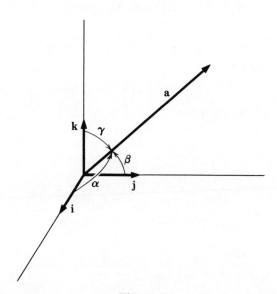

Fig. 18.6.1

***Example 18.6.1.*** Find $\gamma$ if $\alpha = \pi/3$ and $\beta = \pi/4$.

*Solution.* We have $\cos \alpha = 1/2$, $\cos \beta = 1/\sqrt{2}$. Hence, by (18.6.1),

$$\cos^2 \gamma = 1 - \tfrac{1}{4} - \tfrac{1}{2} = \tfrac{1}{4}, \quad \text{and} \quad \cos \gamma = \pm \tfrac{1}{2}.$$

There are thus two possible values of $\gamma$, either $\pi/3$ or $2\pi/3$.

From (18.6.1) we obtain

$$a_1 = |\mathbf{a}| \cos \alpha, \quad a_2 = |\mathbf{a}| \cos \beta, \quad a_3 = |\mathbf{a}| \cos \gamma.$$

Hence the vector $\mathbf{a} = (a_1, a_2, a_3)$ may be expressed as

$$\mathbf{a} = |\mathbf{a}|(\cos \alpha, \cos \beta, \cos \gamma);$$

the vector $(\cos \alpha, \cos \beta, \cos \gamma)$ is a unit vector, by virtue of (18.6.2).

## 496 VECTORS IN THREE DIMENSIONS

Let the vectors **a** and **b** have direction angles $\alpha$, $\beta$, $\gamma$ and $\alpha'$, $\beta'$, $\gamma'$, respectively. If $\theta$ is the angle between **a** and **b**, we may write (18.5.7) as

$$\cos \theta = \frac{a_1}{|\mathbf{a}|} \frac{b_1}{|\mathbf{b}|} + \frac{a_2}{|\mathbf{a}|} \frac{b_2}{|\mathbf{b}|} + \frac{a_3}{|\mathbf{a}|} \frac{b_3}{|\mathbf{b}|}.$$

Now using (18.6.1), and similar formulas for **b**, we obtain

(18.6.3) $\qquad \cos \theta = \cos \alpha \cos \alpha' + \cos \beta \cos \beta' + \cos \gamma \cos \gamma'.$

*Equation* (18.6.3) *is the formula for the angle between two vectors in terms of their direction cosines.*

An immediate consequence of (18.5.7) and (18.6.3) is the following condition that two vectors be perpendicular, since in that case $\theta = \pi/2$ and $\cos \theta = 0$.

**THEOREM 18.6.1.** *The vectors* $\mathbf{a} = (a_1, a_2, a_3)$ *and* $\mathbf{b} = (b_1, b_2, b_3)$ *are perpendicular if and only if*

(18.6.4) $\qquad \mathbf{a} \cdot \mathbf{b} = a_1 b_1 + a_2 b_2 + a_3 b_3 = 0,$

*or, in terms of direction angles,*

(18.6.5) $\qquad \cos \alpha \cos \alpha' + \cos \beta \cos \beta' + \cos \gamma \cos \gamma' = 0.$

***Example 18.6.2.*** Given the points

$$A(2, 4, 3), \qquad B(1, 2, -3), \qquad C(4, 0, 1), \qquad D(2, -2, 2),$$

show that $\overrightarrow{AB}$ is perpendicular to $\overrightarrow{CD}$.

*Solution.* We find $\overrightarrow{AB} = (-1, -2, -6)$, $\overrightarrow{CD} = (-2, -2, 1)$. Then $\overrightarrow{AB}$ is perpendicular to $\overrightarrow{CD}$ by (18.6.4), since

$$(-1)(-2) + (-2)(-2) + (-6)(1) = 0.$$

**EXERCISE GROUP 18.3**

In Exercises 1–6 find (a) $\mathbf{a} \cdot \mathbf{b}$, (b) the projection of **a** on **b**.

1. $\mathbf{a} = (4, 3, 1)$, $\mathbf{b} = (1, -3, 2)$
2. $\mathbf{a} = (0, -2, 4)$, $\mathbf{b} = (-2, 3, 1)$
3. $\mathbf{a} = (5, 0, 2)$, $\mathbf{b} = (2, 3, 0)$
4. $\mathbf{a} = \mathbf{i} + \mathbf{j} - \mathbf{k}$, $\mathbf{b} = 2\mathbf{j} + \mathbf{k}$
5. $\mathbf{a} = 3\mathbf{k}$, $\mathbf{b} = 2\mathbf{i} + \mathbf{j} - \mathbf{k}$
6. $\mathbf{a} = -\mathbf{i} - 2\mathbf{k}$, $\mathbf{b} = 3\mathbf{i} + 5\mathbf{k}$

In Exercises 7–10 find the projection of $\overrightarrow{AB}$ on $\overrightarrow{CD}$.

7. $A(1, 2, 4)$, $B(2, 4, -1)$, $C(-2, 0, 4)$, $D(1, 2, -1)$
8. $A(-1, 3, 0)$, $B(2, -1, 3)$, $C(2, 0, 0)$, $D(3, 2, 0)$
9. $A(0, 0, 3)$, $B(0, 3, 0)$, $C(0, 1, 1)$, $D(2, 0, 0)$
10. $A(-1, -1, 2)$, $B(2, 0, 1)$, $C(-1, 2, -3)$, $D(0, 0, -2)$

In each of Exercises 11–16 describe where the terminal point of a vector must lie if the initial point is at the origin, and $\alpha$, $\beta$, and $\gamma$ are direction angles of the vector.

11. $\alpha = 0$
12. $\gamma = 0$
13. $\alpha = \pi/3$
14. $\beta = 2\pi/3$
15. $\beta = \pi/2$
16. $\gamma = \pi/2$
17. Find $\alpha$ if $\beta = \gamma = \pi/2$.
18. Find $\beta$ if $\alpha = \gamma = \pi/4$.
19. Find the direction cosines of each vector:
    (a) $\mathbf{i} - 2\mathbf{j} + 3\mathbf{k}$    (b) $-4\mathbf{i} - 7\mathbf{j} + 4\mathbf{k}$    (c) $\mathbf{i} - \mathbf{j} - \mathbf{k}$
20. Find the direction cosines of each vector:
    (a) $(3, 3, -1)$    (b) $(-4, 2, 0)$    (c) $(2, 4, -2)$
21. Find the cosines of the angles of the triangle with vertices $(4, 2, 1)$, $(1, 2, 4)$, and $(-2, 1, -3)$.
22. Find the cosines of the angles of the triangle with vertices $(0, 2, 3)$, $(5, 4, -2)$, and $(-1, 3, 2)$.
23. Show with the use of directions that the points $(1, -2, 4)$, $(9, 8, 8)$, and $(10, -8, 1)$ are vertices of a right triangle. (Compare Section 18.3.)
24. Show with the use of directions that the points $(3, 1, 1)$, $(4, 4, -1)$, and $(0, 0, -2)$ are vertices of a right triangle.
25. Prove the commutative property, $\mathbf{a} \cdot \mathbf{b} = \mathbf{b} \cdot \mathbf{a}$.
26. Prove the distributive property, $\mathbf{a} \cdot (\mathbf{b} + \mathbf{c}) = \mathbf{a} \cdot \mathbf{b} + \mathbf{a} \cdot \mathbf{c}$.
27. If $\mathbf{a}$ and $\mathbf{b}$ are nonparallel vectors, show that the vector $\mathbf{c} = |\mathbf{b}|\mathbf{a} + |\mathbf{a}|\mathbf{b}$ makes equal angles with $\mathbf{a}$ and $\mathbf{b}$.

## 18.7 VECTOR PRODUCT OR CROSS PRODUCT

Consider the following problem. We wish to find a vector $\mathbf{u} = (u_1, u_2, u_3)$ which is perpendicular to each of the vectors $\mathbf{a} = (2, -1, 3)$ and $\mathbf{b} = (3, 1, 4)$. By the criterion for perpendicularity (Theorem 18.6.1) applied to each vector, we must have

(18.7.1) $$\begin{cases} 2u_1 - u_2 + 3u_3 = 0, \\ 3u_1 + u_2 + 4u_3 = 0. \end{cases}$$

We have a system of two equations in three unknowns, $u_1, u_2, u_3$. Such a system has infinitely many solutions. In general, we may assign a particular value to one of the unknowns, and then solve the pair of equations for the other two. (This may fail for the choice we make; we may then use a different value or a different unknown for this purpose.) We may also proceed as follows: We rewrite the equations as

$$2u_1 - u_2 = -3u_3,$$
$$3u_1 + u_2 = -4u_3.$$

We now solve for $u_1$ and $u_2$ in terms of $u_3$ and find

$$u_1 = \frac{-7}{5} u_3, \qquad u_2 = \frac{1}{5} u_3.$$

It is now seen that $u_1 = -7$, $u_2 = 1$, $u_3 = 5$ is a solution of (18.7.1). It may be verified that the vector obtained, $(-7, 1, 5)$, is perpendicular to **a** and **b**. Any other vector which is perpendicular to **a** and to **b** must be a multiple of the vector $(-7, 1, 5)$ by a real number.

We now consider the general problem, to find a vector $\mathbf{u} = (u_1, u_2, u_3)$ which is perpendicular to $\mathbf{a} = (a_1, a_2, a_3)$ and to $\mathbf{b} = (b_1, b_2, b_3)$. As above, we must have

$$a_1 u_1 + a_2 u_2 + a_3 u_3 = 0,$$

$$b_1 u_1 + b_2 u_2 + b_3 u_3 = 0.$$

The details are omitted, but the following conclusion which is easily verified may be stated. The statement involves second-order determinants; such a *determinant* is defined by

$$\begin{vmatrix} r_1 & r_2 \\ s_1 & s_2 \end{vmatrix} = r_1 s_2 - r_2 s_1.$$

**THEOREM 18.7.1.** *The vector*

(18.7.2)

$$\left( \begin{vmatrix} a_2 & a_3 \\ b_2 & b_3 \end{vmatrix}, \begin{vmatrix} a_3 & a_1 \\ b_3 & b_1 \end{vmatrix}, \begin{vmatrix} a_1 & a_2 \\ b_1 & b_2 \end{vmatrix} \right) = (a_2 b_3 - a_3 b_2, a_3 b_1 - a_1 b_3, a_1 b_2 - a_2 b_1)$$

*is perpendicular to* $\mathbf{a} = (a_1, a_2, a_3)$ *and to* $\mathbf{b} = (b_1, b_2, b_3)$.

**DEFINITION 18.7.1.** *The vector* (18.7.2) *is called the* **vector product** *or* **cross product** *of* **a** *and* **b**, *and is written as* $\mathbf{a} \times \mathbf{b}$.

It follows from Theorem 18.7.1 that *the vector product of two vectors is perpendicular to each of the vectors.*

It is easily verified that vector multiplication is not commutative. In fact,

(18.7.3) $\qquad \mathbf{a} \times \mathbf{b} = -\mathbf{b} \times \mathbf{a}.$

However, although it is not as easily shown, *vector multiplication is distributive* (Exercise 17),

(18.7.4) $\qquad \mathbf{a} \times (\mathbf{b} + \mathbf{c}) = \mathbf{a} \times \mathbf{b} + \mathbf{a} \times \mathbf{c}.$

An easy way to obtain the vector product $\mathbf{a} \times \mathbf{b}$, with use of third-order determinants, is by the relation

(18.7.5) $\qquad \mathbf{a} \times \mathbf{b} = \begin{vmatrix} \mathbf{i} & \mathbf{j} & \mathbf{k} \\ a_1 & a_2 & a_3 \\ b_1 & b_2 & b_3 \end{vmatrix}$

in which the determinant must be expanded by use of elements of the first row and their cofactors. This follows from the fact that the cofactors of **i**, **j**, and **k** in (18.7.5) are precisely the components of the vector in (18.7.2).

**Example 18.7.1.** Evaluate $(2\mathbf{i} + \mathbf{j} - \mathbf{k}) \times (3\mathbf{i} - 2\mathbf{j} + 4\mathbf{k})$.

*Solution.* By (18.7.5) the cross product is

$$\begin{vmatrix} \mathbf{i} & \mathbf{j} & \mathbf{k} \\ 2 & 1 & -1 \\ 3 & -2 & 4 \end{vmatrix} = \mathbf{i} \begin{vmatrix} 1 & -1 \\ -2 & 4 \end{vmatrix} - \mathbf{j} \begin{vmatrix} 2 & -1 \\ 3 & 4 \end{vmatrix} + \mathbf{k} \begin{vmatrix} 2 & 1 \\ 3 & -2 \end{vmatrix}$$

$$= 2\mathbf{i} - 11\mathbf{j} - 7\mathbf{k}.$$

Formula (18.7.2) may also be used to evaluate the product in the example, directly.

We also present a *schematic* way of evaluating $\mathbf{a} \times \mathbf{b}$. We write the scheme

$$\begin{array}{cccccc} & & \overbrace{\phantom{a_3 \; a_1}}^{②} & & \\ a_1 & a_2 & a_3 & a_1 & a_2 \\ b_1 & b_2 & b_3 & b_1 & b_2 \\ & \underbrace{\phantom{b_2 \; b_3}}_{①} & & \underbrace{\phantom{b_1 \; b_2}}_{③} & \end{array}$$

The second-order determinant whose elements are indicated by ① in the scheme is the first component of $\mathbf{a} \times \mathbf{b}$; the second and third components are indicated in a similar manner. This scheme is valid by a comparison with (18.7.2).

**Example 18.7.2.** Evaluate $(-2, -4, 3) \times (4, -1, -3)$.

*Solution.* Using the above scheme, we write the components of the first vector, and below these we place the components of the second vector. We then repeat the first two columns. The scheme becomes

$$\begin{array}{cccccc} & & \overbrace{\phantom{3 \; -2}}^{②} & & \\ -2 & -4 & 3 & -2 & -4 \\ 4 & -1 & -3 & 4 & -1 \\ & \underbrace{\phantom{-1 \; -3}}_{①} & & \underbrace{\phantom{4 \; -1}}_{③} & \end{array}$$

The components of the vector product are then

$$\begin{vmatrix} -4 & 3 \\ -1 & -3 \end{vmatrix} = 15, \quad \begin{vmatrix} 3 & -2 \\ -3 & 4 \end{vmatrix} = 6, \quad \begin{vmatrix} -2 & -4 \\ 4 & -1 \end{vmatrix} = 18.$$

The vector product is $(15, 6, 18) = 15\mathbf{i} + 6\mathbf{j} + 18\mathbf{k}$.

The magnitude of $\mathbf{a} \times \mathbf{b}$ is related to the angle between $\mathbf{a}$ and $\mathbf{b}$ by the following theorem.

**THEOREM 18.7.2.** *If $\theta$ is the angle between vectors $\mathbf{a}$ and $\mathbf{b}$, then*

(18.7.6) $$|\mathbf{a} \times \mathbf{b}| = |\mathbf{a}||\mathbf{b}| \sin \theta.$$

The proof will be contained in the exercises. An immediate consequence of Theorem 18.7.2 follows.

**THEOREM 18.7.3.** *The vectors* **a** *and* **b** *are parallel if and only if* $\mathbf{a} \times \mathbf{b} = \mathbf{0}$.

This follows from Theorem 18.7.2 since, for parallel vectors, either $\theta = 0$ or $\theta = \pi$.

**THEOREM 18.7.4.** *Let* **a** *and* **b** *be two vectors drawn with the same initial point (Fig. 18.7.1). If K is the area of the parallelogram formed with* **a** *and* **b** *as two adjacent sides, then*

$$K = |\mathbf{a} \times \mathbf{b}|.$$

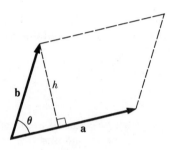

Fig. 18.7.1

*Proof.* Let $h$ be the length of the perpendicular drawn from the terminal point of **b** to **a**. Then $h = |\mathbf{b}| \sin \theta$. The area of the parallelogram is then $K = |\mathbf{a}|h = |\mathbf{a}||\mathbf{b}|\sin \theta = |\mathbf{a} \times \mathbf{b}|$, the last equality arising from (18.7.6).

**Example 18.7.3.** Find the area of the parallelogram with vertices $A(2, 3, 4)$, $B(-1, 3, 2)$, $C(1, -4, 3)$, and $D(4, -4, 5)$.

*Solution.* It is easily verified that $ABCD$ is a parallelogram with $\overrightarrow{AB}$ and $\overrightarrow{AD}$ as adjacent sides. By (18.3.2) we find

$$\overrightarrow{AB} = (-3, 0, -2), \quad \overrightarrow{AD} = (2, -7, 1).$$

We then find

$$\overrightarrow{AB} \times \overrightarrow{AD} = (-3, 0, -2) \times (2, -7, 1) = (-14, -1, 21).$$

Then by Theorem 18.7.4 the area $K$ of the parallelogram is

$$K = |(-14, -1, 21)| = \sqrt{196 + 1 + 441} = \sqrt{638}.$$

We know from Theorem 18.7.1 that $\mathbf{a} \times \mathbf{b}$ is perpendicular to both **a** and **b**. The following is stated without proof.

**THEOREM 18.7.5.** *The vectors* **a**, **b**, *and* **a** × **b** *form a right-handed system in the order given.*

## * 18.8 TRIPLE PRODUCT

Since a vector product is a vector, we may form the scalar product of another vector with it. The product

(18.8.1) $$\mathbf{a} \cdot \mathbf{b} \times \mathbf{c}$$

is called the *triple product* of **a**, **b**, and **c**, in the order given, and is a number. Parentheses are unnecessary in (18.8.1), since only one meaning can be attached to that expression (why does (**a** · **b**) × **c** have no meaning?).

**THEOREM 18.8.1.** *The triple product may be expressed as a determinant*

$$\mathbf{a} \cdot \mathbf{b} \times \mathbf{c} = \begin{vmatrix} a_1 & a_2 & a_3 \\ b_1 & b_2 & b_3 \\ c_1 & c_2 & c_3 \end{vmatrix}.$$

*Proof.* The expansion of the determinant by elements of the first row and their cofactors yields the sum

$$a_1 \begin{vmatrix} b_2 & b_3 \\ c_2 & c_3 \end{vmatrix} + a_2 \begin{vmatrix} b_3 & b_1 \\ c_3 & c_1 \end{vmatrix} + a_3 \begin{vmatrix} b_1 & b_2 \\ c_1 & c_2 \end{vmatrix}.$$

This is precisely the value of the scalar product of **a** and **b** × **c**, and the proof is complete.

The triple product **b** · **c** × **a** may be expressed, by Theorem 18.8.1, as a determinant whose rows are the components of **b**, **c**, and **a**, respectively. Its value, by properties of determinants, is then the same as **a** · **b** × **c**. In this way we get

(18.8.2) $$\mathbf{a} \cdot \mathbf{b} \times \mathbf{c} = \mathbf{b} \cdot \mathbf{c} \times \mathbf{a} = \mathbf{c} \cdot \mathbf{a} \times \mathbf{b}.$$

Furthermore,

(18.8.3) $$\mathbf{a} \times \mathbf{b} \cdot \mathbf{c} = \mathbf{c} \cdot \mathbf{a} \times \mathbf{b} = \mathbf{a} \cdot \mathbf{b} \times \mathbf{c} \quad \text{[by (18.8.2)]}.$$

Combining (18.8.2) and (18.8.3), we may state the following result.

**THEOREM 18.8.2.** *The triple product of* **a**, **b**, *and* **c**, *in that order, has the same value whether the cross is placed between* **a** *and* **b** *or between* **b** *and* **c**. *Furthermore, the value is the same for the vectors in the order* **b**, **c**, *and* **a**, *or* **c**, **a**, *and* **b**.

The arrangements **a**, **b**, **c** or **b**, **c**, **a** or **c**, **a**, **b** are said to be *cyclically equivalent*, since they follow the same order in Fig. 18.8.1. The meaning of Theorem 18.8.2 is then that the triple product of three vectors is unchanged as long as the cyclical order is retained.

Fig. 18.8.1

The following theorem is stated without proof.

**THEOREM 18.8.3.** *Let* **a**, **b**, *and* **c** *be three vectors with the same initial point. The volume of the parallelepiped formed with these vectors as adjacent edges is given by the absolute value of their triple product. The triple product is positive if* **a**, **b**, *and* **c** *form a right-handed system in the order given.*

An alternative proof of Theorem 18.7.1 follows. By Theorem 18.8.1 we have

$$\mathbf{a} \cdot (\mathbf{a} \times \mathbf{b}) = \begin{vmatrix} a_1 & a_2 & a_3 \\ a_1 & a_2 & a_3 \\ b_1 & b_2 & b_3 \end{vmatrix} = 0,$$

using the fact that a determinant with two identical rows is zero. Hence **a** is perpendicular to **a** × **b**. In the same way, **b** is perpendicular to **a** × **b**.

**EXERCISE GROUP 18.4**

1. Given the vectors **a** = (3, −2, 2), **b** = (−1, 1, 3), evaluate directly
   (a) **a** × **b**,  (b) **b** × **a**.

2. Given the vectors **a** = 2**i** − 3**j** + 2**k**, **b** = 3**i** − 2**k**, evaluate directly
   (a) **a** × **b**,  (b) **b** × **a**.

3. Given the vectors **a** = (1, 2, 1), **b** = (2, 0, −1), **c** = (4, −2, 0), find
   (a) **a** × **b**,  (b) **a** · (**b** × **c**),  (c) **a** × (**b** × **c**).

4. Given the vectors **a** = −**i** + **k**, **b** = 2**i** + **j**, **c** = −**i** + 3**j** + 2**k**, find
   (a) **a** × **c**,  (b) (**a** × **b**) · **c**,  (c) (**a** × **b**) × **c**.

5. Show that **i** × **i** = **j** × **j** = **k** × **k** = 0.

6. Show that **i** × **j** = **k**,  **j** × **k** = **i**,  **k** × **i** = **j**.

7. Show that **j** × **i** = −**k**,  **k** × **j** = −**i**,  **i** × **k** = −**j**.

8. Evaluate (a) (**i** × **j**) · **k**, (b) **j** · (**i** × **k**).

9. Find the area of a parallelogram if two adjacent sides are −2**i** + 3**k** and 4**i** − 2**j** − **k**.

10. Find the area of a parallelogram if two adjacent sides are $3\mathbf{i} - 3\mathbf{j} + \mathbf{k}$ and $\mathbf{i} + 2\mathbf{j} - 3\mathbf{k}$.

11. Find the area of the parallelogram with vertices $A(-1, 3, 2)$, $B(3, -3, 4)$, $C(5, 0, 3)$, and $D(1, 6, 1)$.

12. Find the area of the parallelogram with vertices $A(4, 2, 1)$, $B(2, -3, 0)$, $C(1, -1, -2)$, and $D(3, 4, -1)$.

* 13. Find the volume of a parallelepiped if three adjacent edges are $\mathbf{i} - 2\mathbf{j}$, $3\mathbf{i} + 2\mathbf{k}$, and $2\mathbf{j} - \mathbf{k}$.

* 14. Find the volume of a parallelepiped if three adjacent edges are $2\mathbf{i} - 3\mathbf{j} - \mathbf{k}$, $\mathbf{i} - \mathbf{k}$, and $\mathbf{i} + \mathbf{j}$.

15. Prove that $\mathbf{a} \times \mathbf{b} = -\mathbf{b} \times \mathbf{a}$.

16. Prove that $\mathbf{a} \times \mathbf{a} = \mathbf{0}$.

17. Prove that $\mathbf{a} \times (\mathbf{b} + \mathbf{c}) = \mathbf{a} \times \mathbf{b} + \mathbf{a} \times \mathbf{c}$.

18. Prove that $\mathbf{a} \times (\mathbf{b} \times \mathbf{c}) = (\mathbf{a} \cdot \mathbf{c})\mathbf{b} - (\mathbf{a} \cdot \mathbf{b})\mathbf{c}$. [Suggestion: Express each member in terms of the vector components.]

19. Justify each of the following steps:

$$\begin{aligned}|\mathbf{a} \times \mathbf{b}|^2 &= (\mathbf{a} \times \mathbf{b}) \cdot (\mathbf{a} \times \mathbf{b}) = \mathbf{a} \cdot [\mathbf{b} \times (\mathbf{a} \times \mathbf{b})] \\ &= \mathbf{a} \cdot [(\mathbf{b} \cdot \mathbf{b})\mathbf{a} - (\mathbf{b} \cdot \mathbf{a})\mathbf{b}] \\ &= (\mathbf{a} \cdot \mathbf{a})(\mathbf{b} \cdot \mathbf{b}) - (\mathbf{a} \cdot \mathbf{b})^2 \\ &= |\mathbf{a}|^2|\mathbf{b}|^2 - (\mathbf{a} \cdot \mathbf{b})^2.\end{aligned}$$

20. Using $\mathbf{a} \cdot \mathbf{b}$ from (18.5.5) and the result of Exercise 19, obtain the result stated in Theorem 18.7.2.

21. Show that $\mathbf{a} \cdot [\mathbf{b} \times (\mathbf{a} \times \mathbf{b})]$ is never negative.

22. If $\theta$ is the angle between the nonperpendicular vectors $\mathbf{a}$ and $\mathbf{b}$, show that

$$\tan \theta = \frac{|\mathbf{a} \times \mathbf{b}|}{\mathbf{a} \cdot \mathbf{b}}.$$

# 19 Lines and Planes

## 19.1 DIRECTION NUMBERS

Vectors may be defined along a straight line in space in either of two opposite senses. Corresponding direction angles of oppositely directed vectors are supplementary, as for example, the angles that $\overrightarrow{OA}$ and $\overrightarrow{OB}$ make with the $x$ axis in Fig. 19.1.1. If $\alpha$, $\beta$, and $\gamma$ are the direction angles of $\overrightarrow{OA}$, then $\pi - \alpha$, $\pi - \beta$, and $\pi - \gamma$ are the direction angles of $\overrightarrow{OB}$. *We may then take either set of angles as direction angles of the line.* Thus a straight line has two sets of direction angles; the angles of one set are the supplements of the other set; and the line has two sets of direction cosines.

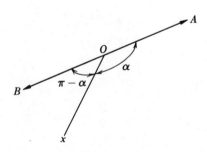

Fig. 19.1.1

In many applications we are concerned with the direction of a line, without distinguishing between oppositely directed vectors on the line. We may describe the direction by introducing *direction numbers* of a line.

**DEFINITION 19.1.1.** *If $\alpha$, $\beta$, and $\gamma$ are direction angles of a line, and for some number $k$ we have*

(19.1.1) $\quad a = k \cos \alpha, \quad b = k \cos \beta, \quad c = k \cos \gamma, \quad k \neq 0,$

the numbers a, b, and c are called a set of **direction numbers** of the line.

Any straight line has infinitely many sets of direction numbers. Any such set is proportional to any other set and to direction cosines of the line. The numbers 0, 0, 0 are not direction numbers for any line.

Direction cosines of a line may be expressed in terms of a set of direction numbers. Let $a$, $b$, and $c$, not all zero, be a set of direction numbers. Then (19.1.1) holds. If both members of each equation in (19.1.1) are squared, and the results added, we obtain

$$a^2 + b^2 + c^2 = k^2(\cos^2 \alpha + \cos^2 \beta + \cos^2 \gamma).$$

The quantity in parentheses equals unity by (18.6.2); hence

$$k = \pm(a^2 + b^2 + c^2)^{1/2},$$

and from (19.1.1) we obtain

(19.1.2) $\quad \cos \alpha = \dfrac{a}{\pm\sqrt{a^2 + b^2 + c^2}}, \quad \cos \beta = \dfrac{b}{\pm\sqrt{a^2 + b^2 + c^2}},$

$$\cos \gamma = \dfrac{c}{\pm\sqrt{a^2 + b^2 + c^2}}.$$

The plus signs yield one set of direction cosines of the straight line, and the minus signs yield the other set.

It is clear that *the components of any vector are a set of direction numbers of any line on which the vector lies*. It follows (see Section 18.3) that if $(x_1, y_1, z_1)$ and $(x_2, y_2, z_2)$ are two distinct points on a line, then *a set of direction numbers is*

$$x_2 - x_1, \quad y_2 - y_1, \quad z_2 - z_1.$$

Conversely, *if a, b, and c are direction numbers of a line, then the vectors $(a, b, c)$ and $(-a, -b, -c)$ are oppositely directed vectors on that line.*

**Example 19.1.1.** Find direction cosines for the straight line which contains the points $(3, -1, 2)$ and $(2, 4, -3)$.

*Solution.* From the above the respective differences of coordinates are a set of direction numbers of the line; thus $-1, 5, -5$ are a set of direction numbers. By (19.1.2) we have

$$\cos \alpha = \dfrac{-1}{\pm\sqrt{1 + 25 + 25}} = -\dfrac{1}{\pm\sqrt{51}}, \quad \cos \beta = \dfrac{5}{\pm\sqrt{51}}, \quad \cos \gamma = \dfrac{-5}{\pm\sqrt{51}}.$$

The two sets of direction cosines for the line are

$$-\dfrac{1}{\sqrt{51}}, \dfrac{5}{\sqrt{51}}, -\dfrac{5}{\sqrt{51}} \quad \text{and} \quad \dfrac{1}{\sqrt{51}}, -\dfrac{5}{\sqrt{51}}, \dfrac{5}{\sqrt{51}}.$$

If we let $k = 1$ in (19.1.1) we see that direction cosines of a line are also a set of direction numbers. In fact, *direction cosines are those direction numbers, among all possible sets, for which the sum of the squares equals unity.*

Various results of Sections 18.5 and 18.6 may be rephrased in terms of direction numbers instead of vector components, including formula (18.6.3) for an angle between two lines. Just as it is proper to speak of the angle between directed line segments whether or not they intersect, so is it proper to speak of an angle between two straight lines whether or not they intersect.

With the use of (19.1.2) and appropriate results from the study of vectors, we may state the following.

**THEOREM 19.1.1.** *Let $a_1, b_1, c_1$ and $a_2, b_2, c_2$ be direction numbers of two lines. Then the lines are perpendicular if and only if*

(19.1.3) $$a_1 a_2 + b_1 b_2 + c_1 c_2 = 0.$$

*The lines are parallel if and only if*

(19.1.4) $$\frac{a_1}{a_2} = \frac{b_1}{b_2} = \frac{c_1}{c_2};$$

*this is to be interpreted to mean, for example, if $b_2 = 0$, then also $b_1 = 0$. The angles $\theta$ between the two lines are given by*

(19.1.5) $$\cos \theta = \pm \frac{a_1 a_2 + b_1 b_2 + c_1 c_2}{\sqrt{a_1^2 + b_1^2 + c_1^2} \sqrt{a_2^2 + b_2^2 + c_2^2}}.$$

*Example 19.1.2.* Find the angles between two lines with direction numbers 1, 2, 2 and $-3, 2, 5$.

*Solution.* Using (19.1.5) we find

$$\cos \theta = \pm \frac{(1)(-3) + (2)(2) + (2)(5)}{\sqrt{1 + 4 + 4}\sqrt{9 + 4 + 25}} = \pm \frac{11}{3\sqrt{38}}.$$

One angle between the lines is $53\frac{1}{2}°$ approximately, and the other angle is accordingly $126\frac{1}{2}°$ approximately.


**EXERCISE GROUP 19.1**

In Exercises 1–8 find a set of direction numbers for the line determined by the given points. Express the result in terms of integers with no common factor (other than $\pm 1$).

1. $(2, 3, -1), (-2, 2, 4)$
2. $(0, -4, 3), (-2, 0, -4)$
3. $(1, 4, -5), (3, 8, -3)$
4. $(2, 9, 5), (8, 3, -4)$
5. $(1, 3/2, 1), (2, -1, 4)$
6. $(1/3, 2, 1/2), (1, -1/3, -1)$
7. $(1/4, 1/2, 1), (-1/3, 1/3, 2)$
8. $(-1, 4, -2/3), (1/2, -1/2, 1)$

In Exercises 9–12 show that the points are vertices of a right triangle.

9. (4, 0, 7), (6, 3, 6), (3, 2, 11)
10. (−5, −9, 13), (−1, −4, 10), (2, −1, 19)
11. (2, 7, −1), (3, 2, 4), (8, 5, 6)
12. $(0, \sqrt{6}, 4), (\sqrt{2}, 2\sqrt{6}, 3), (\sqrt{3}, 1+\sqrt{6}, 4+2\sqrt{6})$

In Exercises 13–16 show that the points are vertices of a parallelogram.

13. (4, −2, 1), (−2, 4, −1), (−1, −4, 4), (−7, 2, 2)
14. (3, 5, −1), (6, −1, 4), (−2, −5, 8), (1, −11, 13)
15. (4.3, 0.25, 1), (1.9, 0.75, −1), (4, 7.5, 3), (1.6, 8, 1)
16. (1/2, 2, −1/3), (3, −2, 1/2), (4, 5, −1), (13/2, 1, −1/6)
17. Show that (−2, 0, 3), (0, 4, 1), (4, 3, 3), and (2, −1, 5) are vertices of a rectangle.
18. Show that (−2, 2, 2), (6, 1, −4), (3, −5, −7), and (−5, −4, −1) are vertices of a rectangle.
19. Find the angles of the triangle with vertices $A(1, 2, 0)$, $B(2, 1, -1)$, and $C(2, 4, 3)$.
20. Find the angles of the triangle with vertices $A(4, 2, 0)$, $B(1, -1, 1)$, and $C(0, -2, 1)$.

## 19.2 THE PLANE

If a point $P$ has coordinates $(x, y, z)$, the vector $\overrightarrow{OP}$ from the origin to $P$, represented by

$$\mathbf{r} = (x, y, z) = x\mathbf{i} + y\mathbf{j} + z\mathbf{k},$$

is called the *position vector* of $P$.

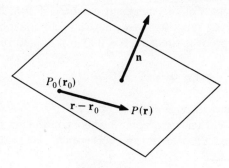

Fig. 19.2.1

Let $P_0(x_0, y_0, z_0)$ be a fixed point in a plane (Fig. 19.2.1), let $P(x, y, z)$ be an arbitrary point in the plane, and let $a$, $b$, $c$ be a set of direction numbers of any line perpendicular to the plane. If $\mathbf{r}_0$ and $\mathbf{r}$ represent the position vectors of $P_0$ and $P$, and $\mathbf{n}$ is the vector $\mathbf{n} = (a, b, c)$, the vector $\overrightarrow{P_0 P}$ lies in the plane and is therefore perpendicular to $\mathbf{n}$. Hence $\mathbf{r} - \mathbf{r}_0$ is perpendicular to $\mathbf{n}$, and by (18.6.4) we have

(19.2.1) $$\mathbf{n} \cdot (\mathbf{r} - \mathbf{r}_0) = 0.$$

Equation (19.2.1) is *the vector form of the equation of the plane which contains the point $P_0(\mathbf{r})$ and is perpendicular to the direction $a, b, c$*. The vector $\mathbf{n}$ is called a *normal* to the plane.

From (19.2.1), and using the distributive property for scalar product, we obtain $\mathbf{n} \cdot \mathbf{r} = \mathbf{n} \cdot \mathbf{r}_0$. If we let $\mathbf{n} \cdot \mathbf{r}_0 = t$, a real number, we may write the equation of the plane (19.2.1) also as

(19.2.2) $$\mathbf{n} \cdot \mathbf{r} = t.$$

If the vectors $\mathbf{n}$, $\mathbf{r}$, and $\mathbf{r}_0$ are expressed in terms of their components, the nonvector forms of (19.2.1) and (19.2.2) become, respectively,

(19.2.3) $$a(x - x_0) + b(y - y_0) + c(z - z_0) = 0,$$

and

(19.2.4) $$ax + by + cz = t.$$

In each of (19.2.3) and (19.2.4) the coefficients $a$, $b$, and $c$ are a set of direction numbers of a *normal* to the plane.

Equation (19.2.3) is called the *standard form* of the equation of a plane, and (19.2.4) is the *general form*.

If $a$, $b$, and $c$ are not all zero. Equation (19.2.4) is a *first degree equation* in $x, y$, and $z$; it is also called a *linear equation* in $x, y$, and $z$.

**THEOREM 19.2.1.** *Any plane may be represented by a linear equation, and conversely.*

*Proof.* The direct part of the statement has been proved in deriving (19.2.4). To prove the converse, suppose that we have a linear equation (19.2.4). We may assume $c \neq 0$ (the proof is similar if $a \neq 0$ or $b \neq 0$), and then write (19.2.3) as

$$a(x - 0) + b(y - 0) + c\left(z - \frac{t}{c}\right) = 0.$$

It follows that the line through $P(x, y, z)$ and $P_0(0, 0, t/c)$ is perpendicular to the fixed direction $a, b, c$. Hence any point $P$ whose coordinates satisfy (19.2.4) must lie in the plane which contains $P_0$ and is perpendicular to the direction $a, b, c$; the proof is complete.

The absence of one or two variables in (19.2.4) is easily interpreted. *If one variable is missing, the plane is parallel to the axis of that variable.* For example, for the plane

$$2x - 3z = 1$$

we have $a = 2$, $b = 0$, $c = -3$. Since $b = 0$, the normal to the plane is parallel to the $xz$ plane, and the given plane is parallel to the $y$ axis.

*If two variables are missing, the plane is parallel to the plane of those variables.* Thus $x = 4$ is a plane parallel to the $yz$ plane.

***Example 19.2.1.*** Find an equation of the plane which contains the point $(2, 3, -1)$, and with $-2, 1, 4$ as direction numbers of the normal.

*Solution.* We may use (19.2.4) with $a = -2$, $b = 1$, $c = 4$, to get

$$-2x + y + 4z = t.$$

Since the coordinates of the point $(2, 3, -1)$ must satisfy this equation, we must have $t = -5$, and a desired equation is

$$-2x + y + 4z = -5 \quad \text{or} \quad 2x - y - 4z - 5 = 0.$$

The same result may be obtained by substituting directly in (19.2.3).

In the following example two methods are presented for finding an equation of a plane which contains three given points.

***Example 19.2.2.*** Find an equation of the plane which contains the points $A(1, -2, 1)$, $B(2, 0, 3)$, and $C(-2, 1, 2)$.

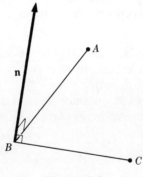

Fig. 19.2.2

*Solution. First method.* The points are indicated schematically in Fig. 19.2.2. The vectors $\overrightarrow{AB}$ and $\overrightarrow{BC}$ are determined as

$$\overrightarrow{AB} = (1, 2, 2), \quad \overrightarrow{BC} = (-4, 1, -1).$$

The vector product of $\overrightarrow{AB}$ and $\overrightarrow{BC}$ is a vector perpendicular to $\overrightarrow{AB}$ and $\overrightarrow{BC}$, and therefore to the plane of $A$, $B$, and $C$. By Section 18.7 we find

$$\overrightarrow{AB} \times \overrightarrow{BC} = (-4, -7, 9).$$

Therefore, $-4, -7, 9$ or $4, 7, -9$ are direction numbers of the normal to the plane. Using the latter set in (19.2.3), with the given point $(1, -2, 1)$, we have

$$4(x - 1) + 7(y + 2) - 9(z - 1) = 0 \quad \text{or} \quad 4x + 7y - 9z + 19 = 0.$$

*Second method.* The coordinates of each point must satisfy (19.2.4); accordingly, we obtain the equations

$$a - 2b + c = t,$$
$$2a + 3c = t,$$
$$-2a + b + 2c = t.$$

These form a system of three linear equations in four unknowns, and may be solved for three of the unknowns in terms of the fourth. In this way we find

$$a = -\frac{4}{19}t, \quad b = -\frac{7}{19}t, \quad c = \frac{9}{19}t.$$

Assigning a convenient nonzero value to $t$, say $t = -19$, we find $a = 4$, $b = 7$, $c = -9$; substitution in (19.2.4) gives the same equation as in the first method.

Other ways in which information may be presented about a plane include the following:

(a) the plane may be perpendicular to a specified line;
(b) the plane may be perpendicular to a specified plane, in which case their normals are perpendicular;
(c) the plane may be perpendicular to two specified planes.

All such conditions may be expressed or utilized in some manner in equation form by our available methods. If they lead to a system of equations, it should be kept in mind that three independent equations in $a$, $b$, $c$, and $t$ are required in order to determine an equation of a plane.

Bear in mind that any condition which involves the direction of a plane may be expressed in terms of a normal to the plane. This will be illustrated.

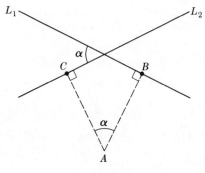

Fig. 19.2.3

## 19.2 THE PLANE

**Example 19.2.3.** Find an angle between the planes

$$2x - 3y + z = 4 \quad \text{and} \quad 3x - 4y + 2 = 0.$$

**Solution.** Let a third plane, which is perpendicular to the intersection of the given planes, intersect them in $L_1$ and $L_2$ (Fig. 19.2.3). Then $\alpha$ is an angle formed by the planes. If $AB$ and $AC$ are drawn in the third plane so that $AB \perp L_1$ and $AC \perp L_2$, then $\alpha$ is also an angle between $AB$ and $AC$. Direction numbers of normals to the given planes are, respectively, 2, $-3$, 1 and 3, $-4$, 0. Then by (19.1.5) we find

$$\cos \alpha = \frac{(2)(3) + (-3)(-4) + (1)(0)}{\sqrt{4+9+1}\sqrt{9+16}} = \frac{18}{5\sqrt{14}} = 0.9621,$$

and $\alpha = 16°$ (approx.).

If a plane intersects a coordinate plane in a straight line, that line is called a *trace* of the plane. If a plane intersects a coordinate axis in a point, the coordinate of the point corresponding to that axis is an *intercept* of the plane.

**Example 19.2.4.** Find the traces and intercepts of the plane $2x - 3y + z = 4$.

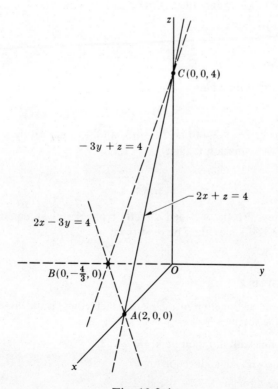

Fig. 19.2.4

*Solution.* We set $z = 0$ to find the $xy$ trace, set $y = 0$ to find the $xz$ trace, and set $x = 0$ to find the $yz$ trace. In this way we get (see Fig. 19.2.4)

$$xy \text{ trace:} \quad 2x - 3y = 4,$$
$$xz \text{ trace:} \quad 2x + z = 4,$$
$$yz \text{ trace:} \quad -3y + z = 4.$$

Each of these equations is an equation of a line in the corresponding coordinate plane in the two-dimensional sense.

To find an intercept we set two of the coordinates equal to zero in the equation of the plane. If we set $y = z = 0$, we find the $x$ intercept to be 2; similarly, the $y$ intercept is $-4/3$, and the $z$ intercept is 4. The intercepts may also be found from the equations of the traces.

Suppose $a$, $b$, $c$, and $t$ are all different from zero in the equation

(19.2.5) $$ax + by + cz = t.$$

Calling the $x$, $y$, and $z$ intercepts $A$, $B$, and $C$, respectively, we find

$$A = \frac{t}{a}, \quad B = \frac{t}{b}, \quad C = \frac{t}{c}.$$

The general form (19.2.5) may then be written in the *intercept form of the equation of a plane*

(19.2.6) $$\frac{x}{A} + \frac{y}{B} + \frac{z}{C} = 1, \quad A, B, C \text{ all} \neq 0.$$

The intercepts of the plane

$$2x - 3y + z = 4$$

were found in Example 19.2.4 to be 2, $-4/3$, and 4, respectively. The intercept form (19.2.6) of the equation is then

$$\frac{x}{2} + \frac{y}{-(4/3)} + \frac{z}{4} = 1.$$

This equation may also be obtained directly from the given equation, and the values of the intercepts may then be read from it.

**EXERCISE GROUP 19.2**

In Exercises 1–4 find an equation of the plane which contains the given point and is perpendicular to the specified line.

**1.** $(-6, 1, -3)$; line with direction numbers $-1, 3, 4$.
**2.** $(3, 4, -5)$; line with equal direction angles.
**3.** $(2, -5, 4)$; line through $(1, 3, -3)$ and $(2, 4, 5)$.
**4.** $(1, -1, \sqrt{2})$; line with $\alpha = \pi/3$, $\beta = 2\pi/3$, $\gamma = \pi/4$.

5. Find an angle between the plane $4x - 3y + z - 7 = 0$ and a line with direction numbers 2, 0, $-3$.

6. Find an angle between the plane $5x + 4y - 2z = 1$ and a line with direction numbers 3, $-4$, 5.

7. Find an angle between the planes $2x - y + 2z = 7$ and $x + 2y - 2z + 3 = 0$.

8. Find an angle between the planes $3x + 2y - z = 0$ and $4 - x + 3y + 5z = 0$.

In Exercises 9–12 find an equation of the plane which contains the given points and is perpendicular to the given plane.

9. $(3, -2, -4), (-1, -1, -8); 3x - 2y - 2z = 0$
10. $(5, -1, -1), (-7, 3, 7); 5x + 3y - z + 2 = 0$
11. $(2, -1, 5), (-1, -3, 3); 3x + 2y + 4z = 1$
12. $(-2, -2, -2), (3, 1, 1); x + y + 2z = 5$

In Exercises 13–16 find an equation of the plane which contains the given point and is perpendicular to the given planes.

13. $(3, 3, -3); 2x - y - z = 4, x + 3y - 2z + 1 = 0$
14. $(-1, 1, -1); 4x - 2y + 3z = 1, 3x - 3y + 2z = 0$
15. $(1, 7, -9); x - 3y + 5z = 4, 5x + 3y - z + 1 = 0$
16. $(7, 1, -1); 3x - 2y + z = 0, x + 3y - 2z = 4$

In Exercises 17–22 find an equation of the plane which contains the given points.

17. $(3, 5, 3), (2, 7, 5), (-1, -3, -1)$
18. $(0, -2, -1), (1, 3, -3), (-1, -5, 3)$
19. $(7, 1, 3), (-2, 4, 7), (3, -2, -1)$
20. $(1, 0, 1), (-1, 7, -3), (2, -2, 2)$
21. $(2, -1, 3), (4, 3, 8), (-3, 2, -3)$
22. $(0, 1, 1), (3, 0, -1), (1, 3, 2)$

23. Show that the four points line in a plane, and find its equation:
$$(2, 2, 5), (-1, 2, -1), (-3, 2, -5), (4, -1, 4).$$

24. Show that the four points lie in a plane, and find its equation:
$$(0, -4, 1), (2, 2, 1), (-3, -3, -1), (5, -4, 4).$$

In Exercises 25–32 write an equation in intercept form, if possible. In other cases find all intercepts which exist.

25. $2x - 3y + 4z = 3$
26. $3x + y - z = -2$
27. $x - 2y - 2z = 0$
28. $3x + 4z = 1$
29. $-3x - y + z = 2$
30. $2x - 2y + 5 = 0$
31. $4x + y = 0$
32. $x - y - 3z - 4 = 0$

In each of the following find equations of all traces which exist.

33. $2x - 5y - 3z = 1$
34. $4x - 3y + z = 0$
35. $x + 2y = 4$
36. $y - 3z + 7 = 0$
37. $x + 4 = 0$
38. $2y - 3 = 0$

## 19.3 THE STRAIGHT LINE

Let $a, b, c$ be a set of direction numbers of a line containing the point $P_1(x_1, y_1, z_1)$. Then $\mathbf{m} = (a, b, c)$ is a vector parallel to the line. If $P(x, y, z)$ is an arbitrary point on the line (Fig. 19.3.1), then the vector $\overrightarrow{P_1P}$ is parallel to $\mathbf{m}$ and must be a real multiple of $\mathbf{m}$. Letting $\mathbf{r} = (x, y, z)$ and $\mathbf{r}_1 = (x_1, y_1, z_1)$ be the position vectors of $P$ and $P_1$, we have $\overrightarrow{P_1P} = \mathbf{r} - \mathbf{r}_1$, and *an equation of the line is, in parametric vector form,*

(19.3.1) $\qquad \mathbf{r} - \mathbf{r}_1 = t\mathbf{m}, \qquad t$ any real number.

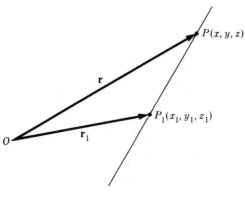

Fig. 19.3.1

We now express $\mathbf{r}$, $\mathbf{r}_1$, and $\mathbf{m}$ in terms of their components. From (19.3.1) and the definition of equality of vectors (Definition 18.2.1), we get the three equations

(19.3.2) $\qquad x = x_1 + at, \qquad y = y_1 + bt, \qquad z = z_1 + ct.$

Equations (19.3.2) are *parametric equations of a line* in terms of a point $(x_1, y_1, z_1)$ on it, and a set of direction numbers $a, b, c$ of the line. *Equations (19.3.2) are valid even when one or two of the direction numbers are zero.*

If none of the quantities $a, b,$ and $c$ is zero, we may solve each of the equations in (19.3.2) for $t$, and obtain

(19.3.3) $\qquad \dfrac{x - x_1}{a} = \dfrac{y - y_1}{b} = \dfrac{z - z_1}{c}.$

We may permit (19.3.3) to include the case where one or two of the denominators are zero, if we understand thereby that the corresponding fraction in (19.3.3) is deleted and is replaced by the equation obtained by setting the numerator equal to zero.

Equations (19.3.3) are called *symmetric equations* of the line which contains the point $(x_1, y_1, z_1)$ and has direction numbers $a, b, c$.

## 19.3 THE STRAIGHT LINE

*Example 19.3.1.* Find equations of the line through the points $(-2, 0, 3)$ and $(1, -2, -4)$, in parametric form and in symmetric form.

*Solution.* A set of direction numbers of the line is
$$a = 1 + 2 = 3, \quad b = -2 - 0 = -2, \quad c = -4 - 3 = -7.$$
Using the first given point in (19.3.2), we obtain *parametric equations*
$$x = -2 + 3t, \quad y = -2t, \quad z = 3 - 7t.$$
Solving each of these equations for $t$ and equating the results, we obtain the *symmetric equations*
$$\frac{x+2}{3} = \frac{y}{-2} = \frac{z-3}{-7}.$$
The use of the second point instead of the first one leads to an equivalent result, that is, symmetric equations representing the same set of points.

*Example 19.3.2.* Express in symmetric form equations of the straight line through the point $(-2, -3, 4)$, with direction numbers 3, 0, 4.

*Solution.* Since the second direction number is zero, we use the first and third fractions in (19.3.3), and set the numerator of the second fraction equal to zero. Symmetric equations are then
$$\frac{x+2}{3} = \frac{z-4}{4}, \quad y = -3.$$

The parametric form (19.3.2) contains three independent equations in $x$, $y$, $z$, and $t$. The symmetric form (19.3.3) contains only two independent equations in $x$, $y$, and $z$, since, if two pairs of ratios are equal, the third pair must also be equal. Either form will yield equivalent equations, whichever point on the line is used.

If $\mathbf{r}_1$ is the position vector of a point on a line, and $\mathbf{m} = (a, b, c)$, where $a, b, c$ are direction numbers of the line, then $\mathbf{r} - \mathbf{r}_1$ is parallel to $\mathbf{m}$, and, by Theorem 18.7.2, we have
$$\mathbf{m} \times (\mathbf{r} - \mathbf{r}_1) = 0 \quad \text{and} \quad \mathbf{m} \times \mathbf{r} = \mathbf{m} \times \mathbf{r}_1.$$
Now letting $\gamma = \mathbf{m} \times \mathbf{r}_1$ we obtain the equation

(19.3.4) $$\mathbf{m} \times \mathbf{r} = \gamma.$$

This is also an equation of the straight line. The vector $\gamma$ in (19.3.4) is a constant vector with the property that $\mathbf{m} \cdot \gamma = 0$, for, using Theorem 18.8.2, we have
$$\mathbf{m} \cdot \gamma = \mathbf{m} \cdot (\mathbf{m} \times \mathbf{r}_1) = (\mathbf{m} \times \mathbf{m}) \cdot \mathbf{r}_1 = 0.$$
We shall now establish the converse if this result.

**THEOREM 19.3.1.** *If $\mathbf{m} \neq 0$ and $\gamma$ are constant vectors such that $\mathbf{m} \cdot \gamma = 0$,* *then the equation*

(19.3.5) $$\mathbf{m} \times \mathbf{r} = \boldsymbol{\gamma}$$

*is an equation of a straight line parallel to* $\mathbf{m}$.

*Proof.* Let $\mathbf{m} = (a, b, c)$, $\boldsymbol{\gamma} = (c_1, c_2, c_3)$. At least one of the numbers $a$, $b$, and $c$ is not zero. If $c \neq 0$, let $\mathbf{r}_1 = (c_2/c, -c_1/c, 0)$. Then

$$\mathbf{m} \times \mathbf{r}_1 = (a, b, c) \times \left(\frac{c_2}{c}, -\frac{c_1}{c}, 0\right) = \left(c_1, c_2, \frac{-ac_1 - bc_2}{c}\right).$$

But $\mathbf{m} \cdot \boldsymbol{\gamma} = 0$. Hence $ac_1 + bc_2 + cc_3 = 0$ and $-ac_1 - bc_2 = cc_3$. Thus the above equation becomes

(19.3.6) $$\mathbf{m} \times \mathbf{r}_1 = (c_1, c_2, c_3) = \boldsymbol{\gamma},$$

so that the vector $\mathbf{r}_1$ satisfies (19.3.5). Using (19.3.5) and (19.3.6), we have

$$\mathbf{m} \times \mathbf{r} = \mathbf{m} \times \mathbf{r}_1 \quad \text{and} \quad \mathbf{m} \times (\mathbf{r} - \mathbf{r}_1) = \mathbf{0}.$$

The latter equation is satisfied for any $\mathbf{r}$ such that

(19.3.7) $$\mathbf{r} - \mathbf{r}_1 = t\mathbf{m}, \quad t \text{ any real number.}$$

Now (19.3.7) is an equation of the straight line through the point with position vector $\mathbf{r}_1$ and parallel to $\mathbf{m}$, and the theorem is proved if $c \neq 0$.

If $c = 0$ and $b \neq 0$ we may take $\mathbf{r}_1 = (-c_3/b, 0, c_1/b)$, while if $b = c = 0$ and $a \neq 0$ we may take $\mathbf{r}_1 = (0, c_3/a, -c_2/a)$. The proof proceeds in the same way in each case.

What has been established in Theorem 19.3.1 and the preceding comments is that the equation (19.3.5) has no solution $\mathbf{r}$ unless $\mathbf{m} \cdot \boldsymbol{\gamma} = 0$; while, if the latter condition is satisfied, then (19.3.5) has infinitely many solutions $\mathbf{r}$, the position vectors of the points on a straight line.

**EXERCISE GROUP 19.3**

In Exercises 1–4 write an equation of the indicated straight line in parametric vector form (19.3.1).

**1.** The line through the points $(-2, 3, 1)$ and $(4, 2, -2)$.

**2.** The line through the point $(2, 0, -3)$ and parallel to the vector $(1, 2, 0)$.

**3.** The line through the point $(4, -2, 1)$ and parallel to the $y$ axis.

**4.** The line through the point $(1, -1, 3)$ and parallel to the line

$$\frac{x}{3} = \frac{y+2}{7} = \frac{z-4}{-2}.$$

In Exercises 5–8 write equations of the indicated straight line in parametric form (19.3.2).

**5.** The line through the point $(5, 2, -3)$ with direction numbers $4, 0, -1$.

6. The line through the points $(3, -2, 2)$ and $(5, -4, 1)$.
7. The line through the point $(-4, 1, -1)$ and parallel to the vector $(6, 3, -3)$.
8. The line through the point $(0, 3, 0)$ and parallel to the $z$ axis.

In Exercises 9–12 write equations of the indicated straight line in symmetric form (19.3.3).

9. The line through the points $(-1, 3, -3)$ and $(0, 2, -1)$.
10. The line through the point $(1, 4, 3)$ and perpendicular to the plane $4x - 3y + 5z + 1 = 0$.
11. The line through the point $(-1, 0, 4)$ and parallel to the vector $(2, -2, 4)$.
12. The line through the origin and parallel to the line $x + 1 = y - 1 = \frac{1}{2}(z + 4)$.

In Exercises 13–16 write an equation of the indicated line in vector form (19.3.5).

13. The line through the point $(4, 1, 1)$ and perpendicular to the plane $7x + y - 3z = 2$.
14. The line through the points $(3, 2, -1)$ and $(5, 2, -1)$.
15. The line through the point $(1, 2, -3)$, with equal direction angles.
16. The line through the point $(-1, -1, 2)$ and parallel to the vector $3\mathbf{i} + \mathbf{j} - 2\mathbf{k}$.

17. Show that the following lines are parallel:

$$x + 4 = 3y = \frac{2z + 1}{3} \quad \text{and} \quad x = 6t, \ y = -1 + 2t, \ z = 4 + 9t.$$

18. Show that the following lines are perpendicular:

$$x = 3t, \ y = 2 - t, \ z = -1 + t \quad \text{and} \quad \frac{2x - 7}{2} = \frac{2y + 3}{4} = \frac{2 - 3z}{3}.$$

19. Rewrite the following equations in symmetric form:

$$\frac{x}{3} = \frac{-y}{2} = \frac{2z}{5}.$$

20. Rewrite the following equations in symmetric form:

$$\frac{2x + 3}{4} = \frac{2 - y}{3} = \frac{3z + 18}{2}.$$

21. Find equations of the line through the point $(2, -1, 4)$, which intersects the line $x = 3 + t, \ y = -5 + t, \ z = 7 - t$ at right angles.
22. Find equations of the line through the point $(-3, 1, 2)$, which intersects the line $x = 4 + 2t, \ y = 4 + t, \ z = 3 + 4t$ at right angles.

## 19.4 INTERSECTION OF TWO PLANES

*For the planes*

(19.4.1) $\quad a_1 x + b_1 y + c_1 z = t_1 \quad$ and $\quad a_2 x + b_2 y + c_2 z = t_2$

to be parallel (*including coincidence*) their normals must be parallel, and the condition

(19.4.2) $$\frac{a_1}{a_2} = \frac{b_1}{b_2} = \frac{c_1}{c_2}$$

must hold where again, *if any denominator is zero, the corresponding member of* (19.4.2) *is removed and replaced by the condition that its numerator is zero.*

**Example 19.4.1.** By the stated condition (19.4.2), the planes

$$4x + 2y - 6z = 3 \quad \text{and} \quad 2x + y - 3z = 3$$

are parallel; so also are the planes

$$x - 4z = 2 \quad \text{and} \quad -2x + 8z = 3.$$

*If the planes* (19.4.1) *are not parallel, they intersect in a straight line which consists of those points whose coordinates* $(x, y, z)$ *satisfy both equations in* (19.4.1). Equations (19.4.1), considered as simultaneous, thus represent a straight line, *if* (19.4.2) does not hold.

In the following example three methods will be presented whereby equations of a line given in the form (19.4.1) may be expressed in the forms of the preceding section. In the first two methods a point on the line and the direction of the line will be found directly. We shall obtain symmetric equations in each case; and the other forms are easily obtained from this one.

**Example 19.4.2.** Obtain equations of the line

(19.4.3) $$x - 4y - 3z = -2, \quad 6x + y + 2z = 23$$

in symmetric form.

*Solution. First method.* We find *any* two points on the line. To find a point we assign a value to one of the unknowns and solve the resulting equations for the other two. For example, setting $z = 1$, the equations become

$$x - 4y = 1, \quad 6x + y = 21.$$

these are satisfied by $x = 17/5$, $y = 3/5$. A point on the line is then $(17/5, 3/5, 1)$.

Setting $z = 0$, we find a second point on the line to be $(18/5, 7/5, 0)$. The line therefore has direction numbers

$$\frac{18}{5} - \frac{17}{5} = \frac{1}{5}, \quad \frac{7}{5} - \frac{3}{5} = \frac{4}{5}, \quad 0 - 1 = -1.$$

To avoid fractions we use the proportional set $1, 4, -5$ as direction numbers. Using the second point obtained above and the direction numbers $1, 4, -5$, we write symmetric equations of the line as

$$\frac{x - 18/5}{1} = \frac{y - 7/5}{4} = \frac{z}{-5}.$$

*Second method.* The line determined by the given planes lies in each plane and is therefore perpendicular to the normals of the planes. The coefficients of $x$, $y$, and $z$ in each equation are components of a normal vector. Thus

$$\mathbf{n}_1 = (1, -4, -3), \qquad \mathbf{n}_2 = (6, 1, 2)$$

are, respectively, normal vectors to the two planes. Their cross product is perpendicular to each. We find $\mathbf{n}_1 \times \mathbf{n}_2 = (-5, -20, 25)$. Hence $-5$, $-20$, and $25$ are a set of direction numbers of the line. We use the proportional set $1, 4, -5$ as direction numbers (these are the same as in the first method). We now find *one* point on the line, as in the first method, and write the corresponding symmetric equations.

*Third method.* We eliminate one of the variables from the two given equations and then eliminate a second one. If we multiply the second equation by 4 and add to the first equation, we eliminate $y$ and get $25x + 5z = 90$ or

(19.4.4) $$5x + z = 18.$$

Multiplying the first equation of (19.4.3) by 2, the second one by 3, and adding, we eliminate $z$ and get $20x - 5y = 65$, or

(19.4.5) $$4x - y = 13.$$

We now solve each of (19.4.4) and (19.4.5) for the common variable $x$, and set the results equal; this gives

$$x = \frac{y + 13}{4} = \frac{18 - z}{5},$$

which we now write in *symmetric form* as

$$\frac{x}{1} = \frac{y + 13}{4} = \frac{z - 18}{-5}.$$

From these equations it is seen that a set of direction numbers is $1, 4, -5$, and a point on the line is $(0, -13, 18)$. The latter may be verified by direct substitution in both equations of (19.4.3).

We may now obtain further insight into the symmetric equations of a line. A straight line lies in infinitely many planes. Any two of them determine the line. By eliminating one of the variables from two such equations we obtain an equation of a plane which contains the line and is perpendicular to a coordinate plane. Thus equation (19.4.4) represents a plane which contains the line of Example 19.4.2, and is perpendicular to the *xz* plane. This is called the *projecting plane* for the given line onto the *xz* plane. The same equation, considered only in the *xz* plane, represents a straight line which is the *projection* of the given line on the *xz* plane. To represent this projection in three-dimensional form, we write equations of the projection of the line of Example 19.4.2 on the *xz* plane as

$$5x + z = 18, \qquad y = 0.$$

Similar considerations apply to the projection of a straight line on any coordinate

520     LINES AND PLANES

plane. *Caution:* The projection of a line on a coordinate plane is obtained by eliminating the proper variable, not by setting it equal to zero.

Any symmetric equations of a line immediately yield the projections of the line on the coordinate planes, by using the three equations determined by the symmetric form.

***Example 19.4.3.*** Find the projections on the coordinate planes of the line

$$\frac{x-3}{2} = y = \frac{z-4}{3}.$$

*Solution.* The line lies in the plane

$$\frac{x-3}{2} = y \quad \text{or} \quad x - 2y = 3.$$

The $xy$ projection is therefore

$$x - 2y = 3, \quad z = 0.$$

The $xz$ projection is obtained from the first and third members above, and is

$$3x - 2z = 1, \quad y = 0.$$

Similarly, the $yz$ projection is

$$3y - z = -4, \quad x = 0.$$

***Example 19.4.4.*** Find the projections on the coordinate planes of the line

$$4x - y - z = 4, \quad 3x + y - 2z = 7.$$

*Solution.* By eliminating $z$ from the two equations we find the $xy$ projection of the line:

$$5x - 3y = 1, \quad z = 0.$$

Eliminating $y$ we have the $xz$ projection:

$$7x - 3z = 11, \quad y = 0.$$

Eliminating $x$ we have the $yz$ projection:

$$7y - 5z = 16, \quad x = 0.$$

**EXERCISE GROUP 19.4**

In Exercises 1–8 determine whether the two planes are parallel. If they are not parallel, obtain symmetric equations of the line of intersection.

1. $3x - 4y + 3z = 3$, $9x - 2y - 6z = 21$
2. $4x + 2y - 2z = 5$, $6x - 3z = 2 - 3y$
3. $x + 2y - z = 8$, $2x - 3y + 3z + 13 = 0$

4. $7x + 3y - 4z = 6$, $4x - 3y - 7z = -6$
5. $4x - 6y + 6z + 1 = 0$, $9y - 2 = 6x + 9z$
6. $5x + y - z = 4$, $2x - y + z = 3$
7. $5x + 6y + 2z = 49$, $8x - 4y - 7z + 10 = 0$
8. $x - y - z + 5 = 0$, $x - 3y + 6z = 23$

In Exercises 9–16 find, for the given line, those projections on the coordinate planes which exist.

9. $\dfrac{2x - 5}{3} = \dfrac{y + 1}{3} = \dfrac{2 - 3z}{2}$

10. $\dfrac{x + 4}{5} = \dfrac{2y + 1}{2} = \dfrac{3z}{4}$

11. $5x + 2y - z = 7$, $x - 2y - 2z = 2$

12. $x + 3y - 8z = 12$, $10x + 3y + 4z = 6$
13. $3x + 6y - 7z = 3$, $4x + 24y - 21z = 8$
14. $x = 2 - 3t$, $y = -1 + 4t$, $z = 7t$
15. $x = 9 + 5t$, $y = 4 - 3t$, $z = -1 - 4t$
16. $x = -3t$, $y = 4 - 9t$, $z = 2 + 4t$

## 19.5 INTERSECTIONS OF LINES AND PLANES

The intersection of two planes was discussed in Section 19.4. In this section we discuss the intersection of a line and a plane, and the intersection of two lines.

A straight line may be parallel to a plane (including the case where it lies in the plane), or it may intersect the plane in exactly one point.

*Example 19.5.1.* Find the intersection of the line

(19.5.1) $$\frac{x - 2}{3} = \frac{y}{2} = \frac{z + 1}{-2}$$

with the plane $3x + 7y - z = 12$.

*Solution. First method.* From the symmetric equations of the line we get two independent equations in $x$, $y$, and $z$. By equating the first and second members we get

$$2x - 3y = 4.$$

From the second and third members we get

$$y + z = -1.$$

Solving simultaneously the system of three equations consisting of these two and the equation of the plane, we find the point of intersection $(13/5, 2/5, -7/5)$.

*Second method.* We introduce a parameter into the equations of the line. This method is also applicable when the line is given in parametric form. Setting each member in (19.5.1) equal to $t$, we obtain parametric equations of the line,

(19.5.2) $$x = 2 + 3t, \quad y = 2t, \quad z = -1 - 2t.$$

Substitution of these in the equation of the plane gives

$$3(2 + 3t) + 7(2t) - (-1 - 2t) = 12, \quad \text{or} \quad 25t = 5.$$

Hence $t = 1/5$, and substitution of this value in (19.5.2) yields the same point of intersection as the first method.

Two straight lines in space may be parallel (including coincidence), or not; which of these is the case is easily determined from their direction numbers. If they are not parallel, they may still have no point of intersection, in which case they are called *skew* lines. We shall illustrate the methods to be used in considering intersections of straight lines for equations in symmetric form and also for equations in parametric form.

**Example 19.5.2.** Find the intersection of the lines

$$\frac{x-2}{3} = \frac{y}{2} = \frac{z+1}{-2} \quad \text{and} \quad \frac{x+1}{4} = \frac{y+2}{5} = \frac{z-1}{-2}.$$

*Solution. First method.* As in Example 19.5.1 we represent the first line by two equations in $x$, $y$, and $z$,

(19.5.3) $\qquad 2x - 3y = 4, \qquad y + z = -1.$

Similarly, the second line is represented by the equations

(19.5.4) $\qquad 5x - 4y = 3, \qquad 2y + 5z = 1.$

The coordinates of any point of intersection must satisfy *all four* equations in (19.5.3) and (19.5.4). These form a system of four equations in three unknowns. We may then solve three of them simultaneously and verify the solution in the fourth equation. We may solve, for example, the system

$$2x - 3y = 4, \qquad y + z = -1, \qquad 5x - 4y = 3.$$

A solution is found as $x = -1$, $y = -2$, $z = 1$. It is easily verified that these values satisfy the remaining equation, $2y + 5z = 1$. Hence $(-1, -2, 1)$ is the point of intersection of the two lines.

*Second method.* We express each line in parametric form. This was done in Example 19.5.1 for the first line. We introduce a *different parameter u* in the second line. In parametric form the two lines are

(19.5.5) $\qquad x = 2 + 3t, \qquad y = 2t, \qquad z = -1 - 2t$

and

(19.5.6) $\qquad x = -1 + 4u, \qquad y = -2 + 5u, \qquad z = 1 - 2u.$

Equations (19.5.5) and (19.5.6), taken together, form a system of six equations in the five unknowns $x$, $y$, $z$, $t$, and $u$, which may be solved as follows. Equating the values of $x$ we have

$$2 + 3t = -1 + 4u \quad \text{or} \quad 3t - 4u = -3.$$

Equating the values of $y$ we have

$$2t = -2 + 5u \quad \text{or} \quad 2t - 5u = -2.$$

The solution of the two equations in $t$ and $u$ is $t = -1$, $u = 0$. Substituting $t = -1$ in (19.5.5), we obtain $x = -1$, $y = -2$, $z = 1$. *Substitution of $u = 0$ in (19.5.6) yields the same values.* Hence $(-1, -2, 1)$ is again obtained as the point of intersection.

If two lines do not intersect, the relevant system of equations has no solution, and the system is *inconsistent*.

The methods given above may clearly be adapted to the case where a line is given as the intersection of two planes (as in Section 19.4).

We now present a method for finding the *undirected* (perpendicular) *distance* between a point and a plane. Fig. 19.5.1 depicts the general situation. A direct

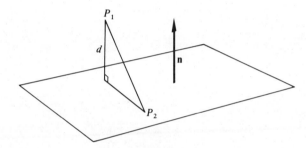

Fig. 19.5.1

method consists of finding equations of the line through $P_1$ and perpendicular to the plane, finding the intersection of this line and the plane, and using the distance formula. The following method is much shorter.

Choose an arbitrary point $P_2$ in the plane. Then $d$ is the absolute value of the projection of $\overrightarrow{P_2 P_1}$ on a normal $\mathbf{n}$ to the plane, and, by (18.5.9), we obtain

(19.5.7) $$d = \frac{|\overrightarrow{P_2 P_1} \cdot \mathbf{n}|}{|\mathbf{n}|}.$$

**Example 19.5.3.** Find the distance between the point $P_1(1, 4, 3)$ and the plane $4x - 3y + 5z + 1 = 0$.

*Solution.* We choose the point $P_2(-1/4, 0, 0)$ in the plane. We obtain

$$\overrightarrow{P_2 P_1} = (5/4, 4, 3), \quad \mathbf{n} = (4, -3, 5), \quad \overrightarrow{P_2 P_1} \cdot \mathbf{n} = 8, \quad |\mathbf{n}| = 5\sqrt{2}.$$

Then by (19.5.7), $d = 8/(5\sqrt{2}) = 4\sqrt{2}/5$.

A general formula for the distance between a point and a plane is requested in Exercise 25.

## EXERCISE GROUP 19.5

In Exercises 1–10 determine whether (a) the line and plane are parallel with no point in common, (b) the line lies in the plane, (c) the line and plane intersect in a single point. If the answer is (c), find the point of intersection.

1. $\dfrac{2x-1}{3} = \dfrac{y-2}{-3} = \dfrac{4-z}{2}$; $7x + 9y - 2z = 1$

2. $2x - y + 1 = 0$, $3x + z + 1 = 0$; $5x - 3y - z = 0$

3. $\dfrac{x}{2} = \dfrac{y + 1/2}{2} = \dfrac{z - 3/2}{-3}$; $7x - y + 4z = 2$

4. $\dfrac{x-2}{-1} = \dfrac{y+7}{3} = \dfrac{z-3}{-1}$; $4x + 3y + 5z = 2$

5. $x = 1 + 3t$, $y = 4 - 2t$, $z = -2 + 4t$; $3x - y + 4z = 3$

6. $x = 1 - 5t$, $y = 2 - 3t$, $z = 4t$; $2x - 2y + z + 2 = 0$

7. $2x - 3y + z = 7$, $4x + 3y + 2z = 8$; $6x + 6y - z = -5$

8. $\dfrac{x - 2/3}{2} = \dfrac{y + 1/2}{3} = \dfrac{z - 1}{-4}$; $4x - 2y + 5z = 25/6$

9. $\dfrac{x+3}{5} = \dfrac{y}{2} = \dfrac{z-2}{-4}$; $2x - y + 2z + 4 = 0$

10. $2x + 3y - 2z = 2$, $x - 4y + z = 4$; $x + 18y - 7z = -8$

In Exercises 11–18 determine whether the two lines (a) are parallel and distinct, (b) coincide, (c) are skew, (d) intersect in a single point. If the answer is (d), find the point of intersection.

11. $x = 1 + 3t$, $y = 4 - 2t$, $z = -2 + 4t$; $x = 4 - 3t$, $y = 2 + 2t$, $z = -8 + t$

12. $x = 16 + 7t$, $y = 2 - 2t$, $z = 4 - t$; $\dfrac{x-1}{2} = \dfrac{y+1}{-3} = \dfrac{z+5}{-4}$

13. $\dfrac{x-1}{4} = \dfrac{y}{2} = \dfrac{z+1}{-4}$; $\dfrac{x+3}{6} = \dfrac{y+2}{3} = \dfrac{3-z}{6}$

14. $x = 3t$, $y = -1 + 3t$, $z = 2 - t$; $x = 1 - 4t$, $y = 4t$, $z = 6 + t$

15. $(1, -1, 1) \times \mathbf{r} = (3, 1, -2)$; $(-2, 2, -2) \times \mathbf{r} = (4, 2, -2)$

16. $(1, -2, 4) \times \mathbf{r} = (16, 8, 0)$; $(3, 0, 5) \times \mathbf{r} = (10, 17, -6)$

17. $3x - 2y + 4z = 6$, $x + y - z = 7$; $2x + 3y - 8z = 17$, $2y + z = 6$

18. $x + y + z = 0$, $2x + y + 2z = 3$; $x - z = 2$, $2x + y - z = 4$

19. Show that the two given lines intersect in a point, and find an equation of the plane containing them:

$x = 4 - t$, $y = 2 + 3t$, $z = 1 - 2t$; $\quad x = 11 + 3t$, $y = 2 - 2t$, $z = 3 + 2t$.

20. Show that the two given lines intersect in a point, and find an equation of the plane containing them:

$\dfrac{x-7}{5} = \dfrac{6-y}{3} = z - 2$; $\quad \dfrac{x}{4} = \dfrac{6-y}{1} = \dfrac{z+6}{3}$.

21. Find equations of the line which intersects at right angles each of the lines

$$\frac{x-1}{1} = \frac{y}{-1} = \frac{z}{-1} \quad \text{and} \quad \frac{x+2}{1} = \frac{y+5}{1} = \frac{z-7}{5}.$$

22. Find equations of the line which intersects at right angles each of the lines

$$\frac{x+3}{-1} = \frac{y-4}{1} = \frac{z-4}{1} \quad \text{and} \quad \frac{x+6}{3} = \frac{y+1}{5} = \frac{z+3}{1}.$$

23. Find the distance between the point $(2, -1, 3)$ and the plane

$$2x - 2y + z + 1 = 0.$$

24. Find the distance between the point $(2, 0, 4)$ and the plane

$$7x - 4y + 4z = 1.$$

25. Use the method of Section 19.5 to show that the distance between the point $(x_1, y_1, z_1)$ and the plane $ax + by + cz = t$ is

$$d = \frac{|ax_1 + by_1 + cz_1 - t|}{\sqrt{a^2 + b^2 + c^2}}.$$

## *19.6  NORMAL FORM OF THE EQUATION OF A PLANE

Let us write the equation of a plane

(19.6.1) $$ax + by + cz = t$$

in vector form as

(19.6.2) $$\mathbf{n} \cdot \mathbf{r} = t,$$

where $\mathbf{n}$ is a vector normal to the plane. We divide both sides of (19.6.2) by $\pm |\mathbf{n}|$, using the sign of $t$ if $t \neq 0$. If $t = 0$ and $c \neq 0$, we use the sign of $c$; if both $t = 0$ and $c = 0$, but $b \neq 0$, we use the sign of $b$. If $t = 0$, $c = 0$, and $b = 0$, we take $\mathbf{N} = (1, 0, 0)$. In all cases, we obtain

(19.6.3) $$\mathbf{N} \cdot \mathbf{r} = p, \quad p \geq 0,$$

where

$$\mathbf{N} = \frac{\mathbf{n}}{\pm|\mathbf{n}|}, \quad p = \frac{t}{\pm|\mathbf{n}|}.$$

It is clear that $\mathbf{N}$ is a unit vector. Equation (19.6.3) is the *vector normal form of the equation of a plane*.

Since $\mathbf{N}$ is a unit vector, it may be written in terms of its direction angles as

$$\mathbf{N} = (\cos \alpha, \cos \beta, \cos \gamma).$$

It is seen that $\mathbf{r}_0 = p\mathbf{N}$ satisfies (19.6.3), since

$$\mathbf{N} \cdot \mathbf{r}_0 = \mathbf{N} \cdot (p\mathbf{N}) = p\mathbf{N} \cdot \mathbf{N} = p.$$

Hence, the corresponding point $P_0(p \cos \alpha, p \cos \beta, p \cos \gamma)$ lies in the plane (19.6.3), $\overrightarrow{OP_0}$ is perpendicular to the plane, and $|\overrightarrow{OP_0}| = p$ (Fig. 19.6.1).

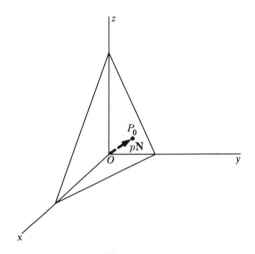

Fig. 19.6.1

In nonvector form equation (19.6.3) becomes

(19.6.4) $\qquad x \cos \alpha + y \cos \beta + z \cos \gamma = p, \qquad p \geq 0.$

This is the *normal form of the equation of a plane* in which $\alpha$, $\beta$, and $\gamma$ are direction angles of the normal vector $\overrightarrow{OP_0}$, and $p = |\overrightarrow{OP_0}|$.

The general form of the equation of a plane, (19.6.1), is converted to normal form by dividing both members by $\pm(a^2 + b^2 + c^2)^{1/2}$, determining the sign as above. If $t = 0$, $c = 0$, and $b = 0$, the normal form is simply $x = 0$.

**Example 19.6.1.** Write the equation

$$2x - 3y - z + 2 = 0$$

in normal form, and find the normal vector $\overrightarrow{OP_0}$ where $P_0$ is in the plane.

*Solution.* We divide both members of the given equation by

$$-(4 + 9 + 1)^{1/2} = -14^{1/2};$$

the minus sign is chosen, since $t < 0$ ($t = -2$). The *normal form* is then

$$\frac{-2}{\sqrt{14}} x + \frac{3}{\sqrt{14}} y + \frac{1}{\sqrt{14}} z = \frac{2}{\sqrt{14}}.$$

We then have $\cos \alpha = -2/\sqrt{14}$, $\cos \beta = 3/\sqrt{14}$, $\cos \gamma = 1/\sqrt{14}$, $p = 2/\sqrt{14}$. To find $\overrightarrow{OP_0}$ we obtain, by the above,

$$\overrightarrow{OP_0} = (p \cos \alpha, p \cos \beta, p \cos \gamma) = (-\tfrac{2}{7}, \tfrac{3}{7}, \tfrac{1}{7}).$$

The point $(-2/7, 3/7, 1/7)$ is in the plane and is the intersection of the normal from $O$ with the plane.

Consider a plane

(19.6.5) $$x \cos \alpha + y \cos \beta + z \cos \gamma = p'$$

parallel to the plane (19.6.4) in normal form. The two planes are on the same side of the origin if $p' > 0$, and on opposite sides if $p' < 0$ (in which case (19.6.5) is not in normal form). In either case, the directed distance from the plane (19.6.4) to the plane (19.6.5) is $p' - p$, *the positive direction along the normal being taken as the direction from the origin to the first plane.*

Now let $P_1(x_1, y_1, z_1)$ be a given point. The plane through $P_1$ and parallel to the plane (19.6.4) is given by (19.6.5) where

$$p' = x_1 \cos \alpha + y_1 \cos \beta + z_1 \cos \gamma.$$

The directed distance from (19.6.4) to the parallel plane through $P_1$ is then

$$p' - p = x_1 \cos \alpha + y_1 \cos \beta + z_1 \cos \gamma - p,$$

even when $p = 0$ or $p' = 0$. Since this directed distance is also the directed distance from (19.6.4) to $P_1$, we have proved the following theorem.

**THEOREM 19.6.1.** *The directed distance from the plane, in normal form,*

$$x \cos \alpha + y \cos \beta + z \cos \gamma = p$$

*to the point $P_1(x_1, y_1, z_1)$ is*

(19.6.6) $$x_1 \cos \alpha + y_1 \cos \beta + z_1 \cos \gamma - p,$$

*the positive normal direction being taken from the origin to the given plane.*

Suppose $p \neq 0$. If the directed distance given by Theorem 19.6.1 is positive, the point and the origin are on opposite sides of the plane. If the distance is negative, the point and the origin are on the same side of the plane.

**Example 19.6.2.** Find the directed distance from the plane

$$2x - 3y - 2z + 4 = 0$$

to the point $(-3, 2, 0)$.

*Solution.* The normal form of the equation is

$$-\frac{2}{\sqrt{17}} x + \frac{3}{\sqrt{17}} y + \frac{2}{\sqrt{17}} z = \frac{4}{\sqrt{17}}.$$

By Theorem 19.6.1 the directed distance from the plane to the point $(-3, 2, 0)$ is

$$\frac{(-2)(-3) + 3(2) + 2(0) - 4}{\sqrt{17}} = \frac{8}{\sqrt{17}}.$$

The point and the origin are on opposite sides of the plane.

From the preceding discussion it is clear that two points are on the same side of the plane (19.6.4) if the directed distances (19.6.6) have the same sign for these points, and are on opposite sides if the directed distances have opposite signs. A corresponding statement follows for an equation in general form.

**THEOREM 19.6.2.** *Two points lie on the same side of the plane*

(19.6.7) $$ax + by + cz - t = 0$$

*if the left member of* (19.6.7) *has the same sign for both points, and are on opposite sides if the left member has opposite signs.*

*Example 19.6.3.* The points $(2, 3, -1)$ and $(-4, 2, 3)$ are on opposite sides of the plane $2x + 4y - 3z = 4$, since

for the first point: $2(2) + 4(3) - 3(-1) - 4 = 15 > 0$;
for the second point: $2(-4) + 4(2) - 3(3) - 4 = -13 < 0$.

⋆ **EXERCISE GROUP 19.6**

In Exercises 1–8 write the equation of the given plane in normal form.

1. $x - 2y - 2z = 4$
2. $2x + y + 2z = 3$
3. $2x + 3y - 6z + 1 = 0$
4. $4x - 7y + 4z = 0$
5. $2y + 3z = 0$
6. $5x - 12y = 0$
7. $7x - 3y = 2$
8. $3x + 4z = 1$

In Exercises 9–12 find the point $P_0$ in the given plane such that $OP_0$ is perpendicular to the plane.

9. $-2x - 2y + z = 2$
10. $x - 3y - 3z + 2 = 0$
11. $5z - 4x + 2 = 0$
12. $2x + 5y = 9$

In Exercises 13–18 find the directed distance from the given plane to the given point. State whether the origin and the point are on the same side or on opposite sides of the plane.

13. $2x - 2y + z + 1 = 0$, $(2, -1, -3)$
14. $2x + y + 2z + 3 = 0$, $(-1, 3, 4)$
15. $7x - 4y + 4z = 1$, $(2, 0, 4)$
16. $3x + 6y - 2z = -3$, $(-3, 0, 2)$
17. $2y + 3z = 0$, $(1, -2, -2)$
18. $5x - 12y = 0$, $(-3, -4, -2)$

In Exercises 19–22 determine which of the given points are on the same side of the given plane as the point $(2, 2, -4)$.

19. $8x - 4y + 3z = 4$; (a) $(1, 2, 1)$, (b) $(-4, -6, 5)$, (c) $(0, 0, 1)$

20. $3x + 7y - z = 3$; (a) $(1, -1, 1)$, (b) $(-3, 1, 2)$, (c) $(1, 0, -1)$
21. $2x - 7y = -3$; (a) $(1, 1, 1)$, (b) $(-4, -1, 7)$, (c) $(6, 2, -3)$
22. $4x - 3z - 5 = 0$; (a) $(1, 1, 1)$, (b) $(1, -1, -1)$, (c) $(5, 7, 6)$
23. In each part find equations of the planes which bisect the angles formed by the two given planes:
    (a) $3x - 2y + 6z = 1$, $2x + 6y + 3z = -4$
    (b) $2x + 2y - 3z = 4$, $2x + 8z = 5$.

## * 19.7  INTERSECTION OF THREE PLANES

The following are the different possibilities that may occur with three planes

(a) *All three planes are parallel.* This includes the cases where (i) all three planes coincide, so that the intersection of the three planes is the common plane, or (ii) exactly two planes coincide and there is no point of intersection of the three planes.
(b) *Two planes are parallel, but the third is not.* If the two parallel planes coincide, there is a *line of intersection* of the three planes. Otherwise, there is no point of intersection of the three planes; however, the third plane intersects the other two in parallel lines.
(c) *No two planes are parallel, but the pairwise lines of intersection are parallel.* This includes the case where the planes intersect in a single line.
(d) *The three planes intersect in a single point.*

Rather than undertake a systematic algebraic analysis of the various possibilities, we shall discuss some of these by means of examples. *The basic procedure in each case is to try to solve the equations simultaneously*, by the method of elimination, and to interpret the procedure and the result.

**Example 19.7.1.** Discuss the intersection of the planes

$$2x - 3y + 6z = 0, \quad 2x + y - 6z = 8, \quad 4x - y - 3z = 10.$$

*Solution.* We note that no two of the planes are parallel. Eliminating $y$, for example, from *any two* of the three equations, we find

$$2x - 3z = 6.$$

The same equation is obtained whichever two of the given equations are used. Hence the intersection of any two of the three planes lies in the plane $2x - 3z = 6$. Since no two planes are parallel, *the planes intersect in a single line* which is represented by any two of the three given equations. This example comes under case (c) above.

**Example 19.7.2.** Discuss the intersection of the planes

$$2x + 3y - z = 2, \quad -x + 2y + 4z = 5, \quad x - 9y - 11z = 1.$$

*Solution.* Elimination of $x$ from the first two equations gives

(19.7.1) $$7y + 7z = 12.$$

Elimination of $x$ from the first and third equations gives

(19.7.2) $$y + z = 0.$$

But (19.7.1) and (19.7.2) represent two distinct parallel planes. Hence the given planes do not intersect. Since no two of the planes are parallel (verify this!), the planes intersect in three distinct parallel lines; each of these lines is represented by a pair of the given equations. This example also falls under case (c).

*Example 19.7.3.* Discuss the intersection of the planes

$$7x + y - 4z = 12, \quad 3x + 5y - 3z = -2, \quad 5x - 3y - 4z = 10.$$

*Solution.* Eliminating $y$ from the first two equations, and then from the first and third equations, we get

$$32x - 17z = 62, \quad 13x - 8z = 23.$$

These have the single solution $x = 3$, $z = 2$. Substitution of these values in one of the given equations gives $y = -1$. Hence the three equations have the *simultaneous, unique solution* $x = 3$, $y = -1$, $z = 2$; and thus $(3, -1, 2)$ is the *unique point of intersection* of the three planes. This example, of course, falls under case (d).

## * EXERCISE GROUP 19.7

In each of the following exercises find the unique point of intersection of the three planes, if such point exists. In other cases describe the relation among the planes, and state which case of Section 19.7 applies.

1.  $3x - 4y + z = -2$
    $5x + y - 2z = 1$
    $7x + 3y + 2z = 19$

2.  $5x - y + 11z = 21$
    $-3x + 4y + 5z = 29$
    $2x - 7y + 9z = 9$

3.  $6x - 10y + 18z = 3$
    $12x + 20y - 3z = 15/2$
    $15x - 13y - 24z = -8/5$

4.  $7x + y - 4z = 8$
    $x + 4y - 7z = -1$
    $2x - y + z = -3$

5.  $3x - y - 2z = 2$
    $7x - 2y - 5z = 5$
    $2x - 9y + 7z = -7$

6.  $x - 5y - 7z = 6$
    $2x + y + 5z = 9$
    $4x - 3y + 9z = 27$

7.  $9x - 8y + 10z = 8$
    $21x + 16y - 35z = -2$
    $-19x + 17y + 40z = 385/12$

8.  $6x + 7y - 10z + 3 = 0$
    $13x - 5y + 17z = 54$
    $-5x - 11y + 13z = 18$

9.  $4x + 6y - 2z = 2$
    $2x + 3y - z = 4$
    $2x + 3y - z = 1$

10. $-5x + 6y + 19z = 12$
    $3x - 4y - 8z = -4$
    $2x + 8y - 11z = -3$

11. $3x - 5y + 6z + 33 = 0$
    $5x + 7y - 2z = 39$
    $7x - 3y - 5z - 25 = 0$

12. $2x - y - 2z = -1$
    $5x - 3y - 4z = 0$
    $3x - 5y + 4z = 16$

13. $4x + 6y - 2z = 5$
    $x - 3y + z = 2$
    $6x + 9y - 3z = 4$

14. $3x - 2y - z = 8$
    $x - 2y + z = -2$
    $5x - 3y - 2z = 10$

## *19.8 SPECIALIZED DISTANCE FORMULAS

In this section we obtain rather specialized formulas for the (perpendicular) distance from a point to a line, and for the (perpendicular) distance between two lines. In both cases the distances are undirected.

To find the perpendicular distance from a point $P_0(\mathbf{r}_0)$ to a line $L$ directly can be cumbersome. We should find equations of the line through $P_0$ which intersects $L$ at right angles (as in Exercises 21 and 22 of Exercise Group 19.3), find the point of intersection, and then use the distance formula (18.3.3) to find $d$ (Fig. 19.8.1). The following alternative method is surprisingly brief.

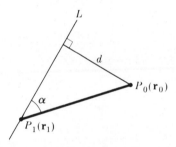

Fig. 19.8.1

Let $P_1(\mathbf{r}_1)$ with position vector $\mathbf{r}_1$ be *any* point on $L$, and let $\alpha$ be an angle $\overrightarrow{P_1 P_0}$ makes with $L$ (Fig. 19.8.1). Let $\mathbf{m}$ be a vector parallel to $L$. Then by (18.7.6) we have

$$|\mathbf{m} \times \overrightarrow{P_0 P_1}| = |\mathbf{m}| |\overrightarrow{P_0 P_1}| \sin \alpha = |\mathbf{m}| d,$$

the last equality since $d = |\overrightarrow{P_0 P_1}| \sin \alpha$ from Fig. 19.8.1. Hence

(19.8.1) $$d = \frac{|\mathbf{m} \times \overrightarrow{P_0 P_1}|}{|\mathbf{m}|}.$$

Equation (19.8.1) gives the perpendicular distance between $P_0$ and $L$. In this formula $P_1(\mathbf{r}_1)$ is *any* point on the line.

**Example 19.8.1.** Find the distance between the point $P_0(2, -1, -3)$ and the line

$$\frac{x-1}{4} = \frac{y+2}{-2} = \frac{z+1}{5}.$$

*Solution.* We take a point $P_1$ on the line as $(1, -2, -1)$ and the vector **m** as $\mathbf{m} = (4, -2, 5)$. Then $\overrightarrow{P_0 P_1} = (-1, -1, 2)$, and, by (19.8.1), we have

$$d = \frac{|(4, -2, 5) \times (-1, -1, 2)|}{|(4, -2, 5)|} = \frac{|(1, -13, -6)|}{|(4, -2, 5)|} = \frac{\sqrt{206}}{\sqrt{45}} = \frac{\sqrt{206}}{3\sqrt{5}}.$$

The use of any other point on the line as $P_1$ leads to the same result.

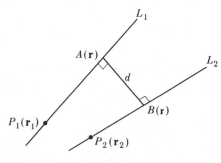

Fig. 19.8.2

Suppose $L_1$ and $L_2$ are two nonparallel lines in space, parallel to vectors $\mathbf{m}_1$ and $\mathbf{m}_2$. If $AB$ is perpendicular to both $L_1$ and $L_2$ (Fig. 19.8.2), where $A$ is on $L_1$ and $B$ on $L_2$, then $\overrightarrow{AB}$ is parallel to $\mathbf{m}_1 \times \mathbf{m}_2$, and $d = |\overrightarrow{AB}|$ is the absolute value of the projection of $\overrightarrow{P_1 P_2}$ on $\overrightarrow{AB}$. Thus by (18.5.9) we have, since $\overrightarrow{P_1 P_2} = \mathbf{r}_2 - \mathbf{r}_1$,

(19.8.2) $$d = |\overrightarrow{AB}| = \frac{|(\mathbf{r}_2 - \mathbf{r}_1) \cdot (\mathbf{m}_1 \times \mathbf{m}_2)|}{|\mathbf{m}_1 \times \mathbf{m}_2|}.$$

*Equation (19.8.2) yields the perpendicular distance between two lines, $L_1$ and $L_2$, which are not parallel, where $\mathbf{m}_1$ and $\mathbf{r}_1$ are, respectively, a vector parallel to, and a position vector of an arbitrary point on, $L_1$, and $\mathbf{m}_2$ and $\mathbf{r}_2$ are similarly defined for $L_2$.*

**Example 19.8.2.** Find the perpendicular distance between the lines

$$\frac{x-1}{2} = \frac{y-3}{3} = \frac{z+1}{1} \quad \text{and} \quad \frac{x+2}{3} = \frac{z-5}{4}, \quad y = 0.$$

*Solution.* We may take

$$\mathbf{m}_1 = (2, 3, 1), \quad \mathbf{r}_1 = (1, 3, -1); \quad \mathbf{m}_2 = (3, 0, 4), \quad \mathbf{r}_2 = (-2, 0, 5).$$

We find
$$\mathbf{r}_2 - \mathbf{r}_1 = (-3, -3, 6), \quad \mathbf{m}_1 \times \mathbf{m}_2 = (12, -5, -9),$$
$$(\mathbf{r}_2 - \mathbf{r}_1) \cdot (\mathbf{m}_1 \times \mathbf{m}_2) = -75, \quad |\mathbf{m}_1 \times \mathbf{m}_2| = \sqrt{250} = 5\sqrt{10}.$$

Then formula (19.8.2) yields, as the distance,
$$d = \frac{|-75|}{5\sqrt{10}} = \frac{15}{\sqrt{10}} = \frac{3\sqrt{10}}{2}.$$

## * EXERCISE GROUP 19.8

In Exercises 1–8 find the distance between the given point and the given line.

1. $(-1, 4, 3)$; $\dfrac{x+3}{2} = \dfrac{y-7}{5} = \dfrac{z+2}{-4}$

2. $(4, 0, 5)$; $\dfrac{x-2}{4} = \dfrac{y+1}{2} = \dfrac{1-z}{3}$

3. $(4, -2, 3)$; $x = 1 - 2t, y = 4 + t, z = -3 + 2t$

4. $(-5, 1, 6)$; $x = 2 - 6t, y = -3t, z = 4 + 2t$

5. $(2, 1, 2)$; $3x + y = 7, 2x - z = 1$

6. $(1, -3, 3)$; $2x - y + z = 4, 3x + 2y - z = 3$

7. $(3, -4, 1)$; the line through the points $(2, 1, 3)$ and $(1, 0, -1)$

8. $(0, 2, 0)$; the line through the points $(3, -1, -1)$ and $(2, 1, 0)$

In Exercises 9–14 find the distance between the given lines.

9. $\dfrac{x-1}{2} = \dfrac{y+1}{3} = z - 2$; $\dfrac{x}{3} = \dfrac{y-2}{4} = \dfrac{z}{-1}$

10. $\dfrac{x+4}{-1} = \dfrac{z}{4}, y = 0$; $\dfrac{x-2}{5} = \dfrac{y+1}{2} = z$

11. $x = t, y = 1 + 2t, z = -2t$; $x = 1 - t, y = 3t, z = 1 + t$

12. $x = 1 + 3t, y = 4t, z = 5 - t$; $x = 2t, y = -1 - t, z = 4 + t$

13. $x - 2y = 3, y - 2z = 1$; $2x + z = 4, x + y + z = 2$

14. $x = 4, y = 2$; $2x + y = 7, x - z = 1$

15. Find the distance between the line $x/2 = (y+3)/2 = -z$ and the intersection of the lines
$$\frac{x-1}{3} = \frac{y-2}{3} = z \quad \text{and} \quad \frac{x}{2} = \frac{y+1}{3} = z + 1.$$

16. Find the distance between the line $x = 1 + 5t, y = 2 + 3t, z = t$ and the intersection of the lines
$$x = 4 - 2t, y = 2 + t, z = 1 - t \quad \text{and} \quad x = t, y = -1 + 2t, z = -2 + t.$$

# 20 Surfaces and Curves

## 20.1 INTRODUCTION

In Chapter 19 we found that a plane can be represented by a *linear equation* in three variables; a plane is one type of surface. In this and following chapters we study more general surfaces. In general, a surface in 3-dimensional space is represented by a single equation in $x$, $y$, and $z$ in the sense that it is the locus or set of points $(x, y, z)$ whose coordinates satisfy the equation.

A curve in space may be considered as an intersection of two surfaces in many different ways. If two surfaces intersect in a curve, that curve is the locus or set of points $(x, y, z)$ whose coordinates satisfy simultaneously equations of the two surfaces.

We shall study certain aspects of surfaces and curves, and then consider special types.

## 20.2 SKETCHING A SURFACE

The sketching of a surface from an equation involves procedures which are different from those used in drawing the graph of an equation in a coordinate plane. For example, plotting of points is generally less helpful. A very useful aid in sketching a surface is finding the intersections of the surface with various planes. In particular, we use *sections* of a surface, which are *curves of intersection of the surface with planes parallel to a coordinate plane*. More particularly, the *traces* of a surface are the *intersections of the surface with the coordinate planes themselves*. In many cases, the knowledge of the traces will suggest at once the sketch of the surface. In other cases, other sections of the surface will be needed, in addition to the traces.

**Example 20.2.1.** Sketch the surface $x^2 + 2y^2 = 2 - z$.

*Solution.* The trace in the $xy$ plane, or the $xy$ trace, is obtained by letting $z = 0$ in the equation, and is

$$z = 0, \quad x^2 + 2y^2 = 2.$$

This curve is represented by two equations; it lies in the plane $z = 0$, and in that plane it is the ellipse $x^2 + 2y^2 = 2$, which is shown in Fig. 20.2.1. Similarly, the trace in the $yz$ plane is the curve with equations

$$x = 0, \quad 2y^2 = 2 - z.$$

This is a parabola in the $yz$ plane, opening in the direction of the negative $z$ axis, with vertex at $(0, 0, 2)$. Finally, the trace in the $xz$ plane is

$$y = 0, \quad x^2 = 2 - z,$$

and is also a parabola. The drawing of the three traces in Fig. 20.2.1 is sufficient to sketch the surface. The surface continues below the $xy$ plane; however, the appearance of this continuation is clear, and the surface is not shown there.

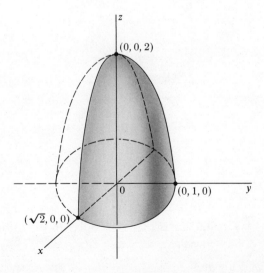

Fig. 20.2.1

In the following example the use of sections other than traces is also illustrated.

**Example 20.2.2.** Sketch the surface $2x^2 + y^2 = 4z$.

*Solution.* The $xy$ trace has an equation $2x^2 + y^2 = 0$, in the $xy$ plane, and is a single point, the origin. The $yz$ trace is

$$x = 0, \qquad y^2 = 4z,$$

which is a parabola in the $yz$ plane with vertex at the origin, and opening upward. The $xz$ trace is

$$y = 0, \qquad x^2 = 2z,$$

which is also a parabola, in the $xz$ plane. When sketched, these traces give a fairly good idea of the appearance of the surface. If we now consider a section of the surface by a plane $z = k > 0$, we have

$$z = k, \qquad 2x^2 + y^2 = 4k, \qquad k > 0.$$

For any positive value of $k$ this section is an ellipse in the plane $z = k$. The surface may now be sketched as in Fig. 20.2.2. Note that if $k < 0$, there is no section at all.

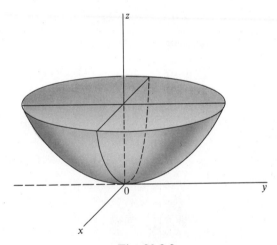

Fig. 20.2.2

The surfaces in the preceding examples have *symmetry* with respect to the $yz$ plane and the $xz$ plane; such symmetries may be utilized in sketching. Tests for symmetry in the coordinate planes are similar to the earlier tests (Section 6.2) for symmetry in the coordinate axes in a plane, and may be stated as follows.

I. *If replacing $z$ by $-z$ yields an equivalent equation, the surface is symmetric in the $xy$ plane.*

II. *If replacing $x$ by $-x$ yields an equivalent equation, the surface is symmetric in the yz plane.*

III. *If replacing $y$ by $-y$ yields an equivalent equation, the surface is symmetric in the xz plane.*

In addition, a test for symmetry in the origin may be stated as follows.

IV. *If replacing $x$ by $-x$, $y$ by $-y$, and $z$ by $-z$ yields an equivalent equation, the surface is symmetric in the origin.*

Tests for symmetry in the coordinate axes may be similarly stated and are left to the exercises.

*Intercepts* of a surface are defined in a similar manner to those of a curve in a plane (Section 6.2), and may also be used as an aid in sketching. It is clear that any intercept of a surface is also an intercept of appropriate traces of the surface.

**EXERCISE GROUP 20.1**

In Exercises 1–8 find equations of the traces of the given surface, and identify each trace.

1. $4x^2 + 4y^2 + 9z^2 = 4$
2. $y^2 + 4z - 2x = 0$
3. $4x^2 - 3y = 0$
4. $2x^2 + y^2 - z = 0$
5. $x^2 + y^2 + 2z^2 + 4x = 4$
6. $x^2 - 4y^2 - 4z^2 = 1$
7. $2x + 3y - 5z + 2 = 0$
8. $x^2 + 4y^2 - z^2 = 0$

In Exercises 9–18 sketch the given surface, using intercepts, traces, sections, and symmetry when appropriate.

9. $x^2 + 4y^2 = 4z$
10. $x^2 + z^2 - 2x = 0$
11. $x^2 + 4y^2 = z^2$
12. $x^2 + 4y^2 + 9z^2 = 4$
13. $x^2 - y^2 + z^2 = 4$
14. $x^2 + y^2 - z^2 + 4 = 0$
15. $x^2 + 2y^2 = 4$
16. $x^2 - y^2 + 2z^2 = 0$
17. $x^2 + y^2 + z^2 = 9$
18. $x^2 - y^2 = 0$

19. State a test for symmetry of a surface in the $x$ axis.
20. State a test for symmetry of a surface in the $y$ axis.
21. State a test for symmetry of a surface in the $z$ axis.

## 20.3 CYLINDRICAL SURFACES

Before considering a particular class of surfaces (Section 20.5), we discuss certain general types of surface.

A *cylinder* or *cylindrical surface* is the locus of points on a family of parallel straight lines such that every line of the family intersects a specified curve. Each line of the family is called an *element* or *ruling* (see also Section 20.6) of the cylinder; the specified curve is the *directrix*.

In this sense a plane may be described as a cylindrical surface. A *right circular cylinder* has rulings which intersect a circle and are perpendicular to the plane of the circle.

We may also say that a cylinder is generated by a straight line which moves parallel to itself in such a way that it always intersects a specified curve.

In general, we shall deal with cylinders whose rulings are parallel to a coordinate axis. Equations of such cylinders are particularly easy to recognize; *an equation of a cylinder with rulings parallel to a coordinate axis lacks the variable corresponding to that axis.* For example, suppose that an equation of a surface contains only $x$ and $y$. If $(x_0, y_0, 0)$ is a point on the surface, then $(x_0, y_0, z)$ satisfies the equation and is on the surface for any value of $z$. Thus the surface contains the straight line $x = x_0$, $y = y_0$, which is parallel to the $z$ axis.

*Example 20.3.1.* Discuss and sketch the surface $y^2 = 4z$.

*Solution.* The surface is cylindrical, with rulings parallel to the $x$ axis. The $yz$ trace is the parabola $y^2 = 4z$, $x = 0$, and any section in a plane $x = a$ parallel to the $yz$ plane is a congruent parabola in that plane. A sketch of the surface appears in Fig. 20.3.1.

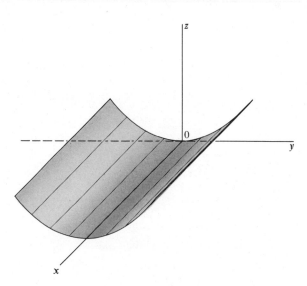

Fig. 20.3.1

Cylinders may be characterized as *parabolic, elliptic* (including circular), *hyperbolic,* and so on, depending on the nature of the directrix. The surface in Example 20.3.1 is a *parabolic cylinder.*

One should not conclude that the graph of an equation with all three variables present is automatically not a cylinder—it may be a cylinder with rulings not parallel to any coordinate axis. One may state, however, that in this case the surface is not a cylinder with rulings parallel to a coordinate axis.

## 20.4 SURFACES OF REVOLUTION

An important type of surface which is easily studied in certain positions is a *surface of revolution*. Such a surface is defined as *the surface obtained by revolving a plane curve about a straight line in the plane of the curve*; this straight line is called the *axis of revolution*. A surface of revolution then has the property that the intersections of the surface with a family of parallel planes which are perpendicular to the axis of revolution are circles whose centers lie on that axis. We shall consider, in particular, surfaces of revolution about a coordinate axis.

Consider a surface of revolution about the $y$ axis, generated by a curve $f(y, z) = 0$. Since the image of the curve in the $y$ axis generates the same surface, we may use $f(y, |z|) = 0$ as the generating curve. As in Fig. 20.4.1 a typical point $P(x, y, z)$ on the surface arises from the rotation of the point $Q(0, Y, Z)$ in the $yz$ plane along a circle which has its center on the $y$ axis and is in a plane perpendicular to the $y$ axis. (The $y$ coordinates of $P$ and $Q$ are equal, but in general the $z$ coordinates are not.) If $P(x, y, z)$ is a point on the circle and $r$ is the radius of the circle, then

(20.4.1) $$r = |Z| = \sqrt{x^2 + z^2},$$

from which

(20.4.2) $$Z = \pm\sqrt{x^2 + z^2} \quad \text{or} \quad Z^2 = x^2 + z^2.$$

This is sufficient to permit us to obtain an equation of a surface of revolution about the $y$ axis.

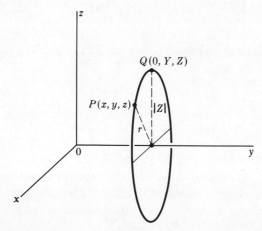

Fig. 20.4.1

*Example 20.4.1.* Find an equation of the surface obtained by revolving the curve $z^2 = 2y$, $x = 0$, about the $y$ axis.

*Solution.* Let $P(x, y, z)$ be a point on the surface which arises from a point $Q(0, Y, Z)$ on the given curve. From the equation of the curve we may write

(20.4.3) $$Z^2 = 2Y.$$

From (20.4.2) we have $Z^2 = x^2 + z^2$. We also have $Y = y$. Substituting these in (20.4.3), we obtain as an equation of the surface of revolution

$$x^2 + z^2 = 2y.$$

The graph of this equation is a surface similar to that of Fig. 20.2.2, but opening along the positive $y$ axis.

*Example 20.4.2.* Find an equation of the surface obtained by revolving the curve of Example 20.4.1 about the $z$ axis.

*Solution.* By analogy with Example 20.4.1 we see that a point $(x, y, z)$ on the surface originates from a point $(0, Y, Z)$ on the given curve, where (since $Y \geq 0$) $Y = \sqrt{x^2 + y^2}$, $Z = z$. Substituting these into Equation (20.4.3), we have

$$z^2 = 2\sqrt{x^2 + y^2}.$$

Now squaring both members, we obtain as the desired equation,

$$z^4 = 4(x^2 + y^2).$$

The following description of the procedure applies, but it should not be used as a substitute for an understanding of the method. *Let a curve be given in a coordinate plane. To obtain an equation of the surface generated by revolving the curve about a coordinate axis in that plane, replace the square of the other coordinate in that plane by the sum of the squares of itself and the third coordinate.*

Furthermore, we may characterize surfaces of revolution about a coordinate axis as follows. *An equation represents a surface of revolution about a coordinate axis if and only if two of the variables appear only in a form which is the sum of their squares. The axis of revolution is then the axis of the third variable.*

For example, the equation

(20.4.4) $$(x^2 + y^2)^2 - (x^2 + y^2) = z$$

represents a surface of revolution about the $z$ axis, since $x$ and $y$ appear only in the form $x^2 + y^2$. On the other hand, the equation

(20.4.5) $$x^2 + y^2 - 2x = z$$

does not represent a surface of revolution about a coordinate axis, since $x$ appears by itself in a term, in addition to its presence in the sum $x^2 + y^2$. However, equation (20.4.5) may represent a surface of revolution about some other axis, and in fact does (can you find this axis?).

Once a surface is identified as a surface of revolution about a coordinate axis, it is an easy matter to find the generating curves. This is done by finding the intersection of the surface with the appropriate coordinate planes.

*Example 20.4.3.* Find the generating curves of the surface (20.4.4).

*Solution.* We have seen that the axis of revolution is the $z$ axis. We may find a generating curve by either setting $x = 0$ or $y = 0$. If we set $x = 0$, we have the generating curve in the $yz$ plane,

$$y^4 - y^2 = z, \quad x = 0.$$

If we set $y = 0$, we have the generating curve in the $xz$ plane,

$$x^4 - x^2 = z, \quad y = 0.$$

Each of these curves is a position assumed by revolving the other one about the $z$ axis.

### EXERCISE GROUP 20.2

In Exercises 1–8 sketch the cylindrical surface represented by the given equation. In each case identify the directrix.

1. $x^2 = 4z$
2. $x^2 + 4z = 4$
3. $x^2 - z^2 = 4$
4. $x^2 - z^2 = 0$
5. $z = e^x$
6. $y = \sin x$
7. $x + 3z = 2$
8. $2y - z + 1 = 0$

In Exercises 9–14 obtain an equation of the surface generated by revolving the given curve about the stated axis.

9. $x = y, z = 0$ about (a) $x$ axis, (b) $y$ axis
10. $x^2 + z^2 = 4, y = 0$ about (a) $x$ axis, (b) $z$ axis
11. $y + z = 1, x = 0$ about (a) $y$ axis, (b) $z$ axis
12. $y + z^2 = 1, x = 0$ about (a) $y$ axis, (b) $z$ axis
13. $y = 3, z = 0$ about (a) $x$ axis, (b) $y$ axis
14. $y^2 = 9, z = 0$ about (a) $x$ axis, (b) $y$ axis

In each of Exercises 15–22 find equations of a generating curve for each equation which represents a surface of revolution about a coordinate axis. Identify the curve by name where feasible.

15. $y^2 + z^2 = 4x$
16. $x^2 + y^2 = 4x + z$
17. $x^2 - y^2 = z^2 + 3$
18. $x^2 + y^2 = 4$
19. $x^2 + z^2 + 2z = 3 - y^2$
20. $x^2 + z^2 = 3 + y^2$
21. $\sin x^2 + \sin y^2 = 2z$
22. $x^4 + y^4 = z - 2x^2y^2$

## 20.5 QUADRIC SURFACES

A linear equation in three variables has been seen to represent a plane in space. The algebraic step up to the *equation of second degree*, or *quadratic equation* in three variables, leads to a class of surfaces known as *quadric surfaces*. In general, an equation of second degree will have as its graph one of the quadric surfaces to be described. We have met some of these in preceding sections. We shall consider such surfaces mainly in certain special positions relative to the coordinate system. A complete analysis of the quadratic equation must be left to a more advanced course in the analytic geometry of three dimensions.

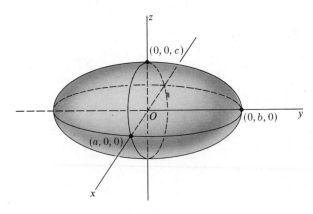

Fig. 20.5.1

*Ellipsoid.* The ellipsoid in the special position considered (see Fig. 20.5.1) has the standard equation

$$\frac{x^2}{a^2} + \frac{y^2}{b^2} + \frac{z^2}{c^2} = 1, \quad a > 0, \quad b > 0, \quad c > 0.$$

It has $x$ intercepts $\pm a$, $y$ intercepts $\pm b$, and $z$ intercepts $\pm c$. Each trace of the ellipsoid is an ellipse or circle, as is also each section which is indeed a curve. In this position of the ellipsoid the *center* is at the origin. It is clear that we must have

$$|x| \leq a, \quad |y| \leq b, \quad |z| \leq c.$$

If two of the numbers $a$, $b$, and $c$ are equal, the surface is an *ellipsoid of revolution*; while if all three are equal, the surface is a *sphere*.

*Elliptic paraboloid.* We present a standard form of the equation of the *elliptic paraboloid* as

$$\frac{x^2}{a^2} + \frac{y^2}{b^2} = cz, \quad a > 0, \quad b > 0.$$

The traces in the $xz$ plane and the $yz$ plane are parabolas, while the trace in the $xy$ plane is a single point, the origin. In Fig. 20.5.2 the surface is sketched for the case $c > 0$. In this case any section by a plane $z = k > 0$ is an ellipse

$$\frac{x^2}{a^2} + \frac{y^2}{b^2} = ck,$$

for which the semi-axes are $a\sqrt{ck}$ and $b\sqrt{ck}$.

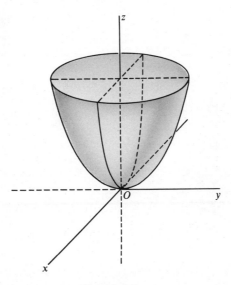

Fig. 20.5.2

If $a = b$, the sections $z = k$ are circles, and the surface is a *paraboloid of revolution* about the $z$ axis.

The cases where $c < 0$, or where the linear term in the equation is the $x$ or $y$ term, are entirely analogous.

*Hyperboloid of one or two sheets.* One of the standard forms of the equation of a *hyperboloid of one sheet* (Fig. 20.5.3) is

(20.5.1) $$\frac{x^2}{a^2} + \frac{y^2}{b^2} - \frac{z^2}{c^2} = 1, \quad a > 0, \quad b > 0, \quad c > 0.$$

Any section $z = k$ is an ellipse or circle. The $xz$ trace and the $yz$ trace are hyperbolas. Let us consider sections $y = k$, with equations

(20.5.2) $$\frac{x^2}{a^2} - \frac{z^2}{c^2} = 1 - \frac{k^2}{b^2}.$$

If $k^2 < b^2$, the hyperbolic section has as its transverse axis a line parallel to the $x$ axis, while if $k^2 > b^2$, the transverse axis is parallel to the $z$ axis. If $k^2 = b^2$ in (20.5.2), the section consists of two straight lines $cx = \pm az$.

If $a^2 = b^2$ the surface is also a surface of revolution about the $z$ axis.

The reason for the designation, "one sheet," is that the surface is a single continuous configuration, in contrast to a hyperboloid of two sheets.

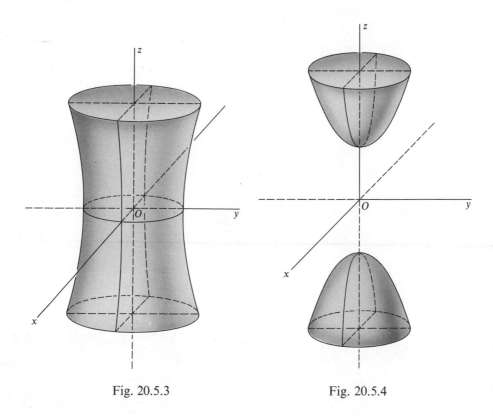

Fig. 20.5.3          Fig. 20.5.4

For a *hyperboloid of two sheets* we consider the equation

(20.5.3) $$\frac{x^2}{a^2} + \frac{y^2}{b^2} - \frac{z^2}{c^2} = -1, \qquad a > 0, \quad b > 0, \quad c > 0.$$

The $xz$ trace and the $yz$ trace are hyperbolas, but there is no $xy$ trace (Fig. 20.5.4). Let $z = k$ in (20.5.3). An equation of the section is

$$\frac{x^2}{a^2} + \frac{y^2}{b^2} = \frac{k^2}{c^2} - 1.$$

In order for a section to exist we must have $k^2 \geq c^2$. If $k^2 = c^2$, the section is a single point; while if $k^2 > c^2$, the section is an ellipse or circle with axes parallel to the $x$ and $y$ axes. Thus there is no point of the surface between the planes $z = c$

and $z = -c$, and the surface consists of two separate parts and hence the designation, "two sheets." Other sections of the surface may be similarly studied. If $a^2 = b^2$, the hyperboloid of two sheets (20.5.3) is also a surface of revolution.

Since the signs of the terms in either (20.5.1) or (20.5.3) also determine the position of the surface in the coordinate system, it is important to notice how these signs determine the type of surface. Rather than state a "rule" for this, we shall illustrate the kind of analysis used.

**Example 20.5.1.** Identify the surface $2x^2 - y^2 - 3z^2 = 1$.

*Solution.* By comparison with (20.5.1) and (20.5.3) it is clear that the surface is a hyperboloid in some position. For a section $x = k$ we have

$$y^2 + 3z^2 = 2k^2 - 1.$$

Hence there is no section for $|k| < 1/\sqrt{2}$, and the surface is a *hyperboloid of two sheets*. The two parts of the surface open along the $x$ axis.

**Example 20.5.2.** The equation $2x^2 - y^2 - 3z^2 = -1$ represents a hyperboloid of one sheet. It is clear that the elliptic sections occur for $x = k$ and have equations

$$y^2 + 3z^2 = 2k^2 + 1.$$

Since such a section occurs for any value of $k$, the surface must consist of one piece (one sheet).

*Cone.* The hyperboloids, in the positions in which they have been considered, have equations which may be written in the form

(20.5.4) $$Ax^2 + By^2 + Cz^2 = D,$$

where two of the constants $A$, $B$, and $C$ are of one sign, and the third one is of the opposite sign. Then if $D > 0$, one type of hyperboloid occurs, that is, of one sheet or two sheets; while if $D < 0$, the other type occurs. If we think of $D$ as varying, then as $D$ changes from positive to negative the hyperboloid will change from one type to the other. In the transition between the two types we have $D = 0$, and the surface is a *cone*, a type of surface which we now consider.

Suppose now that $D = 0$ in (20.5.4), and that $A$ and $B$ are of one sign, while $C$ is of the opposite sign. A standard form of such an equation may be written as

(20.5.5) $$\frac{x^2}{a^2} + \frac{y^2}{b^2} - \frac{z^2}{c^2} = 0, \quad a > 0, \quad b > 0, \quad c > 0.$$

The $xy$ trace is the origin, and the trace in each of the other coordinate planes is a pair of straight lines (Fig. 20.5.5). Any section $z = k \neq 0$ gives an ellipse or a circle; while any section $x = k \neq 0$ or $y = k \neq 0$ is a hyperbola.

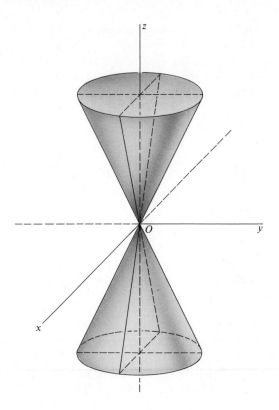

Fig. 20.5.5

The graph of (20.5.5) is an *elliptic cone* (*circular* if $a = b$), and may be said to be generated by a straight line passing through a fixed point (the origin here), and intersecting a specified curve (an ellipse here). A further discussion of such straight lines on a surface appears in Section 20.6.

*Hyperbolic paraboloid.* Perhaps one of the most interesting quadric surfaces (and one of the more difficult ones to sketch) is the *hyperbolic paraboloid.* We shall consider it in the position given by the equation

(20.5.6) $$\frac{y^2}{b^2} - \frac{x^2}{a^2} = cz, \quad a > 0, \quad b > 0, \quad c > 0.$$

The $xy$ trace is $z = 0$, $y^2/b^2 - x^2/a^2 = 0$, which consists of two straight lines through the origin. Refer to Fig. 20.5.6 repeatedly in the course of this discussion. The $yz$ trace is the parabola $x = 0$, $y^2 = b^2 cz$, opening upward with vertex at the origin. The $xz$ trace is the parabola $y = 0$, $x^2 = -a^2 cz$, opening downward with vertex at the origin. The sections of the hyperbolic paraboloid are of particular interest. Consider the sections $z = k$. If $k > 0$, the sections are hyperbolas opening in the $y$ direction. As $k$ decreases these hyperbolas spread out, become two straight lines when $k = 0$, and then open in the $x$ direction when $k < 0$. All sections $x = k$ are parabolas opening upward; while all sections $y = k$ are parabolas opening downward.

The corresponding behavior is similar for other positions of the surface, as for example when $c < 0$ in (20.5.6), or when the roles of $x$, $y$, and $z$ are changed. In such cases it is perhaps more difficult to sketch the surface, but the traces and sections can readily be visualized.

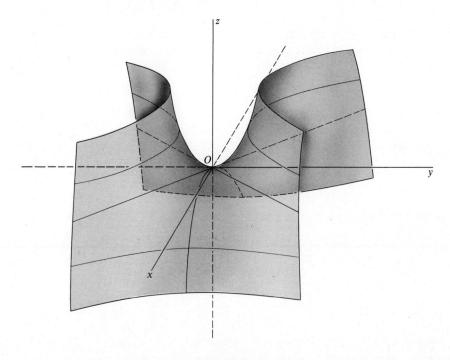

Fig. 20.5.6

*Other quadric surfaces.* The general second-degree equation in $x$, $y$, and $z$ includes cylindrical surfaces (Section 20.3) whose sections perpendicular to the rulings are conic sections. Such equations are easily identified when the rulings are parallel to a coordinate axis. For example, the equation

$$2x^2 - 3y + 2 = 0$$

represents a *parabolic cylinder* with rulings parallel to the $z$ axis; while the equation

$$2x^2 - 4 - z^2 = 0$$

represents a *hyperbolic cylinder* with rulings parallel to the $y$ axis.

In addition to all the quadric surfaces previously discussed there are the *degenerate quadric surfaces*. A degenerate surface of a specified type may be thought of as a surface obtained by deforming that type until in an extreme position the type of surface changes. Thus an elliptic cone may be considered either as a degenerate hyperboloid of one sheet or as a degenerate hyperboloid of two sheets. A pair of intersecting planes may be a degenerate hyperbolic cylinder, and it may result from the equation

$$x^2 - y^2 = c,$$

for example, by letting $c = 0$. Other types will appear in the exercises.

**Summary.** The equations which have been considered in this section are certain second-degree equations in $x$, $y$, and $z$ which contain no terms of the form $axy$, $axz$, or $ayz$. Any equation of the form

(20.5.7) $$Ax^2 + By^2 + Cz^2 + Dx + Ey + Fz + G = 0$$

which has a locus will represent one of the surfaces of this section in a position which is perhaps translated from the position discussed. By completion of squares any equation of the form (20.5.7) can be put in a form which can be associated with one of those studied.

***Example 20.5.3.*** Identify the surface $x^2 + 4y^2 - 8x + 8y + 4z = 0$.

*Solution.* By completing squares in $x$ and $y$ we rewrite the equation as

$$(x - 4)^2 + 4(y + 1)^2 = -4(z - 5).$$

The surface is an elliptic paraboloid opening downward (we must have $z \le 5$), and with the highest point at $(4, -1, 5)$. This point corresponds to the origin in the positions studied earlier.

If a second-degree equation contains terms in $xy$, $xz$, or $yz$, the surface represented is also of one of the types discussed above; however, it is in a rotated position with respect to the coordinate system. These equations are not treated here.


**EXERCISE GROUP 20.3**

In Exercises 1–22 identify the graph of each of the given equations by name, and sketch each surface.

1. $x^2 + y^2 = 7$
2. $x^2 - y^2 + z^2 + 2 = 0$
3. $4y^2 = x^2 + z$
4. $x^2 + y^2 + z^2 - 4 = 0$
5. $2x^2 + y^2 = 3 - z^2$
6. $3x^2 - z^2 = 0$
7. $4x^2 + y = 0$
8. $4x^2 + y^2 + 4z^2 = 4$
9. $4y^2 + x^2 = z$
10. $y^2 + 2z^2 = 4x$
11. $4x^2 + z^2 = y^2 - 1$
12. $x^2 - 4y^2 = z^2$
13. $x^2 - 4y^2 + z^2 = 0$
14. $x^2 + 2y^2 = z - 2$

15. $x^2 + 3y^2 = -z^2$
16. $x^2 + 2xy + y^2 - 4 = 0$
17. $4y^2 - 9 = 0$
18. $x^2 + y^2 - z^2 + 2 = 0$
19. $x^2 + y^2 + z^2 - 2x + 4y = 4$
20. $x^2 + 4y^2 - 2x - 2y + z = 4$
21. $x^2 + y^2 - z^2 - 2x - 4y + 6z = 4$
22. $x^2 + y^2 - z^2 - 2x - 4y + 6z = 5$

23. For the hyperbolic paraboloid (20.5.6) find the coordinates of the foci of the section $z = k$ if (a) $k > 0$, (b) $k < 0$.

24. For the hyperbolic paraboloid (20.5.6) find the coordinates of the focus of the section (a) $x = k$, (b) $y = k$.

25. For the hyperboloid of two sheets (20.5.3), with $a > b$, find the coordinates of the foci of the section (a) $z = k$, where $k^2 > c^2$, (b) $x = k$.

26. For the hyperboloid of one sheet (20.5.1), with $a > b$, find the coordinates of the foci of the section $z = k$.

27. For the hyperboloid of one sheet (20.5.1) find the coordinates of the foci of the section $x = k$ if (a) $k > a$, (b) $k < a$.

## *20.6  RULED SURFACES

A *ruling* of a surface is a straight line which lies entirely in the surface. If every point of a surface lies on a ruling of the surface, the surface is called a *ruled surface*.

It is clear that planes, cones, and cylindrical surfaces are ruled surfaces in this sense. We wish to consider here the other two quadric surfaces which are ruled surfaces—a property which is not immediately apparent for these surfaces.

*Hyperbolic paraboloid.* Consider this surface with the equation

(20.6.1) $$\frac{y^2}{b^2} - \frac{x^2}{a^2} = cz,$$

as in (20.5.6). This equation may be written in the form

$$\left(\frac{y}{b} + \frac{x}{a}\right)\left(\frac{y}{b} - \frac{x}{a}\right) = cz.$$

If $x$, $y$, and $z$ satisfy this equation, they must satisfy the two equations

(20.6.2) $$r\left(\frac{y}{b} - \frac{x}{a}\right) = cz \quad \text{and} \quad \frac{y}{b} + \frac{x}{a} = r$$

for some number $r$. Conversely, if $x$, $y$, and $z$ satisfy both equations in (20.6.2), then they must satisfy (20.6.1). For any value of $r$ Equations (20.6.2) represent a straight line. Hence any straight line (20.6.2) lies in the surface (20.6.1), and is a ruling of (20.6.1). Conversely, any point in (20.6.1) must lie in one of the rulings (20.6.2). *Equations* (20.6.2) *represent, for different values of $r$, a family of rulings of* (20.6.1).

The statements of the preceding paragraph may be repeated if (20.6.2) is replaced by

(20.6.3) $$r'\left(\frac{y}{b} + \frac{x}{a}\right) = cz \quad \text{and} \quad \frac{y}{b} - \frac{x}{a} = r'.$$

Thus Equations (20.6.3) represent, for different values of $r'$, a *second family of rulings* of (20.6.1). *Through each point of (20.6.1) there is a ruling of each of the families* (20.6.2) *and* (20.6.3); moreover, *no ruling can belong to both of the families* (20.6.2) *and* (20.6.3) [see Exercise 8].

*Hyperboloid of one sheet.* We consider the equation of this surface in the form

(20.6.4) $$\frac{x^2}{a^2} + \frac{y^2}{b^2} - \frac{z^2}{c^2} = 1.$$

If we write this equation as

$$\left(\frac{x}{a} + \frac{z}{c}\right)\left(\frac{x}{a} - \frac{z}{c}\right) = \left(1 + \frac{y}{b}\right)\left(1 - \frac{y}{b}\right),$$

we see that any values $x$, $y$, and $z$ which satisfy (20.6.4) must satisfy both of the equations

(20.6.5) $$r\left(\frac{x}{a} + \frac{z}{c}\right) = s\left(1 + \frac{y}{b}\right) \quad \text{and} \quad s\left(\frac{x}{a} - \frac{z}{c}\right) = r\left(1 - \frac{y}{b}\right)$$

for some values of $r$ and $s$, and conversely. Similarly, any values $x$, $y$, and $z$ which satisfy (20.6.4) must satisfy both of the equations

(20.6.6) $$r'\left(\frac{x}{a} + \frac{z}{c}\right) = s'\left(1 - \frac{y}{b}\right) \quad \text{and} \quad s'\left(\frac{x}{a} - \frac{z}{c}\right) = r'\left(1 + \frac{y}{b}\right)$$

for some values of $r'$ and $s'$, and conversely. Then (20.6.5) and (20.6.6) represent two families of rulings of (20.6.4), with properties similar to the rulings of the hyperbolic paraboloid [see Exercise 9].

In the following example we shall use the procedure employed above, for a surface in a different position.

**Example 20.6.1.** Find the rulings of the surface

$$\frac{x^2}{4} - y^2 + \frac{z^2}{9} = 1$$

which contain the point (4, 2, 3) of the surface.

*Solution.* We write the equation of the surface as

(20.6.7) $$\frac{z^2}{9} - y^2 = 1 - \frac{x^2}{4}.$$

The two families of rulings are (in the same way as (20.6.5) and (20.6.6) were derived):

(20.6.8) $\quad r\left(\dfrac{z}{3}+y\right) = s\left(1+\dfrac{x}{2}\right), \quad s\left(\dfrac{z}{3}-y\right) = r\left(1-\dfrac{x}{2}\right),$

and

(20.6.9) $\quad r'\left(\dfrac{z}{3}+y\right) = s'\left(1-\dfrac{x}{2}\right), \quad s'\left(\dfrac{z}{3}-y\right) = r'\left(1+\dfrac{x}{2}\right).$

Substituting (4, 2, 3) for $(x, y, z)$ in either equation of (20.6.8), we find $r = s$. We may take $r = s = 1$; with these values the ruling of family (20.6.8) becomes

$$3x - 6y - 2z = -6, \quad 3x - 6y + 2z = 6,$$

which is a line through the point (4, 2, 3) with direction numbers 2, 1, 0.

In a similar manner the ruling of family (20.6.9) through the point (4, 2, 3) has $3r' = -s'$. We may take $r' = 1$, $s' = -3$. The desired ruling has equations

$$9x - 6y - 2z = 18, \quad x - 6y + 2z = -2,$$

which is a line through the given point with direction numbers 6, 5, 12.

An alternative method for doing Example 20.6.1, more algebraic in nature, is the following. Any line through the point (4, 2, 3) may be written in parametric form as

(20.6.10) $\quad x = 4 + mt, \quad y = 2 + nt, \quad z = 3 + pt.$

On substitution into (20.6.7), the resulting equation reduces to

(20.6.11) $\quad (9m^2 - 36n^2 + 4p^2)t^2 + 24(3m - 6n + p)t = 0.$

If (20.6.10) is a ruling of the surface, equation (20.6.11) must be an identity (valid for any value of $t$), and we must have

(20.6.12) $\quad 9m^2 - 36n^2 + 4p^2 = 0, \quad 3m - 6n + p = 0.$

This is a system of two equations in three unknowns. We may assign a value to one of them and solve for the other two. Choosing $p$ for this purpose, we must consider the two possibilities, $p = 0$ and $p \neq 0$. If $p = 0$ in (20.6.12), the only solution is $m = 2n$. Since $m$, $n$, and $p$ are direction numbers, any solution set may be used. Taking $n = 1$, we have a set of direction numbers 2, 1, 0 for one ruling (20.6.10).

If $p \neq 0$ in (20.6.12), we may choose $p = 3$. Then (20.6.12) becomes

$$m^2 - 4n^2 = -4, \quad m - 2n = -1,$$

which has as the only solution $m = 3/2$, $n = 5/4$. Thus another set of direction numbers is 3/2, 5/4, 3 or 6, 5, 12 for the second ruling. These results agree with those obtained with the first method.

* **EXERCISE GROUP 20.4**

1. Find both rulings of the surface $x^2 - y^2 + z^2 = 1$ which contain the point (a) (1, 0, 0), (b) (1, 1, 1), (c) (4, 4, 1).
2. Find both rulings of the surface $x^2 + y^2/25 - z^2 = 1$ which contain the point (a) (0, 5, 0), (b) (5, 5, 5), (c) (3, 5, 3).
3. Find both rulings of the surface $x^2 - y^2 = 2z$ which contain the point (a) (3, 3, 0), (b) (0, −2, −2), (c) (4, 2, 6).
4. Find both rulings of the surface $y^2/4 - z^2/9 = 4x$ which contain the point (a) (0, 2, 3), (b) (−1, 0, 6), (c) (2, 6, 3).
5. Find both rulings of the surface of Example 20.6.1 which contain the point (0, 0, 3).
* 6. Find equations of any rulings of the surface $x^2/4 - y^2/9 + z^2 = 1$ which have direction numbers 6, 15, 4.
* 7. Find equations of any rulings of the surface $x^2 - y^2/9 + 2z = 0$ which have direction numbers (a) 1, 3, 1, (b) 1, −3, 1.
* 8. Prove that no straight line can belong to both of the families (20.6.2) and (20.6.3).
* 9. Prove that no straight line can belong to both of the families (20.6.5) and (20.6.6).
* 10. Find an equation of the locus of points on the hyperbolic paraboloid $4x^2 - y^2 = 2z$ at which the two rulings are perpendicular.

## 20.7 CURVES IN SPACE

A curve in space lies in infinitely many surfaces. Any two of them may be used to define the curve, in the sense that the curve consists of those points in space which lie in both surfaces.

Without attempting a completely rigorous analysis, we shall say that *a curve is defined as the intersection of two intersecting surfaces.* Since a surface is the graph of a single equation in $x$, $y$, and $z$, *a curve is then the locus of points whose coordinates satisfy simultaneously the equations of two surfaces.*

We have already encountered this situation in the case of a straight line which was represented as the intersection of two planes which contain the line. Furthermore, traces and sections of a surface are curves which are the intersections of the surface with suitable planes.

Of all the surfaces which contain a curve, the *projecting cylinders* are particularly useful. These are cylindrical surfaces, with rulings parallel to a coordinate axis, which contain the curve. In general, an equation of a projecting cylinder for a curve is obtained by eliminating the corresponding variable from the defining equations of the curve. This is illustrated in the following example.

***Example 20.7.1.*** Sketch the curve which is the intersection of the paraboloid $x^2 + y^2 = z$ and the plane $y + z = 1$.

*Solution.* The curve lies in the plane $y + z = 1$, which has rulings parallel to the $x$ axis; this plane therefore is itself one of the projecting cylinders. To find an equation of the projecting cylinder onto the $xy$ plane we *eliminate z* from the two equations, to get $x^2 + y^2 = 1 - y$, or

$$x^2 + (y + \tfrac{1}{2})^2 = \tfrac{5}{4}.$$

This is an equation of a right circular cylinder, whose trace in the $xy$ plane has the same equation as the cylinder, together with $z = 0$. This cylinder is shown in Fig. 20.7.1, in addition to the given surfaces and their intersection.

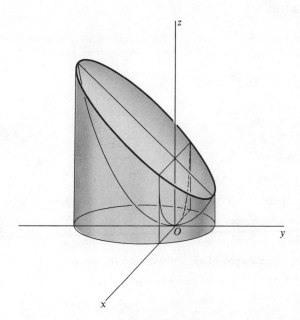

Fig. 20.7.1

An equation of the third projecting cylinder is obtained by eliminating $y$ from the given equations, and is $x^2 + (1 - z)^2 = z$. This equation may be written

$$x^2 + (z - \tfrac{3}{2})^2 = \tfrac{5}{4},$$

and the surface is also a right circular cylinder. However, the graph of this projecting cylinder is not needed for our purpose, and is not shown.

The sketching of a curve may present some difficulties. The following suggestions are offered as guidelines. Sketch each surface reasonably well and determine obvious points of intersection, such as those which lie in coordinate planes. These points may suggest the appearance of the intersection of the surfaces. If not, find one or more of the projecting cylinders and attempt to use these to sketch the curve. Bear in mind that the curve being sought is also the intersection of projecting cylinders. Be prepared to alter the figure as new information is obtained, and to discard a started figure and draw a new one as the appearance of the curve becomes more clear.

In some parts of our study it will be necessary to sketch the solid bounded by given surfaces. This involves the same procedures as in sketching surfaces and curves. For example, Fig. 20.7.1 can readily be used to give a sketch of the solid bounded by the surfaces $x^2 + y^2 = z$ and $y + z = 1$.

**EXERCISE GROUP 20.5**

In each of Exercises 1–10 sketch the curve defined by the given surfaces. Obtain equations of any projecting cylinders which are not given surfaces.

1. $x^2 + z^2 = 4y$, $z = 2y$
2. $z = x^2 + y^2$, $x + y + z = 1$
3. $x^2 + y^2 = 4$, $y - z + 2 = 0$
4. $x^2 + z = 4$, $x + y = 2$
5. $x^2 + z^2 = 9$, $z = x^2 + y^2$
6. $x^2 + z^2 = 4$, $x + y + z = 0$
7. $x^2 + z^2 = a^2$, $y^2 + z^2 = a^2$
8. $x^2 + z^2 = 9$, $y^2 + z^2 = 4$
9. $z^2 = x^2 + y^2$, $x^2 + y^2 + z^2 = 2x$
10. $x^2 + 4y^2 = z - 4$, $x^2 + y^2 = 2y$

In Exercises 11–14 sketch the solid described.

11. Inside the sphere $x^2 + y^2 + z^2 = 9$ and inside the cylinder $x^2 + y^2 = 4$.
12. Bounded by the cylinder $x^2 + y^2 = 4$, the plane $z = x$, and the $xy$ plane.
13. Bounded by the sphere $x^2 + y^2 + z^2 = 9$ and the plane $y + z = 1$ (smaller solid).
14. Bounded by the paraboloid $x^2 + y^2 = z$ and the plane $x + y + z = 1$, in the first octant.

## 20.8 VECTOR FUNCTIONS. PARAMETRIC REPRESENTATION OF A CURVE

The treatment of vector functions in three dimensions follows so closely the treatment in two dimensions that most of the results will merely be stated. A *vector function*

(20.8.1) $$\mathbf{R}(t) = (f(t), g(t), h(t)) = f(t)\mathbf{i} + g(t)\mathbf{j} + h(t)\mathbf{k}$$

is defined in a domain which is common to the domains of $f(t)$, $g(t)$, and $h(t)$.

**DEFINITION 20.8.1.** *The function $\mathbf{R}(t)$ has a* **limit** *$\mathbf{A}$ as $t \to t_0$, and we write*

$$\lim_{t \to t_0} \mathbf{R}(t) = \mathbf{A},$$

*if, given $\epsilon > 0$ there exists $\delta$ such that*

$$|\mathbf{R}(t) - \mathbf{A}| < \epsilon \quad \text{whenever} \quad 0 < |t - t_0| < \delta.$$

**DEFINITION 20.8.2.** *The vector function* $\mathbf{R}(t)$ *is* **continuous** *at* $t_0$ *if*

$$\lim_{t \to t_0} \mathbf{R}(t) = \mathbf{R}(t_0).$$

**DEFINITION 20.8.3.** *The vector function* $\mathbf{R}(t)$ *possesses a* **derivative** $\mathbf{R}'(t)$, *and we write*

$$\mathbf{R}'(t) = \lim_{t \to t_0} \frac{\mathbf{R}(t+h) - \mathbf{R}(t)}{h},$$

*if the indicated limit exists.*

**THEOREM 20.8.1.** *Let* $\mathbf{R}(t)$ *be as in* (20.8.1). *Then*

(a) $\lim_{t \to t_0} \mathbf{R}(t) = a\mathbf{i} + b\mathbf{j} + c\mathbf{k}$ *if and only if all of the following hold:*

$$\lim_{t \to t_0} f(t) = a, \quad \lim_{t \to t_0} g(t) = b, \quad \lim_{t \to t_0} h(t) = c.$$

(b) $\mathbf{R}(t)$ *is continuous at* $t_0$ *if and only if* $f(t)$, $g(t)$, *and* $h(t)$ *are all continuous at* $t_0$.

(c) $\mathbf{R}'(t)$ *exists if and only if* $f'(t)$, $g'(t)$, *and* $h'(t)$ *all exist, in which case*

$$\mathbf{R}'(t) = f'(t)\mathbf{i} + g'(t)\mathbf{j} + h'(t)\mathbf{k}.$$

Assuming that the involved derivatives exist, we have the following.

**THEOREM 20.8.2.**  (a) $\dfrac{d}{dt}[a\mathbf{R}(t)] = a\mathbf{R}'(t)$, $a$ *constant*,

(b) $\dfrac{d}{dt}[\mathbf{R}(t) + \mathbf{U}(t)] = \mathbf{R}'(t) + \mathbf{U}'(t)$,

(c) $\dfrac{d}{dt}[a(t)\mathbf{R}(t)] = a(t)\mathbf{R}'(t) + a'(t)\mathbf{R}(t)$,

(d) $\dfrac{d}{dt}[\mathbf{R}(t) \cdot \mathbf{U}(t)] = \mathbf{R}(t) \cdot \mathbf{U}'(t) + \mathbf{R}'(t) \cdot \mathbf{U}(t)$,

(e) $\dfrac{d}{dt}[\mathbf{R}(t) \times \mathbf{U}(t)] = \mathbf{R}(t) \times \mathbf{U}'(t) + \mathbf{R}'(t) \times \mathbf{U}(t)$.

Only part (e) of Theorem 20.8.2 is new. It is essential that the cross products be written in the correct order. The proof of (e) may be carried out by working with the appropriate components of the vector functions (Exercise 20).

**THEOREM 20.8.3.** *If* $\mathbf{U}(t)$ *is a unit vector function, then either* $\mathbf{U}(t)$ *and* $\mathbf{U}'(t)$ *are orthogonal or* $\mathbf{U}'(t) = \mathbf{0}$.

The proof of Theorem 16.5.3 applies directly to this result.

If $\mathbf{R} = x\mathbf{i} + y\mathbf{j} + z\mathbf{k}$ is the *position vector* of the point $(x, y, z)$, the equation (20.8.1) is equivalent to the three equations

(20.8.2) $\qquad x = f(t), \qquad y = g(t), \qquad z = h(t).$

Equations (20.8.2) are a *parametric representation* of a curve in space, a representation that is very similar to that of a curve in a plane (Section 16.1), but with the addition of a third equation. Thus equation (20.8.1) is a vector equation for the curve given by (20.8.2).

Any equation obtained by eliminating the parameter $t$ from (20.8.2) represents a surface which contains the curve. In particular, *an equation obtained by eliminating t from two of the equations (20.8.2) represents a projecting cylinder of the curve.*

*Example 20.8.1.* Sketch the curve given by the parametric equations

$$x = 4t + 1, \qquad y = 1 - 2t, \qquad z = 4t^2.$$

*Solution.* By eliminating $t$ from the first two equations we obtain the equation $x + 2y = 3$ of a projecting cylinder—a plane. By eliminating $t$ from the last two equations we obtain a second projecting cylinder $z = (y - 1)^2$ which is a parabolic cylinder. These surfaces, and the curve of intersection are shown in Fig. 20.8.1. The third projecting cylinder, $4z = (x - 1)^2$, is not needed for the sketch; it is also a parabolic cylinder.

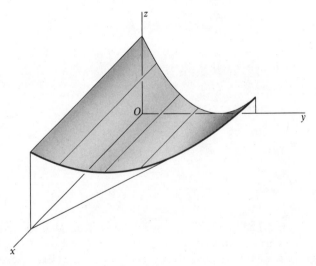

Fig. 20.8.1

Let $P_0$ and $P$ be two points on a curve, as in Fig. 20.8.2. The line joining these points is a *secant* of the curve. As $P$ moves along the curve toward $P_0$ the secant varies, and may approach a certain straight line, which is then the *tangent* to the curve at $P_0$. The secant has direction numbers $x - x_0, y - y_0, z - z_0$, and hence also

$$\frac{x - x_0}{\Delta t}, \quad \frac{y - y_0}{\Delta t}, \quad \frac{z - z_0}{\Delta t},$$

where $t$ corresponds to $P$, $t_0$ to $P_0$, and $\Delta t = t - t_0$. Then $\Delta t \to 0$ as $P$ approaches $P_0$ along the curve, and

$$\lim_{\Delta t \to 0} \frac{x - x_0}{\Delta t}, \quad \lim_{\Delta t \to 0} \frac{y - y_0}{\Delta t}, \quad \lim_{\Delta t \to 0} \frac{z - z_0}{\Delta t}$$

are a set of direction numbers of the tangent at $P_0$, if these limits exist and are not all zero. Thus $dx/dt, dy/dt, dz/dt$ are a set of direction numbers of the tangent to the curve at $P_0$ (and hence of the curve itself). We may then state that $\mathbf{R}'(t)$ *is a vector tangent to the curve* (20.8.1) *at the point corresponding to t*. Any statement involving direction of a curve in space refers to the direction of the tangent to the curve.

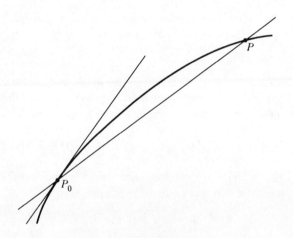

Fig. 20.8.2

**Example 20.8.2.** Find equations of the tangent line to the curve $x = t^3 - t$, $y = 2t - 1$, $z = t^2$ at the point with $t = 2$.

*Solution.* A set of direction numbers at a point of the curve is given by $dx/dt = 3t^2 - 1$, $dy/dt = 2$, $dz/dt = 2t$. Hence at the point with $t = 2$ the vector $\mathbf{A} = (11, 2, 4)$ is parallel to the tangent. The point itself is $(6, 3, 4)$. Symmetric equations of the tangent line are

$$\frac{x - 6}{11} = \frac{y - 3}{2} = \frac{z - 4}{4}.$$

**Example 20.8.3.** Show that the tangent to the curve
$$\mathbf{R} = (2t + 1)\mathbf{i} + (1 - t)\mathbf{j} + t^2\mathbf{k},$$
at the point where it intersects the plane $2x + 2y - z = 5$, lies in the plane.

*Solution.* To find the intersection we substitute
$$x = 2t + 1, \qquad y = 1 - t, \qquad z = t^2$$
in the equation of the plane, and we get
$$2(2t + 1) + 2(1 - t) - t^2 = 5.$$
There is one solution, $t = 1$, and the point of intersection is $(3, 0, 1)$. Since $\mathbf{R}'(t) = 2\mathbf{i} - \mathbf{j} + 2t\mathbf{k}$, a tangent vector at this point is $\mathbf{R}'(1) = 2\mathbf{i} - \mathbf{j} + 2\mathbf{k}$. A normal to the plane is $\mathbf{N} = 2\mathbf{i} + 2\mathbf{j} - \mathbf{k}$. Then
$$\mathbf{N} \cdot \mathbf{R}'(1) = 4 - 2 - 2 = 0.$$
Hence the vectors are perpendicular, and the tangent vector to the curve at $t = 1$ lies in the plane.

The derivation of the formula for arc length of a curve in space parallels that of a curve in a plane (Section 16.4). We shall state the result.

**THEOREM 20.8.4.** *Let $\mathbf{R}(t) = f(t)\mathbf{i} + g(t)\mathbf{j} + h(t)\mathbf{k}$ represent a curve in space. If $\mathbf{R}'(t)$ is continuous in the interval $[t_1, t_2]$, with $t_1 < t_2$, the length, L, of the curve between the points corresponding to $t_1$ and $t_2$ is*

(20.8.3) $$L = \int_{t_1}^{t_2} |\mathbf{R}'(t)|\, dt = \int_{t_1}^{t_2} \sqrt{f'^2(t) + g'^2(t) + h'^2(t)}\, dt.$$

It follows that if $s(t)$ is an arc length function, representing the length of arc from a fixed point to a variable point of the curve, then

(20.8.4) $$ds = |\mathbf{R}'(t)|\, dt = \sqrt{f'^2(t) + g'^2(t) + h'^2(t)}\, dt.$$

**Example 20.8.4.** Find the length of the curve with equation
$$\mathbf{R}(t) = 3t^2\mathbf{i} + 3t\mathbf{j} + 2t^3\mathbf{k}$$
in the interval $0 \le t \le 2$.

*Solution.* We have $\mathbf{R}'(t) = 6t\mathbf{i} + 3\mathbf{j} + 6t^2\mathbf{k}$. Hence
$$|\mathbf{R}'(t)| = \sqrt{36t^2 + 9 + 36t^4} = \sqrt{9(2t^2 + 1)^2} = 3(2t^2 + 1).$$
Then from (20.8.3)
$$L = \int_0^2 3(2t^2 + 1)\, dt = 22.$$

## EXERCISE GROUP 20.6

In each of Exercises 1–10 obtain equations of the projecting cylinders, and sketch the curve.

1. $x = 2t + 1$, $y = 3t$, $z = t - 2$
2. $x = 5t$, $y = 3t + 2$, $z = 2t + 1$
3. $x = t$, $y = t$, $z = t^2 - t$
4. $x = 2t$, $y = t^2$, $z = 2t^2 + 1$
5. $x = 2 + 3t^2$, $y = 2 - t^2$, $z = 2t^2$
6. $x = t$, $y = t^2$, $z = t^3$
7. $x = \sin u$, $y = \cos u$, $z = \sin u$
8. $x = \sinh u$, $y = \cosh u$, $z = \sinh u$
9. $x = t^2$, $y = t + 1$, $z = t^4$
10. $x = 1/t - t$, $y = t$, $z = 1/t + t$

In Exercises 11–14 find equations of the tangent line to the curve at the specified point.

11. $x = 2t$, $y = 3t^2$, $z = t^3$; $t = 1$
12. $x = t^3 - t$, $y = 2t - 1$, $z = t^2$; $t = 2$
13. $\mathbf{R}(u) = u\mathbf{i} - 3u\mathbf{j} + u^{3/2}\mathbf{k}$; $u = 1$
14. $\mathbf{R}(t) = 2t\mathbf{i} + 2\cos t\,\mathbf{j} - \sin 2t\,\mathbf{k}$; $t = 0$
15. Find an angle between the given curves at their point of intersection:
$$x = u^2,\ y = 2u,\ z = u \quad \text{and} \quad x = 2 - t^2,\ y = 1 + t,\ z = t^2$$
16. Find an angle between the given curves at their point of intersection:
$$x = t,\ y = t^2,\ z = t^3 - 3 \quad \text{and} \quad x = u^2 - 3,\ y = u - 1,\ z = -2$$
17. Find angles at which the curve $x = t - t^2$, $y = 2 - t^3$, $z = 2 - t - t^2$ intersects the $xy$ plane.
18. Find an angle that the curve $x = 2 - t^2$, $y = 1 + t$, $z = t^2$ makes with the plane $2x - 5y + 5z = 1$ at each point of intersection.
19. Find an angle that the curve $x = t$, $y = 2t^2$, $z = t^3$ makes with the plane $x - y + z = 0$ at each point of intersection.
20. Prove Theorem 20.8.2, Part (e).

In each of the following find the length of the curve represented by the vector function in the specified interval.

21. $\mathbf{R}(t) = t\mathbf{i} + 2t\mathbf{j} + t^{3/2}\mathbf{k}$, $0 \leq t \leq 1$
22. $\mathbf{R}(t) = (\sinh t)\mathbf{i} + (\cosh t)\mathbf{j} + t\mathbf{k}$, $0 \leq t \leq 1$
23. $\mathbf{R}(t) = (2\cos t)\mathbf{i} + (2\sin t)\mathbf{j} + t\mathbf{k}$, $0 \leq t \leq 2$

## 20.9 UNIT TANGENT AND UNIT NORMAL VECTORS

In Section 20.8 it was seen that $\mathbf{R}'(t)$ is a vector tangent to the curve $\mathbf{R}(t)$ if $\mathbf{R}'(t) \neq \mathbf{0}$. Hence the unit vector

(20.9.1) $$\mathbf{T}(t) = \frac{\mathbf{R}'(t)}{|\mathbf{R}'(t)|}$$

is also tangent to the curve.

**DEFINITION 20.9.1.** *The vector* $\mathbf{T}(t)$ *defined in* (20.9.1) *is the* **unit tangent vector** *for the curve* $\mathbf{R}(t)$.

By Theorem 20.8.3 the vector $\mathbf{T}'(t)$ is orthogonal to $\mathbf{T}(t)$; hence the vector

$$(20.9.2) \qquad \mathbf{N}(t) = \frac{\mathbf{T}'(t)}{|\mathbf{T}'(t)|}$$

is a unit vector orthogonal to $\mathbf{T}(t)$. Of the infinitely many unit vectors perpendicular to $\mathbf{T}(t)$, we select $\mathbf{N}(t)$ for particular attention.

**DEFINITION 20.9.2.** *The vector* $\mathbf{N}(t)$ *defined in* (20.9.2) *is the* **principal unit normal** *for the curve* $\mathbf{R}(t)$.

Since $|\mathbf{R}'(t)| = ds/dt$ (by (20.8.4)), we obtain, from (20.9.1),

$$(20.9.3) \qquad \mathbf{R}'(t) = \frac{ds}{dt} \mathbf{T}(t).$$

Then

$$\mathbf{R}''(t) = \frac{d^2 s}{dt^2} \mathbf{T}(t) + \frac{ds}{dt} \mathbf{T}'(t),$$

and with use of (20.9.2) this becomes

$$(20.9.4) \qquad \mathbf{R}''(t) = \frac{d^2 s}{dt^2} \mathbf{T}(t) + \frac{ds}{dt} |\mathbf{T}'(t)| \mathbf{N}(t).$$

Thus $\mathbf{R}''(t)$ is a linear combination, that is, a sum of scalar multiples of, $\mathbf{T}(t)$ and $\mathbf{N}(t)$, and hence $\mathbf{R}''(t)$ *is parallel to the plane determined by* $\mathbf{T}(t)$ *and* $\mathbf{N}(t)$.

**DEFINITION 20.9.3.** *The plane through the point P of a curve which contains the vectors* $\mathbf{T}(t)$ *and* $\mathbf{N}(t)$ *at P is the* **osculating plane** *of the curve at P.*

We may state the result preceding Definition 20.9.3 as follows: *The vector* $\mathbf{R}''(t)$ *is parallel to the osculating plane at any point of the curve.*

*Example 20.9.1.* Given the curve

$$\mathbf{R}(t) = t^2 \mathbf{i} + (2t^2 - 1)\mathbf{j} + t^3 \mathbf{k},$$

(a) find the unit tangent vector $\mathbf{T}(t)$ and unit normal vector $\mathbf{N}(t)$.
(b) find an equation of the osculating plane at $t = 1$.

*Solution.* (a) We find

$$\mathbf{R}'(t) = 2t\mathbf{i} + 4t\mathbf{j} + 3t^2 \mathbf{k},$$

and

$$|\mathbf{R}'(t)| = \sqrt{4t^2 + 16t^2 + 9t^4} = |t|\sqrt{20 + 9t^2}.$$

Hence

$$\mathbf{T}(t) = \frac{\mathbf{R}'(t)}{|\mathbf{R}'(t)|} = \frac{2t\mathbf{i} + 4t\mathbf{j} + 3t^2\mathbf{k}}{|t|\sqrt{20 + 9t^2}} = \pm \frac{2\mathbf{i} + 4\mathbf{j} + 3t\mathbf{k}}{\sqrt{20 + 9t^2}}, \quad \begin{cases} + \text{ if } t > 0 \\ - \text{ if } t < 0 \end{cases}$$

since $t/|t| = 1$ if $t > 0$ and $= -1$ if $t < 0$.

To find $\mathbf{N}(t)$ we obtain

$$\mathbf{T}'(t) = \pm \frac{-18t\mathbf{i} - 36t\mathbf{j} + 60\mathbf{k}}{(20 + 9t^2)^{3/2}}, \quad (+ \text{ if } t > 0, - \text{ if } t < 0).$$

We then find

$$|\mathbf{T}'(t)| = \frac{6(45t^2 + 100)^{1/2}}{(20 + 9t^2)^{3/2}},$$

and hence

$$\mathbf{N}(t) = \frac{\mathbf{T}'(t)}{|\mathbf{T}'(t)|} = \pm \frac{-3t\mathbf{i} - 6t\mathbf{j} + 10\mathbf{k}}{\sqrt{45t^2 + 100}}.$$

(b) At $t = 1$ we have, from the results in (a),

$$\mathbf{T}(1) = \frac{2\mathbf{i} + 4\mathbf{j} + 3\mathbf{k}}{\sqrt{29}}, \quad \mathbf{N}(1) = \frac{-3\mathbf{i} - 6\mathbf{j} + 10\mathbf{k}}{\sqrt{145}}.$$

Then $\mathbf{T}(1) \times \mathbf{N}(1)$ is a vector perpendicular to the osculating plane, and so is the vector

$$(2\mathbf{i} + 4\mathbf{j} + 3\mathbf{k}) \times (-3\mathbf{i} - 6\mathbf{j} + 10\mathbf{k}) = 29(2\mathbf{i} - \mathbf{j}),$$

which is the cross product of the numerators in $\mathbf{T}(1)$ and $\mathbf{N}(1)$. Thus $2, -1, 0$ are direction numbers of a line perpendicular to the osculating plane. The point of the curve at which $t = 1$ is $(1, 1, 1)$. An equation of the osculating plane is, accordingly,

$$2(x - 1) - (y - 1) = 0 \quad \text{or} \quad 2x - y - 1 = 0.$$

## 20.10 VELOCITY AND ACCELERATION. CURVATURE. THE MOVING TRIHEDRAL

With the use of the chain rule for a vector function we have

(20.10.1) $$\frac{d\mathbf{T}}{dt} = \frac{d\mathbf{T}}{ds}\frac{ds}{dt},$$

where $\mathbf{T}$ is the unit tangent vector at a point $P$ of a curve, and $s$ is an arc length function $s = s(t)$. We define curvature as in (16.6.8).

**DEFINITION 20.10.1.** *The **curvature** $k$ of a curve is defined as*

$$k = \left|\frac{d\mathbf{T}}{ds}\right|.$$

Then (20.10.1) yields the relation (since $ds/dt$ is assumed positive)

$$\left|\frac{d\mathbf{T}}{dt}\right| = k\frac{ds}{dt},$$

and (20.9.4) becomes

(20.10.2) $$\mathbf{R}''(t) = \frac{d^2s}{dt^2}\mathbf{T}(t) + k\left(\frac{ds}{dt}\right)^2\mathbf{N}(t).$$

We now define a vector $\mathbf{B}(t)$ as

(20.10.3) $$\mathbf{B}(t) = \mathbf{T}(t) \times \mathbf{N}(t).$$

Then $|\mathbf{B}(t)| = |\mathbf{T}(t)||\mathbf{N}(t)| \sin \pi/2$ (since $\mathbf{T}$ and $\mathbf{N}$ are orthogonal), and since $\mathbf{T}$ and $\mathbf{N}$ are also unit vectors it follows that $\mathbf{B}(t)$ is a unit vector. Furthermore, as a cross product of $\mathbf{T}$ and $\mathbf{N}$, $\mathbf{B}(t)$ is orthogonal to both $\mathbf{T}$ and $\mathbf{N}$.

**DEFINITION 20.10.2.** *The unit vector $\mathbf{B}(t)$ in (20.10.3) is called the* **binormal vector** *at P.*

The vectors $\mathbf{T}$, $\mathbf{N}$, and $\mathbf{B}$ are a set of three mutually orthogonal unit vectors, forming a right-handed system in the order given, and are designated as the *moving trihedral*. The moving trihedral plays an important role in the detailed study of curves, and is treated at length in Differential Geometry. We shall return to this shortly.

Suppose now that in the vector, or parametric, representation of a curve, the parameter $t$ denotes time. The curve becomes the path of a moving particle. The *velocity* of the particle is defined as $\mathbf{v}(t) = \mathbf{R}'(t)$, and the *acceleration* is defined as

$$\mathbf{a}(t) = \mathbf{v}'(t) = \mathbf{R}''(t).$$

The *speed* of the particle is defined as

$$v(t) = |\mathbf{v}(t)| = \frac{ds}{dt}.$$

From (20.9.3) and (20.10.2) we obtain

(20.10.4) $$\mathbf{v}(t) = \frac{ds}{dt}\mathbf{T}(t), \qquad \mathbf{a}(t) = \frac{d^2s}{dt^2}\mathbf{T}(t) + k\left(\frac{ds}{dt}\right)^2\mathbf{N}(t).$$

The latter relation shows that *the acceleration vector always lies in the plane of $\mathbf{T}$ and $\mathbf{N}$, that is, in the osculating plane, even for motion along a curve in space.*

Let us write the relation for $\mathbf{a}(t)$ in (20.10.4) as

(20.10.5) $$\mathbf{a}(t) = a_T\mathbf{T} + a_N\mathbf{N}, \quad \text{where} \quad a_T = \frac{d^2s}{dt^2}, \quad a_N = k\left(\frac{ds}{dt}\right)^2;$$

$a_T$ and $a_N$ are called the *tangential* and *normal components* of $\mathbf{a}(t)$. Since $\mathbf{T}$ and $\mathbf{N}$ are perpendicular, unit vectors, it follows from (20.10.5) that

(20.10.6) $$|\mathbf{a}(t)|^2 = a_T^2 + a_N^2.$$

A use of (20.10.6) appears in the following example.

**Example 20.10.1.** The path of a moving particle is defined by the vector function

(20.10.7) $$\mathbf{R}(t) = \ln t\, \mathbf{i} + t^2 \mathbf{j} - 2t\, \mathbf{k}.$$

Find the tangential and normal components of the acceleration at any value of $t$.

*Solution.* We find
$$\mathbf{v}(t) = \mathbf{R}'(t) = \frac{1}{t}\mathbf{i} + 2t\mathbf{j} - 2\mathbf{k},$$

$$\frac{ds}{dt} = |\mathbf{v}(t)| = \sqrt{\frac{1}{t^2} + 4t^2 + 4} = 2t + \frac{1}{t}, \qquad (t > 0).$$

Hence by (20.10.5),
$$a_T = \frac{d^2 s}{dt^2} = \frac{d}{dt}\left(\frac{ds}{dt}\right) = 2 - \frac{1}{t^2}.$$

For the acceleration we have
$$\mathbf{a}(t) = \mathbf{v}'(t) = -\frac{1}{t^2}\mathbf{i} + 2\mathbf{j}, \qquad |\mathbf{a}(t)|^2 = \frac{1}{t^4} + 4.$$

Now using (20.10.6), we obtain
$$a_N^2 = |\mathbf{a}(t)|^2 - a_T^2 = \frac{1}{t^4} + 4 - \left(2 - \frac{1}{t^2}\right)^2 = \frac{4}{t^2}, \qquad a_N = \frac{2}{t}.$$

We have found
$$a_T = 2 - \frac{1}{t^2}, \qquad a_N = \frac{2}{t}, \quad \text{if} \quad t > 0.$$

The methods used in Example 20.10.1 provide an effective way of finding the curvature of a curve, *whether the parameter denotes time or not.* Using the above notation, we have from (20.10.5),
$$k = \frac{a_N}{(ds/dt)^2}.$$

With the values of $a_N$ and $ds/dt$ from Example 20.10.1, we obtain

(20.10.8) $$k = \frac{2/t}{(2t + 1/t)^2} = \frac{2t}{(2t^2 + 1)^2}.$$

**Example 20.10.2.** Find the curvature of the curve (20.10.7) by use of Definition 20.10.1.

*Solution.* We find
$$\mathbf{T}(t) = \frac{\mathbf{R}'(t)}{|\mathbf{R}'(t)|} = \frac{(1/t)\mathbf{i} + 2t\mathbf{j} - 2\mathbf{k}}{2t + 1/t} = \frac{\mathbf{i} + 2t^2\mathbf{j} - 2t\mathbf{k}}{2t^2 + 1},$$

$$\mathbf{T}'(t) = \frac{1}{(2t^2 + 1)^2} [-4t\mathbf{i} + 4t\mathbf{j} + 2(1 - 2t^2)\mathbf{k}],$$

$$\frac{d\mathbf{T}}{ds} = \frac{d\mathbf{T}/dt}{ds/dt} = \frac{t}{(2t^2 + 1)^3} [-4t\mathbf{i} + 4t\mathbf{j} + 2(1 - 2t^2)\mathbf{k}].$$

Finally, if $t > 0$,

$$k = \left|\frac{d\mathbf{T}}{ds}\right| = \frac{t}{(2t^2 + 1)^3} \sqrt{16t^2 + 16t^2 + 4(1 - 2t^2)^2} = \frac{2t}{(2t^2 + 1)^2}.$$

This is the same result as is (20.10.8), but the method used in obtaining (20.10.8) seems to be simpler.

\*Frenet formulas. As a matter of some interest here, and of much importance in Differential Geometry, we obtain formulas for the derivatives of $\mathbf{T}$, $\mathbf{N}$, and $\mathbf{B}$ with respect to $s$, as linear combinations of $\mathbf{T}$, $\mathbf{N}$, and $\mathbf{B}$ themselves.

In the definition of $\mathbf{N}$ in (20.9.2) any parameter may be used. If the parameter is arc length $s$, that relation yields

(20.10.9) $$\frac{d\mathbf{T}}{ds} = \left|\frac{d\mathbf{T}}{ds}\right| \mathbf{N} = k\mathbf{N}.$$

Since $\mathbf{N}$ is a unit vector, $d\mathbf{N}/ds$ is orthogonal to $\mathbf{N}$, by Theorem 20.8.3; hence $d\mathbf{N}/ds$ lies in the plane of $\mathbf{T}$ and $\mathbf{B}$, and we may write, for some values $a$ and $\tau$,

$$\frac{d\mathbf{N}}{ds} = a\mathbf{T} + \tau\mathbf{B}.$$

Now from $\mathbf{B} = \mathbf{T} \times \mathbf{N}$ we obtain (using $\mathbf{T} \times \mathbf{T} = \mathbf{0}$, $\mathbf{T} \times \mathbf{B} = -\mathbf{N}$),

(20.10.10) $$\frac{d\mathbf{B}}{ds} = \mathbf{T} \times \frac{d\mathbf{N}}{ds} = \mathbf{T} \times (a\mathbf{T} + \tau\mathbf{B}) = \tau\mathbf{T} \times \mathbf{B} = -\tau\mathbf{N}.$$

From $\mathbf{N} = \mathbf{B} \times \mathbf{T}$ we derive

$$\frac{d\mathbf{N}}{ds} = \mathbf{B} \times \frac{d\mathbf{T}}{ds} + \frac{d\mathbf{B}}{ds} \times \mathbf{T},$$

and, using (20.10.9) and (20.10.10), we get

(20.10.11) $$\frac{d\mathbf{N}}{ds} = \mathbf{B} \times (k\mathbf{N}) - \tau\mathbf{N} \times \mathbf{T} = k\mathbf{B} \times \mathbf{N} - \tau\mathbf{N} \times \mathbf{T} = -k\mathbf{T} + \tau\mathbf{B}.$$

The *Frenet formulas*, in (20.10.9), (20.10.10), and (20.10.11), are collected as follows:

(20.10.12) $$\frac{d\mathbf{T}}{ds} = k\mathbf{N}, \quad \frac{d\mathbf{N}}{ds} = -k\mathbf{T} + \tau\mathbf{B}, \quad \frac{d\mathbf{B}}{ds} = -\tau\mathbf{N}.$$

The coefficients of $\mathbf{T}$, $\mathbf{N}$, and $\mathbf{B}$ in these formulas involve the curvature $k$, and a function $\tau$ which is defined as the *torsion* of the curve at a point. The torsion may be construed as a measure of the degree by which a curve deviates from a plane curve at a point—it, too, is of considerable importance in the study of Differential Geometry. (See Exercise 12 for a formula for torsion.)

**EXERCISE GROUP 20.7**

In Exercises 1–5 the path of a moving particle is given, where $t$ denotes time.

1. $x = 2t^2 + t$, $y = t^2 - 2t$, $z = t^2 - 1$
   (a) Find **T**, **N**, $a_T$, and $a_N$ at $t = 0$
   (b) Find an equation of the osculating plane at $t = 0$

2. $x = 2 \cos t$, $y = 2 \sin t$, $z = t$
   (a) Find **T**, **N**, $a_T$, and $a_N$ for any $t$
   (b) Find an equation of the osculating plane at $t = \pi/2$

3. $\mathbf{R}(t) = (\sin t - t \cos t)\mathbf{i} + (\cos t + t \sin t)\mathbf{j} + t^2\mathbf{k}$
   (a) Find $a_T$ and $a_N$ for any $t$
   (b) Find the curvature for any $t$

4. $\mathbf{R}(t) = 2t\mathbf{i} + 2t\mathbf{j} + t^2\mathbf{k}$
   (a) Find **T** and **N** for any $t$
   (b) Find an equation of the osculating plane at $t = 1$
   (c) Find **B** at $t = 1$

5. $\mathbf{R}(t) = t^2\mathbf{i} + (1 + t^2)\mathbf{j} + (1 - t)\mathbf{k}$
   (a) Find **T** and **N** at $t = 0$
   (b) Find the curvature at $t = 0$
   (c) Find **B** at $t = 0$

6. By differentiating $\mathbf{T} = \mathbf{R}'(t)/|\mathbf{R}'(t)|$ with respect to $s$, and then using Definition 20.10.1, derive the following formula for curvature:

$$k = \frac{\sqrt{|\mathbf{R}'(t)|^2 |\mathbf{R}''(t)|^2 - (\mathbf{R}'(t) \cdot \mathbf{R}''(t))^2}}{|\mathbf{R}'(t)|^3}.$$

7. Use Exercise 6 to find the curvature of the curve in Exercise 1 at $t = 0$.

8. Use Exercise 6 to find the curvature of the curve in Exercise 4 at $t = 1$.

9. By using the result of Exercise 6 show that $k = |\mathbf{R}''(s)|$, where the parameter is $s$, arc length.

10. With use of (20.10.2) and (20.10.3) show that

    $\mathbf{R}'(t) \times \mathbf{R}''(t) = k|\mathbf{R}'(t)|^3 \mathbf{B}$, and hence $k = |\mathbf{R}'(t) \times \mathbf{R}''(t)|/|\mathbf{R}'(t)|^3$.

11. Use the result in Exercise 10 to find the curvature for each curve:
    (a) $\mathbf{R}(t) = \mathbf{i} \cos t^2 + \mathbf{j} \sin t^2$  (b) $\mathbf{R}(t) = 2\mathbf{i} \cos t + 2\mathbf{j} \sin t + 3\mathbf{k}$

12. If $\mathbf{R} = \mathbf{R}(t)$ represents a curve in space, it can be shown that the torsion is given by

$$\tau = \frac{(\mathbf{R}', \mathbf{R}'', \mathbf{R}''')}{|\mathbf{R}' \times \mathbf{R}''|^2},$$

where $(\mathbf{R}', \mathbf{R}'', \mathbf{R}''') = \mathbf{R}' \cdot \mathbf{R}'' \times \mathbf{R}'''$ is a triple scalar product. Use this result to find the torsion in each of the following:
(a) $\mathbf{R}(t) = (3 \cos t)\mathbf{i} + (3 \sin t)\mathbf{j} + 4t\mathbf{k}$
(b) $\mathbf{R}(t) = t^3\mathbf{i} + t^2\mathbf{j} + t\mathbf{k}$

13. It can be shown that if $\tau = 0$ for each point of a curve, the curve is a plane curve, that is, the curve lies in a plane. Use Exercise 12 to show that $\tau = 0$ for each of the following curves, and find an equation of the plane which contains the curve.

   (a) $\mathbf{R}(t) = \left(t - \dfrac{1}{t}\right)\mathbf{i} + \left(1 + \dfrac{1}{t}\right)\mathbf{j} + t\mathbf{k}$

   (b) $\mathbf{R}(t) = (\cos 2t)\mathbf{i} + t^2\mathbf{j} + (t^2 + \cos^2 t)\mathbf{k}$

* 14. Prove that if the curve $\mathbf{R}(t) = x(t)\mathbf{i} + y(t)\mathbf{j} + z(t)\mathbf{k}$ lies in a plane, then $\mathbf{T}(t)$ and $\mathbf{N}(t)$ are parallel to that plane. [*Suggestion*: If $ax + by + cz = d$ is an equation of the plane, then $a\,x(t) + b\,y(t) + c\,z(t) = d$.]

15. Derive formula (20.10.6).

## 20.11 CYLINDRICAL AND SPHERICAL COORDINATES

While a rectangular coordinate system is perhaps the most widely used in mathematics and its applications, two other coordinate systems are of sufficient interest and importance to merit consideration here. In some connections, in fact, one or the other of these systems may be more advantageous than a rectangular system.

In a rectangular system all three coordinates represent linear measure; in the systems to be introduced at least one coordinate represents angular measure.

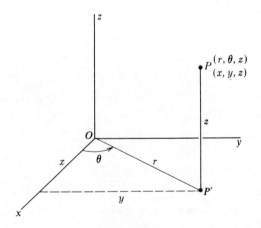

Fig. 20.11.1

*Cylindrical coordinates.* If $P$ is a point in space (Fig. 20.11.1), *cylindrical coordinates* of $P$ are defined as the polar coordinates $r$ and $\theta$ of the projection of $P$ on the $xy$ plane, and the $z$ coordinate of $P$. We designate the point as $P(r, \theta, z)$. As in polar coordinates, we permit $r$ to be negative. It is clear that a point has infinitely many representations in cylindrical coordinates.

The equations connecting cylindrical and rectangular coordinates are obtained directly from the corresponding equations for polar coordinates, and may be written as

(20.11.1) $$x = r \cos \theta, \quad y = r \sin \theta, \quad z = z$$

for the *rectangular coordinates of P in terms of its cylindrical coordinates*, and as

(20.11.2) $$r = \pm \sqrt{x^2 + y^2}, \quad \theta = \arctan \frac{y}{x}, \quad z = z$$

for the *cylindrical coordinates in terms of the rectangular coordinates*. The value of $\theta$ and the sign of $r$ in (20.11.2) must be consistent with (20.11.1).

**Example 20.11.1.** Find cylindrical coordinates of the point $P$ with rectangular coordinates $(-2, -2, 3)$.

*Solution.* Using (20.11.2), we may take $r = \sqrt{4 + 4} = 2\sqrt{2}$ and $\theta = \arctan 1 = 5\pi/4$, where an angle in the third quadrant is used since $x$ and $y$ are both negative. We may then give $(2\sqrt{2}, 5\pi/4, 3)$ as cylindrical coordinates of $P$.

If $c$ is a constant, the surfaces representing the equations $x = c$, $y = c$, and $z = c$ are planes parallel to the coordinate planes. Corresponding surfaces in cylindrical coordinates are $r = c$, a family of right circular cylinders with the $z$ axis as axis, $\theta = c$, a family of planes containing the $z$ axis, and again $z = c$.

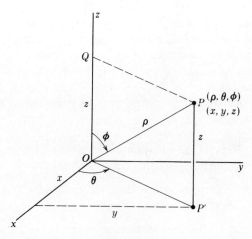

Fig. 20.11.2

**Spherical coordinates.** In this system one coordinate represents linear measure, and the other two represent angular measure. If $P$ is a point in space (Fig. 20.11.2), *spherical coordinates* of $P$ are the length $\rho$ of $OP$ (hence $\rho \geq 0$), the angle $\theta$ as used in cylindrical coordinates, and the angle $\phi$ that $OP$ makes with the positive $z$ axis. We designate the point as $P(\rho, \theta, \phi)$, and restrict $\phi$ to the interval $0 \leq \phi \leq \pi$. From the figure we obtain

$$x = |OP'| \cos \theta, \qquad y = |OP'| \sin \theta, \qquad |OP'| = |QP| = \rho \sin \phi.$$

With these relations we find

(20.11.3) $\qquad x = \rho \sin \phi \cos \theta, \qquad y = \rho \sin \phi \sin \theta, \qquad z = \rho \cos \phi$

for the *rectangular coordinates of P in terms of its spherical coordinates*. If we solve (20.11.3) for $\rho$, $\theta$, and $\phi$, we obtain

(20.11.4) $\quad \rho = \sqrt{x^2 + y^2 + z^2}, \qquad \theta = \arctan \dfrac{y}{x}, \qquad \phi = \arccos \dfrac{z}{\sqrt{x^2 + y^2 + z^2}}$

for the *spherical coordinates of P in terms of its rectangular coordinates*. The quadrant for $\theta$ is determined by the signs of $x$ and $y$.

**Example 20.11.2.** Find rectangular coordinates of the point $P(2, \pi/6, \pi/4)$ in spherical coordinates.

*Solution.* Using (20.11.3), we find

$$x = 2 \cdot \dfrac{\sqrt{2}}{2} \cdot \dfrac{\sqrt{3}}{2} = \dfrac{\sqrt{6}}{2}, \qquad y = 2 \cdot \dfrac{\sqrt{2}}{2} \cdot \dfrac{1}{2} = \dfrac{\sqrt{2}}{2}, \qquad z = 2 \cdot \dfrac{\sqrt{2}}{2} = \sqrt{2}.$$

The rectangular coordinates of $P$ are $(\sqrt{6}/2, \sqrt{2}/2, \sqrt{2})$.

If $c$ is constant, the surfaces $\rho = c$ are concentric spheres with center at the origin; the surfaces $\theta = c$ are half-planes bounded by the $z$ axis; and the surfaces $\phi = c$ are nappes of right circular cones with vertex at the origin and axis on the $z$ axis. The spherical coordinates $\theta$ and $\phi$ correspond to *longitude* and *colatitude*, respectively, on the earth.

Equations in one coordinate system may be transformed into equations in another coordinate system.

**Example 20.11.3.** Express the equation of the sphere $x^2 + y^2 + z^2 - 2x = 0$ in (a) cylindrical coordinates, (b) spherical coordinates.

*Solution.* (a) In cylindrical coordinates we have $x^2 + y^2 = r^2$, $x = r \cos \theta$. The equation becomes $r^2 + z^2 - 2r \cos \theta = 0$.

(b) In spherical coordinates we have $x^2 + y^2 + z^2 = \rho^2$, and we also have $x = \rho \sin \phi \cos \theta$. The equation becomes $\rho^2 - 2\rho \sin \phi \cos \theta = 0$. We may divide out $\rho$, and write an equation of the sphere in spherical coordinates as

(20.11.5) $\qquad\qquad \rho - 2 \sin \phi \cos \theta = 0.$

No point of the surface is lost in dividing out $\rho$, since $\rho = 0$ corresponds to the origin, and the origin is on the graph of (20.11.5).

## EXERCISE GROUP 20.8

In Exercises 1–14 express the point, whose coordinates are given in the first specified coordinate system, in terms of coordinates in the second specified coordinate system.

1. $(1, -1, 1)$, rectangular; cylindrical
2. $(6, 3, 3)$, rectangular; cylindrical
3. $(-2, \sqrt{3}, -2)$, rectangular; cylindrical
4. $(2, -3, -3)$, rectangular; cylindrical
5. $(-4, \pi/2, -1)$, cylindrical; rectangular
6. $(3, -\pi/3, 4)$, cylindrical; rectangular
7. $(2, 5\pi/4, -3)$, cylindrical; rectangular
8. $(1, -1, 1)$, rectangular; spherical
9. $(\sqrt{5}, 2, 4)$, rectangular; spherical
10. $(2, \pi/2, \pi/2)$, spherical; rectangular
11. $(3, -\pi/3, 3\pi/4)$, spherical; rectangular
12. $(2, \pi/4, 2\pi/3)$, spherical; cylindrical
13. $(5, -\pi/4, 3\pi/4)$, spherical; cylindrical
14. $(-2, \pi/3, 2)$, cylindrical; spherical

In Exercises 15–26 obtain an equation of the given surface in (a) cylindrical coordinates, (b) in spherical coordinates.

15. $x^2 + y^2 = 2z$
16. $x^2 + y^2 = 2y$
17. $x^2 + y^2 = z^2$
18. $x^2 + y^2 - z^2 = 4$
19. $x^2 + y^2 + 3z^2 = 3$
20. $x^2 + y^2 = 5$
21. $x^2 + y^2 + z^2 - 2y = 0$
22. $x^2 + yz = 1$
23. $x^2 + y^2 + z^2 = 4$
24. $x^2 + y^2 = 9x$
25. $x^2 + 2y = 4$
26. $x^2 - y^2 = 2z$

# 21 Partial Differentiation

## 21.1 FUNCTIONS OF SEVERAL VARIABLES

Our previous work was devoted primarily to functions of a single real variable, that is, functions for which the domain consisted of a set of real numbers. We now extend the concept so that it applies to functions of several variables, with a domain as a set of pairs of real numbers, or a set of triples of real numbers, and so on. Much of the discussion will be directed to functions of two variables, but the extension to more than two variables will be clear. For a function of two variables we shall interchangeably refer to the domain as a set of pairs of real numbers, or a set of points in a plane. The following definition is similar to Definition 4.7.1.

**DEFINITION 21.1.1.** *Let S be a set of points in a plane. If to each point p in S there corresponds a unique real number z, the set of ordered pairs (p, z) defines a* **function**. *The set S is the* **domain** *of the function; the set of numbers z, for p in S, is the* **range** *of the function.*

If the function is denoted by $f$, then for any point $p = (x, y)$ in $S$ we may write

$$z = f(p) \quad \text{or} \quad z = f(x, y).$$

The number $z$ is then called the *value* of the function $f$ at the point $p$ or at $(x, y)$.
If the values of a function are given by a mathematical expression with no

domain specified, the domain is understood to be the set of points for which that expression is defined.

**Example 21.1.1.** The function $f$ defined by

$$f(x, y) = \frac{x + y}{x - y}$$

has as its domain all points for which $x - y \neq 0$, that is, the entire $xy$ plane except for the points on the line $x = y$.

The *distance* between two points $p$ and $q$ was obtained in Section 18.3 and is now denoted by $|p - q|$. If $p = (x, y)$, $q = (x_2, y_2)$, then

$$|p - q| = \sqrt{(x_2 - x_1)^2 + (y_2 - y_1)^2}.$$

Limit and continuity are defined in terms of distance.

**DEFINITION 21.1.2.** *Let the domain of $f$ include the interior of some circle with $(a, b)$ as center, except possibly $(a, b)$ itself. Then*

$$\lim_{(x, y) \to (a, b)} f(x, y) = A,$$

*if, given $\epsilon > 0$, there exists $\delta$ such that*

$$|f(x, y) - A| < \epsilon \quad \text{whenever} \quad 0 < |(x, y) - (a, b)| < \delta.$$

According to this definition, the value $f(a, b)$, or even its existence, is not involved in the limit. Only values of $f$ for points near, but different from $(a, b)$, appear in the definition.

The properties of limits, as stated in Section 3.5 for functions of one variable, are preserved and are not restated here.

**Example 21.1.2.** Show that

$$\lim_{(x, y) \to (0, 0)} \frac{x^2 y^2}{x^2 + y^2} = 0.$$

*Solution.* The limit theorem for a quotient is not applicable, since the denominator approaches zero. However, we have $x^2 \leq x^2 + y^2$, $y^2 \leq x^2 + y^2$. Hence $x^2 y^2 < (x^2 + y^2)^2$ [equality cannot hold since $(x, y) \neq (0, 0)$], and

(21.1.1) $$\left| \frac{x^2 y^2}{x^2 + y^2} \right| < |x^2 + y^2|.$$

Given $\epsilon > 0$, choose $\delta = \epsilon^{1/2}$. Then, from (21.1.1)

$$\left| \frac{x^2 y^2}{x^2 + y^2} \right| < \epsilon \quad \text{when} \quad 0 < \sqrt{x^2 + y^2} < \delta,$$

and the desired limit has been established.

**DEFINITION 21.1.3.** *A function $f(x, y)$ is* **continuous** *at $(a, b)$ if*

$$\lim_{(x, y) \to (a, b)} f(x, y) = f(a, b).$$

We shall assume the continuity properties of the familiar functions. Thus a polynomial in $x$ and $y$ is everywhere continuous; a rational function, the quotient of two polynomials, is continuous wherever the denominator is not zero; composite functions are continuous if the component functions are; and so on.

The function in Example 21.1.1 is continuous at $(a, b)$ if $a - b \neq 0$; it is discontinuous at any point $(a, a)$, since the function is then undefined.

The function in Example 21.1.2 is continuous if $(x, y) \neq (0, 0)$, and is undefined at $(0, 0)$. If we extend the domain by defining the value of the function at $(0, 0)$ as zero, the limit of $f$ as $(x, y) \to (0, 0)$, the new function is continuous also at $(0, 0)$ and hence at each point in the plane. At $(0, 0)$ there is a *removable discontinuity*, according to the following definition.

**DEFINITION 21.1.4.** *If a function, discontinuous at $(a, b)$, may be made continuous at $(a, b)$ by defining or redefining its value there, the point $(a, b)$ is a* **removable discontinuity** *of the function.*

In some cases the following result may be used to prove the nonexistence of a limit.

**THEOREM 21.1.1.** *Let $y = g(x)$ be a curve which passes through $(a, b)$, with $g(x)$ continuous at $a$. If $\lim\limits_{(x, y) \to (a, b)} f(x, y) = A$, then*

$$\lim_{x \to a} f(x, g(x)) = A.$$

*Proof.* We must show that, given $\epsilon > 0$, there exists $\delta$ such that

$$|f(x, g(x)) - A| < \epsilon \quad \text{when} \quad 0 < |x - a| < \delta.$$

Now, given $\epsilon > 0$ we may choose $\delta'$ so that

(21.1.2) $\quad |f(x, y) - A| < \epsilon \quad \text{when} \quad 0 < |(x, y) - (a, b)| < \delta'.$

Then by the continuity of $g$ we may choose $\delta''$ so that

$$|g(x) - b| < \delta'/2 \quad \text{when} \quad |x - a| < \delta''.$$

The required $\delta$ is chosen as the smaller of $\delta'/2$ and $\delta''$. Now suppose that $0 < |x - a| < \delta$. Then

$$|(x, g(x)) - (a, b)| = \sqrt{(x - a)^2 + (g(x) - b)^2} \leq |x - a| + |g(x) - b|$$
$$< \delta + \delta'/2 \leq \delta'/2 + \delta'/2 = \delta'.$$

Then, by (21.1.2), $|f(x, g(x)) - A| < \epsilon$, and the proof is complete.

The use of Theorem 21.1.1 to prove the nonexistence of a limit consists in finding two paths along which the limits are not the same.

**Example 21.1.3.** Discuss the existence of the limit

$$\lim_{(x,y)\to(0,0)} \frac{xy}{x^2+y^2}.$$

*Solution.* Let the function be $f$. Along the curve $y = x$ we have $(x, y) = (x, x)$ and

$$\lim_{(x,x)\to(0,0)} \frac{xy}{x^2+y^2} = \lim_{x\to 0} \frac{x^2}{2x^2} = 2.$$

Along the curve $y = 0$ we have $(x, y) = (x, 0)$ and

$$\lim_{(x,0)\to(0,0)} \frac{xy}{x^2+y^2} = \lim_{x\to 0} \frac{0}{x^2} = 0.$$

Since different limits are obtained along different paths, $\lim f(x, y)$ cannot exist as $(x, y) \to (0, 0)$, by Theorem 21.1.1.

A statement similar to Theorem 21.1.1 can also be given for a curve $x = h(y)$ through the point $(a, b)$, or for a curve $x = x(t)$, $y = y(t)$ through the point.

It has been seen that the proof of the existence of a limit can be based on the use of limit theorems or on the definition of limit. Theorem 21.1.1 cannot be used for this purpose.

### EXERCISE GROUP 21.1

In Exercises 1–6 describe the domain of the function.

1. $e^{x-y}$
2. $\sqrt{x^2+y^2-4}$
3. $x\sqrt{x^2+y^2-4}$
4. $\ln(x^2+y^2-1)$
5. $\sqrt{x^2-y^2-1}$
6. $\ln(2-xy)$

In each of Exercises 7–12 prove that the limit does not exist.

7. $\lim_{(x,y)\to(0,0)} \dfrac{x}{x^2+y^2}$

8. $\lim_{(x,y)\to(0,0)} \dfrac{x^2-y^2}{x^2+y^2}$

9. $\lim_{(x,y)\to(0,0)} \ln(x^2+y^2)$

10. $\lim_{(x,y)\to(0,0)} \dfrac{x}{x-y}$

11. $\lim_{(x,y)\to(0,0)} \dfrac{x^2}{x^2+y^2}$

12. $\lim_{(x,y)\to(0,0)} \dfrac{x^4 y^4}{(x^2+y^4)^3}$

13. Each of the following functions has a removable discontinuity at $(0, 0)$. How should $f(0, 0)$ be defined to remove the discontinuity?

(a) $f(x, y) = \dfrac{x^2 y^2}{x^2 + y^2}$

(b) $\begin{cases} f(x, y) = \dfrac{x(1 + y^2)}{\sin x}, & x \neq 0 \\ f(0, y) = 1 + y^2 \end{cases}$

14. Let $f(x, y) = \dfrac{x^2 y^2}{x^2 y^2 + (x - y)^2}$. Find

(a) $\lim\limits_{x \to 0} f(x, x)$  (b) $\lim\limits_{x \to 0} f(x, 2x)$  (c) $\lim\limits_{y \to 0} f(y^2, y)$

A *level curve* of a function $f(x, y)$ is the set of points for which $f(x, y) = c$, constant. Describe the level curves for each of the following functions. Sketch several level curves for different values of $c$.

15. $\dfrac{x - y}{x + y}$

16. $\dfrac{x}{x^2 + y^2}$

17. $\sqrt{x^2 + y^2 - 4}$

18. $\ln(2 - xy)$

## 21.2 DIRECTIONAL DERIVATIVE. PARTIAL DERIVATIVE

The derivative of a function of one variable is basically the rate of change of the function with respect to the variable. We wish to retain this idea as much as possible in working with a function $f(x, y)$ of two variables; but when the domain is a set of points in a plane, a point may move along any curve through a given point $(x_0, y_0)$, and the rate of change of the function depends on the curve. To narrow down the possibilities, we consider straight line paths through the point, and define a derivative of $f$ related to such a path. Let $\mathbf{u}$ be a unit vector in the $xy$ plane in the direction $\theta$, so that

$$\mathbf{u} = \lambda \mathbf{i} + \mu \mathbf{j}, \qquad \lambda = \cos \theta, \qquad \mu = \sin \theta,$$

and consider a line in this direction, as in Fig. 21.2.1. A point on this line at a distance $h$ from $P_0$ is $P(x_0 + \lambda h, y_0 + \mu h)$. The change in the value of the function from $P_0$ to $P$ is $f(x_0 + \lambda h, y_0 + \mu h) - f(x_0, y_0)$; the distance $\overline{P_0 P}$, directed along the line, is $h$; and the limit of the ratio of these numbers as $h \to 0$ is called the directional derivative of $f$ at the point $(x_0, y_0)$ in the direction of $\mathbf{u}$.

**DEFINITION 21.2.1.** *The* **directional derivative** *of a function $f(x, y)$ at the point $(x_0, y_0)$ in the direction of the unit vector $\mathbf{u} = \lambda \mathbf{i} + \mu \mathbf{j}$ is*

$$D_{\mathbf{u}} f(x_0, y_0) = \lim_{h \to 0} \frac{f(x_0 + \lambda h, y_0 + \mu h) - f(x_0, y_0)}{h},$$

*if the limit exists.*

If $\mathbf{u}$ makes an angle $\theta$ with the positive $x$ axis, the directional derivative $D_{\mathbf{u}} f$ is also denoted by $D_\theta f$.

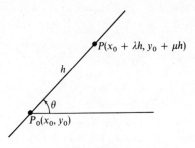

Fig. 21.2.1

**Example 21.2.1.** Find the directional derivative of $f(x, y) = x^2 y$ in the direction with $\theta = \pi/6$.

*Solution.* We have $\mathbf{u} = \mathbf{i}(\sqrt{3}/2) + \mathbf{j}/2$, $\lambda = \sqrt{3}/2$, $\mu = 1/2$. We may omit subscripts. Then

$$f\left(x + \frac{\sqrt{3}}{2}h, y + \frac{1}{2}h\right) - f(x, y) = \left(x + \frac{\sqrt{3}}{2}h\right)^2 \left(y + \frac{1}{2}h\right) - x^2 y$$

$$= \left(\sqrt{3}xy + \frac{1}{2}x^2\right)h + \left(\frac{\sqrt{3}}{2}x + \frac{3}{4}y\right)h^2 + \frac{3}{8}h^3.$$

We now divide by $h$, and with Definition 21.2.1 we get

$$D_{\pi/6} f = \lim_{h \to 0}\left[\left(\sqrt{3}xy + \frac{1}{2}x^2\right) + \left(\frac{\sqrt{3}}{2}x + \frac{3}{4}y\right)h + \frac{3}{8}h^2\right] = \sqrt{3}xy + \frac{1}{2}x^2.$$

**Example 21.2.2.** Find the directional derivative of $f(x, y) = x^2 y$ in the direction of the unit vector $\mathbf{u} = \lambda \mathbf{i} + \mu \mathbf{j}$.

*Solution.* We have

$$\frac{f(x + \lambda h, y + \mu h) - f(x, y)}{h} = \frac{(x + \lambda h)^2 (y + \mu h) - x^2 y}{h}$$

$$= \frac{(2\lambda xy + \mu x^2)h + (\lambda^2 y + 2\lambda\mu x)h^2 + \lambda^2 \mu h^3}{h}$$

$$= (2\lambda xy + \mu x^2) + (\lambda^2 y + 2\lambda\mu x)h + \lambda^2 \mu h^2.$$

Hence

$$D_\mathbf{u} f = \lim_{h \to 0} \frac{f(x + \lambda h, y + \mu h) - f(x, y)}{h} = 2\lambda xy + \mu x^2.$$

If $\lambda = \sqrt{3}/2$ and $\mu = 1/2$, this agrees with the result in Example 21.2.1.

Of particular interest are the directional derivatives of a function $f(x, y)$ in the directions of **i** and **j**. If $\mathbf{u} = \mathbf{i}$, then $\lambda = 1$, $\mu = 0$, and Definition 21.2.1 gives

$$(21.2.1) \qquad D_\mathbf{i} f = \lim_{h \to 0} \frac{f(x+h, y) - f(x, y)}{h}.$$

Since both functional values in the numerator involve the same value of $y$, the value of $D_\mathbf{i} f$ is precisely what would be obtained if the original definition of derivative for a function of one variable were applied to $f(x, y)$ with $y$ held constant. For this reason, the limit in (21.2.1) is also called the *partial derivative* of $f$ with respect to $x$. Two commonly used notations are $\partial f / \partial x$ and $f_x(x, y)$. Similar considerations apply to $D_\mathbf{j} f$, in which case $x$ is fixed and $y$ varies.

**DEFINITION 21.2.2.** *For $f(x, y)$ we define the* **partial derivatives**

$$\frac{\partial f}{\partial x} = f_x(x, y) = \lim_{h \to 0} \frac{f(x+h, y) - f(x, y)}{h}$$

$$\frac{\partial f}{\partial y} = f_y(x, y) = \lim_{h \to 0} \frac{f(x, y+h) - f(x, y)}{h}$$

*if these limits exist.*

It then follows that for $f(x, y)$,

$$\frac{\partial f}{\partial x} = D_\mathbf{i} f, \qquad \frac{\partial f}{\partial y} = D_\mathbf{j} f.$$

The formulas and techniques of ordinary differentiation carry over to partial differentiation. To differentiate with respect to $x$ we treat $y$ as constant, and to differentiate with respect to $y$ we treat $x$ as constant. In general, partial derivatives of a function are functions of the same variables as in the original.

**Example 21.2.3.** For the function $f(x, y) = x^2 y$ we have

$$f_x = 2xy, \qquad f_y = x^2.$$

**Example 21.2.4.** If $u(x, y) = xf(y) + yg(x)$, find $u_x$ and $u_y$.

*Solution.* Treating $y$ as a constant to get $u_x$, and then $x$ as a constant to get $u_y$, we find

$$u_x = f(y) + yg'(x), \qquad u_y = xf'(y) + g(x),$$

where each of $f'(y)$ and $g'(x)$ is an ordinary derivative of a function of a single variable.

Much of the preceding discussion can be extended to functions of more than 2 variables. For a function $f(x, y, z)$ of 3 variables, we consider a unit vector

$$\mathbf{u} = \lambda \mathbf{i} + \mu \mathbf{j} + \nu \mathbf{k}.$$

If $\alpha$, $\beta$, and $\gamma$ are direction angles of $\mathbf{u}$, then

$$\lambda = \cos \alpha, \qquad \mu = \cos \beta, \qquad \nu = \cos \gamma.$$

The directional derivative of $f(x, y, z)$ in the direction of the unit vector $\mathbf{u}$ is *defined* as

$$D_\mathbf{u} f = \lim_{h \to 0} \frac{f(x + \lambda h, y + \mu h, z + \nu h) - f(x, y, z)}{h}.$$

The partial derivatives of $f$ with respect to $x$, $y$, and $z$ are the directional derivatives of $f$ in the directions of the unit vectors $\mathbf{i}$, $\mathbf{j}$, and $\mathbf{k}$, respectively. Thus

$$\frac{\partial f}{\partial x} = f_x(x, y, z) = \lim_{h \to 0} \frac{f(x + h, y, z) - f(x, y, z)}{h},$$

with similar representations for $f_y$ and $f_z$. In the case of a function of more than 3 variables, we dispense here with the idea of directional derivative, but still define a partial derivative as a rate of change of the function with respect to a single variable. For example,

$$f_t(x, y, z, t) = \lim_{h \to 0} \frac{f(x, y, z, t + h) - f(x, y, z, t)}{h}.$$

In order to determine a partial derivative with respect to some variable, all the other variables are held constant.

**Example 21.2.5.** Find the partial derivatives of

$$s(u, v, w) = u^2 v w + \ln(uw) + e^{vw}.$$

*Solution.* Thinking of $v$ and $w$ as constant, we obtain

$$\frac{\partial s}{\partial u} = 2uvw + \frac{w}{uw} = 2uvw + \frac{1}{u}.$$

Similarly,

$$\frac{\partial s}{\partial v} = u^2 w + w e^{vw}, \qquad \frac{\partial s}{\partial w} = u^2 v + \frac{1}{w} + v e^{vw}.$$

**Example 21.2.6.** If $F(x, y, z) = zf(x, y) + xg(y, z) + yh(x, z)$, find $F_x$, $F_y$, and $F_z$.

*Solution.* We obtain $F_x = zf_x(x, y) + g(y, z) + yh_x(x, z)$

$$F_y = zf_y(x, y) + xg_y(y, z) + h(x, z)$$
$$F_z = f(x, y) + xg_z(y, z) + yh_z(x, z).$$

It may be helpful to see a geometric interpretation of some of the derivatives discussed, in the case of a function of two variables. Fig. 21.2.2 shows a surface $ABC$, the graph of the function $f(x, y)$ or of the equation $z = f(x, y)$. The curve $RP_0 S$ is the intersection of the surface with the plane $x = x_0$, and has the tangent $P_0 T$ at $P_0(x_0, y_0, z_0)$. The slope of this tangent line in the plane $x = x_0$ (that is, the tangent of the angle which $P_0 T$ makes with a line through the point $K$ and parallel to the $y$ axis) is $f_y(x_0, y_0)$. Similarly, the curve $MP_0 Q$ is the intersection of the surface with the plane $y = y_0$ and has the tangent $P_0 T'$ at $P_0$, with slope $f_x(x_0, y_0)$ in that plane. Finally, the curve $UP_0 V$ is the intersection of the surface with the plane $(x - x_0) \sin \theta - (y - y_0) \cos \theta = 0$ and has the tangent $P_0 T''$ at $P_0$. The slope of $P_0 T''$ in this plane is the directional derivative of $f$ at $P_0$ in the direction of $\theta$.

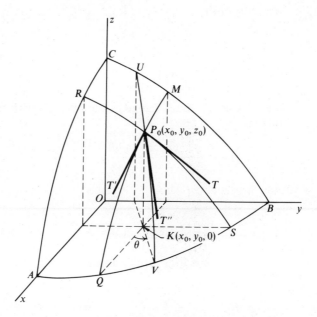

Fig. 21.2.2

**EXERCISE GROUP 21.2**

In each of Exercises 1–14 find the partial derivatives with respect to each variable in the function. The letters $x$, $y$, $z$, $u$, $v$, and $w$ represent variables.

1. $f(x, y) = x^2 + xy + 3y^2$
2. $f(x, y) = xy^2 - 2x^3 y + y^3$
3. $f(x, y) = \ln \dfrac{x}{x - y}$
4. $f(x, y) = x \cos(xy)$
5. $f(x, y) = \tan y/x$
6. $h(u, v) = e^{uv}$
7. $f(x, z) = \ln \sqrt{x^2 + z^2}$
8. $f(z, u) = z^u$
9. $f(x, y) = \cos^2(xy)$
10. $g(u, v) = \sin^{-1} u/v$
11. $r(x, y) = \sin^3(x + 2y)$
12. $r(x, y, u) = x^2 y + y^2 u + u^2 x$
13. $h(u, v, w) = \dfrac{u}{v + w}$
14. $f(x, y, z) = \sqrt{x^2 + y^2 + z^2}$

15. If $f(u, v) = u^2v - 2uv^2$, find $f_u(2, 3)$, $f_v(-1, 2)$.
16. If $g(x, y) = 4x^3y - \sin xy$, find $g_x(1, \pi/3)$, $g_y(-2, \pi/2)$.
17. If $f(x, y) = x^2 + y^3$ and $F(x, y) = f(\sin x, y)$, find $\partial F/\partial x$, $\partial F/\partial y$.
18. If $f(x, y) = x^2 + xy + y^2$ and $F(r, \theta) = f(r \cos \theta, r \sin \theta)$, find $\partial F/\partial r$, $\partial F/\partial \theta$.

In each of Exercises 19–22 show that $r_u = s_v$ and $r_v = -s_u$.

19. $r = u^2 - v^2$, $s = 2uv$
20. $r = u^3 - 3uv^2 - u^2 + v^2$, $s = 3u^2v - v^3 - 2uv$
21. $r = e^u \cos v$, $s = e^u \sin v$
22. $r = \sin u \cosh v$, $s = \cos u \sinh v$

23. If $u(x, y) = \sqrt{x} - \sqrt{y}$ show that $xu_x + yu_y = u/2$.
24. If $u(x, y) = \sin(x^2 + y^2)$ show that $yu_x - xu_y = 0$.
25. If $u(x, y) = -\frac{1}{2} + (x - \frac{3}{4}y)e^{y/2}$ show that $3u_x + 4u_y - 2u = 1$.
26. If $f(x, y) = y^2/x + x(\ln x - \ln y)$ show that $xf_x + yf_y = f$.
27. If $f(x, y, z) = \sqrt{x} \tan^{-1} y/z$ show that $xf_x + yf_y + zf_z = f/2$.
28. If $u = \dfrac{x^3 + y^3}{(x + y)^{1/2}}$ show that $x\dfrac{\partial u}{\partial x} + y\dfrac{\partial u}{\partial y} = \dfrac{5u}{2}$.
29. If $u = x^2y + y^2z + z^2x$ show that $u_x + u_y + u_z = (x + y + z)^2$.
30. If $u(x, y) = -y \cos x + f(y)$ find $u_x$ and $u_y$.
31. If $u(x, y) = xf(x, y) + yg(x, y)$ find $u_x$ and $u_y$.
32. If $u(x, y) = x^2/2 + xf(y) + g(y)$ find $u_x$ and $u_y$.
33. If $u(x, y, z) = x^2/2 + xf(y, z) + g(y, z)$ find $u_x$, $u_y$, and $u_z$.
34. If $F(x, y, z) = f(x, y) - z$ find $F_x$, $F_y$, and $F_z$.

Find the designated directional derivative in Exercises 35–38 by use of Definition 21.2.1.

35. $f(x, y) = x + 3y$; $D_{\pi/4} f$
36. $f(x, y) = xy$; $D_{\pi/2} f$
37. $f(x, y) = 2x - 3y$; $D_\pi f$, $D_{\mathbf{u}} f$
38. $f(x, y) = x^2y$; $D_{-\pi} f$, $D_{\mathbf{u}} f$
39. Find equations of the tangent line to the curve of intersection of the sphere $x^2 + y^2 + z^2 = 9$ and the plane $x = 1$ at the point $(1, 2, 2)$.
40. Find equations of the tangent line to the curve of intersection of the surface $x^2 + y^2 = 4z$ and the plane $y = 2$ at the point $(2, 2, 2)$.

## 21.3 PARTIAL DERIVATIVES OF HIGHER ORDER

The partial derivatives previously considered were *first-order* derivatives of a function of several variables. Such derivatives are themselves functions of the original variables. Thus for a function of two variables $f(x, y)$ we may consider the derivatives of $f_x$ with respect to $x$ and with respect to $y$, and similarly for $f_y$. For example, the derivative of $f_x$ with respect to $y$ is written as $f_{xy}$ and is *defined* as

$$f_{xy} = \lim_{h \to 0} \frac{f_x(x, y+h) - f_x(x, y)}{h},$$

*if the limit exists*. The definitions of the other *second-order partial* derivatives of $f(x, y)$ are similar, and need not be written here.

The notation, however, should be examined carefully. In the following we also give the notation corresponding to $\partial f/\partial x$. We have:

$$f_{xx} = (f_x)_x = \frac{\partial}{\partial x}\left(\frac{\partial f}{\partial x}\right) = \frac{\partial^2 f}{\partial x^2}$$

$$f_{xy} = (f_x)_y = \frac{\partial}{\partial y}\left(\frac{\partial f}{\partial x}\right) = \frac{\partial^2 f}{\partial y\,\partial x}$$

$$f_{yx} = (f_y)_x = \frac{\partial}{\partial x}\left(\frac{\partial f}{\partial y}\right) = \frac{\partial^2 f}{\partial x\,\partial y}$$

$$f_{yy} = (f_y)_y = \frac{\partial}{\partial y}\left(\frac{\partial f}{\partial y}\right) = \frac{\partial^2 f}{\partial y^2}$$

Note that in the notation for $f_{xy}$ the order of differentiation is from left to right, that is, first with respect to $x$ and then with respect to $y$. *This order is reversed* in the symbol $\partial^2 f/\partial y\,\partial x$, where the order of differentiation is from right to left.

There are accordingly four second-order partial derivatives of a function of two variables, but it will turn out later (Theorem 21.3.1) under the conditions commonly met that the *mixed derivatives* $f_{xy}$ and $f_{yx}$ are equal.

Partial derivatives of order higher than the second may be similarly defined for a function of two variables, as well as for a function of more than two variables. The techniques and formulas for finding higher-order derivatives are the same as for first-order derivatives.

*Example 21.3.1.* Find the second-order derivatives of

$$f(x, y) = \sin(xy) + xe^y.$$

*Solution.* We find

$$f_x = y\cos(xy) + e^y, \qquad f_y = x\cos(xy) + xe^y.$$

Then

$$f_{xx} = (f_x)_x = -y^2 \sin(xy),$$

$$f_{xy} = (f_x)_y = -xy \sin(xy) + \cos(xy) + e^y,$$

$$f_{yx} = (f_y)_x = -xy \sin(xy) + \cos(xy) + e^y,$$

$$f_{yy} = (f_y)_y = -x^2 \sin(xy) + xe^y.$$

Note that $f_{xy}$ and $f_{yx}$ turn out the same; but $f_{xy}$ was obtained by differentiating $f_x$ with respect to $y$, while $f_{yx}$ was obtained by differentiating $f_y$ with respect to $x$.

**Example 21.3.2.** Show that the function
$$u(x, t) = e^{-a^2kt}(A \cos ax + B \sin ax),$$
where $A$, $B$, and $a$ are constants, and $k$ is also constant, satisfies the relation
$$u_t = ku_{xx}.$$

*Solution.* We find
$$u_t = -a^2 k e^{-a^2kt}(A \cos ax + B \sin ax),$$
$$u_x = ae^{-a^2kt}(-A \sin ax + B \cos ax),$$
$$u_{xx} = a^2 e^{-a^2kt}(-A \cos ax - B \sin ax).$$
It is clear that $u_t = ku_{xx}$.

**Example 21.3.3.** Show that $w_{xzx} = w_{xxz}$ for the function
$$w = x^2 y^2 z + \sin(2x - 3z).$$

*Solution.* We obtain $w_{xzx}$ as follows:
$$w_x = 2xy^2 z + 2 \cos(2x - 3z),$$
$$w_{xz} = 2xy^2 + 6 \sin(2x - 3z),$$
$$w_{xzx} = 2y^2 + 12 \cos(2x - 3z).$$
To obtain $w_{xxz}$ we use $w_x$ as already obtained. Then
$$w_{xx} = 2y^2 z - 4 \sin(2x - 3z),$$
$$w_{xxz} = 2y^2 + 12 \cos(2x - 3z),$$
and $w_{xzx} = w_{xxz}$.

We now present a theorem which specifies a set of conditions under which mixed partial derivatives are equal. More specialized sets of conditions are considered in more advanced work.

**THEOREM 21.3.1.** *If $f(x, y), f_x, f_y, f_{xy}$ and $f_{yx}$ are continuous in a neighborhood of $(a, b)$, then $f_{xy}(a, b) = f_{yx}(a, b)$.*

*Proof.* A neighborhood of $(a, b)$ may be considered as the set of points within a circle with $(a, b)$ as center. Any point $(x, y)$ in the proof is assumed to lie within this circle.

Define functions $\phi(y)$ and $\psi(x)$ as follows
$$\phi(y) = f(a + h, y) - f(a, y),$$
$$\psi(x) = f(x, b + k) - f(x, b).$$

Then

$$\phi(b+k) - \phi(b) = [f(a+h, b+k) - f(a, b+k)] - [f(a+h, b) - f(a, b)],$$

$$\psi(a+h) - \psi(a) = [f(a+h, b+k) - f(a+h, b)] - [f(a, b+k) - f(a, b)],$$

and hence

(21.3.1) $$\phi(b+k) - \phi(b) = \psi(a+h) - \psi(a).$$

We now apply the Mean-Value Theorem to both members of (21.3.1), and get, for some numbers $\alpha$ and $\beta$,

(21.3.2.) $$k\phi'(b + \alpha k) = h\psi'(a + \beta h), \quad 0 < \alpha < 1, 0 < \beta < 1.$$

Now,

$$\phi'(y) = f_y(a+h, y) - f_y(a, y), \quad \psi'(x) = f_x(x, b+k) - f_x(x, b).$$

Hence (21.3.2) becomes

$$k[f_y(a+h, b+\alpha k) - f_y(a, b+\alpha k)] = h[f_x(a+\beta h, b+k) - f_x(a+\beta h, b)].$$

Since both values of $f_y$ on the left are for the same value of $y$, we may consider $f_y(x, b + \alpha k)$ as a function of $x$ and apply the Mean-Value Theorem to the difference. Similar considerations apply to the difference on the right. In this way we get, for some numbers $s$ and $t$,

$$khf_{yx}(a + sh, b + \alpha k) = hkf_{xy}(a + \beta h, b + tk), \quad 0 < s < 1, 0 < t < 1.$$

We now divide by $hk$, and then let $h \to 0$, $k \to 0$. By the continuity of $f_{yx}$ and $f_{xy}$, we have at once

$$f_{yx}(a, b) = f_{xy}(a, b),$$

and the proof is complete.

With the use of Theorem 21.3.1 it may be shown that, for a function of any number of variables, the value of a mixed derivative with respect to any variables is unchanged by a change in the order of differentiation. We have, for example,

$$f_{xxyy} = f_{xyxy} = f_{yxxy}.$$

To show the first equality we differentiate $f_x$ with respect to $x$ and $y$, or with respect to $y$ and $x$, and we have

$$(f_x)_{xy} = (f_x)_{yx};$$

we now differentiate both sides with respect to $y$. To show the second equality we start with the relation $f_{xy} = f_{yx}$, and differentiate with respect to $x$ and $y$.

**EXERCISE GROUP 21.3**

In each of Exercises 1–14 find the specified partial derivatives.

**1.** $f(x, y) = x^2 + xy + 3y^2; f_{xx}, f_{yy}$  **2.** $f(x, y) = xy^2 - 2x^3y + y^3; f_{xxy}, f_{xyy}$

3. $g(x, y) = x/y$; $g_{xy}$, $g_{yxy}$
4. $u(x, y) = \ln(x^2 + y)$; $u_{xy}$, $u_{xyy}$
5. $f(x, y) = \dfrac{x}{x^2 - y^2}$; $f_{yy}$, $f_{xy}$
6. $g(x, z) = \dfrac{1}{\sqrt{x^2 + z^2}}$; $g_{xx}$
7. $h(r, s) = 3r^2 s - s^3 - 2rs$; $h_{rrs}$, $h_{ssr}$
8. $f(x, y, z) = \sqrt{x^2 + y^2 + z^2}$; $f_{xz}$, $f_{xxz}$
9. $f(x, y, z) = x\sin(xyz)$; $f_{xy}$, $f_{yy}$, $f_{xz}$
10. $r(u, v) = e^u \cos v$; $r_{uv}$
11. $f(x, y) = \sin^2 x + y^3$; $f_{xy}$, $f_{xxx}$
12. $h(u, v) = e^{uv}$; $h_{uv}$, $h_{vv}$
13. $r(u, v) = \cos u \sinh v$; $r_{uv}$
14. $g(u, v) = \tan^{-1} u/v$; $g_{uu}$, $g_{uv}$

15. Show that each of the functions $f(x, y) = x^3 - 3xy^2$ and $f(x, y) = 3x^2 y - y^3$ satisfies the equation $f_{xx} + f_{yy} = 0$.
16. If $f(x, y) = \ln(x^2 + y^2) + \tan^{-1} y/x$ show that $f_{xx} + f_{yy} = 0$.
17. If $u = \sqrt{x} - \sqrt{y}$ show that $x^2 u_{xx} + 2xy u_{xy} + y^2 u_{yy} = -u/4$.
18. If $u = x^3 - 4x^2 y + 2y^3$ show that
    (a) $x^2 u_{xx} + 2xy u_{xy} + y^2 u_{yy} = 6u$
    (b) $x^3 u_{xxx} + 3x^2 y u_{xxy} + 3xy^2 u_{xyy} + y^3 u_{yyy} = 6u$
19. Show that $u(x, y) = e^x + \ln(x^2 + y^2) - 2$ satisfies the equation
$$y^2 u_{xy} - xy u_{yy} + x u_y = 0.$$
20. If $u(x, y) = (x^3 + y^3)^{2/3}$ show that $x^2 u_{xx} + 2xy u_{xy} + y^2 u_{yy} = 2u$.
21. Show that each of the functions $u(x, y) = x^2 - y^2$ and $u(x, y) = e^x \cos y$, satisfies the equation $u_{xx} + u_{yy} = 0$.

In Exercises 22–26, $f$, $g$, and $h$ are arbitrary functions, with any needed differentiability properties.

22. Show that $u(x, y) = f(x)g(y)$ satisfies the equation $uu_{xy} - u_x u_y = 0$.
23. Show that $u(x, y) = f(y)e^x + g(y)e^{-x}$ satisfies the equation $u_{xx} - u = 0$.
24. Show that $u(x, y) = \frac{1}{2}x^2 y + f(x) + xg(y)$ satisfies the equation $u_{xxy} = 1$.
25. Show that both $u(x, y) = \frac{1}{2}x^2 + xf(y) + g(y)$ and $u(x, y, z) = \frac{1}{2}x^2 + xf(y, z) + g(y, z)$ satisfy the equation $u_{xx} = 1$.
26. Show that $u(x, y, z) = f(y, z) + g(y, z)\sin x + h(y, z)\cos x$ satisfies the equation $u_{xxx} + u_x = 0$.
27. Assuming the necessary continuity and differentiability properties, prove each of the following
    (a) $f_{xxy} = f_{yxx} = f_{xyx}$    (b) $f_{xxxy} = f_{yxxx}$
    (c) $f_{xyyx} = f_{xxyy}$    (d) $f_{xxyyy} = f_{yyxxy}$

## 21.4 THE CHAIN RULE

The chain rule of differentiation for functions of a single variable (Section 4.5) provided a means for finding the derivative of a composite function in terms of the derivatives of the component functions. A corresponding formula applies to composite functions where more than one variable is involved. First we derive a formula for the increment of a function. The details will be carried out for a function of two variables, but they can easily be extended to a function of more than two variables. If $x$ is increased by an amount $h$, and $y$ by $k$, the *increment* $\Delta f$ of $f$, or change in $f$, is

(21.4.1) $$\Delta f = f(x+h, y+k) - f(x, y).$$

**THEOREM 21.4.1.** *If $f$ is continuous in a neighborhood of $(x, y)$, and $f_x$ and $f_y$ exist in this neighborhood and are continuous at $(x, y)$, then there are functions $r_1(h, k)$ and $r_2(h, k)$ for which*

(21.4.2) $$\lim_{(h,k)\to(0,0)} r_1(h,k) = 0, \qquad \lim_{(h,k)\to(0,0)} r_2(h,k) = 0$$

*and such that*

(21.4.3) $$\Delta f = h f_x(x, y) + k f_y(x, y) + h r_1(h, k) + k r_2(h, k).$$

*Proof.* By subtracting and adding $f(x, y+k)$, we write (21.4.1) as

(21.4.4) $$\Delta f = [f(x+h, y+k) - f(x, y+k)] + [f(x, y+k) - f(x, y)].$$

The first difference is the increment of the function $f(x, y+k)$, considered as a function of $x$ alone with $y$ fixed, and the second difference is the increment of the function $f(x, y)$ as a function of $y$ with $x$ fixed. The hypotheses make the Mean-Value Theorem applicable to each difference, with the derivative in each case becoming the appropriate partial derivative of $f$. We get

(21.4.5) $$\Delta f = h f_x(x + \alpha h, y + k) + k f_y(x, y + \beta k), \qquad 0 < \alpha < 1, \quad 0 < \beta < 1.$$

By the continuity of $f_x$ and $f_y$ at $(x, y)$, we may write

$$f_x(x + \alpha h, y + k) = f_x(x, y) + r_1(h, k), \qquad f_y(x, y + \beta k) = f_y(x, y) + r_2(h, k),$$

with (21.4.2) holding. Substitution of these values in (21.4.5) yields the desired result.

**THEOREM 21.4.2. Chain rule of differentiation.** *Let $f(x, y)$, $f_x$, and $f_y$ be continuous. Let $x(u, v)$, $y(u, v)$ be continuous and let $x_u$, $x_v$, $y_u$, $y_v$ exist. Then if*

$$F(u, v) = f(x(u, v), y(u, v)),$$

*we have*

$$F_u = f_x x_u + f_y y_u, \qquad F_v = f_x x_v + f_y y_v.$$

*Proof.* We derive the result for $F_u$; the derivation for $F_v$ is similar. We hold $v$ constant ($\Delta v = 0$) and then have

$$\Delta F = F(u + \Delta u, v) - F(u, v) = f(x+h, y+k) - f(x, y),$$

where

$$h = x(u + \Delta u, v) - x(u, v) = \Delta x, \qquad k = y(u + \Delta u, v) - y(u, v) = \Delta y.$$

Since $\Delta f = \Delta F$ for corresponding values of the variables, we have from (21.4.3),

$$\Delta F = f_x(x, y)\, \Delta x + f_y(x, y)\, \Delta y + r_1\, \Delta x + r_2\, \Delta y,$$

and

(21.4.6) $$\frac{\Delta F}{\Delta u} = f_x(x, y)\frac{\Delta x}{\Delta u} + f_y(x, y)\frac{\Delta y}{\Delta u} + r_1 \frac{\Delta x}{\Delta u} + r_2 \frac{\Delta y}{\Delta u}.$$

As $\Delta u \to 0$,

$$\lim_{\Delta u \to 0} \frac{\Delta F}{\Delta u} = F_u, \qquad \lim_{\Delta u \to 0} \frac{\Delta x}{\Delta u} = x_u, \qquad \lim_{\Delta u \to 0} \frac{\Delta y}{\Delta u} = y_u.$$

Furthermore, $h \to 0$, $k \to 0$, and hence $r_1 \to 0$, $r_2 \to 0$. We get from (21.4.6)

$$F_u = f_x x_u + f_y y_u,$$

the result we wanted.

The relation

(21.4.7) $$F_u = f_x x_u + f_y y_u$$

may also be written in the form

$$\frac{\partial F}{\partial u} = \frac{\partial f}{\partial x}\frac{\partial x}{\partial u} + \frac{\partial f}{\partial y}\frac{\partial y}{\partial u},$$

which, in more complete form, reads

$$\frac{\partial F(u, v)}{\partial u} = \frac{\partial f(x(u, v), y(u, v))}{\partial x}\frac{\partial x(u, v)}{\partial u} + \frac{\partial f(x(u, v), y(u, v))}{\partial y}\frac{\partial y(u, v)}{\partial u}.$$

The first derivative on the right means the partial derivative of $f(x, y)$ with respect to $x$, evaluated at $x = x(u, v)$, $y = y(u, v)$. The meanings of the other derivatives are then clear. All of this is implied in the meaning of (21.4.7).

**Example 21.4.1.** Given $f(x, y) = x^3 - 3xy^2$, $x = 2u + 3v$, $y = u - 2v$, find $\partial f(x(u, v), y(u, v))/\partial u$ and $\partial f(x(u, v), y(u, v))/\partial v$.

**Solution.** Let $f(x(u, v), y(u, v)) = F(u, v)$; then $F_u$ and $F_v$ are the derivatives we want. With all the functions given, we may find $F(u, v)$ explicitly, and differentiate. Thus

$$F(u, v) = (2u + 3v)^3 - 3(2u + 3v)(u - 2v)^2$$
$$= 2u^3 + 51u^2 v + 66uv^2 - 9v^3,$$

and by direct differentiation we find

$$F_u = 6u^2 + 102uv + 66v^2, \qquad F_v = 51u^2 + 132uv - 27v^2.$$

The same derivatives may be found by the chain rule. By this method we have

$$F_u = f_x x_u + f_y y_u$$
$$= (3x^2 - 3y^2) \cdot 2 + (-6xy) \cdot 1 = 6x^2 - 6xy - 6y^2;$$
$$F_v = f_x x_v + f_y y_v$$
$$= (3x^2 - 3y^2) \cdot 3 + (-6xy)(-2) = 9x^2 + 12xy - 9y^2.$$

These derivatives may also be expressed in terms of $u$ and $v$, by substituting for $x$ and $y$ their values in terms of $u$ and $v$, and the same results are obtained as above.

Because of the different variables and several functions involved in any application of the chain rule, the matter of notation is important. The notation as used thus far is correct. However, it is not uncommon in practice to refer to the derivatives $F_u$ and $F_v$ in Example 21.4.1 as $f_u$ and $f_v$, respectively. In a strict sense this is incorrect since the variables in $f$ are $x$ and $y$, not $u$ and $v$. The confusion arises because of the relation $F(u, v) = f(x, y)$; that is, the *values* of $F$ and $f$ are equal for corresponding values of the variables, but the manner of dependence on the variables is different.

The situation in Example 21.4.1 is also sometimes described as
$$z = x^3 - 3xy^2, \qquad x = 2u + 3v, \qquad y = u - 2v,$$
and then the derivatives $\partial z/\partial x$, $\partial z/\partial y$, $\partial z/\partial u$, and $\partial z/\partial v$ are also used. This must then be interpreted to mean that we are thinking of $z$ *either* as a function of $x$ and $y$ *or* as a function of $u$ and $v$, and the above derivatives are for the appropriate function.

If we have
$$z = f(x, y), \qquad x = x(u, v), \qquad y = y(u, v),$$
the statement of Theorem 21.4.2 applies. With the interpretation as above, we may also write
$$\frac{\partial z}{\partial u} = \frac{\partial z}{\partial x}\frac{\partial x}{\partial u} + \frac{\partial z}{\partial y}\frac{\partial y}{\partial u},$$
$$\frac{\partial z}{\partial v} = \frac{\partial z}{\partial x}\frac{\partial x}{\partial v} + \frac{\partial z}{\partial y}\frac{\partial y}{\partial v}.$$

As another illustration of the Chain Rule, with different variables, consider the situation (for a function of three variables)
$$f(x, y, z), \qquad x = x(t), \qquad y = y(t), \qquad z = z(t).$$
The composite function is a function of $t$ alone,
$$f(x(t), y(t), z(t)) = F(t),$$
and the Chain Rule gives
$$F'(t) = f_x x'(t) + f_y y'(t) + f_z z'(t),$$
where the primes indicate ordinary derivatives of the corresponding functions of

$t$, and the functions $f_x$, $f_y$, and $f_z$ are evaluated for $x(t)$, $y(t)$, and $z(t)$. If, in addition, we write

$$w = f(x, y, z),$$

then we also write

$$\frac{dw}{dt} = \frac{\partial w}{\partial x}\frac{dx}{dt} + \frac{\partial w}{\partial y}\frac{dy}{dt} + \frac{\partial w}{\partial z}\frac{dz}{dt}.$$

It is absolutely essential to understand which function is intended in each derivative.

***Example 21.4.2.*** Find the first partial derivatives of $\sin(x^2 + 2xy)$ with respect to $x$ and $y$.

*Solution.* Let $f(x, y) = \sin(x^2 + 2xy)$. We may write

$$g(s) = \sin s, \qquad s = x^2 + 2xy, \qquad f(x, y) = g(s(x, y)).$$

Of course, $g$ is a function of one variable, and in this case the Chain Rule gives

$$f_x = g'(s)s_x, \qquad f_y = g'(s)s_y,$$

and we get, since $g'(s) = \cos s$,

$$f_x = (\cos s)(2x + 2y) = 2(x + y)\cos(x^2 + 2xy),$$
$$f_y = (\cos s)(2x) = 2x \cos(x^2 + 2xy).$$

**EXERCISE GROUP 21.4**

In each of Exercises 1–6 express the given function as a composite function by the introduction of suitable variables. For example, if $f(x, y) = (x - 3)^2 + (y + 2)^2$, set $u = x - 3$, $v = y + 2$, and get $f(x, y) = F(u, v) = u^2 + v^2$. Different answers are possible; try to give that answer which is most "obvious."

1. $f(x, y) = \ln(x^2 - y^2)$
2. $f(x, y) = e^{2x + 3y}$
3. $f(x, y) = (2x - 3y)^{10}$
4. $f(x, y) = \sin x^2 + \cos y^2$
5. $f(t) = \cosh 3t + \ln(2t - 3)$
6. $f(x, y, z) = (x - 2y)^2 + e^{3x - 2z} + \ln(x + y + z)$

Use the Chain Rule for each of the following. These exercises may also be done directly; this may be used as a check.

7. If $u = x^3 - 3xy^2$, $x = r + s$, $y = r - s$, find $\partial u/\partial r$, $\partial u/\partial s$.
8. If $u = x^3 - 3xy^2$, $x = e^r$, $y = e^{-r}$, find $du/dr$.
9. If $f(t) = \cosh 3t + \ln(2t - 3)$, find $f'(t)$.
10. If $z = \sin x^2 y$, $x = e^u$, $y = u + v$, find $\partial z/\partial u$, $\partial z/\partial v$.

11. If $v = 4x^2 - y^2$, $x = r \tan \theta$, $y = r \sec \theta$, find $\partial v/\partial r$, $\partial v/\partial \theta$.
12. If $u = e^{s^2 - t^2}$ find $\partial u/\partial s$, $\partial u/\partial t$.
13. If $w = (2x - y)^2 - (3x + 2y)^2$, find $\partial w/\partial x$, $\partial w/\partial y$.
14. If $z = x^2 + y^2$, $x = u \cosh v$, $y = u \sinh v$, find $\partial z/\partial u$, $\partial z/\partial v$.
15. If $f(x, y) = x^2 + y^3$, $F(x, y) = f(x, \sin y)$, find $\partial F/\partial x$, $\partial F/\partial y$.
16. If $f(x, y) = x^3 - 2xy + y^2$, $F(u, v) = f(u^2, uv)$, find $F_u$, $F_v$.

## 21.5 USE OF THE CHAIN RULE

In the preceding section the Chain Rule was applied to composite functions where the individual functions were explicitly given. The greater value of the rule lies perhaps in its application where one or more of the functions are not so given. For example, suppose we are to differentiate $f(y/x)$ with respect to $x$ or $y$. Here $f$ is a function of a single variable $y/x$. We may consider $f(y/x)$ as a function of the variables $x$ and $y$, but then it should be renamed, say as $F(x, y)$, and we are to find $F_x$ and $F_y$. If we set $u = y/x$, we have

$$F(x, y) = f(y/x) = f(u), \qquad u = y/x.$$

To find $F_x$ and $F_y$ we apply the Chain Rule, to give

$$F_x = f'(u)u_x, \qquad F_y = f'(u)u_y,$$

and hence

$$F_x = -\frac{y}{x^2} f'(u), \qquad F_y = \frac{1}{x} f'(u).$$

These results may be written in terms of the given function as

$$\frac{\partial f(y/x)}{\partial x} = -\frac{y}{x^2} f'(y/x), \qquad \frac{\partial f(y/x)}{\partial y} = \frac{1}{x} f'(y/x).$$

It is understood that $f'(y/x)$ means the value of the derivative of $f$ with respect to its (single) variable, when that variable has the value $y/x$.

As another illustration consider the function

$$f(2x + 3y, x - 2y).$$

Here $f$ is a function of two variables, $2x + 3y$ and $x - 2y$ (not of $x$ and $y$). If we set

$$F(x, y) = f(2x + 3y, x - 2y),$$

then $F$ is a function of $x$ and $y$. Let us write

$$F(x, y) = f(u, v), \qquad u = 2x + 3y, \qquad v = x - 2y.$$

We apply the Chain Rule, and get

(21.5.1)
$$F_x = f_u u_x + f_v v_x = 2f_u + f_v,$$
$$F_y = f_u u_y + f_v v_y = 3f_u - 2f_v.$$

The desire to be able to write these without introducing the new variables $u$ and $v$ (or other letters) leads to a modified notation. If $f$ is a function of several variables, then $f_1$ is used for the partial derivative of $f$ with respect to its first variable, $f_2$ is used for the partial derivative of $f$ with respect to its second variable, and so on. Then (21.5.1) may be written as

$$\frac{\partial f(2x + 3y, x - 2y)}{\partial x} = 2f_1(2x + 3y, x - 2y) + f_2(2x + 3y, x - 2y)$$

with no ambiguity as to the meaning.

For higher order partial derivatives of $f(x, y)$ we also have the notation

$$f_{11} = \frac{\partial^2 f}{\partial x^2}, \qquad f_{12} = \frac{\partial^2 f}{\partial y \, \partial x}, \qquad \text{and so on.}$$

**Example 21.5.1.** If $w(x, y) = f(2x - 3y)$, show that $9w_{11} - 4w_{22} = 0$.

**Solution.** We find
$$w_1(x, y) = 2f'(2x - 3y), \qquad w_2 = -3f'(2x - 3y)$$
$$w_{11}(x, y) = 4f''(2x - 3y), \qquad w_{22} = 9f''(2x - 3y).$$

The desired relation follows at once. Although not needed for this problem, it is clear that $w_{12}(x, y) = -6f''(2x - 3y)$.


### EXERCISE GROUP 21.5

1. If $u(x, y) = f(x^2 + y^2)$, show that $yu_x - xu_y = 0$.
2. If $u(x, y) = -\frac{1}{2} + e^{y/2} f(x - \frac{3}{4}y)$, show that $3u_x + 4u_y - 2u = 1$.
3. If $F(x, y) = f(x^2 - y, x + y^2)$, find $F_x$, $F_y$.
4. If $F(x, y) = f(x - y, y - x)$, show that $F_x + F_y = 0$.
5. Find $\dfrac{d}{dx} f\left(\dfrac{g(x)}{h(x)}\right)$.
6. If $F(x, y) = \ln f(y, g(x, y))$, find $F_x$, $F_y$.
7. If $u = \phi(x - y, y - z, z - x)$ show that $u_x + u_y + u_z = 0$.
8. If $u = f(x + at, y + bt)$, show that $\partial u/\partial t = af_1 + bf_2$.
9. If $u = y^2 f(xy)$, find $\partial u/\partial x$, $\partial u/\partial y$.
10. If $u = f(x, y)$, $x = r \cos \theta$, $y = r \sin \theta$, show that $u_x^2 + u_y^2 = u_r^2 + \dfrac{1}{r^2} u_\theta^2$.

Consider carefully the meaning of $u$ for each derivative. [*Hint*: Find $u_r$ and $u_\theta$ in terms of $u_x$ and $u_y$.]

11. If $x = \dfrac{1}{x} f\left(\dfrac{y}{x}\right)$, show that $xz_x + yz_y + z = 0$.

12. If $u = f\left(\dfrac{x-y}{x+y}\right)$, show that $x\dfrac{\partial u}{\partial x} + y\dfrac{\partial u}{\partial y} = 0$.

13. If $u = xf(y/x) + yg(x/y)$, show that
   (a) $xu_x + yu_y = u$,     *(b) $x^2 u_{xx} + 2xy u_{xy} + y^2 u_{yy} = 0$.

14. If $u = xf\left(\dfrac{x-y}{x+y}\right)$, show that
   (a) $xu_x + yu_y = u$,     *(b) $x^2 u_{xx} + 2xy u_{xy} + y^2 u_{yy} = 0$.

## 21.6 TANGENT PLANES

In general, an equation of the form

(21.6.1) $$F(x, y, z) = 0,$$

where $F$ is a function of three variables, has as its graph a surface $S$ in space (Fig. 21.6.1). A curve in space may be represented by parametric equations

(21.6.2)     $C:$     $x = x(t),$     $y = y(t),$     $z = z(t).$

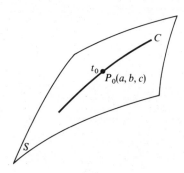

Fig. 21.6.1

Let the curve $C$ of (21.6.2) lie in the surface (21.6.1). We may then write

$$\phi(t) = F(x(t), y(t), z(t)) = 0.$$

It follows that $\phi'(t) = 0$. By the Chain Rule, we have

$$\phi'(t) = F_x x'(t) + F_y y'(t) + F_z z'(t) = 0.$$

If $P_0(a, b, c)$, corresponding to $t = t_0$, is a point on $C$, then

(21.6.3)     $F_x(a, b, c) x'(t_0) + F_y(a, b, c) y'(t_0) + F_z(a, b, c) z'(t_0) = 0.$

Now $F_x$, $F_y$, and $F_z$ in (21.6.3) depend only on $F$ and $P_0$ and not on the curve $C$; they may be defined as the components of a vector which is defined as the *gradient* of $F$, and denoted **grad** $F$:

(21.6.4) $$\mathbf{grad}\ F = F_x\mathbf{i} + F_y\mathbf{j} + F_z\mathbf{k}.$$

A tangent vector to $C$ at $P_0$ is given by

$$\boldsymbol{\tau} = \mathbf{i}x'(t_0) + \mathbf{j}y'(t_0) + \mathbf{k}z'(t_0).$$

Since the left member of (21.6.3) is $(\mathbf{grad}\ F) \cdot \boldsymbol{\tau}$, we have

$$(\mathbf{grad}\ F) \cdot \boldsymbol{\tau} = 0,$$

and $\boldsymbol{\tau}$ is perpendicular to **grad** $F$. Thus *the tangent to any curve on S and passing through $P_0$ is perpendicular to* **grad** $F$, *and hence all such tangent lines must lie in a single plane.* The following definitions and theorem have now been justified.

**DEFINITION 21.6.1.** *The* **tangent plane** *to a surface S at a point $P_0$ is the plane which contains the tangent lines to all curves on S passing through $P_0$.*

**DEFINITION 21.6.2.** *The* **normal** *to a surface S at a point $P_0(a, b, c)$ is the line perpendicular to the tangent plane to S at $P_0$.*

**THEOREM 21.6.1.** *If $F(x, y, z) = 0$ is an equation of a surface S, then*

$$\mathbf{grad}\ F\big|_{P_0} = \mathbf{i}F_x(a, b, c) + \mathbf{j}F_y(a, b, c) + \mathbf{k}F_z(a, b, c)$$

*is a vector perpendicular to the tangent plane to S at $(a, b, c)$.*

*Example 21.6.1.* Find equations of the tangent plane and normal line to the ellipsoid $2x^2 + 3y^2 + z^2 = 13$ at the point $(2, -1, -\sqrt{2})$.

*Solution.* If we set $F(x, y, z) = 2x^2 + 3y^2 + z^2 - 13$, we have

$$F_x = 4x, \quad F_y = 6y, \quad F_z = 2z.$$

A set of direction numbers of the normal at the given point is $8, -6, -2\sqrt{2}$ or $4, -3, -\sqrt{2}$; an equation of the tangent plane is then

$$4(x - 2) - 3(y + 1) - \sqrt{2}(z + \sqrt{2}) = 0 \quad \text{or} \quad 4x - 3y - \sqrt{2}\,z = 13.$$

Symmetric equations of the normal are

$$\frac{x-2}{4} = \frac{y+1}{-3} = \frac{z+\sqrt{2}}{-\sqrt{2}}.$$

An equation of a surface is often given in the form

$$z = f(x, y),$$

where $f$ is a function of two variables. This may be written as

$$F(x, y, z) = f(x, y) - z = 0.$$

Then $F_x = f_x$, $F_y = f_y$, $F_z = -1$, and we may state the following corollary to Theorem 21.6.1.

**COROLLARY.** *If $z = f(x, y)$ is an equation of a surface $S$, then*

$$\mathbf{i}f_x(a, b) + \mathbf{j}f_y(a, b) - \mathbf{k}$$

*is a vector perpendicular to the tangent plane to $S$ at $(a, b, f(a, b))$.*

***Example 21.6.2.*** Find equations of the tangent plane and normal line to the surface $z = 2x^2 - 3y^2 + x - 2y + 3$ at the point for which $x = 2$, $y = 1$.

*Solution.* We find that $z = 8$. A set of direction numbers of the normal at the given point is

$$\frac{\partial z}{\partial x} = 4x + 1 = 9, \qquad \frac{\partial z}{\partial y} = -6y - 2 = -8, \qquad -1.$$

An equation of the tangent plane is

$$9(x - 2) - 8(y - 1) - (z - 8) = 0 \quad \text{or} \quad 9x - 8y - z = 2.$$

Symmetric equations of the normal are

$$\frac{x - 2}{9} = \frac{y - 1}{-8} = \frac{z - 8}{-1}.$$

Besides being given in a parametric representation in terms of a single parameter, a curve in space may be given as the intersection of two surfaces,

$$F(x, y, z) = 0, \qquad G(x, y, z) = 0.$$

At a point $P_0(a, b, c)$ of the curve, the tangent line must lie in the tangent plane to each of the surfaces $F = 0$, $G = 0$, and hence is the intersection of the two tangent planes. We then know from Section 19.6 that the tangent line is perpendicular to the normal of each tangent plane, and hence *the cross product*

$$(\mathbf{grad}\ F) \times (\mathbf{grad}\ G)$$

*is a vector tangent to the curve.*

***Example 21.6.3.*** Find a unit vector tangent to the curve

$$x^2 - 2y^2 + z^2 = 9, \qquad x^2 - 3x + z^2 = y + 16$$

at the point $(1, -2, 4)$.

*Solution.* We set

$$F(x, y, z) = x^2 - 2y^2 + z^2 - 9, \qquad G(x, y, z) = x^2 - 3x + z^2 - y - 16.$$

We find
$$\text{grad } F = 2x\mathbf{i} - 4y\mathbf{j} + 2z\mathbf{k}, \quad \text{grad } G = (2x-3)\mathbf{i} - \mathbf{j} + 2z\mathbf{k}.$$
At the point $(1, -2, 4)$ a tangent vector to the curve is
$$(2\mathbf{i} + 8\mathbf{j} + 8\mathbf{k}) \times (-\mathbf{i} - \mathbf{j} + 8\mathbf{k}) = 72\mathbf{i} - 24\mathbf{j} + 6\mathbf{k}.$$
Dividing by 6, we may say that the vector $12\mathbf{i} - 4\mathbf{j} + \mathbf{k}$ is also tangent to the curve. Finally, a unit tangent vector is
$$\frac{12\mathbf{i} - 4\mathbf{j} + \mathbf{k}}{\pm\sqrt{144 + 16 + 1}} = \pm\frac{12\mathbf{i} - 4\mathbf{j} + \mathbf{k}}{\sqrt{161}}.$$

**EXERCISE GROUP 21.6**

In Exercises 1–6 find equations of (a) the tangent plane, and (b) the normal line, to the given surface at the stated point.

1. $z = 2x^2 + 3y^2$, $(-1, 2, 14)$
2. $z = \tan^{-1} y/x$, $(1, 1, \pi/4)$
3. $x^2yz = 1$, $(1/2, 2, 2)$
4. $x^2 - y^2 + 2xy - 3xz - 1 = 0$, $(-1, 1, 1)$
5. $2x^2 - 3y^2 + z^2 - 2x + 11 = 0$, $(1, 2, -1)$
6. $z = \ln(x^2 + y^2)$, $(0, 1, 0)$

In Exercises 7–10 find a unit vector tangent to the curve of intersection of the given surfaces at the stated point.

7. $x^2 + y^2 + z^2 = 30$, $z = x^2 + y^2$; $(-2, 1, 5)$
8. $x^2 + y^2 + z^2 = 14$, $2x + y = 0$; $(1, -2, 3)$
9. $x^3 = z$, $y^3 = z^2$; $(1, 1, 1)$
10. $9x^2 + 4y^2 = 36$, $3x = y \cot z$; $(0, 3, \pi/2)$
11. Show that the surfaces
$$x^2 + y^2 + z^2 = 14, \quad 9x^2 + 9y^2 - 5z^2 = 0, \quad 2x + y = 0$$
intersect at $(1, -2, 3)$ in curves which are mutually orthogonal.
12. Show that the surfaces
$$2x^2 + 4y = 1, \quad 2x^2 - 2z^2 = 2y + 1, \quad y^2 - 3xz = 0$$
intersect at $(\sqrt{2}/2, 0, 0)$ in curves which are mutually orthogonal.
13. Prove that if the tangent planes to three surfaces at a point of intersection are mutually orthogonal, then the curves of intersection are mutually orthogonal.

## 21.7 DIRECTIONAL DERIVATIVE AND GRADIENT

The directional derivative of a function $f(x, y)$, in the direction of the unit vector $\mathbf{u} = \lambda\mathbf{i} + \mu\mathbf{j}$, was defined in Section 21.2 as
$$D_{\mathbf{u}} f = \lim_{h \to 0} \frac{f(x + \lambda h, y + \mu h) - f(x, y)}{h}.$$

By Theorem 21.4.1, we may write, with $\lambda h$ in place of $h$, and $\mu h$ in place of $k$,

$$f(x + \lambda h, y + \mu h) - f(x, y)$$
$$= \lambda h f_x(x, y) + \mu h f_y(x, y) + \lambda h r_1(\lambda h, \mu h) + \mu h r_2(\lambda h, \mu h).$$

If we divide by $h$ and then let $h \to 0$, we obtain, since $r_1 \to 0$, $r_2 \to 0$,

$$D_{\mathbf{u}} f = \lambda f_x(x, y) + \mu f_y(x, y).$$

We have proved the following result.

**THEOREM 21.7.1.** *Under the conditions of Theorem 21.4.1, and if*

$$\mathbf{u} = \lambda \mathbf{i} + \mu \mathbf{j} = \cos \theta \, \mathbf{i} + \sin \theta \, \mathbf{j}$$

*is a unit vector, then*

(21.7.1.) $\quad D_{\mathbf{u}} f(x, y) = \lambda f_x(x, y) + \mu f_y(x, y) = f_x(x, y) \cos \theta + f_y(x, y) \sin \theta.$

***Example 21.7.1.*** (a) Find the directional derivative of $f(x, y) = 2x^2 - 3xy$ at the point $(2, 1)$, in the direction $\theta$. (b) Find the direction in which the directional derivative at $(2, 1)$ is a maximum.

*Solution.* (a) We find $f_x = 4x - 3y$, $f_y = -3x$. Then $f_x(2, 1) = 5$, $f_y(2, 1) = -6$, and by (21.7.1),

$$D_{\theta} f(2, 1) = 5 \cos \theta - 6 \sin \theta.$$

(b) The directional derivative of $f$ at $(2, 1)$ was found in (a), and is a function of $\theta$,

$$\phi(\theta) = 5 \cos \theta - 6 \sin \theta.$$

We may maximize $\phi(\theta)$ in the usual way. Thus

$$\phi'(\theta) = -5 \sin \theta - 6 \cos \theta.$$

To satisfy $\phi'(\theta) = 0$, we find $\tan \theta = -6/5$, for which $\theta$ may be an appropriate angle in the second or fourth quadrant. We find

$$\phi''(\theta) = -5 \cos \theta + 6 \sin \theta.$$

It is clear that when $\theta$ is in the fourth quadrant, $\cos \theta > 0$ and $\sin \theta < 0$, and therefore $\phi''(\theta) < 0$. Hence $\phi(\theta)$ has a maximum, if $\tan \theta = -6/5$ with $\theta$ in the fourth quadrant. We may also describe the angle by writing $\cos \theta = 5/\sqrt{61}$, $\sin \theta = -6/\sqrt{61}$. With these values we find the maximum value of $\phi(\theta)$ to be

$$5\left(\frac{5}{\sqrt{61}}\right) - 6\left(\frac{-6}{\sqrt{61}}\right) = \sqrt{61}.$$

The corresponding result for a function of three variables may be proved in the same way as Theorem 21.7.1, and is stated as follows.

**THEOREM 21.7.2.** *Let* **u** *be a unit vector*

$$\mathbf{u} = \lambda\mathbf{i} + \mu\mathbf{j} + \nu\mathbf{k} = \cos\alpha\,\mathbf{i} + \cos\beta\,\mathbf{j} + \cos\gamma\,\mathbf{k}.$$

*Then the directional derivative of* $f(x, y, z)$ *in the direction of* **u**, *at the point* $P(x, y, z)$, *is given by*

(21.7.2.) $\quad D_{\mathbf{u}}f = \lambda f_x + \mu f_y + \nu f_z = f_x \cos\alpha + f_y \cos\beta + f_z \cos\gamma,$

*where* $f_x, f_y,$ *and* $f_z$ *are evaluated at P.*

The gradient of $f$ was defined in (21.6.4) as $\mathbf{grad}\,f = f_x\mathbf{i} + f_y\mathbf{j} + f_z\mathbf{k}$. From (21.7.2) we then see that

(21.7.3) $\quad\quad\quad\quad\quad\quad D_{\mathbf{u}}f = \mathbf{u}\cdot\mathbf{grad}\,f;$

hence *the directional derivative of f in the direction of a vector is the component of* **grad** *f in the direction of that vector*. Furthermore, since the component of a vector in a given direction is a maximum in the direction of the vector itself, the following statement characterizes the gradient of a function.

**THEOREM 21.7.3.** 1. $D_{\mathbf{u}}f = \mathbf{u}\cdot\mathbf{grad}\,f.$

2. $D_{\mathbf{u}}f$ *is a maximum when* **u** *is in the direction of* **grad** $f$.

3. *The maximum value of* $D_{\mathbf{u}}f$ *is* $|\mathbf{grad}\,f|$.

*Example 21.7.2.* Find the direction and value of the maximum directional derivative for $f(x, y, z) = e^{yz} + e^{xz} + e^{xy}$ at the point $(1, 0, 2)$.

*Solution.* We have

$$f_x = ze^{xz} + ye^{xy}, \quad f_y = ze^{yz} + xe^{xy}, \quad f_z = ye^{yz} + xe^{xz}.$$

At the point $(1, 0, 2)$ these derivatives have the values $2e^2$, 3, $e^2$, respectively. Hence

$$\mathbf{grad}\,f(1, 0, 2) = 2e^2\mathbf{i} + 3\mathbf{j} + e^2\mathbf{k}.$$

By Theorem 21.7.3 the maximum directional derivative has the value

$$\sqrt{4e^4 + 9 + e^4} = \sqrt{9 + 5e^4},$$

and the unit vector in the direction in which the maximum occurs is

$$\mathbf{u} = \frac{2e^2\mathbf{i} + 3\mathbf{j} + e^2\mathbf{k}}{\sqrt{9 + 5e^4}}.$$

Theorem 21.7.3 may also be used for a function of two variables, since such a function may be interpreted as a function of three variables with the third variable missing. In Example 21.7.1 the point $(2, 1)$ may be interpreted as the point $(2, 1, 0)$ for the function $F(x, y, z) = 2x^2 - 3xy$. Then, at $(2, 1, 0)$,

$$\mathbf{grad}\,F = 5\mathbf{i} - 6\mathbf{j} + 0\mathbf{k}.$$

By Theorem 21.7.3, the maximum directional derivative occurs in the direction of the unit vector

$$\mathbf{u} = \frac{5\mathbf{i} - 6\mathbf{j}}{\sqrt{61}},$$

and has the value $|\operatorname{grad} F| = \sqrt{25 + 36} = \sqrt{61}$. These results conform to those of Example 21.7.1.

## EXERCISE GROUP 21.7

1. (a) Find the directional derivative of the function
$$f(x, y) = x^2 - 2xy - 2y^2 + x - 3$$
at the point (2, 3) in each of the following directions:
   i. $\mathbf{u} = -\mathbf{j}$, ii. $\mathbf{u} = \dfrac{3\mathbf{i} - 4\mathbf{j}}{5}$, iii. $\theta = \pi/3$, iv. in the direction of the vector drawn from the point (2, 3) to the point (−1, 2).
   (b) Find the largest value of the directional derivative of $f$ at (2, 3).

2. (a) Find the directional derivative of the function
$$f(x, y, z) = x^2 - yz + xz^2$$
at the point (1, 1, 2) in each of the following directions:
   i. $\mathbf{u} = -\mathbf{k}$, ii. $\mathbf{u} = \dfrac{\mathbf{i} - 2\mathbf{j} - 2\mathbf{k}}{3}$, iii. in the direction of the vector drawn from the point (1, 1, 2) to the point (2, 0, −1).
   (b) Find the largest value of the directional derivative of $f$ at (1, 1, 2).

3. For a function $f(x, y)$ find
   (a) $D_{-\mathbf{i}}f$, (b) $D_{-\mathbf{j}}f$, (c) $D_{\frac{\mathbf{i}+\mathbf{j}}{\sqrt{2}}}f$, (d) $D_{\tan^{-1}\sqrt{3}}f$.

4. For a function $f(x, y)$ find (a) $D_{\theta + \pi/2}f$, (b) $D_{\theta - \pi/2}f$, (c) $D_{\theta - \pi}f$.

5. Find the directional derivative of the function $f(x, y) = x^2 - 3xy^2$ at the point (2, 1) in the direction of the curve $y^2 = x - 1$, $x$ increasing.

6. Find the directional derivative of the function $f(x, y) = \sin xy$ in the direction of the curve $x = t$, $y = t^2$ when $t = 2$, $t$ increasing.

7. Find the directional derivative of the function $f(x, y, z) = x^2 - yz$ in the direction of the curve $x = t$, $y = t^2$, $z = t^3$ when $t = 1$, $t$ decreasing.

8. (a) Find a unit vector in the direction of the largest directional derivative for the function $xy - \sin yz$ at the point $(2, 1, \pi/2)$.
   (b) Same as (a), for the smallest directional derivative.

9. For a function $f(x, y)$ show that the sum of the squares of the directional derivatives in *any* two perpendicular directions in the $xy$ plane is equal to $|\operatorname{grad} f|^2$.

10. For a function $f(x, y, z)$ it can be shown that the sum of the squares of the directional derivatives in the directions of any three mutually vectors is equal to $|\operatorname{grad} f|^2$. Verify this statement for the vectors
$$\mathbf{u} = \mathbf{i} + \mathbf{j} + \mathbf{k}, \quad \mathbf{v} = \mathbf{i} + 2\mathbf{j} - 3\mathbf{k}, \quad \mathbf{w} = 5\mathbf{i} - 4\mathbf{j} - \mathbf{k}.$$

* 11. If $\mathbf{u}_1$ and $\mathbf{u}_2$ are unit vectors in two nonorthogonal directions $\alpha$ and $\beta$ in the $xy$ plane and $\mathbf{u}$ is a unit vector in the direction $\theta$, show that the directional derivative of $f(x, y)$ in the direction of $\mathbf{u}$ is given by

$$D_\mathbf{u} f = \frac{(D_{\mathbf{u}_1} f)\sin(\beta - \theta) - (D_{\mathbf{u}_2} f)\sin(\alpha - \theta)}{\sin(\beta - \alpha)}.$$

## 21.8  TOTAL DIFFERENTIAL

The development in this section will be carried out for functions of two variables. The extension to other numbers of variables will be immediately clear, and will be used in illustrations.

The increment

$$\Delta f = f(x + h, y + k) - f(x, y)$$

of the function $f(x, y)$ was obtained in Section 21.4 as

(21.8.1)  $\quad \Delta f = h f_x(x, y) + k f_y(x, y) + h r_1(h, k) + k r_2(h, k),$

where $r_1$ and $r_2$ approach zero as $(h, k) \to (0, 0)$. In the terms $h r_1(h, k)$ and $k r_2(h, k)$ each factor approaches zero. Hence these terms approach zero faster, in a sense, than the terms $h f_x(x, y) + k f_y(h, k)$, since a product of small numbers is smaller than each number. In this sense the first two terms of $\Delta f$ in (21.8.1) constitute the "principal part" of $\Delta f$. We define the principal part of $\Delta f$ as the *total differential* of $f$.

**DEFINITION 21.8.1.**  *The **total differential** of $f(x, y)$ is*

(21.8.2)  $\quad df(x, y) = h f_x(x, y) + k f_y(x, y),$

*where $x$, $y$, $h$, and $k$ are all independent variables.*

Equation (21.8.1) may now be written as

$$\Delta f = df + h r_1(h, k) + k r_2(h, k)$$

from which it can be shown that

(21.8.3)  $\quad \displaystyle\lim_{(h, k) \to (0, 0)} \frac{\Delta f - df}{|h| + |k|} = 0.$

Equation (21.8.3) may be interpreted as stating that the difference between $\Delta f$ and $df$ is small in comparison with $|h| + |k|$. This statement expresses again the fact that $df$ is the principal part of $\Delta f$.

A consequence of the above is that for computational purposes the total differential $df$ can serve as an approximation to the increment $\Delta f$ of a function, for small changes in the variables. One type of application is illustrated in the following example.

***Example 21.8.1.***  The area of a triangle is computed by measuring the altitude and base as $x = 2.3$, $y = 4.7$, subject to errors in measurement of 0.1 for $x$ and 0.2 for $y$. Find an approximate value for the largest possible error in the area.

*Solution.* The area of the triangle is

$$A = \tfrac{1}{2}xy.$$

The maximum error in the area may be interpreted as the maximum *increment* $\Delta A$ when $|h| < 0.1$, $|k| < 0.2$, where $h$ and $k$ are the increments of $x$ and $y$, respectively. We approximate the increment $\Delta A$ by the total differential $dA$. We find

$$dA = \tfrac{1}{2}(hy + kx),$$

and we see that max $dA$ occurs when $h$ and $k$ are a maximum. Hence

$$\max dA = \tfrac{1}{2}(0.1 \times 4.7 + 0.2 \times 2.3) = 0.465.$$

This value approximates the maximum error in the area.

We may compare this value with the maximum increment in $A$, which is $2.4 \times 4.9 - 2.3 \times 4.7 = 0.475$.

We present another example, different in form, which involves the same ideas.

**Example 21.8.2** Find an approximate value of

$$\sqrt[3]{(1.98)^2 + (11.03)^2}.$$

*Solution.* The given number may be computed from the function

$$f(x, y) = \sqrt[3]{x^2 + y^2}$$

as

$$f(1.98, 11.03) = f(2, 11) + \Delta f,$$

where $\Delta f$ is computed with $h = -0.02$, $k = 0.03$. Now, $f(2, 11) = 5$, and we approximate $\Delta f$ by $df$:

$$df = f_x h + f_y k$$
$$= \frac{2xh + 2yk}{3\sqrt[3]{(x^2 + y^2)^2}}.$$

Thus with $x = 2$, $y = 11$, $h = -0.02$, and $k = 0.03$ we have

$$df = \frac{4(-0.02) + 22(0.03)}{3 \times 25} = \frac{0.58}{75} \approx 0.0077.$$

Hence approximately $\sqrt[3]{(1.98)^2 + (11.03)^2} \approx 5 + 0.0077 = 5.0077$. This value is actually correct to 4 decimals.

If we set $h = dx$, $k = dy$, (21.8.2) becomes

$$df(x, y) = f_x(x, y)\, dx + f_y(x, y)\, dy.$$

Similarly, for a function $f(x, y, z)$ of three variables, the definition of total differential becomes

$$df(x, y, z) = f_x(x, y, z)\, dx + f_y(x, y, z)\, dy + f_z(x, y, z)\, dz.$$

### EXERCISE GROUP 21.8

In Exercises 1–6 find the total differential of each function.

1. (a) $x^2 + y^2$, (b) $xy$, (c) $x/y$
2. (a) $\sin^{-1} x/y$, (b) $x \cos y$, (c) $x^3 + 3xy^2 - 6x^2y$
3. (a) $\ln y/x$, (b) $\tan^{-1} y/x$, (c) $\ln(x^2 + y^2)$
4. (a) $e^{xy} \sin xy$, (b) $xye^{xyz}$
5. (a) $\dfrac{xy}{z}$, (b) $xyz^2$, (c) $\dfrac{x-y}{z^2}$
6. (a) $x^2z + xy^2 - 2xz + y^2$, (b) $\ln \dfrac{x-z}{y^2}$

7. Find $du$ if $u = f(x/y)$.
\* 8. Find $du$ if $u = f(xy, x/y)$.
\* 9. If $\psi$ is the angle defined in Section 17.7, from the extended radial line to the tangent line at a point of a curve, show that

$$\tan \psi = \frac{x\, dy - y\, dx}{x\, dx + y\, dy} = \frac{2x^2\, d(y/x)}{d(x^2 + y^2)}.$$

10. The area of a triangle is $K = \tfrac{1}{2}bc \sin A$. With $A = \pi/6$, $b = 4$, and $c = 5$, find the approximate error and relative error in $K$ if angle $A$ is subject to an error of 0.01 rad, and $b$ and $c$ are subject to errors of 0.2 and 0.3, respectively.

11. The crushing load of a certain pillar is $L = kr^4/h^2$, where $k$ is constant, $r$ is the radius, and $h$ is the height. What is the maximum error (approx.) in $L$ if $r = 0.25$ and $h = 30$, subject to errors of 0.01 and 1, respectively?

12. The kinetic energy of a body is $E = \tfrac{1}{2}mv^2$ where $m$ is the mass and $v$ is the speed. Find the approximate percentage error in $E$ due to errors of 1% in $m$ and 2% in $v$.

13. If a projectile is fired at an angle $\alpha$ with the horizontal with initial speed $v_0$, the maximum height reached is

$$H = \frac{v_0^2 \sin^2 \alpha}{2g}, \qquad g \text{ constant}.$$

Find an approximate error in $H$ corresponding to small errors in $v_0$ and $\alpha$.

14. If $z = x^a + y^a$ and $x$ and $y$ are subject to the same relative error $r$, find the relative error in $z$.

## 21.9 USE OF THE TOTAL DIFFERENTIAL

The definition of total differential (Definition 21.8.1) may also be written as

(21.9.1) $$df(x, y) = f_x(x, y)\, dx + f_y(x, y)\, dy,$$

where again $x$, $y$, $dx$, and $dy$ are independent variables. We have simply set $h = dx$ and $k = dy$ in Definition 21.8.1.

Suppose now that $x$ and $y$ are functions of $u$ and $v$,

(21.9.2) $$x = x(u, v), \qquad y = y(u, v).$$

Then
$$f(x, y) = f(x(u, v), y(u, v)) = F(u, v).$$

Now, by Definition 21.8.1,

(21.9.3) $\quad dF = F_u\, du + F_v\, dv, \quad dx = x_u\, du + x_v\, dv, \quad dy = y_u\, du + y_v\, dv,$

where $u$, $v$, $du$, and $dv$ are independent. Is (21.9.1) still valid when $x$, $y$, $dx$, and $dy$ are governed by (21.9.2) and (21.9.3)? Substituting from (21.9.3) into (21.9.1), we obtain

(21.9.4) $$df = f_x(x_u\, du + x_v\, dv) + f_y(y_u\, du + y_v\, dv)$$
$$= (f_x x_u + f_y y_u)\, du + (f_x x_v + f_y y_v)\, dv.$$

By the Chain Rule we know that
$$F_u = f_x x_u + f_y y_u, \qquad F_v = f_x x_v + f_y y_v.$$

Hence (21.9.4) becomes
$$df = F_u\, du + F_v\, dv,$$

which is the same as $dF$ in (21.9.3). In (21.9.1) $x$ and $y$ are independent variables, whereas in (21.9.3) and (21.9.4) $x$ and $y$ depend on $u$ and $v$. We have shown that the total differentials of the corresponding functions are equal. We state the result as follows.

**THEOREM 21.9.1.** *For a function $f(x, y)$ of two variables we have*
$$df = f_x\, dx + f_y\, dy,$$
*whether $x$ and $y$ are independent variables or depend on other variables.*

The significance of Theorem 21.9.1 is that for a function $f(x, y)$ of two variables we may write
$$df = f_x\, dx + f_y\, dy$$
without knowing whether $x$ and $y$ are independent variables, dependent on other variables, or related to each other. Corresponding statements apply to other sets of variables.

We illustrate the application of the above ideas. Suppose we have again $f(x, y) = F(u, v)$, where

(21.9.5) $$x = x(u, v), \quad y = y(u, v).$$

We may then write

(21.9.6) $$f_x\, dx + f_y\, dy = F_u\, du + F_v\, dv.$$

In any application of this relation we must keep clearly in mind which variables are independent and which are dependent. We may divide (21.9.6) by the differential of a variable. Then, if we consider all other independent variables as constant, any quotient of differentials is interpreted as an appropriate derivative, ordinary or partial as the case may be. For example, suppose we divide (21.9.6) by $du$, where $u$ and $v$ are independent. Then $dv = 0$, we interpret $dx/du$ as $\partial x/\partial u$ and $dy/du$ as $\partial y/\partial u$, and we get

$$f_x \frac{\partial x}{\partial u} + f_y \frac{\partial y}{\partial u} = F_u$$

which we know to be valid by the Chain Rule.

It may be possible to solve (21.9.5) for $u$ and $v$ in terms of $x$ and $y$. Then $x$ and $y$ are independent, while $u$ and $v$ are dependent. Now divide (21.9.6) by $dx$; then $dy = 0$ and with the proper interpretation of quotients of differentials, we get

$$f_x = F_u \frac{\partial u}{\partial x} + F_v \frac{\partial v}{\partial x},$$

which we again know to be valid by the Chain Rule.

Now suppose that in addition to the above, $v$ is a function of $u$. Then (21.9.6) is still valid, where $u$ is treated as independent, while $x$, $y$, and $v$ are functions of $u$. Dividing (21.9.6) by $du$, we obtain the valid relation

(21.9.7) $$f_x \frac{dx}{du} + f_y \frac{dy}{du} = F_u + F_v \frac{dv}{du},$$

where the quotients of differentials are ordinary derivatives.

Equation (21.9.7) may be verified in the following situation, as a special case.

$$f(x, y) = x^2 + 3xy, \quad x = u + v, \quad y = u - v, \quad v = u^2.$$

Then we obtain

$$F(u, v) = (u + v)^2 + 3(u + v)(u - v) = 4u^2 + 2uv - 2v^2.$$

We now find

(21.9.8) $\quad f_x = 2x + 3y, \quad f_y = 3x, \quad F_u = 8u + 2v, \quad F_v = 2u - 4v.$

In terms of $u$ we have $x = u + u^2$, $y = u - u^2$. Hence

(21.9.9) $$\frac{dx}{du} = 1 + 2u, \quad \frac{dy}{du} = 1 - 2u, \quad \frac{dv}{du} = 2u.$$

With the derivatives obtained in (21.9.8) and (21.9.9), and with the relations connecting $x$, $y$, and $u$, equation (21.9.7) may be verified.

An equivalent way of applying the above ideas is the following: An equation in any number of variables may be *differentiated implicitly* with respect to a variable if it is kept in mind that any other independent variables are held constant, and derivatives of dependent variables are interpreted properly.

In any application of the methods being considered here, the dependent or independent nature of any variables must be either stated or inferrable from the problem.

**Example 21.9.1.** Find $dy/dz$ from the equations

$$x^2 + y^2 + z^2 = 2, \qquad x^2 - y^2 + 2z^2 = 7.$$

*Solution.* In general, two equations can be solved for two variables, which are then dependent. Since there are three variables in the equations, one variable is independent. If $dy/dz$ is sought, the independent variable is $z$, and derivatives of $x$ and $y$ with respect to $z$ are ordinary derivatives. Differentiating in each equation with respect to $z$, we get

$$2x\frac{dx}{dz} + 2y\frac{dy}{dz} + 2z = 0, \qquad 2x\frac{dx}{dz} - 2y\frac{dy}{dz} + 4z = 0.$$

This pair of *linear* equations in $dx/dz$ and $dy/dz$ is solved, and we get, in particular,

$$\frac{dy}{dz} = \frac{z}{2y}.$$

We can also obtain $dx/dz = -3z/(2x)$.

**Example 21.9.2.** Find $u_x$ and $v_y$ if

$$ue^v - xy + v = 0, \qquad ve^y - xv + u = 0.$$

*Solution.* Here $u$ and $v$ are dependent, $x$ and $y$ are independent. Differentiating in each equation with respect to $x$, we have

$$ue^v v_x + e^v u_x - y + v_x = 0,$$

$$e^y v_x - xv_x - v + u_x = 0.$$

Collecting terms in $u_x$ and $v_x$ in each equation, and solving, we find (we can also find $v_x$ from these equations),

$$u_x = \frac{y(e^y - x) - v(ue^v + 1)}{e^v(e^y - x - u) - 1}.$$

To find $v_y$ we repeat the process with respect to $y$, getting

$$ue^v v_y + e^v u_y - x + v_y = 0,$$

$$e^y v_y + ve^y - xv_y + u_y = 0,$$

and

$$v_y = \frac{x + ve^{y+v}}{e^v(u - e^y + x) + 1}.$$

We shall now do the same kind of thing as in Example 21.9.2, but with a pair of equations where explicit functions are not given.

*Example 21.9.3.* Find $u_x$ and $v_x$ if

$$f(x, y, u, v) = 0, \qquad g(x, y, u, v) = 0.$$

*Solution.* The variables $u$ and $v$ are dependent, $x$ and $y$ are independent. Differentiating with respect to $x$ (holding $y$ fixed), we get

$$f_x + f_u u_x + f_v v_x = 0,$$
$$g_x + g_u u_x + g_v v_x = 0,$$

from which we find, treating these equations as a linear system in $u_x$ and $v_x$,

$$u_x = \frac{f_v g_x - f_x g_v}{f_u g_v - f_v g_u}, \qquad v_x = \frac{f_x g_u - f_u g_x}{f_u g_v - f_v g_u}.$$

The derivatives $u_y$ and $v_y$ may be found in a similar manner.

*Note*: If the question in Example 21.9.3 is merely to find $u_x$, an ambiguity exists since it is not clear what function is meant. If the equations are solved for $u$ and $v$ in terms of $x$ and $y$, the functions are as treated in Example 21.9.3. However, presumably the equations may be solved for $u$ and $x$ in terms of $v$ and $y$, or for $u$ and $y$ in terms of $v$ and $x$, in which cases the functions are different.

### EXERCISE GROUP 21.9

1. Find $dy/dx$ if $y \ln x - x \ln y = 0$.
2. Find $dy/dx$ if $x \tan y + x^2 y = 2$.
3. Find $\partial u/\partial x$, $\partial u/\partial y$ if $u^2 + v^2 - y^3 + 2x = 0, \qquad u - v - x^3 - 3y = 0$.
4. If $w + \ln w = xy$ show that $xw_x - yw_y = 0$.
5. Find $z_x$ and $z_y$ if $x^2 y - \cos z + z = 0$.
6. Find $u_x$, $u_y$, $v_x$, and $v_y$ if $u^2 + x^2 + y^2 = 3, \qquad u - v^3 + 3x = 4$.
7. Find $u_x$ and $v_x$ if $u + \ln v = xy$, $v + \ln u = x - y$.
8. Find $\partial u/\partial y$ and $\partial v/\partial y$ if $x = u^2 - v^2, \qquad y = uv$.
9. Find $\partial z/\partial x$ and $\partial z/\partial y$ if $f(x, y, z) = 0$.
10. Find $dx/du$ if $u = f(x, u - x)$.
11. Find $\partial u/\partial x$ and $\partial u/\partial y$ if $u = f(x, y, u)$.
12. Find $du/dx$ if $u = f(x - u)$.
13. Find $\partial x/\partial u$ if $u = f(x, y, u - x)$.

14. If $f(x, y, z) = 0$ prove that

(a) $\dfrac{\partial z}{\partial x} \dfrac{\partial x}{\partial z} = 1$, (b) $\dfrac{\partial x}{\partial y} \dfrac{\partial y}{\partial z} \dfrac{\partial z}{\partial x} = -1$.

Note carefully which function is involved in each derivative.

15. If $f(x, y, z, u) = 0$ prove that $\dfrac{\partial x}{\partial y} \dfrac{\partial y}{\partial z} \dfrac{\partial z}{\partial u} \dfrac{\partial u}{\partial x} = 1$.

16. Find $dz/dy$ if $f(x, y) = 0$ and $g(x, z) = 0$.

17. The equations $u = f(x, y)$, $g(x, y) = 0$ determine $u = u(x)$ as a function of $x$. Find $u'(x)$.

18. Find $u_x$ and $v_x$ if $x = x(u, v)$, $y = y(u, v)$.

19. Find $du/dx$ if $u = f(x, y, u)$, $u = g(x, y, u)$.

20. If $z = f(x, y)$, $x = x(u, v)$, $y = y(u, v)$, $u = \phi(t)$, $v = \psi(t)$, find $dz/dt$.

## 21.10 MORE ON HIGHER ORDER DERIVATIVES

Higher order derivatives were defined and obtained in Section 21.3 for functions given explicitly. We illustrate the use of our methods to obtain such derivatives for *functions defined implicitly* by one or more equations.

**Example 21.10.1.** Find $w_{xy}$ if $w + \ln w = xy$.

**Solution.** We obtain for $w_x$ and $w_y$:

$$w_x = \frac{wy}{w+1}, \qquad w_y = \frac{wx}{w+1};$$

these should have been obtained in Exercise 4 of Exercise Group 21.9. Differentiating $w_x$ with respect to $y$, we get

$$w_{xy} = \frac{(w+1)(w + yw_y) - wyw_y}{(w+1)^2} = \frac{w(w+1) + yw_y}{(w+1)^2}.$$

Substituting for $w_y$, we have

$$w_{xy} = \frac{w(w+1) + \dfrac{wxy}{w+1}}{(w+1)^2} = \frac{w(w+1)^2 + wxy}{(w+1)^3} = \frac{w}{w+1} + \frac{wxy}{(w+1)^3}.$$

In finding higher derivatives it often becomes necessary to apply the Chain Rule to a function which is itself a partial derivative. For example, suppose that, with

$$x = x(u, v), \qquad y = y(u, v),$$

we have, on substituting into a function $f(x, y)$,

(21.10.1) $$f(x, y) = F(u, v).$$

We may refer to derivatives of any order of $f$ with respect to $x$ and $y$, or of $F$ with respect to $u$ and $v$, if they exist; any other partial derivatives should be expressed in terms of these. From (21.10.1) we get

(21.10.2) $$F_u = f_x x_u + f_y y_u.$$

Let us now find $F_{uu}$. We obtain, from (21.10.2),

(21.10.3) $$F_{uu} = f_x x_{uu} + x_u (f_x)_u + f_y y_{uu} + y_u (f_y)_u.$$

We wish to express $(f_x)_u$ and $(f_y)_u$ in terms of derivatives of $f$ with respect to $x$ and $y$. We apply the Chain Rule to $f_x$ and find

$$(f_x)_u = (f_x)_x x_u + (f_x)_y y_u = f_{xx} x_u + f_{xy} y_u.$$

In the same way we find

$$(f_y)_u = (f_y)_x x_u + (f_y)_y y_u = f_{yx} x_u + f_{yy} y_u.$$

Then $F_{uu}$ from (21.10.3) becomes

$$F_{uu} = f_x x_{uu} + (f_{xx} x_u + f_{xy} y_u) x_u + f_y y_{uu} + (f_{yx} x_u + f_{yy} y_u) y_u,$$

and, since $f_{xy} = f_{yx}$, we get

$$F_{uu} = f_x x_{uu} + f_y y_{uu} + f_{xx} x_u^2 + 2 f_{xy} x_u y_u + f_{yy} y_u^2.$$

**Example 21.10.2.** If $x = e^u \cos v$, $y = e^u \sin v$, show that

$$z_{xx} + z_{yy} = e^{-2u}(z_{uu} + z_{vv}).$$

*Solution.* It is assumed that $z$ is a function of $x$ and $y$ so that

$$z = f(x, y) = f(e^u \cos v, e^u \sin v) = F(u, v).$$

It is then understood that derivatives of $z$ with respect to $x$ and $y$ denote derivatives of $f(x, y)$; while derivatives of $z$ with respect to $u$ and $v$ denote derivatives of $F(u, v)$. We get, using the Chain Rule,

$$z_u = z_x x_u + z_y y_u = z_x e^u \cos v + z_y e^u \sin v,$$

and
$$\begin{aligned} z_{uu} &= z_x e^u \cos v + (z_{xx} e^u \cos v + z_{xy} e^u \sin v) e^u \cos v \\ &\quad + z_y e^u \sin v + (z_{yx} e^u \cos v + z_{yy} e^u \sin v) e^u \sin v \\ &= z_x e^u \cos v + z_y e^u \sin v + z_{xx} e^{2u} \cos^2 v + 2 z_{xy} e^{2u} \sin v \cos v \\ &\quad + z_{yy} e^{2u} \sin^2 v. \end{aligned}$$

In a similar way we get

$$z_v = z_x x_v + z_y y_v = -z_x e^u \sin v + z_y e^u \cos v,$$

and
$$\begin{aligned} z_{vv} &= -z_x e^u \cos v - (-z_{xx} e^u \sin v + z_{xy} e^u \cos v) e^u \sin v \\ &\quad - z_y e^u \sin v + (-z_{yx} e^u \sin v + z_{yy} e^u \cos v) e^u \cos v \\ &= -z_x e^u \cos v - z_y e^u \sin v + z_{xx} e^{2u} \sin^2 v - 2 z_{xy} e^{2u} \sin v \cos v \\ &\quad + z_{yy} e^{2u} \cos^2 v. \end{aligned}$$

Addition of $z_{uu}$ and $z_{vv}$ gives the following equation, equivalent to the desired one:

$$z_{uu} + z_{vv} = e^{2u}(z_{xx} + z_{yy}).$$

**EXERCISE GROUP 21.10**

1. If $u = f(xy)$ verify that $u_{xy} = u_{yx}$.
2. If $u = f\left(\dfrac{x+y}{z}\right)$ verify that $u_{zx} = u_{xz}$.
3. If $u = f(x + at, y + bt)$ show that $u_{tt} = a^2 f_{11} + 2ab f_{12} + b^2 f_{22}$.
4. If $u(x, y) = f(x^2 + y^2) + g(x)$ show that $y^2 u_{xy} - xy u_{yy} + x u_y = 0$.
5. If $x = e^u \cos v$, $y = e^u \sin v$, find $u_{xx}$.
6. If $F(x, y) = f(x^2 - y, x + y^2)$ find $F_{xy}$.
7. If $f(x, y) = 0$ find $d^2y/dx^2$ in terms of derivatives of $f$.
8. If $u = f(g(t), h(t))$ find $d^2u/dt^2$.
9. If $u = f(x, y)$, $x = r \cos \theta$, $y = r \sin \theta$, show that

$$u_{xx} + u_{yy} = u_{rr} + \frac{1}{r^2} u_{\theta\theta} + \frac{1}{r} u_r.$$

* 10. Show that the curvature of a curve with equation $f(x, y) = 0$ is given by

$$k = \frac{|f_y^2 f_{xx} - 2 f_x f_y f_{xy} + f_x^2 f_{yy}|}{(f_x^2 + f_y^2)^{3/2}}.$$

Use the result in Exercise 10 in Exercises 11–14.

11. Find the curvature of $x^3 + y^3 = 3xy$ at $(3/2, 3/2)$.
12. Find the curvature of $2x^2 - 3y^2 = 5$ at $(2, -1)$.
13. Find the curvature of $x^2 + y + \sin xy = 0$ at $(0, 0)$.
14. Find the curvature of $x^3 - 3xy^2 = 2$ at $(-1, 1)$.

* 15. A function $f(x, y)$ is said to be *homogeneous of degree n* if $f(tx, ty) = t^n f(x, y)$. For such a function prove:

    (a) $x f_1 + y f_2 = nf$, (b) $x^2 f_{11} + 2xy f_{12} + y^2 f_{22} = n(n-1)f$.

    [*Suggestion*. Differentiate in the above equation with respect to $t$, and then set $t = 1$.]

## 21.11 EXTREMA OF FUNCTIONS OF TWO VARIABLES

In the earlier treatment of extrema, in Chapter 5, the quantity under consideration was ultimately expressed as a function of one variable, either directly or by virtue of auxiliary conditions. It may also happen that we want a maximum or minimum of a quantity which depends on more than one independent variable. This leads us to the question of extrema of functions of several variables.

## 21.11 EXTREMA OF FUNCTIONS OF TWO VARIABLES

**DEFINITION 21.11.1.** *A function $f(x, y)$ of two variables has a **relative minimum** at $(a, b)$ if, for all $(x, y)$ in some neighborhood of $(a, b)$,*

(21.11.1) $$f(x, y) \geq f(a, b).$$

*There is a **relative maximum** at $(a, b)$ if, for all $(x, y)$ in some neighborhood of $(a, b)$,*

(21.11.2) $$f(x, y) \leq f(a, b).$$

*Either one is called an **extremum**.*

If $f(x, y)$ possesses partial derivatives it is easy to determine *necessary conditions* for an extremum. Thus, if (21.11.1) holds in a neighborhood of $(a, b)$, then it holds when $y = b$, and for $x$ in a neighborhood of $a$ we have

$$f(x, b) \geq f(a, b).$$

Hence $f(a, b)$ is a relative minimum of $f(x, b)$, considered as a function of $x$ alone, and $f_x(a, b) = 0$. Similarly, we get $f_y(a, b) = 0$. The same conclusions result, if (21.11.2) holds. Hence *if $f_x$ and $f_y$ exist, then for an extremum at $(a, b)$ we must have*

(21.11.3) $$f_x(a, b) = 0, \quad f_y(a, b) = 0.$$

Solutions of Equations (21.11.3) should be considered as potentially yielding extrema. However, some kind of test is required to provide a definite answer. These matters are treated in the examples, and later in this section.

***Example 21.11.1.*** Find any extrema of the function

$$f(x, y) = x^2 + 3xy - 2y^2 + 4x - 11y + 7.$$

*Solution.* The necessary conditions (21.11.3) give the equations

$$2a + 3b + 4 = 0, \quad 3a - 4b - 11 = 0.$$

The solution $a = 1$, $b = -2$ provides the only possible extremum. For a second degree function in two variables an algebraic test is available. Let $x = 1 + h$, $y = -2 + k$. Then $f(1, -2) = 20$, and

$$f(x, y) - f(1, -2) = f(1 + h, -2 + k) - 20 = h^2 + 3hk - 2k^2.$$

The discriminant of this quadratic function in $h$ and $k$ is $9 - 4(-2) = 17 > 0$; because the discriminant is positive, the difference $f(x, y) - f(1, -2)$ is positive for some values of $h$ and $k$ near zero, and negative for others. Hence there cannot be a relative maximum or relative minimum [see Exercise 8(a)]. The function is said to have a *saddle point* at $(1, -2)$.

***Example 21.11.2.*** Find any extrema of the function

$$f(x, y) = -2x^2 + 5xy - 4y^2 - 4x - 2y.$$

*Solution.* The conditions $f_x = 0$ and $f_y = 0$ become

$$-4x + 5y - 4 = 0, \qquad 5x - 8y - 2 = 0,$$

and have the one solution $a = -6$, $b = -4$. Setting $x = -6 + h$, $y = -4 + k$, we find

(21.11.4) $\qquad f(x, y) - f(-6, -4) = -2h^2 + 5hk - 4k^2.$

The discriminant of this quadratic function in $h$ and $k$ is $25 - 32 = -7 < 0$; hence the value of this quadratic is always of the same sign when $(h, k) \neq (0, 0)$. Because the coefficients of $h^2$ and $k^2$ are negative, the difference in (21.11.4) is always negative when $(h, k) \neq (0, 0)$, and hence $f(x, y)$ has a relative maximum at $(-6, -4)$. [See Exercise 8(b).]

*The use of constraints.* Sometimes a function which is to be maximized or minimized is expressed initially in terms of variables which are related by one or more equations, and therefore are not all independent. Such equations are called *constraints* or *auxiliary conditions*. For example, we may wish to find extrema of a function $f(x, y, z)$ with the constraint $g(x, y, z) = 0$. The constraint may then be solved for one of the variables in terms of the other two; and when the resulting function is substituted into $f$, a new function of only two independent variables is obtained, to which we may apply the available methods.

We illustrate an alternative procedure in which the substitution is performed, in effect, via derivatives.

**Example 21.11.3.** A box is to be constructed in the shape of a rectangular parallelepiped, open at the top, and with a fixed volume. Find the dimensions that will minimize the surface area of the box.

*Solution.* Let $x$ and $y$ be the dimensions of the rectangular base, and let $z$ be the height. Then the constraint is

(21.11.5) $\qquad\qquad xyz = k$ (constant),

and the function to be minimized is (there is no top!)

(21.11.6) $\qquad\qquad S = xy + 2yz + 2xz.$

We may solve (21.11.5) for $z = k/(xy)$, substitute into (21.11.6), and express $S$ directly in terms of $x$ and $y$. Instead of carrying this out explicitly, we proceed as though it has been done. A necessary condition is $\partial S/\partial x = 0$ and $\partial S/\partial y = 0$. We differentiate (21.11.6) implicitly and find

(21.11.7)
$$\partial S/\partial x = y + 2z + 2(x + y)\, \partial z/\partial x = 0,$$
$$\partial S/\partial y = x + 2z + 2(x + y)\, \partial z/\partial y = 0.$$

From (21.11.5) we find, on differentiating in turn with respect to $x$ and with respect to $y$,

$$yz + xy\, \partial z/\partial x = 0, \qquad xz + xy\, \partial z/\partial y = 0.$$

Since neither $x$ nor $y$ can be zero, by virtue of (21.11.5), we find $\partial z/\partial x = -z/x$, $\partial z/\partial y = -z/y$. On substitution into (21.11.7) and simplifying, we get $x = 2z$, $y = 2z$. Thus $x = y$, the base is square, and the altitude is half of a side of the base. The dimensions may be found from (21.11.5) by substituting $x = 2z$, $y = 2z$, and we have

$$x = y = 2\sqrt[3]{k/4}, \quad z = \sqrt[3]{k/4}.$$

It is easy to show that we have found a minimum. For example, we can choose $x$ and $y$ large, and then find $z$ from (21.11.5). Thus $S$ is large for many possible choices of dimensions, and a minimum must exist. Since the necessary conditions yielded only one possible choice of dimensions, this choice furnishes the minimum.

As a second example consider the following, for which an answer can be obtained by geometric methods (see Exercise 11).

**Example 21.11.4.** Find the point on the sphere

(21.11.8) $$x^2 + y^2 + z^2 = 4$$

which is closest to the point $(1, 2, 3)$.

*Solution.* It is enough to minimize the square of the distance between a point on the sphere and the given point:

$$Q = (x - 1)^2 + (y - 2)^2 + (z - 3)^2,$$

subject to (21.11.8) which we view as solved for $z$ in terms of $x$ and $y$. We have

(21.11.9) $$\begin{cases} \partial Q/\partial x = 2(x - 1) + 2(z - 3)\, \partial z/\partial x = 0 \\ \partial Q/\partial y = 2(y - 2) + 2(z - 3)\, \partial z/\partial y = 0. \end{cases}$$

From (21.11.8) we find $\partial z/\partial x = -x/z$, $\partial z/\partial y = -y/z$. Substitution into (21.11.9) gives $y = 2x$, $z = 3x$, and then with (21.11.8) we get

$$x = \sqrt{14}/7, \quad y = 2\sqrt{14}/7, \quad z = 3\sqrt{14}/7.$$

It is clear that this point furnishes a minimum.

We proceed to develop a test for an extremum of a function $f(x, y)$. We set $x = a + \lambda t$, $y = b + \mu t$, where $\lambda$ and $\mu$ are constants and $t$ is variable, and we write

(21.11.10) $$f(x, y) = f(a + \lambda t, b + \mu t) = \phi(t).$$

By Exercise 14 of Exercise Group 9.1 we may write

(21.11.11) $$\phi(t) = \phi(0) + t\phi'(0) + t^2 \frac{\phi''(c)}{2}, \quad c \text{ in } (0, t).$$

By use of the Chain Rule in (21.11.10) we obtain

$$\phi'(t) = \lambda f_x + \mu f_y,$$
$$\phi''(t) = \lambda^2 f_{xx} + 2\lambda\mu f_{xy} + \mu^2 f_{yy}.$$

Now let $(a, b)$ be a point for which both $f_x$ and $f_y$ are zero. Then $\phi(0) = f(a, b)$, $\phi'(0) = 0$, and from (21.11.11) we obtain

$$f(x, y) - f(a, b) = \tfrac{1}{2}[\lambda^2 t^2 f_{xx} + 2\lambda\mu t^2 f_{xy} + \mu^2 t^2 f_{yy}](\xi, \eta),$$

where $(\xi, \eta)$ is a point on the line segment joining $(a, b)$ and $(x, y)$. Now let $\lambda t = h$, $\mu t = k$; then $h$ and $k$ are small in absolute value if $(x, y)$ is near $(a, b)$, and we may write

(21.11.12) $\quad f(x, y) - f(a, b) = \tfrac{1}{2}[h^2 f_{xx} + 2hk f_{xy} + k^2 f_{yy}](a + \theta h, b + \theta k)$

where $0 < \theta < 1$. Now suppose that

(21.11.13) $\qquad\qquad f_{xy}^2(a, b) - f_{xx}(a, b) f_{yy}(a, b) < 0.$

If the derivatives are continuous, it follows that a similar inequality holds at $(a + \theta h, b + \theta k)$ for $h$ and $k$ small enough. Then the expression on the right of (21.12.10) is of the same sign for $h$ and $k$ small and $(h, k) \neq (0, 0)$ [Exercise 9] and is positive if $f_{xx}(a, b) > 0$ (or $f_{yy}(a, b) > 0$), and negative if $f_{xx}(a, b) < 0$ (or $f_{yy}(a, b) < 0$). In the former case, $f$ has a minimum at $(a, b)$, and $f$ has a maximum in the latter case. If the left member of (21.11.13) is positive, the left side of (21.11.12) is both positive and negative in any neighborhood of $(a, b)$, and there can be no extremum. These results are included in the following statement.

**Second-derivative test for an extremum.** *Let $f(x, y)$, its first derivatives and its second derivatives be continuous in a neighborhood of $(a, b)$. Let*

$$f_x(a, b) = 0, \qquad f_y(a, b) = 0.$$

*Let*

$$\Delta = f_{xy}^2(a, b) - f_{xx}(a, b) f_{yy}(a, b).$$

*Then:* (a) *if $\Delta < 0$, there is an extremum at $(a, b)$, which is a relative maximum if $f_{xx}(a, b) < 0$, and a relative minimum if $f_{xx}(a, b) > 0$.*
(b) *if $\Delta > 0$, there is a saddle point at $(a, b)$, neither a relative maximum nor a relative minimum.*
(c) *if $\Delta = 0$, the test fails. In this case the question regarding the existence of an extremum at $(a, b)$ remains unanswered.*

Let us apply this test to Examples 21.11.1 and 21.11.2. In Example 21.11.1 we found $a = 1$, $b = -2$ for a potential extremum. We now find $f_{xx} = 2, f_{xy} = 3$, $f_{yy} = -4$ (all constant). Then $\Delta = 3^2 - 2(-4) = 17 > 0$, and by part (b) of the above test there is a saddle point (no extremum).

In Example 21.11.2 we find $f_{xx} = -4$, $f_{xy} = 5$, and $f_{yy} = -8$. Then $\Delta = 5^2 - (-4)(-8) = -7 < 0$. The potential extremum was at $(-6, -4)$. Since $f_{xx}(-6, -4) < 0$, there is a relative maximum at $(-6, -4)$, by (a) of the above test.

## EXERCISE GROUP 21.11

1. Find the extrema of the function $x^2 + y^2 - x + y - 3$.
2. Find the extrema of the function $3x^2 + 2xy + y^2 - 4x + 2y + 7$.
3. Find the extrema of the function $x^4 + y^4 + 4x - 32y - 7$.
4. Find the extrema of the function $x^3 + y^3 - 3xy$.
5. Find the extrema of the function $2x^2 - 4xy + y^4$.
6. Use the methods of Section 21.11 to find the distance from the point $(x_0, y_0, z_0)$ to the plane $ax + by + cz + d = 0$.
7. Find the point $P(x, y)$ such that the sum of the squares of the distances from $P$ to the points $(2, 3)$, $(-1, 4)$, $(3, -3)$ is a minimum.
8. (a) Show that $h^2 + 3hk - 2k^2$ is positive if $h = k \neq 0$, and is negative if $h = -k \neq 0$.
   (b) Show that $-2h^2 + 5hk - 4k^2$ may be written as $-2(h - 5k/4)^2 - 7k^2/8$, and is negative unless $h = k = 0$.
9. (a) Show that if $a \neq 0$,
$$ax^2 + bx + c = a\left[\left(x + \frac{b}{2a}\right)^2 - \frac{b^2 - 4ac}{4a^2}\right].$$

   From this relation show that
   (i) if $b^2 - 4ac < 0$ then $ax^2 + bx + c$ is always of the same sign, that of $a$;
   (ii) if $b^2 - 4ac > 0$ then $ax^2 + bx + c$ is positive for some values of $x$, and negative for others.
   (b) Interpret the results of (a) for the expression $ah^2 + bhk + ck^2$ for small $|h|$ and $|k|$.
10. Find the point on the surface $z = x^2 + y^2$ which is closest to the point $(2, -2, 3/2)$.
11. Prove: If $P_1(x_1, y_1, z_1)$ is the point on the surface $f(x, y, z) = 0$ which is closest to a point $P_0(x_0, y_0, z_0)$, then the line through $P_0$ and $P_1$ is normal to the surface at $P_1$.
12. Given $n$ points $P_i(x_i, y_i)$, $i = 1, 2, \ldots, n$, find a point $P(x, y)$ for which
$$|PP_1|^2 + |PP_2|^2 + \cdots + |PP_n|^2$$
is a minimum.

* 13. Find the (perpendicular) distance between the two lines
$$x = 2 + t, y = 3 - 2t, z = t \quad \text{and} \quad x = 6 - u, y = -u, z = 3 + u$$
by considering a suitable extremal problem.

* 14. The maximum value of the directional derivative of a function $f(x, y, z)$ at a fixed point was obtained in Section 21.7. Try to obtain the same result by maximizing $f_x \cos \alpha + f_y \cos \beta + f_z \cos \gamma$ ($f_x, f_y, f_z$ are constant at a fixed point) subject to the condition
$$\cos^2 \alpha + \cos^2 \beta + \cos^2 \gamma = 1.$$

* 15. Find the shortest distance between the surfaces
$$x^2 + y^2 + z^2 = 1 \quad \text{and} \quad 2x + 2y + z = 6. \qquad \text{(Cont.)}$$

*Note.* This problem can be done geometrically. However, try the following method. Minimize the square of the distance between a point $(u, v, w)$ on the first surface and a point $(x, y, z)$ on the second surface, subject to two constraints; that is, minimize

$$K = (u-x)^2 + (v-y)^2 + (w-z)^2,$$

subject to $u^2 + v^2 + w^2 = 1$, $2x + 2y + z = 6$. If the latter equations are considered as solved for $w$ and $z$, then $K$ becomes a function of $u, v, x$, and $y$, and we must have $K_u = 0$, $K_v = 0$, $K_x = 0$, $K_y = 0$.

## 21.12 EXACT DIFFERENTIAL

The total differential of a function $f(x, y)$ was expressed in Section 21.9 as

$$df = f_x(x, y)\, dx + f_y(x, y)\, dy.$$

For example, if $f(x, y) = x^3 y + y^2$, then

$$df = 3x^2 y\, dx + (x^3 + 2y)\, dy.$$

We may also state in the latter case that a function $f(x, y)$ exists (since we started with it) such that

$$f_x = 3x^2 y, \qquad f_y = x^3 + 2y.$$

We raise the following question: given functions $P(x, y)$ and $Q(x, y)$, does a function exist such that

$$f_x(x, y) = P(x, y), \qquad f_y(x, y) = Q(x, y)?$$

Before giving a complete answer, we show that this is not always possible. Let $P(x, y) = 3x^2 y$, $Q(x, y) = x^2$. If $f_x = 3x^2 y$, then

(21.12.1) $$f(x, y) = x^3 y + g(y),$$

where $3x^2 y$ was integrated with respect to $x$ ($y$ being constant), and $g(y)$, a function of $y$ alone, acts as a constant of integration. If $f_y = Q$ is to hold, then by differentiating (21.12.1) we must have

$$x^3 + g'(y) = x^2 \quad \text{and} \quad g'(y) = x^2 - x^3.$$

However, $g'(y)$ may not contain $x$, and hence the original assumption that $f(x, y)$ exists such that $f_x = 3x^2 y$, $f_y = x^2$ is false.

We return to the original question. Suppose $f(x, y)$ exists such that $f_x = P$, $f_y = Q$. If $P_y$ and $Q_x$ exist, we then have

$$f_{xy} = P_y, \qquad f_{yx} = Q_x.$$

If $P_y$ and $Q_x$ are continuous, it follows that $f_{xy} = f_{yx}$, and hence $P_y = Q_x$ is a necessary condition. With suitable differentiability properties it is also sufficient, as stated in the following.

**THEOREM 21.12.1.** *If $P(x, y)$, $P_y(x, y)$, $Q(x, y)$, and $Q_x(x, y)$ are continuous in a rectangular region R bounded by straight lines parallel to the coordinate axes, a necessary and sufficient condition that $f(x, y)$ exist such that*

(21.12.2)    $f_x(x, y) = P(x, y), \quad f_y(x, y) = Q(x, y), \quad \text{in } R,$

*is that*

(21.12.3)    $P_y(x, y) = Q_x(x, y) \quad \text{in } R.$

*Proof.* We have previously shown the necessity of (21.12.3). We establish the sufficiency by actually producing a function $f$ which satisfies (21.12.2) when (21.12.3) holds. We define $f$ as

(21.12.4)    $f(x, y) = \int_a^x P(u, b)\, du + \int_b^y Q(x, v)\, dv,$

where $(a, b)$ is any point in $R$, and prove that (21.12.2) holds for this function. In the definition of $f$ the variable $x$ appears twice, once as a limit in the first integral, and again in the integrand of the second integral. The variable $y$ appears only in a limit of the second integral. Differentiating (21.12.4) with respect to $y$, we get

$$f_y = \frac{\partial}{\partial y} \int_b^y Q(x, v)\, dv = Q(x, y)$$

by Theorem 9.10.1. Differentiating (21.12.4) with respect to $x$, we have

(21.12.5)    $f_x = \dfrac{\partial}{\partial x} \int_a^x P(u, b)\, du + \dfrac{\partial}{\partial x} \int_b^y Q(x, v)\, dv.$

Now, again by Theorem 9.10.1,

(21.12.6)    $\dfrac{\partial}{\partial x} \int_a^x P(u, b)\, du = P(x, b).$

On the other hand, it can be shown that

(21.12.7)    $\dfrac{\partial}{\partial x} \int_b^y Q(x, v)\, dv = \int_b^y Q_x(x, v)\, dv.$

We now use (21.12.3) for the first time, in the form

$$Q_x(x, v) = P_v(x, v).$$

Continuing (21.12.7), we get

(21.12.8)    $\dfrac{\partial}{\partial x} \int_b^y Q(x, v)\, dv = \int_b^y P_v(x, v)\, dv = P(x, v)\Big|_b^y = P(x, y) - P(x, b).$

Combining (21.12.5), (21.12.6), and (21.12.8), we have

$$f_x(x, y) = P(x, b) + [P(x, y) - P(x, b)] = P(x, y),$$

and the proof is complete.

The definition of $f$ in (21.12.4) is a ready-made formula for determining $f$. We illustrate its use in the following example. In the subsequent examples we present alternative methods that may be employed. We note that once a function $f(x, y)$ is found to satisfy (21.12.2), then $f(x, y) + C$ is also such a function, where $C$ is a constant. The choice of the point $(a, b)$ in (21.12.4) can affect only this constant.

**Example 21.12.1.** Show that $f(x, y)$ exists such that

$$f_x = 3x^2 y + 2x, \qquad f_y = x^3 + 3y^2,$$

and find such a function.

**Solution.** If we set

$$P(x, y) = 3x^2 y + 2x, \qquad Q(x, y) = x^3 + 3y^2,$$

we see that $P_y = 3x^2 = Q_x$, and, since $P$, $Q$, $P_y$, and $Q_x$ are continuous, a function exists as desired.

To find $f$, we use (21.12.4) with the choice $a = 0$, $b = 0$. Then $P(u, b) = 2u$, $Q(x, v) = x^3 + 3v^2$, and we get

$$f(x, y) = \int_0^x 2u \, du + \int_0^y (x^3 + 3v^2) \, dv = u^2 \Big|_0^x + (x^3 v + v^3)\Big|_0^y = x^2 + x^3 y + y^3.$$

It is easy to verify that this function fulfills the conditions. It is also clear that the same is true of the function

$$x^2 + x^3 y + y^3 + C, \qquad C \text{ constant.}$$

If $f$ exists such that $f_x = P$ and $f_y = Q$, we may write

$$P \, dx + Q \, dy = f_x \, dx + f_y \, dy = df;$$

in other words, $P \, dx + Q \, dy$ is the total differential of $f$, and $P \, dx + Q \, dy$ is said to be an *exact differential*. Theorem 21.12.1 may then be rephrased in an obvious way so as to state a necessary and sufficient condition that $P \, dx + Q \, dy$ be an exact differential.

**Example 21.12.2.** Show that

$$\frac{1}{y} dx + \frac{y - x}{y^2} dy$$

is an exact differential, and find a function for which the expression is the total differential.

**Solution.** We write

$$P = \frac{1}{y}, \qquad Q = \frac{y - x}{y^2}, \qquad P_y = -\frac{1}{y^2} = Q_x.$$

Thus the conditions of Theorem 21.12.1 are met if $y \neq 0$, and the given expression is an exact differential. We shall now find a function $f$ by the method attempted at the beginning of this section (in that case such a function did not exist). We desire a function $f$ such that

(21.12.9) $$f_x = \frac{1}{y}, \quad f_y = \frac{y-x}{y^2}.$$

Integrating the first equation with respect to $x$, we have

(21.12.10) $$f(x, y) = \frac{x}{y} + g(y).$$

To satisfy the second equation of (21.12.9), we differentiate (21.12.10) with respect to $y$ and get

$$f_y = -\frac{x}{y^2} + g'(y) = \frac{y-x}{y^2}.$$

This yields the equation

$$g'(y) = \frac{1}{y}, \quad \text{hence} \quad g(y) = \ln|y|,$$

and substituting in (21.12.10) we get a function as desired,

$$f(x, y) = \frac{x}{y} + \ln|y|.$$

*Comments*: (1) A constant may be added to $f(x, y)$ as obtained.
(2) This method worked here, in contrast to the previous attempt, because we came out with $g'(y)$ in terms of $y$ alone. In fact, the method may be employed without previously establishing the existence of $f$, and, if it is successful, the function $f$ will satisfy the requirements.

*Example 21.12.3.* Find $f(x, y)$ such that
$$df = \left(e^y - \frac{1}{x^2}\right) dx + (xe^y + 1) \, dy.$$

*Solution.* We present here a type of trial-and-error method. If $f$ exists, we must have

(21.12.11) $$f_x = e^y - \frac{1}{x^2}, \quad f_y = xe^y + 1.$$

We integrate the first with respect to $x$ and the second with respect to $y$, writing the results separately:

$$xe^y + \frac{1}{x}, \quad xe^y + y.$$

We use those terms common to both expressions, as well as those terms appearing separately in each,

$$f(x, y) = xe^y + \frac{1}{x} + y.$$

We may now prove that this is a desired function by directly verifying (21.12.11). The direct verification is necessary because we have not previously established the existence of a function, nor does the procedure itself establish it.

**EXERCISE GROUP 21.12**

In each of Exercises 1–12 test for exactness; if the form is exact, find a function for which the given form is the total differential.

1. $(1 + y) \, dx + x \, dy$
2. $x \, dy - y \, dx$
3. $(y - 14x) \, dx + x \, dy$
4. $\dfrac{x+y}{x^2} \, dx - \dfrac{x+y}{xy} \, dy$
5. $\dfrac{2x \, dx}{y} + \dfrac{x^2 \, dy}{y^2}$
6. $\dfrac{x \, dx}{y^2} - \dfrac{x^2 \, dy}{y^3}$
7. $[2xy + y \cos(xy)] \, dx + [x^2 + x \cos(xy)] \, dy$
8. $(3x^2 y^2 - y) \, dx + (2x^3 y - x + 7) \, dy$
9. $\dfrac{x \, dy - y \, dx}{x^2}$
10. $\dfrac{-y \, dx + x \, dy}{x^2 + y^2}$
11. $\dfrac{1 + y^2}{x^2} \, dx - \dfrac{2y - xy^2}{x} \, dy$
12. $e^{xy}[(1 + xy) \, dx + x^2 \, dy]$

In each of the following extend the method of Example 21.12.2 to find a function $f(x, y, z)$ such that the given form is the total differential of $f$. Bear in mind that in any integration there is, in place of a constant of integration, an arbitrary function of the remaining variables.

13. $2xy^3 z \, dx + 3x^2 y^2 z \, dy + (x^2 y^3 + 2z) \, dz$
14. $(yz + 1) \, dx + (xz + 1) \, dy + (xy + 1) \, dz$
15. $\dfrac{dx}{z} - \dfrac{dy}{z} - \dfrac{x - y}{z^2} \, dz$
16. $z \, dx + z \cos(yz) \, dy + [x + y \cos(yz)] \, dz$

17. If $P \, dx + Q \, dy$ is exact, show that $\mathbf{F}(x, y) = P(x, y)\mathbf{i} + Q(x, y)\mathbf{j}$ is the gradient of some function.

## *21.13 LINE INTEGRALS

The definite integral $\int_a^b f(x) \, dx$ was defined in Section 9.9. Its value is completely determined by the values of $f(x)$ for $x$ in the interval $[a, b]$. For the concept of *line integral* in a plane, which we will consider here, the underlying principle is that the value depends on the values that a function of two variables has along a curve in the plane. For a line integral in space the value depends on the values of a function of three variables along a curve in space.

The formal definition of a line integral is similar to that of a definite integral. We dispense with the definition, and consider the properties and evaluation of line integrals.

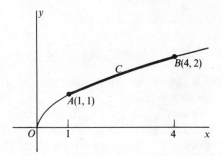

Fig. 21.13.1

Let $C$ be the part of the parabola $x = y^2$, between the points $A(1, 1)$ and $B(4, 2)$, directed from $A$ to $B$ (Fig. 21.13.1). An example of a line integral along $C$ is

$$\int_C (x - 2y)\, dx,$$

where $x$ and $y$ are such that $x = y^2$, and the integration is with respect to $x$. The evaluation can be carried out by substituting $y = x^{1/2}$, and we get

$$\int_C (x - 2y)\, dx = \int_1^4 (x - 2x^{1/2})\, dx = \left(\frac{x^2}{2} - \frac{4x^{3/2}}{3}\right)\bigg|_1^4$$

$$= \left(8 - \frac{32}{3}\right) - \left(\frac{1}{2} - \frac{4}{3}\right) = -\frac{11}{6}.$$

In this way the line integral is expressed as an ordinary definite integral in $x$, between limits 1 and 4 for $x$ as determined by the way the curve is traced.

The relation between $x$ and $y$ may also be used to obtain an integral in $y$, if we use the relation $dx = 2y\, dy$. The limits for the integral will be values of $y$ corresponding to $A$ and $B$. Thus

$$\int_C (x - 2y)\, dx = \int_1^2 (y^2 - 2y)2y\, dy = \int_1^2 (2y^3 - 4y^2)\, dy$$

$$= \left(\frac{y^4}{2} - \frac{4y^3}{3}\right)\bigg|_1^2 = -\frac{11}{6}.$$

We may also use the relation between $x$ and $y$ by using parametric equations of $C$. If we write equations of the curve as $x = t^2$, $y = t$, then $t = 1$ at $A$ and $t = 2$ at $B$. Then we have, since $dx = 2t\, dt$,

$$\int_C (x - 2y)\, dx = \int_1^2 (t^2 - 2t)2t\, dt = -\frac{11}{6}.$$

The example just considered is typical of a line integral of the form

$$\int_C f(x, y)\, dx.$$

Similarly, we may treat line integrals of the forms

$$\int_C f(x, y)\, dy \quad \text{and} \quad \int_C f(x, y)\, ds,$$

where $ds = \sqrt{dx^2 + dy^2}$ is the element of arc and may be used for any type of representation of the equation of $C$.

The curve used in the above illustration is a simple one, but it is typical of the kind of curve permitted. In fact, the type of curve permitted, in general, is made up of arcs, like the one above, joined to form a continuous curve, such as in Fig. 21.13.2. A line integral along such a curve is defined as the sum of line integrals along the parts, traced in a continuous manner.

Fig. 21.13.2

For reasons which will appear later, line integrals of the type

$$\int_C [P(x, y)\, dx + Q(x, y)\, dy]$$

are used. This is a sum of two types used above. In the evaluation of such a line integral any of the methods illustrated may be used, when applicable, for either of the parts.

**Example 21.13.1.** Evaluate the line integral

$$\int_C (xy\, dx + x^2\, dy)$$

along the path $C$ indicated in Fig. 21.13.3, consisting of three line segments $C_1, C_2$, and $C_3$.

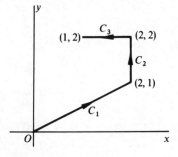

Fig. 21.13.3

*Solution.* We may add the values of the line integral along the separate segments. We obtain, for the respective segments:

$C_1: y = x/2, dy = dx/2, \int_{C_1} (xy\, dx + x^2\, dy) = \int_0^2 \left(\frac{x^2}{2} dx + x^2 \frac{dx}{2}\right) = \int_0^2 x^2\, dx = \frac{8}{3};$

$C_2: x = 2, dx = 0, \int_{C_2} (xy\, dx + x^2\, dy) = \int_1^2 4\, dy = 4;$

$C_3: y = 2, dy = 0, \int_{C_3} (xy\, dx + x^2\, dy) = \int_2^1 2x\, dx = x^2 \Big|_2^1 = -3.$

Hence
$$\int_C (xy\, dx + x^2\, dy) = \frac{8}{3} + 4 - 3 = \frac{11}{3}.$$

A line integral may be taken along a *closed curve* which does not intersect itself. The endpoints of $C$ are then the same point (any point on $C$). However, the value of a variable may change after a complete circuit of $C$. We illustrate some of the techniques and considerations involved in the evaluation of a line integral around a closed curve.

*Example 21.13.2.* Evaluate the integral
$$\int_C (y\, dx - x\, dy),$$
where $C$ is the circle $x^2 + y^2 = 4$, traced in a counterclockwise direction.

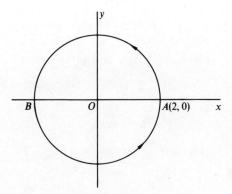

Fig. 21.13.4

*Solution.* The parametric form of $C$ [Fig. 21.13.4],
$$x = 2 \cos \theta, \qquad y = 2 \sin \theta,$$
is particularly advantageous for this purpose since, if we start at the point

(2, 0), we may trace $C$ by letting $\theta$ vary from 0 to $2\pi$. We get

$$\int_C (y\,dx - x\,dy) = \int_0^{2\pi} [2\sin\theta(-2\sin\theta\,d\theta) - 2\cos\theta(2\cos\theta\,d\theta)]$$

$$= -4\int_0^{2\pi} (\sin^2\theta + \cos^2\theta)\,d\theta = -4\int_0^{2\pi} d\theta = -8\pi.$$

The advantage of the method just used will be clear in comparison with the method we shall now employ to evaluate the same integral by use of the Cartesian equation of the circle. The purpose in doing this is to point out the kind of analysis used which, in other cases, may be necessary to use. We write

(21.13.1) $$\int_C (y\,dx - x\,dy) = \int_C y\,dx - \int_C x\,dy$$

and evaluate the integrals separately. To evaluate $\int_C y\,dx$ we treat $C$ as a sum of two arcs, $C = C_1 + C_2$, where $C_1$ is the upper half of the circle from $A$ to $B$, while $C_2$ is the lower half from $B$ to $A$. Then

$$y = \sqrt{4 - x^2} \quad \text{on} \quad C_1, \qquad y = -\sqrt{4 - x^2} \quad \text{on} \quad C_2$$

and

$$\int_C y\,dx = \int_{C_1} y\,dx + \int_{C_2} y\,dx = \int_2^{-2} \sqrt{4-x^2}\,dx + \int_{-2}^{2} (-\sqrt{4-x^2})\,dx$$

$$= -2\int_{-2}^{2} \sqrt{4-x^2}\,dx \quad \left(\text{since } \int_2^{-2} f(x)\,dx = -\int_{-2}^{2} f(x)\,dx\right).$$

We may now set $x = 2\sin\theta$, $dx = 2\cos\theta\,d\theta$, and get

$$\int_C y\,dx = -2\int_{-\pi/2}^{\pi/2} 4\cos^2\theta\,d\theta = -4\int_{-\pi/2}^{\pi/2} (1 + \cos 2\theta)\,d\theta = -4\pi.$$

To evaluate $\int_C x\,dy$ we use the right and left halves of $C$, on which we have, respectively,

$$x = \sqrt{4 - y^2} \quad \text{and} \quad x = -\sqrt{4 - y^2}.$$

Then, starting at the point $(0, -2)$, we get

$$\int_C x\,dy = \int_{-2}^{2} \sqrt{4-y^2}\,dy + \int_{2}^{-2} (-\sqrt{4-y^2})\,dy = 2\int_{-2}^{2} \sqrt{4-y^2}\,dy = 4\pi.$$

With (21.13.1) the result becomes

$$\int_C (y\,dx - x\,dy) = -4\pi - 4\pi = -8\pi,$$

the same result as with the first method, whose advantage is now apparent.

*Example 21.13.3.* Evaluate the line integral $\int_C xy\,ds$, where $C$ is the parabola $y = x^2$ from $(0, 0)$ to $(\sqrt{3}/2, 3/4)$.

*Solution.* It is convenient to express the integral in terms of $y$. We find

$$x = y^{1/2}, \quad dx = \frac{1}{2} y^{-1/2} \, dy, \quad ds = \sqrt{1 + \frac{1}{4y}} \, dy.$$

Then

$$\int_C xy \, ds = \int_0^{3/4} \sqrt{y} \cdot y \sqrt{1 + \frac{1}{4y}} \, dy = \frac{1}{2} \int_0^{3/4} y \sqrt{4y + 1} \, dy.$$

To evaluate this definite integral we substitute $4y + 1 = u^2$; continuing the evaluation, we have, as the value of the original line integral,

$$\frac{1}{16} \int_0^2 (u^4 - u^2) \, du = \frac{1}{16} \left( \frac{u^5}{5} - \frac{u^3}{3} \right) \Big|_0^2 = \frac{7}{30}.$$

A line integral in space may be written in the form

(21.13.2) $$\int_C [P(x, y, z) \, dx + Q(x, y, z) \, dy + R(x, y, z) \, dz],$$

where $x, y, z$ as well as $dx, dy, dz$ are connected by equations of the curve $C$. In particular, if we have in parametric form

$$C: x = x(t), \quad y = y(t), \quad z = z(t),$$

and substitute into (21.13.2), we get an ordinary definite integral in $t$.

*Example 21.13.4.* Evaluate the line integral

$$\int_C [xy \, dx + x^2 z \, dy - (x + y) \, dz],$$

with $C: x = t, y = t^2, z = t^3, 0 \le t \le 2$.

*Solution.* We obtain directly

$$\int_C [xy \, dx + x^2 z \, dy - (x + y) \, dz] = \int_0^2 [t^3 \, dt + 2t^6 \, dt - 3t^2(t + t^2) \, dt]$$

$$= \int_0^2 (2t^6 - 3t^4 - 2t^3) \, dt = \left( \frac{2t^7}{7} - \frac{3t^5}{5} - \frac{t^4}{2} \right) \Big|_0^2 = \frac{328}{35}.$$

*Work.* The work done as a particle moves along a straight line subject to a constant force was defined in Section 13.2. A direct application of line integrals is in finding the work done as a particle moves along a curve subject to a variable force.

Let $C$ be a curve lying in a region of the $xy$ plane in which the force acting at any point $(x, y)$ is given by

$$\mathbf{F}(x, y) = P(x, y)\mathbf{i} + Q(x, y)\mathbf{j}.$$

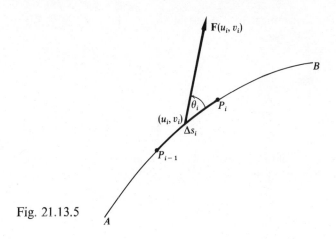

Fig. 21.13.5

Subdivide the arc from $A$ to $B$ in the usual manner. Fig. 21.13.5 shows one subarc. Choose a point $(u_i, v_i)$ on this subarc and let $\mathbf{F}(u_i, v_i)$ be the force acting at this point. The work done as the particle moves from $P_{i-1}$ to $P_i$ is, approximately,

$$|\mathbf{F}(u_i, v_i)|\sqrt{(\Delta x_i)^2 + (\Delta y_i)^2}\cos\theta_i.$$

Summing this over the subarcs from $i = 1$ to $i = n$, and letting $n \to \infty$, we get the definition of work done, $W$, as

$$W = \lim_{n\to\infty} \sum_{i=1}^{n} |\mathbf{F}(u_i, v_i)|\sqrt{(\Delta x_i)^2 + (\Delta y_i)^2}\cos\theta_i.$$

It can be shown that this leads to the line integral

(21.13.3) $$W = \int_C |\mathbf{F}(x, y)|\cos\theta\, ds,$$

where $\theta$ is the angle between $\mathbf{F}$ and the tangent to $C$ at $(x, y)$.

Now if $\mathbf{R}$ is the position vector of $P$, we have

$$\mathbf{R} = x\mathbf{i} + y\mathbf{j}, \qquad d\mathbf{R} = \mathbf{i}\, dx + \mathbf{j}\, dy, \qquad |d\mathbf{R}| = ds,$$

and

$$\mathbf{F} \cdot d\mathbf{R} = |\mathbf{F}|\cos\theta\, ds.$$

Hence (21.13.3) may be written as

(21.13.4) $$W = \int_C \mathbf{F} \cdot d\mathbf{R} = \int_C P(x, y)\, dx + Q(x, y)\, dy.$$

Thus if a particle moves along a curve $C$ subject to a force $\mathbf{F}$, the work done is given by (21.13.4), where $P$ and $Q$ are the $x$ and $y$ components of $\mathbf{F}$, respectively. It is sufficient if $P$ and $Q$ are continuous in a region containing the curve $C$, where $C$ is given parametrically by $x = x(t)$, $y = y(t)$ with $x'(t)$ and $y'(t)$ continuous, or if $C$ consists of a finite number of arcs for each of which the stated conditions hold.

The extension to a curve in space is easily made, so that if the force is

$$\mathbf{F}(x, y, z) = P(x, y, z)\mathbf{i} + Q(x, y, z)\mathbf{j} + R(x, y, z)\mathbf{k},$$

the work done is

(21.13.5) $$W = \int_C \mathbf{F} \cdot d\mathbf{R} = \int_C P\,dx + Q\,dy + R\,dz.$$

**Example 21.13.5.** If the force acting on a particle is

$$\mathbf{F} = x\mathbf{i} + y\mathbf{j} + z\mathbf{k},$$

find the work done as the particle moves from $(1, 0, 0)$ to $(1, 0, 2\pi)$ along the curve $x = \cos t$, $y = \sin t$, $z = t$.

*Solution.* The work done is given by (21.13.5), where

$$P = x, \quad Q = y, \quad R = z, \quad dx = -\sin t\, dt, \quad dy = \cos t\, dt, \quad dz = dt.$$

Then, since $t$ varies from 0 to $2\pi$,

$$W = \int_C \mathbf{F} \cdot d\mathbf{R} = \int_0^{2\pi} [\cos t(-\sin t\, dt) + \sin t(\cos t\, dt) + t\, dt]$$

$$= \int_0^{2\pi} t\, dt = 2\pi^2.$$

★ **EXERCISE GROUP 21.13**

1. Evaluate $\int_C y\,dx + (x - 14y)\,dy$ along each of the following curves from $(1, 0)$ to $(0, 1)$.
   (a) the straight line joining the given points
   (b) the parabola $x^2 = 1 - y$
   (c) the straight line from $(1, 0)$ to $(1, 1)$, followed by the straight line from $(1, 1)$ to $(0, 1)$
   (d) the circle $x^2 + y^2 = 1$ in the first quadrant

2. Evaluate each line integral along the curve $x^2 = y$ from $(1, 1)$ to $(3, 9)$.
   (a) $\int_C 2xy\,dx + x^2\,dy$  (b) $\int_C xy\,dx + xy\,dy$  (c) $\int_C ds$

3. Evaluate $\int_C 2\,dx + x\,dy + y\,dz$, $C$: $x = \cos t$, $y = \sin t$, $z = t$, $0 \le t \le 2\pi$.

4. Evaluate $\int_{(1,0,1)}^{(2,3,2)} x^2\,dx - xz\,dy + y^2\,dz$ along the straight line which joins the two points.

5. If $C$ is the circle $x^2 + y^2 = 4$ traced in a counterclockwise direction, evaluate each of the following line integrals:
   (a) $\int_C y^2\,ds$  (b) $\int_C (x^2 + y^2)\,ds$  (c) $\int_C (x^2 + 2y)\,ds$

6. Evaluate $\int_C x^2yz\,ds$, $C$: $x = \cos t$, $y = \cos t$, $z = \sqrt{2}\sin t$, $0 \le t \le \pi/2$.

* 7. Evaluate $\int_C \sin yz\, dx + xz \cos yz\, dy + xy \cos yz\, dz$ along the curve $C$: $x = t^2$, $y = t^3$, $z = t$, $0 \leq t \leq 1$.

8. Find the work done as a particle moves from $(0, 0)$ to $(4, 8)$ along the curve $y^2 = x^3$, subject to a force $\mathbf{F} = 2\mathbf{i} + \mathbf{j}$.

9. Find the work done as a particle moves around the circle $x^2 + y^2 = 4$ in a counterclockwise direction subject to a force $\mathbf{F} = x\mathbf{i} + x\mathbf{j}$.

10. Find the work done as a particle moves along the curve $x = t$, $y = t^2$, $z = t^3$ from $t = 0$ to $t = 1$ subject to a force $2xyz\mathbf{i} + x^2z\mathbf{j} + x^2y\mathbf{k}$.

## * 21.14  INDEPENDENCE OF PATH

The name "line integral," or "curvilinear integral," as it is also called, carries with it the implication that the value depends on the curve or path which joins two points. In other words, the value of a line integral is, in general, different along different paths joining two given points (Fig. 21.14.1).

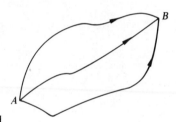

Fig. 21.14.1

There are line integrals, however, for which the same value is obtained regardless of which path is followed from $A$ to $B$. Consider, for example, the line integral

(21.14.1) $$\int_C 2xy\, dx + x^2\, dy,$$

where $C$ is any curve joining two fixed points $A$ and $B$ (Fig. 21.14.1). It is easily verified that

$$d(x^2 y) = 2xy\, dx + x^2\, dy,$$

and this holds along $C$ whether $C$ is given in the form $y = f(x)$ or $x = g(y)$ or $F(x, y) = 0$ or $x = x(t)$, $y = y(t)$. Thus whatever the equation of $C$ we have

$$\int_C 2xy\, dx + x^2\, dy = \int_C d(x^2 y) = x^2 y \Big|_A^B,$$

where the notation on the right means the difference of the values of the function $x^2 y$ at $B$ and at $A$. The value of (21.14.1) depends only on the values of $x^2 y$ at $A$ and $B$, and not on the path $C$. The line integral (21.14.1) is said to be *independent of the path*.

From the above it appears that the independence of path is directly related to the question of whether the integrand is an exact differential. This is indeed the case, and the same reasoning is used in the general proof.

**THEOREM 21.14.1.** *Let $P(x, y)\, dx + Q(x, y)\, dy$ be an exact differential of a function $f(x, y)$ in some region. Then the line integral of $P\, dx + Q\, dy$ is independent of the path in the region, and*

(21.14.2) $$\int_C P\, dx + Q\, dy = f(x, y)\Big|_A^B,$$

*where C is any path joining A to B.*

*Proof.* We need merely write

$$\int_C P\, dx + Q\, dy = \int_C df = f(x,y)\Big|_A^B.$$

*Example 21.14.1.* Evaluate the line integral

$$\int_C (2x + y)\, dx + (x - y^2)\, dy$$

along the curve $C\colon x = t,\ y = t^2$, from the point $(1, 1)$ to the point $(2, 4)$.

*Solution.* The evaluation may be carried out by the methods of Section 21.13, for the specific path given. However, if we set

$$P(x, y) = 2x + y, \qquad Q(x, y) = x - y^2,$$

we see that $P_y = Q_x = 1$ and hence the integrand is an exact differential. A function $f(x, y)$ such that $df = P\, dx + Q\, dy$ is easily found to be $f(x, y) = x^2 + xy - y^3/3$. Hence

$$\int_C (2x + y)\, dx + (x - y^2)\, dy = (x^2 + xy - y^3/3)\Big|_{(1,1)}^{(2,4)}$$

$$= (4 + 8 - 64/3) - (1 + 1 - 1/3) = -11.$$

If the point $A$ in (21.14.2) is a fixed point $(a, b)$, and $B$ is the point $(x, y)$, then (21.14.2) may be used to find $f(x, y)$ in cases where $P\, dx + Q\, dy$ is an exact differential. To avoid confusion concerning variables used, we may write (21.14.2) as

(21.14.3) $$\int_C P(u, v)\, du + Q(u, v)\, dv = f(u, v)\Big|_{(a, b)}^{(x, y)}$$

where $C$ is any path joining $(a, b)$ to $(x, y)$.

*Example 21.14.2.* Find $f(x, y)$ such that

$$f_x = 2x + y, \qquad f_y = x - y^2.$$

*Solution.* It was shown in Example 21.14.1 that $f$ exists, and, in fact, $f$ was found. We wish to find it here by using (21.14.3). We choose $A$ as the point $(0, 0)$. Then

(21.14.4) $$\int_C (2u + v)\, du + (u - v^2)\, dv = f(u, v)\Big|_{(0,0)}^{(x,y)},$$

where $C$ is any path from $(0, 0)$ to $(x, y)$. We now choose the path shown in Fig. 21.14.2 consisting of the line segments $v = 0$ and $u = x$. Along the segment $v = 0$ we have $dv = 0$, and along the segment $u = x$ (constant) we have $du = 0$. Hence for the chosen path the line integral in (21.14.4) becomes

$$\int_0^x 2u\, du + \int_0^y (x - v^2)\, dv = u^2 \Big|_0^x + (xv - v^3/3)\Big|_0^y = x^2 + xy - y^3/3.$$

We may take $f(x, y) = x^2 + xy - y^3/3$, as in Example 21.14.1.

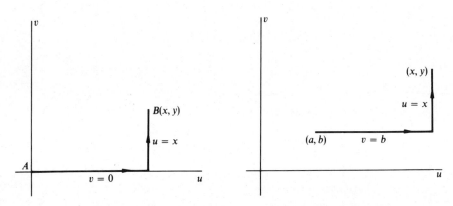

Fig. 21.14.2    Fig. 21.14.3

The method of Example 21.14.2 may be applied generally. If $P\, dx + Q\, dy$ is an exact differential we may use (21.14.3), and the path consisting of $v = b$ and $u = x$ (Fig. 21.14.3). We then get

$$\int_a^x P(u, b)\, du + \int_b^y Q(x, v)\, dv = f(u, v)\Big|_{(a,b)}^{(x,y)} = f(x, y) - f(a, b).$$

This equation helps explain the source of the function $f(x, y)$ used in the proof of Theorem 21.12.1.

We close this section with a discussion of an example which provides a deeper insight into the relation between exact differential and independence of path.

**Example 21.14.3.** Consider the line integral

$$\int_C \frac{-y\,dx + x\,dy}{x^2 + y^2},$$

where $C$ is the unit circle traced in the counterclockwise sense. Along the curve $C_1$ of Fig. 21.14.4, the upper half of the unit circle, we have

$$x = \cos t, \qquad y = \sin t, \qquad 0 \leq t \leq \pi.$$

We have $x^2 + y^2 = 1$, $dx = -\sin t\,dt$, $dy = \cos t\,dt$, and the integral becomes

$$\int_{C_1} = \int_0^\pi [-\sin t(-\sin t\,dt) + \cos t(\cos t\,dt)] = \int_0^\pi dt = \pi.$$

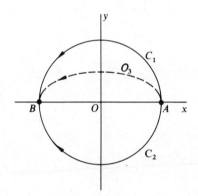

Fig. 21.14.4

Along $C_2$, the lower half of the unit circle, we have $-\pi \leq t \leq 0$, and the line integral becomes

$$\int_{C_2} = \int_0^{-\pi} dt = -\pi.$$

We have obtained different values along two different curves joining the same points, and therefore the line integral *depends on the path*.

Now, the line integral may be written as

$$\int_C P\,dx + Q\,dy, \quad \text{where} \quad P = \frac{-y}{x^2 + y^2}, \quad Q = \frac{x}{x^2 + y^2}.$$

It is easily verified that $P_y = Q_x$, if $(x, y) \neq (0, 0)$. This situation does not contradict Theorems 21.14.1 and 21.12.1. In Theorem 21.12.1 the region was restricted to the type which lies interior to a simple closed curve. Since the origin must be excluded, the hypotheses of Theorem 21.12.1 are not satisfied, and the theorem does not apply. If, however, we chose a path like $C_3$, we would find the same value for the line integral as for $C_1$.

In Exercise 10 of Exercise Group 21.12, a function was sought for which the above integrand is the exact differential, and the function found was $\tan^{-1} y/x$. It is important to note that $\tan^{-1} y/x$ is a function in the mathematical sense only in a region which does not surround the origin. Thus it is not a function in the region interior to the unit circle, but it is a function in the region interior to the curve formed by $C_1$ and $C_3$. In more rigorous treatments, these considerations involve the concept of *simply-connected regions*.

## ★ EXERCISE GROUP 21.14

In Exercises 1–8 verify that the line integral is independent of the path, and evaluate the integral. The following notation may be used in such cases.

$$\int_{(a,b)}^{(c,d)} P\,dx + Q\,dy.$$

1. $\displaystyle\int_{(1,2)}^{(-3,1)} x\,dx + y\,dy$

2. $\displaystyle\int_{(1,-1)}^{(2,1)} 2xy\,dx + x^2\,dy$

3. $\displaystyle\int_{(1,1)}^{(e,e)} (1-xy)\left(\frac{dx}{x}+\frac{dy}{y}\right)$

4. $\displaystyle\int_{(1,3)}^{(0,-2)} \frac{xy\,dx - x^2\,dy}{y^3}$

5. $\displaystyle\int_{(1,3)}^{(u,v)} \frac{y\,dx - x\,dy}{x^2}$

6. $\displaystyle\int_{(1,3)}^{(u,v)} \frac{y\,dx - x\,dy}{y^2}$

7. $\displaystyle\int_{(0,0)}^{(1,2)} (y^2 - 2xy - 2)\,dx + (2xy - x^2)\,dy$

8. $\displaystyle\int_{(0,1)}^{(\sqrt{\pi},1/2)} 2xy\cos x^2 y\,dx + x^2 \cos x^2 y\,dy$

9. (a) Show that the following line integral is independent of the path in a suitable region.
$$\int_c \frac{x\,dy - y\,dx}{x\sqrt{x^2 - y^2}}.$$
   (b) Evaluate the integral from $(1, 0)$ to $(2, 1)$ along a path formed by the lines $x = 1, y = 1$.
   (c) Find a function for which the integrand is an exact differential.

10. (a) Show that if a particle moves subject to a force
$$\mathbf{F} = \frac{x\mathbf{i} + y\mathbf{j}}{(x^2 + y^2)^{3/2}},$$
then the work done is independent of the path.
   (b) Find the work done in moving from $(a, b)$ to $(c, d)$.

11. Prove that if a line integral is independent of path in some region, then the value of the line integral around a closed curve is zero.

# 22 Multiple Integration

## 22.1 INTRODUCTION

The definite integral of a function of one variable in an interval was defined in Section 9.9. In a broad sense the definition involved a process of summation together with a type of limit operation. The application of the definite integral to finding certain geometric or physical quantities depended on approximating the quantity in question for a small part of some figure. It was generally characteristic of such a "small part" that in the limit process one of its dimensions approached zero.

In the present chapter the concept of definite integral is extended to a function of two variables or a function of three variables. It again involves a process of summation and a type of limit operation. This definition, and the application of *multiple integrals* as thus defined, also involve a "small part" of some figure, but it is now characteristic of this small part that in the limit process the part approaches zero in more than one dimension. For example, in the case of a *double integral* the small part may be a rectangle in which both dimensions become smaller and smaller.

Additionally, we recall that the evaluation of a definite integral was effected by use of the Fundamental Theorem of Integral Calculus, and not by a direct application of the definition. Correspondingly, it will be seen that the evaluation of multiple integrals can be carried out by successive applications of that theorem, and thus the definition need not be used directly for this purpose.

## 22.2 DOUBLE INTEGRALS

A *smooth arc* is the graph of equations $x = f(t)$, $y = g(t)$ in an interval, with $f'$ and $g'$ continuous and not both zero for the same value of $t$. We consider a *regular closed curve* in a plane, consisting of arcs which are smooth, and such that the curve does not cross itself. We specify initially that a line $x = a$ may intersect the curve in not more than two points, except that such a line may intersect the curve in a straight line segment if $x = a$ is the smallest value of $x$ on the curve, or if $x = a$ is the largest value of $x$ on the curve. A similar specification applies to lines parallel to the $x$ axis. A *simple region* is now defined as the region interior to such a curve. The regions in Fig. 22.2.1(a) and (b) are simple, but the one in Fig. 22.2.1(c) is not.

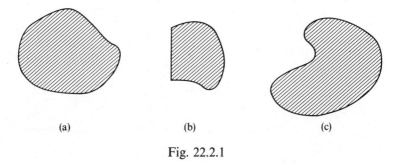

(a)  (b)  (c)

Fig. 22.2.1

On the boundary of a simple region there are a smallest value and largest value of $x$, say $a$ and $b$, respectively. We subdivide the interval $[a, b]$ by points $x_i$, $i = 0, 1, \ldots, n$, such that (Fig. 22.2.2)

$$a = x_0 < x_1 < \cdots < x_n = b; \qquad \Delta x_i = x_i - x_{i-1}, \quad i = 1, 2, \ldots, n.$$

A similar subdivision is made of the interval $[c, d]$ between the smallest and largest values of $y$:

$$c = y_0 < y_1 < \cdots < y_m = d; \qquad \Delta y_i = y_i - y_{i-1}, \quad i = 1, 2, \ldots, m.$$

At the subdivision points, lines are drawn parallel to the coordinate axes, yielding a rectangular network. We consider all rectangles formed in this manner which, except possibly for points which lie on the boundary of a rectangle, lie in the region, and number these from 1 to $N \leq mn$. Let the area of the $i$th rectangle be $\Delta A_i$. Now choose any point $(u_i, v_i)$ in this rectangle, and form the sum

$$(22.2.1) \quad \sum_{i=1}^{N} f(u_i, v_i) \, \Delta A_i = f(u_1, v_1) \, \Delta A_1 + f(u_2, v_2) \, \Delta A_2 + \cdots + f(u_N, v_N) \, \Delta A_N.$$

The idea we wish to pursue is that this sum approximates a quantity which is called the *double integral* of the function $f$ in the region $A$ and is written as

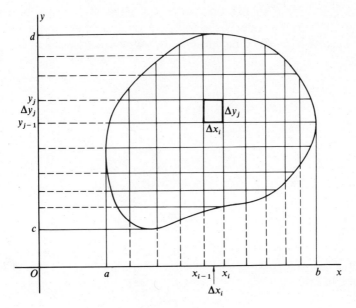

Fig. 22.2.2

$\iint_A f(x, y)\, dA$. The smaller the dimensions of the rectangles become (which implies that $N$ is increasing and that $N \to \infty$), the better approximation becomes. Implicit in this idea is the existence of a unique number which is approximated by sums of the type (22.2.1), provided only that $\Delta x_i$ and $\Delta y_i$ are sufficiently small. The following formal definition incorporates the above ideas.

**DEFINITION 22.2.1.** *Suppose that a number $I$ exists with the following property: given $\epsilon = 0$ there exists $\delta > 0$ such that*

$$\left| I - \sum_{i=1}^{N} f(u_i, v_i)\, \Delta A_i \right| < \epsilon, \quad \text{whenever} \quad \max \Delta x_i < \delta, \quad \max \Delta y_i < \delta,$$

*where $(u_i, v_i)$ is any point in the ith rectangle, and the sum extends over all rectangles which lie in the region, as above. Then we write*

$$I = \iint_A f(x, y)\, dA,$$

*and we call $I$ the* **double integral** *of $f$ in the region $A$.*

We shall be concerned primarily with the evaluation and application of double integrals. Proofs of relevant theorems are rather involved and will generally be omitted. The first main theorem follows.

**THEOREM 22.2.1.** *If $f$ is continuous in (and on the boundary of) a region $A$ as defined above, then $\iint_A f(x, y)\, dA$ exists.*

If a region is not simple, but can be assembled by joining simple regions, Definition 22.2.1 can be extended to such a region by adding the double integrals for the separate parts. Such regions are illustrated in Fig. 22.2.3.

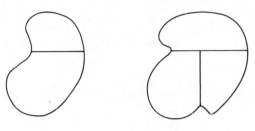

Fig. 22.2.3

In the following example we illustrate the evaluation of sums of the type appearing in Definition 22.2.1.

***Example 22.2.1.*** Approximate the double integral $\iint_R x^2 y\, dA$, where $R$ is the rectangle bounded by $x = 1$, $x = 3$, $y = 0$, and $y = 2$. Use equal subintervals for $x$ and for $y$, and let $(u_i, v_i)$ be the midpoint of the $i$th rectangle. Consider the following cases

(a) $n = m = 2$, (b) $n = m = 4$.

*Solution.* The subdivisions are shown in Fig. 22.2.4, with a specific choice of the numbering of the rectangles formed.

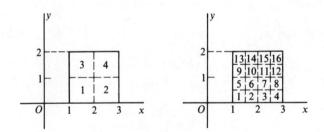

Fig. 22.2.4

In Case (a) the midpoints of the rectangles are, following the numbering shown in the figure,

$$\left(\frac{3}{2}, \frac{1}{2}\right), \left(\frac{5}{2}, \frac{1}{2}\right), \left(\frac{3}{2}, \frac{3}{2}\right), \left(\frac{5}{2}, \frac{3}{2}\right).$$

The areas of the rectangles are equal, $\Delta A_i = 1$, $i = 1, \ldots, 4$. The desired approximation is given by

$$\sum_{i=1}^{N} f(u_i, v_i) \Delta A_i = \frac{1}{8}(9 + 25 + 27 + 75) \cdot 1 = 17.$$

In Case (b) the midpoints are, in order,

$$\left(\frac{5}{4}, \frac{1}{4}\right), \left(\frac{7}{4}, \frac{1}{4}\right), \left(\frac{9}{4}, \frac{1}{4}\right), \left(\frac{11}{4}, \frac{1}{4}\right), \left(\frac{5}{4}, \frac{3}{4}\right), \left(\frac{7}{4}, \frac{3}{4}\right), \left(\frac{9}{4}, \frac{3}{4}\right), \left(\frac{11}{4}, \frac{3}{4}\right),$$

$$\left(\frac{5}{4}, \frac{5}{4}\right), \left(\frac{7}{4}, \frac{5}{4}\right), \left(\frac{9}{4}, \frac{5}{4}\right), \left(\frac{11}{4}, \frac{5}{4}\right), \left(\frac{5}{4}, \frac{7}{4}\right), \left(\frac{7}{4}, \frac{7}{4}\right), \left(\frac{9}{4}, \frac{7}{4}\right), \left(\frac{11}{4}, \frac{7}{4}\right).$$

In this case, $\Delta A_i = 1/4$, $i = 1, \ldots, 16$, and the approximation is

$$\frac{1}{64}(25 + 49 + 81 + 121 + 75 + 147 + 243 + 363 + 125 + 245 + 405 + 605$$

$$+ 175 + 343 + 567 + 847) \cdot \frac{1}{4} = \frac{69}{4} = 17\tfrac{1}{4}.$$

The values thus obtained may be compared with the true value, $17\tfrac{1}{3}$, which can be evaluated by the methods to be presented in Section 22.3. (Exercise 5 of Exercise Group 22.2.)

In general the function $f$ in Definition 22.2.1 has no special significance for the region involved except that it is defined there and, in most cases, is continuous. In each application to be made of double integrals the function $f$ will assume a specific form appropriate to that application. We direct our attention now to the evaluation of double integrals.

**EXERCISE GROUP 22.1**

In Exercises 1–4 approximate the given integral for the region and subdivisions as described in Example 22.2.1. Specifically, (a) use the subdivision in Case (a) of that example, and the midpoints of the rectangles; (b) use the subdivision in Case (b) of that example, and the midpoints of the rectangles.

1. $\iint_R (x + 3) \, dA$

2. $\iint_R y^2 \, dA$

3. $\iint_R xy \, dA$

4. $\iint_R (2x - 3y) \, dA$

5. Carry out the approximations for the double integral in Example 22.2.1, in both cases as described, choosing $(u_i, v_i)$ as the upper right vertex in each rectangle.

6. Consider the integral $\iint_A xy^2 \, dA$, where $A$ is the region $x^2 + y^2 \leq 9$. Subdivide the region $A$ by lines $x = -2, -1, 0, 1, 2$ and $y = -2, -1, 0, 1, 2$. Approximate the double integral using (a) the lower left vertex of each included rectangle, and (b) the midpoint of each included rectangle.

## 22.3 EVALUATION OF DOUBLE INTEGRALS. ITERATED INTEGRALS

Definition 22.2.1 clearly does not lend itself readily to the evaluation of double integrals. For this purpose we may equate a double integral to an *iterated integral* or *repeated integral*, whereby two integrations are performed successively by ordinary integration, first with respect to one variable, and then with respect to the other one. To form an intuitive understanding of this procedure we return to the sum (22.2.1). We must add terms of the type $f(u_i, v_i) \Delta A_i$ over all rectangles of the subdivision that lie in the region. Suppose we systematize the addition by first adding these terms over each horizontal strip of such rectangles. In this way we have a number, the sum, over each horizontal strip. We then add these numbers, one for each horizontal strip, and we have the sum over all the rectangles. As the dimensions of the rectangles approach zero, the first summation suggests an integration with respect to $x$, with $y$ constant, and with the limits for $x$ depending on $y$; the second summation suggests an integration with respect to $y$. This, in fact, is the case; a proof is not given here. A similar discussion can be given with the first integration with respect to $y$, and the second one with respect to $x$.

Fig. 22.3.1 shows a typical region, the same one in both parts of the figure. In (a) the boundary is such that the left and right parts can be represented, respectively, by functions $x = x_1(y)$ and $x = x_2(y)$, and the double integral can be represented as an *iterated integral* as

(22.3.1) $$\iint_A f(x, y)\, dA = \int_c^d dy \int_{x_1(y)}^{x_2(y)} f(x, y)\, dx = \int_c^d \int_{x_1(y)}^{x_2(y)} f(x, y)\, dx\, dy.$$

There are two forms of the iterated integral in (22.3.1); in each of them the first integration is with respect to $x$, the limits are functions of $y$ (which may be constant), and an ordinary definite integral in $y$ results after the first integration.

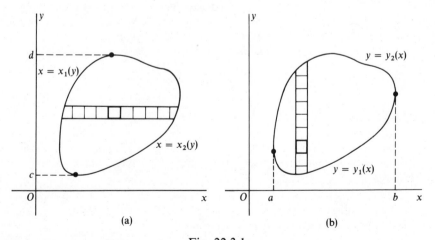

Fig. 22.3.1

In Fig. 22.3.1(b) the lower and upper parts of the boundary are given, respectively, by functions $y = y_1(x)$ and $y = y_2(x)$. In a similar manner as above, the double integral may be expressed as an iterated integral in two forms

(22.3.2) $$\iint_A f(x, y) \, dA = \int_a^b dx \int_{y_1(x)}^{y_2(x)} f(x, y) \, dy = \int_a^b \int_{y_1(x)}^{y_2(x)} f(x, y) \, dy \, dx.$$

In the second form of the iterated integral in either order of integration, the first integration is associated with the inner differential and the inner limits of integration; while the limits for the second integration are constants.

**Example 22.3.1.** Evaluate the integral $\iint_A x \, dA$ for the region bounded by the curves $x = y^2$ and $x = y$.

*Solution.* The given curves, a parabola and a straight line, determine the region shown in Fig. 22.3.2. For the evaluation by an iterated integral with the $x$ integration first, Fig. 22.3.2(a) shows a typical horizontal strip, extending from the parabola on the left to the straight line on the right. The functions of $y$ on these boundaries are, respectively, $y^2$ and $y$, which are the limits for the first integration. The smallest and largest values of $y$ in the region, 0 and 1, respectively, are the limits for the second integration, and we have

$$\iint_A x \, dA = \int_0^1 \int_{y^2}^y x \, dx \, dy = \int_0^1 \frac{1}{2} x^2 \bigg|_{y^2}^y dy = \frac{1}{2} \int_0^1 (y^2 - y^4) \, dy$$

$$= \frac{1}{2} \left( \frac{1}{3} y^3 - \frac{1}{5} y^5 \right) \bigg|_0^1 = \frac{1}{2} \left( \frac{1}{3} - \frac{1}{5} \right) = \frac{1}{15}.$$

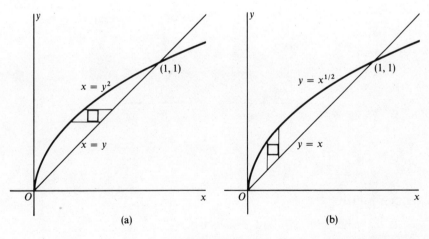

Fig. 22.3.2

If the first integration is with respect to $y$, Fig. 22.3.2(b) shows that a typical vertical strip extends from the straight line below to the parabola above, and the corresponding functions of $x$ are $x$ and $\sqrt{x}$. In this case we have

$$\iint_A x \, dA = \int_0^1 \int_x^{\sqrt{x}} x \, dy \, dx = \int_0^1 xy \Big|_x^{\sqrt{x}} dx = \int_0^1 (x^{3/2} - x^2) \, dx$$

$$= \left( \frac{2}{5} x^{5/2} - \frac{1}{3} x^3 \right) \Big|_0^1 = \frac{2}{5} - \frac{1}{3} = \frac{1}{15}.$$

***Example 22.3.2.*** Write an equivalent integral, with the order of integration reversed, for the integral

$$\int_0^2 \int_0^{2x-x^2} dy \, dx.$$

*Solution.* From the inner limits we see that the lower boundary of the region is $y = 0$ and the upper boundary is $y = 2x - x^2$, a parabola. The region of integration is shown in Fig. 22.3.3. To reverse the order of integration we think of a horizontal strip and note that both the left and right boundaries of the region are on the parabola. If we solve the equation of the parabola for $x$, we get $x = 1 \pm \sqrt{1 - y}$, with the minus sign applying to the left part and the plus sign to the right part. We then get, for the desired integral,

$$\int_0^1 \int_{1-\sqrt{1-y}}^{1+\sqrt{1-y}} dx \, dy.$$

The evaluation of either form yields the value 4/3 (Exercise 17).

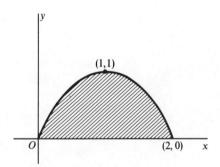

Fig. 22.3.3

**EXERCISE GROUP 22.2**

Evaluate the iterated integrals in Exercises 1–4.

**1.** $\int_0^1 \int_{x^2}^1 x \, dy \, dx$

**2.** $\int_0^1 \int_{2y}^{2\sqrt{y}} xy \, dx \, dy$

**3.** $\int_0^{\pi/2} \int_{2\sin\theta}^{4\sin\theta} r \, dr \, d\theta$

**4.** $\int_0^{\pi/3} \int_1^{2\cos\theta} r \sin\theta \, dr \, d\theta$

In Exercises 5–10 evaluate the double integral for the region $A$ bounded by the given curves.

5. $\iint_A x^2 y \, dA;\ x = 1,\ x = 3,\ y = 0,\ y = 2$

6. $\iint_A (x^2 + y^2)\, dA;\ x = -1,\ x = 2,\ y = 1,\ y = 2$

7. $\iint_A dA;\ y = x,\ y = 0,\ 2x - y = 2$

8. $\iint_A dA;\ y = x,\ y = -x,\ 2x - y = 2$

9. $\iint_A y\, dA;\ y^2 = x + 1,\ 2y = x + 1$

10. $\iint_A dA;\ x = 0,\ x = \pi,\ y = 0,\ y = 2 - \cos x$

In Exercises 11–16 rewrite the given iterated integral with the reverse order of integration, and evaluate either form.

11. $\int_1^4 \int_1^{x^2} dy\, dx$

12. $\int_0^1 \int_{\sqrt{x}}^{\frac{x+1}{2}} dy\, dx$

13. $\int_0^1 \int_{-\sqrt{1-y}}^{1-y} dx\, dy$

14. $\int_0^2 \int_1^{\sqrt{y^2+1}} dx\, dy$

15. $\int_0^2 \int_0^{\sqrt{y^2+1}} dx\, dy$

16. $\int_a^{2a} \int_0^{\sqrt{y^2-a^2}} (x - y)\, dx\, dy$

17. Evaluate the given integral in Example 22.3.2, and the one obtained with the order of integration reversed.

18. Express the sum

$$\int_0^1 dy \int_0^y f(x, y)\, dx + \int_1^2 dy \int_{2y-2}^y f(x, y)\, dx$$

as a single iterated integral in the reverse order of integration.

## 22.4 DOUBLE INTEGRALS IN POLAR COORDINATES

Many of the ideas developed in previous sections for a double integral in rectangular coordinates carry over to the concept of a double integral in polar coordinates. We consider a region in the plane bounded by a curve such that a radial line $\theta = \theta_0$ (constant) intersects the curve in not more than two points, except that such a line may intersect the curve in a straight line segment if $\theta = \theta_0$ is the smallest value of $\theta$ on the curve, or if $\theta = \theta_0$ is the largest value of $\theta$ on the curve. If $\alpha$ and $\beta$ are the smallest and largest values of $\theta$ on the boundary of the region (Fig. 22.4.1), we subdivide the interval $[\alpha, \beta]$,

$$\alpha = \theta_0 < \theta_1 < \cdots < \theta_n = \beta; \quad \Delta\theta_i = \theta_i - \theta_{i-1}, \quad i = 1, 2, \ldots, n,$$

and draw the lines $\theta = \theta_i$. If $a$ and $b$ are the smallest and largest values of $r$ in the region, we subdivide the interval $[a, b]$,

$$a = r_0 < r_1 < \cdots < r_m = b; \quad \Delta r_i = r_i - r_{i-1}, \quad i = 1, 2, \ldots, m,$$

and draw the circular arcs $r = r_i$. In this manner we form a set of *pseudo-rectangles*, and we consider all such figures which, except possibly for points which lie on the boundary of such a figure, lie in the region. These pseudo-rectangles are numbered from 1 to $N$. If $\Delta A_i$ is the area of the $i$th figure, we choose any point $(r_i^*, \theta_i^*)$ in this figure, and form the sum

$$(22.4.1) \quad \sum_{i=1}^{N} f(r_i^*, \theta_i^*) \Delta A_i = f(r_1^*, \theta_1^*) \Delta A_1 + f(r_2^*, \theta_2^*) \Delta A_2 + \cdots + f(r_N^*, \theta_N^*) \Delta A_N.$$

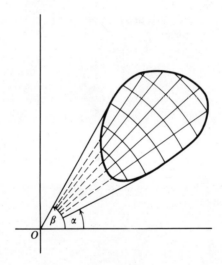

Fig. 22.4.1

**DEFINITION 22.4.1.** *Suppose a number $I$ exists with the following property: given $\epsilon > 0$, there exists $\delta > 0$ such that*

$$\left| I - \sum_{i=1}^{N} f(r_i^*, \theta_i^*) \Delta A_i \right| < \epsilon, \quad \text{whenever} \quad \max \Delta r_i < \delta, \quad \max \Delta \theta_i < \delta,$$

*where $(r_i^*, \theta_i^*)$ is any point in the $i$th pseudo-rectangle, and the sum extends over all pseudo-rectangles which lie in the region, as above. Then we write*

$$I = \iint_A f(r, \theta) \, dA,$$

*and we call $I$ the* **double integral** *of $f$ in the region $A$.*

As in the case of rectangular coordinates, the definition of double integral may be extended to a region which is formed by joining together regions of the type described.

The evaluation of a double integral in polar coordinates can be effected by an iterated integral in the form

(22.4.2) $$\iint_A f(r, \theta)\, dA = \int_\alpha^\beta \int_{r_1(\theta)}^{r_2(\theta)} f(r, \theta)\, r\, dr\, d\theta,$$

where $r$ varies from $r_1(\theta)$ to $r_2(\theta)$ for any value of $\theta$ in the region, and $\alpha$ and $\beta$ are the extreme values of $\theta$ for the region (Fig. 22.4.2).

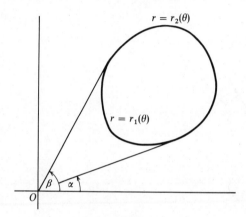

Fig. 22.4.2

To justify (22.4.2) we have, for the area of the pseudo-rectangle in Fig. 22.4.3,
$$\Delta A_i = \tfrac{1}{2} r_i^2\, \Delta \theta_i - \tfrac{1}{2} r_{i-1}^2\, \Delta \theta_i = \tfrac{1}{2}(r_i^2 - r_{i-1}^2)\, \Delta \theta_i$$
$$= \tfrac{1}{2}(r_i + r_{i-1})(r_i - r_{i-1})\, \Delta \theta_i = \tfrac{1}{2}(r_i + r_{i-1})\, \Delta r_i\, \Delta \theta_i.$$

Now $r_i + r_{i-1} = 2r_i^*$, where $r_i^*$ is a value of $r$ between $r_{i-1}$ and $r_i$. Hence $\Delta A_i = r_i^*\, \Delta r_i\, \Delta \theta_i$. We now choose a point in the above pseudo-rectangle with $r_i^*$ as thus determined. Then the sum (22.4.1) is of the same type described in Definition 22.2.1, now with variables $r$ and $\theta$, for the function $rf(r, \theta)$, and the representation (22.4.2) follows as in Section 22.3.

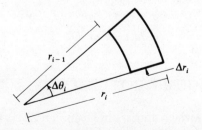

Fig. 22.4.3

***Example 22.4.1.*** Express $\iint_A f(r, \theta)\, dA$ as an iterated integral for the loop in the first quadrant formed by the lemniscate $r^2 = a^2 \sin 2\theta$.

*Solution.* The curve appears in Fig. 22.4.4. The extreme values of $\theta$ for the region are 0 and $\pi/2$. The functions of $\theta$ on the boundary are $r_1(\theta) = 0$, $r_2(\theta) = a\sqrt{\sin 2\theta}$. Hence

$$\iint_A f(r, \theta)\, dA = \int_0^{\pi/2} \int_0^{a\sqrt{\sin 2\theta}} f(r, \theta)\, r\, dr\, d\theta.$$

The order of integration used has the effect of adding the terms of (22.4.1) along sectors (Fig. 22.4.4(a)), and then adding the values for the sectors.

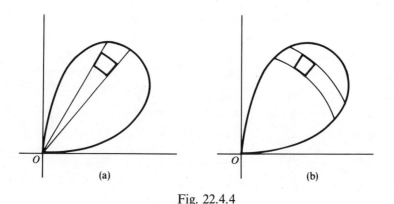

Fig. 22.4.4

The reverse order of integration is less often used in polar coordinates, but it is possible. Its effect is to add the terms of (22.4.1) in a ring first (Fig. 22.4.4(b)), and then to add the values for the rings. For the integral in Example 22.4.1 we get

$$\iint_A f(r, \theta)\, dA = \int_0^a \int_{\frac{1}{2}\sin^{-1} r^2/a^2}^{\pi/2 - \frac{1}{2}\sin^{-1} r^2/a^2} f(r, \theta)\, r\, d\theta\, dr.$$

This representation as an iterated integral suggests some of the reasons for the less frequent use of this order of integration.

It is important to note that *in the representation of a double integral in polar coordinates as an iterated integral, the element of area is*

(22.4.3) $\qquad\qquad dA = r\, dr\, d\theta.$

## 22.5 AREA BY DOUBLE INTEGRALS

If $f(x, y) = 1$, the sum which appears in Definition 22.2.1 becomes $\sum_{i=1}^{N} \Delta A_i$, and is an approximation to the area of the region. It is then reasonable to *define the area of a region* as the double integral of the function $f(x, y) = 1$ over that region, and we have

$$\text{Area} = \iint_A dA.$$

The previous discussion about the evaluation by use of an iterated integral and the determination of the limits of integration is applicable.

***Example 22.5.1.*** Find the area of the region bounded by the curves

$$y^2 - 2y + x = 0, \qquad y = 2x.$$

*Solution.* The curves and the region bounded by them are shown in Fig. 22.5.1. Clearly it is simpler to integrate with respect to $x$ first, since the left boundary is the same curve throughout (the straight line), and the right boundary is the same curve throughout (the parabola). The intersection (3/4, 3/2) is obtained by solving the two equations simultaneously. We get

$$\text{Area} = \int_0^{3/2} \int_{y/2}^{2y-y^2} dx\, dy = \int_0^{3/2} (3y/2 - y^2)\, dy = 9/16.$$

It is of interest to set up the iteration in the opposite order. We separate the region into two parts by a line $x = 3/4$, and then use a separate iterated integral for each part. We get (as the student should verify)

$$\text{Area} = \int_0^{3/4} \int_{1-\sqrt{1-x}}^{2x} dy\, dx + \int_{3/4}^{1} \int_{1-\sqrt{1-x}}^{1+\sqrt{1-x}} dy\, dx.$$

This evaluates as $19/48 + 1/6 = 9/16$. The functions of $x$ representing the lower and upper parts of the parabola were obtained by solving for $y$ the equation of the parabola.

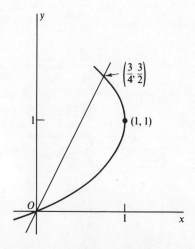

Fig. 22.5.1

In polar coordinates the area of a region of the plane is given by

$$\text{Area} = \iint_A dA = \int_\alpha^\beta \int_{r_1(\theta)}^{r_2(\theta)} r\, dr\, d\theta,$$

with the iterated integral set up as in the discussion of Section 22.4.

**Example 22.5.2.** Find the area of the loop in the first quadrant formed by the curve $r^2 = a^2 \sin 2\theta$.

*Solution.* This is the region of Example 22.4.1, to which Fig. 22.4.4 applies. Accordingly, we have

$$\text{Area} = \int_0^{\pi/2} \int_0^{a\sqrt{\sin 2\theta}} r\, dr\, d\theta = \frac{a^2}{2} \int_0^{\pi/2} \sin 2\theta\, d\theta$$

$$= \frac{a^2}{2} \left(-\frac{\cos 2\theta}{2}\right)\bigg|_0^{\pi/2} = \frac{a^2}{2}\left(\frac{1}{2} + \frac{1}{2}\right) = \frac{a^2}{2}.$$

**Example 22.5.3.** Find the area of the region inside the circle $r = 2\cos\theta$ and outside the circle $r = 1$.

*Solution.* The desired region is the shaded part of Fig. 22.5.2. We make use of the symmetry of the region in the $x$ axis, and determine the area for the upper half. The intersection of the circles in the first quadrant is at $(1, \pi/3)$. Thus, if $K$ is the total area desired,

$$\frac{K}{2} = \int_0^{\pi/3} \int_1^{2\cos\theta} r\, dr\, d\theta = \frac{1}{2}\int_0^{\pi/3} r^2 \bigg|_1^{2\cos\theta} d\theta$$

$$= \frac{1}{2}\int_0^{\pi/3}(2\cos 2\theta + 1)\, d\theta = \frac{1}{2}(\sin 2\theta + \theta)\bigg|_0^{\pi/3} = \frac{3\sqrt{3} + 2\pi}{12}.$$

The desired area is $K = (3\sqrt{3} + 2\pi)/6$.

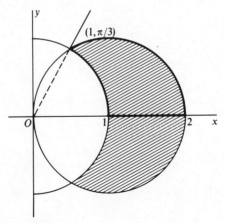

Fig. 22.5.2

## EXERCISE GROUP 22.3

In Exercises 1–22 use a double integral to find the area of the region described.

1. The region bounded by $y = x$, $y = 2x$, $x = 4$
2. The region bounded by $y = 4 - x^2$, $y = 3x$
3. The region bounded by $y = x^2 + 1$, $y = 0$, $x = 0$, $x = 2$
4. The region bounded by $y^2 = 4x + 4$, $y^2 + 2x = 6$
5. The region bounded by $y = x + 1$, $x = 2y^2 + y - 3$
6. The region bounded by $y = 2x^2 - 2x - 1$, $y = x^2 - 2x + 3$
7. The region bounded by $x^2 = 2 - y$, $y = x^2$
8. The region bounded by $y^2 = x$, $x - 2y + 1 = 0$, $x = 0$
9. The region bounded by $x^2 = 2 - y$, $x^2 - 2x = 2y + 1$
10. The interior of the ellipse $b^2 x^2 + a^2 y^2 = a^2 b^2$
11. The region bounded by $y = \ln x$, $y = e^x$, $y = 1$, $y = 2$
* 12. The region bounded by $y = 0$, $x = 0$, $y^2 - x^2 = 1$, $x + y = 3$, $x^2 - y^2 = 1$
* 13. The loop enclosed by the curve $y^2 = x^3 - x^4$
* 14. The loop enclosed by the curve $x^{2/3} + y^2 = 1$
* 15. The region in the first quadrant bounded by $xy^2 = 1$, $y^3 = x^2$, $x = 0$
16. The smaller region bounded by $r = 2a \cos \theta$, $\theta = \pi/6$
17. One loop enclosed by the curve $r = a \cos 2\theta$
18. The region bounded by $r = 5$, $r \sin \theta = 5$, $\theta = \pi/6$ which lies outside the circle
19. The region interior to the cardioid $r = a(1 + \cos \theta)$
20. The region interior to the limaçon $r = 2 - \sin \theta$
21. The region in the first quadrant inside the circle $r = a(\sin \theta + \cos \theta)$ and outside the cardioid $r = a(1 - \sin \theta)$
22. The smaller loop formed by the limaçon $r = 1 - 2 \sin \theta$
23. Express the area of the upper half of the region enclosed by the cardioid $r = 1 + \cos \theta$ as an iterated integral using the order of integration $d\theta\, dr$. (The evaluation is challenging.)

## 22.6 VOLUME BY DOUBLE INTEGRALS

A second geometric application of the double integral concerns the volume of a solid, or 3-dimensional region. We consider a solid with lower boundary in the $xy$ plane, with lateral boundary a cylindrical surface whose elements are parallel to the $z$ axis, and with upper boundary in a surface with equation $z = f(x, y)$ (Fig. 22.6.1). The lower boundary is assumed here to be a simple region $A$, as described in Section 22.2. We subdivide the region $A$ and project each rectangle thus formed parallel to the $z$ axis so as to intersect the surface. In this way we

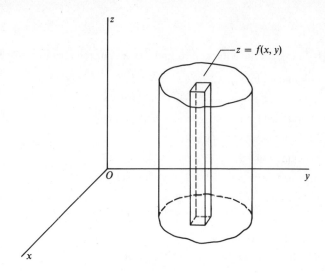

Fig. 22.6.1

subdivide the solid into a set of columns, the sum of whose volumes approximates the volume of the solid. If $(u_i, v_i)$ is a point in the $i$th rectangle, with area $\Delta A_i$, the product $f(u_i, v_i) \Delta A_i$ approximates the volume of the column formed on that rectangle, and the sum

$$\sum_{i=1}^{N} f(u_i, v_i) \Delta A_i$$

approximates the volume of the solid. As the dimensions of the rectangles approach zero, the approximation gets better, and it is reasonable to *define the volume of the solid* as

(22.6.1) $$V = \iint_A f(x, y) \, dA,$$

where *the integration is performed over the region A in the xy plane*. The region $A$ may be viewed as the projection of the upper boundary of the solid, or of the solid itself, on the $xy$ plane.

**Example 22.6.1.** Find the volume of the solid bounded by the surfaces

$$x^2 + y^2 = 9, \quad 2x + 3y + 4z = 12, \quad z = 0.$$

*Solution.* The solid is bounded above and below by planes, and laterally by a cylinder (Fig. 22.6.2). From the equation of the upper surface we have

$$z = f(x, y) = \tfrac{1}{4}(12 - 2x - 3y).$$

The region in the $xy$ plane, the region of integration, is circular. We then have

$$V = \int_{-3}^{3} \int_{-\sqrt{9-x^2}}^{\sqrt{9-x^2}} \frac{1}{4}(12 - 2x - 3y)\, dy\, dx$$

$$= \frac{1}{4}\int_{-3}^{3} \left(12y - 2xy - \frac{3}{2}y^2\right)\bigg|_{-\sqrt{9-x^2}}^{\sqrt{9-x^2}} dx$$

$$= \frac{1}{4}\int_{-3}^{3} (24\sqrt{9-x^2} - 4x\sqrt{9-x^2})\, dx$$

$$= 27\pi$$

The integral of the second term, between $-3$ and $3$, is zero. The remaining integration can be effected by a trigonometric substitution. In this manner the result was obtained, and $V = 27\pi$.

**Example 22.6.2.** Find the volume bounded by the surfaces

$$x^2 + y^2 = 1, \quad z = y^2, \quad z = 0, \quad x = 0.$$

*Solution.* The bounding surfaces are two coordinate planes, a right circular cylinder, and a parabolic cylinder (Fig. 22.6.3). The upper boundary is the surface $z = y^2$, and the projection of the solid on the $xy$ plane is bounded by $x = 0$, $y = 0$, and the circle $x^2 + y^2 = 1$. We obtain, for the volume $V$,

$$V = \int_0^1 \int_0^{\sqrt{1-x^2}} y^2 \, dy \, dx = \frac{\pi}{16}.$$

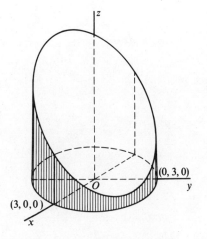

Fig. 22.6.2

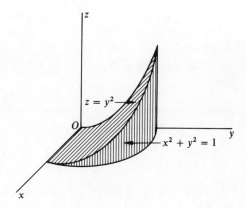

Fig. 22.6.3

The foregoing discussion is easily adapted to a similar solid which can be projected onto one of the other coordinate planes. To illustrate the relevant considerations, we reexamine the problem of Example 22.6.2. We may view the solid (Fig. 22.6.3) as bounded by a cylinder with elements parallel to the $x$ axis. The boundary in front is in the surface $x^2 + y^2 = 1$, which projects into the region in the $yz$ plane bounded by the curves $z = y^2$, $z = 0$, and $y = 1$. By analogy we have

$$V = \int_0^1 \int_0^{y^2} \sqrt{1 - y^2}\, dz\, dy = \frac{\pi}{16},$$

where the function $\sqrt{1 - y^2}$ is obtained from the equation $x^2 + y^2 = 1$.

The representation (22.6.1) for volume may also be used with the *polar representation* of area in the plane. The equation of the surface $z = f(x, y)$ is then converted to polar coordinates by the relations

$$x = r \cos \theta, \qquad y = r \sin \theta,$$

and we have

$$z = f(x, y) = f(r \cos \theta, r \sin \theta) = F(r, \theta).$$

In effect we are using *cylindrical coordinates* (see Section 20.11), and are representing the equation of the upper bounding surface in this coordinate system. For the element of area we use $dA = r\, dr\, d\theta$, and an iterated integral is set up for the region in the plane which is the projection of the solid.

*Example 22.6.3.* Let us find the volume of Example 22.6.1 by the use of polar coordinates. From the equation of the upper plane we find

$$z = \tfrac{1}{4}(12 - 2x - 3y) = \tfrac{1}{4}(12 - 2r \cos \theta - 3r \sin \theta) = F(r, \theta).$$

The volume is

$$V = \frac{1}{4} \int_0^{2\pi} \int_0^3 (12 - 2r \cos \theta - 3r \sin \theta) r\, dr\, d\theta$$

$$= \frac{1}{4} \int_0^{2\pi} \left( 6r^2 - \frac{2}{3} r^3 \cos \theta - r^3 \sin \theta \right) \bigg|_0^3 d\theta$$

$$= \frac{1}{4} \int_0^{2\pi} (54 - 18 \cos \theta - 27 \sin \theta)\, d\theta$$

$$= 27\pi.$$

### EXERCISE GROUP 22.4

1. Find the volume of the tetrahedron formed by the plane $2x + 3y + 4z = 12$ and the coordinates planes.

2. Find the volume of the solid in the first octant bounded by the surfaces $z = 2x$, $x^2 = y$, $y = 1$.

3. Find the volume of the solid bounded by the surfaces $z = 2 - x^2 - y^2$, $x^2 + y^2 - 2y = 0$, $z = 0$.

4. Find the volume of the solid bounded by the surfaces $x^2 + y^2 = 9$, $y = \sqrt{z}$, $z = 0$.

5. Find the volume of the solid bounded by the surfaces $z = 2x + 6$, $x^2 + y^2 = 9$, $z = 0$.

6. Find the volume of that portion of the solid described in Exercise 5 for which $x \geq 0$.

7. Find the volume of the solid bounded by the surfaces $x^2 + z^2 = a^2$, $x^2 + y^2 = a^2$.

8. Find the volume of the solid which lies below the plane $2x - 3y + 3z = 5$ and above the square bounded by $x = \pm 1$, $y = \pm 1$.

9. Find the volume of the solid which lies below the plane $2x - 3y + 3z = 5$ and above the square bounded by $|x| + |y| = 1$.

10. Find the volume of the solid bounded by the planes $x - y - 2z = 1$, $y + z = 3$, and the coordinate planes.

11. Find the volume of the solid which lies below the plane $2x - 3y + 3z = 5$ and above the circle $x^2 + y^2 = 1$. [*Hint*: Use polar coordinates in the $xy$ plane.]

12. Find the volume of the solid bounded by the surfaces $z = x^2 + y^2 + 1$, $x^2 + y^2 = 4$, $z = 0$. [*Hint*: Use polar coordinates in the $xy$ plane.]

13. Find the volume of the solid bounded by the surfaces $y = 4 - x^2 - z^2$, $x^2 + z^2 = 1$, $y = 0$. [*Hint*: Use polar coordinates in the $xz$ plane.]

## 22.7  PHYSICAL APPLICATIONS OF DOUBLE INTEGRALS

Moment with respect to an axis, and moment of inertia with respect to an axis, have been defined for a particle and for a system of particles. They have also been obtained for certain plane regions. These applications are particularly easy to set up with the use of double integrals. We think of a region in the plane as subdivided into small regions of area $\Delta A_i$ in the fashion described previously for a double integral; this subdivision may be in rectangular or polar coordinates. Let $\rho(x, y)$ denote a density function (in units of mass per unit area). If $(u_i, v_i)$ is a point in the $i$th subregion, $\rho(u_i, v_i) \Delta A_i$ approximates the mass of this part, and we may write, where $m$ denotes mass,

$$\Delta m_i = \rho(u_i, v_i) \Delta A_i.$$

Summing over all included subregions, we have an approximation to the total mass, and the limit of such a sum gives the mass. Thus we find

(22.7.1) $$m = \iint dm = \iint \rho(x, y) \, dA,$$

and we also write

$$dm = \rho(x, y) \, dA.$$

Formula (22.7.1) expresses the mass of a plane region as a double integral. The integration is performed by an iterated integral, with $dA$ expressed in either of the forms

$$dA = dx \, dy \quad \text{or} \quad dA = r \, dr \, d\theta.$$

In the latter case $\rho(x, y)$ is also expressed in polar coordinates. The limits for the iterated integral are determined entirely from the region, as for any double integral.

**Example 22.7.1.** Find the mass of a semi-circular plate if the density is proportional to the square of the distance from the straight boundary.

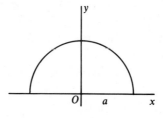

Fig. 22.7.1

*Solution.* If the plate is placed as in Fig. 22.7.1, the boundary consists of the semi-circle $x^2 + y^2 = a^2$, $y \geq 0$, and the line $y = 0$. The density is $\rho(x, y) = y^2$. Hence

$$m = \iint y^2 \, dA = \int_{-a}^{a} \int_{0}^{\sqrt{a^2-x^2}} y^2 \, dy \, dx = \pi a^4 / 8.$$

In polar coordinates the density is $\rho(x, y) = r^2 \sin^2 \theta$, and

$$m = \int_{0}^{\pi} \int_{0}^{a} (r^2 \sin^2 \theta) r \, dr \, d\theta.$$

If the density function in a plane region is constant, $\rho(x, y) = \rho$, then (22.7.1) gives $m = \rho \iint dA = \rho \cdot$ (area of $A$), as expected. In particular, if $\rho = 1$, then mass and area are numerically equal (but with different units).

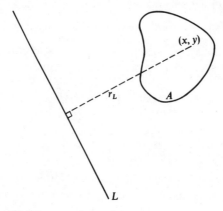

Fig. 22.7.2

## 22.7 PHYSICAL APPLICATIONS OF DOUBLE INTEGRALS

Let $r_L(x, y)$ be a directed distance from a point in a plane region $A$ to a line $L$ in the plane of $A$ (Fig. 22.7.2). If $(x, y)$ is a point in an element of area $dA$, the *moment of the element with respect to $L$* is approximately

$$dM_L = r_L(x, y)\, dm,$$

and the *moment* (also called *first moment*) of the region with respect to $L$ is

(22.7.2) $$M_L = \iint_A r_L(x, y)\, dm = \iint_A r_L(x, y)\rho(x, y)\, dA.$$

In particular, the moments with respect to the coordinate axes are

(22.7.3) $$M_x = \iint y\, dm, \qquad M_y = \iint x\, dm.$$

As in earlier discussions, the *center of mass*, or *center of gravity*, is that point such that the moment about any axis through it is zero. It turns out, if $(\bar{x}, \bar{y})$ denotes the center of mass, that

(22.7.4) $$\bar{x} = \frac{M_y}{m}, \qquad \bar{y} = \frac{M_x}{m}.$$

With constant density, $\bar{x}$ and $\bar{y}$ are independent of the density $\rho$, and we may set $\rho = 1$. In the latter case we also refer to the center of gravity as the *centroid* of the region.

**Example 22.7.2.** Find the center of gravity of the region bounded by the curves $x - y + 2 = 0$ and $y = x^2 + 2x$.

*Solution.* The region is shown in Fig. 22.7.3. We take $\rho = 1$. Then

$$m = A = \int_{-2}^{1} \int_{x^2+2x}^{x+2} dy\, dx = 9/2,$$

$$M_y = \int_{-2}^{1} \int_{x^2+2x}^{x+2} x\, dy\, dx = -9/4, \qquad M_x = \int_{-2}^{1} \int_{x^2+2x}^{x+2} y\, dy\, dx = 27/10.$$

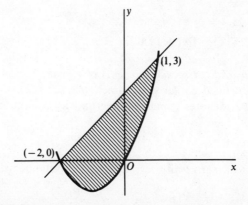

Fig. 22.7.3

We then obtain

$$\bar{x} = \frac{M_y}{m} = -1/2, \qquad \bar{y} = \frac{M_x}{m} = 3/5.$$

*Comments.* 1. If a region possesses an axis of symmetry, and the density is constant, the center of gravity lies on this axis.

2. The center of gravity of a region need not lie within the region. (See Exercise 4).

3. The much earlier comments about finding the center of gravity of a composite body still apply (Section 13.6).

### EXERCISE GROUP 22.5

1. Find the centroid of the region bounded by $y^2 = x^3$, $x = 0$, $y = 1$.
2. Find the centroid of the region bounded by $y^2 = 12x$, $y = 2x$.
3. Find the centroid of the region bounded by $y^2 = 6x$, $3y = 4x^2$.
4. Find the centroid of the region bounded by $y = x^2$, $4y = 4 + 3x^2$.
5. Find the centroid of the region bounded by $x = 4y - y^2$, $x = y^2$.
6. Find the centroid of the region bounded by $y = \sin x$, $y = 0$, for $0 \leq x \leq \pi$.
7. Find $M_x$ and $M_y$ for the region bounded by $x^2 = 2 - y$, $x^2 - 2x = 2y + 1$.
8. A plate is bounded by $x^2 = 2y$, $y = 2$, with density proportional to the distance from the straight line. Find the center of gravity.
9. A rectangular plate is bounded by $x = 0$, $x = a$, $y = 0$, $y = b$. If the density is proportional to the square of the distance from the origin, find the center of gravity.
10. Find the centroid of the region enclosed by the cardioid $r = a(1 + \sin \theta)$.
11. Find the centroid of the region interior to the curve $r = a \cos 2\theta$ for $-\pi/4 \leq \theta \leq \pi/4$.

## 22.8 MOMENT OF INERTIA

The *moment of inertia* (or *second moment*) of a plane region with respect to a line in its plane is defined similarly to moment (see (22.7.2)), but with the distance squared. Thus, if $I_L$ denotes moment of inertia with respect to a line $L$, we have

$$I_L = \iint r_L^2(x, y)\, dm,$$

and the moments of inertia with respect to the coordinate axes are

$$I_x = \iint y^2\, dm, \qquad I_y = \iint x^2\, dm.$$

There is also occasion to refer to the *polar moment of inertia* of a region, that is, with respect to a line through the origin and perpendicular to the plane. For this moment of inertia we have $r_L^2 = x^2 + y^2$, and hence, if $I_0$ denotes the polar moment,

$$I_0 = \iint (x^2 + y^2)\, dm = I_x + I_y.$$

*Moment of inertia is additive*, in the sense that the moment of inertia of a body with respect to a line is equal to the sum of such moments for the various parts making up the complete body.

*Example 22.8.1.* Find the moment of inertia of a uniform circular plate of radius $a$, about a diameter.

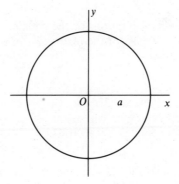

Fig. 22.8.1.

*Solution.* For the circular plate shown in Fig. 22.8.1, with constant density $\rho$, and taking the $x$ axis as the diameter, we may find $I_x$ as four times the moment of inertia of the sector in the first quadrant. We have

$$I_x/4 = \rho \iint y^2\, dA.$$

Polar coordinates may be conveniently used for the evaluation, and we have

$$I_x/4 = \rho \int_0^{\pi/2} \int_0^a (r^2 \sin^2 \theta) r\, dr\, d\theta = \rho \pi a^4/16.$$

Hence $I_x = \pi a^4 \rho/4$, which is the desired moment of inertia.

When a particle rotates about an axis with angular acceleration $\alpha$, the tangential component of the linear acceleration is $r\alpha$, the tangential component of the force acting on the particle is $mr\alpha$, and the moment of the total force about the axis is $mr^2\alpha$ (the radial component of force has zero moment). If a system of rigidly connected particles with mass $m_i$ and distance $r_i$ for the $i$th particle rotates about an axis, the total moment of force is

$$\sum m_i r_i^2 \alpha = \alpha \sum m_i r_i^2 = I\alpha,$$

where $I$ is the moment of inertia of the system about the axis. Finally, for a rotation of a continuous body about an axis with angular acceleration $\alpha$, the moment of force can be defined to be $I\alpha$, where $I$ is the moment of inertia of the body with respect to that axis. Thus for a rotation the moment of inertia has a role similar to that of mass for linear motion, and is a measure of the torque required to produce a given angular acceleration.

If the mass of a body rotating about an axis were to be concentrated at a point, how far should it be from the axis to produce the same moment of inertia? Let $k$ be this distance, called the *radius of gyration*. Then

$$k^2 m = I \quad \text{and} \quad k = \sqrt{I/m};$$

this value for $k$ expresses the radius of gyration in terms of $I$ and $m$.

Let us find the radius of gyration for the circular plate of Example 22.8.1, about a diameter. Since the mass of the plate is $\pi a^2 \rho$, we have, for the radius of gyration $k$ about a diameter,

$$k^2 = \frac{I}{m} = \frac{\pi a^2 \rho / 4}{\pi a^2 \rho} = \frac{a^2}{4}, \quad k = \frac{a}{2}.$$

**Example 22.8.2.** Find the moment of inertia of a uniform circular plate about a tangent to the boundary.

*Solution.* We refer again to Fig. 22.8.1, and let the tangent line be $y = a$. Then for the distance of any point in the plate from the line $y = a$ we have $r_{y=a}^2 = (a - y)^2$, and hence

$$I_{y=a} = \rho \iint (a - y)^2 \, dA.$$

This becomes, in polar coordinates,

$$I_{y=a} = \rho \int_0^{2\pi} \int_0^a (a - r\sin\theta)^2 r \, dr \, d\theta$$

$$= \rho \int_0^{2\pi} \int_0^a (a^2 r - 2ar^2 \sin\theta + r^3 \sin^2\theta) \, dr \, d\theta$$

$$= \frac{\rho a^4}{12} \int_0^{2\pi} (6 - 8\sin\theta + 3\sin^2\theta) \, d\theta$$

$$= \frac{5\pi a^4 \rho}{4}.$$

Note that the symmetry of the figure may not be used in the same way as in Example 22.8.1, since the moment of inertia about the line $y = a$ is not the same for all quadrants. However, the moment of inertia of the left and right halves are equal; hence the integration with respect to $\theta$ could have been done between $-\pi/2$ and $\pi/2$, and the result doubled.

**TRANSFER OF AXIS THEOREM.** *Let $L$ be any line and let $L_G$ be a parallel line through the center of gravity of a body. If $s$ is the distance between $L$ and $L_G$, then*

$$I_L = I_G + s^2 m.$$

*Proof.* If $r_L$ and $r_G$ are the distances from $L$ and $L_G$, respectively, of a point in the body (Fig. 22.8.2), then $r_L = s \pm r_G$, and

$$I_L = \iint_A r_L^2 \, dm = \iint_A (s \pm r_G)^2 \, dm$$

$$= s^2 \iint_A dm \pm 2s \iint_A r_G \, dm + \iint_A r_G^2 \, dm.$$

The first integral in the last line is the mass $m$, and the last integral is $I_G$. The middle integral is zero, since the first moment of a body about any line through its center of gravity is zero. The result follows.

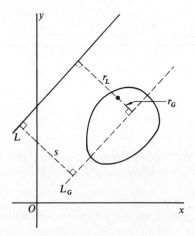

Fig. 22.8.2

In Example 22.8.1 the moment of inertia of a circular plate about a diameter was found to be $\pi a^4 \rho/4$. In Example 22.8.2 the moment of inertia about a tangent line to the boundary was obtained as $5\pi a^4 \rho/4$. Either can be found from the other, since the mass of the plate is $\pi a^2 \rho$, the center of gravity is the center of the circle, and $s = a$. Thus, if $I_T$ is the moment of inertia about a tangent line,

$$I_T = I_G + s^2 m = \frac{\pi a^4 \rho}{4} + a^2 \cdot \pi a^2 \rho = \frac{5\pi a^4 \rho}{4}.$$

The above theorem, and essentially the same type of proof, is valid for any body, even 3-dimensional, and for any line $L$ and a parallel line $L_G$ through the center of gravity. It is a consequence of the theorem that *of all the lines of a family of parallel lines, the moment of inertia is least about that line of the family which passes through the center of gravity.*

**EXERCISE GROUP 22.6**

1. Find $I_x$ and $I_y$ for the region bounded by $y^2 = x^3$, $x = 0$, $y = 1$.
2. Find the moment of inertia of a rectangular plate of sides $a$ and $b$ about (a) a line through the center parallel to a side of length $a$, (b) about a side of length $a$.
3. Find the moment of inertia of a rectangular plate of sides $a$ and $b$ (a) about a line through the center of the rectangle and perpendicular to its plane; (b) about a line perpendicular to the plane of the rectangle at a vertex.
4. Find the polar moment of inertia for the region bounded by $y^2 = 2x$, $y = 4x^2$.
5. Find the moment of inertia of a circular plate of radius $a$ about a diameter if the density varies as the distance from the center.
6. Find $I_x$, $I_y$, and $I_0$ for the region bounded by the ellipse $x^2/a^2 + y^2/b^2 = 1$.
* 7. Find the polar moment of inertia of the region in the first quadrant bounded by $y = 0$, $x^2 - y^2 = 1$, $xy = 1$, $x = y$. (*Hint:* Use polar coordinates.)
8. Find the moment of inertia about the line $x = 1$ of the region bounded by $y = 2x^2$, $y = 2x$.
9. Find the polar moment of inertia of the region interior to one loop of the curve $r^2 = a^2 \sin 2\theta$.
10. Find $I_x$ and $I_0$ for the region interior to the cardioid $r = a(1 - \cos \theta)$.
11. Find the moment of inertia about the polar axis of the region bounded by one loop of the curve $r = \sin 2\theta$.
12. Let $L_1$ and $L_2$ be two perpendicular lines in a plane, and let $L$ be a line perpendicular to the plane at their point of intersection. If $R$ is a region in the plane, prove that

$$I_L(R) = I_{L_1}(R) + I_{L_2}(R).$$

13. Let $L_1$ and $L_2$ be two parallel lines in a plane, at distances $r_1$ and $r_2$, respectively, from the center of gravity of a region. If $m$ is the mass of the region, prove that

$$I_{L_1} - I_{L_2} = m(r_1^2 - r_2^2).$$

## 22.9 TRIPLE OR VOLUME INTEGRAL

Multiple integrals were introduced in Section 22.2 with the double integral. Now a further generalization leads to the concept of a *triple* or *volume integral*. The region $V$ over which the integration is performed is 3-dimensional, a solid lying within a closed surface. We assume initially that a straight line, parallel to a coordinate axis and passing through an interior point, intersects the bounding surface in at most two points. Planes are drawn parallel to the coordinate planes, forming a set of rectangular parallelepipeds, one of which is shown in Fig. 22.9.1. We consider those parallelepipeds which have no points outside the region or its boundary, and number them from 1 to $N$. If $f(x, y, z)$ is a function defined in some domain containing $V$ and its boundary, we choose a point $(u_k, v_k, w_k)$ in the $k$th parallelepiped, with volume $\Delta V_k$, and form the sum

$$(22.9.1) \qquad \sum_{k=1}^{N} f(u_k, v_k, w_k) \, \Delta V_k.$$

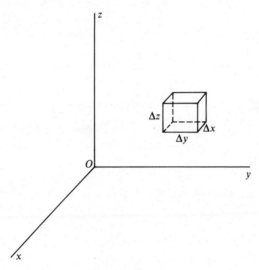

Fig. 22.9.1

As in the case of a double integral, this sum approximates a number, called the *triple* or *volume integral* of $f$ in $V$, written as $\iiint_V f(x, y, z) \, dV$. A formal definition of triple integral parallels Definition 22.2.1 and thus is not stated here. Again the underlying idea is that the smaller the dimensions of the parallelepipeds, the more closely does the sum (22.9.1) approximate the triple integral of $f$. The following theorem is stated without proof.

**THEOREM 22.9.1.** *If $f$ is continuous in and on the boundary of a region $V$ as defined above, then $\iiint_V f(x, y, z) \, dV$ exists.*

For purposes of evaluation a triple integral is usually expressed as an *iterated integral* with a suitable order of integration (six are generally possible), and suitable limits for each integration. In general these observations apply:

1. The limits for the first integration are obtained from two bounding surfaces, and are functions of the remaining variables. They depend only on the region $V$.

2. The remaining limits are obtained from the region in the plane of the remaining variables which is the projection of the solid on that plane. These limits are determined entirely from this plane region.

A typical iteration may appear as

$$\iiint_V f(x, y, z)\, dV = \int_a^b \int_{f(x)}^{g(x)} \int_{\phi(x,y)}^{\psi(x,y)} f(x, y, z)\, dz\, dy\, dx,$$

where $z = \phi(x, y)$ is obtained from the equation of the lower bounding surface, and $z = \psi(x, y)$ is obtained from the equation of the upper bounding surface; $y = f(x)$ and $y = g(x)$ are obtained from the equations of the left and right boundaries of the projection of $V$ on the $xy$ plane; $x = a$ and $x = b$ are the extreme values of $x$ in the $xy$ projection.

One of the parallelepipeds in the definition of triple integral is called an *element of volume*, and we have for its volume

$$dV = dx\, dy\, dz \text{ (in any order)}.$$

If $\rho(x, y, z)$ is the density of the solid at the point $(x, y, z)$, the *element of mass*, that is, the mass of an element of volume, is

$$dm = \rho(x, y, z)\, dV,$$

and the *mass* of the body is the triple integral

$$m = \iiint_V dm = \iiint_V \rho(x, y, z)\, dV.$$

If $\rho(x, y, z) = 1$, the mass is numerically equal to the volume of the solid, and

$$V = \iiint_V dV = \iiint_V dx\, dy\, dz.$$

The *moment* of a body with respect to a plane $K$ becomes

$$M_K = \iiint_V r_K\, dm,$$

where $r_K$ is a directed distance from the plane to a point of the body. In particular, the moments with respect to the coordinate planes are

$$M_{yz} = \iiint_V x\, dm, \qquad M_{xz} = \iiint_V y\, dm, \qquad M_{xy} = \iiint_V z\, dm.$$

These moments play a role in finding the center of gravity of a solid similar to that of $M_x$ and $M_y$ in finding the center of gravity of a plate (Section 22.7). Thus if $(\bar{x}, \bar{y}, \bar{z})$ is the center of gravity of a solid, then

$$\bar{x} = \frac{M_{yz}}{m}, \qquad \bar{y} = \frac{M_{xz}}{m}, \qquad \bar{z} = \frac{M_{xy}}{m}.$$

It may be proved that *the moment of a body about any plane through the center of* gravity of a solid is zero. If $\rho = 1$, we may refer to the center of gravity or centroid of the geometric figure.

The *moment of inertia* or *second moment* of a body with respect to a line $L$ is

$$I_L = \iiint_V r_L^2 \, dm,$$

where $r_L$ is the distance from the line to a point of the body. For example, the moment of inertia of a body with respect to the $x$ axis is, since $r_L^2 = y^2 + z^2$,

$$I_x = \iiint_V (y^2 + z^2) \, dm = \iiint_V (y^2 + z^2) \rho(x, y, z) \, dV.$$

**Example 22.9.1.** Find the moment of inertia of a solid sphere about a diameter.

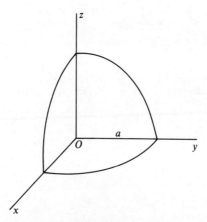

Fig. 22.9.2

*Solution.* Let us take the surface of the sphere as $x^2 + y^2 + z^2 = a^2$. In Fig. 22.9.2 one-eighth of the solid is shown. If we choose the $z$ axis as the diameter, then by symmetry the moment of inertia in the first octant is one-eighth of the total, and

$$I_z/8 = \iiint_V (x^2 + y^2) \, dV,$$

where the integration is performed over the first octant. Indicating the first integration with respect to $z$, and using $dA$ for the element of area in the $xy$ plane, we have

$$I_z/8 = \iint_A \int_0^{\sqrt{a^2 - x^2 - y^2}} (x^2 + y^2) \, dz \, dA = \iint_A (x^2 + y^2)\sqrt{a^2 - x^2 - y^2} \, dA.$$

We now convert to polar coordinates in the $xy$ plane,

$$x = r \cos \theta, \qquad y = r \sin \theta, \qquad dA = r\, dr\, d\theta,$$

and the above becomes

$$I_z/8 = \int_0^{\pi/2} \int_0^a r^2 \sqrt{a^2 - r^2}\, r\, dr\, d\theta.$$

The integration with respect to $\theta$ may actually be performed first—it merely introduces a factor $\pi/2$—and we get

$$I_z/8 = \frac{\pi}{2} \int_0^a r^3 \sqrt{a^2 - r^2}\, dr.$$

The final integration may be done by a change of variable,

$$r = a \sin u, \qquad dr = a \cos u\, du,$$

to give

$$I_z/8 = \frac{\pi a^5}{2} \int_0^{\pi/2} \sin^3 u \cos^2 u\, du = \frac{\pi a^5}{15},$$

and we obtain, finally, $I_z = 8\pi a^5/15$.

***Example 22.9.2.*** Find the volume of the solid in the first octant bounded by the cylinder $z = 2 - x^2$, the plane $x + y = 1$, and the coordinate planes.

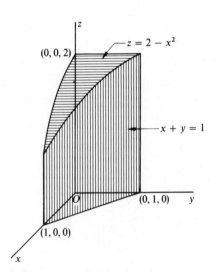

Fig. 22.9.3

*Solution.* The solid is shown in Fig. 22.9.3. The triple integral is easily set up with the $z$ integration first. The remaining limits are obtained from the triangular region in the $xy$ plane. We get

$$V = \int_0^1 \int_0^{1-x} \int_0^{2-x^2} dz\, dy\, dx = \int_0^1 \int_0^{1-x} (2-x^2)\, dy\, dx$$

$$= \int_0^1 (2 - 2x - x^2 + x^3)\, dx = 11/12.$$

*Note:* The effect of the order of integration may be emphasized by a comparison with Exercise 8.

**Example 22.9.3.** Set up an iterated integral for the volume integral of a function $f(x, y, z)$, over the solid in the first octant bounded by the surfaces

$$x^2 + z^2 = 1, \quad 2x + 2y + z = 2, \quad x = 0, \quad z = 0, \quad y = 3.$$

*Solution.* The solid is shown in Fig. 22.9.4. An inspection of the figure indicates that the best choice is to integrate first with respect to $y$. The subsequent order of integration is immaterial. The left boundary is in the plane $2x + 2y + z = 2$, and the right boundary is in the plane $y = 3$. We obtain

$$\int_0^1 \int_0^{\sqrt{1-x^2}} \int_{(2-2x-z)/2}^3 f(x, y, z)\, dy\, dz\, dx.$$

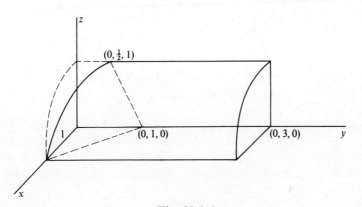

Fig. 22.9.4

## EXERCISE GROUP 22.7

1. Find the volume of the solid in the first octant bounded by $x + 2y + 3z = 6$ and the coordinate planes.
2. Find the volume of the solid bounded by $x^2 = z$, $x^2 + y^2 = a^2$, $z = 0$.
3. Find the volume of the solid bounded by the surfaces $x^2 + y^2 + z^2 = 8$, $2z = x^2 + y^2$.

4. Find the volume of the solid in the first octant bounded by the surfaces $x^2 + z^2 = 1$, $x + y = 1$, $x = 0$, $z = 0$, $y = 3$.

5. Find the volume of the solid in the first octant bounded by the surfaces $z^2 - y^2 = 1$, $x^2 - y^2 = 1$, $y = 2$.

6. Find the volume of the solid bounded by $x^2 + y^2 = 4z$, $z = y$.

7. Express the volume of the solid in the first octant bounded by the surfaces $x^2 + y^2 = 9$, $y = \sqrt{z}$, $z = 0$ as a triple integral using the order $dy\,dz\,dx$.

8. Express the volume described in Example 22.9.2 as a triple integral using the order $dx\,dz\,dy$.

9. (a) Find the volume of the solid bounded by the surfaces $y^2 + z^2 = 9$, $x = 1$, $x + z = 5$.
   ★ (b) Express the volume described in (a) as a triple integral in the order $dz\,dx\,dy$.

10. (a) Find the volume of the solid in the first octant bounded by the surfaces $z = x + y + 2$, $x^2 + y = 1$.
    (b) Express the volume described in (a) as a triple integral in the order $dx\,dz\,dy$.

11. (a) Find the volume of the solid in the first octant bounded by the surfaces $x^2 + y^2 = 36$, $x^2 + 4z = 36$.
    (b) Find $\bar{y}$ for the solid in (a).

12. Find the moment of inertia of a solid right circular cylinder of radius of base $a$ and altitude $b$ about a diameter of the base.

13. Use the result of Example 22.9.1 to find the moment of inertia about a diameter of a spherical shell of outer radius $a$ and inner radius $b$.

★ 14. Find the volume of the solid bounded by the surfaces $x^2 + y^2 = 2z$, $y - z + 1 = 0$.
[*Hint:* Integrate with respect to $x$ first.]

## 22.10 TRIPLE INTEGRALS IN CYLINDRICAL AND SPHERICAL COORDINATES

A process similar to that for rectangular coordinates is used in defining a triple integral in cylindrical or spherical coordinates. We recall that the surfaces which established the subdivisions in rectangular coordinates were planes $x = $ constant, $y = $ constant, $z = $ constant. In cylindrical and spherical coordinates some of the corresponding surfaces are curved rather than plane.

*Cylindrical coordinates.* Let $f(r, \theta, z)$ be a function expressed in terms of cylindrical coordinates $r$, $\theta$, $z$, and defined in a 3-dimensional region $V$. We subdivide $V$ by surfaces $r = $ constant, $\theta = $ constant, $z = $ constant. A typical element of volume is shown in Fig. 22.10.1.

The elements of volume which lie in $V$ are numbered from 1 to $N$. Let $\Delta V_i$ be the volume of the $i$th element, in which an arbitrary point $(r_i{}^*, \theta_i{}^*, z_i{}^*)$ is chosen, and the sum

$$\sum_{i=1}^{N} f(r_i{}^*, \theta_i{}^*, z_i{}^*)\,\Delta V_i$$

is formed. If the limit of this sum exists as the maximum dimension of any

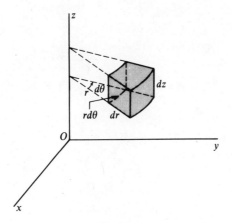

Fig. 22.10.1

element of volume approaches zero, in a similar sense to that of Definition 22.2.1, the limit is *defined* as the value of the triple integral $\iiint_V f(r, \theta, z)\, dV$.

It can be shown that a triple integral in cylindrical coordinates exists if $f$ is continuous in $V$, and that it may be expressed as an iterated integral in the form

$$\iiint_V f(r, \theta, z)\, dV = \int_\alpha^\beta \int_{r_1(\theta)}^{r_2(\theta)} \int_{z_1(r, \theta)}^{z_2(r, \theta)} f(r, \theta, z)\, r\, dz\, dr\, d\theta.$$

The *element of volume* in cylindrical coordinates may be taken as

(22.10.1) $$dV = r\, dz\, dr\, d\theta.$$

We note that each cross section of an element of volume (see Fig. 22.10.1) by a plane perpendicular to the $z$ axis is the same, and is the type of pseudo-rectangle used in polar coordinates, with area $dA = r\, dr\, d\theta$. The volume of the element is the product of the area of a cross section with the altitude, which is consistent with the relation in (22.10.1).

***Example 22.10.1.*** Set up an iterated integral in cylindrical coordinates for the function $f(r, \theta, z)$ over the solid in the first octant bounded by the surfaces

$$x^2 + y^2 = z, \qquad x^2 + y^2 = 2y, \qquad z = 0.$$

*Solution.* The first two surfaces are a paraboloid of revolution and a circular cylinder; the solid is shown in Fig. 22.10.2. The equation of the paraboloid in cylindrical coordinates is $z = r^2$. The limits for the integration with respect to $z$ are obtained from the equations of the lower and upper bounding surfaces of the solid, and are 0 and $r^2$. The limits for the remaining integrations are obtained from the projection of the solid on the $xy$ plane. This projection is a semicircular region, bounded by a circle $r = 2\sin\theta$ in polar coordinates and the $y$ axis. Hence

$$\iiint_V f(r, \theta, z)\, dV = \int_0^{\pi/2} \int_0^{2\sin\theta} \int_0^{r^2} f(r, \theta, z)\, r\, dz\, dr\, d\theta.$$

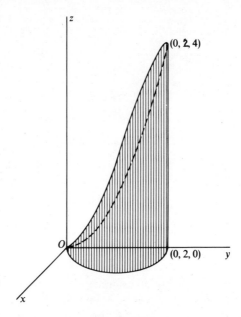

Fig. 22.10.2

*Spherical coordinates.* In spherical coordinates (Section 20.11) a subdivision of a region is produced by surfaces of the form $\rho =$ constant, $\theta =$ constant $\phi =$ constant. In this system an element of volume is bounded (see Fig. 22.10.3) by two spheres with center at the origin, two planes which contain the $z$ axis, and two cones with the $z$ axis as axis. These bounding surfaces intersect at right angles, and it is plausible to use for the volume of the element the product of three perpendicular dimensions. From the figure we see that the dimensions are, approximately, $d\rho$, $\rho \sin \phi \, d\theta$, and $\rho \, d\phi$. Hence

(22.10.2) $$dV = \rho^2 \sin \phi \, d\rho \, d\theta \, d\phi,$$

and a triple integral in spherical coordinates is expressed as an iterated integral in the form

$$\iiint_V f(\rho, \theta, \phi) \, dV = \int_\alpha^\beta \int_{\theta_1(\phi)}^{\theta_2(\phi)} \int_{\rho_1(\theta, \phi)}^{\rho_2(\theta, \phi)} f(\rho, \theta, \phi) \rho^2 \sin \phi \, d\rho \, d\theta \, d\phi.$$

*Example 22.10.2.* Find the volume of the solid which lies inside the upper nappe of the cone $x^2 + y^2 = z^2$, and the sphere $x^2 + y^2 + z^2 = 4$.

*Solution.* Spherical coordinates are particularly suitable here. The volume is given by a triple integral of the function $f(\rho, \theta, \phi) = 1$. The solid is shown in Fig. 22.10.4. Throughout the solid $\rho$ varies from 0 to 2, and $\theta$ varies from 0 to $2\pi$. The equation of the cone in spherical coordinates is $\phi = \pi/4$ so that $\phi$ varies from 0 to $\pi/4$. All limits of integration are thus constant, and we have

$$V = \int_0^{\pi/4} \int_0^{2\pi} \int_0^2 \rho^2 \sin\phi \, d\rho \, d\theta \, d\phi = \frac{8}{3} \int_0^{\pi/4} \int_0^{2\pi} \sin\phi \, d\theta \, d\phi$$

$$= \frac{16\pi}{3} \int_0^{\pi/4} \sin\phi \, d\phi = \frac{8\pi(2-\sqrt{2})}{3}.$$

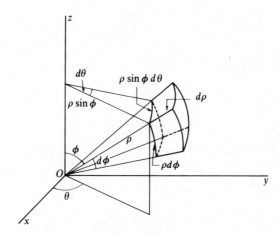

Fig. 22.10.3

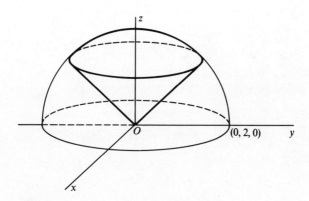

Fig. 22.10.4

The formulas for determining centers of gravity, moments of inertia, and so on, of a solid have been expressed in terms of the element of volume, $dV$. Such formulas remain valid in cylindrical and spherical coordinates; it is necessary merely to express $dV$ appropriately, using (22.10.1) or (22.10.2), and then to set up the corresponding iterated integral.

**Example 22.10.3.** Find the moment of inertia about the $z$ axis of the solid described in Example 22.10.1, assuming constant density $\mu$.

*Solution.* We use cylindrical coordinates. The squared distance of a point in the solid from the $z$ axis is $r^2$. Hence

$$I_z = \iiint r^2 \, dm = \mu \iiint r^2 \, dV = \mu \int_0^{\pi/2} \int_0^{2\sin\theta} \int_0^{r^2} r^3 \, dz \, dr \, d\theta$$

$$= \mu \int_0^{\pi/2} \int_0^{2\sin\theta} r^5 \, dr \, d\theta = \frac{32\mu}{3} \int_0^{\pi/2} \sin^6\theta \, d\theta$$

$$= \frac{5\pi\mu}{3}.$$

The value of the last integral may be obtained by the methods of Section 14.2.

**Example 22.10.4.** Find the coordinates of the center of gravity of the solid described in Example 22.10.2.

*Solution.* The density may be taken as unity. By symmetry of the solid we have $\bar{x} = \bar{y} = 0$, and it is necessary only to find $\bar{z}$. Now,

$$\bar{z} = \frac{M_{xy}}{V} = \frac{\iiint z \, dV}{V},$$

with the volume $V$ as found in Example 22.10.2. Using spherical coordinates, we find

$$M_{xy} = \iiint z \, dV = \iiint \rho \cos\phi \cdot \rho^2 \sin\phi \, d\rho \, d\theta \, d\phi$$

$$= \int_0^{\pi/4} \int_0^{2\pi} \int_0^2 \rho^3 \sin\phi \cos\phi \, d\rho \, d\theta \, d\phi$$

$$= 4 \int_0^{\pi/4} \int_0^{2\pi} \sin\phi \cos\phi \, d\theta \, d\phi = 2\pi.$$

Hence

$$\bar{z} = \frac{2\pi}{8\pi(2-\sqrt{2})/3} = \frac{3(2+\sqrt{2})}{8}.$$

### EXERCISE GROUP 22.8

1. Find the volume of the solid sphere $x^2 + y^2 + z^2 \leq a^2$ (a) with the use of cylindrical coordinates, (b) with the use of spherical coordinates.

2. Use cylindrical coordinates to find the volume of the solid interior to both the sphere $x^2 + y^2 + z^2 = a^2$ and the cylinder $x^2 + y^2 = b^2$, $0 < b < a$.

3. Find the volume of the solid bounded by the paraboloid $x^2 + y^2 + z = 3$ and the upper nappe of the cone $x^2 + y^2 = 2z^2$.

4. Find the volume of the solid bounded by the surfaces $x^2 + y^2 = 2y$, $x^2 + y^2 - z^2 = 0$.

5. Find the volume of the solid described in Example 22.10.1.

6. Find the volume of the solid interior to the surface $x^2 + y^2 + 4z^2 = 9$.

7. Find the volume of the solid bounded by the lower nappe of the cone $x^2 + y^2 = (z-3)^2$ and the paraboloid $4z = x^2 + y^2$.

8. Find the volume of the solid bounded by the upper nappes of the cones $x^2 + y^2 = z^2$, $x^2 + y^2 = 4z^2$ and the sphere $x^2 + y^2 + z^2 = 4$.

9. Find (a) the center of gravity, (b) $I_y$ and $I_z$, for the solid bounded by the cone $z^2 = h^2(x^2 + y^2)$ and the plane $z = h$.

\* 10. Find the volume of the solid in the first octant bounded by the surfaces $x^2 + y^2 + z^2 = a^2$, $y = b_1 x$, $y = b_2 x$, $x^2 + y^2 = (\tan^2 \phi_1) z^2$, $x^2 + y^2 = (\tan^2 \phi_2) z^2$, $\phi_2 > \phi_1$, $b_2 > b_1 > 0$.

11. (a) Find the volume of the solid bounded by the surfaces $x^2 + y^2 = 9$, $y + z = 3$, $z = 0$.
    (b) Find the coordinates of the centroid of the solid in (a).

12. Find the centroid of the solid bounded by the surfaces $x^2 + y^2 = 9$, $y + z = 3$, $y + 3z = 3$. [*Hint:* This can be found by making appropriate use of Exercise 11, without carrying out all the integrations.]

## 22.11 AREA OF A CURVED SURFACE

In finding the area of a curved surface we have previously been restricted to the case of a surface of revolution (Section 12.6). We now extend our methods to apply to more general surfaces. We consider, at first, a portion $S$ of a surface (Fig. 22.11.1) whose equation may be written in the form $z = f(x, y)$. If $C$ is the

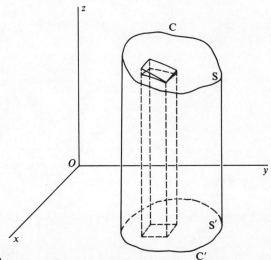

Fig. 22.11.1

boundary curve of $S$, then the projection of $S$ on the $xy$ plane is a region $S'$ whose boundary $C'$ is the projection of $C$. The region $S'$ in the $xy$ plane is subdivided by lines parallel to the coordinate axes (as in Section 22.2); one of the rectangles thus formed is shown in the figure. A point $(u_i, v_i)$ is chosen in this rectangle, and at the corresponding point $(u_i, v_i, f(u_i, v_i))$ the tangent plane to the surface is drawn. The part of the tangent plane which projects into the rectangle of the $xy$ plane is a parallelogram, whose area is taken as an approximation of the area $\Delta S_i$ of the part of $S$ above the rectangle.

It can be shown that the area of the parallelogram described is $|\sec \gamma_i| \, \Delta A_i$, where $\Delta A_i = \Delta x_i \Delta y_i$, and $\alpha_i, \beta_i, \gamma_i$ are the direction angles of a unit normal to the surface. We know that a normal to the surface $z = f(x, y)$ is $\mathbf{i} f_x + \mathbf{j} f_y - \mathbf{k}$. We may then take

$$|\sec \gamma| = (1 + f_x^2 + f_y^2)^{1/2},$$

and hence, if $dS$ is an *element of area* on the surface,

(22.11.1) $\qquad dS = |\sec \gamma| \, dA = (1 + f_x^2 + f_y^2)^{1/2} \, dA,$

where $dA$ is the corresponding element of area in the $xy$ plane.

Finally, the area $S$ of the surface itself is given by

(22.11.2) $\qquad S = \iint_{S'} |\sec \gamma| \, dA = \iint_{S'} (1 + f_x^2 + f_y^2)^{1/2} \, dA;$

the double integral is evaluated over the region $S'$, the projection of the surface $S$ on the $xy$ plane.

We shall illustrate the use of (22.11.2) by an example which can also be treated by earlier methods (Section 12.6).

**Example 22.11.1.** Find the area of that portion of the hemisphere

(22.11.3) $\qquad\qquad x^2 + y^2 + z^2 = a^2, \qquad z \geq 0,$

which lies inside the cylinder $x^2 + y^2 = b^2$, $0 < b \leq a$.

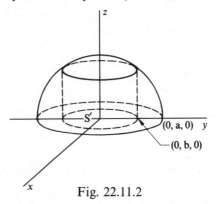

Fig. 22.11.2

*Solution.* The surface described appears in Fig. 22.11.2 as a cap of the sphere.

The projection of the surface on the $xy$ plane is a circle of radius $b$. The function $f$ for the surface is obtained by solving (22.11.3) for $z$, and is

$$f(x, y) = (a^2 - x^2 - y^2)^{1/2},$$

from which we may get $f_x$ and $f_y$. We may also differentiate (22.11.3) implicitly, and we have

$$2x + 2z f_x = 0, \qquad 2y + 2z f_y = 0,$$

and hence

$$f_x = -x/z, \qquad f_y = -y/z.$$

Then $1 + f_x^2 + f_y^2 = 1 + x^2/z^2 + y^2/z^2 = (x^2 + y^2 + z^2)/z^2$. Since $(x, y, z)$ is a point on the sphere we get $1 + f_x^2 + f_y^2 = a^2/z^2$, and (22.11.2) yields the representation

(22.11.4) $$S = \iint_{S'} \frac{a}{z} \, dA = \iint_{S'} \frac{a \, dA}{\sqrt{a^2 - x^2 - y^2}}.$$

We may now express this as an iterated integral in rectangular coordinates over the region bounded by $x^2 + y^2 = a^2$. But the evaluation is much simpler in polar coordinates which we now elect to use. Thus $x = r \cos \theta$, $y = r \sin \theta$, $dA = r \, dr \, d\theta$, and (22.11.4) becomes

$$S = \int_0^{2\pi} \int_0^b \frac{ar \, dr \, d\theta}{\sqrt{a^2 - r^2}} = 2\pi a(a - \sqrt{a^2 - b^2}).$$

The integration with respect to $r$ involved the power rule.

Formulas corresponding to (22.11.2) may be readily obtained for a surface which is projected on one of the other coordinate planes. For example, if the surface is $x = h(y, z)$, then

$$S = \iint \sqrt{1 + h_y^2 + h_z^2} \, dA,$$

where the integration is over a region in the $yz$ plane which is the projection of the surface. In this case we have

$$dS = |\sec \alpha| \, dA,$$

and the element of area $dA$ refers to the $yz$ plane.

If a surface is given by an equation

$$F(x, y, z) = 0,$$

and the surface is projected on the $xy$ plane, the expression for the area of the surface may be written as

(22.11.5) $$S = \iint_{S'} \frac{\sqrt{F_x^2 + F_y^2 + F_z^2}}{|F_z|} \, dA.$$

This may be obtained by noting that a normal to the surface is

(22.11.6) $$\text{grad } F = \mathbf{i}F_x + \mathbf{j}F_y + \mathbf{k}F_z$$

and hence

$$|\sec \gamma| = \sqrt{F_x^2 + F_y^2 + F_z^2}/|F_z|.$$

We may also obtain the expressions for $|\sec \alpha|$ and $|\sec \beta|$ from (22.11.6), and use them, if the surface is projected onto the $yz$ or $xz$ planes, respectively. Thus

$$|\sec \alpha| = \sqrt{F_x^2 + F_y^2 + F_z^2}/|F_x|, \qquad |\sec \beta| = \sqrt{F_x^2 + F_y^2 + F_z^2}/|F_y|.$$

In every expression for surface area presented in this section it is implied that the function being integrated must be expressed in terms of the variables involved in the $dA$. For example, the expression in (22.11.5) which multiplies $dA$ must be in terms of $x$ and $y$, with use of the equation of the surface, if necessary, to eliminate $z$. Remember that $(x, y, z)$ is a point in the surface.

In those cases where the equation of a surface cannot be expressed in the form $z = f(x, y)$ or $y = g(x, z)$ or $x = h(y, z)$, it may be possible to consider the surface as made up of separate parts, for each of which one of the above representations does apply. The area of each part may then be expressed by an applicable integral, and the total area is the sum of the areas of the parts.

*Surface integrals.* We mention briefly that it is possible to define a *surface integral* for a function $G(x, y, z)$ which is defined at the points of a surface $S$; this integral is designated as

$$\iint_S G(x, y, z) \, dS.$$

It is emphasized that the variables $x$, $y$, and $z$ are not independent, but are related by the equation of the surface.

A surface integral may now be used to express moments of inertia, etc., for a curved surface (to which a density function $\mu(x, y, z)$ may be assigned). We shall illustrate some of these ideas in an example which is chosen because it can readily be carried out completely.

**Example 22.11.2.** Find the coordinates of the center of gravity of the triangle cut from the following plane by the coordinate planes. Assume a density of 1.

$$\frac{x}{a} + \frac{y}{b} + \frac{z}{c} = 1, \qquad a > 0, \, b > 0, \, c > 0.$$

*Solution.* The triangle is shown in Fig. 22.11.3. We find the area from (22.11.5) with

$$F(x, y, z) = \frac{x}{a} + \frac{y}{b} + \frac{z}{c} - 1.$$

Then $F_x = 1/a$, $F_y = 1/b$, $F_z = 1/c$, and

$$\sec \gamma = \frac{\sqrt{1/a^2 + 1/b^2 + 1/c^2}}{1/c} = \frac{\sqrt{a^2b^2 + a^2c^2 + b^2c^2}}{ab}.$$

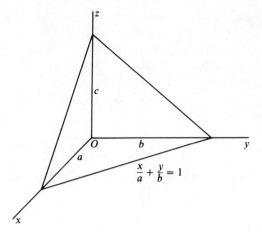

Fig. 22.11.3

Thus $\sec \gamma$ is constant (which we knew in advance), $dS = \sec \gamma\, dA$ and the area of the triangle is

$$S = \sec \gamma \iint dA = \tfrac{1}{2}ab \sec \gamma;$$

we have used the fact that $\iint dA$ is the area of the triangle in the $xy$ plane, and is $ab/2$.

For the moment $M_{yz}$ we have

$$M_{yz} = \iint x\, dS = \sec \gamma \int_0^a \int_0^{b(1-x/a)} x\, dy\, dx;$$

the limits for the iterated integral are obtained from the triangle in the $xy$ plane. The value of the iterated integral is $a^2b/6$. Hence $M_{yz} = (a^2b \sec \gamma)/6$, and

$$\bar{x} = \frac{M_{yz}}{S} = \frac{\tfrac{1}{6}a^2b \sec \gamma}{\tfrac{1}{2}ab \sec \gamma} = \frac{a}{3}.$$

By analogy, it follows that $\bar{y} = b/3$, $\bar{z} = c/3$, and hence the center of gravity is at $(a/3, b/3, c/3)$.

## EXERCISE GROUP 22.9

1. Find the area of that part of the surface $z = x^2 + y^2$ which lies interior to the cylinder $x^2 + y^2 = 2$.

2. Find the area of that part of the cone $x^2 + y^2 = z^2$ in the first octant which projects into the region in the $xy$ plane bounded by $x = 0$, $x = 3$, $y = 0$, $y = 3$.

3. Find the area of that part of one nappe of the cone $x^2 + y^2 = z^2$ which lies inside the cylinder $x^2 + y^2 = 2y$.

4. Find the total area of the surface which bounds the solid interior to both cylinders $x^2 + z^2 = a^2$ and $x^2 + y^2 = a^2$.

5. Find the area of that part of the cylinder $y^{2/3} + z^{2/3} = a^{2/3}$ which is in the first octant and inside the cylinder $x^{2/3} + y^{2/3} = a^{2/3}$.

* 6. Find the area of that part of the sphere $x^2 + y^2 + z^2 = a^2$ in the first octant which is bounded by the surfaces $y = b_1 x$, $y = b_2 x$, $x^2 + y^2 = (\tan^2 \phi_1) z^2$, $x^2 + y^2 = (\tan^2 \phi_2) z^2$, $\phi_2 > \phi_1$, $b_2 > b_1 > 0$.

7. If $z = f(r, \theta)$ is an equation of a surface in cylindrical coordinates, show that the formula for surface area becomes

$$S = \iint_A \left(1 + f_r^2 + \frac{1}{r^2} f_\theta^2\right)^{1/2} r\, dr\, d\theta,$$

where $A$ is the projection of the surface into the $r\theta$ plane. (See Exercise 10 of Exercise Group 21.5.)

# 23  Infinite Series

## 23.1  SEQUENCES

The development of infinite series is based on *sequences*, which were introduced in Section 9.8; we will pursue this topic further here. A sequence will be defined in terms of a function whose domain is the set of positive integers. Such a function, for example, is,

$$(23.1.1) \qquad f(n) = \frac{1}{1 + n^2}, \qquad n = 1, 2, 3, \ldots.$$

We may also need to refer to the function

$$f(x) = \frac{1}{1 + x^2}, \qquad x \text{ real},$$

whose values coincide with (23.1.1) when $x$ is a positive integer.

Another example is the *factorial function*, defined by

$$(23.1.2) \qquad \begin{cases} f(n) = (n - 1)! = 1 \cdot 2 \cdot 3 \cdots (n - 1), & n = 2, 3, 4, \ldots \\ f(1) = 0! = 1. \end{cases}$$

This function is not defined unless $n$ is a positive integer.

The values of (23.1.1), written in order for $n = 1, 2, 3, \ldots$ are

$$(23.1.3) \qquad \frac{1}{2}, \frac{1}{5}, \frac{1}{10}, \frac{1}{17}, \ldots.$$

The values of (23.1.1) when $n$ is an *even* integer are, in order,

$$(23.1.4) \qquad \frac{1}{5}, \frac{1}{17}, \frac{1}{37}, \ldots.$$

Both (23.1.3) and (23.1.4) are examples of infinite sequences. Although (23.1.4) was obtained by writing alternate values of the numbers in (23.1.3), it may also be considered as the set of values of the function

$$g(n) = \frac{1}{1 + 4n^2},$$

written in succession for $n = 1, 2, 3, \ldots$.

**DEFINITION 23.1.1.** *A* **sequence** *of numbers is the set of values of a function $f(n)$, whose domain in the set of positive integers, taken in succession for $n = 1, 2, 3, \ldots$. Each number in the sequence is called an* **element** *or* **term** *of the sequence.*

It is common practice to use a subscript notation for the elements of a sequence. Thus, corresponding to (23.1.1), we write

$$a_n = \frac{1}{1 + n^2}, \quad n = 1, 2, 3, \ldots,$$

and we designate the sequence as $\{a_n\}$, $n = 1, 2, 3, \ldots$.

If we write the values of $f(n) = 1/(n - 1)$, we must exclude $n = 1$, and we have

$$1, \tfrac{1}{2}, \tfrac{1}{3}, \tfrac{1}{4}, \ldots;$$

however, this is a sequence of numbers corresponding to

$$g(n) = \frac{1}{n}, \quad n = 1, 2, 3, \ldots$$

In fact, any collection of numbers, exhibited in such a manner that one of them is first, one is second, and so on, so as to include every number of the collection, is a sequence. We need only define $f(1)$ as the first number, $f(2)$ as the second number, and so on.

It is sometimes required to determine an expression for $a_n$ when a few elements are given at the beginning of a sequence. In the absence of more definite information, this determination is not unique. For example, if

$$a_1 = 1, \quad a_2 = 2, \quad a_3 = 3, \quad a_4 = 4,$$

two representations are

$$a_n = n \quad \text{and} \quad a_n = n + (n - 1)(n - 2)(n - 3)(n - 4).$$

In general, the writing of a few terms at the beginning of a sequence without specifying $a_n$ is a matter of convenience, and it is implied that $a_n$ is the "simplest" function for which the given elements are consistent. Thus we should take $a_n = n$ in the above.

If the signs of the terms of a sequence are alternately plus and minus, the expression for $a_n$ will generally include a power of $(-1)$. For example, for the sequence

(23.1.5) $\quad\quad\quad \sin 1, -\sin 2, \sin 3, -\sin 4, \ldots,$

we may take $a_n = (-1)^{n+1} \sin n$. Note that we are not claiming that the terms in (23.1.5) are themselves alternately positive and negative; in this sequence they are not!

## EXERCISE GROUP 23.1

In Exercises 1–12 write the first five terms of the sequence for which $a_n$ is given.

1. $a_n = \dfrac{1}{3n+1}$

2. $a_n = \dfrac{3n}{2n+3}$

3. $a_n = \dfrac{(-1)^{n+1}}{2^n}$

4. $a_n = \dfrac{n}{2n^2-1}$

5. $a_n = \dfrac{n+1}{n!}$

6. $a_n = \dfrac{2^n}{3^n-1}$

7. $a_n = \dfrac{(2n)!}{n!}$

8. $a_n = \dfrac{(-1)^{n(n-1)/2}}{n^2}$

9. $a_n = (-2)^{n(n-1)/2}$

10. $a_n = (-1)^{n(n-1)(n-2)/6}$

11. $a_n = \dfrac{x^{2n}}{n}$

12. $a_n = (-1)^n x^n$

In Exercises 13–18 find (a) a (simple) expression for $a_n$ for the given sequence, (b) an expression for $a_n$ for the sequence which starts with the second term.

13. $2, 5, 8, 11, \ldots$

14. $\dfrac{2}{3}, \dfrac{3}{4}, \dfrac{4}{5}, \dfrac{5}{6}, \ldots$

15. $\dfrac{2}{3 \cdot 4}, \dfrac{-3}{4 \cdot 5}, \dfrac{4}{5 \cdot 6}, \dfrac{-5}{6 \cdot 7}, \ldots$

16. $\dfrac{1}{2}, \dfrac{1 \cdot 2}{5}, \dfrac{1 \cdot 2 \cdot 3}{8}, \dfrac{1 \cdot 2 \cdot 3 \cdot 4}{11}, \ldots$

17. $\dfrac{1}{3}, -\dfrac{1}{8}, \dfrac{1}{15}, -\dfrac{1}{24}, \ldots$

18. $\dfrac{x}{1 \cdot 2}, \dfrac{x^3}{3 \cdot 4}, \dfrac{x^5}{5 \cdot 6}, \dfrac{x^7}{7 \cdot 8}, \ldots$

In Exercises 19–24 find an expression for $b_n$, where $b_n = a_{n+1} - a_n$ for the given $a_n$, and also write the first three terms for the sequence $\{b_n\}$.

19. $a_n = 1/n$

20. $a_n = 2n+1$

21. $a_n = \ln n$

22. $a_n = (-1)^n$

23. $a_n = (-1)^{n+1} 2^n$

24. $a_n = \sin 2n$

25. If $f(x) = \dfrac{1}{1+x^2}$, obtain an expression for $a_n$ if (a) $a_n = f(2n)$, (b) $a_n = f(n^2)$, (c) $a_n = f\left(\dfrac{n+1}{n}\right)$.

26. If $f(x) = \dfrac{x-1}{x+1}$, obtain an expression for $a_n$ if (a) $a_n = f(1/n)$, (b) $a_n = f(2n+1)$, (c) $a_n = f(\tan n)$.

## 23.2 THE LIMIT OF A SEQUENCE

The behavior of a sequence of numbers $a_n$ as $n$ increases indefinitely is expressed in the idea of limit. The definition which follows is clearly analogous to the definition of the limit of a function $f(x)$ as $x \to \infty$.

**DEFINITION 23.2.1.** $\lim_{n \to \infty} a_n = A$ *if for each $\epsilon > 0$ a positive number $N$ exists such that*
$$|a_n - A| < \epsilon \quad \text{whenever} \quad n > N.$$

*If* $\lim_{n \to \infty} a_n$ *exists, the sequence $\{a_n\}$ is said to* **converge**. *Otherwise, it* **diverges**.

The existence of a limit implies that the limit is a real number. We also define infinite limit for a sequence.

**DEFINITION 23.2.2.** (a) $\lim_{n \to \infty} a_n = \infty$ *if for each $B > 0$, a positive number $N$ exists such that*
$$a_n > B \quad \text{whenever} \quad n > N.$$

(b) $\lim_{n \to \infty} a_n = -\infty$ *if for each $B > 0$, a positive number $N$ exists such that*
$$a_n < -B \quad \text{whenever} \quad n > N.$$

In the event $\lim_{n \to \infty} a_n = \infty$ or $\lim_{n \to \infty} a_n = -\infty$, the sequence $\{a_n\}$ still diverges. However, the knowledge that a divergent sequence has an infinite limit tells us more about the behavior of the sequence than the simple fact that it diverges.

Most of the properties of limits of functions, as well as the proofs, carry over to limits of sequences. We then state the following without proof.

**THEOREM 23.2.1.** *If* $\lim_{n \to \infty} a_n = A$ *and* $\lim_{n \to \infty} b_n = B$, *then*:

1. $\lim_{n \to \infty} (ka_n) = kA$, $k$ constant
2. $\lim_{n \to \infty} (a_n + b_n) = A + B$
3. $\lim_{n \to \infty} a_n b_n = AB$
4. $\lim_{n \to \infty} \dfrac{a_n}{b_n} = \dfrac{A}{B}$, *provided $B \neq 0$ and $b_n \neq 0$*.

Other properties will appear in the course of the work. We single out the property that if $\lim_{n \to \infty} a_n = A$, then also

$$\lim_{n \to \infty} a_{n-1} = A, \quad \lim_{n \to \infty} a_{n+3} = A, \quad \text{and so on.}$$

We illustrate some of the techniques used in the evaluation of limits.

**Example 23.2.1.** Evaluate $\lim\limits_{n \to \infty} \dfrac{2n^2 + 1}{3(n + 1)^2 + 1}$.

**Solution.** $\lim\limits_{n \to \infty} \dfrac{2n^2 + 1}{3(n + 1)^2 + 1} = \lim\limits_{n \to \infty} \dfrac{2n^2 + 1}{3n^2 + 6n + 4} = \lim\limits_{n \to \infty} \dfrac{2 + 1/n^2}{3 + 6/n + 4/n^2} = \dfrac{2}{3}.$

We have used the fact that $\lim\limits_{n \to \infty} 1/n^k = 0$ if $k > 0$.

The following example includes an illustration of a limit of a quotient of two sequences, where Theorem 23.2.1 does not apply since $B = 0$.

**Example 23.2.2.** Evaluate $\lim\limits_{n \to \infty} \dfrac{n^3 + 2}{2n^2 + 1}$.

**Solution.** We write

$$\lim_{n \to \infty} \frac{n^3 + 2}{2n^2 + 1} = \lim_{n \to \infty} \frac{1 + 2/n^3}{2/n + 1/n^3}.$$

We see that $\lim\limits_{n \to \infty} (2/n + 1/n^3) = 0$, by Theorem 23.2.1, part 2, and hence we have a limit of the type $a_n/b_n$, where $\lim\limits_{n \to \infty} b_n = 0$. Since $2/n + 1/n^3 > 0$, and $\lim\limits_{n \to \infty} (1 + 2/n^3) = 1$, it is reasonable to expect, and it can be shown, that the limit of the quotient is infinite, and we write

$$\lim_{n \to \infty} \frac{n^3 + 2}{2n^2 + 1} = \infty.$$

An example of a sequence with no limit, finite or infinite, is $a_n = (-1)^{n-1}$, for which the terms are $1, -1, 1, -1, \ldots$. The values of $a_n$ are alternately $+1$ and $-1$, and there is no number that the $a_n$ get close to and remain close to. On the other hand, if $a_n = (-1)^{n-1}/n$, the sequence becomes

$$1, -\tfrac{1}{2}, \tfrac{1}{3}, -\tfrac{1}{4}, \ldots.$$

Although the terms are again alternately positive and negative, they are all getting and remaining close to zero, and we have

$$\lim_{n \to \infty} (-1)^{n-1}/n = 0.$$

Another example of a divergent sequence, one for which there is not an infinite limit, is given by (cf. Example 23.2.2)

$$a_n = (-1)^{n-1} \frac{n^3 + 2}{2n^2 + 1}.$$

The terms alternate in sign, and in Example 23.2.2 we saw that they increase in absolute value. Hence there is no limit.

We now illustrate a type of limit that will arise later.

**Example 23.2.3.** Evaluate $\lim_{n\to\infty} a_{n+1}/a_n$ if $a_n = 2^n/n!$.

**Solution.** We have $a_{n+1} = 2^{n+1}/(n+1)!$. Hence

$$\frac{a_{n+1}}{a_n} = \frac{2^{n+1}}{(n+1)!} \cdot \frac{n!}{2^n} = \frac{2}{n+1} \to 0 \quad \text{as} \quad n \to \infty.$$

In the subsequent work we shall on occasion be in the position of having to show that a limit exists, without being able to produce the limit. Such demonstrations can often be based on the following fundamental principle, which is one form of the *completeness property of the real number system*.

If $\{a_n\}$ is a nondecreasing, bounded sequence, that is, if

$$a_n \geq a_m \quad \text{whenever} \quad n > m,$$

and if, for some number $A$,

$$a_n < A \quad \text{for all positive integers } n,$$

then $\lim_{n\to\infty} a_n$ exists.

Similarly, $\lim_{n\to\infty} a_n$ exists if $a_n$ is a nonincreasing, bounded sequence, that is, if

$$a_n \leq a_m \quad \text{whenever} \quad n > m,$$

and if, for some number $B$,

$$a_n > B \quad \text{for all positive integers } n.$$

These follow from the *least upper bound axiom* (Section 1.1). For example, in the first case $A$ is an upper bound, and hence there exists a least upper bound $a \leq A$. We prove that

$$\lim_{n\to\infty} a_n = a.$$

Given $\epsilon > 0$ there exists a positive integer $N$ such that $a - \epsilon < a_N$, otherwise $a - \epsilon$ is an upper bound which is less than $a$. Since $\{a_n\}$ is nondecreasing, we have

$$a - \epsilon < a_N \leq a_n, \quad n > N.$$

But $a_n \leq a$ for any positive integer $n$ since $a$ is an upper bound. Thus

$$a - \epsilon < a_n \leq a, \quad n > N,$$

from which it follows that

$$|a_n - a| < \epsilon, \quad n > N,$$

and Definition 23.2.1 is satisfied.

**EXERCISE GROUP 23.2**

In Exercises 1–10 evaluate $\lim_{n\to\infty} a_n$ for the given $a_n$, if the limit exists.

**1.** $a_n = \dfrac{n^2}{2n^2 - 1}$

**2.** $a_n = \dfrac{n}{\sqrt{n^2 + 1}}$

**3.** $a_n = \dfrac{(n+1)^3 - 1}{(n-1)^3 + 1}$

**4.** $a_n = \dfrac{2n^2 + 3n + 1}{4n^3 - 2}$

**5.** $a_n = \dfrac{2^n}{3^n}$

**6.** $a_n = \dfrac{2^n - 1}{2^n + 1}$

**7.** $a_n = 1 + (-1)^n$

**8.** $a_n = \dfrac{2n^3}{(n+1)^3}$

**9.** $a_n = \dfrac{n^2(n+1)}{2n^3 + n^2 + n - 3}$

**10.** $a_n = \dfrac{2n^{3/2}}{\sqrt{n^2 + 1}}$

In Exercises 11–20 evaluate $\lim\limits_{n \to \infty} a_{n+1}/a_n$ if the limit exists.

**11.** $a_n = \dfrac{n(2n+1)}{3^n}$

**12.** $a_n = \dfrac{4^n}{n^3}$

**13.** $a_n = nk^n,\ 0 \neq k$

**14.** $a_n = \dfrac{n!}{10^n}$

**15.** $a_n = \dfrac{n^2}{n!}$

**16.** $a_n = \dfrac{n!}{n^n}$

**17.** $a_n = \dfrac{1 \cdot 3 \cdot 5 \cdots (2n-1)}{3 \cdot 6 \cdot 9 \cdots (3n)}$

**18.** $a_n = \dfrac{4 \cdot 8 \cdot 12 \cdots (4n)}{3 \cdot 8 \cdot 13 \cdots (5n-2)}$

**19.** $a_n = \dfrac{1 \cdot 3 \cdot 5 \cdots (2n-1)}{2 \cdot 4 \cdot 6 \cdots (2n)}$

**20.** $a_n = \dfrac{n!}{3 \cdot 5 \cdot 7 \cdots (2n+1)} \cdot \dfrac{1}{n^2}$

## 23.3 INFINITE SERIES

The name *infinite series* is used to describe the expression

(23.3.1) $$a_1 + a_2 + a_3 + \cdots + a_n + \cdots,$$

that is, the indicated sum of the terms of an infinite sequence, in order.

The following are examples of infinite series:

(23.3.2) $$1 + \frac{1}{2} + \frac{1}{3} + \cdots + \frac{1}{n} + \cdots$$

(23.3.3) $$\frac{1}{1 \cdot 2} - \frac{1}{2 \cdot 3} + \frac{1}{3 \cdot 4} - \cdots + \frac{(-1)^{n+1}}{n(n+1)} + \cdots$$

For brevity we employ the sigma notation, by which (23.3.1), (23.3.2), and (23.3.3) are written, respectively, as

$$\sum_{n=1}^{\infty} a_n, \quad \sum_{n=1}^{\infty} \frac{1}{n}, \quad \text{and} \quad \sum_{n=1}^{\infty} \frac{(-1)^{n+1}}{n(n+1)}.$$

Each number $a_n$ in (23.3.1) is called a *term* of the series, and $a_n$ is the $n$th term. Associated with the series (23.3.1) are various sequences, of which the sequence of *partial sums* is particularly important.

**DEFINITION 23.3.1.** *The $n$th **partial sum** $s_n$ is the sum of the first $n$ terms of the series, that is,*

(23.3.4) $$s_n = \sum_{k=1}^{n} a_k = a_1 + a_2 + \cdots + a_n.$$

Whereas the summation indicated in (23.3.1) cannot be performed directly, since there are always additional terms to be added, each partial sum $s_n$ is a *finite* sum, and clearly defined. The following definition gives a meaning that can be assigned to an infinite series in certain cases.

**DEFINITION 23.3.2.** *The series $\sum_{n=1}^{\infty} a_n$ **converges** if the sequence of its partial sums $\{s_n\}$ converges. If $\lim_{n\to\infty} s_n = S$ (finite), we write*

$$\sum_{n=1}^{\infty} a_n = a_1 + a_2 + \cdots + a_n + \cdots = S,$$

*and $S$ is called the **sum** or **value** of the series. If a series does not converge, it is called **divergent**.*

From Definitions 23.2.1 and 23.3.2 it follows that the partial sums of a convergent series can be made to differ from the sum of the series $S$ by as little as we choose, by taking $n$ sufficiently large.

Some basic properties of series are now presented.

**THEOREM 23.3.1.** *Let $k$ be a constant. If $\sum_{n=1}^{\infty} a_n$ converges, then so does the series $\sum_{n=1}^{\infty} ka_n$, and we have*

$$\sum_{n=1}^{\infty} ka_n = k \sum_{n=1}^{\infty} a_n.$$

*If $\sum_{n=1}^{\infty} a_n$ diverges and $k \neq 0$, then $\sum_{n=1}^{\infty} ka_n$ diverges.*

*Proof.* Let $s_n = a_1 + a_2 + \cdots + a_n$ be the partial sum of the given series, and $s_n'$ the partial sum of the second series. Then

$$s_n' = ka_1 + ka_2 + \cdots + ka_n = k(a_1 + a_2 + \cdots + a_n) = ks_n.$$

Hence if $\lim_{n\to\infty} s_n = s$, then $\lim_{n\to\infty} s_n' = ks$, and the first statement is proved. If $\lim_{n\to\infty} s_n$ does not exist and $k \neq 0$, then $\lim_{n\to\infty} (ks_n)$ cannot exist, hence $\lim_{n\to\infty} s_n'$ does not exist, and the second statement is proved.

For example, the convergence or divergence of the series

$$\frac{1}{2} + \frac{1}{4} + \cdots + \frac{1}{2n} + \cdots$$

is the same as for the series

$$1 + \frac{1}{2} + \cdots + \frac{1}{n} + \cdots.$$

Both series actually diverge (see Section 23.4).

**THEOREM 23.3.2.** *If* $\sum_{n=1}^{\infty} a_n$ *and* $\sum_{n=1}^{\infty} b_n$ *converge, then the series* $\sum_{n=1}^{\infty} (a_n + b_n)$ *converges, and*

$$\sum_{n=1}^{\infty} (a_n + b_n) = \sum_{n=1}^{\infty} a_n + \sum_{n=1}^{\infty} b_n.$$

*Proof.* Let $s_n$, $s_n'$, and $s_n''$ be the partial sums of $\sum_{n=1}^{\infty} a_n$, $\sum_{n=1}^{\infty} b_n$, and $\sum_{n=1}^{\infty} (a_n + b_n)$, respectively. Then

$$s_n'' = s_n + s_n'$$

and the statement follows from Theorem 23.2.1 on taking limits as $n \to \infty$.

**THEOREM 23.3.3.** *The series* $\sum_{n=1}^{\infty} a_n$ *and* $\sum_{n=k}^{\infty} a_n$ *converge or diverge together.*

This means that the convergence or divergence of a series is not affected by dropping a fixed number of terms (no matter how many) at the beginning of the series.

*Proof.* Let the partial sums be, respectively,

$$s_n = a_1 + a_2 + \cdots + a_n, \qquad s_n' = a_k + a_{k+1} + \cdots + a_{k+n-1}.$$

Then $s_n' = s_{k+n-1} - \sum_{j=1}^{k-1} a_j$. Since the latter sum is a fixed number, independent of $n$, the sequences $s_n'$ and $s_{k+n-1}$ converge or diverge together as $n \to \infty$, and hence so do $s_n'$ and $s_n$.

The use of this property will appear later. A further property will be stated in two equivalent forms.

**THEOREM 23.3.4.** 1. *If* $\sum_{n=1}^{\infty} a_n$ *converges, then* $\lim_{n \to \infty} a_n = 0$.

2. *If* $\lim_{n \to \infty} a_n \neq 0$, *the series* $\sum_{n=1}^{\infty} a_n$ *diverges.*

*Proof* of 1. If $s_n = a_1 + a_2 + \cdots + a_n$, we may write

$$a_n = s_n - s_{n-1}, \qquad n > 1.$$

Since $\{s_n\}$ converges, say to $S$, we have

$$\lim_{n\to\infty} a_n = \lim_{n\to\infty} s_n - \lim_{n\to\infty} s_{n-1} = S - S = 0,$$

and the proof is complete.

Form 2 of Theorem 23.3.4 is logically equivalent to Form 1, and follows directly from it; it is included for greater emphasis. Theorem 23.3.4 expresses a *necessary condition* for convergence of an infinite series, which may also be stated as follows: *in order that the series* $\sum_{n=1}^{\infty} a_n$ *converge it is necessary that* $\lim_{n\to\infty} a_n = 0$. The theorem can never be used to prove convergence of a series. However, it can be used to prove divergence if it can be shown that $\lim_{n\to\infty} a_n \neq 0$, and hence *it may be considered as a test for divergence.*

*Note*: The statement $\lim_{n\to\infty} a_n \neq A$ is considered to apply if the sequence $\{a_n\}$ has no limit, or if it has a limit different from $A$.

Theorem 23.3.4 will be used in part of the proof of the following result.

**THEOREM 23.3.5.** *The* **geometric series**

$$\sum_{n=1}^{\infty} ar^{n-1} = a + ar + ar^2 + \cdots + ar^{n-1} + \cdots, \qquad a \neq 0,$$

*converges if* $|r| < 1$, *to the sum* $a/(1-r)$. *The series diverges if* $|r| \geq 1$.

*Proof.* If $r \neq 1$ the partial sum $s_n$ is

$$s_n = a + ar + \cdots + ar^{n-1} = a\frac{1-r^n}{1-r} = \frac{a}{1-r} - \frac{ar^n}{1-r}.$$

If $|r| < 1$, then $\lim_{n\to\infty} r^n = 0$, and hence

$$\lim_{n\to\infty} s_n = \frac{a}{1-r}.$$

The series then converges, by Definition 23.2.2, and the sum is $a/(1-r)$.

If $|r| \geq 1$, the limit of $a_n$ as $n \to \infty$ is not zero, and the series diverges, by Theorem 23.3.4.

The definition of convergence, which may be considered in a sense as a test for convergence, is of very limited practical use. In some cases, however, the convergence of a series may be determined by a direct examination of the partial sums.

***Example 23.3.1.*** Find $s_n$ for the given series, and test the series for convergence:

$$\frac{1}{1\cdot 2}+\frac{1}{2\cdot 3}+\cdots+\frac{1}{n(n+1)}+\cdots.$$

*Solution.* Each term may be written as a difference of two fractions (by partial fractions, for example), and we have

$$s_n = \left(1-\frac{1}{2}\right)+\left(\frac{1}{2}-\frac{1}{3}\right)+\cdots+\left(\frac{1}{n-1}-\frac{1}{n}\right)+\left(\frac{1}{n}-\frac{1}{n+1}\right).$$

This is an example of a *telescoping sum*. The terms cancel in pairs, and we are left with

$$s_n = 1 - \frac{1}{n+1}.$$

Hence $\lim_{n\to\infty} s_n = 1$, the series therefore converges, and the sum of the series is 1.

**Example 23.3.2.** If $s_n = 2n/(n+2)$ for a series, (a) does the series converge? (b) obtain the series.

*Solution.* (a) We have at once

$$\lim_{n\to\infty}\frac{2n}{n+2} = \lim_{n\to\infty}\frac{2}{1+2/n} = 2.$$

Hence the series converges, to the sum 2.

(b) From (23.3.4) we have

$$a_1 = s_1, \qquad a_n = s_n - s_{n-1}, \qquad n > 1.$$

Hence $a_1 = 2/3$, and if $n > 1$,

$$a_n = \frac{2n}{n+2} - \frac{2(n-1)}{n+1} = \frac{4}{(n+1)(n+2)}.$$

Since this expression for $a_n$ gives the correct value also for $n=1$, the series is

$$\sum_{n=1}^{\infty} a_n = \sum_{n=1}^{\infty} \frac{4}{(n+1)(n+2)}.$$

**EXERCISE GROUP 23.3**

In Exercises 1–7 show that the given series diverges.

1. $\sum_{n=1}^{\infty} \dfrac{n}{2n-1}$

2. $\tfrac{2}{3} - \tfrac{3}{5} + \tfrac{4}{7} - \cdots$

3. $\sum_{n=1}^{\infty} (-1)^n \sin n$

4. $\sum_{n=1}^{\infty} (-1)^{n(n-1)/2}$

5. $\sum_{n=1}^{\infty} n^2\left(1 - \cos\dfrac{1}{n}\right)$

6. $\sin 1 + 2\sin\tfrac{1}{2} + 3\sin\tfrac{1}{3} + \cdots$

7. $\dfrac{1}{1+\sqrt{2}} + \dfrac{1}{1+\sqrt[3]{2}} + \dfrac{1}{1+\sqrt[4]{2}} + \cdots$

In Exercises 8–17 test the series for convergence or divergence by finding $s_n$. If the series converges, find its sum.

8. $\dfrac{1}{1\cdot 3} + \dfrac{1}{3\cdot 5} + \cdots + \dfrac{1}{(2n-1)(2n+1)} + \cdots$

9. $\sum\limits_{n=1}^{\infty} \dfrac{2}{(3n+1)(3n-2)}$

10. $\sum\limits_{n=1}^{\infty} \dfrac{1}{2^n}$

11. $\sum\limits_{n=1}^{\infty} \dfrac{2-n}{2^n}$

12. $\sum\limits_{n=1}^{\infty} \dfrac{2n^2-1}{n^2(n+1)^2}$

13. $1 - \sum\limits_{n=2}^{\infty} \dfrac{2n-1}{n^2(n-1)^2}$

14. $\sum\limits_{n=1}^{\infty} (-1)^{n+1} \dfrac{2n+1}{n(n+1)}$

15. $\sum\limits_{n=1}^{\infty} \ln \dfrac{n+1}{n}$

16. $\ln 2 + \sum\limits_{n=2}^{\infty} \ln \dfrac{n^2-1}{n^2}$

17. $-\ln 3 + \sum\limits_{n=2}^{\infty} \ln \dfrac{2n^2-n}{2n^2-n-1}$

In Exercises 18–23 find the general term $a_n$ of a series for which the partial sum $s_n$ is given.

18. $s_n = 3n - 1$

19. $s_n = 1/n$

20. $s_n = 2^n$

21. $s_n = \ln n$

22. $s_n = (-1)^{n+1}$

23. $s_n = \cos 4n$

24. Find the sum of each geometric series:
    (a) $4 - 2 + 1 - \cdots$
    (b) $1 + 1/3 + 1/9 + \cdots$
    (c) $6 + 2 + 2/3 + \cdots$
    (d) $3 - 2 + 4/3 - \cdots$

25. Express each of the repeating decimals as a quotient of two integers with no common factor. [*Hint*: Express the decimal as a geometric series.]
    (a) $1.51515\ldots$
    (b) $0.363636\ldots$
    (c) $0.603603603\ldots$
    (d) $2.34234234\ldots$

## 23.4 THE INTEGRAL TEST FOR SERIES

Of prime importance in work with series is the determination of whether a series converges or not. (It is quite another matter to find the sum of a convergent series.) No single test can be used for all series. In this and subsequent sections we present several tests for convergence of an infinite series. The integral test and comparison tests apply to series of positive terms. The test to be presented here depends on the concept of improper integral (Section 15.5), in statement as well as in application.

**THEOREM 23.4.1.** *Integral test for series.* Let $\sum\limits_{n=1}^{\infty} a_n$ be a series whose terms are positive and nonincreasing,

$$a_n > 0, \qquad a_{n+1} \leq a_n, \qquad n = 1, 2, 3, \ldots.$$

*If $f(x)$ is a continuous nonincreasing function for $x \geq 1$, such that*

$$f(n) = a_n, \qquad n = 1, 2, 3, \ldots,$$

*then the improper integral $\int_1^\infty f(x)\, dx$ and the series $\sum_{n=1}^\infty a_n$ either both converge or both diverge.*

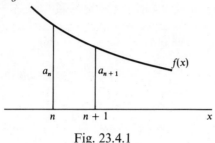

Fig. 23.4.1

*Proof.* Consider the function $f(x)$ in an interval $[n, n+1]$ (Fig. 23.4.1). We have

$$a_{n+1} \leq f(x) \leq a_n, \qquad n \leq x \leq n+1.$$

Hence with use of Theorem 9.9.3, part 4, we have

(23.4.1) $\quad a_{n+1} = \int_n^{n+1} a_{n+1}\, dx \leq \int_n^{n+1} f(x)\, dx \leq \int_n^{n+1} a_n\, dx = a_n.$

Summing (23.4.1) for $n = 1, 2, \ldots, k$, we obtain

$$\sum_{n=1}^k a_{n+1} \leq \sum_{n=1}^k \int_n^{n+1} f(x)\, dx \leq \sum_{n=1}^k a_n,$$

and hence

(23.4.2) $\quad s_{k+1} - a_1 \leq \int_1^{k+1} f(x)\, dx \leq s_k.$

Suppose now that the improper integral converges. Then the limit of the integral in (23.4.2) exists as $k \to \infty$, and since $f(x) \geq 0$, we have

$$s_{k+1} - a_1 \leq \int_1^{k+1} f(x)\, dx \leq \int_1^\infty f(x)\, dx.$$

Hence the $s_k$ are bounded, and since they are nondecreasing, the sequence $\{s_k\}$ has a limit as $k \to \infty$ (Section 23.2). It follows that the series $\sum_{n=1}^\infty a_n$ converges.

Now suppose that $\sum_{n=1}^\infty a_n$ converges. Then $\lim_{k\to\infty} s_k$ exists, with value $S$, say, and since the $s_k$ are nondecreasing, we have from (23.4.2),

$$\int_1^{k+1} f(x)\, dx \leq s_k \leq \lim_{k\to\infty} s_k = S, \qquad k \times 1, 2, \ldots.$$

Since $f(x) \geq 0$ it follows that

(23.4.3) $$\int_1^t f(x)\,dx \leq S, \quad \text{for any} \quad t \geq 2,$$

the limit of the integral in (23.4.3) exists as $t \to \infty$, and the improper integral converges.

We have shown that if either the improper integral or the infinite series converges, then so does the other. It is a logical consequence of this that if one of them diverges, so must the other. The proof is complete.

We shall illustrate the integral test in establishing the following important result.

**THEOREM 23.4.2.** *The* **hyperharmonic** *series, or p-series,*

$$\sum_{n=1}^{\infty} \frac{1}{n^p} = 1 + \frac{1}{2^p} + \frac{1}{3^p} + \cdots + \frac{1}{n^p} + \cdots$$

*converges if $p > 1$, and diverges if $p \leq 1$.*

*Proof.* If $p \leq 0$ the series diverges, by Theorem 23.3.4, since $\lim_{n \to \infty} 1/n^p \neq 0$.

If $p > 0$, let $f(x) = 1/x^p$. Then $f(x)$ is positive and nonincreasing for $x \geq 1$, and $f(n) = 1/n^p$. We have

$$\int_1^t \frac{dx}{x^p} = \int_1^t x^{-p}\,dx = \begin{cases} \left.\dfrac{x^{1-p}}{1-p}\right|_1^t & \text{if } p \neq 1 \\[6pt] \left.\ln x\right|_1^t & \text{if } p = 1 \end{cases}$$

$$= \begin{cases} \dfrac{t^{1-p} - 1}{1-p} & \text{if } p \neq 1 \\[6pt] \ln t & \text{if } p = 1 \end{cases}$$

If $p = 1$, the integral diverges since $\ln t \to \infty$ as $t \to \infty$. If $p \neq 1$, we have

$$\lim_{t \to \infty} \frac{t^{1-p} - 1}{1 - p} = \begin{cases} 1/(p-1) & \text{if } p > 1 \\ \infty & \text{if } 0 < p < 1. \end{cases}$$

Hence the integral converges when $p > 1$, and diverges when $p < 1$, the same is true of the series, and the proof is complete.

When $p = 1$ the hyperharmonic series becomes

$$\sum_{n=1}^{\infty} \frac{1}{n} = 1 + \frac{1}{2} + \frac{1}{3} + \cdots + \frac{1}{n} + \cdots;$$

this series is called the *harmonic* series, and is divergent. Note that for this divergent series $\lim_{n \to \infty} a_n = \lim_{n \to \infty} 1/n = 0$.

It may be instructive to see a direct proof of the divergence of the harmonic series. If $s_n$ is the $n$th partial sum, we have

$$s_{2n} - s_n = \frac{1}{n+1} + \frac{1}{n+2} + \cdots + \frac{1}{2n}.$$

If $0 < k \leq n$, then $n + k \leq 2n$, each term in the sum on the right is greater than or equal to $1/(2n)$, and since there are $n$ terms we obtain

(23.4.4) $$s_{2n} - s_n \geq 1/2.$$

Now, if $\lim_{n \to \infty} s_n$ exists, then $\lim_{n \to \infty} s_{2n}$ exists and has the same value, and hence $\lim_{n \to \infty} (s_{2n} - s_n) = 0$. But this is impossible by (23.4.4). Thus $\lim_{n \to \infty} s_n$ cannot exist, and the harmonic series diverges.

We see that when $a_n$ is positive and nonincreasing, we merely set $f(n) = a_n$. It is then required that $f(x)$ be defined also when $x$ is not an integer, that $f(x)$ be positive and nonincreasing, and that we be able to test the convergence of $\int_1^\infty f(x)\, dx$. Thus it is not feasible to apply the integral test, for example, when $a_n = 1/n!$, since $x!$ is defined only when $x$ is a nonnegative integer. Moreover, we would not apply the test to something like $a_n = 1/\sqrt{n^3 + 1}$, since we cannot find an antiderivative of $1/\sqrt{x^3 + 1}$. (In the study of more advanced mathematics, tests of convergence are devised for improper integrals which often circumvent the need to find an antiderivative.)

It is clear that the lower limit of the improper integral need not be 1 but may be any integer $\geq 1$, or in fact any number $\geq 1$. It is the behavior of the integrand for large $x$ that is significant.

To illustrate the use of the test further, we give another example, which can perhaps be done better by other tests.

**Example 23.4.1.** Test the series $\sum_{n=1}^{\infty} \frac{1}{\sqrt{1+n^2}}$ for convergence.

*Solution.* The requirements of the integral test are clearly met. We have (perhaps using the substitution $x = \tan \theta$ for the integration)

$$\int_1^t \frac{dx}{\sqrt{1+x^2}} = \ln(x + \sqrt{1+x^2})\Big|_1^t$$
$$= \ln(t + \sqrt{1+t}\,) - \ln(1 + \sqrt{2}).$$

This function approaches infinity as $t \to \infty$, and the series diverges.

**EXERCISE GROUP 23.4**

Test each of the following series for convergence or divergence. The integral test applies in each case, but it is not necessarily the best test to use.

1. $\sum_{n=1}^{\infty} \frac{1}{2n}$  2. $\sum_{n=1}^{\infty} \frac{1}{2n-1}$  3. $\sum_{n=1}^{\infty} \frac{1}{n(n+1)}$

4. $\sum_{n=2}^{\infty} \dfrac{1}{n^2 - 1}$

5. $\sum_{n=2}^{\infty} \dfrac{1}{\sqrt{n^2 - 1}}$

6. $\sum_{n=1}^{\infty} \dfrac{n}{n^2 + 1}$

7. $\sum_{n=2}^{\infty} \dfrac{1}{n \ln n}$

8. $\sum_{n=1}^{\infty} \dfrac{1}{\sqrt{2n - 1}}$

9. $\sum_{n=1}^{\infty} 2^{-n}$

10. $\sum_{n=1}^{\infty} n \cdot 2^{-n}$

11. $\sum_{n=1}^{\infty} \dfrac{\ln n}{n}$

12. $\sum_{n=1}^{\infty} \dfrac{\ln n}{n^2}$

13. $\sum_{n=1}^{\infty} \cot^{-1} n$

14. $\sum_{n=1}^{\infty} n^2 \cdot 3^{-n}$

## 23.5 COMPARISON TESTS FOR SERIES

The convergence or divergence of a series of positive terms can often be determined by relating its terms to those of a known series. The following tests involve a direct comparison of the magnitudes of corresponding terms.

**THEOREM 23.5.1. Comparison tests.** Let $\sum_{n=1}^{\infty} a_n$ and $\sum_{n=1}^{\infty} b_n$ be two series such that

$$0 \leq a_n \leq b_n, \quad n = 1, 2, 3, \ldots.$$

1. If $\sum_{n=1}^{\infty} b_n$ converges, then $\sum_{n=1}^{\infty} a_n$ converges.

2. If $\sum_{n=1}^{\infty} a_n$ diverges, then $\sum_{n=1}^{\infty} b_n$ diverges.

*Proof.* Let $s_k$ and $s_k'$ be partial sums of $\sum_{n=1}^{\infty} a_n$ and $\sum_{n=1}^{\infty} b_n$, respectively. We then have

(23.5.1) $$s_k = \sum_{n=1}^{k} a_n \leq \sum_{n=1}^{k} b_n = s_k'.$$

Now suppose that $\sum_{n=1}^{\infty} b_n$ converges, to a sum $S$. Then $\lim_{k \to \infty} s_k' = S$ and, since $s_k'$ is increasing, $s_k' \leq S$. It follows that $s_k \leq S$, by (23.5.1). Hence $\{s_k\}$ is a bounded, nondecreasing sequence, and $\lim_{k \to \infty} s_k$ exists, by Section 23.2. The series $\sum_{n=1}^{\infty} a_n$ thus converges, and part 1 is proved.

The proof of part 2 consists of noting that this statement is logically equivalent to the statement of part 1. This completes the proof.

The use of the comparison tests depends on a knowledge of the convergence or divergence of various series. Whenever the convergence or divergence of a series of positive terms has been established by any method, that series becomes available as a comparison series. Perhaps the most useful series for this purpose are the geometric series and *p*-series, whose behavior is completely known.

**Example 23.5.1.** Test the series

$$\frac{1}{1\cdot 2}+\frac{1}{2\cdot 3}+\frac{1}{3\cdot 4}+\cdots+\frac{1}{n(n+1)}+\cdots.$$

*Solution.* We establish a direct comparison with the convergent $p$-series with $p=2$,

$$1+\frac{1}{2^2}+\frac{1}{3^2}+\cdots+\frac{1}{n^2}+\cdots,$$

since

$$\frac{1}{n(n+1)}<\frac{1}{n^2}, \qquad n=1,2,3,\ldots$$

The given series then converges by Theorem 23.5.1, part 1. (See Example 23.3.1 for the same series.)

The comparisons in Theorem 23.5.1 were stated for $n \geq 1$. We know from Theorem 23.3.3 that convergence or divergence of a series is not affected by disregarding a finite number of the terms at the beginning of a series. Hence the relation $a_n \leq b_n$ need be established only for $n$ greater than some positive integer $N$, and the proof of the comparison tests goes through with almost no change.

**Example 23.5.2.** Test the series

(23.5.2) $\quad 1+\dfrac{1}{\sqrt{3}}+\dfrac{1}{\sqrt{8}}+\dfrac{1}{\sqrt{17}}+\cdots+\dfrac{1}{\sqrt{n^2-1}}+\cdots, \qquad n \geq 2.$

*Solution.* Since $n^2 - 1 < n^2$, it follows that

$$\frac{1}{\sqrt{n^2-1}} > \frac{1}{n}, \qquad n \geq 2.$$

The series $\sum\limits_{n=1}^{\infty} 1/n$ is known to diverge. By the comparison test, part 2, the given series also diverges.

We emphasize that each comparison test applies in only one way. *In order to prove convergence we must show that the terms of the given series are term by term less than or equal to the terms of a convergent series; in order to prove divergence we must show that the terms of the given series are term by term greater than or equal to the terms of a divergent series.* For example, if $a_n = 1/[n(n-1)]$, $n = 2, 3, \ldots,$ and $b_n = 1/n^2$, we have

$$a_n = \frac{1}{n(n-1)} > \frac{1}{n^2} = b_n, \qquad n = 2, 3, \ldots.$$

However, $\sum\limits_{n=1}^{\infty} 1/n^2$ converges, and neither comparison test yields any conclusion. On the other hand, we may also make the comparison

$$\frac{1}{n(n-1)} < \frac{1}{(n-1)^2}, \qquad n = 2, 3, \ldots,$$

and in this way conclude that $\sum_{n=2}^{\infty} a_n$ converges.

## *23.6  AN EXTENDED COMPARISON TEST

An analysis of $a_n$ in an infinite series from a point of view to be presented can often determine whether a comparison test can be used. For example, consider the series $\sum_{n=1}^{\infty} a_n$ and $\sum_{n=1}^{\infty} b_n$ where

$$a_n = \frac{1}{n}, \qquad b_n = \frac{n}{n^2 + 1}.$$

We see that $b_n < a_n$; however, since $\sum_{n=1}^{\infty} a_n$ diverges, the usual comparison test fails. We note that when $n$ is large the denominator of $b_n$ differs from $n^2$ by relatively little, so that $b_n$ differs by little from $n/n^2 = 1/n$. It would then appear that $\sum_{n=1}^{\infty} b_n$ should behave about the same as $\sum_{n=1}^{\infty} a_n$, and should diverge. If this reasoning is valid, a similar type of discussion would indicate that the series $\sum_{n=1}^{\infty} (n+1)/[(n+2)(n+3)]$ diverges (it does!). These ideas are made precise in the following test.

**THEOREM 23.6.1.  Extended comparison test.** *Suppose*

$$a_n > 0, \qquad b_n > 0, \qquad \text{and} \qquad \lim_{n \to \infty} \frac{a_n}{b_n} = c \neq 0.$$

*Then the series* $\sum_{n=1}^{\infty} a_n$ *and* $\sum_{n=1}^{\infty} b_n$ *converge or diverge together.*

*Proof.* Since we must have $c > 0$, we may choose a number $d$ such that $0 < d < c$. There is then an integer $N$ such that

$$c - d < \frac{a_n}{b_n} < c + d, \qquad n > N,$$

or

(23.6.1) $\qquad (c-d)b_n < a_n < (c+d)b_n, \qquad n > N.$

If $\sum_{n=1}^{\infty} b_n$ converges, so does $\sum_{n=1}^{\infty} (c+d)b_n$, and by (23.6.1) and Theorem 23.5.1, $\sum_{n=1}^{\infty} a_n$ converges. If $\sum_{n=1}^{\infty} b_n$ diverges, so does $\sum_{n=1}^{\infty} (c-d)b_n$, and hence by (23.6.1) $\sum_{n=1}^{\infty} a_n$ diverges. This completes the proof.

***Example 23.6.1.*** Test the series $\sum_{n=1}^{\infty} \dfrac{n+1}{(n+2)(n+3)}$ for convergence.

***Solution.*** In the discussion preceding Theorem 23.6.1 it appeared that the series behaves like $\sum_{n=1}^{\infty} 1/n$. We have

$$\lim_{n \to \infty} \frac{n+1}{(n+2)(n+3)} \div \frac{1}{n} = \lim_{n \to \infty} \frac{n(n+1)}{(n+2)(n+3)} = 1.$$

By Theorem 23.6.1 the given series diverges, since $\sum_{n=1}^{\infty} 1/n$ diverges.

In Section 23.5 we attempted a direct comparison between the two series with

$$a_n = \frac{1}{n(n-1)}, \quad b_n = \frac{1}{n^2},$$

a comparison which failed. With the extended comparison test we now see that the two series behave the same, and both converge.

The method of proof of Theorem 23.6.1 is valid when $\sum_{n=1}^{\infty} b_n$ converges even if $c = 0$, and we may state the following.

**COROLLARY.** *Suppose that* $a_n > 0$, $b_n > 0$, *and* $\lim_{n \to \infty} a_n/b_n = 0$; *then if* $\sum_{n=1}^{\infty} b_n$ *converges, so does* $\sum_{n=1}^{\infty} a_n$.

**EXERCISE GROUP 23.5**

In Exercises 1–20 test the series for convergence or divergence by a comparison test.

1. $\sum_{n=1}^{\infty} \dfrac{1}{(n+1)(n+2)}$

2. $\sum_{n=1}^{\infty} \dfrac{1}{n^2 + 1}$

3. $\sum_{n=2}^{\infty} \dfrac{n}{n^2 - 1}$

4. $\sum_{n=1}^{\infty} \dfrac{n}{(n+1)3^n}$

5. $\sum_{n=1}^{\infty} \dfrac{1}{n \cdot 2^n}$

6. $\sum_{n=1}^{\infty} \dfrac{1}{n!}$

7. $\sum_{n=1}^{\infty} \dfrac{1}{\sqrt{n!}}$

8. $\sum_{n=2}^{\infty} \dfrac{1}{\ln n}$

9. $\sum_{n=1}^{\infty} \dfrac{n+2}{n(n+1)}$

10. $\sum_{n=1}^{\infty} \dfrac{1}{\sqrt{n^2 + 1}}$

11. $\sum_{n=1}^{\infty} \dfrac{1}{\sqrt{n^2 - 1/4}}$

12. $\sum_{n=2}^{\infty} \dfrac{1}{n\sqrt{n^2 - 1}}$

13. $\sum_{n=1}^{\infty} \dfrac{\sqrt{n}}{n^2 + 1}$

14. $\sum_{n=1}^{\infty} \dfrac{1}{\sqrt{n+1}}$

15. $\displaystyle\sum_{n=1}^{\infty} \frac{n+1}{n \cdot 2^n}$

16. $\displaystyle\sum_{n=1}^{\infty} \frac{1}{n+\sqrt{n^2+1}}$

17. $\displaystyle\sum_{n=1}^{\infty} \frac{n!}{3 \cdot 5 \cdot 7 \cdots (2n+1)}$

18. $\displaystyle\sum_{n=1}^{\infty} \frac{1}{n^n}$

* 19. $\displaystyle\sum_{n=1}^{\infty} \sin \frac{1}{n}$

* 20. $\displaystyle\sum_{n=1}^{\infty} \sin^2 \frac{1}{n}$

* 21. Suppose $a_n > 0$, $b_n > 0$. (a) If $\dfrac{a_{n+1}}{a_n} \leq \dfrac{b_{n+1}}{b_n}$ for $n$ sufficiently large, $n \geq N$, and if $\displaystyle\sum_{n=1}^{\infty} b_n$ converges, prove that $\displaystyle\sum_{n=1}^{\infty} a_n$ converges. (b) If $\dfrac{a_{n+1}}{a_n} \geq \dfrac{b_{n+1}}{b_n}$ for $n$ sufficiently large, $n \geq N$, and if $\displaystyle\sum_{n=1}^{\infty} b_n$ diverges, prove that $\displaystyle\sum_{n=1}^{\infty} a_n$ diverges. [*Hint* for (a). We write

$$\frac{a_N}{b_N} \geq \frac{a_{N+1}}{b_{N+1}} \geq \frac{a_{N+2}}{b_{N+2}} \geq \cdots.$$

Let $a_N = k b_N$. Then $a_{N+1} \leq k b_{N+1}$, $a_{N+2} k \leq b_{N+2}$, ....]

* 22. Let $\displaystyle\sum_{n=1}^{\infty} a_n$ and $\displaystyle\sum_{n=1}^{\infty} b_n$ be two series with $a_n > 0$, $b_n > 0$, such that $\displaystyle\lim_{n \to \infty} a_n/b_n = 1$.

(a) If $\displaystyle\sum_{n=1}^{\infty} b_n$ converges, and $r$ is any number $r > 1$, prove that there exists $N$ such that there is a direct comparison of the form $a_n < r b_n$, $n \geq N$.

(b) If $\displaystyle\sum_{n=1}^{\infty} b_n$ diverges, and $r$ is any number $0 < r < 1$, prove that there exists $N$ such that there is a direct comparison of the form $a_n > r b_n$, $n \geq N$.

## 23.7 THE RATIO TEST FOR SERIES

We present one additional test for series of positive terms.

**THEOREM 23.7.1. The Ratio Test.** *Let $a_n > 0$ be the general term of an infinite series, and let*

$$\lim_{n \to \infty} \frac{a_{n+1}}{a_n} = L.$$

*Then:*

*if $L < 1$, the series converges;*

*if $L > 1$, the series diverges;*

*if $L = 1$, there is no conclusion—the series may converge or it may diverge.*

*Proof.* Suppose $L < 1$. Choose $r$ so that $L < r < 1$. Then an integer $N$ exists such that

$$\frac{a_{n+1}}{a_n} < r \quad \text{whenever} \quad n > N,$$

and we may write

$$\frac{a_n}{a_N} = \frac{a_{N+1}}{a_N} \cdot \frac{a_{N+2}}{a_{N+1}} \cdots \frac{a_n}{a_{n-1}} < \underbrace{r \cdot r \cdots r}_{(n-N) \text{ factors}} = r^{n-N}, \quad n > N.$$

Hence $a_n < a_N r^{n-N}$, $n > N$. The series with general term $a_N r^{n-N}$ is a convergent geometric series since $r < 1$. Hence by the comparison test the series $\sum_{n=1}^{\infty} a_n$ converges. Note that the comparison of terms has been shown only for $n > N$.

Now suppose that $L > 1$. An integer $N$ exists such that

$$\frac{a_{n+1}}{a_n} > 1 \quad \text{whenever} \quad n > N.$$

Thus $a_{n+1} > a_n$, $n > N$, and the terms of the series eventually increase. Thus we cannot have $\lim_{n \to \infty} a_n = 0$, and by Theorem 23.3.4 the series diverges.

We now consider the case $L = 1$. For the $p$-series, with $a_n = 1/n^p$, we have

$$\frac{a_{n+1}}{a_n} = \frac{n^p}{(n+1)^p} = \left(\frac{n}{n+1}\right)^p \to 1 \quad \text{as} \quad n \to \infty.$$

We know that the series converges when $p > 1$, and diverges when $p \leq 1$, but in both cases the limit of $a_{n+1}/a_n$ is 1. Hence, if $L = 1$, the test fails, since the same limit occurs for some convergent series as well as some divergent series.

***Example 23.7.1.*** Test the series $\sum_{n=1}^{\infty} \frac{2n+1}{3^n}$.

*Solution.* Applying the ratio test, we have

$$\frac{a_{n+1}}{a_n} = \frac{2n+3}{3^{n+1}} \cdot \frac{3^n}{2n+1} = \frac{1}{3} \cdot \frac{2n+3}{2n+1} \to \frac{1}{3} < 1 \quad \text{as} \quad n \to \infty.$$

Since $L = 1/3 < 1$, the series converges.

***Example 23.7.2.*** Test the series $\sum_{n=1}^{\infty} \frac{n^n}{n!}$.

*Solution.* We have

$$\frac{a_{n+1}}{a_n} = \frac{(n+1)^{n+1}}{(n+1)!} \cdot \frac{n!}{n^n} = \left(\frac{n+1}{n}\right)^n.$$

Now, with use of the expression for $e$ in (11.3.6),

$$\left(\frac{n+1}{n}\right)^n = \left(1 + \frac{1}{n}\right)^n \to e \quad \text{as} \quad n \to \infty.$$

Hence $\lim_{n \to \infty} a_{n+1}/a_n = e > 1$, and the given series diverges, by the ratio test.

## EXERCISE GROUP 23.6

Examine each of the following series for convergence or divergence.

1. $\sum_{n=1}^{\infty} \dfrac{1}{n!}$

2. $\sum_{n=1}^{\infty} \dfrac{1}{\sqrt{n!}}$

3. $\sum_{n=1}^{\infty} \dfrac{n^2}{3^n}$

4. $\sum_{n=1}^{\infty} \dfrac{n!}{2^n}$

5. $\sum_{n=1}^{\infty} \dfrac{4^n}{n!}$

6. $\sum_{n=1}^{\infty} \dfrac{1}{(2n)!}$

7. $\sum_{n=1}^{\infty} \dfrac{1}{n^n}$

8. $\sum_{n=1}^{\infty} \dfrac{2^n}{n^n}$

9. $\sum_{n=1}^{\infty} \dfrac{n!}{n^n}$

10. $\sum_{n=1}^{\infty} \dfrac{10^{n/2}}{n!}$

11. $\sum_{n=1}^{\infty} \dfrac{7^{2n}}{(2n)!}$

12. $\sum_{n=1}^{\infty} \dfrac{4^n}{(2n-1)!}$

13. $1 + \dfrac{1\cdot 2}{1\cdot 3} + \dfrac{1\cdot 2\cdot 3}{1\cdot 3\cdot 5} + \dfrac{1\cdot 2\cdot 3\cdot 4}{1\cdot 3\cdot 5\cdot 7} + \cdots$

14. $2 + \dfrac{2\cdot 5}{1\cdot 3}\cdot\dfrac{1}{3} + \dfrac{2\cdot 5\cdot 8}{1\cdot 3\cdot 5}\cdot\dfrac{1}{3^2} + \cdots$

## 23.8 TESTING A SERIES OF POSITIVE TERMS

No single test for convergence will serve for all series. In the preceding sections a number of very useful tests have been presented; other tests are developed in more advanced treatments of the subject. Many series can be tested by more than one of the tests presented earlier. However, for certain types of series, one test may be more easily applied than another.

The use of the integral test was discussed earlier (Section 23.4). The ratio test is generally applicable to a series in which $a_n$ contains a factorial, or an exponential (in which $n$ appears in an exponent), or both. The discussion in Section 23.6 may serve as a guide to series for which a comparison test is feasible. In the last analysis, however, no complete criteria can be given for examining a series for convergence, and there is no substitute for the experience gained in applying the various tests to specific series.

An infinite series of positive terms converges if the general term $a_n$ approaches zero "rapidly enough," and diverges if either $a_n$ does not approach zero or else approaches zero too slowly. No series exists which may be considered as a dividing line between convergent and divergent series. Given a convergent series of positive terms, it is always possible to find another convergent series with larger terms; given a divergent series, one can devise another divergent series with smaller terms.

The following exercises should be approached with an eye toward the choice of test to make, as well as the actual application of the test.

**EXERCISE GROUP 23.7**

In each of the following test the series for convergence or divergence.

1. $\sum_{n=1}^{\infty} \dfrac{1}{n+p}, p > 0$
2. $\sum_{n=1}^{\infty} \dfrac{3^{2n}}{n!}$
3. $\sum_{n=2}^{\infty} \dfrac{n}{n^2 - 1}$
4. $\sum_{n=1}^{\infty} \dfrac{1}{(n^2 + 1)^{1/3}}$
5. $\sum_{n=1}^{\infty} \dfrac{3^n}{n(2^n + 1)}$
6. $\sum_{n=1}^{\infty} n^2 e^{-n}$
7. $\sum_{n=1}^{\infty} \dfrac{n!}{(2n)!}$
8. $\sum_{n=2}^{\infty} \dfrac{1}{n(\ln n)^2}$
9. $\sum_{n=0}^{\infty} \dfrac{2^n}{4^n + 1}$
10. $\sum_{n=1}^{\infty} \dfrac{n}{(n+1)^{5/2}}$
11. $\sum_{n=1}^{\infty} \dfrac{2n-1}{n(n+2)}$
12. $\sum_{n=1}^{\infty} \left(1 + \dfrac{1}{n}\right)^{-n}$
13. $\sum_{n=1}^{\infty} \left(2 + \dfrac{2}{n}\right)^n$
14. $\sum_{n=1}^{\infty} \dfrac{n}{\sqrt{n^3 + 1}}$
15. $\sum_{n=1}^{\infty} \dfrac{1}{n^{(n+1)/n}}$
16. $\sum_{n=1}^{\infty} \dfrac{1 \cdot 3 \cdot 5 \cdots (2n-1)}{2 \cdot 4 \cdot 6 \cdots (2n)}$
17. $\sum_{n=1}^{\infty} \left(1 - \dfrac{1}{n}\right)^n$
18. $\sum_{n=1}^{\infty} \dfrac{1}{\sqrt[3]{n!}}$

\* 19. $\sum_{n=1}^{\infty} \dfrac{1}{\sqrt[n]{n!}}$

\* 20. $\sum_{n=2}^{\infty} \dfrac{1 \cdot 3 \cdot 5 \cdots (2n-3)}{4 \cdot 6 \cdot 8 \cdots (2n)}$

## 23.9 ALTERNATING SERIES

Except for such general considerations as definitions and Theorem 23.3.4, we have hitherto been concerned with series of all positive terms (or series with all negative terms). We turn our attention to a special type of series with terms of varying signs, a series in which the signs alternate. Such a series is called an *alternating series*, and may be written as

(23.9.1) $\quad b_1 - b_2 + b_3 - \cdots + (-1)^{n+1} b_n + \cdots, \quad b_n > 0.$

The following test applies specifically to such series.

**THEOREM 23.9.1. Alternating series test.** Let $\sum_{n=1}^{\infty} (-1)^{n+1} b_n$, $b_n > 0$, be an alternating series for which

(a) $b_1 \geq b_2 \geq b_3 \geq \cdots$

(b) $\lim_{n \to \infty} b_n = 0.$

*Then the series converges.*

*Proof.* If $n$ is even, with $n = 2m$, we write the partial sum $s_n$ in the two forms:

(23.9.2) $\quad s_{2m} = (b_1 - b_2) + (b_3 - b_4) + \cdots + (b_{2m-1} - b_{2m}),$

(23.9.3) $\quad s_{2m} = b_1 - (b_2 - b_3) - \cdots - (b_{2m-2} - b_{2m-1}) - b_{2m}.$

Since each difference in (23.9.2) is nonnegative, $s_{2m}$ is nonnegative and nondecreasing. From (23.9.3) we have $s_{2m} \leq b_1$. Then by the completeness property of Section 23.2 the sequence $s_{2m}$ converges, say to $S$. Given $\epsilon > 0$, we can find $N_1$ such that

(23.9.4) $\qquad |s_{2m} - S| < \epsilon/2 \quad \text{whenever} \quad m > N_1.$

Now $s_{2m+1} - s_{2m} = b_{2m+1}$, so that by hypothesis (b) we can find $N_2$ such that

(23.9.5) $\qquad |s_{2m+1} - s_{2m}| < \epsilon/2 \quad \text{whenever} \quad m > N_2.$

If $N$ is the larger of $N_1$ and $N_2$, we have, with (23.9.4) and (23.9.5),

(23.9.6) $\quad |s_{2m+1} - S| = |(s_{2m+1} - s_{2m}) + (s_{2m} - S)| < \epsilon/2 + \epsilon/2 = \epsilon, \quad m > N.$

Combining (23.9.4) and (23.9.6) we see that $|s_n - S| < \epsilon$ for $n$ large enough, regardless of whether $n$ is odd or even, and the series must converge.

**Example 23.9.1.** Test the series

$$1 - \tfrac{1}{2} + \tfrac{1}{3} - \tfrac{1}{4} + \cdots$$

*Solution.* This is of the form (23.9.1) with $b_n = 1/n$. We see that $b_n$ decreases with $n$ increasing, and also $\lim_{n \to \infty} b_n = 0$. By the alternating series test the series converges.

It is clear that hypothesis (a) may be disregarded for a fixed number of terms at the beginning of a series, but the test cannot be applied unless (a) holds for $n$ sufficiently large. It should be noted that condition (b) is not sufficient for the convergence of an arbitrary series, but it is sufficient for an alternating series for which (a) also holds.

For an alternating series satisfying the hypotheses of the above test a simple criterion is available to estimate how closely a partial sum approximates the sum of the series. Let us write

(23.9.7) $\qquad S = b_1 - b_2 + b_3 - \cdots.$

Then

$$0 \leq S - s_{2m} = b_{2m+1} - (b_{2m+2} - b_{2m+3}) - \cdots \leq b_{2m+1}$$

and also

$$0 \leq s_{2m+1} - S = b_{2m+2} - (b_{2m+3} - b_{2m+4}) - \cdots \leq b_{2m+2}.$$

We may now state that *the difference between the sum of the series* (23.9.7) *and any partial sum does not exceed in absolute value the magnitude of the first neglected term:*

$$|S - s_n| \leq b_{n+1}.$$

Also, the partial sums $s_n$ oscillate, being greater than $S$ if $n$ is odd, and less than $S$ if $n$ is even. Fig. 23.9.1 shows the relation between the partial sums and the sum of a series (23.9.7).

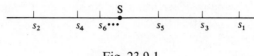

Fig. 23.9.1

**Example 23.9.2.** In order to have 2-decimal accuracy in a partial sum approximation to the sum of the series in Example 23.9.1 we must have

$$|S - s_n| \leq 0.005.$$

By the criterion previously stated, this can be attained by having

$$\frac{1}{n+1} \leq 0.005 \quad \text{or} \quad n \geq 199.$$

Hence $s_{199}$ yields the required accuracy, and $s_{199} > S$. The series converges very slowly and is not well suited for computational purposes. Of course, it is possible that fewer than 199 terms will also yield the required accuracy.

## 23.10 ABSOLUTE CONVERGENCE OF SERIES

The convergence of the series $\sum_{n=1}^{\infty} (-1)^{n+1}/n$ and the divergence of the series $\sum_{n=1}^{\infty} 1/n$ demonstrate that for some series the signs of the terms have a bearing on the question of convergence, apart from the sum of the series. We shall establish a criterion for convergence to be independent of the signs of the terms.

**THEOREM 23.10.1.** *If the series $\sum_{n=1}^{\infty} |u_n|$ converges, the series $\sum_{n=1}^{\infty} u_n$ also converges.*

*Proof.* Let the partial sums of the two series be

$$s_n = u_1 + u_2 + \cdots + u_n,$$
$$s_n' = |u_1| + |u_2| + \cdots + |u_n|.$$

By hypothesis the sequence $\{s_n'\}$ converges, say to $S$. Let $v_n$ be the sum of the terms of $s_n$ which are positive, and $-w_n$ the sum of the terms of $s_n$ which are negative. We then have

(23.10.1) $$s_n = v_n - w_n, \quad s_n' = v_n + w_n.$$

Now $s_n'$, $v_n$, and $w_n$ are nonnegative, and each of the corresponding sequences is nondecreasing, since the terms in each of these sums are nonnegative. Since $\lim_{n \to \infty} s_n' = S$, we have

$$v_n + w_n = s_n' \leq S, \quad \text{hence} \quad v_n \leq S, \quad w_n \leq S.$$

Then $\{v_n\}$ and $\{w_n\}$, being nondecreasing, bounded sequences, have limits, say $v$ and $w$. It follows from (23.10.1) that

$$\lim_{n \to \infty} s_n = v - w;$$

hence $\sum_{n=1}^{\infty} u_n$ converges, and the proof is complete.

**DEFINITION 23.10.1.** *A series* $\sum_{n=1}^{\infty} u_n$ *is called* **absolutely convergent** *if* $\sum_{n=1}^{\infty} |u_n|$ *converges.*

*A series* $\sum_{n=1}^{\infty} u_n$ *is called* **conditionally convergent** *if the series itself converges, while* $\sum_{n=1}^{\infty} |u_n|$ *diverges, i.e., if the series converges, but not absolutely.*

Theorem 23.10.1 can now be reworded to state that *an absolutely convergent series is convergent.*

**COROLLARY.** *If* $\sum_{n=1}^{\infty} u_n$ *is absolutely convergent, then*

$$\left| \sum_{n=1}^{\infty} u_n \right| \leq \sum_{n=1}^{\infty} |u_n|.$$

*Proof.* With the notation used in the proof of Theorem 23.10.1 we have

$$\left| \sum_{n=1}^{\infty} u_n \right| = |v - w| \leq v + w = \sum_{n=1}^{\infty} |u_n|.$$

The previous discussion for the series $\sum_{n=1}^{\infty} (-1)^{n+1}/n$ proves that the series is conditionally convergent. It is also apparent that *any convergent series of positive terms is automatically absolutely convergent.*

In testing a series for absolute convergence we must test a series of positive terms; the previous tests for such series may be used. An alternating series which is not absolutely convergent may be tested for convergence by the alternating series test, when applicable. If the alternating series test is not applicable, or if the terms of a series do not alternate in sign (and are not of the same sign), other specialized methods are required; such tests do exist but they are not considered here.

***Example 23.10.1.*** Test the following series for absolute convergence:

$$1 - \frac{1}{2^2} + \frac{1}{3^2} - \cdots + (-1)^{n+1} \frac{1}{n^2} + \cdots.$$

*Solution.* The series of absolute values to be tested is

$$1 + \frac{1}{2^2} + \frac{1}{3^2} + \frac{1}{4^2} + \cdots.$$

This is a convergent *p*-series, with $p = 2$, and is convergent. The given series is therefore *absolutely convergent* (and hence *convergent*).

Absolute convergence of a series is stronger than conditional convergence, and absolutely convergent series possess properties that conditionally convergent series do not. We shall indicate some of these properties without proofs.

1. The terms of an absolutely convergent series may be rearranged in any manner without affecting the convergence or the sum of the series. On the other hand, if $u_n \neq 0$, the sum of the positive terms of a conditionally convergent series approaches infinity, and the sum of the negative terms approaches negative infinity. It then can be shown that the terms of a conditionally convergent series may be rearranged so as to make the new series have any sum we like (see Exercise 25).

2. Absolutely convergent series may be multiplied or divided in the same general way as finite sums. We do not give a formal statement of this, but the method is illustrated in Example 23.17.5.

In effect, the arithmetic operations on absolutely convergent series may be performed according to the same rules as for finite sums. The same need not be true of conditionally convergent series.

**EXERCISE GROUP 23.8**

Test each of the series in Exercises 1–8 for convergence or divergence.

1. $\dfrac{1}{1^2} - \dfrac{1}{2^2} + \dfrac{1}{3^2} - \dfrac{1}{4^2} + \cdots$

2. $\dfrac{1}{2 \cdot 3} - \dfrac{2}{3 \cdot 4} + \dfrac{3}{4 \cdot 5} - \dfrac{4}{5 \cdot 6} + \cdots$

3. $\dfrac{1}{3} - \dfrac{2}{3^2} + \dfrac{3}{3^3} - \dfrac{4}{3^4} + \cdots$

4. $\ln 2 - \dfrac{\ln 3}{2} + \dfrac{\ln 4}{3} - \dfrac{\ln 5}{4} + \cdots$

5. $1 - \dfrac{2}{3} + \dfrac{3}{5} - \dfrac{4}{7} + \cdots$

6. $\dfrac{1}{\ln 2} - \dfrac{1}{\ln 4} + \dfrac{1}{\ln 6} - \dfrac{1}{\ln 8} + \cdots$

7. $\dfrac{1}{4} - \dfrac{3}{7} + \dfrac{5}{10} - \dfrac{7}{13} + \cdots$

*8. $\displaystyle\sum_{n=1}^{\infty} (-1)^{n-1} \dfrac{1 \cdot 3 \cdot 5 \cdots (2n-1)}{2 \cdot 4 \cdot 6 \cdots (2n)}$

9. How many terms will insure 2-decimal-place accuracy in computing the sum of the series in Exercise 1?

10. How many terms will insure 2-decimal-place accuracy in computing the sum of the series in Exercise 2?

11. Approximate the error in Exercise 3 incurred by using $s_4$ as the sum of the series.

12. Approximate the error in Exercise 4 incurred by using $s_{20}$ as the sum of the series.

Determine whether each of the series in Exercises 13–24 is absolutely convergent, conditionally convergent, or divergent.

13. The series in Exercise 1
14. The series in Exercise 2
15. The series in Exercise 3
16. The series in Exercise 4
17. The series in Exercise 5
18. The series in Exercise 6

19. $\sum_{n=1}^{\infty} \dfrac{(-1)^{n+1}}{1+2^{1/n}}$

20. $\sum_{n=1}^{\infty} \dfrac{1-\cos n}{n^2}$

21. $\sum_{n=1}^{\infty} \dfrac{(-1)^n}{\sqrt{n^3+1}}$

22. $\sum_{n=1}^{\infty} (-1)^n \dfrac{n}{n^2+1}$

*23. $1 + \dfrac{1}{2^2} - \dfrac{1}{2} - \dfrac{1}{2^3} + \dfrac{1}{2^4} + \dfrac{1}{2^6} - \dfrac{1}{2^5} - \dfrac{1}{2^7} + \cdots$

*24. $\dfrac{1}{2} - \dfrac{1}{3} + \dfrac{1}{2^2} - \dfrac{1}{3^2} + \dfrac{1}{2^3} - \dfrac{1}{3^3} + \cdots$

*25. Justify this argument. Let $\sum_{n=1}^{\infty} u_n$ be a conditionally convergent series with $u_n \neq 0$.
If $A$ is any number, the terms of the series may be rearranged to yield a series with sum $A$ as follows. Start with just enough positive terms of the series to yield a partial sum greater than $A$. Then take just enough negative terms so that the new partial sum is less than $A$. Add just enough additional positive terms to bring the partial sum to a number greater than $A$, and continue this process. Then this rearrangement of the original series converges to $A$.

## 23.11 POWER SERIES

The infinite series which were studied previously were series with constant terms. We may also study a series of variable terms,

(23.11.1) $\quad \sum_{n=0}^{\infty} u_n(x) = u_0(x) + u_1(x) + \cdots + u_n(x) + \cdots,$

in which each term is a function of $x$, all defined in a common domain. For each value of $x$ in the common domain of the $u_n(x)$, the series above is a series of constant terms, and all previous considerations apply; the series may converge or diverge, and any of the earlier tests for convergence may be used when applicable. In particular, the series converges for any value of $x$ for which it converges absolutely. Note that the series starts with $n = 0$. The general term $u_n(x)$ is now the $(n+1)$st term.

We shall be concerned specifically with *power series*, to be considered next.

The most general power series is of the form

(23.11.2)

$\sum_{n=0}^{\infty} a_n(x-a)^n = a_0 + a_1(x-a) + a_2(x-a)^2 + \cdots + a_n(x-a)^n + \cdots,$

in which the general term, $u_n(x) = a_n(x - a)^n$, is the $(n + 1)$st term. The series (23.11.2) is called a *power series in* $x - a$, or a *power series about* $x = a$.

If $a = 0$ in (23.11.2) we have a *power series in* $x$, or a *power series about* $x = 0$,

$$(23.11.3) \qquad \sum_{n=0}^{\infty} a_n x^n = a_0 + a_1 x + a_2 x^2 + \cdots + a_n x^n + \cdots.$$

Since a power series in $x - a$ can be transformed into a power series in $u$ by setting $u = x - a$, any result which may be established for (23.11.3) will furnish a corresponding result for (23.11.2). Power series are particularly useful, and they have special properties.

**THEOREM 23.11.1.** *If the series* (23.11.3) *converges for a value* $x_0$, *it converges absolutely for all values* $x$ *such that* $|x| < |x_0|$.

*Proof.* Let $x_0 \neq 0$ be a number for which (23.11.3) converges, and let $x$ be a number such that $|x| < |x_0|$. We write

$$(23.11.4) \qquad a_n x^n = a_n x_0^n \cdot \frac{x^n}{x_0^n},$$

and set $|x/x_0| = r$; then $r < 1$. Since $\sum_{n=0}^{\infty} a_n x_0^n$ converges, it follows that $\lim_{n \to \infty} a_n x_0^n = 0$ (Theorem 23.3.4), so that for $n$ large enough, say $n > N$, we have $|a_n x_0^n| < 1$. Then from (23.11.4),

$$|a_n x^n| = |a_n x_0^n| \left| \frac{x}{x_0} \right|^n < r^n, \qquad n > N,$$

and by the comparison test, $\sum_{n=0}^{\infty} |a_n x^n|$ converges; this in turn means that (23.11.3) converges absolutely.

An implication of Theorem 23.11.1 is that if the series (23.11.3) converges for any $x \neq 0$, either (i) there must be a number $R > 0$ such that the series converges for $|x| < R$ and diverges for $|x| > R$, or (ii) the series converges for all real numbers $x$. In the latter case we say $R = \infty$. If the series converges only for $x = 0$, then $R = 0$.

**DEFINITION 23.11.1.** *The number* $R$ *just described is the* **radius of convergence** *of the series* (23.11.3). *The interval (finite or infinite, open, closed, or semi-open) in which the series converges is the* **interval of convergence.**

In many cases the radius of convergence can be found by applying the ratio test to the series of absolute values, as will be illustrated. If $R$ is finite, the convergence of the series (23.11.3) for $x = R$ and $x = -R$ must then be investigated separately.

**Example 23.11.1.**  Find all values of $x$ for which this series converges:
$$\sum_{n=0}^{\infty} \frac{(-1)^n n}{3^n(n+1)} x^n.$$

*Solution.*  Applying the ratio test to the series of absolute values, we have
$$\frac{|u_{n+1}|}{|u_n|} = \frac{(n+1)|x|^{n+1}}{3^{n+1}(n+2)} \cdot \frac{3^n(n+1)}{n|x|^n} = \frac{(n+1)^2}{n(n+2)} \cdot \frac{|x|}{3} \to \frac{|x|}{3}, \text{ as } n \to \infty.$$

By the ratio test the series converges absolutely and hence converges, if $|x|/3 < 1$ or $|x| < 3$; the series diverges if $|x| > 3$. The values $x = -3$ and $x = 3$, for which $|x| = 3$, are now examined separately.

If $x = -3$, direct substitution gives the series $\sum_{n=0}^{\infty} n/(n+1)$ for which
$$\lim_{n \to \infty} a_n = \lim_{n \to \infty} \frac{n}{n+1} = 1,$$
and the series diverges by Theorem 23.3.4. If $x = 3$, the series takes the form $\sum_{n=0}^{\infty} (-1)^n n/(n+1)$, and diverges for the same reason. Hence the complete interval of convergence is $-3 < x < 3$, or $(-3, 3)$. The radius of convergence is $R = 3$.

**Example 23.11.2.**  Find all values of $x$ for which the series $\sum_{n=0}^{\infty} x^{2n}/n!$ converges.

*Solution.*  We find
$$\frac{|u_{n+1}|}{|u_n|} = \frac{|x|^{2n+2}}{(n+1)!} \cdot \frac{n!}{|x|^{2n}} = \frac{n!}{(n+1)!}|x|^2 = \frac{1}{n+1}|x|^2 \to 0, \text{ as } n \to \infty.$$

The limit of the ratio is zero for any value of $x$. Hence the series converges (absolutely) for all real $x$, and $R = \infty$. The interval of convergence is $(-\infty, \infty)$.

A power series of the general form (23.11.2) may be treated in a very similar manner.

**Example 23.11.3.**  Find all values of $x$ for which this series converges:
$$\sum_{n=0}^{\infty} \frac{(x+2)^n}{n+1}.$$

*Solution.*  To use the ratio test, we find
$$\frac{|u_{n+1}|}{|u_n|} = \frac{|x+2|^{n+1}}{n+2} \cdot \frac{n+1}{|x+2|^n} = \frac{n+1}{n+2}|x+2| \to |x+2|, \text{ as } n \to \infty.$$

The series converges absolutely, by the ratio test, if
$$|x+2| < 1 \quad \text{or} \quad -3 < x < -1,$$

and diverges outside this interval. Writing the series for $x = -3$ and $x = -1$ separately, we have

$$\text{for } x = -3: \sum_{n=0}^{\infty} \frac{(-1)^n}{n+1} = 1 - \frac{1}{2} + \frac{1}{3} - \frac{1}{4} + \cdots$$

$$\text{for } x = -1: \sum_{n=0}^{\infty} \frac{1}{n+1} = 1 + \frac{1}{2} + \frac{1}{3} + \frac{1}{4} + \cdots.$$

The first is a convergent alternating series, and the second is the divergent harmonic series. Hence the complete interval of convergence is $-3 \leq x < -1$, or $[-3, 1)$. We still say that the radius of convergence is $R = 1$.

In general, *the radius of convergence of a power series is one-half the length of the interval of convergence.*

### EXERCISE GROUP 23.9

Determine the interval of convergence for each series in Exercises 1–17.

1. $\sum_{n=1}^{\infty} (-1)^{n+1} \frac{x^n}{n}$

2. $\sum_{n=1}^{\infty} (-1)^{n+1} \frac{(2n+1)x^n}{n(n+1)}$

3. $\sum_{n=1}^{\infty} \frac{x^{2n}}{n^2}$

4. $\sum_{n=1}^{\infty} \frac{x^n}{n \cdot 2^n}$

5. $\sum_{n=1}^{\infty} \left(1 + \frac{1}{n}\right)^n x^n$

6. $\sum_{n=2}^{\infty} \frac{x^n}{\ln n}$

7. $\sum_{n=0}^{\infty} (-1)^n (x+1)^n$

8. $\sum_{n=0}^{\infty} \frac{(x-2)^n}{4^n}$

9. $\sum_{n=1}^{\infty} \frac{x^n}{n(2^n + 1)}$

10. $\sum_{n=0}^{\infty} \frac{x^{3n}}{n+1}$

11. $\frac{2!}{3} x + \frac{3!}{3 \cdot 6} x^2 + \frac{4!}{3 \cdot 6 \cdot 9} x^3 + \cdots$

12. $1 + (2x) + (2x)^2 + (2x)^3 + \cdots$

13. $2x - \frac{(2x)^3}{3} + \frac{(2x)^5}{5} - \frac{(2x)^7}{7} + \cdots$

14. $\frac{2}{1} x + \frac{2 \cdot 3}{1 \cdot 3} x^2 + \frac{2 \cdot 3 \cdot 4}{1 \cdot 3 \cdot 5} x^3 + \cdots$

★ 15. $\sum_{n=2}^{\infty} \frac{1 \cdot 3 \cdot 5 \cdots (2n-3)}{2^n n!} (x-1)^n$

★ 16. $\sum_{n=1}^{\infty} \frac{n!}{n^n} x^n$

★ 17. $\sum_{n=1}^{\infty} (-1)^{n-1} \frac{1 \cdot 3 \cdot 5 \cdots (2n-1)}{2 \cdot 4 \cdot 6 \cdots (2n)} x^n$

18. If $m$ is not an odd integer, find the radius of convergence of the series

$$\sum_{n=2}^{\infty} (-1)^{n-1} \frac{(m^2 - 1)(m^2 - 9) \cdots [m^2 - (2n-3)^2]}{(2n-1)!} x^{2n-1}.$$

19. If $m$ is not zero or a positive integer, find the radius of convergence of the series

$$1 + mx + \frac{m(m-1)}{2!} x^2 + \cdots + \frac{m(m-1) \cdots (m-k+1)}{k!} x^k + \cdots.$$

The methods of Section 23.11 can often be used for series of functions which are not power series. Determine all values of $x$ for which each of the following series converges.

20. $\sum_{n=0}^{\infty} (-1)^n x^n (1+x)^n$

21. $\sum_{n=0}^{\infty} (-1)^n x^n (1-x)^n$

22. $\sum_{n=1}^{\infty} \frac{x^n}{x^{2n}+1}$

23. $\sum_{n=1}^{\infty} \frac{x^n}{(x+1)^n}$

24. $\sum_{n=1}^{\infty} \frac{1}{2n-1} \left( \frac{x-1}{x+1} \right)^{2n-1}$

## 23.12 SUMS OF POWER SERIES

In the last section we considered the convergence of a power series. In the present section we shall study the sum of a convergent power series. Since any general result for a power series in $x$ transfers immediately to a power series in $x - a$, we shall give a detailed development for a power series in $x$.

A power series represents a function in the interval of convergence. If the radius of convergence is $R$, we may write

(23.12.1) $$f(x) = \sum_{n=0}^{\infty} a_n x^n, \quad |x| < R.$$

We should like to be able to differentiate and integrate a power series in the same way as a finite sum, term by term. For example, for $f(x)$ given in (23.12.1) we should like to be able to write, since $D(a_n x^n) = n a_n x^{n-1}$,

(23.12.2) $$f'(x) = \sum_{n=1}^{\infty} n a_n x^{n-1} = a_1 + 2a_2 x + 3a_3 x^2 + \cdots.$$

Such a statement requires proof—the validity for a finite sum does not carry over automatically to an infinite series. Fortunately, it is valid to do what we should like to do. The results will be given with proofs, to show some of the flavor of proofs involving infinite series. However, the theorems are easily understood without the proofs, and are to the effect that term-by-term integration and differentiation of a convergent power series are valid in the interval of convergence.

The first result tells us that the series in (23.12.2) converges.

**THEOREM 23.12.1.** *If the series* $\sum_{n=0}^{\infty} a_n x^n$ *converges for* $|x| < R$, *then the series* $\sum_{n=1}^{\infty} n a_n x^{n-1}$ *also converges for* $|x| < R$, *and absolutely.*

*Proof.* The second series clearly converges when $x = 0$. Now if $|x| < R$, $x \neq 0$, choose $x_0$ such that $|x| < |x_0| < R$. Letting $|x/x_0| = r < 1$, we have

$$|a_n x^n| = |a_n x_0^n| |x/x_0|^n = |a_n x_0^n| r^n.$$

Since the first series converges for $x_0$, there is an integer $N$ such that $|a_n x_0^n| < 1$, $n > N$. Hence $|a_n x^n| < r^n$ for $n > N$, and

$$|na_n x^{n-1}| = \frac{n}{|x|}|a_n x^n| < \frac{n}{|x|}r^n, \quad n > N.$$

Now $\sum_{n=1}^{\infty} nr^n/|x|$ converges by the ratio test, since $r < 1$. Hence the series $\sum_{n=1}^{\infty} na_n x^{n-1}$ converges absolutely, and therefore converges. Thus the second series converges for any $x$ such that $|x| < R$.

Theorem 23.12.1 enables us to prove the *continuity* of the sum of a convergent power series.

**THEOREM 23.12.2.** *If* $f(x) = \sum_{n=0}^{\infty} a_n x^n$ *converges for* $|x| < R$, *then* $f(x)$ *is continuous for* $|x| < R$.

*Proof.* Let $x_0$ be any number for which $|x_0| < R$. We wish to show that

(23.12.3) $$\lim_{x \to x_0} |f(x) - f(x_0)| = 0.$$

We obtain

(23.12.4) $$f(x) - f(x_0) = \sum_{n=1}^{\infty} a_n(x^n - x_0^n).$$

Now,

(23.12.5) $$|x^n - x_0^n| = |x - x_0||x^{n-1} + x^{n-2}x_0 + \cdots + x_0^{n-1}|.$$

We may choose $r$ such that $0 < r < R$ and $|x| < r$, $|x_0| < r$. Then

(23.12.6) $$|x^{n-1} + x^{n-2}x_0 + \cdots + x_0^{n-1}| < nr^n,$$

and from (23.12.4), (23.12.5), and (23.12.6),

(23.12.7) $$|f(x) - f(x_0)| < \sum_{n=1}^{\infty} |a_n||x^n - x_0^n|$$

$$< |x - x_0| \sum_{n=1}^{\infty} n|a_n|r^n = r|x - x_0| \sum_{n=1}^{\infty} n|a_n|r^{n-1}.$$

Since $0 < r < R$, and since $\sum_{n=1}^{\infty} na_n x^{n-1}$ converges absolutely by the previous theorem, the series on the right of (23.12.7) converges, with a sum, say $A$. Then from (23.12.7)

$$|f(x) - f(x_0)| < Ar|x - x_0|.$$

Hence (23.12.3) must hold, and the proof is complete.

We state the term-by-term *integrability* of (23.12.1) in the following form.

**THEOREM 23.12.3.** *If* $f(x) = \sum_{n=0}^{\infty} a_n x^n$ *converges for* $|x| < R$, *and c and d are two numbers in this interval, then*

(23.12.8) $$\int_c^d f(x)\,dx = \sum_{n=0}^{\infty} \int_c^d a_n x^n\,dx = \sum_{n=0}^{\infty} a_n \frac{d^{n+1} - c^{n+1}}{n+1}.$$

*Note.* The entire point of the theorem is that in (23.12.8) the second member can be written as the sum of the integrals of the individual terms in the series for $f(x)$.

*Proof.* The integral of $f$ exists, by the continuity of $f$. Let $s_n$ be the partial sum of the series in (23.12.8),

$$s_n = \sum_{k=0}^{n} \int_c^d a_k x^k\,dx.$$

The proof will be complete once we show that

$$\lim_{n \to \infty} s_n = \int_c^d f(x)\,dx \quad \text{or} \quad \lim_{n \to \infty} \left[ \int_c^d f(x)\,dx - s_n \right] = 0.$$

Let us write the series for $f(x)$ as

$$f(x) = \sum_{k=0}^{n} a_k x^k + R_n(x), \quad \text{where} \quad R_n(x) = \sum_{k=n+1}^{\infty} a_k x^k.$$

Then

$$\int_c^d f(x)\,dx = s_n + \int_c^d R_n(x)\,dx,$$

and it is sufficient to show that $\lim_{n \to \infty} \int_c^d R_n(x)\,dx = 0$. If $c < d$, we have

$$\left| \int_c^d R_n(x)\,dx \right| \leq \int_c^d |R_n(x)|\,dx \leq \int_c^d \sum_{k=n+1}^{\infty} |a_k x^k|\,dx.$$

Now choose $x_0$ such that $|c| < |x_0| < R$ and $|d| < |x_0| < R$. Then, as in the proof of Theorem 23.12.1, there is an integer $N$ such that

$$|a_k x^k| < r^k, \qquad k > N, \qquad c \leq x \leq d, \qquad r < 1.$$

Hence

$$\sum_{k=N+1}^{\infty} |a_k x^k| < \sum_{k=N+1}^{\infty} r^k = \frac{r^{N+1}}{1-r},$$

and

$$\int_c^d \sum_{k=N+1}^{\infty} |a_k x^k|\,dx < \frac{r^{N+1}}{1-r}(d-c).$$

Since the number on the right approaches zero as $N \to \infty$, it follows that

$$\lim_{n \to \infty} \int_c^d R_n(x)\, dx = 0.$$

If $c > d$ the same conclusion follows, and the proof is complete.

We are now ready to state the term-by-term *differentiability* of a power series in the following form.

**THEOREM 23.12.4.** *If* $f(x) = \sum_{n=0}^{\infty} a_n x^n$ *converges for* $|x| < R$, *then*

$$f'(x) = \sum_{n=1}^{\infty} n a_n x^{n-1}, \qquad |x| < R.$$

*Proof.* Let us write

(23.12.9) $$g(x) = \sum_{n=1}^{\infty} n a_n x^{n-1}.$$

By Theorem 23.12.1, we know that this series converges for $|x| < R$. We apply Theorem 23.12.3 to (23.12.9), and obtain

(23.12.10) $$\int_0^t g(x)\, dx = \sum_{n=1}^{\infty} \int_0^t n a_n x^{n-1}\, dx = \sum_{n=1}^{\infty} a_n t^n.$$

Now

$$f(t) = \sum_{n=0}^{\infty} a_n t^n = a_0 + \sum_{n=1}^{\infty} a_n t^n.$$

Hence we get, from (23.12.10),

(23.12.11) $$\int_0^t g(x)\, dx = f(t) - a_0.$$

We now differentiate (23.12.10) and apply Theorem 9.10.1. We get

$$g(t) = f'(t),$$

and the proof is complete.

The preceding theorems can be restated easily so as to apply to the series $f(x) = \sum_{n=0}^{\infty} a_n(x-a)^n$ in the interval $|x-a| < R$, where $R$ is the radius of convergence of the series.

Each of the theorems above expresses the interchangeability of two operations. Thus the continuity of a power series can be stated as

$$\lim_{x \to x_0} \left( \sum_{k=0}^{\infty} a_k x^k \right) = \sum_{k=0}^{\infty} \left( \lim_{x \to x_0} a_k x^k \right),$$

in which the limit and summation operations are interchanged.

The term-by-term differentiability becomes

$$\frac{d}{dx} \sum_{k=0}^{\infty} a_k x^k = \sum_{k=0}^{\infty} \left( \frac{d}{dx} a_k x^k \right),$$

and the term-by-term integrability becomes

$$\int_a^b \left( \sum_{k=0}^{\infty} a_k x^k \right) dx = \sum_{k=0}^{\infty} \int_a^b a_k x^k \, dx.$$

Each of these relations has previously been rather easily established where a finite sum was involved. The extension to the case of an infinite sum is by no means automatic and requires a special proof for each result. No additional restriction was needed in order to make this extension to a convergent power series. However, when one works with a convergent series of other functions, even absolute convergence is not generally sufficient, and it is necessary to impose further restrictions before relations similar to the above can be proved.

Some applications of the results of this section will appear in succeeding sections.

We add a property of power series which will be useful later.

**THEOREM 23.12.5.** *If two power series are equal in value in an interval $|x| < R$, they must be identical; that is, if*

(23.12.12) $\quad a_0 + a_1 x + a_2 x^2 + \cdots = b_0 + b_1 x + b_2 x^2 + \cdots,$

*when $|x| < R$, then $a_i = b_i$, $i = 0, 1, 2, \ldots$.*

*Proof.* Since (23.12.12) holds when $x = 0$, we get $a_0 = b_0$. Differentiating (23.12.12) we have

$$a_1 + 2a_2 x + \cdots = b_1 + 2b_2 x + \cdots;$$

setting $x = 0$ we find $a_1 = b_1$. Successive differentiations and setting $x = 0$ yield the conclusion of the theorem.

**EXERCISE GROUP 23.10**

1. If $f(x) = 1 + x + x^2 + x^3 + \cdots$, prove that $(1 - x)f'(x) = f(x)$.

2. If $f(x) = \sum_{n=0}^{\infty} \frac{m^n x^n}{n!}$, prove that $f'(x) = mf(x)$.

3. If $f(x) = \sum_{n=0}^{\infty} \frac{x^{2n+1}}{(2n+1)!}$ and $g(x) = \sum_{n=0}^{\infty} \frac{x^{2n}}{(2n)!}$, prove that $f'(x) = g(x)$ and $g'(x) = f(x)$.

4. If $f(x) = \sum_{n=0}^{\infty} (-1)^n \frac{x^{2n+1}}{(2n+1)!}$ and $g(x) = \sum_{n=0}^{\infty} (-1)^n \frac{x^{2n}}{(2n)!}$, prove the following:
    (a) $f'(x) = g(x)$  (b) $g'(x) = -f(x)$
    (c) $f''(x) + f(x) = 0$  (d) $g''(x) + g(x) = 0$

5. If $f(x) = \sum_{n=0}^{\infty} (-1)^n \frac{x^{2n}}{(2n)!}$ and $F(x) = \int_0^x f(t)\,dt$ prove that

$$\int_0^x F(t)\,dt = 1 - f(x).$$

6. If $f(x) = \sum_{n=0}^{\infty} \frac{x^n}{n!}$, prove each of the following:
   (a) $f'(x) = f(x)$
   (b) $f''(x) + f'(x) - 2f(x) = 0$
   (c) $\int_0^x f(t)\,dt = f(x) - 1$
   (d) $g(x) = xf(x)$ satisfies the equation $g''(x) - 2g'(x) + g(x) = 0$

7. If $f(x) = 1 + \sum_{n=1}^{\infty} \frac{m(m-1)\cdots(m-n+1)}{n!} x^n$, prove that $(1+x)f'(x) = mf(x)$.

## 23.13 TAYLOR SERIES

We have seen that a convergent power series, with radius of convergence $R$,

$$(23.13.1) \qquad f(x) = \sum_{n=0}^{\infty} a_n(x-a)^n,$$

represents a function $f$ which is continuous, differentiable term-by-term, and integrable term-by-term, in the interval $|x - a| < R$. We now obtain the relation between $f$ and the coefficients $a_n$.

We may differentiate (23.13.1) as often as we like, and we get, in general,

$$f^{(k)}(x) = \sum_{n=k}^{\infty} n(n-1)(n-2)\cdots(n-k+1)a_n(x-a)^{n-k}$$
$$= \sum_{n=k}^{\infty} \frac{n!}{(n-k)!} a_n(x-a)^{n-k}.$$

The only term in this series which does not contain a power of $x - a$ is the first one, for $n = k$. Thus setting $x = a$ in the series we get

$$(23.13.2) \qquad a_k = \frac{f^{(k)}(a)}{k!}, \qquad k = 0, 1, 2, \ldots.$$

Note that (23.13.2) includes the case $k = 0$, which is permissible since $0! = 1$ and $f^{(0)}(x)$ is *defined* as $f(x)$. Inserting (23.13.2) in (23.13.1), we obtain

$$(23.13.3) \qquad f(x) = \sum_{n=0}^{\infty} \frac{f^{(n)}(a)}{n!} (x-a)^n.$$

Series (23.13.3) is called the *Taylor series* of the function $f$ about $x = a$. In what has preceded, we have proved the following.

**THEOREM 23.13.1.** *If a function $f$ is defined as the sum of a convergent power series* (23.13.1), *then $f$ may be represented by its Taylor series* (23.13.3) *for $|x| < R$.*

Even if we do not know that the series on the right of (23.13.3) has $f(x)$ as its sum, that series is still called the Taylor series for the function $f$, and we write

$$f(x) \sim \sum_{n=0}^{\infty} \frac{f^{(n)}(a)}{n!} (x-a)^n.$$

The symbol $\sim$ signifies that $f(x)$ has associated with it the series on the right.

The particular form of a Taylor series (23.13.3) in the case that $a = 0$ is

(23.13.4) $$f(x) \sim \sum_{n=0}^{\infty} \frac{f^{(n)}(0)}{n!} x^n,$$

and is called the *Maclaurin series* for $f$.

**Example 23.13.1.** Find the Taylor series for $f(x) = \ln x$ about $x = 2$.

*Solution.* We arrange the work as follows, with $a = 2$:

$f(x) = \ln x$ $\qquad$ $f(2) = \ln 2$
$f'(x) = 1/x$ $\qquad$ $f'(2) = 1/2$
$f''(x) = -1/x^2$ $\qquad$ $f''(2) = -1/2^2$
$f'''(x) = 2/x^3$ $\qquad$ $f'''(2) = 2/2^3$
$f^{(4)}(x) = -3!/x^4$ $\qquad$ $f^{(4)}(2) = -3!/2^4$
$\vdots$ $\qquad$ $\vdots$
$f^{(n)}(x) = (-1)^{n+1}(n-1)!/x^n$ $\qquad$ $f^{(n)}(2) = (-1)^{n+1}(n-1)!/2^n, \quad n \geq 1.$

We differentiated enough times to be able to determine a general formula for $f^{(n)}(x)$. From (23.13.2) we get

$$a_n = \frac{f^{(n)}(2)}{n!} = \frac{(-1)^{n+1}(n-1)!}{2^n n!} = \frac{(-1)^{n+1}}{2^n n}, \quad n \geq 1,$$

and the Taylor series (23.13.3) may be written as

$$\ln 2 + \sum_{n=1}^{\infty} (-1)^{n+1} \frac{(x-2)^n}{2^n n}.$$

It is clear from (23.13.3) that one of the requirements for a function to be represented by its Taylor series about $x = a$ is that all orders of derivative of $f$ exist at $x = a$. In the case of a function *defined* as the sum of a convergent power series, this requirement is met automatically by Theorem 23.13.1 and its consequences. However, this condition is not sufficient in itself (see Exercise 28). Moreover, what can be stated for a function that can be differentiated at $x = a$ perhaps only a limited number of times? We turn to this question in the next theorem, a very important result for many purposes.

**THEOREM 23.13.2. Taylor's Formula with Remainder.** *Let $f$ be a function such that $f, f', f'', \ldots, f^{(n)}$ are continuous in a closed interval $[c, d]$, and $f^{(n+1)}$ exists in the open interval $(c, d)$. Let $a$ be a number in $(c, d)$. If $x$ is any number in $(c, d)$, let us define $R_n(x)$ by the equation*

(23.13.5) $$f(x) = \sum_{k=0}^{n} \frac{f^{(k)}(a)}{k!}(x-a)^k + R_n(x).$$

Then there exists a number $\xi$ between $a$ and $x$ such that

(23.13.6) $$R_n(x) = \frac{f^{(n+1)}(\xi)}{(n+1)!}(x-a)^{n+1}.$$

*Proof.* The proof is somewhat contrived, but there seems to be no feasible way of avoiding this. Let $x \neq a$ be fixed in $(c, d)$, and define a function $\phi(u)$, $u$ in $[c, d]$, as

(23.13.7) $$\phi(u) = f(x) - f(u) - (x-u)f'(u) - \frac{(x-u)^2}{2!}f''(u) - \cdots - \frac{(x-u)^n}{n!}f^{(n)}(u),$$

and then define

(23.13.8) $$\psi(u) = \phi(u) - \frac{(x-u)^{n+1}}{(x-a)^{n+1}}\phi(a).$$

Now $\phi(u)$ is continuous for $u$ in $[c, d]$ and differentiable in $(c, d)$; the same is true of $\psi(u)$. Substituting $u = a$ and $u = x$ in (23.13.8) we find, since $\phi(x) = 0$,

$$\psi(a) = 0, \quad \psi(x) = \phi(x) = 0.$$

Hence Rolle's Theorem applies in the interval $[a, x]$, and there exists a number $\xi$ in $(a, x)$ such that $\psi'(\xi) = 0$, so that from (23.13.8)

(23.13.9) $$\phi'(\xi) + (n+1)\frac{(x-\xi)^n}{(x-a)^{n+1}}\phi(a) = 0.$$

From (23.13.7) we get, differentiating as a product where needed,

$$\phi'(u) = -f'(u) + [f'(u) - (x-u)f''(u)] + \left[(x-u)f''(u) - \frac{(x-u)^2}{2!}f'''(u)\right]$$
$$+ \cdots + \left[\frac{(x-u)^{n-1}}{(n-1)!}f^{(n)}(u) - \frac{(x-u)^n}{n!}f^{(n+1)}(u)\right].$$

The terms in this expression cancel in pairs, and we are left with

$$\phi'(u) = -\frac{(x-u)^n}{n!}f^{(n+1)}(u).$$

Substitution in (23.13.9) gives

$$-\frac{(x-\xi)^n}{n!}f^{(n+1)}(\xi) + (n+1)\frac{(x-\xi)^n}{(x-a)^{n+1}}\phi(a) = 0,$$

from which we get

(23.13.10) $$\phi(a) = \frac{(x-a)^{n+1}}{(n+1)!}f^{(n+1)}(\xi).$$

Now letting $u = a$ in (23.13.7), and using $\phi(a)$ from (23.13.10), we may write $f(x)$ in the form

$$f(x) = f(a) + f'(a)(x - a) + \frac{f''(a)}{2!}(x - a)^2 + \cdots + \frac{f^{(n)}(a)}{n!}(x - a)^n$$
$$+ \frac{f^{(n+1)}(\xi)}{(n + 1)!}(x - a)^{n+1}, \qquad \xi \text{ in } (a, x).$$

This is the same as (23.13.5) and (23.13.6), and the proof is complete.

The right member of (23.13.5), apart from the term $R_n(x)$, is a polynomial in $x - a$ and hence in $x$, whose first $n$ derivatives coincide with those of $f(x)$ at $x = a$. This polynomial, called a *Taylor polynomial* for $f(x)$, may be considered as an approximating polynomial of degree $n$ to the function $f(x)$ in the interval between $a$ and $x$. The remainder $R_n(x)$ is then the difference between the function and its Taylor polynomial, and is a measure of the accuracy of the approximation. In a way, the problem of evaluating $f(x)$ is shifted to the evaluation of $R_n(x)$. This is not of immediate benefit, since all that is known about $\xi$ is that it lies between $a$ and $x$. We shall see, however, that in many cases we are able to restrict $R_n(x)$ between two numbers, and this will furnish valuable information about $f(x)$. Furthermore, (23.13.5) is of theoretical value (see Theorem 23.13.3).

*Example 23.13.2.* Obtain the Taylor formula with remainder for $\ln x$ about $x = 2$.

*Solution.* In Example 23.13.1 we found

$$f^{(n)}(x) = (-1)^{n+1} \frac{(n - 1)!}{x^n}.$$

Taylor's formula may then be written as

$$(23.13.11) \quad \ln x = \ln 2 + \sum_{k=1}^{n} (-1)^{k+1} \frac{(x - 2)^k}{2^k k} + \frac{(-1)^n}{\xi^{n+1}(n + 1)}(x - 2)^{n+1}$$

for some number $\xi$ between 2 and $x$. The remainder $R_n(x)$ is the last term in (23.13.11).

In the following example we shall find only a few terms in a Maclaurin series because it is not feasible to find a general formula for the $n$th derivative of the function.

*Example 23.13.3.* Find the Maclaurin series for $\sec x$ through the term in $x^4$.

*Solution.* We obtain, with $a = 0$,

$f(x) = \sec x$ $\qquad\qquad\qquad\qquad f(0) = 1$

$f'(x) = \sec x \tan x$ $\qquad\qquad\qquad f'(0) = 0$

$$f''(x) = \sec^3 x + \sec x \tan^2 x \qquad\qquad f''(0) = 1$$
$$f'''(x) = 5 \sec^3 x \tan x + \sec x \tan^3 x \qquad\qquad f'''(0) = 0$$
$$f^{(4)}(x) = 5 \sec^5 x + 18 \sec^3 x \tan^2 x + \sec x \tan^4 x \qquad f^{(4)}(0) = 5$$

Substituting into the Maclaurin series, we get

$$1 + \frac{1}{2} x^2 + \frac{5}{24} x^4 + \cdots.$$

We have obtained three nonzero terms, but they include the first five terms of the expansion, through $n = 4$.

A theoretical use of the remainder in Taylor's formula is in proving that the sum of the Taylor series of a function is the function itself. Let us write Taylor's formula as

$$f(x) = s_n(x) + R_n(x).$$

For any value of $x$, if it can be shown that $\lim_{n \to \infty} R_n|x| = 0$, then

$$f(x) = \lim_{n \to \infty} s_n(x) + \lim_{n \to \infty} R_n(x) = \lim_{n \to \infty} s_n(x) = \sum_{k=0}^{\infty} \frac{f^{(k)}(a)}{k!} (x - a)^k,$$

and, for that value of $x$, $f(x)$ equals the sum of its Taylor series.

We have proved the following result.

**THEOREM 23.13.3.** *If* $\lim_{n \to \infty} R_n(x) = 0$ *in* (23.13.5), *then* $f(x)$ *is equal to the sum of its Taylor series,*

$$f(x) = \sum_{k=0}^{\infty} \frac{f^{(k)}(a)}{k!} (x - a)^k.$$

## 23.14   THE BINOMIAL SERIES

The binomial formula gives the expansion of $(a + b)^m$ when $m$ is a positive integer. The important *binomial series* is an extension of the binomial formula to any real value of $m$, and the appropriate result may be stated in the following form.

**THEOREM 23.14.1.** *If $m$ is any real number, the Maclaurin expansion*

(23.14.1) $\qquad (1 + x)^m = 1 + mx + \dfrac{m(m-1)}{2!} x^2$

$$+ \cdots + \frac{m(m-1) \cdots (m-k+1)}{k!} x^k + \cdots$$

*is valid for* $|x| < 1$; *that is, the series on the right converges to* $(1 + x)^m$.

*Proof.* The ratio test may be used to show that the series on the right of (23.14.1) converges for $|x| < 1$ (Exercise 19 of Exercise Group 23.9). Let $f(x)$ be the sum:

(23.14.2)
$$f(x) = 1 + mx + \frac{m(m-1)}{2!}x^2 + \cdots + \frac{m(m-1)\cdots(m-k+1)}{k!}x^k + \cdots.$$

We wish to prove $f(x) = (1 + x)^m$. The method is to show from (23.14.2) that

(23.14.3) $$(1 + x)f'(x) = mf(x).$$

If equation (23.14.3) is true, then it is true that

(23.14.4) $$(1 + x)^{-m}f'(x) - m(1 + x)^{-m-1}f(x) = 0.$$

The left side of (23.14.4) is the derivative of $(1 + x)^{-m}f(x)$. Hence we have $D[(1 + x)^{-m}f(x)] = 0$, from which it follows that $(1 + x)^{-m}f(x) = c$, constant, or

(23.14.5) $$f(x) = c(1 + x)^m.$$

To find $c$, we see from (23.14.2) that $f(0) = 1$, and from (23.14.5) that $f(0) = c$. Hence $c = 1$ and $f(x) = (1 + x)^m$, which is what we wished to prove.

To complete the proof we must obtain (23.14.3). By Theorem 23.12.4, we get

$$f'(x) = m + m(m-1)x + \cdots + \frac{m(m-1)\cdots(m-k+1)}{(k-1)!}x^{k-1}$$
$$+ \frac{m(m-1)\cdots(m-k)}{k!}x^k + \cdots,$$

and, multiplying by $x$,

$$xf'(x) = mx + \cdots + \frac{m(m-1)\cdots(m-k+1)}{(k-1)!}x^k + \cdots.$$

On adding the last two equations we get $(1 + x)f'(x)$ on the left; the first two terms on the right are $m$ and $m^2x$, and the general term in $x^k$ has the coefficient

$$\frac{m(m-1)\cdots(m-k)}{k!} + \frac{m(m-1)\cdots(m-k+1)}{(k-1)!}$$
$$= \frac{m(m-1)\cdots(m-k+1)}{(k-1)!}\left[\frac{m-k}{k} + 1\right]$$
$$= \frac{m(m-1)\cdots(m-k+1)}{(k-1)!}\frac{m}{k}$$
$$= m\frac{m(m-1)\cdots(m-k+1)}{k!}.$$

This is $m$ times the coefficient of $x^k$ in $f(x)$, and hence

$$(1 + x)f'(x) = m\left[1 + mx + \cdots + \frac{m(m-1)\cdots(m-k+1)}{k!}x^k + \cdots\right] = mf(x).$$

Thus (23.14.3) has been proved, and the proof is now complete.

If $m$ is a positive integer, the expansion in (23.14.1) terminates with the term in $x^m$ and becomes the finite binomial expansion of $(1 + x)^m$.

The binomial series (23.14.1) was not obtained by use of (23.13.4). However, as a valid series representation it coincides with the series on the right of (23.13.4).

## EXERCISE GROUP 23.11

In Exercises 1–12 obtain the Taylor series of the given function about the specified value $a$. If $a = 0$, the series is also a Maclaurin series.

1. $e^x$, $a = 0$
2. $e^x$, $a = 1$
3. $\sin x$, $a = 0$
4. $\sin x$, $a = \pi/4$
5. $\sin(x + \pi/4)$, $a = 0$
6. $\cos x$, $a = 0$
7. $\ln(x + 1)$, $a = 0$
8. $\ln x$, $a = 1$
9. $\dfrac{1}{1-x}$, $a = 0$
10. $\dfrac{1}{1-x}$, $a = 2$
11. $(1 + x)^{-2}$, $a = 0$
12. $(1 - x)^{-1/2}$, $a = 0$

In Exercises 13–18 find the remainder $R_n$ in the form (23.13.6) for those of the above exercises which are specified.

13. Exercise 2
14. Exercise 3
15. Exercise 6
16. Exercise 7
17. Exercise 10
18. Exercise 11

In Exercises 19–24 find the terms of the Maclaurin series through the term in $x^4$.

19. $\tan x$
20. $e^x(x - 1)$
21. $\dfrac{x}{e^x}$
22. $\ln \sec x$
23. $\tan^{-1} x$
24. $\sin^{-1} x$

25. If $f$ is a function such that $(1 + x)f'(x) = mf(x)$,
    (a) show directly that
    $$f^{(k)}(x) = m(m - 1) \cdots (m - k + 1) \frac{f(x)}{(1 + x)^k}, \quad k \geq 1.$$
    (b) Then show that the Maclaurin series for $f(x)$ is
    $$f(0)\left[1 + \sum_{k=1}^{\infty} \frac{m(m - 1) \cdots (m - k + 1)}{k!} x^k\right].$$
    (c) Show directly that the Maclaurin series for $(1 + x)^m$ is the series in brackets in (b).

26. In the result of Exercise 12 replace $x$ by $x^2$ to get the expansion of $(1 - x^2)^{-1/2}$. Then integrate from 0 to $x$ to get the Maclaurin series for $\sin^{-1} x$. (Compare with Exercise 24.)

**27.** If $f$ has a continuous second derivative in a neighborhood of $x = a$, prove that
$$\lim_{h \to 0} \frac{f(a+h) + f(a-h) - 2f(a)}{h^2} = f''(a).$$

* **28.** Let $f(x) = e^{-1/x^2}$, $x \neq 0$.
   (a) Prove that $\lim_{x \to 0} f(x) = 0$. Hence if we define $f(0) = 0$, then $f$ is continuous for all $x$.
   (b) Prove that $\lim_{x \to 0} f'(x) = 0$. Hence if we define $f'(0) = 0$, then $f'$ is continuous for all $x$.
   (c) Prove that $f^{(n)}(x) = e^{-1/x^2} P_n(1/x)$ if $x \neq 0$, where $P_n(u)$ is a polynomial in $u$. Hence show that $\lim_{x \to 0} f^{(n)}(x) = 0$.
   The Maclaurin series for $f(x)$ then is simply a sum of zeros, which cannot equal $f(x) = e^{-1/x^2}$ in any interval. The above shows that we may have a function $f$ with $f(0) = 0$, $f^{(n)}(0) = 0$, $n = 1, 2, 3, \ldots$, for which the Maclaurin series does not converge to the function.

* **29.** (a) Let $f$ be an even function. Prove that if $f'$ exists, then it is an odd function, and $f'(0) = 0$. (See Exercise 28 of Exercise Group 6.2.)
   (b) Let $g$ be an odd function. Prove that if $g'$ exists, then it is an even function.
   It then follows that the Maclaurin series of an odd function has only odd powers of $x$, and that the Maclaurin series of an even function has only even powers of $x$.

## 23.15 USE OF THE REMAINDER IN TAYLOR'S FORMULA

We shall discuss two main uses of the remainder term in the Taylor formula.

For example, suppose we wish to compute $e^{0.4}$. We write Taylor's formula for $e^x$ with $a = 0$,

$$e^x = 1 + x + \frac{x^2}{2!} + \cdots + \frac{x^n}{n!} + \frac{e^\xi x^{n+1}}{(n+1)!}, \quad \xi \text{ between } 0 \text{ and } x.$$

If we use the approximating polynomial of degree 5, with $x = 0.4$, we have

(23.15.1) $\quad e^{0.4} \approx 1 + 0.4 + \frac{(0.4)^2}{2!} + \frac{(0.4)^3}{3!} + \frac{(0.4)^4}{4!} + \frac{(0.4)^5}{5!}.$

What error do we incur by neglecting the remainder? We know that $0 < \xi < 0.4$; hence $0 < e^\xi < e^{0.4}$, and

(23.15.2) $\quad 0 < R_5 = \frac{e^\xi (0.4)^6}{6!} < \frac{e^{0.4}(0.4)^6}{6!}.$

Since we do not know the value of $e^{0.4}$, we can accentuate the error and be certain that our estimate is still valid. It can be shown independently that $e < 3$; we then have $e^{0.4} < e < 3$, and from (23.15.2),

$$0 < R_5 < \frac{3(0.4)^6}{6!} = \frac{3(0.004096)}{720} < 0.0000171.$$

This tells us that if the expression on the right of (23.15.1) is evaluated exactly, the value approximates $e^{0.4}$ with 4-decimal-place accuracy. When we do the actual computation, errors may arise due to the process of rounding off. To assure that the accumulation of such round-off errors does not affect the first four decimal places, we shall carry the computation correct to six decimal places:

$$
\begin{aligned}
1 &= 1.000000 \\
0.4 &= .400000 \\
(0.4)^2/2! &= .080000 \\
(0.4)^3/3! &= .010667 \\
(0.4)^4/4! &= .001067 \\
(0.4)^5/5! &= .000085
\end{aligned}
$$

$$\text{Sum} \quad 1.491819$$

We may now say that

$$1.491819 < e^{0.4} < 1.491819 + 0.000017,$$

or

$$1.491819 < e^{0.4} < 1.491836.$$

Hence $e^{0.4} \approx 1.4918$ correct to 4 decimal places.

As a practical procedure, if we use additional terms in the approximating polynomial until there is no contribution in the first 6 decimal places, we can be fairly certain that the result is correct to 4 decimal places. The next term is $(0.4)^6/6! = 0.000006$, while the following one has zeros in the first 6 decimal places. Adding 0.000006 to the above sum we get 1.491825, which is 1.4918 to 4 decimal places. This value is correct.

The use of the remainder in representing a function by its Taylor series was expressed in Theorem 23.13.3. We shall illustrate this use of the remainder for the expansion (see Example 23.13.2)

$$\ln x = \ln 2 + \sum_{k=1}^{n} (-1)^{k+1} \frac{(x-2)^k}{2^k k} + \frac{(-1)^n}{\xi^{n+1}(n+1)} (x-2)^{n+1}.$$

The Taylor series with $u_k = (-1)^{k+1}(x-2)^k/(2^k k)$ converges when $|x-2| < 2$, by use of the Ratio Test. We can show directly that the remainder approaches zero as $n \to \infty$ if $0 < x - 2 \leq 2$. We have $|x - 2| \leq 2$ and $\xi > 2$ (Fig. 23.15.1). Hence

$$|R_n(x)| = \frac{|x-2|^{n+1}}{\xi^{n+1}(n+1)} < \frac{2^{n+1}}{2^{n+1}(n+1)} = \frac{1}{n+1}.$$

Hence $\lim\limits_{n \to \infty} R_n(x) = 0$, and $\ln x$ equals its Taylor series about $x = 2$, at least when $2 < x \leq 4$. The given proof cannot be used when $0 < x < 2$, but the result can be proved by other methods in this case (see Section 23.17).

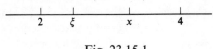

Fig. 23.15.1

We present an example where a complete answer can be given by considering the remainder directly.

**Example 23.15.1**  Prove that for any real $x$,

$$e^x = \sum_{n=0}^{\infty} \frac{x^n}{n!}.$$

*Proof.* The Taylor formula with $a = 0$ was used at the beginning of this section and is

$$e^x = \sum_{k=0}^{n} \frac{x^k}{k!} + \frac{e^\xi x^{n+1}}{(n+1)!},$$

with $\xi$ between 0 and $x$. If $x > 0$, then $e^\xi < e^x$, and we have

$$0 < R_n(x) < e^x \frac{x^{n+1}}{(n+1)!}.$$

Now, by the Ratio Test the series $\sum_{n=0}^{\infty} x^{n+1}/(n+1)!$ converges for all $x$, from which it follows, by Theorem 23.3.4, that $\lim_{n \to \infty} x^{n+1}/(n+1)! = 0$. Hence $\lim_{n \to \infty} R_n(x) = 0$, and the proof is complete if $x > 0$. If $x < 0$, then $\xi < 0$ and $0 < e^\xi < 1$, so that

$$|R_n(x)| < \frac{|x|^{n+1}}{(n+1)!}.$$

We have shown that the right member approaches zero. Hence $\lim_{n \to \infty} R_n(x) = 0$, also if $x < 0$, and the proof is complete.

We shall not carry out the proofs that the Taylor series for each function we work with converges to that function (however, see Exercises 11-13). But we will assume that henceforth, *when a Taylor series is used for a specific function, the representation is valid in the interior of the interval of convergence.* We shall avoid cases where this is not true.

## 23.16   COMPUTATION WITH POWER SERIES

In Section 23.15 we used the Maclaurin series for $e^x$ to compute $e^{0.4}$. This was an illustration of one computational use of the power series representation of a function, that is, the evaluation of specific values of the function. In that computation we used the remainder term to estimate the error. In the present section the computations will be carried out without an estimate of the error, but in such a way that the conclusions are reasonable and also can be validated.

Any series that has been obtained earlier, either in the text or in exercises, may be used for the present purpose.

***Example 23.16.1.*** Evaluate $\sin^{-1} 1/2$ to 4 decimals.

*Solution.* We have (see Exercise 26 of Exercise Group 23.11)

$$\sin^{-1} x = x + \frac{1}{2}\frac{x^3}{3} + \frac{1 \cdot 3}{2 \cdot 4}\frac{x^5}{5} + \frac{1 \cdot 3 \cdot 5}{2 \cdot 4 \cdot 6}\frac{x^7}{7} + \cdots, \quad |x| < 1.$$

With $x = 1/2$ we obtain, term by term,

| | |
|---|---|
| 1st term | .500000 |
| 2nd " | .020833 |
| 3rd " | .002344 |
| 4th " | .000349 |
| 5th " | .000059 |
| 6th " | .000011 |
| 7th " | .000002 |
| Sum | .523598 |

To 4 decimals we may say that $\sin^{-1} 1/2 = 0.5236$.

We know that $\sin^{-1} 1/2 = \pi/6$. We may use the decimal result to approximate $\pi$. Rounding off the above sum to 5 decimals minimizes the chance for error. Thus

$$\pi/6 \approx 0.52360, \quad \pi \approx 6(0.52360) = 3.14160,$$

and to 4 decimals $\pi = 3.1416$. (This is actually correct.)

***Example 23.16.2.*** Compute $10^{1/3}$ correct to 4 decimal places.

*Solution.* We may use the binomial series by writing

$$10^{1/3} = (8 + 2)^{1/3} = 2(1 + .25)^{1/3}$$

$$= 2\left[1 + \frac{1}{3}(.25) + \frac{(1/3)(-2/3)}{2!}(.25)^2 + \frac{(1/3)(-2/3)(-5/3)}{3!}(.25)^3 + \cdots\right]$$

$$= 2[1 + .083333 - .006944 + .000965 - .000161$$

$$+ .000030 - .000006 + .000001 - \cdots]$$

$$\approx 2(1.077218) = 2.154436$$

To 4 decimal places $10^{1/3} = 2.1544$. The series which was used converges slowly.

A second computational use of power series expansions is the evaluation of certain definite integrals for which an antiderivative either cannot be found or is difficult to work with (but the method is not limited to such cases).

**Example 23.16.3.** Evaluate $\int_0^{0.3} \dfrac{dx}{\sqrt{1-x^3}}$.

*Solution.* We cannot find an explicit expression for an antiderivative of $(1-x^3)^{-1/2}$. Instead, we use the binomial series

$$(1-x^3)^{-1/2} = 1 + \frac{1}{2}x^3 + \frac{1\cdot 3}{2^2 2!}x^6 + \frac{1\cdot 3\cdot 5}{2^3 3!}x^9 + \cdots, \quad |x| < 1.$$

We then integrate between 0 and 0.3 (Theorem 23.12.3) and get

$$\int_0^{0.3} (1-x^3)^{-1/2}\, dx = 0.3 + \frac{1}{2}\frac{(.3)^4}{4} + \frac{1\cdot 3}{2^2 2!}\frac{(.3)^7}{7} + \frac{1\cdot 3\cdot 5}{2^3 3!}\frac{(.3)^{10}}{10} + \cdots.$$

The computation follows.

|         |         |
|---------|---------|
| 1st term | .300000 |
| 2nd "    | .001012 |
| 3rd "    | .000012 |
| Sum     | .301024 |

Hence to 4 decimal places the value of the integral is 0.3010. Only 3 terms of the above expansion were used. Two terms, in fact, yield the same result.

**EXERCISE GROUP 23.12**

In Exercises 1–10 compute the value of the given expression correct to 4 decimal places. (Assume the validity of any needed expansion.)

1. $\sin 0.1$
2. $e^{0.2}$
3. $\cos 0.3$
4. $\int_0^{0.2} \cos x^2 \, dx$
5. $\int_0^1 e^{-x^2}\, dx$
6. $\int_0^{0.1} (1+x^2)^{-1/3}\, dx$
7. $31^{1/5}$
8. $30^{1/3}$
9. $70^{1/2}$
10. $19^{1/4}$

11. Prove, for any real $x$, the validity of the equation

$$\sin x = \sum_{n=0}^{\infty} (-1)^n \frac{x^{2n+1}}{(2n+1)!}$$

12. Prove, for any real $x$, the validity of the equation

$$\cos x = \sum_{n=0}^{\infty} (-1)^n \frac{x^{2n}}{(2n)!}$$

13. (a) Obtain the Taylor formula with remainder for $\ln(1+x)$ with $a=0$, and use the method of proof preceding Example 23.15.1 to show that the remainder approaches zero for $0 < x \leq 1$.
    (b) Using the result of (a), obtain a series representation for $\ln 2$. Compute a few partial sums to convince yourself that this series is not the one to use practically to compute $\ln 2$.

14. (a) Find the series for $\ln(1-x)$ from the result of Exercise 13(a).
    (b) Subtract the series for $\ln(1-x)$ from that for $\ln(1+x)$ to obtain the series
    $$\ln\frac{1+x}{1-x} = 2\left(x + \frac{x^3}{3} + \frac{x^5}{5} + \cdots\right).$$
    (c) Use the series in (b) to compute $\ln 2$. Compare the efficiency of this method with that of Exercise 13(b).

## 23.17 FINDING SERIES FROM KNOWN SERIES

The great advantage of power series is that in almost all ways they may be treated as finite sums for values of the variable in a suitable interval. That they may be added, subtracted, and multiplied term-by-term by a constant follows from Section 23.3. That they may be differentiated and integrated term-by-term was proved in Section 23.12. Power series may also be multiplied, divided, and even used as exponents. These properties can often be used to find a number of terms in a Taylor or Maclaurin expansion of a function, where the Taylor formula might not be very practicable. It should be kept in mind that whenever a valid power series expansion of a function is obtained, that expansion is a Taylor (or Maclaurin) series for the function, whether the Taylor formula was used or not.

In both the examples and the exercises that follow the series that are asked for will be obtainable from the following series.

(23.17.1) $$e^x = \sum_{n=0}^{\infty} \frac{x^n}{n!} = 1 + x + \frac{x^2}{2!} + \frac{x^3}{3!} + \cdots$$

(23.17.2) $$\sin x = \sum_{n=0}^{\infty} (-1)^n \frac{x^{2n+1}}{(2n+1)!} = x - \frac{x^3}{3!} + \frac{x^5}{5!} - \cdots$$

(23.17.3) $$\cos x = \sum_{n=0}^{\infty} (-1)^n \frac{x^{2n}}{(2n)!} = 1 - \frac{x^2}{2!} + \frac{x^4}{4!} - \cdots$$

(23.17.4)
$$(1+x)^m = 1 + mx + \frac{m(m-1)}{2!}x^2 + \cdots + \frac{m(m-1)\cdots(m-k+1)}{k!}x^k + \cdots$$

*Example 23.17.1.* Obtain the Maclaurin series for $\cos(x^2)$.

*Solution.* Replacing $x$ by $x^2$ in the Maclaurin series for $\cos x$ (23.17.3), we have at once
$$\cos(x^2) = 1 - \frac{x^4}{2!} + \frac{x^8}{4!} - \cdots + (-1)^{n+1} \frac{x^{4n-4}}{(2n-2)!} + \cdots, \quad n \geq 1.$$

This is clearly simpler than evaluating derivatives of $\cos(x^2)$ at $x = 0$.

*Example 23.17.2.* Obtain the Maclaurin series for $\cos^2 x$.

*Solution.* Using the identity $\cos^2 x = (1 + \cos 2x)/2$, together with (23.17.3) with $x$ replaced by $2x$, we have

$$\cos^2 x = \frac{1}{2}\left[1 + 1 - \frac{(2x)^2}{2!} + \frac{(2x)^4}{4!} - \frac{(2x)^6}{6!} + \cdots\right]$$

$$= 1 - \frac{2x^2}{2!} + \frac{2^3 x^4}{4!} - \frac{2^5 x^6}{6!} + \cdots.$$

We can often obtain the series for a function by integrating the series for the derivative of the function.

***Example 23.17.3.*** Obtain the Maclaurin series for $\ln(1 + x)$.

*Solution.* This series was found directly in Exercise 13 of Exercise Group 23.12. We give another method. We have, either as a binomial series or as the sum of a geometric series,

(23.17.5) $\qquad (1 + t)^{-1} = 1 - t + t^2 - t^3 + \cdots, \quad |t| < 1.$

Integrating between $t = 0$ and $t = x$, we obtain

$$\int_0^x (1 + t)^{-1}\, dt = \ln(1 + x) = x - \frac{x^2}{2} + \frac{x^3}{3} - \frac{x^4}{4} + \cdots, \quad |x| < 1.$$

We may now prove the validity of the Taylor expansion of $\ln x$ about $x = 2$ for $0 < x < 2$, a result which was not covered by the proof in Section 23.15.

By Example 23.17.3 the Maclaurin series for $\ln(1 + y)$ converges to the function for $|y| < 1$. Let $1 + y = x/2$. Then $\ln(1 + y) = \ln x/2 = \ln x - \ln 2$, the powers of $y$ become powers of $(x - 2)/2$, and the expansion of $\ln x$ in powers of $x - 2$ is obtained. Now $|y| < 1 \Leftrightarrow |x - 2| < 2$. Hence the expansion of $\ln x$ in powers of $x - 2$ is also valid for $0 < x < 2$.

***Example 23.17.4.*** Obtain the Maclaurin series for $x(1 + x^2)^{-2}$.

*Solution.* We employ a method using derivatives. We know that

$$\frac{d}{dx}(1 + x^2)^{-1} = -2x(1 + x^2)^{-2}.$$

Letting $t = x^2$ in (23.17.5) we have

$$(1 + x^2)^{-1} = 1 - x^2 + x^4 - x^6 + \cdots, \quad |x| < 1.$$

Differentiating, and dividing by $-2$, we get the result:

$$x(1 + x^2)^{-2} = x - 2x^3 + 3x^5 - 4x^7 + \cdots$$

The same result can be obtained by using the binomial series for $(1 + x^2)^{-2}$, and then multiplying by $x$.

We state the results for *multiplying and dividing series* in the following forms, omitting proofs.

**THEOREM 23.17.1.** *If the series*

$$f(x) = a_0 + a_1 x + a_2 x^2 + a_3 x^3 + \cdots,$$
$$g(x) = b_0 + b_1 x + b_2 x^2 + b_3 x^3 + \cdots,$$

*converge for* $|x| < R$, *then*

(a) *the Maclaurin series for* $f(x)\,g(x)$ *is*

$$f(x)\,g(x) = a_0 b_0 + (a_0 b_1 + a_1 b_0)x + (a_0 b_2 + a_1 b_1 + a_2 b_0)x^2 + \cdots$$
$$+ \left(\sum_{k=0}^{n} a_k b_{n-k}\right) x^n + \cdots,$$

*and it is valid for* $|x| < R$;

(b) *if* $b_0 \neq 0$, *the Maclaurin series for* $f(x)/g(x)$ *is*

(23.17.6) $\qquad f(x)/g(x) = c_0 + c_1 x + c_2 x^2 + c_3 x^3 + \cdots,$

*where* $c_0, c_1, c_2, \ldots$ *can be obtained from the equations*

$$b_0 c_0 = a_0$$
$$b_0 c_1 + b_1 c_0 = a_1$$
$$b_0 c_2 + b_1 c_1 + b_2 c_0 = a_2$$
$$\cdots \cdots \cdots \cdots$$
$$b_0 c_n + b_1 c_{n-1} + \cdots + b_n c_0 = a_n.$$

*These equations are equivalent to the long-division algorithm. If* $g(x) \neq 0$ *for* $|x| < R' < R$, *then the series* (23.17.6) *is valid at least in the interval* $|x| < R'$.

The seemingly involved statement of Theorem 23.17.1 states, in brief, that power series may be multiplied, and divided if $b_0 \neq 0$, by the procedures of elementary algebra. The procedures will be illustrated.

*Example 23.17.5.* Expand $e^x \sin x$ as a power series in $x$.

*Solution.* We write

$$e^x = 1 + x + \frac{x^2}{2} + \frac{x^3}{6} + \frac{x^4}{24} + \frac{x^5}{120} + \cdots$$

$$\sin x = \quad x \quad\quad - \frac{x^3}{6} \quad\quad + \frac{x^5}{120} - \cdots$$

and multiply each term of $e^x$ by each term of $\sin x$, collecting coefficients of like powers of $x$:

$$e^x \sin x = 0 + (1 + 0)x + (0 + 1 + 0)x^2 + \left(-\frac{1}{6} + 0 + \frac{1}{2} + 0\right)x^3$$
$$+ \left(0 - \frac{1}{6} + 0 + \frac{1}{6} + 0\right)x^4 + \left(\frac{1}{120} + 0 - \frac{1}{12} + 0 + \frac{1}{24} + 0\right)x^5 + \cdots$$
$$= x + x^2 + \frac{x^3}{3} - \frac{x^5}{30} + \cdots, \quad \text{all } x.$$

Note that we cannot always expect to get a simple expression for the general term in the expansion.

**Example 23.17.6.** Find the Maclaurin series of $\tan x$ up to and including the term in $x^5$.

*Solution.* Since $\tan x = \sin x / \cos x$, we use the long-division algorithm with the series for $\sin x$ and $\cos x$:

$$
\begin{array}{r}
x + \dfrac{x^3}{3} + \dfrac{2x^5}{15} + \cdots \\[4pt]
1 - \dfrac{x^2}{2} + \dfrac{x^4}{24} - \cdots \overline{\smash{\big)}\, x - \dfrac{x^3}{6} + \dfrac{x^5}{120} - \cdots} \\[4pt]
\underline{x - \dfrac{x^3}{2} + \dfrac{x^5}{24}} \\[4pt]
\dfrac{x^3}{3} - \dfrac{x^5}{30} \\[4pt]
\underline{\dfrac{x^3}{3} - \dfrac{x^5}{6}} \\[4pt]
\dfrac{2}{15} x^5
\end{array}
$$

In this manner we get

$$\tan x = x + \frac{x^3}{3} + \frac{2x^5}{15} + \cdots.$$

To get an additional term it would be necessary to carry out the complete procedure with an additional term used in the series for $\sin x$ and $\cos x$.

*Comment*: It might occur to the student to consider the possibility of using the long-division algorithm for the division of $\cos x$ by $\sin x$ to find $\cot x$ or, similarly, to use it for $1/\tan x$ with the series just found for $\tan x$. If we attempt the first suggestion, we come out with

$$\cot x = \frac{1}{x} - \frac{x}{3} - \frac{4x^3}{45} + \cdots.$$

What meaning can this have? It *cannot* be valid for $x = 0$. By more advanced methods, which we will not consider here, the expansion can be shown to be valid if $0 < x < \pi/2$. The feature to notice, however, is that *the series on the right is not a power series* (why?), and hence not a Maclaurin series.

### EXERCISE GROUP 23.13

In Exercises 1–14 find the Maclaurin series for the given function, making use of the known series for $e^x$, $\sin x$, $\cos x$, $(1 + x)^m$.

1. sinh $x$
2. cosh $x$
3. $\dfrac{\sin x}{x}$
4. $\sin^2 x$
5. $(1+x)e^x$
6. $e^x \cos x$
7. $\dfrac{\sin x}{1+x}$
8. tanh $x$
9. sech $x$
10. $\displaystyle\int_0^x \dfrac{e^t - 1}{t}\, dt$
11. $\cos (\sin x)$
12. $e^{\sin x}$
13. $e^{\tan x}$
14. $\tan^{-1} x$

15. It can be shown that the expansion in Exercise 14 is valid for $x = 1$. Let $x = 1$ in that series and obtain a series expansion for $\pi$.

16. Determine the coefficient of $x^{3n+1}$, $n = 0, 1, 2, \ldots$, in the expansion of $(1+x+x^2)^{-1}$ in powers of $x$. [*Hint*: The given function may be expressed as $(1-x)/(1-x^3)$.]

*  17. Expand $\tan^{-1} \dfrac{2x+1}{\sqrt{3}}$ in a Maclaurin series. [*Hint*: Expand the derivative of the function, and integrate.]

## 23.18 INDETERMINATE FORMS VIA INFINITE SERIES

In Section 15.3 L'Hospital's rule was used to evaluate indeterminate forms of the type 0/0. It is often necessary to apply the rule more than once and, conceivably, the resulting form may become more difficult to work with than the preceding one. The use of infinite series may be effective for the purpose of evaluating indeterminate forms which are otherwise difficult, as well as some where L'Hospital's rule works easily. We illustrate the method.

**Example 23.18.1.** Evaluate $\displaystyle\lim_{x \to 0} \dfrac{\tan x - \sin x}{x^2 \ln(1+x)}$.

*Solution.* All we need are the following series, obtained previously, to the extent written:

$$\tan x = x + \frac{x^3}{3} + \cdots, \qquad \sin x = x - \frac{x^3}{6} + \cdots, \qquad \ln(1+x) = x - \frac{x^2}{2} + \cdots.$$

Then we get

$$\tan x - \sin x = \frac{x^3}{2} + \cdots, \qquad x^2 \ln(1+x) = x^3 - \cdots$$

and

$$\lim_{x \to 0} \frac{\tan x - \sin x}{x^2 \ln(1+x)} = \lim_{x \to 0} \frac{x^3/2 + \cdots}{x^3 - \cdots} = \lim_{x \to 0} \frac{1/2 + \cdots}{1 - \cdots} = \frac{1}{2}.$$

The three dots in each sum indicate terms of higher degree than the preceding term. In the next to last step the numerator and denominator were divided by $x^3$; in the last step the terms indicated by the dots approach zero as $x \to 0$, and the limit 1/2 is obtained.

We may look upon this procedure in the following way. The given form is indeterminate, because numerator and denominator are zero at $x = 0$; this is the case, because numerator and denominator have powers of $x$ as a factor, in some manner. The use of Maclaurin series permitted us to divide out a suitable power of $x$, leaving a form which was no longer indeterminate.

**EXERCISE GROUP 23.14**

Evaluate each of the following indeterminate forms by the use of infinite series. (Some of these may be easily done by L'Hospital's Rule, but this is not the case in all of them. It may be instructive to try the use of that rule.)

1. $\lim\limits_{x \to 0} \dfrac{\sin x}{e^x - \cos x}$

2. $\lim\limits_{x \to 0} \dfrac{e^x - e^{-x} - 2x}{x - \sin x}$

3. $\lim\limits_{x \to 0} \dfrac{\tan x - x}{x^2 \ln(1 + x)}$

4. $\lim\limits_{x \to 0} \dfrac{x^3}{2 \tan x - \tan 2x}$

5. $\lim\limits_{x \to 0} \dfrac{2 \sin x - \sin 2x}{x(\cos x - \cos 2x)}$

6. $\lim\limits_{x \to 0} \dfrac{e^x - e^{-x} + 2 \sin x - 4x}{x^5}$

7. $\lim\limits_{x \to 1} \dfrac{\ln x}{1 - x}$

8. $\lim\limits_{x \to 0} \left( \cot x - \dfrac{1}{x} \right)$

9. $\lim\limits_{x \to 0} \left( \dfrac{1}{x^2} - \csc^2 x \right)$

10. $\lim\limits_{x \to 0} \dfrac{e^x - e^{\tan x}}{x - \tan x}$

11. $\lim\limits_{x \to 0} \dfrac{1 - \cos x \cosh x}{x^4}$

12. $\lim\limits_{x \to 0} \dfrac{2 \sinh x - 2x - x^2 \cos x}{x^4}$

13. $\lim\limits_{x \to a} \dfrac{f(x) - f(a) - f'(a)(x - a)}{(x - a)^2}$

14. $\lim\limits_{x \to a} \dfrac{f(x) - \sum_{k=0}^{n} \dfrac{f^{(k)}(a)}{k!}(x - a)^k}{(x - a)^{n+1}}$

## 23.19 TAYLOR'S FORMULA FOR A FUNCTION OF TWO VARIABLES

Taylor's formula for a function $f(x)$ about a value $x = a$ provides an expansion of $f$ in powers of $x - a$. The extension to a function $f(x, y)$ about a point $(a, b)$ is an expansion of $f$ in powers of $x - a$ and $y - b$.

Suppose we desire an expansion of $f(x, y)$. We set

$$x = a + \lambda t, \quad y = b + \mu t,$$

where $\lambda$ and $\mu$ are constant and $t$ is variable. We may then set

(23.19.1) $\qquad \phi(t) = f(a + \lambda t, b + \mu t),$

and expand $\phi(t)$ by Maclaurin's formula,

(23.19.2) $\qquad \phi(t) = \sum\limits_{k=0}^{n} \dfrac{\phi^{(k)}(0)}{k!} t^k + \dfrac{\phi^{(n+1)}(\theta)}{(n+1)!} t^{n+1}, \quad 0 < \theta < t.$

We require the values $\phi^{(k)}(0)$. From (23.19.1) we get
$$\phi'(t) = \lambda f_x(a + \lambda t, b + \mu t) + \mu f_y(a + \lambda t, b + \mu t).$$
Then,
$$\phi''(t) = \lambda[\lambda f_{xx}(a + \lambda t, b + \mu t) + \mu f_{xy}(a + \lambda t, b + \mu t)]$$
$$+ \mu[\lambda f_{yx}(a + \lambda t, b + \mu t) + \mu f_{yy}(a + \lambda t, b + \mu t)]$$
$$= \lambda^2 f_{xx}(a + \lambda t, b + \mu t) + 2\lambda\mu f_{xy}(a + \lambda t, b + \mu t)$$
$$+ \mu^2 f_{yy}(a + \lambda t, b + \mu t).$$
In particular, if $t = 0$,
$$\phi''(0) = \lambda^2 f_{xx}(a, b) + 2\lambda\mu f_{xy}(a, b) + \mu^2 f_{yy}(a, b).$$
We may abbreviate this with the notation
$$\phi''(0) = \left(\lambda \frac{\partial}{\partial x} + \mu \frac{\partial}{\partial y}\right)^2 f(a, b),$$
if, in squaring the binomial, we make the following interpretation:
$$\left(\frac{\partial}{\partial x}\right)^r \left(\frac{\partial}{\partial y}\right)^s f(a, b) = \frac{\partial^{r+s}}{\partial x^r \, \partial y^s} f(a, b).$$
By mathematical induction it can be shown that
$$\phi^{(k)}(t) = \left(\lambda \frac{\partial}{\partial x} + \mu \frac{\partial}{\partial y}\right)^k f(a + \lambda t, b + \mu t),$$
and hence

(23.19.3) $$\phi^{(k)}(0) = \left(\lambda \frac{\partial}{\partial x} + \mu \frac{\partial}{\partial y}\right)^k f(a, b).$$

Then
$$\phi^{(k)}(0) t^k = t^k \left(\lambda \frac{\partial}{\partial x} + \mu \frac{\partial}{\partial y}\right)^k f(a, b) = \left[(\lambda t) \frac{\partial}{\partial x} + (\mu t) \frac{\partial}{\partial y}\right]^k f(a, b)$$
$$= \left[(x - a) \frac{\partial}{\partial x} + (y - b) \frac{\partial}{\partial y}\right]^k f(a, b).$$
Similarly, if $0 < \theta < t$,

(23.19.4) $$t^n \phi^{(n)}(\theta) = \left[(x - a) \frac{\partial}{\partial x} + (y - b) \frac{\partial}{\partial y}\right]^n f(\xi, \eta),$$

where $(\xi, \eta)$ is a point (other than an endpoint) on the line segment joining $(a, b)$ and $(x, y)$. We get the desired expansion by substituting (23.19.3) and (23.19.4) in (23.19.2). The statement follows.

**THEOREM 23.19.1. Taylor's Formula.** *If $f$ and its partial derivatives up to and including those of order $n$ are continuous in a neighborhood of $(a, b)$, and $f^{(n+1)}$ exists in a neighborhood of $(a, b)$, then*

(23.19.5) $\quad f(x, y) = \sum_{k=0}^{n} \frac{1}{k!} \left[ (x-a)\frac{\partial}{\partial x} + (y-b)\frac{\partial}{\partial y} \right]^k f(a, b) + R_n,$

where the remainder $R_n$ may be expressed as

$$R_n = \frac{1}{(n+1)!} \left[ (x-a)\frac{\partial}{\partial x} + (y-b)\frac{\partial}{\partial y} \right]^{n+1} f(\xi, \eta),$$

with $(\xi, \eta)$ an interior point of the line segment joining $(a, b)$ and $(x, y)$.

It is understood that the term in (23.19.5) for $k = 0$ is $f(a, b)$. The binomial power in (23.19.5) may itself be written as a sum,

$$\left[ (x-a)\frac{\partial}{\partial x} + (y-b)\frac{\partial}{\partial y} \right]^k f = \sum_{j=0}^{k} \frac{k!}{(k-j)!j!} \frac{\partial^k f}{\partial x^{k-j} \partial y^j} (x-a)^{k-j} (y-b)^j,$$

which may then be inserted into (23.19.5). However, the form (23.19.5) appears to be as easy to remember as any.

If $R_n$ can be shown to approach zero as $n \to \infty$, a double series expansion of $f(x, y)$ results, analogous in many ways to the Taylor expansion of a function of one variable.

It can be shown that an expansion of the form (23.19.5), no matter how it may be obtained, is the Taylor expansion of the function. For example, in the exponential expansion

$$e^t = 1 + t + \frac{t^2}{2!} + \cdots + \frac{t^n}{n!} + e^{\xi} \frac{t^{n+1}}{(n+1)!}, \qquad \xi \text{ between } 0 \text{ and } t,$$

we may substitute $t = x - y$ and get

$$e^{x-y} = 1 + (x-y) + \frac{(x-y)^2}{2!} + \cdots + \frac{(x-y)^n}{n!} + e^{\xi} \frac{(x-y)^{n+1}}{(n+1)!}.$$

This is the same expansion one would obtain by applying (23.19.5) to $e^{x-y}$ with $a = 0, b = 0$.

We now illustrate the direct use of (23.19.5).

**Example 23.19.1.** Expand the function

$$f(x, y) = 2x^2 + 3xy + xy^2$$

in a Taylor expansion with remainder, about the point $(-2, 1)$, with (a) $n = 1$, (b) $n = 2$.

*Solution.* (a) With $n = 1$ we need first and second derivatives. Thus

$$f(-2, 1) = 0$$
$$f_x(-2, 1) = (4x + 3y + y^2)_{(-2, 1)} = -4$$
$$f_y(-2, 1) = (3x + 2xy)_{(-2, 1)} = -10$$
$$f_{xx} = 4, \qquad f_{xy} = 3 + 2y, \qquad f_{yy} = 2x.$$

We then get, using (23.19.5),
$$2x^2 + 3xy + xy^2 = 0 - 4(x+2) - 10(y-1) + R_1,$$
where
$$R_1 = \tfrac{1}{2}[4(x+2)^2 + 2(3+2\eta)(x+2)(y-1) + 2\xi(y-1)^2],$$
with $(\xi, \eta)$ as some point on the line segment joining $(-2, 1)$ and $(x, y)$.

(b) With $n = 2$ we need the second derivatives at $(-2, 1)$,
$$f_{xx}(-2, 1) = 4, \quad f_{xy}(-2, 1) = 5, \quad f_{yy}(-2, 1) = -4,$$
and the third derivatives,
$$f_{xxx} = 0, \quad f_{xxy} = 0, \quad f_{xyy} = 2, \quad f_{yyy} = 0.$$
All the third derivatives are constant since $f$ is a polynomial of degree 3. The expansion (23.19.5) becomes

(23.19.6) $\quad 2x^2 + 3xy + xy^2 = -4(x+2) - 10(y-1)$
$$+ \tfrac{1}{2}[4(x+2)^2 + 10(x+2)(y-1) - 4(y-1)^2] + R_2,$$
where
$$R_2 = \tfrac{1}{6}[6(x+2)(y-1)^2] = (x+2)(y-1)^2.$$

With $R_2$, Equation (23.19.6) is an identity in $x$ and $y$. The reason that $R_2$ does not contain $\xi$ or $\eta$ is that $f$ is a polynomial of degree 3 in $x$ and $y$, and $R_2$ is also of degree 3.

**EXERCISE GROUP 23.15**

1. Expand $7 - 3x + 4y^2$ in powers of $x + 4$ and $y - 3$.
2. Expand $x^2 - 3xy + 2y^2$ in powers of $x - 1$ and $y - 2$.
3. Expand $3x^2 - 7xy$ in powers of $x + 2$ and $y - 3$.
4. Expand $2x^2 - 4xy + 3y^2 - 2x - 7y + 8$ in powers of $x + 1$ and $y + 1$.
5. Expand $2 - 12x + 4y - 3x^2 + 4xy + x^2y$ in powers of $x + 2$ and $y - 3$.
6. Expand $(x+2)^2(y-3)$ in powers of $x$ and $y + 2$.
7. Obtain the Taylor expansion for $e^x \cos y$ in powers of $x$ and $y$ through terms of third degree.
8. Obtain the Taylor expansion for $e^{x-2y}$ in powers of $x$ and $y$ through terms of third degree.
9. Obtain the Taylor expansion for $\dfrac{y}{1+x}$ in powers of $x - 1$ and $y - 1$ through terms of degree 2.
10. Write Taylor's Formula with remainder for $n = 1$.

# Trigonometry Review and Formulas

**Radian and degree measure of an angle**

$1 \text{ rad} = \dfrac{180°}{\pi} \approx 57.296° \approx 57°17'45''$

$1° = \dfrac{\pi}{180} \text{ rad} \approx 0.017453 \text{ rad}$

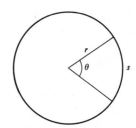

**Circle relations**

$s$ = arc length,   $A$ = area of sector

If $\theta$ is measured in radians,   $s = r\theta$,   $A = \tfrac{1}{2}r^2\theta$

**Trigonometric functions**

$\sin \theta = \dfrac{y}{r}, \quad \cos \theta = \dfrac{x}{r}, \quad \tan \theta = \dfrac{y}{x}$

$\cot \theta = \dfrac{x}{y}, \quad \sec \theta = \dfrac{r}{x}, \quad \csc \theta = \dfrac{r}{y}$

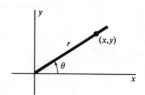

**Trigonometric values for special angles**

| $\theta$ | | $\sin \theta$ | $\cos \theta$ | $\tan \theta$ | $\cot \theta$ | $\sec \theta$ | $\csc \theta$ |
|---|---|---|---|---|---|---|---|
| 30° | $\dfrac{\pi}{6}$ | $\dfrac{1}{2}$ | $\dfrac{\sqrt{3}}{2}$ | $\dfrac{1}{\sqrt{3}} = \dfrac{\sqrt{3}}{3}$ | $\sqrt{3}$ | $\dfrac{2}{\sqrt{3}} = \dfrac{2\sqrt{3}}{3}$ | 2 |
| 45° | $\dfrac{\pi}{4}$ | $\dfrac{1}{\sqrt{2}} = \dfrac{\sqrt{2}}{2}$ | $\dfrac{1}{\sqrt{2}} = \dfrac{\sqrt{2}}{2}$ | 1 | 1 | $\sqrt{2}$ | $\sqrt{2}$ |
| 60° | $\dfrac{\pi}{3}$ | $\dfrac{\sqrt{3}}{2}$ | $\dfrac{1}{2}$ | $\sqrt{3}$ | $\dfrac{1}{\sqrt{3}} = \dfrac{\sqrt{3}}{3}$ | 2 | $\dfrac{2}{\sqrt{3}} = \dfrac{2\sqrt{3}}{3}$ |

**Trigonometric values for quadrantal angles**

| $\theta$ | | $\sin \theta$ | $\cos \theta$ | $\tan \theta$ | $\cot \theta$ | $\sec \theta$ | $\csc \theta$ |
|---|---|---|---|---|---|---|---|
| 0° | 0 | 0 | 1 | 0 | — | 1 | — |
| 90° | $\dfrac{\pi}{2}$ | 1 | 0 | — | 0 | — | 1 |
| 180° | $\pi$ | 0 | −1 | 0 | — | −1 | — |
| 270° | $\dfrac{3\pi}{2}$ | −1 | 0 | — | 0 | — | −1 |

## Variation of the trigonometric functions

|            | $\theta$ in $Q_1$ | $\theta$ in $Q_2$ | $\theta$ in $Q_3$ | $\theta$ in $Q_4$ |
|------------|-------------------|-------------------|-------------------|-------------------|
| $\sin \theta$ | $0 \nearrow 1$      | $1 \searrow 0$      | $0 \searrow -1$     | $-1 \nearrow 0$     |
| $\cos \theta$ | $1 \searrow 0$      | $0 \searrow -1$     | $-1 \nearrow 0$     | $0 \nearrow 1$      |
| $\tan \theta$ | $0 \nearrow \infty$ | $-\infty \nearrow 0$| $0 \nearrow \infty$ | $-\infty \nearrow 0$|
| $\cot \theta$ | $\infty \searrow 0$ | $0 \searrow -\infty$| $\infty \searrow 0$ | $0 \searrow -\infty$|
| $\sec \theta$ | $1 \nearrow \infty$ | $-\infty \nearrow -1$| $-1 \searrow -\infty$| $\infty \searrow 1$|
| $\csc \theta$ | $\infty \searrow 1$ | $1 \nearrow \infty$ | $-\infty \nearrow -1$| $-1 \searrow -\infty$|

### Reciprocal identities

$$\cot \theta = \frac{1}{\tan \theta}, \quad \sec \theta = \frac{1}{\cos \theta}, \quad \csc \theta = \frac{1}{\sin \theta}$$

### Ratio identities

$$\tan \theta = \frac{\sin \theta}{\cos \theta}, \quad \cot \theta = \frac{\cos \theta}{\sin \theta}$$

### Pythagorean or squares identities

$$\sin^2 \theta + \cos^2 \theta = 1, \quad \sec^2 \theta = 1 + \tan^2 \theta, \quad \csc^2 \theta = 1 + \cot^2 \theta$$

### Periodicity of the trigonometric functions

$\sin(\theta + 2\pi) = \sin \theta, \quad \cos(\theta + 2\pi) = \cos \theta$
$\sec(\theta + 2\pi) = \sec \theta, \quad \csc(\theta + 2\pi) = \csc \theta$ } The sine, cosine, secant, and cosecant functions have period $2\pi$.

$\tan(\theta + \pi) = \tan \theta, \quad \cot(\theta + \pi) = \cot \theta$ } The tangent and cotangent functions have period $\pi$.

**Reduction formulas** If $f(x)$ is any of the six basic trigonometric functions and $\operatorname{co} f(x)$ is its cofunction, then

$$f(\pm x + n\pi/2) = \begin{cases} \pm f(x) & \text{if } n \text{ is even} \\ \pm \operatorname{co} f(x) & \text{if } n \text{ is odd} \end{cases}$$

where, for a given integer $n$, the appropriate sign in the right member applies for a' values of $x$ for which the functions are defined.

### Addition formulas (upper signs are taken together)

$\sin(A \pm B) = \sin A \cos B \pm \cos A \sin B$
$\cos(A \pm B) = \cos A \cos B \mp \sin A \sin B$
$\tan(A \pm B) = \dfrac{\tan A \pm \tan B}{1 \mp \tan A \tan B}$

# Graphs of the trigonometric functions

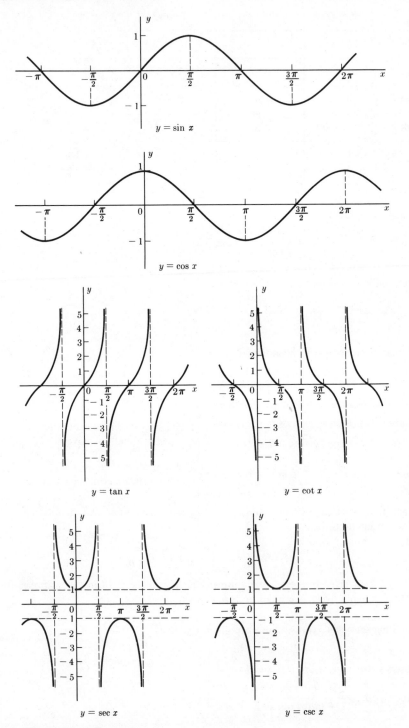

**Double-angle formulas**

$\sin 2A = 2 \sin A \cos A$

$\cos 2A = \cos^2 A - \sin^2 A = 1 - 2 \sin^2 A = 2 \cos^2 A - 1$

$\tan 2A = \dfrac{2 \tan A}{1 - \tan^2 A}$

**Half-angle formulas**

$\sin \dfrac{A}{2} = \pm \sqrt{\dfrac{1 - \cos A}{2}} \qquad \cos \dfrac{A}{2} = \pm \sqrt{\dfrac{1 + \cos A}{2}}$

$\tan \dfrac{A}{2} = \pm \sqrt{\dfrac{1 - \cos A}{1 + \cos A}} = \dfrac{\sin A}{1 + \cos A} = \dfrac{1 - \cos A}{\sin A}$

**Product formulas**

$\sin A \cos B = \tfrac{1}{2}[\sin(A + B) + \sin(A - B)]$

$\cos A \cos B = \tfrac{1}{2}[\cos(A + B) + \cos(A - B)]$

$\sin A \sin B = \tfrac{1}{2}[\cos(A - B) - \cos(A + B)]$

**Factor formulas**

$\sin x + \sin y = 2 \sin \tfrac{1}{2}(x + y) \cos \tfrac{1}{2}(x - y)$

$\sin x - \sin y = 2 \sin \tfrac{1}{2}(x - y) \cos \tfrac{1}{2}(x + y)$

$\cos x + \cos y = 2 \cos \tfrac{1}{2}(x + y) \cos \tfrac{1}{2}(x - y)$

$\cos x - \cos y = -2 \sin \tfrac{1}{2}(x + y) \sin \tfrac{1}{2}(x - y)$

**Relations in a triangle**  Sides are $a$, $b$, $c$; opposite angles are $A$, $B$, $C$.

Law of Sines: $\quad \dfrac{a}{\sin A} = \dfrac{b}{\sin B} = \dfrac{c}{\sin C}$

Law of Cosines: $\quad a^2 = b^2 + c^2 - 2bc \cos A$

Law of Tangents: $\quad \dfrac{a - b}{a + b} = \dfrac{\tan \tfrac{1}{2}(A - B)}{\tan \tfrac{1}{2}(A + B)}$

**Half-angle formulas**  If $2s = a + b + c$, then

$$\tan \dfrac{A}{2} = \dfrac{1}{s - a} \sqrt{\dfrac{(s - a)(s - b)(s - c)}{s}}$$

**Area of a triangle**

$$A = \dfrac{1}{2} bc \sin A = \dfrac{1}{2} c^2 \dfrac{\sin A \sin B}{\sin C} = \sqrt{s(s - a)(s - b)(s - c)}$$

## Table 1  Squares and Square Roots

| $n$ | $n^2$ | $\sqrt{n}$ | $\sqrt{10n}$ | $n$ | $n^2$ | $\sqrt{n}$ | $\sqrt{10n}$ |
|---|---|---|---|---|---|---|---|
| 1 | 1 | 1.000 | 3.162 | 51 | 2601 | 7.141 | 22.583 |
| 2 | 4 | 1.414 | 4.472 | 52 | 2704 | 7.211 | 22.804 |
| 3 | 9 | 1.732 | 5.477 | 53 | 2809 | 7.280 | 23.022 |
| 4 | 16 | 2.000 | 6.325 | 54 | 2916 | 7.348 | 23.238 |
| 5 | 25 | 2.236 | 7.071 | 55 | 3025 | 7.416 | 23.452 |
| 6 | 36 | 2.449 | 7.746 | 56 | 3136 | 7.483 | 23.664 |
| 7 | 49 | 2.646 | 8.367 | 57 | 3249 | 7.550 | 23.875 |
| 8 | 64 | 2.828 | 8.944 | 58 | 3364 | 7.616 | 24.083 |
| 9 | 81 | 3.000 | 9.487 | 59 | 3481 | 7.681 | 24.290 |
| 10 | 100 | 3.162 | 10.000 | 60 | 3600 | 7.746 | 24.495 |
| 11 | 121 | 3.317 | 10.488 | 61 | 3721 | 7.810 | 24.698 |
| 12 | 144 | 3.464 | 10.954 | 62 | 3844 | 7.874 | 24.900 |
| 13 | 169 | 3.606 | 11.402 | 63 | 3969 | 7.937 | 25.100 |
| 14 | 196 | 3.742 | 11.832 | 64 | 4096 | 8.000 | 25.298 |
| 15 | 225 | 3.873 | 12.247 | 65 | 4225 | 8.062 | 25.495 |
| 16 | 256 | 4.000 | 12.649 | 66 | 4356 | 8.124 | 25.690 |
| 17 | 289 | 4.123 | 13.038 | 67 | 4489 | 8.185 | 25.884 |
| 18 | 324 | 4.243 | 13.416 | 68 | 4624 | 8.246 | 26.077 |
| 19 | 361 | 4.359 | 13.784 | 69 | 4761 | 8.307 | 26.268 |
| 20 | 400 | 4.472 | 14.142 | 70 | 4900 | 8.367 | 26.458 |
| 21 | 441 | 4.583 | 14.491 | 71 | 5041 | 8.426 | 26.646 |
| 22 | 484 | 4.690 | 14.832 | 72 | 5184 | 8.485 | 26.833 |
| 23 | 529 | 4.796 | 15.166 | 73 | 5329 | 8.544 | 27.019 |
| 24 | 576 | 4.899 | 15.492 | 74 | 5476 | 8.602 | 27.203 |
| 25 | 625 | 5.000 | 15.811 | 75 | 5625 | 8.660 | 27.386 |
| 26 | 676 | 5.099 | 16.125 | 76 | 5776 | 8.718 | 27.568 |
| 27 | 729 | 5.196 | 16.432 | 77 | 5929 | 8.775 | 27.749 |
| 28 | 784 | 5.292 | 16.733 | 78 | 6084 | 8.832 | 27.928 |
| 29 | 841 | 5.385 | 17.029 | 79 | 6241 | 8.888 | 28.107 |
| 30 | 900 | 5.477 | 17.321 | 80 | 6400 | 8.944 | 28.284 |
| 31 | 961 | 5.568 | 17.607 | 81 | 6561 | 9.000 | 28.460 |
| 32 | 1024 | 5.657 | 17.889 | 82 | 6724 | 9.055 | 28.636 |
| 33 | 1089 | 5.745 | 18.166 | 83 | 6889 | 9.110 | 28.810 |
| 34 | 1156 | 5.831 | 18.439 | 84 | 7056 | 9.165 | 28.983 |
| 35 | 1225 | 5.916 | 18.708 | 85 | 7225 | 9.220 | 29.155 |
| 36 | 1296 | 6.000 | 18.974 | 86 | 7396 | 9.274 | 29.326 |
| 37 | 1369 | 6.083 | 19.235 | 87 | 7569 | 9.327 | 29.496 |
| 38 | 1444 | 6.164 | 19.494 | 88 | 7744 | 9.381 | 29.665 |
| 39 | 1521 | 6.245 | 19.748 | 89 | 7921 | 9.434 | 29.833 |
| 40 | 1600 | 6.325 | 20.000 | 90 | 8100 | 9.487 | 30.000 |
| 41 | 1681 | 6.403 | 20.248 | 91 | 8281 | 9.539 | 30.166 |
| 42 | 1764 | 6.481 | 20.494 | 92 | 8464 | 9.592 | 30.332 |
| 43 | 1849 | 6.557 | 20.736 | 93 | 8649 | 9.644 | 30.496 |
| 44 | 1936 | 6.633 | 20.976 | 94 | 8836 | 9.695 | 30.659 |
| 45 | 2025 | 6.708 | 21.213 | 95 | 9025 | 9.747 | 30.822 |
| 46 | 2116 | 6.782 | 21.448 | 96 | 9216 | 9.798 | 30.984 |
| 47 | 2209 | 6.856 | 21.679 | 97 | 9409 | 9.849 | 31.145 |
| 48 | 2304 | 6.928 | 21.909 | 98 | 9604 | 9.899 | 31.305 |
| 49 | 2401 | 7.000 | 22.136 | 99 | 9801 | 9.950 | 31.464 |
| 50 | 2500 | 7.071 | 22.361 | 100 | 10000 | 10.000 | 31.623 |

## Table 2  Common Logarithms

| n | 0 | 1 | 2 | 3 | 4 | 5 | 6 | 7 | 8 | 9 |
|---|---|---|---|---|---|---|---|---|---|---|
| 1.0 | .0000 | .0043 | .0086 | .0128 | .0170 | .0212 | .0253 | .0294 | .0334 | .0374 |
| 1.1 | .0414 | .0453 | .0492 | .0531 | .0569 | .0607 | .0645 | .0682 | .0719 | .0755 |
| 1.2 | .0792 | .0828 | .0864 | .0899 | .0934 | .0969 | .1004 | .1038 | .1072 | .1106 |
| 1.3 | .1139 | .1173 | .1206 | .1239 | .1271 | .1303 | .1335 | .1367 | .1399 | .1430 |
| 1.4 | .1461 | .1492 | .1523 | .1553 | .1584 | .1614 | .1644 | .1673 | .1703 | .1732 |
| 1.5 | .1761 | .1790 | .1818 | .1847 | .1875 | .1903 | .1931 | .1959 | .1987 | .2014 |
| 1.6 | .2041 | .2068 | .2095 | .2122 | .2148 | .2175 | .2201 | .2227 | .2253 | .2279 |
| 1.7 | .2304 | .2330 | .2355 | .2380 | .2405 | .2430 | .2455 | .2480 | .2504 | .2529 |
| 1.8 | .2553 | .2577 | .2601 | .2625 | .2648 | .2672 | .2695 | .2718 | .2742 | .2765 |
| 1.9 | .2788 | .2810 | .2833 | .2856 | .2878 | .2900 | .2923 | .2945 | .2967 | .2989 |
| 2.0 | .3010 | .3032 | .3054 | .3075 | .3096 | .3118 | .3139 | .3160 | .3181 | .3201 |
| 2.1 | .3222 | .3243 | .3263 | .3284 | .3304 | .3324 | .3345 | .3365 | .3385 | .3404 |
| 2.2 | .3424 | .3444 | .3464 | .3483 | .3502 | .3522 | .3541 | .3560 | .3579 | .3598 |
| 2.3 | .3617 | .3636 | .3655 | .3674 | .3692 | .3711 | .3729 | .3747 | .3766 | .3784 |
| 2.4 | .3802 | .3820 | .3838 | .3856 | .3874 | .3892 | .3909 | .3927 | .3945 | .3962 |
| 2.5 | .3979 | .3997 | .4014 | .4031 | .4048 | .4065 | .4082 | .4099 | .4116 | .4133 |
| 2.6 | .4150 | .4166 | .4183 | .4200 | .4216 | .4232 | .4249 | .4265 | .4281 | .4298 |
| 2.7 | .4314 | .4330 | .4346 | .4362 | .4378 | .4393 | .4409 | .4425 | .4440 | .4456 |
| 2.8 | .4472 | .4487 | .4502 | .4518 | .4533 | .4548 | .4564 | .4579 | .4594 | .4609 |
| 2.9 | .4624 | .4639 | .4654 | .4669 | .4683 | .4698 | .4713 | .4728 | .4742 | .4757 |
| 3.0 | .4771 | .4786 | .4800 | .4814 | .4829 | .4843 | .4857 | .4871 | .4886 | .4900 |
| 3.1 | .4914 | .4928 | .4942 | .4955 | .4969 | .4983 | .4997 | .5011 | .5024 | .5038 |
| 3.2 | .5051 | .5065 | .5079 | .5092 | .5105 | .5119 | .5132 | .5145 | .5159 | .5172 |
| 3.3 | .5185 | .5198 | .5211 | .5224 | .5237 | .5250 | .5263 | .5276 | .5289 | .5302 |
| 3.4 | .5315 | .5328 | .5340 | .5353 | .5366 | .5378 | .5391 | .5403 | .5416 | .5428 |
| 3.5 | .5441 | .5453 | .5465 | .5478 | .5490 | .5502 | .5514 | .5527 | .5539 | .5551 |
| 3.6 | .5563 | .5575 | .5587 | .5599 | .5611 | .5623 | .5635 | .5647 | .5658 | .5670 |
| 3.7 | .5682 | .5694 | .5705 | .5717 | .5729 | .5740 | .5752 | .5763 | .5775 | .5786 |
| 3.8 | .5798 | .5809 | .5821 | .5832 | .5843 | .5855 | .5866 | .5877 | .5888 | .5899 |
| 3.9 | .5911 | .5922 | .5933 | .5944 | .5955 | .5966 | .5977 | .5988 | .5999 | .6010 |
| 4.0 | .6021 | .6031 | .6042 | .6053 | .6064 | .6075 | .6085 | .6096 | .6107 | .6117 |
| 4.1 | .6128 | .6138 | .6149 | .6160 | .6170 | .6180 | .6191 | .6201 | .6212 | .6222 |
| 4.2 | .6232 | .6243 | .6253 | .6263 | .6274 | .6284 | .6294 | .6304 | .6314 | .6325 |
| 4.3 | .6335 | .6345 | .6355 | .6365 | .6375 | .6385 | .6395 | .6405 | .6415 | .6425 |
| 4.4 | .6435 | .6444 | .6454 | .6464 | .6474 | .6484 | .6493 | .6503 | .6513 | .6522 |
| 4.5 | .6532 | .6542 | .6551 | .6561 | .6571 | .6580 | .6590 | .6599 | .6609 | .6618 |
| 4.6 | .6628 | .6637 | .6646 | .6656 | .6665 | .6675 | .6684 | .6693 | .6702 | .6712 |
| 4.7 | .6721 | .6730 | .6739 | .6749 | .6758 | .6767 | .6776 | .6785 | .6794 | .6803 |
| 4.8 | .6812 | .6821 | .6830 | .6839 | .6848 | .6857 | .6866 | .6875 | .6884 | .6893 |
| 4.9 | .6902 | .6911 | .6920 | .6928 | .6937 | .6946 | .6955 | .6964 | .6972 | .6981 |
| 5.0 | .6990 | .6998 | .7007 | .7016 | .7024 | .7033 | .7042 | .7050 | .7059 | .7067 |
| 5.1 | .7076 | .7084 | .7093 | .7101 | .7110 | .7118 | .7126 | .7135 | .7143 | .7152 |
| 5.2 | .7160 | .7168 | .7177 | .7185 | .7193 | .7202 | .7210 | .7218 | .7226 | .7235 |
| 5.3 | .7243 | .7251 | .7259 | .7267 | .7275 | .7284 | .7292 | .7300 | .7308 | .7316 |
| 5.4 | .7324 | .7332 | .7340 | .7348 | .7356 | .7364 | .7372 | .7380 | .7388 | .7396 |
| n | 0 | 1 | 2 | 3 | 4 | 5 | 6 | 7 | 8 | 9 |

**Table 2** *Continued*

| n | 0 | 1 | 2 | 3 | 4 | 5 | 6 | 7 | 8 | 9 |
|---|---|---|---|---|---|---|---|---|---|---|
| 5.5 | .7404 | .7412 | .7419 | .7427 | .7435 | .7443 | .7451 | .7459 | .7466 | .7474 |
| 5.6 | .7482 | .7490 | .7497 | .7505 | .7513 | .7520 | .7528 | .7536 | .7543 | .7551 |
| 5.7 | .7559 | .7566 | .7574 | .7582 | .7589 | .7597 | .7604 | .7612 | .7619 | .7627 |
| 5.8 | .7634 | .7642 | .7649 | .7657 | .7664 | .7672 | .7679 | .7686 | .7694 | .7701 |
| 5.9 | .7709 | .7716 | .7723 | .7731 | .7738 | .7745 | .7752 | .7760 | .7767 | .7774 |
| 6.0 | .7782 | .7789 | .7796 | .7803 | .7810 | .7818 | .7825 | .7832 | .7839 | .7846 |
| 6.1 | .7853 | .7860 | .7868 | .7875 | .7882 | .7889 | .7896 | .7903 | .7910 | .7917 |
| 6.2 | .7924 | .7931 | .7938 | .7945 | .7952 | .7959 | .7966 | .7973 | .7980 | .7987 |
| 6.3 | .7993 | .8000 | .8007 | .8014 | .8021 | .8028 | .8035 | .8041 | .8048 | .8055 |
| 6.4 | .8062 | .8069 | .8075 | .8082 | .8089 | .8096 | .8102 | .8109 | .8116 | .8122 |
| 6.5 | .8129 | .8136 | .8142 | .8149 | .8156 | .8162 | .8169 | .8176 | .8182 | .8189 |
| 6.6 | .8195 | .8202 | .8209 | .8215 | .8222 | .8228 | .8235 | .8241 | .8248 | .8254 |
| 6.7 | .8261 | .8267 | .8274 | .8280 | .8287 | .8293 | .8299 | .8306 | .8312 | .8319 |
| 6.8 | .8325 | .8331 | .8338 | .8344 | .8351 | .8357 | .8363 | .8370 | .8376 | .8382 |
| 6.9 | .8388 | .8395 | .8401 | .8407 | .8414 | .8420 | .8426 | .8432 | .8439 | .8445 |
| 7.0 | .8451 | .8457 | .8463 | .8470 | .8476 | .8482 | .8488 | .8494 | .8500 | .8506 |
| 7.1 | .8513 | .8519 | .8525 | .8531 | .8537 | .8543 | .8549 | .8555 | .8561 | .8567 |
| 7.2 | .8573 | .8579 | .8585 | .8591 | .8597 | .8603 | .8609 | .8615 | .8621 | .8627 |
| 7.3 | .8633 | .8639 | .8645 | .8651 | .8657 | .8663 | .8669 | .8675 | .8681 | .8686 |
| 7.4 | .8692 | .8698 | .8704 | .8710 | .8716 | .8722 | .8727 | .8733 | .8739 | .8745 |
| 7.5 | .8751 | .8756 | .8762 | .8768 | .8774 | .8779 | .8785 | .8791 | .8797 | .8802 |
| 7.6 | .8808 | .8814 | .8820 | .8825 | .8831 | .8837 | .8842 | .8848 | .8854 | .8859 |
| 7.7 | .8865 | .8871 | .8876 | .8882 | .8887 | .8893 | .8899 | .8904 | .8910 | .8915 |
| 7.8 | .8921 | .8927 | .8932 | .8938 | .8943 | .8949 | .8954 | .8960 | .8965 | .8971 |
| 7.9 | .8976 | .8982 | .8987 | .8993 | .8998 | .9004 | .9009 | .9015 | .9020 | .9025 |
| 8.0 | .9031 | .9036 | .9042 | .9047 | .9053 | .9058 | .9063 | .9069 | .9074 | .9079 |
| 8.1 | .9085 | .9090 | .9096 | .9101 | .9106 | .9112 | .9117 | .9122 | .9128 | .9133 |
| 8.2 | .9138 | .9143 | .9149 | .9154 | .9159 | .9165 | .9170 | .9175 | .9180 | .9186 |
| 8.3 | .9191 | .9196 | .9201 | .9206 | .9212 | .9217 | .9222 | .9227 | .9232 | .9238 |
| 8.4 | .9243 | .9248 | .9253 | .9258 | .9263 | .9269 | .9274 | .9279 | .9284 | .9289 |
| 8.5 | .9294 | .9299 | .9304 | .9309 | .9315 | .9320 | .9325 | .9330 | .9335 | .9340 |
| 8.6 | .9345 | .9350 | .9355 | .9360 | .9365 | .9370 | .9375 | .9380 | .9385 | .9390 |
| 8.7 | .9395 | .9400 | .9405 | .9410 | .9415 | .9420 | .9425 | .9430 | .9435 | .9440 |
| 8.8 | .9445 | .9450 | .9455 | .9460 | .9465 | .9469 | .9474 | .9479 | .9484 | .9489 |
| 8.9 | .9494 | .9499 | .9504 | .9509 | .9513 | .9518 | .9523 | .9528 | .9533 | .9538 |
| 9.0 | .9542 | .9547 | .9552 | .9557 | .9562 | .9566 | .9571 | .9576 | .9581 | .9586 |
| 9.1 | .9590 | .9595 | .9600 | .9605 | .9609 | .9614 | .9619 | .9624 | .9628 | .9633 |
| 9.2 | .9638 | .9643 | .9647 | .9652 | .9657 | .9661 | .9666 | .9671 | .9675 | .9680 |
| 9.3 | .9685 | .9689 | .9694 | .9699 | .9703 | .9708 | .9713 | .9717 | .9722 | .9727 |
| 9.4 | .9731 | .9736 | .9741 | .9745 | .9750 | .9754 | .9759 | .9763 | .9768 | .9773 |
| 9.5 | .9777 | .9782 | .9786 | .9791 | .9795 | .9800 | .9805 | .9809 | .9814 | .9818 |
| 9.6 | .9823 | .9827 | .9832 | .9836 | .9841 | .9845 | .9850 | .9854 | .9859 | .9863 |
| 9.7 | .9868 | .9872 | .9877 | .9881 | .9886 | .9890 | .9894 | .9899 | .9903 | .9908 |
| 9.8 | .9912 | .9917 | .9921 | .9926 | .9930 | .9934 | .9939 | .9943 | .9948 | .9952 |
| 9.9 | .9956 | .9961 | .9965 | .9969 | .9974 | .9978 | .9983 | .9987 | .9991 | .9996 |
| n | 0 | 1 | 2 | 3 | 4 | 5 | 6 | 7 | 8 | 9 |

## Table 3
### Natural Logarithms

| $x$ | $\ln x$ |
|---|---|
| 0.2 | 8.391 − 10 |
| 0.4 | 9.084 − 10 |
| 0.6 | 9.489 − 10 |
| 0.8 | 9.777 − 10 |
| 1.0 | .0000 |
| 1.2 | .1823 |
| 1.4 | .3365 |
| 1.6 | .4700 |
| 1.8 | .5878 |
| 2.0 | .6931 |
| 2.2 | .7885 |
| 2.4 | .8755 |
| 2.6 | .9555 |
| 2.8 | 1.0296 |
| 3.0 | 1.0986 |
| 3.2 | 1.1632 |
| 3.4 | 1.2238 |
| 3.6 | 1.2809 |
| 3.8 | 1.3350 |
| 4.0 | 1.3863 |
| 4.2 | 1.4351 |
| 4.4 | 1.4816 |
| 4.6 | 1.5261 |
| 4.8 | 1.5686 |
| 5.0 | 1.6094 |
| 5.2 | 1.6487 |
| 5.4 | 1.6864 |
| 5.6 | 1.7228 |
| 5.8 | 1.7579 |
| 6.0 | 1.7918 |
| 6.5 | 1.8718 |
| 7.0 | 1.9459 |
| 7.5 | 2.0149 |
| 8.0 | 2.0794 |
| 8.5 | 2.1401 |
| 9.0 | 2.1972 |
| 9.5 | 2.2513 |
| 10 | 2.3026 |
| 15 | 2.7081 |
| 20 | 2.9957 |
| 25 | 3.2189 |
| 30 | 3.4012 |
| 35 | 3.5553 |
| 40 | 3.6889 |
| 45 | 3.8067 |
| 50 | 3.9120 |
| 55 | 4.0073 |
| 60 | 4.0943 |
| 65 | 4.1744 |
| 70 | 4.2485 |
| 75 | 4.3175 |
| 80 | 4.3820 |
| 85 | 4.4427 |
| 90 | 4.4998 |
| 95 | 4.5539 |
| 100 | 4.6052 |

## Table 4  Powers of $e$

| $x$ | $e^x$ | $e^{-x}$ |
|---|---|---|
| 0.00 | 1.00000 | 1.00000 |
| 0.01 | 1.01005 | 0.99005 |
| 0.02 | 1.02020 | 0.98020 |
| 0.03 | 1.03045 | 0.97045 |
| 0.04 | 1.04081 | 0.96079 |
| 0.05 | 1.05127 | 0.95123 |
| 0.06 | 1.06184 | 0.94176 |
| 0.07 | 1.07251 | 0.93239 |
| 0.08 | 1.08329 | 0.92312 |
| 0.09 | 1.09417 | 0.91393 |
| 0.10 | 1.10517 | 0.90484 |
| 0.11 | 1.11628 | 0.89583 |
| 0.12 | 1.12750 | 0.88692 |
| 0.13 | 1.13883 | 0.87810 |
| 0.14 | 1.15027 | 0.86936 |
| 0.15 | 1.16183 | 0.86071 |
| 0.16 | 1.17351 | 0.85214 |
| 0.17 | 1.18530 | 0.84366 |
| 0.18 | 1.19722 | 0.83527 |
| 0.19 | 1.20925 | 0.82696 |
| 0.20 | 1.22140 | 0.81873 |
| 0.30 | 1.34986 | 0.74082 |
| 0.40 | 1.49182 | 0.67032 |
| 0.50 | 1.64872 | 0.60653 |
| 0.60 | 1.82212 | 0.54881 |
| 0.70 | 2.01375 | 0.49659 |
| 0.80 | 2.22554 | 0.44933 |
| 0.90 | 2.45960 | 0.40657 |
| 1.00 | 2.71828 | 0.36788 |
| 1.10 | 3.00417 | 0.33287 |
| 1.20 | 3.32012 | 0.30119 |
| 1.30 | 3.66930 | 0.27253 |
| 1.40 | 4.05520 | 0.24660 |
| 1.50 | 4.48169 | 0.22313 |
| 1.60 | 4.95303 | 0.20190 |
| 1.70 | 5.47395 | 0.18268 |
| 1.80 | 6.04965 | 0.16530 |
| 1.90 | 6.68589 | 0.14957 |
| 2.00 | 7.38906 | 0.13534 |
| 2.10 | 8.16617 | 0.12246 |
| 2.20 | 9.02501 | 0.11080 |
| 2.30 | 9.97418 | 0.10026 |
| 2.40 | 11.02318 | 0.09072 |
| 2.50 | 12.18249 | 0.08208 |
| 2.60 | 13.46374 | 0.07427 |
| 2.70 | 14.87973 | 0.06721 |
| 2.80 | 16.44465 | 0.06081 |
| 2.90 | 18.17415 | 0.05502 |
| 3.00 | 20.08554 | 0.04979 |
| 3.50 | 33.11545 | 0.03020 |
| 4.00 | 54.95815 | 0.01832 |
| 4.50 | 90.01713 | 0.01111 |
| 5.00 | 148.41316 | 0.00674 |
| 5.50 | 224.69193 | 0.00409 |
| 6.00 | 403.42879 | 0.00248 |
| 6.50 | 665.14163 | 0.00150 |
| 7.00 | 1096.63316 | 0.00091 |
| 7.50 | 1808.04241 | 0.00055 |
| 8.00 | 2980.95799 | 0.00034 |
| 8.50 | 4914.76884 | 0.00020 |
| 9.00 | 8103.08393 | 0.00012 |
| 9.50 | 13359.72683 | 0.00007 |
| 10.00 | 22026.46579 | 0.00005 |

## Table 5  Trigonometric Values For Angles In Degrees

| Angle | Sin | Tan | Cot | Cos | |
|---|---|---|---|---|---|
| 0° | .0000 | .0000 | —— | 1.0000 | 90° |
| 1° | .0175 | .0175 | 57.29 | .9998 | 89° |
| 2° | .0349 | .0349 | 28.64 | .9994 | 88° |
| 3° | .0523 | .0524 | 19.08 | .9986 | 87° |
| 4° | .0698 | .0699 | 14.30 | .9976 | 86° |
| 5° | .0872 | .0875 | 11.43 | .9962 | 85° |
| 6° | .1045 | .1051 | 9.514 | .9945 | 84° |
| 7° | .1219 | .1228 | 8.144 | .9925 | 83° |
| 8° | .1392 | .1405 | 7.115 | .9903 | 82° |
| 9° | .1564 | .1584 | 6.314 | .9877 | 81° |
| 10° | .1736 | .1763 | 5.671 | .9848 | 80° |
| 11° | .1908 | .1944 | 5.145 | .9816 | 79° |
| 12° | .2079 | .2126 | 4.705 | .9781 | 78° |
| 13° | .2250 | .2309 | 4.331 | .9744 | 77° |
| 14° | .2419 | .2493 | 4.011 | .9703 | 76° |
| 15° | .2588 | .2679 | 3.732 | .9659 | 75° |
| 16° | .2756 | .2867 | 3.487 | .9613 | 74° |
| 17° | .2924 | .3057 | 3.271 | .9563 | 73° |
| 18° | .3090 | .3249 | 3.078 | .9511 | 72° |
| 19° | .3256 | .3443 | 2.904 | .9455 | 71° |
| 20° | .3420 | .3640 | 2.747 | .9397 | 70° |
| 21° | .3584 | .3839 | 2.605 | .9336 | 69° |
| 22° | .3746 | .4040 | 2.475 | .9272 | 68° |
| 23° | .3907 | .4245 | 2.356 | .9205 | 67° |
| 24° | .4067 | .4452 | 2.246 | .9135 | 66° |
| 25° | .4226 | .4663 | 2.145 | .9063 | 65° |
| 26° | .4384 | .4877 | 2.050 | .8988 | 64° |
| 27° | .4540 | .5095 | 1.963 | .8910 | 63° |
| 28° | .4695 | .5317 | 1.881 | .8829 | 62° |
| 29° | .4848 | .5543 | 1.804 | .8746 | 61° |
| 30° | .5000 | .5774 | 1.732 | .8660 | 60° |
| 31° | .5150 | .6009 | 1.664 | .8572 | 59° |
| 32° | .5299 | .6249 | 1.600 | .8480 | 58° |
| 33° | .5446 | .6494 | 1.540 | .8387 | 57° |
| 34° | .5592 | .6745 | 1.483 | .8290 | 56° |
| 35° | .5736 | .7002 | 1.428 | .8192 | 55° |
| 36° | .5878 | .7265 | 1.376 | .8090 | 54° |
| 37° | .6018 | .7536 | 1.327 | .7986 | 53° |
| 38° | .6157 | .7813 | 1.280 | .7880 | 52° |
| 39° | .6293 | .8098 | 1.235 | .7771 | 51° |
| 40° | .6428 | .8391 | 1.192 | .7660 | 50° |
| 41° | .6561 | .8693 | 1.150 | .7547 | 49° |
| 42° | .6691 | .9004 | 1.111 | .7431 | 48° |
| 43° | .6820 | .9325 | 1.072 | .7314 | 47° |
| 44° | .6947 | .9657 | 1.036 | .7193 | 46° |
| 45° | .7071 | 1.000 | 1.000 | .7071 | 45° |
| | Cos | Cot | Tan | Sin | Angle |

## Table 6  Trigonometric Values For Angles In Radians

| Rad | Sin | Tan | Cot | Cos |
| --- | --- | --- | --- | --- |
| .00 | .00000 | .00000 | —— | 1.0000 |
| .05 | .04998 | .05004 | 19.983 | .99875 |
| .10 | .09983 | .10033 | 9.9666 | .99500 |
| .15 | .14944 | .15114 | 6.6166 | .98877 |
| .20 | .19867 | .20271 | 4.9332 | .98007 |
| .25 | .24740 | .25534 | 3.9163 | .96891 |
| .30 | .29552 | .30934 | 3.2327 | .95534 |
| .35 | .34290 | .36503 | 2.7395 | .93937 |
| .40 | .38942 | .42279 | 2.3652 | .92106 |
| .45 | .43497 | .48306 | 2.0702 | .90045 |
| .50 | .47943 | .54630 | 1.8305 | .87758 |
| .55 | .52269 | .61311 | 1.6310 | .85252 |
| .60 | .56464 | .68414 | 1.4617 | .82534 |
| .65 | .60519 | .76020 | 1.3154 | .79608 |
| .70 | .64422 | .84229 | 1.1872 | .76484 |
| .75 | .68164 | .93160 | 1.0734 | .73169 |
| .80 | .71736 | 1.0296 | .97121 | .69671 |
| .85 | .75128 | 1.1383 | .87848 | .65998 |
| .90 | .78333 | 1.2602 | .79355 | .62161 |
| .95 | .81342 | 1.3984 | .71511 | .58168 |
| 1.00 | .84147 | 1.5574 | .64209 | .54030 |
| 1.05 | .86742 | 1.7433 | .57362 | .49757 |
| 1.10 | .89121 | 1.9648 | .50897 | .45360 |
| 1.15 | .91276 | 2.2345 | .44753 | .40849 |
| 1.20 | .93204 | 2.5722 | .38878 | .36236 |
| 1.25 | .94898 | 3.0096 | .33227 | .31532 |
| 1.30 | .96356 | 3.6021 | .27762 | .26750 |
| 1.35 | .97572 | 4.4552 | .22446 | .21901 |
| 1.40 | .98545 | 5.7979 | .17248 | .16997 |
| 1.45 | .99271 | 8.2381 | .12139 | .12050 |
| 1.50 | .99749 | 14.101 | .07091 | .07074 |
| 1.55 | .99978 | 48.078 | .02080 | .02079 |
| 1.60 | .99957 | −34.233 | −.02921 | −.02920 |
| 1.65 | .99687 | −12.599 | −.07937 | −.07912 |
| 1.70 | .99166 | −7.6966 | −.12993 | −.12884 |
| 1.75 | .98399 | −5.5204 | −.18115 | −.17825 |
| 1.80 | .97385 | −4.2863 | −.23330 | −.22720 |
| 1.85 | .96128 | −3.4881 | −.28669 | −.27559 |
| 1.90 | .94630 | −2.9271 | −.34164 | −.32329 |
| 1.95 | .92896 | −2.5095 | −.39849 | −.37018 |
| 2.00 | .90930 | −2.1850 | −.45766 | −.41615 |
| Rad | Sin | Tan | Cot | Cos |

## Table 7  Table of Integrals

Add an arbitrary constant C to each.

1. $\displaystyle\int x^n \, dx = \frac{x^{n+1}}{n+1}, \quad n \neq -1$

2. $\displaystyle\int \frac{dx}{x} = \ln|x|$

3. $\displaystyle\int e^{ax} \, dx = \frac{e^{ax}}{a}$

4. $\displaystyle\int \ln x \, dx = x \ln x - x$

5. $\displaystyle\int a^x \, dx = \frac{a^x}{\ln a}$

6. $\displaystyle\int \frac{dx}{a^2 + x^2} = \frac{1}{a} \tan^{-1} \frac{x}{a}$

7. $\displaystyle\int \frac{dx}{a^2 - x^2} = \frac{1}{2a} \ln \left| \frac{a+x}{a-x} \right|, \quad \text{or} \quad \frac{1}{a} \tanh^{-1} \frac{x}{a}$

8. $\displaystyle\int \frac{dx}{\sqrt{a^2 - x^2}} = \sin^{-1} \frac{x}{a}$

9. $\displaystyle\int \frac{dx}{\sqrt{x^2 \pm a^2}} = \ln|x + \sqrt{x^2 \pm a^2}|$

10. $\displaystyle\int \frac{dx}{x\sqrt{x^2 - a^2}} = \frac{1}{a} \sec^{-1} \frac{x}{a}$

11. $\displaystyle\int \frac{dx}{x\sqrt{a^2 \pm x^2}} = -\frac{1}{a} \ln \left| \frac{a + \sqrt{a^2 \pm x^2}}{x} \right|$

12. $\displaystyle\int \sqrt{x^2 \pm a^2} \, dx = \tfrac{1}{2}[x\sqrt{x^2 \pm a^2} \pm a^2 \ln |x + \sqrt{x^2 \pm a^2}|]$

13. $\displaystyle\int \sqrt{a^2 - x^2} \, dx = \tfrac{1}{2}\left[ x\sqrt{a^2 - x^2} + a^2 \sin^{-1} \frac{x}{a} \right]$

14. $\displaystyle\int \sin x \, dx = -\cos x$

15. $\displaystyle\int \cos x \, dx = \sin x$

16. $\displaystyle\int \tan x \, dx = \ln|\sec x|, \quad \text{or} \quad -\ln|\cos x|$

17. $\displaystyle\int \cot x \, dx = \ln|\sin x|, \quad \text{or} \quad -\ln|\csc x|$

18. $\displaystyle\int \sec x \, dx = \ln|\sec x + \tan x|, \quad \text{or} \quad \ln\left| \tan\left(\frac{\pi}{4} + \frac{x}{2}\right) \right|$

19. $\displaystyle\int \csc x \, dx = \ln|\csc x - \cot x|, \quad \text{or} \quad \ln\left| \tan \frac{x}{2} \right|$

20. $\displaystyle\int \sec^2 x \, dx = \tan x$

21. $\int \csc^2 x \, dx = -\cot x$

22. $\int \sec^3 x \, dx = \frac{1}{2}[\sec x \tan x + \ln|\sec x + \tan x|]$

23. $\int \sin mx \sin nx \, dx = \frac{\sin(m-n)x}{2(m-n)} - \frac{\sin(m+n)x}{2(m+n)}, \quad m^2 \neq n^2$

24. $\int \sin mx \cos nx \, dx = -\frac{\cos(m-n)x}{2(m-n)} - \frac{\cos(m+n)x}{2(m+n)}, \quad m^2 \neq n^2$

25. $\int \cos mx \cos nx \, dx = \frac{\sin(m-n)x}{2(m-n)} + \frac{\sin(m+n)x}{2(m+n)}, \quad m^2 \neq n^2$

26. $\int \tan^2 x \, dx = \tan x - x$

27. $\int \cot^2 x \, dx = -\cot x - x$

28. $\int x \sin x \, dx = \sin x - x \cos x$

29. $\int x \cos x \, dx = \cos x + x \sin x$

30. $\int \sin^{-1} x \, dx = x \sin^{-1} x + \sqrt{1-x^2}$

31. $\int \cos^{-1} x \, dx = x \cos^{-1} x - \sqrt{1-x^2}$

32. $\int \tan^{-1} x \, dx = x \tan^{-1} x - \frac{1}{2}\ln(1+x^2)$

33. $\int \sec^{-1} x \, dx = x \sec^{-1} x - \ln|x + \sqrt{x^2-1}|$

34. $\int x \sin^{-1} x \, dx = \frac{1}{4}[(2x^2-1)\sin^{-1} x + x\sqrt{1-x^2}]$

35. $\int x \ln x \, dx = \frac{x^2}{2} \ln x - \frac{x^2}{4}$

36. $\int xe^{ax} \, dx = \frac{e^{ax}}{a^2}(ax-1)$

37. $\int e^{ax} \sin bx \, dx = \frac{e^{ax}(a \sin bx - b \cos bx)}{a^2 + b^2}$

38. $\int e^{ax} \cos bx \, dx = \frac{e^{ax}(a \cos bx + b \sin bx)}{a^2 + b^2}$

39. $\int \sinh x \, dx = \cosh x$

40. $\int \cosh x \, dx = \sinh x$

41. $\int \tanh x \, dx = \ln \cosh x$

42. $\int \coth x \, dx = \ln|\sinh x|$

43. $\int \text{sech } x \, dx = 2 \tan^{-1} e^x$

44. $\int \text{csch } x \, dx = \ln\left|\tanh \frac{x}{2}\right|$

# Answers To Odd-Numbered Exercises

**EXERCISE GROUP 1.1**

1. $x < 7/3$  3. $x > -7/3$  5. $x > -7/2$  7. $x < -6$  9. $x > -1$
11. $-3 < x < 2$  13. $x < -4/3, x > 2$  15. $-2 < x < 5/2$  17. $x < -2, 1/2 < x < 7/3$
19. $x > 2$  21. $x < -5/2, -2/3 < x < 3/7$  23. $x \geq 1/2$  25. $x \leq -1, x \geq 1$
27. $x < 0, x \geq 1$  29. $-4 < x < 4$  31. $x < -5, x > 5$  33. $-4 < x < 2$
35. $x < 7/4, x > 9/4$  37. $-1 < x < 1, 5 < x < 7$  39. $-4 < x < -2, -2 < x < 0$
41. $x < (3 - \sqrt{65})/4, x > (3 + \sqrt{65})/4$  43. $(5 - \sqrt{65})/4 < x < (5 + \sqrt{65})/4$
45. $x < (7 - \sqrt{17})/4, x > (7 + \sqrt{17})/4$

**EXERCISE GROUP 1.2**

1. $(-4, -1)$  3. $(1, 4]$  5. $(-\infty, 2)$  7. $(-1.4, 1.4)$  9. $(-4, 0)$
11. $(-\infty, -b - 4), (b - 4, \infty)$  13. $(-\infty, 0), (2/3, \infty)$  15. $[-3, -1], [3, 5]$
17. $[-2, -5/3), (-5/3, -4/3]$  19. $\left(\dfrac{-5 - \sqrt{37}}{6}, \dfrac{-5 + \sqrt{37}}{6}\right)$
21. $(1, 2)$  23. $(1/2, \infty)$  25. $(2/3, 3/2), (5/2, \infty)$

**EXERCISE GROUP 1.3**

1. (a) On a straight line 3 units to the left of the $y$ axis
   (b) On a straight line 2 units above the $x$ axis
3. On a straight line $1\frac{1}{2}$ units above the $x$ axis
5. Three solutions: $(4, 1), (-2, -1), (-4, 5)$
7. $(11/2, 4), (11/2, -4)$  11. $2; -3$  13. $10; -14$  15. $4; -3$

**EXERCISE GROUP 1.4**

1. (a) $(-1, -3)$, (b) $(5, -7)$, (c) $(-4, -1)$
3. (a) $\sqrt{29}$, (b) $\sqrt{13}$, (c) $\sqrt{10}$  5. (a) $(15, -1)$, (b) $\sqrt{226}$
7. (a) $\pm(2, -5)/\sqrt{29}$, (b) $\pm(-3, 2)/\sqrt{13}$, (c) $\pm(-1, -3)/\sqrt{10}$
9. $(-2, 5)$  11. $(7/2, -1)$  13. $\frac{11}{2}(-1, 1)$
15. (a) $\sqrt{2a^2 + 2a + 1}$, (b) $\sqrt{2(a^2 + 1)}$  21. (b) $-c|a|$

**EXERCISE GROUP 1.5**

1. $-6\mathbf{i} + 15\mathbf{j}$  3. $12\mathbf{i} + 38\mathbf{j}$  5. $30\mathbf{i} - 24\mathbf{j}$  7. $\dfrac{2}{\sqrt{29}}\mathbf{i} - \dfrac{5}{\sqrt{29}}\mathbf{j}$  9. $\mathbf{a} - \mathbf{b}$
11. $\mathbf{a} - 2\mathbf{b}$  13. $\dfrac{5}{13}\mathbf{a} - \dfrac{1}{13}\mathbf{b}$  15. $\dfrac{9}{13}\mathbf{a} + \dfrac{6}{13}\mathbf{b}$

## EXERCISE GROUP 1.6

**5.** (a) $(-2, -12)$, (b) $(-1, -6)$, (c) $(3, 1)$
**7.** (a) $\mathbf{a} + \mathbf{b}/2$, (b) $\mathbf{a}/2 + \mathbf{b}$, (c) $(\mathbf{b} - \mathbf{a})/2$

## EXERCISE GROUP 2.1

**1.** (a) $\sqrt{10}$, (b) $(1/2, 7/2)$, (c) $\sqrt{58}/2$  **3.** (a) $\sqrt{34}$, (b) $(5/2, 3/2)$, (c) $5\sqrt{2}/2$

## EXERCISE GROUP 2.2

**1.** (a) 1, (b) 0, (c) 25  **3.** (a) 8, (b) 13  **5.** 0  **7.** 17  **9.** $-16$  **11.** $\pi/4$
**13.** $\pi/2 < \cos^{-1}(-\sqrt{2}/10) > \pi$  **15.** $\pi/2 < \cos^{-1}(-\sqrt{65}/65) < \pi$  **17.** $\pi/2$
**21.** $x^2 + y^2 - 4 = 0$

## EXERCISE GROUP 2.3

**1.** $1/3$  **3.** $3/4$  **5** $\dfrac{y-3}{x-1}, x \neq 1$  **7** $\dfrac{y}{x-a}, x \neq a$  **19.** (a) $73/7$, (b) $13/5$  **21.** $11/3$
**23.** 6

## EXERCISE GROUP 2.4

**1.** 0.28 rad (16°20′)  **3.** 0.32 rad (18°30′)  **5.** 1.37 rad (78°40′)  **7.** 1.25 rad (71°30′)
**9.** 27°20′, 40°20′, 112°20′  **11.** 14°30′, 21°, 144°30′  **13.** 1 or 18  **17.** 0.57 rad (32°30′)
**19.** 1.18 rad (67°40′)

## EXERCISE GROUP 2.6

**1.** $3x + y - 13 = 0$  **3.** $3x - 2y - 6 = 0$  **5.** $2x + y - 2a = 0$  **7.** $x - 2y - 2 = 0$
**9.** $2x + y + 4 = 0$  **11.** $mx - y - am = 0$  **13.** $7x - 2y + 31 = 0$
**15.** $2x + 7y - 29 = 0$  **17.** $3x + 8y - 16 = 0$  **19.** $3x - 2y = 0$, $x + 2y - 8 = 0$
**21.** $2x - 5y - 1 = 0$  **23.** $5x + y - 7 = 0$  **25.** $2x - y - 3 = 0$  **27.** $x + 4y - 7 = 0$
**29.** $x + y - 3 = 0$

## EXERCISE GROUP 2.7

**1.** $(2, 3)$  **3.** $(3/2, 2/3)$  **5.** $(-3, -5)$  **7.** $(3/2, -2/3)$  **9.** $(-5/2, 3/2)$
**11.** $(5/11, 23/11)$  **13.** $(13/5, -7/5)$

## EXERCISE GROUP 2.8

In each of the following answers the general shape of the region to be shaded is indicated.

**1.**    **3.**    **5.**

**7.** None   **9.**    **11.**

## EXERCISE GROUP 3.1

**1.** (b) $f(2) = 2, f(5) = 2$, (c) $x = 3$  **3.** $x \neq 0$  **5.** All real $x$  **7.** $|x| < 2$
**17.** (a) $|x|\sqrt{1 + x^2}$, (b) $\sqrt{x^4 - 5x^2 + 4x + 13}$  **19.** $2y = f(2x)$  **21.** $2y = f(x) + g(x)$

## EXERCISE GROUP 3.2

**1.** (a) [(0, 4), (1, 6), (2, 8), (3, 10)], (b) [(0, 1), (1, 4), (2, −1), (3, 11)],
  (c) [(0, 3), (1, 2), (2, 9), (3, −1)], (d) [(0, 4), (1, 9), (2, 16), (3, 25)]
**3.** 5  **5.** $a^2 + a - 1$  **7.** (a) $x^2 - x$, (b) $(x + h)^2 + (x + h)$, (c) $2x + 1 + h$
**9.** (a) $f[g(x)]$: [(1, −5), (2, 3), (3, 4)]; Domain [1, 2, 3],
  (b) $g[f(x)]$: [(2, −1), (4, −1)]; Domain [2, 4]
**11.** (a) $f[g(x)] = x + 1, x \neq -1$, (b) $g[f(x)] = x/(x + 1), x \neq 0, -1$
**13.** (a) $f[g(x)] = x$, all real $x$, (b) $g[f(x)] = x$, all real $x$
**15.** $g(x + 2) = \dfrac{x + 5}{x}, g(x^2 + 1) = \dfrac{x^2 + 4}{x^2 - 1}, g\left(\dfrac{x - 1}{x + 1}\right) = -\dfrac{4x + 2}{x + 3}$
**17.** $g[g(1)] = 64, g[g(2)] = 128$  **19.** $f(x) = x^{17}, g(x) = 2x^2 + 3x - 1$
**21.** $f(x) = 1/x, g(x) = x^2 + x$  **23.** $f(x) = \sqrt[3]{x}, g(x) = (x - 1)/(x + 1)$

## EXERCISE GROUP 3.3

**1.** 7  **3.** 5/3  **5.** 1/3  **7.** 7/6  **9.** 3/10  **11.** 0  **13.** 1/2  **15.** 1/3  **17.** 1/48  **19.** 1
**21.** 3  **23.** $4x - 1$  **25.** $1/(2\sqrt{x})$  **27.** $3x^2$  **29.** $2/(x + 1)^2$

## EXERCISE GROUP 3.4

**1.** −6  **3.** −10/3  **5.** 10/3  **7.** $\sqrt{2}$  **9.** −5  **11.** $\sqrt{2}/4$  **13.** 0

## EXERCISE GROUP 3.5

**9.** $x \neq 1$  **11.** $x \neq -1, 2$  **13.** $x \neq \pm 2$  **15.** $x \neq 0, 2/3$  **17.** $x \neq 0$  **19.** $x \neq -1$
**21.** max 1/2, min 0
**25.** $f$ is continuous at $a$ if, given $\epsilon > 0$, there exists $\delta$ such that $|f(x) - f(a)| < \epsilon$ whenever $|x - a| < \delta$.

## EXERCISE GROUP 4.1

**1.** 5  **3.** $2x$  **5.** $-3/x^2$  **7.** $-1/(x - 1)^2$  **9.** $1/(2\sqrt{x})$  **11.** $-1/[2\sqrt{x}(1 + \sqrt{x})^2]$
**13.** 2  **15.** −1  **17.** −1/2  **19.** $f'(2), f(x) = x^2$  **21.** $f'(4), f(x) = \sqrt{x}$
**23.** $f'(1), f(x) = x^3$  **25.** $f'(-2), f(x) = x^3 - 3x$

## EXERCISE GROUP 4.2

**1.** (a) $2x + y + 4 = 0$, (b) $x - 2y - 3 = 0$
**3.** (a) $x + 4y - 4 = 0$, (b) $8x - 2y - 15 = 0$
**5.** (a) $x - 4y + 4 = 0$, (b) $4x + y - 18 = 0$
**7.** (a) $2x + y - 3 = 0$, (b) $x - 2y + 1 = 0$  **9.** (1/2, 1/4)  **11.** (−1/3, −13/27), (3, −9)
**13.** None  **17.** $\tan^{-1} 4/3$ at $(\pm 1, 1)$
**19.** $\tan^{-1} 2$ at (0, 0), $\tan^{-1} 1/3$ at (1, 0), $\tan^{-1} 2/49$ at (2, 4)

## EXERCISE GROUP 4.3

1. $6x^2 - 3$  3. $-3/x^4$  5. $2 - 12x^2$  7. $1/(2\sqrt{x}) + 2x - 3$  9. $1$
11. $3x^2 - 2x$  13. $1, x \ne -1$  15. $f'(2) = 15, f'(-1/2) = 25/2$
17. (a) $12, -2$, (b) $2/3, 3/2$, (c) $1, 7/6$

## EXERCISE GROUP 4.4

1. $18x(x^2 - 1)^8$  3. $16(2w - 3)^7$  5. $x/\sqrt{x^2 - 1}$  7. $x(1 - x^2)^{-3/2}$
9. $\frac{8}{5}v^3(v^4 - 1)^{-3/5}$  11. $\frac{4}{3}(1 - 2s)(s - s^2)^{1/3}$  13. $-x^{-1/2}(a^{1/2} - x^{1/2})$
15. $\frac{3}{2}(x + 1)(x^{1/2} - x^{-1/2})^2 x^{-3/2}$  17. $-p(1 - x)$  19. $p(2x)/\sqrt{\sin 2x}$
21. $-p(x) \sin x/\sqrt{1 - \sin^2 x}$

## EXERCISE GROUP 4.5

1. $2t(2 + 3t)^3(2 + 9t)$  3. $4(2x + 1)(x - 2)^3(3x - 1)$  5. $\frac{4}{3}(x^2 - 2)x^{-1/3}(x^2 - 4)^{-2/3}$
7. $(1 - x)^{-2}$  9. $16x(x^2 + 4)^{-2}$  11. $-2x(x + 3)^{-3}$  13. $-6x(1 + x^2)^{-2}$
15. $\frac{1}{2}(5 + 3x)(2 + x)^{-1/2}$  17. $x^{-1/2}(1 - \sqrt{x})^{-2}$  19. $x(2 - x)(1 + x)^{-4}$
21. $-(1 - x)^{-1/2}(1 + x)^{-3/2}$  23. $\frac{1}{2}(1 - u^2)u^{-1/2}(u^2 + 1)^{-3/2}$
25. $\frac{1}{2}x(4 + 3\sqrt{x})(1 + \sqrt{x})^{-2}$  27. $\frac{2}{3}(1 + x)^{-2/3}(1 - x)^{-4/3}$
29. $x(12 - x^2)(4 - x^2)^{-3/2}$  31. $-x(4 - x^2)^{-1/2}$
33. $2(1 - \sqrt{1 - x^4})x^{-3}(1 - x^4)^{-1/2}$  35. $xy' + y$  37. $\frac{1}{2}x(xy' + 4y)y^{-1/2}$
39. $(y - xy')(x + y)^{-2}$  41. $2[(x + y)y' + y]$  43. $(x^3y' + y^2 - x^2y)(x^2 + y)^{-2}$

## EXERCISE GROUP 4.6

1. $4a/y$  3. $x/y$  5. $-y^{1/2}/x^{1/2}$  7. $(x^2 - y)/(x - y^2)$  9. $x(2 - 3x)/(2y)$
11. $-y^3/x^3$  13. $x - 2y - 5 = 0$  15. $x + 2y - 6 = 0$  17. $x + y - 6 = 0$
19. $4x + 5y - 17 = 0$

## EXERCISE GROUP 4.7

1. $f'(x) = 8x^3 - 3x^2 + 6x + 2, f''(x) = 24x^2 - 6x + 6, f'''(x) = 48x - 6, f^{(4)}(x) = 48,$
$f^{(n)}(x) = 0, n > 4$
3. $f'(x) = 4x^3 + 4x, f''(x) = 12x^2 + 4, f'''(x) = 24x, f^{(4)}(x) = 24, f^{(n)}(x) = 0, n > 4$
5. $-\frac{1}{4}(x + 4)(x + 1)^{-5/2}$  7. $4(x - 1)^{-3}$  9. $(2x + 1)(x + 1)^{-3/2}(x - 1)^{-5/2}$
11. $(-1)^n n!(x - 1)^{-n-1}$  13. $(-1)^n \cdot 1 \cdot 3 \cdot 5 \cdots (2n - 1)2^{-n}x^{-(2n+1)/2}$
15. $(-1)^{n+1} \cdot 2 \cdot 5 \cdots (3n - 4)3^{-n}x^{-(3n-1)/3}, n \ge 2$  17. $-4y^{-3}$  19. $-\frac{8}{9}y^{-3}$
21. $2(4x^3 + 3x^4y - y^3 - 2)(x^2 + y)^{-3}$

## EXERCISE GROUP 4.8

1. $3x(x - 2) \, dx$  3. $\frac{2}{3}(3x^{1/3} - 2)x^{-2/3} \, dx$  5. $t(t^2 + 1)^{-1/2} \, dt$  7. $-5(3t - 1)^{-2} \, dt$
9. $\frac{1}{2}(t + 2)(t + 1)^{-3/2} \, dt$  11. $4x(x^2 + 1)^{-2} \, dx$  13. (a) $2u \, du$, (b) $2uu' \, dx$
15. (a) $(u^2 + v^2)^{-1/2}(u \, du + v \, dv)$, (b) $(u^2 + v^2)^{-1/2}(uu' + vv') \, dx$
17. (a) $u \, dv + v \, du$, (b) $(uv' + vu') \, dx$  19. $-xy^{-1}$  21. $3x(2x - 3y)^{-1}$
23. $-t(u + 2t + 2)^{-1}$

**EXERCISE GROUP 4.9**

1. 1.995   3. 1.01   5. 238.95   7. 10%   9. (a) $2h/r$, (b) $3h/r$   11. $(3x+h-1)h^2$
13. $[x^2(x+h)]^{-1}h^2$   15. $-[(x+h+1)(x+1)^2]^{-1}h^2$   17. $a_1 = 3/2$, $a_2 = 17/12$

**EXERCISE GROUP 5.1**

1. $y=0$   3. $x-9y+7=0$   5. $x-y+1=0$   7. Tangent $x+y=0$; Normal $x-y=0$   9. (2, 10)   11. (1, −1), (−1, 1)

**EXERCISE GROUP 5.2**

1. (a) $-\infty < x < \infty$, (b) none   3. (a) $x<-1$, $x>2$, (b) $-1<x<2$
5. (a) None, (b) $x<0$, $x>0$   7. (a) $x<0$, (b) $x>0$   9. (a) $x>4$, (b) $2<x<4$
11. (a) $x<-1/2$, $x>0$, (b) $-1/2<x<0$   13. (a) $x>-1$, (b) $x<-1$
15. (a) $x>-1/2$, (b) None

**EXERCISE GROUP 5.3**

1. Rel max 41/8 at $x=-3/4$   3. Rel max 3 at $x=-1$, rel min $-15/4$ at $x=1/2$
5. Rel min 0 at $x=-2$   7. Rel max 1/2 at $x=\sqrt{2}/2$, rel min $-1/2$ at $x=-\sqrt{2}/2$
9. Rel min $-\sqrt{2}$ at $x=-1$   11. Rel max $3^{13}/5^5$ at $x=1/5$, rel min 0 at $x=2$
13. Rel max $2\sqrt{3}/9$ at $x=-\sqrt{3}/3$, rel min $-2\sqrt{3}/9$ at $x=\sqrt{3}/3$
15. Rel max $-2$ at $x=-1$, rel min 2 at $x=1$
17. Rel max 1/6 at $x=3$, rel min $-1/2$ at $x=-1$   19. Rel max $-6$ at $x=-2$
21. Rel max 16 at $x=2$, rel min $-16$ at $x=-2$
23. Rel min $-59/16$ at $x=-3/2$   25. Rel max 4 at $x=1/2$, rel min $-1$ at $x=-2$
27. Rel max 0 at $x=0$, rel min $-4/27$ at $x=8/27$   29. Rel max $-\sqrt{6}/6$ at $x=5/6$

**EXERCISE GROUP 5.4**

1. Max 1, min $-27$   3. Max 10, min 6   5. Max 1, min $-\sqrt[3]{3}$   7. Max 5/2, min 2
9. Max 12, min 0   11. Max 1/4, min 0,   13. Max 4, min 0   15. Max 1, min $-1$

**EXERCISE GROUP 5.5**

1. 12 in. × 12 in. × 3 in.   3. Side of base = twice altitude
5. 5 mi downstream   9. $2\sqrt{6}\pi/3$   11. Alt $=4a/3$, base rad $=2\sqrt{2}a/3$
13. $2a^2$   15. $|ax_0+by_0+c|/\sqrt{a^2+b^2}$   17. (a) $a^2/2$, (b) $a^2$   21. $(1/2, \sqrt{2}/2)$
23. $(-2, \pm 1)$   25. $(a^{4/3}/\sqrt{a^{2/3}+b^{2/3}}, b^{4/3}/\sqrt{a^{2/3}+b^{2/3}})$   27. $5\sqrt{5}$ ft

**EXERCISE GROUP 5.6**

1. Rel max (2, 16), rel min $(-2, -16)$, point of inflection (0, 0), concave up $x<0$, conc down $x>0$

3. Rel max $(-2, 8)$, rel min $(2/3, -40/27)$, point of infl $(-2/3, 88/27)$, conc up $x>-2/3$, conc down $x<-2/3$

5. No rel ext, point of infl (0, 3), conc up $x>0$, conc down $x<0$

7. Rel min (0, 0), conc down $x \neq 0$

9. Rel max (0, 0), rel min $(\pm \sqrt{2}, -4)$, points of infl $(\pm \sqrt{6}/3, -20/9)$, conc up $|x| > \sqrt{6}/3$, conc down $|x| < \sqrt{6}/3$

11. Rel max $(-2, 0)$, rel min $(-4/5, -2^2 3^8/5^5)$, points of infl $(1, 0)$ and at $x = (-8 \pm 3\sqrt{6})/10$, conc up $(-8 - 3\sqrt{6})/10 < x < (-8 + 3\sqrt{6})/10$, $x > 1$, conc down $x < (-8 - 3\sqrt{6})/10$, $(-8 + 3\sqrt{6})/10 < x < 1$

13. Rel max (0, 1), rel min $(\pm 1, 0)$. point of infl $(\pm \sqrt{3}/3, 4/9)$, conc up $|x| > \sqrt{3}/3$, conc down $|x| < \sqrt{3}/3$

### EXERCISE GROUP 5.8

1. $s$ starts at 0, increases to a max of 225/4 at $t = 15/8$, then decreases to 36 at $t = 3$
3. $s$ starts at 4, increases to 5 at $t = 1$, and continues to increase to 6 at $t = 2$
5. $s$ starts at 19, decreases to a min of 8 at $t = 2$, then increases to 49/4 at $t = 4$
7. (a) $t = 1/9$, (b) $-7$
9. $s = 0$ at $t = 0$, then $s$ increases to a max of $v_0^2/(2g)$ at $t = v_0/g$, then $s$ decreases
11. $x/4$   13. $(1/2, 1/4)$   15. (a) $(-2, 0)$, $(\pm \sqrt{2}, 1)$, (b) $(-1, 3/2)$, $(\pm \sqrt{2}, 1)$

### EXERCISE GROUP 5.9

1. $352\sqrt{5}/5 \approx 157$ (ft/sec)   3. $3\sqrt{5}/20 \approx 0.34$ (ft/sec)   5. 160 ft$^3$/min
7. $27\sqrt[3]{\pi}$ ft$^3$/min   9. $172/\sqrt{34} \approx 29.5$ (mph)   11. $19\sqrt{10}/5$
13. (a) $6\sqrt{5}/5$, (b) $[(7 + \sqrt{7})/14, (-7 + \sqrt{7})/14]$   15. $3\sqrt{2}/5$ ft/sec   17. 4 ft/sec
19. $(a^2 + b^2)t/\sqrt{h^2 + (a^2 + b^2)t^2}$ ft/sec

### EXERCISE GROUP 6.1

11. Rel max (1, 1/2), rel min $(-1, -1/2)$, points of infl (0, 0), $(\sqrt{3}, \sqrt{3}/4)$, $(-\sqrt{3}, -\sqrt{3}/4)$, conc up $-\sqrt{3} < x < 0$, $x > \sqrt{3}$, conc down $x < -\sqrt{3}$, $0 < x < \sqrt{3}$

13. No rel ext, no point of infl, conc up $x < -1$, conc down $x > -1$
15. Rel max $(0, -1)$, no point of infl, conc up $|x| > 1$, conc down $|x| < 1$

### EXERCISE GROUP 6.2

1. (a) $-2x$, (b) $2x$   3. (a) $x^2 + x$, (b) $-x^2 - x$   5. (a) $-x^3 + x^2$, (b) $x^3 - x^2$
7. (a) $1/(1 + x^2)$, (b) $-1/(1 + x^2)$   9. $y < -1/8$   11. $|x| > 1$, $|y| > 2$
13. $|x| > 2\sqrt{6}/3$, $|y| > 2\sqrt{6}/3$   15. $y \leq 0$, $y > 1$
17. If replacing $x$ by $2h - x$ yields an equivalent equation
19. If replacing $x$ by $2h - x$ and $y$ by $2k - y$ yields an equivalent equation
21. If replacing $x$ by $-y$ and $y$ by $-x$ yields an equivalent equation
23. (a) $(-1, 8)$

### EXERCISE GROUP 6.3

1. 1   3. 0   5. 3/2   7. $-\infty$   9. $\infty$   11. 4   13. 1   15. $-1$

**EXERCISE GROUP 6.4**

1. (0, 0); None; $y = 1$, $x = -1$  3. (0, 0); None; $x = 3$
5. (0, 0); $y$ axis; $y = 1$, $x = \pm 3$  7. (0, 0); $y$ axis; $y = 1$
9. $(\pm 1, 0)$, (0, 1/4); $y$ axis; $y = 1$, $x = \pm 2$  11. $(-1, 0)$; $x$ axis; $y = \pm 2$, $x = 0$
13. None; none; $y = 0$, $y = 4$; $x = 0$  15. None; origin; $y = 0$; $x = 0$, $x = \pm 2$
17. $(-1, 0)$; None; $y = 0$; $x = 0$, $x = -2$
19. $(\pm 2, 0)$, (0, 4/9); None; $y = 1/2$; $x = -3/2$, $x = 3$
21. (0, 0), $(-1, 0)$; None; $x = 1$  23. (0, 0); None; $x = 2$
25. Given $\epsilon > 0$ there exists $\delta$ such that $|f(x) - A| < \epsilon$ whenever
   (a) $-\delta < x - a < 0$, (b) $0 < x - a < \delta$

**EXERCISE GROUP 6.5**

1. $x = 0$, endpt min; $x = 2$, endpt min  3. None  5. None  7. $x = \pm 1$, endpt min
9. $x = 1$, endpt max  11. $x = \pm 3$  13. $x = \pm 2\sqrt{5}/5$  15. $x = -1$, $x = 3$
17. $x = -2$, $x = 10/7$  19. $x = 0$  21. None

**EXERCISE GROUP 6.7**

1. $y = 2x$  3. $y = 2 - x$  5. $3x - 2y + 1 = 0$  7. $y = x + 3$  9. $y = x$
11. $y = x$  13. $y = x^2$  15. $y = x^2 + 1$  17. $y = x^2 - x + 1$

**EXERCISE GROUP 7.1**

1. $x^2 - 2y - 4 = 0$, excluding $(-2, 0)$ and (2, 0)
3. $x^2 + y^2 - x - y = 0$, excluding (1, 0) and (0, 1)  5. $y^2 - 4x + 4 = 0$
7. $3x^2 + 16y = 0$  9. $y = 0$, excluding $(-2, 0)$ and (2, 0); $x = -6$
11. $4x^2 = y$  13. $2x^2 + 4y^2 = 1$  15. $y = \dfrac{a^2 - ax}{\sqrt{a^2 - 2ax}}$, $0 \le x < a/2$

**EXERCISE GROUP 7.2**

1. Center $(5, -7)$, rad $= 8$  3. Point $(2, -3)$  5. Center $(3/4, -2/3)$, rad $= 5/4$
7. Center $(-3/4, 7/4)$, rad $= 2$  9. $x^2 + y^2 - 4y - 5 = 0$
11. $x^2 + y^2 + 6x - 14y + 42 = 0$  13. $x^2 + y^2 - 4x - 6y = 0$
15. $x^2 + y^2 - 2x - 3y - 1 = 0$  17. $x - 2y + 5 = 0$  19. $5x + 9y - 51 = 0$
21. $2x + 11y - 25 = 0$, $x - 2y - 5 = 0$
23. $4x + 5y - 28 = 0$, $32x - y + 104 = 0$  27. $\sqrt{2}$

**EXERCISE GROUP 7.3**

1. $x^2 + y^2 - x - 2y - 5 = 0$  3. $x^2 + y^2 - 4x - 6y + 8 = 0$
5. $4x^2 + 4y^2 - 9x + 3y - 5 = 0$  7. $x^2 + y^2 - 4x + 6y - 37 = 0$
9. $x^2 + y^2 - 2x + 4y - 21 = 0$
11. $x^2 + y^2 - 8x - y + 10 = 0$; $x^2 + y^2 - 4x - 3y = 0$
13. $x^2 + y^2 - 8x \pm 6y + 16 = 0$  15. $2x^2 + 2y^2 + 2x - 13y + 6 = 0$
17. $x^2 + y^2 - x - 8y + 15 = 0$

## EXERCISE GROUP 7.4

**1.** $(2, 2)$, $(-22/13, 58/13)$  **3.** $(3, -2)$  **5.** $(3, -1)$,  **7.** $(3, 1)$, $(1/6, -11/6)$
**9.** (a) $(0, 0)$, $(-1, 1)$, (b) $x + y = 0$  **11.** (a) $(-1, -1)$, $(59/13, -22/13)$,
   (b) $x + 8y + 9 = 0$  **13.** (a) None, (b) $2x + y - 10 = 0$
**15.** (a) None, (b) $x - 2y - 9 = 0$  **19.** Yes, at $(3, 3/2)$

## EXERCISE GROUP 7.5

**1.** $y^2 = 8x$  **3.** $x^2 = -16y$  **5.** $3x^2 = -16y$  **7.** $3y^2 = -8x$  **9.** $x^2 = -16y$
**11.** $F(3/2, 0)$; $D: x = -3/2$  **13.** $F(0, -1/6)$; $D: y = 1/6$  **15.** $F(0, -5/8)$; $D: y = 5/8$
**17.** $3x^2 = -16y$  **19.** $x^2 = 4y - 12$

## EXERCISE GROUP 7.6

**1.** (a) $(0, 0)$, (b) $(-4, -2)$, (c) $(-3, 7)$  **3.** $2x' - 3y' + 17 = 0$
**5.** $x'^2 + y'^2 + 4x' - 6y' + 11 = 0$  **7.** $y^2 + 16x - 6y + 25 = 0$
**9.** $y^2 - 8x + 4y + 4 = 0$  **11.** $V(-1/2, 0)$; $F(0, 0)$; $D: x = -1$
**13.** $V(-1, -2/3)$; $F(-1, -7/24)$; $D: y = -25/24$
**15.** $V(-1/3, -13/3)$; $F(-1/3, -17/4)$; $D: y = -53/12$
**17.** $V(-37/12, -3/4)$; $F(-139/48, -3/4)$; $D: x = -157/48$
**19.** $V(-1, -5)$; $F(-1, -39/8)$; $D: y = -41/8$  **23.** $y = ax^2 + x - a$
**25.** $x + 4y + 5 = 0$  **27.** $8x + 3y - 20 = 0$

## EXERCISE GROUP 7.7

**1.** $16x^2 + 25y^2 = 400$. Focal radii $19/5$, $31/5$  **3.** $9x^2 + 8y^2 = 144$
**5.** $4x^2 + y^2 = 12$  **7.** $16(x - 2)^2 + 25(y - 3)^2 = 400$
**9.** $21(x - 2)^2 + 25(y + 3)^2 = 525$  **11.** Foc $(0, \pm\sqrt{5})$, Vert $(0, \pm 3)$
**13.** Foc $(0, \pm 1)$, Vert $(0, \pm 2)$  **15.** Foc $(\pm 3, 0)$, Vert $(\pm 5, 0)$
**17.** Foc $(\pm\sqrt{210}/10, 0)$, Vert $(\pm\sqrt{14}/2)$  **19.** Foc $(-1, \pm 3/2)$, Vert $(-1, \pm 3\sqrt{2}/2)$
**21.** Foc $(-2, -2)$, $(4, -2)$, Vert $(-4, -2)$, $(6, -2)$
**23.** Foc $(4, \pm 4)$, Vert $(4, \pm 4\sqrt{2})$  **25.** Foc $(4 \pm \sqrt{3}, -2)$, Vert $(4 \pm \sqrt{5}, -2)$
**27.** $9x^2 + 5y^2 = 45$  **29.** $8(x + 2)^2 + 9(y - 4)^2 = 72$

## EXERCISE GROUP 7.8

**1.** $16x^2 - 9y^2 = 144$. Focal radii $7$, $13$  **3.** $y^2 - 7x^2 = 14$  **5.** $2y^2 - 27x^2 = 6$
**7.** $9x^2 - 4y^2 = 36$  **9.** $16(x - 1)^2 - 9(y - 3)^2 = 144$  **11.** $(x - 4)^2 - 4(y - 4)^2 = 36$
**13.** $(x - 3)^2 - 4(y - 1)^2 = 16$  **15.** Foc $(\pm\sqrt{17}, 0)$, Vert $(\pm 1, 0)$, Asymp $y = \pm 4x$
**17.** Foc $(0, \pm\sqrt{13})$, Vert $(0, \pm 2)$, Asymp $3y = \pm 2x$
**19.** Foc $(0, \pm 2\sqrt{2})$, Vert $(0, \pm 2)$, Asymp $y = \pm x$
**21.** Foc $(\pm\sqrt{5}, 0)$, Vert $(\pm\sqrt{3}, 0)$, Asymp $\sqrt{3}y = \pm\sqrt{2}x$
**23.** Foc $(3 \pm \sqrt{6}, -2)$, Vert $(1, -2)$, $(5, -2)$, Asymp $\sqrt{2}(y + 2) = \pm(x - 3)$
**25.** Foc $(-1 \pm \sqrt{5}, 0)$, Vert $(-2, 0)$, $(0, 0)$, Asymp $y = \pm 2(x + 1)$

**27.** Foc $(1/2 \pm \sqrt{17}, 1/2)$, Vert $(1/2 \pm \sqrt{34}/2, 1/2)$, Asymp $y - 1/2 = \pm(x - 1/2)$
**29.** Foc $(0, -2), (0, 4)$, Vert $(0, -1), (0, 3)$, Asymp $\sqrt{5}(y - 1) = \pm 2x$
**31.** $3y^2 - x^2 = 3$  **33.** $3(y + 1)^2 - (x - 4)^2 = 3$

## EXERCISE GROUP 7.9

**1.** $2x^2 - 3y^2 = 0$  **3.** $2(x + 1)^2 - (y - 3)^2 = 0$  **5.** $2(x + 1)^2 - 3y^2 = 0$
**7.** $3(x + 1)^2 - 4(y - 1)^2 = 0$  **9.** $4y^2 - x^2 = 15$  **11.** $25x^2 - 4y^2 = K^2, K \neq 0$
**13.** $(x - 1)^2 - 16(y + 3)^2 = K, K \neq 0$

## EXERCISE GROUP 8.1

**1.** $x = \frac{1}{2}(\sqrt{3}x' + y'), y = \frac{1}{2}(-x' + \sqrt{3}y')$  **3.** $(2, -2\sqrt{3})$  **5.** $(-\sqrt{3}, 1)$
**7.** $(-2\sqrt{2}, -4\sqrt{2})$  **9.** $(-3/2, -3\sqrt{3}/2)$  **11.** $[\frac{1}{2}(\sqrt{3} - 1), \frac{1}{2}(1 + \sqrt{3})]$
**13.** $(3\sqrt{2}/2, \sqrt{2}/2)$  **15.** $(3, 2)$  **17.** $5y' + 7 = 0$
**19.** $194x'^2 - 120x'y' + 313^2y' = 169$  **21.** $2x'y' = a^2$  **23.** $2x'^2 - 3y'^2 + 4 = 0$
**29.** $2abx'y' + (a^2 - b^2)y'^2 + a^2b^2 = 0$

## EXERCISE GROUP 8.2

**1.** $7x'^2 - 3y'^2 = 10$  **3.** $x'^2 + 14y'^2 = 84$  **5.** $26x'^2 - 15y'^2 = 9$
**7.** $100y'^2 + 9\sqrt{10}x' - 3\sqrt{10}y' = 30$  **9.** $30x'^2 - 95y'^2 + 6x' + 17y' = 60$

## EXERCISE GROUP 8.3

**1.** (a) Two straight lines: $x = y, x = -y$, (b) Hyperbola
**3.** (a) Two straight lines: $x = -3, x = 1$, (b) Parabola
**5.** (a) The point $(2, -1)$, (b) Ellipse  **9.** (b) Parabola
**11.** Hyperbola if $|A| < 1$, parabola if $|A| = 1$, ellipse if $|A| > 1$

## EXERCISE GROUP 8.4

**1.** $3x^2 + 4y^2 = 108$  **3.** $5(x - 4)^2 + 9(y + 1)^2 = 20$  **5.** $4x^2 + 3y^2 = 108$
**7.** $45(x - 4)^2 - 36(y + 1)^2 = 80$  **9.** $18x^2 - 2y^2 = 45$  **11.** $3(y - 4)^2 - x^2 = 48$
**17.** $e = \sqrt{2}$  **19.** $e = \sqrt{2}/2$  **23.** (a) $e = \sqrt{1 + m^2}$, (b) $e = \sqrt{1 + m^2}/m$

## EXERCISE GROUP 8.5

**1.** $4x + y - 7 = 0$  **3.** $\sqrt{3}x + 2y - 4\sqrt{3} = 0$  **5.** $7x - 6y + 5 = 0$
**7.** $x - 2y + 7 = 0$  **9.** $8x - y - 5 = 0$  **11.** $11x - 9y + 5 = 0$
**13.** $8x - 5y - 1 = 0$  **15.** $3x - 4y - 7 = 0$  **17.** $y = ax^2 + x$

## EXERCISE GROUP 9.1

**1.** 2  **3.** $\sqrt{21}/3$  **5.** $\sqrt[3]{30}$  **7.** $\sqrt{(-8 + \sqrt{85})/3} \approx 0.64$  **9.** $-\sqrt{2}/2$
**11.** $\frac{1}{2}(x_0 + x_1)$  **13.** Derivative does not exist at $x = 1$; 7/8  **15.** 5/3

## EXERCISE GROUP 9.2

**1.** $\frac{2}{3}x^3 - \frac{3}{2}x^2 + C$  **3.** $\frac{1}{2}x^2 + \frac{2}{3}x^{3/2} + \frac{3}{4}x^{4/3} + C$  **5.** $2\sqrt{2}t^{3/2}/3 + C$

**7.** $2\sqrt{a}u^{3/2}/3 + C$  **9.** $\frac{4}{5}x^5 - \frac{4}{3}x^3 + x + C$  **11.** $\frac{1}{2}x^2 + \frac{4}{3}x^{3/2} + x + C$
**13.** $\frac{3}{5}t^{3/2} - 2t + 3t^{1/3} + C$  **15.** $\frac{3}{7}u^{7/3} - 2u - 3u^{-1/3} + C$  **17.** $3x^{1/3} + \frac{3}{4}x^{4/3} + C$
**19.** $2x^{1/2} + \frac{4}{3}x^{3/2} + \frac{2}{5}x^{5/2} + C$  **21.** $\frac{3}{7}y^{7/3} + \frac{3}{2}y^{4/3} + 3y^{1/3} + C$
**23.** $x^2 + 1$  **25.** $x - \frac{2}{3}x^{3/2} + \frac{4}{3}\sqrt{2}$  **27.** $2x^{1/2} - x + 2$  **29.** $2t^2 - 3t + 9$
**31.** $3y = 4 - x^2$  **33.** $(2t^{3/2} + 5)/7$

### EXERCISE GROUP 9.3

**1.** $\frac{2}{3}(2+x)^{3/2} + C$  **3.** $-\frac{1}{3}(3-2x)^{3/2} + C$  **5.** $\frac{1}{3}(2x+3)^{3/2} + C$
**7.** $-\frac{1}{10}(3-y^2)^5 + C$  **9.** $\frac{1}{10}(u^2 - 2u - 3)^5 + C$  **11.** $-\frac{1}{6}(3-2x^2)^{3/2} + C$
**13.** $\frac{1}{2}(2+x^3)^{2/3} + C$  **15.** $-\frac{1}{2}(y^2 - 2y + 2)^{-1} + C$  **17.** $\pm(t^2 - 1)^{1/2} + C$
**19.** $\frac{1}{2}(1+x^{3/2})^{4/3} + C$  **21.** $\frac{1}{4}(2x^{4/3} - 3)^{3/2} + C$  **23.** $(x^2 + 4x)^{1/2} + C$
**25.** $\frac{4}{3}(x^{1/2} - a^{1/2})^{3/2} + C$  **27.** $\frac{1}{2}(x^{4/3} + 8)^{3/2} + C$  **29.** $-\frac{2}{3}(1+y^{-1})^{3/2} + C$
**31.** $2(x^{1/3} + 2x)^{3/2} + C$  **33.** Yes; 1  **35.** Yes; 2  **37.** Yes; 3  **39.** No  **41.** Yes; 3
**43.** No  **45.** No

### EXERCISE GROUP 9.4

**1.** $-\frac{2}{9}(4-3x)^{3/2} + C$  **3.** $-\frac{1}{40}(3-2y^2)^{10} + C$  **5.** $\frac{1}{3}(x^2 - 2x)^{3/2} + C$
**7.** $\frac{1}{2}(3+2t^2)^{1/2} + C$  **9.** $-\frac{4}{3}(4-t^{1/2})^{3/2} + C$
**11.** (a) $\frac{1}{4}\int(u+5)u^{1/2}\,du$, (b) $\frac{1}{10}(2x-3)^{5/2} + \frac{5}{6}(2x-3)^{3/2} + C$
**13.** (a) $\frac{1}{2}\int(u-2)u^{-1/2}\,du$, (b) $\frac{1}{3}(x^2+1)^{3/2} - 2(x^2+1)^{1/2} + C$
**15.** (a) $\frac{3}{4}\int v^{1/2}(v-1)\,dv$, (b) $\frac{3}{10}(1-2u^{1/3})^{5/2} - \frac{1}{2}(1-2u^{1/3})^{3/2} + C$
**17.** (a) $\frac{1}{2}\int(z^2+1)z^{-2}\,dz$, (b) $x(2x+1)^{-1/2} + C$
**19.** (a) $\frac{1}{9}\int(u-2)u^{-1/2}\,du$, (b) $\frac{2}{27}(3x+2)^{3/2} - \frac{4}{9}(3x+2)^{1/2} + C$

### EXERCISE GROUP 9.5

**1.** 285  **3.** $-150$  **5.** $-1/7$  **7.** $122/105$  **9.** 30  **11.** 0  **13.** 360  **15.** 69
**17.** $-1176$  **19.** 193  **21.** 1255  **23.** (a) $n(4n^2 + 12n + 11)/3$, (b) $n(2n+1)(2n-1)/3$
**25.** $r^{n+1} - r$  **29.** (b) $n/(n+1)$

### EXERCISE GROUP 9.6

**1.** 11/6  **3.** 31/8  **5.** (a) $(n-1)/(2n)$, (b) $(n+1)/(2n)$, (c) 1/2
**7.** (a) $(7n-3)/(2n)$, (b) $(7n+3)/(2n)$, (c) 7/2
**9.** (a) $(5n^2 - 6n + 1)/(6n^2)$, (b) $(5n^2 + 6n + 1)/(6n^2)$, (c) 5/6

### EXERCISE GROUP 9.7

**3.** $\pm\sqrt{3}/3$  **5.** $\sqrt{7}$  **7.** (a) 2/3, (b) $(3+\sqrt{3})/3$  **9.** (a) 0, (b) $\sqrt{2}$

### EXERCISE GROUP 9.8

**1.** 28/3  **3.** 64/3  **5.** 3  **7.** 119/6  **9.** 4  **11.** 3  **13.** $2(\sqrt{2}-1)/\sqrt{a}$
**15.** $(3 - 2\sqrt{2})a/3$  **17.** $\sqrt{21} - \sqrt{5}$  **19.** $\{[f(b)]^{n+1} - [f(a)]^{n+1}\}/(n+1)$  **21.** 3
**23.** $7\sqrt{21}/9$  **25.** $\frac{1}{3}(\sqrt{3} + \sqrt{35})$

**EXERCISE GROUP 9.9**

1. $2 - \sqrt{3}$   3. $a^6/6$   5. $2/15$   7. $175/64$   9. $9/100$   11. $2/5$   13. $106/15$
15. $(16 - 9\sqrt{3})/3$   17. $61/24$

**EXERCISE GROUP 9.10**

1. $36$   3. $2$   5. $27/4$   7. $1/4$   9. $16/3$   11. $a^2/6$   13. $8$   15. $1/10$   17. $125/24$
19. $9/2$   21. $27/8$   23. $8/15$   25. $4096/105$   27. (a) $2 - 2(t+1)^{-1/2}$, (b) $2$

**EXERCISE GROUP 10.1**

1. $1/2$   3. $1/2$   5. $1/2$   7. $0$   9. $0$   11. $-1/\pi$   13. $0$   15. $1/2$   17. $3/2$

**EXERCISE GROUP 10.2**

1. $2 \cos 2x$   3. $1 + 3 \cos 3x$   5. $\cos 2x / \sqrt{\sin 2x}$   7. $\sec^2 x / (2\sqrt{\tan x})$
9. $-6 \csc^2 3x \cot 3x$   11. $3 \sec 3x \tan 3x$   13. $2x \cos x - (x^2 - 2) \sin x$
15. $\sec^2 x - \sec x \tan x$   17. $\sin(x/2) \cos(x/2)$   19. $2 \sin x / (1 + \cos x)^2$
21. $\sec x / (1 + \sec x)$   23. $\sec^2 x / (1 + \tan x)^2$   25. $(x^2 + 3) \cos x$
33. Rel max $\sqrt{2}$ at $x = \pi/4$, rel min $-\sqrt{2}$ at $x = 5\pi/4$
35. Rel max $\sqrt{13}$ where $\tan x = -2/3$ in 2nd quad, rel min $-\sqrt{13}$ where $\tan x = -2/3$ in 4th quad
37. Rel max $3\sqrt{3}/2$ at $x = \pi/3$, rel min $-3\sqrt{3}/2$ at $x = 5\pi/3$

**EXERCISE GROUP 10.3**

1. $-\frac{1}{3} \cos 3x + C$   3. $-\frac{1}{2} \sin(1 - 2x) + C$   5. $\frac{1}{3} \tan(3x - 1) + C$
7. $\frac{1}{6} \sin^2 3x + C$   9. $-\frac{1}{3} \sin^{-3} x + C$   11. $-2\sqrt{\cos x} + C$   13. $-2 \cos \sqrt{x} + C$
15. $\frac{2}{3} \sec^{3/2} x + C$   17. $\frac{1}{3} \tan^{3/2} 2x + C$   19. $-\frac{1}{9} \cot^3 3x + C$   21. $\frac{1}{12}(4 - \sqrt{2})$
23. $15/4$   25. $\frac{1}{3} \tan^3 x + C$   27. $\frac{1}{2} \sin^4 x/2 + C$   29. $-2\sqrt{1 + \cot x} + C$
31. $\sqrt{1 + \sin^2 x} + C$   33. $\frac{1}{2}(2 - \sin^2 x)^{-2} + C$   35. $-2\sqrt{2 - \tan x} + C$   37. $2$
39. $2\sqrt{5}$

**EXERCISE GROUP 10.4**

1. (a) $-\pi/6$, (b) $2\pi/3$, (c) $\pi/6$, (d) $2\pi/3$, (e) $\pi/2$, (f) $\pi/2$
17. $\sqrt{1 - x^2}$   19. $2x\sqrt{1 - x^2}/(1 - 2x^2)$   21. $2x/(1 + x^2)$   23. $(1 - x^2)/(1 + x^2)$

**EXERCISE GROUP 10.5**

1. $-2/\sqrt{1 - 4x^2}$   3. $2x/\sqrt{1 - x^4}$   5. $\sin^{-1} x + x/\sqrt{1 - x^2}$   7. $1/(1 + x^2)$
9. $-(x + \sqrt{1 - x^2} \cos^{-1} x)/(x^2 \sqrt{1 - x^2})$   11. $1/(2x\sqrt{x - 1})$   13. $1/(2\sqrt{x - x^2})$
15. $\pm 2$   17. $1/(2\sqrt{2ay - y^2})$   19. $1/(1 + x^2)$   21. $\pm 1/\sqrt{1 - x^2}$   23. $\sin^{-1} x$
25. $2\sqrt{2x - x^2}$   27. $2x^2/\sqrt{4 - x^2}$   29. $\sqrt{(1 + x)/(1 - x)}$   31. $-2x^2/\sqrt{a^2 - x^2}$

**EXERCISE GROUP 10.6**

1. $\sin^{-1} x/2 + C$   3. $\frac{1}{6} \tan^{-1} 3u/2 + C$   5. $\frac{1}{2} \sin^{-1} 2x + C$   7. $\frac{1}{2}(\tan^{-1} x)^2 + C$

**9.** $-\frac{1}{2}(\cos^{-1} x)^2 + C$  **11.** $\frac{1}{2} \sin^{-1} 2u/3 + C$  **13.** $\frac{1}{2} \tan^{-1} x^2 + C$
**15.** $\tan^{-1}(\sin x) + C$  **17.** $\sin^{-1}[(\sin x)/\sqrt{2}] + C$  **19.** $\sin^{-1}(\sqrt{3}/3)$
**21.** $2 \sin^{-1} \sqrt{x} + C$  **23.** $a \sin^{-1} x/a - \sqrt{a^2 - x^2} + C$  **25.** $3 \tan^{-1} x^{1/3} + C$

## EXERCISE GROUP 11.1

**1.** 2.370, 2.488, 2.546  **3.** (a) $x > 3/2$, (b) $x > 2$, (c) $-3 < x < 2$, (d) $x > 1$
**13.** $2x f(x^2)$

## EXERCISE GROUP 11.3

**1.** $(1 + 2x)e^{2x}$  **3.** $(1 + 2x^2)e^{x^2}$  **5.** $(2 \ln x)/x$  **7.** $\cot x$  **9.** $(\sin x + \cos x)e^x$
**11.** $2 \cos x \, e^{2 \sin x}$  **13.** $13 e^{2x} \sin 3x$  **15.** $e^x[1/(x-1) + \ln(x-1)]$  **17.** $1/(1 + e^x)$
**19.** $2e^{2x}/\sqrt{1 - e^{4x}}$  **21.** $-a/(x\sqrt{x^2 + a^2})$  **23.** $2x/(1 - x^4)$  **25.** $-x \csc^2 x$
**27.** $\tan^{-1} x$  **29.** $\sec x$  **41.** $\frac{4}{3} x(1 + x^2)^{-2/3}(1 - x^2)^{-4/3}$
**43.** $(1 + 1/x)^x[\ln(1 + 1/x) - (x + 1)^{-1}]$  **45.** $(\ln x)^x[\ln(\ln x) + (\ln x)^{-1}]$
**47.** $(\sin x)^x[x \cot x + \ln \sin x]$  **49.** $e^x x^{e^x}(1/x + \ln x)$

## EXERCISE GROUP 11.4

**1.** $\frac{1}{2}\ln |2u - 3| + C$  **3.** $\frac{1}{2}\ln |x^2 - 4x - 5| + C$  **5.** $\frac{1}{3}\ln |x^3 - 2| + C$
**7.** $1/\ln 3$  **9.** $2e^{\sqrt{x}} + C$  **11.** $\ln|1 + \tan x| + C$  **13.** $-\ln|\cos x| + C$
**15.** $\frac{1}{2}(\ln \sin x)^2 + C$  **17.** $\frac{1}{2}(1 + \ln x)^2 + C$
**19.** $(\ln x)^{n+1}/(n + 1) + C$ if $n + 1 \neq 0$, $\ln|\ln x| + C$ if $n + 1 = 0$
**21.** $2\sqrt{e^t - 1} + C$  **23.** $-\frac{2}{3}\ln|a^{3/2} - x^{3/2}| + C$  **25.** $5\pi^2/288$  **27.** $2\sqrt{\tan^{-1} x} + C$
**29.** $-\sin^{-1}[(\cos x)/\sqrt{3}] + C$  **31.** $\ln|1 + \ln x| + C$  **33.** $(2^x - 2)/\ln 2$  **35.** $1/2$

## EXERCISE GROUP 11.5

**17.** $3 \cosh(3x - 2)$  **19.** $(\sinh \sqrt{x})/(2\sqrt{x})$  **21.** $x \cosh x$  **23.** $e^x \cosh e^x$
**25.** $-3a \, \text{sech}^3 ax \tanh ax$  **27.** $(\sinh x)^x(x \coth x + \ln \sinh x)$
**29.** $\sinh x/\sqrt{1 - \cosh^2 x}$  **31.** $a \coth ax$  **33.** $\text{sech } x$

## EXERCISE GROUP 11.6

**1.** $\frac{2}{3} \cosh 3x/2 + C$  **3.** $2 \sinh \sqrt{x} + C$  **5.** $x - \tanh x + C$
**7.** $\frac{1}{3} \sinh^3 x + \frac{1}{5} \sinh^5 x + C$  **9.** $\ln \cosh x + C$  **11.** $\cosh 1 - 1$  **13.** $\ln \cosh 2$

## EXERCISE GROUP 11.8

**1.** $2/\sqrt{4x^2 + 1}$  **3.** $1/[2\sqrt{x}(1 - x)]$  **5.** $-1/[(2x - 1)\sqrt{x - x^2}]$
**7.** $e^{\sinh^{-1} x}/\sqrt{x^2 + 1}$  **9.** $x/\sqrt{x^2 + 1} + \sinh^{-1} x$  **11.** $-\text{csch } x$

## EXERCISE GROUP 12.1

**1.** $40/3$  **3.** $2\pi$  **5.** $16/3$  **7.** $\pi/2$  **9.** $\pi/2$  **11.** $16a^3/3$

## EXERCISE GROUP 12.2

**1.** $26\pi/3$  **3.** $64\sqrt{2}\pi/15$  **5.** $\pi$  **7.** $2\pi/3$  **9.** $2\pi$  **11.** $128\pi/15$  **13.** $\pi(\sinh 2 - 2)/4$
**15.** $2(2 - \ln 3)\pi$  **17.** $3\pi/10$  **19.** $32\pi/3$

## EXERCISE GROUP 12.3

**1.** $16\pi/3$  **3.** $512(\sqrt{2}-1)\pi/5$  **5.** $4a^3\pi/3$  **7.** $11\pi/30$  **9.** $2\pi^2 a^2 b$

## EXERCISE GROUP 12.4

**1.** $8(10\sqrt{10}-1)/27$  **3.** $(80\sqrt{10}-13\sqrt{13})/27$  **5.** $\sinh 1$  **7.** $123/32$  **9.** $59/24$
**11.** $6a$  **13.** $\ln(e^6-1) - \ln(e^2-1) - 2$  **17.** $-15/(64\pi)$ in./sec

## EXERCISE GROUP 12.5

**1.** $13\pi/3$  **3.** $8(2\sqrt{2}-1)\pi a^2/3$  **5.** $47\pi/16$  **7.** $(16e^2 - e^{-2} + 1)\pi/16$

## EXERCISE GROUP 13.1

**1.** 51,000 in. lb  **3.** 88 in. lb  **5.** $200w$  **7.** $15\pi w/4$  **9.** $280w/3$

## EXERCISE GROUP 13.2

**1.** $15w$  **3.** $32w/3$  **5.** $125w$  **7.** $b/2$  **9.** $45w/4$

## EXERCISE GROUP 13.3

**1.** $(15/4, -17/8)$  **3.** $(-1/7, 3/7)$  **5.** (a) $(10/3, 7/3)$, $(9/11, 14/11)$,
(b) $(19/14, 3/2)$  **7.** (a) $(29/9, 4/9)$, $(-4/7, 30/7)$, (b) $(25/16, 17/8)$
**9.** (a) $(-7/5, 38/15)$, $(21/4, -1/4)$, (b) $(0, 37/19)$
**11.** (a) $(1, 11/3)$, $(10/3, 4/3)$, $(2, 2/3)$, (b) $(74/33, 68/33)$

## EXERCISE GROUP 13.4

**1.** $(5/9, 7/9)$  **3.** $(3/8, 2/5)$  **5.** $(5/7, 64/35)$  **7.** $(0, 7/5)$

## EXERCISE GROUP 13.5

**1.** On the center line, 31/11 units above base of square
**3.** On center line, $112/(9\pi)$ units above common diameter
**5.** $(-12/5, 0)$  **7.** $(-1/2, 2/\pi)$  **9.** (a) $(0, 23/21)$, (b) $(23/21, 23/21)$

## EXERCISE GROUP 13.6

**1.** $(4/3, 0)$  **3.** $(0, 5/16)$  **5.** $(2a/5, 2a/5)$  **7.** $2L/3$ from end mentioned
**9.** $(0, 2a/(\pi+2))$  **13.** On center line, $2a/\pi$ units from diameter  **15.** $4\pi^2 ab$

## EXERCISE GROUP 13.7

**1.** (a) $bh^3/12$, (b) $\sqrt{6}h/6$  **3.** $I_x = 128/21$, $I_y = 32/5$
**5.** (a) $\pi(a^4 - b^4)/4$, (b) $\pi(a^4 + b^4)/4$  **7.** (a) $L^3/3$, (b) $\sqrt{3}L/3$  **11.** $4\pi/3$

## EXERCISE GROUP 14.1

**1.** $\frac{1}{8}(2x-1)^4 + C$  **3.** $\frac{4}{5}x^{5/2} - \frac{2}{3}x^{3/2} + 6x^{1/2} + C$  **5.** $\frac{1}{2}\sqrt{1+4x} + C$
**7.** $\ln(x^2 - x + 1) + C$  **9.** $-\frac{1}{2}\cos x^2 + C$  **11.** $\sin^{-1} x/2 + C$
**13.** $\frac{1}{2}(\sin^{-1} x)^2 + C$  **15.** $\sec x + C$  **17.** $5^x/\ln 5 + C$  **19.** $\frac{1}{2}(\ln x)^2 + C$
**21.** $\frac{1}{2}(1 + \ln x)^2 + C$  **23.** $x - \ln|1+x| + C$  **25.** $\ln(2 - \cos x) + C$
**27.** $-(\ln|3 - 2^x|)/\ln 2 + C$  **29.** $2 - \sqrt{2}$  **31.** $(3\sqrt{3} - 2\sqrt{2})/72$  **35.** $\ln(2 + \sqrt{3})$

## EXERCISE GROUP 14.2

1. $-\cos x - \frac{2}{3}\cos^{3/2}x + C$  3. $\frac{1}{3}\sin^3 x - \frac{2}{5}\sin^5 x + \frac{1}{7}\sin^7 x + C$
5. $\frac{1}{8}(4x + \sin 4x) + C$  7. $-\cos x + \cos^3 x - \frac{3}{5}\cos^5 x + \frac{1}{7}\cos^7 x + C$
9. $-\cos x + \frac{1}{2}\cos^2 x + C$  11. $(4x - \sin 4x)/32 + C$
13. $(180x + 48 \sin 6x - 4 \sin^3 6x + 9 \sin 12x)/576 + C$
15. $-\csc x - \sin x + C$  17. $\sec x + \cos x + C$  19. $\frac{1}{6}(3 \sin x + \sin 3x) + C$
21. $\frac{1}{9}[9 \cos x/2 - \cos 9x/2] + C$  23. 0
25. 0 if $m \neq n$ or $m = n = 0$; $\pi$ if $m = n \neq 0$

## EXERCISE GROUP 14.3

1. $\frac{1}{2}\sec(2x + 4) + C$  3. $\frac{1}{5}\ln|\sin(5x - 8)| + C$  5. $\tan x - \cot x + C$
7. $-\frac{1}{3}\csc^3 x + C$  9. $-\frac{1}{5}\csc^5 x + \frac{1}{3}\csc^3 x + C$  11. $\frac{1}{7}\tan^7 x + \frac{1}{5}\tan^5 x + C$
13. $\frac{2}{5}\sec^{5/2} x - 2 \sec^{1/2} x + C$  15. $-\frac{2}{9}\cot^{9/2} x - \frac{2}{5}\cot^{5/2} x + C$
17. $\frac{1}{5}\sec^5 x - \frac{1}{3}\sec^3 x + C$  19. $x + \cot x - \frac{1}{3}\cot^3 x + C$  21. $\frac{1}{6}\sec^6 x - \frac{1}{4}\sec^4 x + C$

## EXERCISE GROUP 14.4

1. $-x/\sqrt{4x^2 - 1} + C$  3. $\frac{1}{2}[a^2 \sin^{-1} x/a + x\sqrt{a^2 - x^2}] + C$
5. $\sqrt{x^2 - 9} - 3 \sec^{-1} x/3 + C$
7. $\frac{1}{8}[a^4 \sin^{-1} x/a - x(a^2 - 2x^2)\sqrt{a^2 - x^2}] + C$  9. $\pi$  11. $\frac{1}{2}[\tan^{-1} x - x/(1 + x^2)] + C$
13. 162/5  15. $\sqrt{16 - x^2} + 4 \ln|(4 - \sqrt{16 - x^2})/x| + C$
17. $(2a^2 + u^2)/\sqrt{a^2 + u^2} + C$  19. $\frac{1}{32}[4x/(4 - x^2) + \ln|(2 + x)/(2 - x)|] + C$
21. $\frac{1}{5}(1 + x^2)^{5/2} - \frac{1}{3}(1 + x^2)^{3/2} + C$  23. $\frac{1}{2}\ln(e^{2x} + \sqrt{e^{4x} - 1}) + C$
25. $\sin^{-1}\sqrt{x} - \sqrt{x - x^2} + C$  27. $\pi\{4\sqrt{17} - \sqrt{2} + \ln[(4 + \sqrt{17})/(1 + \sqrt{2})]\}$
31. $\pi a^3$  33. $\pi a b^3/4$

## EXERCISE GROUP 14.5

1. $\sin^{-1}(x + 2)/3 + C$  3. $\frac{1}{2}[\ln(x^2 - 6x + 13) + 3 \tan^{-1}(x - 3)/2] + C$
5. $\frac{1}{2}\sin^{-1}(2x - 5) - \sqrt{5x - 6 - x^2} + C$
7. $-\frac{1}{2}\sqrt{7 - 12x - 2x^2} - 3\sqrt{2} \sin^{-1}[\sqrt{2}(x + 3)/5] + C$
9. $\sqrt{x^2 - 2x} + \ln|x - 1 + \sqrt{x^2 - 2x}| + C$
11. $\frac{3}{4}\sqrt{4x^2 - 4x - 3} + \frac{7}{4}\ln|2x - 1 + \sqrt{4x^2 - 4x - 3}| + C$  13. $\ln(1 + \sqrt{2})$

## EXERCISE GROUP 14.6

1. $\ln|(x - 2)^4(x + 1)^3| + C$  3. $\ln|(x^2 - 1)/x| + C$
5. $\frac{1}{6}\ln|(x + 2)(x - 1)^2/x^3| + C$  7. $-\ln 18$  9. $1/x + \ln|(x - 1)/x| + C$
11. $-\frac{1}{8}x/(x^2 - 4) + \frac{1}{32}\ln|(x + 2)/(x - 2)| + C$  13. $-1/x^2 + \ln|x^3/(x + 1)| + C$
15. $x^2/2 - 1/(x - 2) + \ln|x/(x - 2)^2| + C$  17. $6 \ln 3 - \frac{26}{3}\ln 2$
19. $\sqrt{x}/(1 - x) + \frac{1}{2}\ln|(1 - \sqrt{x})/(1 + \sqrt{x})| + C$

## EXERCISE GROUP 14.7

1. $\ln|x-1| - \tfrac{1}{2}\ln(x^2+x+1) - \sqrt{3}\tan^{-1}[(2x+1)/\sqrt{3}] + C$
3. $-1/(x-1) - (2/\sqrt{3})\tan^{-1}[(2x+1)/\sqrt{3}] + C$
5. $\ln|x+1| - \tfrac{1}{2}\ln(x^2+1) + C$  7. $\ln|x| - (2/\sqrt{3})\tan^{-1}[(2x+1)/\sqrt{3}] + C$
9. $2/(x+1) + \ln|x+1| + (\sqrt{2}/4)\ln|(x-\sqrt{2})/(x+\sqrt{2})| + C$
11. $\tfrac{1}{4}\tan^{-1}x/2 - \tfrac{1}{2}x/(x^2+4) + C$  13. $2/(x^2+4) + \tfrac{1}{2}\ln(x^2+4) + C$
15. $-1/(x^2-x+1) + (4/\sqrt{3})\tan^{-1}[(2x-1)/\sqrt{3}] + C$
17. $(\sqrt{2}/32)[\tan^{-1}x/\sqrt{2} + \sqrt{2}x(x^2-2)/(x^2+2)^2] + C$

## EXERCISE GROUP 14.8

1. $\tfrac{1}{9}(\cos 3x + 3x\sin 3x) + C$  3. $\tfrac{1}{4}(2x-1)e^{2x} + C$  5. $(x^3 - 3x^2 + 6x - 6)e^x + C$
7. $\tfrac{1}{2}[(2v-1)\sin^{-1}\sqrt{v} + \sqrt{v-v^2}] + C$  9. $2e - 4$
11. $x[2 - 2\ln x + (\ln x)^2] + C$  13. $x\ln(4+x^2) - 2x + 4\tan^{-1}x/2 + C$
15. $e^{-x}(3\sin 3x - \cos 3x)/10 + C$  17. $x\cot^{-1}x + \tfrac{1}{2}\ln(1+x^2) + C$
19. $x\sec^{-1}x - \ln|x + \sqrt{x^2-1}| + C$  21. $\tfrac{1}{2}[(x^2+1)\tan^{-1}x - x] + C$
23. $\tfrac{1}{3}(2\cos x \sin 2x - \sin x \cos 2x) + C$  25. $\tfrac{1}{8}[3\sin x \sin 3x + \cos x \cos 3x] + C$
27. $e^x(x+2)/x^3 + C$
29. $\tfrac{1}{8}[2\sec^3 x \tan x + 3\sec x \tan x + 3\ln|\sec x + \tan x|] + C$
31. $\tfrac{1}{8}[2\sec^3 x \tan x - 5\sec x \tan x + 3\ln|\sec x + \tan x|] + C$
37. $1$   39. $\pi$   41. $\pi$   43. $\tfrac{1}{16}[16\sqrt{257} + \ln(16+\sqrt{257})]$

## EXERCISE GROUP 14.10

1. $x - \tan x/2 + C$  3. $(\sqrt{2}/2)\ln|(1+\sqrt{2}+\tan x/2)/(1-\sqrt{2}+\tan x/2)| + C$
5. $(2/\sqrt{3})\tan^{-1}[(2\tan x/2 - 1)/\sqrt{3}] + C$  7. $\ln|\tan x/2/(1+\tan x/2)| + C$
9. $\tfrac{1}{3}\ln|(3+\tan x/2)/(3-\tan x/2)| + C$
13. (a) $\sin x = 2w/(w^2+1)$, $\cos x = (w^2-1)/(w^2+1)$, $dx = -2\,dw/(w^2+1)$

## EXERCISE GROUP 14.11

1. $-2\sqrt{x} - 2\ln|1 - \sqrt{x}| + C$  3. $x/2 + 2/x + \tfrac{1}{4}\ln|2x-1| + C$
5. $\tfrac{1}{2}\sin^{-1}x^2/2 + C$  7. $a\sin^{-1}[(y-a)/a] - \sqrt{2ay - y^2} + C$
9. $\sqrt{u^2+4} - 2\ln|(2+\sqrt{u^2+4})/u| + C$  11. $\ln(x + \sqrt{x^2+1}) - x/\sqrt{x^2+1} + C$
13. $\tfrac{1}{16}[\ln|(2+\sqrt{t^2+4})/t| - 2\sqrt{t^2+4}/t^2] + C$
15. $\tfrac{1}{2}[\tan^{-1}(x-1) + (x-1)/(x^2 - 2x + 2)] + C$
17. $\sec\theta + \tfrac{1}{2}\ln|(1-\cos\theta)/(1+\cos\theta)| + C$
19. $\tfrac{1}{2}[(\cot x - \csc x)(\cot x - 2) - \ln|\csc x + \cot x|] + C$
21. $-\sec^{-1}x - 1/\sqrt{x^2-1} + C$  23. $\tfrac{1}{9}[3x^3 \sin^{-1}x - (1-x^2)^{3/2} + 3\sqrt{1-x^2}] + C$
25. $\tfrac{1}{2}(x + \sqrt{1-x^2})e^{\sin^{-1}x} + C$  27. $3\ln|x^{1/3}/(1-x^{1/3})| + C$
29. $\tfrac{3}{8}[\sin^{-1}x^{1/3} + (2x - x^{1/3})\sqrt{1-x^{2/3}}] + C$  31. $\tfrac{1}{2}\ln|(1+e^x)/(1-e^x)| + C$

**33.** $\frac{1}{2}\ln|(1-e^x)/(1+e^x)| + C$   **35.** $(e^x - 1)e^{e^x} + C$   **39.** $2\ln 3 - 1$
**41.** $\frac{1}{2}(4 + \sinh 4)\pi$   **43.** $\frac{1}{512}[1026\sqrt{257} - \frac{1}{8}\ln(16 + \sqrt{257})]$

## EXERCISE GROUP 15.1

**1.** 1/4   **3.** 1/4   **5.** 2/3   **7.** $-\infty$   **9.** 0   **11.** 0   **13.** $-\infty$   **15.** $\infty$   **17.** 2/3
**19.** 3/2   **21.** 0   **23.** $e^{-3}$

## EXERCISE GROUP 15.2

**1.** 2/5   **3.** $\sqrt{2}$   **5.** 2   **7.** $-\infty$   **9.** 0   **11.** 1/3   **13.** 0   **15.** $-\infty$   **17.** $-2$   **19.** $\infty$
**21.** 0   **23.** $-2$   **25.** 1/2   **27.** 1/3   **29.** 1/2   **31.** 0   **33.** 2   **35.** 1   **37.** 1/6   **39.** 0
**41.** $f''(x)$

## EXERCISE GROUP 15.3
**1.** $e^2$   **3.** $e^{-1}$   **5.** 1   **7.** 1   **9.** 1   **11.** $e^2$   **13.** 1

## EXERCISE GROUP 15.4

**1.** 2   **3.** Div   **5.** $\pi/32$   **7.** $\pi/2$   **9.** Div   **11.** $1/\ln 2$   **13.** 6   **15.** $\pi/(2a)$   **17.** Div
**19.** Div   **21.** $\pi$   **23.** $\frac{1}{4}(4-\pi)$   **25.** Div   **27.** $\frac{1}{2}\ln 2$   **29.** 1   **31.** 4   **33.** 6
**35.** (a) Div, (b) Div, (c) $(b-a)^{1-k}/(1-k)$   **37.** $-k/a$

## EXERCISE GROUP 16.1

**1.** $x + 2y - 3 = 0$   **3.** $x^2 - 24x + 16y = 0$   **5.** $x^2 + xy - 2x - y + 2 = 0$
**7.** $x^2 - 2xy + y^2 - y = 0$   **9.** $x(y-3)^2 = 4$   **11.** $x^3 - y^2 + 3xy + y = 0$
**13.** $x^2 + y^2 = 1$ (circle)   **15.** $n^2(x-a)^2 - m^2(y-b)^2 + m^2n^2 = 0$ (hyperbola)
**17.** $x^{2/3} + y^{2/3} = a^{2/3}$   **19.** $x = a\cos^{-1}\dfrac{a-y}{a} \pm \sqrt{2ay - y^2}$   **21.** $(x^2 + y^2)^3 = 4x^2y^2$
**23.** $y = 2$   **25.** $y = 1$   **27.** $6[(2+\sqrt{3})x - y] = a[(2+\sqrt{3})\pi - 12]$
**29.** (a) $y = 9$, (b) None   **31.** (a) $y = -1/4$, (b) $x = -1$
**33.** (a) $y = 3x$, (b) $x = -1/4$, (c) $t = -1, 2$, (d) $y = 2$, $3x - y - 4 = 0$

## EXERCISE GROUP 16.2

**1.** $(26, -15)$   **3.** $(49/4, 27/2)$   **5.** $x = -\dfrac{2t}{1+t^2}, y = -\dfrac{2t^2}{1+t^2}$
**7.** $x = \dfrac{3}{t-1}, y = \dfrac{3t}{t-1}$   **9.** $x = \dfrac{3at}{1+t^3}, y = \dfrac{3at^2}{1+t^3}$
**11.** $x = 3\sin\theta$, $y = 3\cos\theta$ or $y = -3\cos\theta$
**13.** $x = 2\sin\theta$ or $x = -2\sin\theta$, $y = 3\cos\theta$
**15.** $y^2 = ax$   **17.** $x = a\cot\theta \pm b\cos\theta$, $y = a \pm b\sin\theta$
**23.** (a) $y = 10\sin\theta - 8\tan\theta$, $y = 10\cos\theta$, $0 \leq \theta \leq \cos^{-1}4/5$
   (b) $x^2y^2 = (100 - y^2)(y-8)^2$
**25.** Part of the line $b^2x + a^2my = 0$   **27.** $3\pi a^2$   **29.** $2\sqrt{2}/3$

## EXERCISE GROUP 16.3

**1.** $2\pi a$   **3.** $\frac{1}{4}[2\sqrt{5} + \ln(2 + \sqrt{5})]$   **5.** (a) $8a$, (b) $64\pi a^2/3$, (c) $16\pi^2 a^2$

7. (a) $6a$, (b) $6\pi a^2/5$  9. $2\pi a\{a + (b^2/\sqrt{a^2-b^2})\ln[(a+\sqrt{a^2-b^2})/b]\}$
11. (a) $\sqrt{2}(e^{\pi/2}-1)$, (b) $2\sqrt{2}\pi(e^\pi-2)/5$, (c) $2\sqrt{2}\pi(2e^\pi+1)/5$,
    (d) $x^2+y^2 = e^{2\tan^{-1}x/y}$

**EXERCISE GROUP 16.4**

1. $4\mathbf{j}$  3. $\mathbf{i}/2$  5. (a) $2t\mathbf{i}+2t\mathbf{j}$, (b) $2\mathbf{i}+2\mathbf{j}$, (c) $8t^3$
7. (a) $-2\sin t\,\mathbf{i}+3\cos t\,\mathbf{j}$, (b) $-2\cos t\,\mathbf{i}-3\sin t\,\mathbf{j}$, (c) $10\sin t\cos t$

**EXERCISE GROUP 16.5**

1. (a) $\mathbf{v}=12\mathbf{i}-4\mathbf{j}$, $v=4\sqrt{10}$, $\mathbf{a}=12\mathbf{i}-2\mathbf{j}$, $a_T=19\sqrt{10}/5$, $a_N=3\sqrt{10}/5$,
   (b) $3\sqrt{10}/800$
3. (a) $\mathbf{v}=\frac{1}{4}(4\mathbf{i}+3\mathbf{j})$, $v=5/4$, $\mathbf{a}=-\frac{1}{16}(2\mathbf{i}-\mathbf{j})$, $a_T=-1/16$, $a_N=1/8$, (b) $2/25$
5. (a) $\mathbf{v}=2(-\sin 2t\,\mathbf{i}+\cos 2t\,\mathbf{j})$, $v=2$, $\mathbf{a}=-4(\cos 2t\,\mathbf{i}+\sin 2t\,\mathbf{j})$, $a_T=0$, $a_N=4$,
   (b) $k=1$
7. (a) $\mathbf{v}=e^{\pi/2}(\mathbf{i}-\mathbf{j})$, $v=\sqrt{2}e^{\pi/2}$, $\mathbf{a}=-2e^{\pi/2}\mathbf{j}$, $a_T=\sqrt{2}e^{\pi/2}$, $a_N=\sqrt{2}e^{\pi/2}$,
   (b) $k=1/(\sqrt{2}e^{\pi/2})$
9. (a) $\mathbf{v}=2\mathbf{i}+8\mathbf{j}$, $v=2\sqrt{17}$, $\mathbf{a}=8\mathbf{j}$. $a_T=32/\sqrt{17}$, $a_N=8/\sqrt{17}$, (b) $k=2/17^{3/2}$
11. (b) $(\cos t/2)\mathbf{i}-(\sin t/2)\mathbf{j}$

**EXERCISE GROUP 16.6**

1. $6/13^{3/2}$  3. $b/a^2$  5. $1/[a(\sec^2 t+\tan^2 t)^{3/2}]$  7. $16/17^{3/2}$  9. $\sqrt{10}/400$
11. $2\sqrt{5}/25$  13. $2/(3a)$  15. $\cos x$  19. 2

**EXERCISE GROUP 16.7**  1. $x=a\cos(s/a)$, $y=a\sin(s/a)$

**EXERCISE GROUP 17.1**

1. $(2\sqrt{3}, 2)$  3. $(-1, -\sqrt{3})$  5. $(6.44, 2.74)$  7. $(2.30, -1.93)$
9. $(0, 8)$  11. $(7\sqrt{3}/2, 7/2)$  13. Circle of radius 2, center at origin
15. Line through the origin, making an angle of 30° with positive $x$ axis
17. Same as Ex. 15, excluding part in third quadrant
19. Quarter of circle in first quadrant, radius 3, center at origin, including points $(0, 3)$ and $(3, 0)$
21. Points interior to and on circle with radius 2, center at origin
23. Points in first and third quadrants between and on lines $y=0$, $y=x$

**EXERCISE GROUP 17.2**  11. $\sqrt{3}x-y=0$  13. $3x-2y=1$
15. $(x^2+y^2-x)^2 = x^2+y^2$  17. $y^2+16x-64=0$  19. $x^2+y^2=2y+3x$
21. $3x^2-y^2-16x+16=0$  23. $r(2\cos\theta-5\sin\theta)+4=0$
25. $r=2\cos\theta-4\sin\theta$  27. $r=\pm(r\cos\theta+4)$  29. $r=\cos\theta/2$

**EXERCISE GROUP 17.4**

1. Parabola, vertex $(-1, 0)$  3. Hyperbola, center $(32/7, 0)$, slopes $\pm\sqrt{7}/3$

**5.** Hyperbola, center (0, 8/3), slopes $\pm\sqrt{3}/3$   **7.** Ellipse, center (4/3, 0)
**9.** Parabola, vertex (3/4, $3\sqrt{3}/4$)   **15.** $r = 2\sin\theta$
**17.** $r(4 - \cos^2\theta) = 12\cos\theta$   **19.** $r = 10 - 8\csc\theta$, $\sin^{-1}4/5 \leq \theta \leq \pi/2$

### EXERCISE GROUP 17.5

The following coordinates do not necessarily satisfy both equations.

**1.** Pole, $(1, \pi/2)$, $(1, -\pi/2)$   **3.** Pole, $(\sqrt{3}/2, \pi/3)$, $(-\sqrt{3}/2, -\pi/3)$
**5.** (1, 0 rad)   **7.** Pole, $(3/2, \pi/3)$, $(3/2, -\pi/3)$   **9.** $(2, \pi/3)$
**11.** $(2/5, 7\pi/6)$, $(2/5, 11\pi/6)$   **13.** Pole, $(a, 0 \text{ rad})$, $(a/5, \cos^{-1}(-3/5))$

### EXERCISE GROUP 17.6

**1.** $\pi/2, \tan^{-1}2$   **3.** $3\pi/4, 2\pi/3$   **5.** $\pi - \tan^{-1}(\sqrt{3}/6), 0$   **7.** $\pi/2, \pi - \tan^{-1}\frac{1}{3}$
**15.** $0, -1$   **17.** $-(2+\sqrt{3}), -1$
**19.** (a) $(3/2, \pm\pi/3)$, $(0, \pi)$, (b) $(2, 0)$, $(1/2, \pm 2\pi/3)$
**21.** (a) $(0, 0)$, $(\pm\sqrt[4]{3}, \pi/3)$, (b) $(0, \pi/2)$, $(\pm\sqrt[4]{3}, \pi/6)$
**23.** (a) $(e^{3\pi/4}, 3\pi/4)$, $(e^{7\pi/4}, 7\pi/4)$, (b) $(e^{\pi/4}, \pi/4)$, $(e^{5\pi/4}, 5\pi/4)$

### EXERCISE GROUP 17.7

**1.** $a^2/2$   **3.** $\pi a^2/12$   **5.** 1   **7.** $(8\pi + 3\sqrt{3})/3$   **9.** $3\pi a^2/8$   **11.** $(4\pi - 3\sqrt{3})a^2/6$
**13.** $20\pi/3$

### EXERCISE GROUP 17.8

**1.** $2\pi$   **3.** $3\sqrt{3}$   **5.** (a) $(\sqrt{6} - \sqrt{2})/2$, (b) $\sqrt{2}/2$   **7.** $2\pi^2 + 4\pi$
**9.** $2\sqrt{2}\pi(e^{2\pi}+1)/5$   **11.** $32\pi a^2/5$   **13.** $\frac{3}{4}|\csc\theta/2|/a$   **15.** $3\sqrt{\sin 2\theta}/a$

### EXERCISE GROUP 17.9

**1.** $\mathbf{v} = 2\mathbf{u}_r + 2\mathbf{u}_\theta$, $\mathbf{a} = -4\mathbf{u}_r + 8\mathbf{u}_\theta$   **3.** $\mathbf{v} = -\frac{3}{2}(\mathbf{u}_r - 2\mathbf{u}_\theta)$, $\mathbf{a} = -\frac{9}{2}(\mathbf{u}_r + 2\mathbf{u}_\theta)$
**5.** $a_r = -5 - 4\sqrt{3}$, $a_\theta = 2 - \sqrt{3}$

### EXERCISE GROUP 18.1   **1.** (a) $15\mathbf{i} - 2\mathbf{j} + 3\mathbf{k}$, (b) $\sqrt{238}$

**3.** (a) $\pm(2\mathbf{i} - 5\mathbf{j} + \mathbf{k})/\sqrt{30}$, (b) $\pm(-3\mathbf{i} + 3\mathbf{j} - 2\mathbf{k})/\sqrt{22}$, (c) $\pm(-\mathbf{i} - 2\mathbf{j} - \mathbf{k})/\sqrt{6}$
**5.** $-10\mathbf{i} + 5\mathbf{j} - 3\mathbf{k}$   **7.** $(-11\mathbf{i} + 14\mathbf{j} - 7\mathbf{k})/2$   **19.** $26\mathbf{i} - 6\mathbf{j} + 36\mathbf{k}$
**27.** Center $(-1/2, 3/4, -1/4)$, rad $= 1$

### EXERCISE GROUP 18.2   **1.** $\sqrt{38}/2$, $\sqrt{77}/2$, $\sqrt{41}/2$

**5.** (a) $(7/3, -8/3, 7/3)$, (b) $(4, -1, 4)$, (c) $(3/2, -7/2, 3/2)$, (d) $(3, -2, 3)$
**7.** $(0, 5/3, 10/3)$, $(1, 1/3, 8/3)$   **9.** (a) $(7/3, -2/3, -1/3)$, $(8/3, 5/3, -2/3)$

### EXERCISE GROUP 18.3

**1.** (a) $-3$, (b) $-3\sqrt{14}/14$   **3.** (a) 10, (b) $10\sqrt{13}/13$   **5.** (a) $-3$, (b) $-\sqrt{6}/2$
**7.** $16\sqrt{38}/19$   **9.** 0   **11.** On positive $x$ axis

**13.** On a cone with positive $x$ axis as axis   **15.** In $xz$ plane   **17.** $0, \pi$
**19.** (a) $\sqrt{14}/14, -\sqrt{14}/7, 3\sqrt{14}/14$, (b) $-4/9, -7/9, 4/9$, (c) $\sqrt{3}/3, -\sqrt{3}/3, -\sqrt{3}/3$
**21.** $2/\sqrt{106}, 47/\sqrt{59 \cdot 53}, 4/\sqrt{118}$

## EXERCISE GROUP 18.4

**1.** (a) $(-8, -11, 1)$, (b) $(8, 11, -1)$   **3.** (a) $(-2, 3, -4)$, (b) $-14$, (c) $(-4, 2, 0)$
**9.** $2\sqrt{38}$   **11.** $8\sqrt{10}$   **13.** 10

## EXERCISE GROUP 19.1

**1.** $4, 1, -5$   **3.** $1, 2, 1$   **5.** $2, -5, 6$   **7.** $7, 2, -12$
**19.** $A: 128°10', B: 36°0', C: 15°50'$

## EXERCISE GROUP 19.2

**1.** $x - 3y - 4z = 3$   **3.** $x + y + 8z = 29$   **5.** $\sin^{-1} 5\sqrt{2}/26$
**7.** $\cos^{-1} 4/9$   **9.** $2x + 4y - z = 2$   **11.** $2x - 3y = 7$   **13.** $5x + 3y + 7z = 3$
**15.** $6x - 13y - 9z = -4$   **17.** $2x - 3y + 4z = 3$   **19.** $4y - 3z = -5$
**21.** $3x + y - 2z = -1$   **23.** $6x + 5y - 3z = 7$   **25.** $\dfrac{x}{3/2} + \dfrac{y}{-1} + \dfrac{z}{3/4} = 1$
**27.** Intercepts: $0, 0, 0$   **29.** $\dfrac{x}{-2/3} + \dfrac{y}{-2} + \dfrac{z}{2} = 1$   **31.** Intercepts: $0, 0, z$ axis
**33.** $xy: 2x - 5y = 1$, $yz: 5y + 3z = -1$, $xz: 2x - 3z = 1$
**35.** $xy: x + 2y = 4$, $yz: y = 2$, $xz: x = 4$   **37.** $xy: x + 4 = 0$, $yz:$ None, $xz: x + 4 = 0$

## EXERCISE GROUP 19.3

**1.** $(x, y, z) - (-2, 3, 1) = t(6, -1, -3)$   **3.** $(x, y, z) - (4, -2, 1) = t(0, 1, 0)$
**5.** $x = 5 + 4t, y = 2, z = -3 - t$   **7.** $x = -4 + 2t, y = 1 + t, z = -1 - t$
**9.** $x + 1 = \dfrac{y - 3}{-1} = \dfrac{z + 3}{2}$   **11.** $x + 1 = \dfrac{y}{-1} = \dfrac{z - 4}{2}$
**13.** $(7, 1, -3) \times (x, y, z) = (4, -19, 3)$   **15.** $(1, 1, 1) \times (x, y, z) = (-5, 4, 1)$
**19.** $\dfrac{x}{3} = \dfrac{y}{-2} = \dfrac{z}{5/2}$   **21.** $x = 2 + 3t, y = -1 - 2t, z = 4 + t$

## EXERCISE GROUP 19.4

**1.** $\dfrac{x}{2} = \dfrac{y + 27/10}{3} = \dfrac{z + 13/5}{2}$   **3.** $\dfrac{x - 1}{3} = \dfrac{y - 2}{-5} = \dfrac{z + 3}{-7}$
**5.** Parallel   **7.** $\dfrac{x - 3}{2} = \dfrac{y - 5}{-3} = \dfrac{z - 2}{4}$
**9.** $xy: 2x - y = 6$, $yz: 2y + 9z = 4$, $xz: 4x + 9z = 16$
**11.** $xy: 3x + 2y = 4$, $yz: 4y + 3z = -1$, $xz: 2x - z = 3$
**13.** $xy: 5x - 6y = 1$, $yz: 48y - 35z = 12$, $xz: 8x - 7z = 4$
**15.** $xy: 3x + 5y = 47$, $yz: 4y - 3z = 19$, $xz: 4x + 5z = 31$

### EXERCISE GROUP 19.5

**1.** (c), (2, −1, 2)   **3.** (a)   **5.** (c), (7/3, 28/9, −2/9)
**7.** (c), (1/2, −2/3, 4)   **9.** (a)   **11.** (d), (−2, 6, −6)
**13.** (b)   **15.** (a)
**17.** (d), (4, 3, 0)   **19.** $2x - 4y - 7z = -7$   **21.** $\dfrac{x-1}{2} = \dfrac{y}{3} = \dfrac{z}{-1}$   **23.** 10/3

### EXERCISE GROUP 19.6

**1.** $\frac{1}{3}x - \frac{2}{3}y - \frac{2}{3}z = \frac{4}{3}$   **3.** $-\frac{2}{7}x - \frac{3}{7}y + \frac{6}{7}z = \frac{1}{7}$   **5.** $\dfrac{2}{\sqrt{13}}y + \dfrac{3}{\sqrt{13}}z = 0$

**7.** $\dfrac{7}{\sqrt{58}}x - \dfrac{3}{\sqrt{58}}y = \dfrac{2}{\sqrt{58}}$   **9.** (−4/9, −4/9, 2/9)   **11.** (8/41, 0, −10/41)
**13.** −4/3; same   **15.** 29/9; opposite   **17.** $-10/\sqrt{13}$; does not apply
**19.** (a), (c)   **21.** (a)
**23.** (a) $x - 8y + 3z = 5$, $5x + 4y + 9z = -3$,
   (b) $2x + 4y - 14z = 3$, $6x + 4y + 2z = 13$

### EXERCISE GROUP 19.7

**1.** (1, 2, 3)   **3.** (1/3, 1/5, 1/6)
**5.** Three planes intersect in line $x = y = z + 1$. Case (c)   **7.** (2/3, 3/4, 4/5)
**9.** Two planes coincide and are parallel to third plane. Case (a)   **11.** (2, 3, −4)
**13.** Two planes parallel, third plane intersects them in parallel lines. Case (b)

### EXERCISE GROUP 19.8

**1.** $\sqrt{3745}/15$   **3.** 9   **5.** $\sqrt{35}/7$   **7.** $\sqrt{22}$   **9.** $8\sqrt{3}/5$   **11.** $2\sqrt{10}/5$   **13.** $3\sqrt{110}/55$
**15.** $10\sqrt{2}/3$

### EXERCISE GROUP 20.1

**1.** $xy$: $x^2 + y^2 = 1$, circle; $yz$: $4y^2 + 9z^2 = 4$, ellipse; $xz$: $4x^2 + 9z^2 = 4$, ellipse
**3.** $xy$: $4x^2 - 3y = 0$, parabola; $yz$: $y = 0$, $z$ axis; $xz$: $x = 0$, $z$ axis
**5.** $xy$: $x^2 + y^2 + 4x = 4$, circle; $yz$: $y^2 + 2z^2 = 4$, ellipse; $xz$: $x^2 + 2z^2 + 4x = 4$, ellipse
**7.** $xy$: $2x + 3y + 2 = 0$, line; $yz$: $3y - 5z + 2 = 0$, line; $xz$: $2x - 5z + 2 = 0$, line
**19.** If replacing $y$ by $-y$ and $z$ by $-z$ yields an equivalent equation
**21.** If replacing $x$ by $-x$ and $y$ by $-y$ yields an equivalent equation

### EXERCISE GROUP 20.2

**9.** (a) $x^2 = y^2 + z^2$, (b) $x^2 + z^2 = y^2$
**11.** (a) $x^2 + z^2 = (1-y)^2$, (b) $x^2 + y^2 = (1-z)^2$
**13.** (a) $y^2 + z^2 = 9$, (b) $y = 3$   **15.** $y^2 = 4x$, $z = 0$; parabola
**17.** $x^2 = y^2 + 3$, $z = 0$; hyperbola   **19.** $x^2 + z^2 + 2z = 3$, $y = 0$; circle
**21.** Not a surface of specified type

## EXERCISE GROUP 20.3

1. Right circular cylinder  3. Hyperbolic paraboloid  5. Ellipsoid
7. Parabolic cylinder  9. Elliptic paraboloid  11. Hyperboloid of 2 sheets
13. Right circular cone  15. Point (origin)  17. Two planes  19. Sphere
21. Right circular cone
23. (a) $(0, \pm\sqrt{ck(a^2+b^2)}, k)$, (b) $(\pm\sqrt{-ck(a^2+b^2)}, 0, k)$
25. (a) $(\pm\sqrt{(a^2-b^2)(k^2-c^2)}/c, 0, k)$, (b) $(k, 0, \pm\sqrt{(b^2+c^2)(a^2+k^2)}/a)$
27. (a) $(k, 0, \pm\sqrt{(b^2+c^2)(k^2-a^2)}/a)$, (b) $(k, \pm\sqrt{(b^2+c^2)(a^2-k^2)}/a, 0)$

## EXERCISE GROUP 20.4

1. (a) $x=1, y=z; x=1, y=-z$, (b) $x=1, y=z; z=1, x=y$,
(c) $\dfrac{x-4}{15} = \dfrac{y-4}{17} = \dfrac{z-1}{8}; z=1, x=y$

3. (a) $z=0, x=y; x-3 = \dfrac{y-3}{-1} = \dfrac{z}{6}$, (b) $x=y+2 = \dfrac{z+2}{2}; x = \dfrac{y+2}{-1} = \dfrac{z+2}{-1}$,
(c) $x=y+2 = \dfrac{z+2}{2}; x-4 = \dfrac{y-2}{-1} = \dfrac{z-6}{6}$

5. $z=3, x=2y; z=3, x=-2y$

7. (a) $x = \dfrac{y-3}{3} = \dfrac{2z-1}{2}$, (b) $x = \dfrac{y+3}{-3} = \dfrac{2z-1}{2}$

## EXERCISE GROUP 20.5

1. $x^2+4y^2=4y, x^2+z^2=2z$  3. $x^2+(z-2)^2=4$
5. $x^2+(x^2+y^2)^2=9, z^2-y^2+z=9$  7. $x^2=y^2$
9. $x^2+y^2=x, z^4-z^2+y^2=0, z^2=x$

## EXERCISE GROUP 20.6

1. $3x-2y=3, y-3z=6, x-2z=5$  3. $x=y, z=y^2-y, z=x^2-x$
5. $x+3y=8, 2y+z=4, 2x-3z=4$  7. $x^2+y^2=1, y^2+z^2=1, x=z$
9. $x=(y-1)^2, z=(y-1)^4, z=x^2$  11. $(x-2)/2 = (y-3)/6 = (z-1)/3$
13. $x-1 = -(y+3)/3 = 2(z-1)/3$  15. $\pi/2$
17. $\sin^{-1}3/\sqrt{19}$ at $(0, 1, 0)$, $\sin^{-1}3/\sqrt{178}$ at $(-6, 10, 0)$
19. $\sin^{-1}1/\sqrt{3}$ at $(0, 0, 0)$, $0$ at $(1, 2, 1)$  21. $(29\sqrt{29}-40\sqrt{5})/27$  23. $2\sqrt{5}$

## EXERCISE GROUP 20.7

1. (a) $\mathbf{T}=(\mathbf{i}-2\mathbf{j})/\sqrt{5}, \mathbf{N}=(2\mathbf{i}+\mathbf{j}+\mathbf{k})/\sqrt{6}, a_T=0, a_N=2\sqrt{6}$, (b) $2x+y-5z=5$
3. (a) $a_T=\pm\sqrt{5}, a_N=|t|$, (b) $1/(5|t|)$
5. (a) $\mathbf{T}=-\mathbf{k}, \mathbf{N}=(\mathbf{i}+\mathbf{j})/\sqrt{2}$, (b) $2\sqrt{2}$, (c) $(\mathbf{i}-\mathbf{j})/\sqrt{2}$  7. $2\sqrt{6}/5$
11. (a) 1, (b) 1/2  13. (a) $x+y-z=1$, (b) $x+2y-2z=-1$

## EXERCISE GROUP 20.8

1. $(\sqrt{2}, -\pi/4, 1)$   3. $(\sqrt{7}, \pi - \tan^{-1}\sqrt{3}/2, -2)$   5. $(0, -4, -1)$
7. $(-\sqrt{2}, -\sqrt{2}, -3)$   9. $(5, \tan^{-1}2/\sqrt{5}, \cos^{-1}4/5)$
11. $(3\sqrt{2}/4, -3\sqrt{6}/4, -3\sqrt{2}/2)$
13. $(5\sqrt{2}/2, -\pi/4, -5\sqrt{2}/2)$   15. (a) $r^2 = 2z$, (b) $\rho = 2 \csc \phi \cot \phi$
17. (a) $r = z$, (b) $\tan^2 \phi = 1$
19. (a) $r^2 + 3z^2 = 3$, (b) $\rho^2(1 + 2 \cos^2 \phi) = 3$
21. (a) $r^2 + z^2 - 2r \sin \theta = 0$, (b) $\rho = 2 \sin \phi \sin \theta$
23. (a) $r^2 + z^2 = 4$, (b) $\rho = 2$
25. (a) $r^2 \cos^2\theta + 2r \sin \theta = 4$, (b) $\rho^2 \sin^2 \phi \cos^2 \theta = 2\rho \sin \phi \sin \theta = 4$

## EXERCISE GROUP 21.1

1. All points in the plane
3. All points outside or on the circle $x^2 + y^2 = 4$
5. $x^2 - y^2 \geq 1$, that is, all points on the hyperbola $x^2 - y^2 = 1$ or on the concave side of the curve
13. (a) $f(0, 0) = 0$, (b) $f(0, 0) = 1$
15. Straight lines through the origin, other than $x + y = 0$
17. Circles with center at origin and radius $\geq 2$

## EXERCISE GROUP 21.2

1. $f_x = 2x + y, f_y = x + 6y$   3. $f_x = -y/[x(x - y)], f_y = 1/(x - y)$
5. $f_x = -(y/x^2) \sec^2 y/x, f_y = (1/x) \sec^2 y/x$   7. $f_x = x/(x^2 + z^2), f_z = z/(x^2 + z^2)$
9. $f_x = -2y \sin(xy) \cos(xy), f_y = -2x \sin(xy) \cos(xy)$
11. $f_x = 3 \sin^2(x + 2y) \cos(x + 2y), f_y = 6 \sin^2(x + 2y) \cos(x + 2y)$
13. $h_u = 1/(v + w), h_v = h_w = -u/(v + w)^2$   15. $f_u(2, 3) = -6, f_v(-1, 2) = 9$
17. $\partial F/\partial x = 2 \sin x \cos x, \partial F/\partial y = 3y^2$
31. $u_x = f + xf_x + yg_x, u_y = g + xf_y + yg_y$
33. $u_x = x + f(y, z), u_y = xf_y(y, z) + g_y(y, z), u_z = xf_z(y, z) + g_z(y, z)$
35. $2\sqrt{2}$   37. $D_\pi f = -2, D_\mathbf{u} f = 2\lambda - 3\mu$   39. $y + z = 4, x = 1$

## EXERCISE GROUP 21.3

1. $f_{xx} = 2, f_{yy} = 6$   3. $g_{xy} = -1/y^2, g_{yxy} = 2/y^3$
5. $f_{yy} = 2x(x^2 + 2y^2)/(x^2 - y^2)^3, f_{xy} = -2y(3x^2 + y^2)/(x^2 - y^2)^3$
7. $h_{rrs} = 6, h_{ssr} = 0$   9. $f_{xy} = xz(2 \cos xyz - xyz \sin xyz), f_{yy} = -x^3z^2 \sin xyz,$
$f_{xz} = xy(2 \cos xyz - xyz \sin xyz)$
11. $f_{xy} = 0, f_{xxx} = -8 \sin x \cos x$   13. $r_{uv} = -\sin u \cosh v$

## EXERCISE GROUP 21.4

1. $\ln u, u = x^2 - y^2$   3. $u^{10}, u = 2x - 3y$   5. $\cosh u + \ln v, u = 3t, v = 2t - 3$

7. $\partial u/\partial r = 3(x^2 - 2xy - y^2)$, $\partial u/\partial s = 3(x^2 + 2xy - y^2)$
9. $f'(t) = 3 \sinh 3t + 2/(2t - 3)$
11. $\partial v/\partial r = 2(4x \tan \theta - y \sec \theta)$, $\partial v/\partial \theta = 2r \sec \theta(4x \sec \theta - y \tan \theta)$
13. $\partial w/\partial x = -10x - 16y$, $\partial w/\partial y = -16x - 6y$
15. $\partial F/\partial x = 2x$, $\partial F/\partial y = 3 \sin^2 y \cos y$

**EXERCISE GROUP 21.5**

3. $F_x = 2xf_1(x^2 - y, x + y^2) + f_2(x^2 - y, x + y^2)$,
   $F_y = -f_1(x^2 - y, x + y^2) + 2yf_2(x^2 - y, x + y^2)$
5. $[h(x)g'(x) - g(x)h'(x)]f'(g(x)/h(x))/h^2(x)$
9. $\partial u/\partial x = y^3 f'(xy)$, $\partial u/\partial y = 2yf(xy) + xy^2 f'(xy)$

**EXERCISE GROUP 21.6**

1. (a) $4x - 12y + z = -14$, (b) $(x + 1)/4 = -(y - 2)/12 = z - 14$
3. (a) $8x + y + z = 8$, (b) $(2x - 1)/16 = y - 2 = z - 2$
5. (a) $x - 6y - z = -10$, (b) $x - 1 = -(y - 2)/6 = -(z + 1)$
7. $\pm(\mathbf{i} + 2\mathbf{j})/\sqrt{5}$  9. $\pm(\mathbf{i} + 2\mathbf{j} + 3\mathbf{k})/\sqrt{14}$

**EXERCISE GROUP 21.7**

1. (a) i. 16, ii. 61/5, iii. $-\frac{1}{2} - 8\sqrt{3}$, iv. $19/\sqrt{10}$, (b) $\sqrt{257}$
3. (a) $-f_x$, (b) $-f_y$, (c) $(f_x + f_y)/\sqrt{2}$, (d) $\frac{1}{2}(f_x + \sqrt{3} f_y)$
5. $-2\sqrt{5}$  7. $3/\sqrt{14}$

**EXERCISE GROUP 21.8**

1. (a) $2xh + 2yk$, (b) $yh + xk$, (c) $(yh - xk)/y^2$
3. (a) $(-yh + xk)/(xy)$, (b) $(-yh + xk)/(x^2 + y^2)$, (c) $2(xh + yk)/(x^2 + y^2)$
5. (a) $(yz\, dx + xz\, dy - xy\, dz)/z^2$, (b) $yz^2\, dx + xz^2\, dy + 2xyz\, dz$,
   (c) $[z\, dx - z\, dy - 2(x - y)\, dz]/z^3$  7. $(y\, dx - x\, dy)f'(x/y)/y^2$  11. $10^{-6}k$
13. $v_0 \sin \alpha (\sin \alpha\, dv_0 + v_0 \cos \alpha\, d\alpha)/g$

**EXERCISE GROUP 21.9**

1. $y(x \ln y - y)/[x(y \ln x - x)]$
3. $\partial u/\partial x = (3x^2 v - 1)/(u + v)$, $\partial u/\partial y = 3(y^2 - 2u)/[2(u + v)]$
5. $z_x = -2xy/(1 + \sin z)$, $z_y = -x^2/(1 + \sin z)$
7. $u_x = u(vy - 1)/(uv - 1)$, $v_x = v(u - y)/(uv - 1)$
9. $\partial z/\partial x = -f_x/f_z$, $\partial z/\partial y = -f_y/f_z$  11. $\partial u/\partial x = f_x/(1 - f_u)$, $\partial u/\partial y = f_y/(1 - f_u)$
13. $\partial x/\partial u = (1 - f_3)/(f_1 - f_3)$  17. $u'(x) = (f_x g_y - f_y g_x)/g_y$
19. $du/dx = (f_2 g_1 - f_1 g_2)/[f_2(1 - g_3) - g_2(1 - f_3)]$

**EXERCISE GROUP 21.10**

5. $u_{xx} = -e^{-2u} \cos 2v$  7. $d^2y/dx^2 = -(f_y^2 f_{xx} - 2f_x f_y f_{xy} + f_x^2 f_{yy})/f_y^3$
11. $8\sqrt{2}/3$  13. 2

## EXERCISE GROUP 21.11

**1.** Min $-7/2$ at $(1/2, -1/2)$  **3.** Min $-58$ at $(-1, 2)$
**5.** Min $-1$ at $(1, 1)$, min $-1$ at $(-1, -1)$, saddle point at $(0, 0)$
**7.** $(4/3, 4/3)$  **13.** $\sqrt{14}/2$  **15.** 1

## EXERCISE GROUP 21.12

**1.** $x + xy + C$  **3.** $xy - 7x^2 + C$  **5.** Not exact  **7.** $x^2y + \sin(xy) + C$
**9.** $y/x + C$  **11.** $(xy^3 - 3y^2 - 3)/(3x) + C$  **13.** $x^2y^3z + z^2 + C$
**15.** $(x - y)/z + C$

## EXERCISE GROUP 21.13

**1.** (a), (b), (c), (d) $-7$  **3.** $\pi$  **5.** (a) $8\pi$, (b) $16\pi$, (c) $8\pi$  **7.** $\sin 1$  **9.** $4\pi$

## EXERCISE GROUP 21.14

**1.** $5/2$  **3.** $3 - e^2$  **5.** $(3u - v)/u$  **7.** 0  **9.** (b) $\pi/6$, (c) $\sin^{-1}y/x + C$

## EXERCISE GROUP 22.1

**1.** (a) 20, (b) 20  **3.** (a) 8, (b) 8  **5.** (a) 39, (b) 215/8

## EXERCISE GROUP 22.2

**1.** $1/4$  **3.** $3\pi/2$  **5.** $52/3$  **7.** 1  **9.** $4/3$  **11.** $\int_1^{16} \int_{\sqrt{y}}^4 dx\, dy$, 18
**13.** $\int_{-1}^0 \int_0^{1-x^2} dy\, dx + \int_0^1 \int_0^{1-x} dy\, dx$, $7/6$
**15.** $\int_0^1 \int_0^2 dy\, dx + \int_1^{\sqrt{5}} \int_{\sqrt{x^2-1}}^2 dy\, dx$, $\frac{1}{2}[2\sqrt{5} + \ln(2 + \sqrt{5})]$  **17.** $4/3$

## EXERCISE GROUP 22.3

**1.** 8  **3.** $14/3$  **5.** $8/3$  **7.** $8/3$  **9.** $128/27$  **11.** $e^2 - e + 1 - 2\ln 2$
**13.** $\pi/8$  **15.** $7/5$  **17.** $\pi a^2/8$  **19.** $3\pi a^2/2$  **21.** $(12 - \pi)a^2/8$
**23.** $\int_0^2 \int_2^{\cos^{-1}(r-1)} r\, d\theta\, dr$

## EXERCISE GROUP 22.4

**1.** 12  **3.** $\pi/2$  **5.** $54\pi$  **7.** $16a^3/3$  **9.** $10/3$  **11.** $5\pi/3$  **13.** $7\pi/2$

## EXERCISE GROUP 22.5

**1.** $(5/14, 5/8)$  **3.** $(27/40, 27/20)$  **5.** $(2, 1)$  **7.** $M_x = 896/405$, $M_y = 128/81$
**9.** $[a(3a^2 + 2b^2)/(4(a^2 + b^2))$, $b(2a^2 + 3b^2)/(4(a^2 + b^2))]$
**11.** $\bar{x} = 128\sqrt{2}\,a/(105\pi)$, $\bar{y} = 0$

## EXERCISE GROUP 22.6

**1.** $I_x = 3/11$, $I_y = 1/9$  **3.** (a) $ab(a^2 + b^2)/12$, (b) $ab(a^2 + b^2)/3$  **5.** $\pi a^5 k/5$  **7.** $1/2$
**9.** $\pi a^4/16$  **11.** $3\pi/128$

## EXERCISE GROUP 22.7

**1.** 6  **3.** $4(8\sqrt{2} - 7)\pi/3$  **5.** $14/3$  **7.** $\int_0^3 \int_0^{9-x^2} \int_{\sqrt{z}}^{\sqrt{9-x^2}} dy\, dz\, dx$

9. (a) $36\pi$, (b) $\int_{-3}^{3}\int_{1}^{5-\sqrt{9-y^2}}\int_{-\sqrt{9-y^2}}^{\sqrt{9-y^2}} dz\,dx\,dy + \int_{-3}^{3}\int_{5-\sqrt{9-y^2}}^{5+\sqrt{9-y^2}}\int_{-\sqrt{9-y^2}}^{5-x} dz\,dx\,dy$
11. (a) $243\pi/4$, (b) $128/(15\pi)$   13. $8\pi(a^5 - b^5)/15$

### EXERCISE GROUP 22.8

1. $4\pi a^3/3$   3. $8\pi/3$   5. $3\pi/4$   7. $14\pi/3$
9. (a) $(0, 0, 3h/4)$, (b) $I_y = h(1 + 4h^2)\pi/20$, $I_z = \pi h/10$
11. (a) $27\pi$, (b) $(0, -3/4, 15/8)$

### EXERCISE GROUP 22.9

1. $13\pi/3$   3. $\sqrt{2}\pi$   5. $3a^2/5$

### EXERCISE GROUP 23.1

1. 1/4, 1/7, 1/10, 1/13, 1/16   3. 1/2, $-1/4$, 1/8, $-1/16$, 1/32
5. 2, 3/2, 2/3, 5/24, 1/20   7. 2, 12, 120, 1680, 30240   9. 1, $-2$, $-8$, 64, 1024
11. $x^2$, $x^4/2$, $x^6/3$, $x^8/4$, $x^{10}/5$   13. (a) $3n - 1$, (b) $3n + 2$
15. (a) $(-1)^{n-1}(n+1)/[(n+2)(n+3)]$, (b) $(-1)^n(n+2)/[(n+3)(n+4)]$
17. (a) $(-1)^{n-1}/[n(n+2)]$, (b) $(-1)^n/[(n+1)(n+3)]$
19. $b_n = -1/[n(n+1)]$; $-1/2$, $-1/6$, $-1/12$
21. $b_n = \ln[(n+1)/n]$; $\ln 2$, $\ln 3/2$, $\ln 4/3$   23. $b_n = 3(-2)^n$; $-6$, 12, $-24$
25. (a) $1/(1 + 4n^2)$, (b) $1/(1 + n^4)$, (c) $n^2/(2n^2 + 2n + 1)$

### EXERCISE GROUP 23.2

1. 1/2   3. 1   5. 0   7. No limit   9. 1/2   11. 1/3   13. $k$   15. 0   17. 2/3   19. 1

### EXERCISE GROUP 23.3

9. $s_n = 2n/(3n + 1)$, conv, sum $= 2/3$   11. $s_n = n/2^n$, conv, sum $= 0$
13. $s_n = 1/n^2$, conv, sum $= 0$   15. $s_n = \ln(n+1)$, div
17. $s_n = \ln[n/(2n+1)]$, conv, sum $= -\ln 2$   19. $a_1 = 1$; $a_n = -1/[n(n-1)]$, $n > 1$
21. $a_1 = 0$; $a_n = \ln[n/(n-1)]$, $n > 1$   23. $a_1 = \cos 4$; $a_n = -2\sin 2\sin(4n-2)$, $n > 1$
25. (a) 50/33, (b) 4/11, (c) 67/111, (d) 260/111

### EXERCISE GROUP 23.4

1. Div   3. Conv   5. Div   7. Div   9. Conv   11. Div   13. Div

### EXERCISE GROUP 23.5

1. Conv   3. Div   5. Conv   7. Conv   9. Div   11. Div   13. Conv   15. Conv
17. Conv   19. Div

### EXERCISE GROUP 23.6

1. Conv   3. Conv   5. Conv   7. Conv   9. Conv   11. Conv   13. Conv

### EXERCISE GROUP 23.7

1. Div   3. Div   5. Div   7. Conv   9. Conv   11. Div   13. Div   15. Div
17. Div   19. Div

## EXERCISE GROUP 23.8

1. Conv  3. Conv  5. Div  7. Div  9. 14  11. 0.021  13. Abs conv
15. Abs conv  17. Div  19. Div  21. Abs conv  23. Abs conv

## EXERCISE GROUP 23.9

1. $-1 < x \leq 1$  3. $-1 \leq x \leq 1$  5. $-1 < x < 1$  7. $-2 < x < 0$
9. $-2 \leq x < 2$  11. $-3 < x < 3$  13. $-\frac{1}{2} \leq x \leq \frac{1}{2}$  15. $0 \leq x \leq 2$
17. $-1 < x \leq 1$  19. 1  21. $\frac{1}{2}(1 - \sqrt{5}) < x < \frac{1}{2}(1 + \sqrt{5})$  23. $x > -\frac{1}{2}$

## EXERCISE GROUP 23.11

1. $\sum_{k=0}^{\infty} \frac{x^k}{k!}$  3. $\sum_{k=0}^{\infty} (-1)^k \frac{x^{2k+1}}{(2k+1)!}$  5. $\frac{\sqrt{2}}{2} \sum_{k=0}^{\infty} (-1)^{k(k-1)/2} \frac{x^k}{k!}$  7. $\sum_{k=1}^{\infty} (-1)^{k+1} \frac{x^k}{k}$
9. $\sum_{k=0}^{\infty} x^k$  11. $\sum_{k=0}^{\infty} (-1)^k k x^k$  13. $e^{\xi}(x-1)^{n+1}/(n+1)!$
15. $\cos[\xi + (n+1)\pi/2]x^{n+1}/(n+1)!$  17. $(x-2)^{n+1}/(1-\xi)^{n+2}$  19. $x + x^3/3$
21. $x - x^2 + x^3/2! - x^4/3! + x^5/4!$  23. $x - x^3/3$

## EXERCISE GROUP 23.12

1. 0.0998  3. 0.9553  5. 0.7468  7. 1.9873  9. 8.3666
13. (a) $\sum_{k=1}^{n} (-1)^{k+1} \frac{x^k}{k} + (-1)^n \frac{x^{n+1}}{(n+1)(1+\xi)^{n+1}}$, (b) $1 - \frac{1}{2} + \frac{1}{3} - \frac{1}{4} + \cdots$

## EXERCISE GROUP 23.13

1. $\sum_{k=0}^{\infty} \frac{x^{2k+1}}{(2k+1)!}$  3. $\sum_{k=0}^{\infty} (-1)^k \frac{x^{2k}}{(2k+1)!}$  5. $\sum_{k=0}^{\infty} \frac{(k+1)x^k}{k!}$
7. $x - x^2 + 5x^3/6 - 5x^4/6 + 101x^5/120 + \cdots$
9. $1 - x^2/2 + 5x^4/24 - 61x^6/720 + \cdots$  11. $1 - x^2/2 + 5x^4/24 - \cdots$
13. $1 + x + x^2/2 + x^3/2 + 3x^4/8 + \cdots$  15. $4(1 - \frac{1}{3} + \frac{1}{5} - \frac{1}{7} + \cdots)$
17. $\pi/6 + (\sqrt{3}/2)[x - x^2/2 + x^4/4 - x^5/5 + x^7/7 - x^8/8 + \cdots]$

## EXERCISE GROUP 23.14

1. 1  3. 1/3  5. 2/3  7. $-1$  9. $-1/3$  11. 1/6  13. $\frac{1}{2}f''(a)$

## EXERCISE GROUP 23.15

1. $55 - 3(x+4) + 24(y-3) + 4(y-3)^2$
3. $54 - 33(x+2) + 14(y-3) + 3(x+2)^2 - 7(x+2)(y-3)$
5. $14 + (x+2)^2(y-3)$  7. $1 + x + x^2/2 - y^2/2 + x^3/6 - xy^2/2$
9. $1/2 - (x-1)/4 + (y-1)/2 + (x-1)^2/8 - (x-1)(y-1)/4$

# Index

Absolute value, 7
Acceleration, 141, 441, 442, 561
  normal component of, 444, 562
  in polar coordinates, 479
  radial component of, 480
  tangential component of, 444, 562
  transverse component of, 480
Alternating series test, 693
Antiderivative, 238
Approximations, 108
Arc length, 336, 338, 433, 476, 558
  as parameter, 449
Area, 269
  by double integrals, 640
  in polar coordinates, 472
  of a (curved) surface, 665
  of a parallelogram, 500
  of a plane region, 268
  of a region, 269, 270, 641
  of a surface of revolution, 339, 433, 476
  polar representation of, 646
  under a curve, 431
Associative property, 2
  of vector addition, 16
Asymptote, horizontal, 159
  oblique, 172
  of a hyperbola, 206, 210
  vertical, 160, 161
Auxiliary condition, 128, 608
Auxiliary variable, 128

Base vectors, 18, 485
Binomial series, 711
Binormal vector, 562

Cardioid, 455, 461
Center of gravity, 649
Center of mass, 352, 353, 649
  of a composite body, 361
  of a lamina, 356, 357
Centroid, 358, 649
  of an arc, 364
  of a solid of revolution, 365
  of a surface of revolution, 366
Chain rule, 88, 89, 104, 583, 584, 588
  in integration, 242
Change of variable, in a definite integral, 266
  in an integral, 246
  in a sum, 251
Chord, 192
Circle, 179, 180, 196
Cissoid, 420, 433
Closure, 2
Colatitude, 568
Commutative property, 2
  of scalar product, 29, 491
  of vector addition, 16
Comparison tests for series, 686, 688

Completeness property, 3, 676
Completing the square, 8, 180, 202
Composite body, center of mass, 361
Composite function, 57, 72, 88
Concavity, 132, 133
Conditional convergence, 696
Cone, 545
  circular, 546
  element, 339
  elliptic, 546
  frustum, 331, 339
Conics, 175
  directrix, 462
  focus, 462
  focus-directrix property, 225, 226
  identification of, 222
  in polar form, 462
  optical properties, 229
  tangents of, 228
Constant of integration, 239
Constraint, 608
Continuity, of a function, 70, 258
  function of two variables, 572
  of a power series, 703
  of a vector function, 438, 555
  of trigonometric functions, 277
Convergence, 415, 416, 674, 678
  absolute, 695, 696
  interval of, 699
  radius of, 699
Convex set, 52
Coordinates, cylindrical, 566, 646, 660
  rectangular, 12, 483
  spherical, 566, 567, 662
Critical relation, 129
Critical value, 121
Cross product, 497, 498
Curvature, 446, 561
  in polar coordinates, 478
Curve, 534
  area under a, 431
  asymptotic, 173
  closed, 619
  direction of a, 112
  equation of a, 175
  extent of a, 154
  in space, 552
  level, 574
  normal to, 83
  parametric representation, 554, 556
  rectifiable, 434
  regular closed, 630
  rotation of a, 460
  slope of a, 83, 425
  tangent to a, 81
Curvilinear integral, 624
Cyclically equivalent, 501
Cycloid, 430, 431
Cylinder, 537, 538
  directrix of a, 538
  elliptic, 538
  hyperbolic, 538, 547
  parabolic, 538, 547
  projecting, 552, 556
  right circular, 538
Cylindrical discs, method of, 329
Cylindrical shells, 332

Definite integral, 234, 248, 257, 258
  applications of a, 324, 345
  change of variable in a, 266
Density, 356
Derivative, 77, 79
  of a composite function, 88
  directional, 574, 593
  first-order, 579
  higher order, 604
  mixed, 580, 581
  partial, 574, 576
  of a polynomial, 86, 87
  of a product, 93
  of a quotient, 93
  of a scalar product, 440
  second, 100
  of a sum, 87
  of a vector function, 438, 555
  of a vector product, 555
Derivatives, applications of, 112
  Mean-Value Theorem for, 235
  of exponential functions, 308
  of higher order, 100
  of inverse hyperbolic functions, 321
  of inverse trigonometric functions, 291
  of logarithmic functions, 308
  of trigonometric functions, 279
Determinant, 498
Differential, 103, 106, 108
  exact, 612, 614
  total, 597, 600
Differential equation, 281
Differentiation, chain rule of, 88, 89, 104, 583, 584
  implicit, 97, 602
  logarithmic, 309
  power rule of, 86, 90, 95, 243
  product formula of, 94
  quotient formula of, 94
Differentiation formulas, 85
Directed distance from a plane to a point, 527
Directed line segment, 13
Direction, 20
  in space, 491
  of a curve, 112, 469
  of a vector, 30
Direction angle, 494
Direction cosine, 494
Direction number, 504, 505
Discontinuous, 70
Discriminant, 223
Displacement vector, 442
Distance, 25, 454, 485, 523, 531, 571
Domain, 54, 97, 570
Dot product, 29, 491
Double integral, 629, 630, 638
  applications of, 647
  area by a, 640
  evaluation of a, 634
  in polar coordinates, 637
  volume by a, 643
Double point, 427
Dummy index (variable), 249

$e$, 414
Eccentricity, 226
Ellipse, 198, 223, 224
  axes, 200
  center of an, 198
  equation of an, 199, 200, 201
  focal radius of an, 200
  focal vector of an, 200
  foci of an, 198
  horizontal, 201
  vertical, 201, 202
  vertices of an, 199
Ellipsoid, 542
Enabling factor, 243
Endpoint maximum, 125
Equation, degenerate, 272
  graph of an, 41, 455
  linear, 43, 508
  locus of an, 455
  Newton's method for solving, 110
  of a circle, 179, 180
  of a conic in polar form, 462
  of a curve, 175
  of a hyperbola, 205, 207
  of a locus, 177
  of a parabola, 191, 192, 195
  of a plane, 508, 512, 525, 526
  of a sphere, 487
  of a straight line, 43, 462
  of an ellipse, 199, 200, 201
  of first-degree, 508
  of second-degree, 542
  solution of an, 40
Equations, dependent, 47
  equivalent, 151
  in parametric form, 427
  in polar form, 455
  inconsistent, 48, 523
  of rotation, 215
  of transformation, 454
  of translation, 194
  parametric, 421
Exponential function, 297, 299, 311
  derivative of an, 308
  graph of an, 304
Extended comparison test, 688
Extended Mean-Value Theorem, 237
Extent, 459
  of a curve, 154
Extremum, 120
  extended criterion for an, 137
  first-derivative test for an, 121
  of a function of two variables, 606
  second-derivative test, 123, 610

Factorial, 101
Factorial function, 671
First-derivative test, 121
First moment, 368, 649
Fluid pressure, 349
Focal vector, 200
Focus-directrix property, 225, 226
Frenet formulas, 564
Function, 53, 54, 56, 285
  algebraic, 275
  average rate of change, 77
  composite, 57, 72
  continuous, 70, 258
  decreasing, 115, 117, 286
  defined implicitly, 98, 604
  derivative, 76
  derived, 76
  differentiable, 78
  domain of a, 54, 97, 570

even, 152
exponential, 297, 299, 311
factorial, 671
graph, 54
homogeneous, 606
hyperbolic, 314
increasing, 115, 117, 286
inverse, 285, 286
inverse trigonometric, 288
linear, 55
logarithmic, 297, 311
maximum (minimum), 72
odd, 152
position, 141
range, 54, 95, 570
rate of change, 77, 139
rational, 72, 393, 401
transcendental, 275
value of a, 570
vector, 421, 437, 554
Function of several variables, 570
Function of two variables, 97
  continuity of a, 572
  extrema of a, 606
  limit of a, 571
  Taylor formula for a, 724
Fundamental Theorem of Algebra, 235
Fundamental Theorem of Integral Calculus, 262, 264
Fundamental trigonometric limit, 278

Generalized Mean-Value Theorem, 408, 425
Gradient, 591, 593
Graph, of a function, 54
  of a hyperbolic function, 315
  of a logarithmic function, 304
  of a polar equation, 457
  of a relation, 40
  of an equation, 41, 455
  of an exponential function, 304
Gudermannian, 321
Gyration, radius of, 371, 652

Hollow-disc method, 330
Hooke's Law, 346
Hyperbola, 178, 204, 223, 224
  asymptote of a, 206, 210
  center of a, 204
  conjugate axis of a, 206
  equation of a, 205, 207
  equilateral, 227
  focal radii of a, 206
  foci of a, 204
  horizontal, 207
  transverse axis of a, 206
  vertical, 207
  vertices of a, 206
Hyperbolic function, 314, 315, 318
Hyperboloid, of one sheet, 543, 550
  of two sheets, 543, 545
Hypocycloid, 436

Image, 150
Implicit differentiation, 97, 602
Improper integral, 406, 415, 416
Incident ray, 229
Increment, 108, 584

Indefinite integral, 238, 239
Independence of path, 624
Indeterminate form, 406, 413, 722
Index of summation, 249, 251
Inequality, 3, 5, 49
Infinite series, 671, 677
  absolute convergence of, 695, 696
Inscribed polygon, 336, 433
Integrable, 258
Integral, 311
  change of variable in an, 246
  convergent, 415, 416
  curvilinear, 624
  definite, 234, 248, 257, 258
  divergent, 415
  double, 629, 630, 638
  improper, 406, 415
  indefinite, 238, 239
  iterated (repeated), 634, 655
  Law of the Mean, 261
  multiple, 629
  of a hyperbolic function, 318
  of a trigonometric function, 282
  particular, 241
  substitution in an, 246
  surface, 668
  triple, 654, 655, 660
  volume, 654
Integral test for series, 682
Integration, 234, 239, 374
  chain rule in, 242
  constant of, 239
  by parts, 394
  reduction formulas, 398, 399, 400
Intercept, 45, 150, 511, 537
Intercept form, 45, 512
Intermediate Value Theorem, 73
Intersection, of straight lines, 47
  of curves in polar coordinates, 466
  of planes, 517, 528, 530
Interval, 10
  excluded, 154
  maximum (minimum), 125
Invariant, 223
Inverse function, 285, 286
Inverse hyperbolic function, 319, 321
Inverse trigonometric function, 288, 291
Iterated integral, 634, 655

Lamina, center of mass, 356, 357
  mass, 357
  moment, 357
  moment of inertia, 368
Latus rectum, 192
Least upper bound, 2, 3, 676
Lemniscate, 459
Level curve, 574
L'Hospital's Rule, 408, 409, 417
Limaçon, 459
Limit, 59, 63, 155
  infinite, 155
  of a function of two variables, 571
  of a sequence, 256, 674
  of a vector function, 438, 554
  one-sided, 160
  properties of, 66
  trigonometric, 276
  uniqueness of a, 69
Limits of summation, 249

Line integral, 616, 624
Linear combinations, 18
Linear motion, 141
Lines, angle between, 37, 506
  skew, 522
Locus, 56, 176, 177, 455
Logarithmic function, 297, 311
  derivative of a, 308
  graph of a, 304
Longitude, 568

Maclaurin series, 708
Mass, 647, 656
  of a lamina, 357
Maximum and minimum, 119, 127
  absolute, 119
  endpoint, 125
  in an interval, 125
  of a function, 72
  relative, 120, 607
Mean-Value Theorem for derivatives, 235
Midpoint formula, 25, 27, 489
Minkowski inequality, 24
Moment, 352, 647, 649, 656
  of a lamina, 357
Moment of inertia, 368, 647, 650, 657
  of a lamina, 368
  polar, 372, 651
Moving trihedral, 561, 562

Naperian logarithm, 302
Natural logarithm, 302
Newton's method, 110
Norm of a subdivision, 257
Normal, 508, 591
  to a curve, 83
Normal component, 444, 562
Normal form of equation, of a plane, 525, 526
  of a straight line, 462
Normal vector, 445, 560
Normalize a vector, 17, 485

Octant, 484
Optical properties of conics, 229
Osculating plane, 560, 562

$p$-series, 684
Pappus, Theorems of, 367
Parabola, 190, 194, 223, 224
  axis of a, 191
  directrix of a, 190
  equation of a, 191, 192, 195
  focus of a, 190
  horizontal, 191, 194
  vertex of a, 191
  vertical, 191
Paraboloid, elliptic, 542
  hyperbolic, 546, 549
  of revolution, 543
Parallelogram law, 20, 487
Parameter, 421, 424, 428
Parametric equations, 421, 427
  of a cycloid, 431
  of a line, 514
Parametric representation, 421, 554, 556
Partial derivative, 574, 576, 580

Partial fractions, linear factors, 388
  quadratic factors, 391
Partial sum, 678
Plane, 507, 508, 525
  normal form of equation, 525, 526
  osculating, 560, 562
  projecting, 519
  tangent, 590, 591
Planes, intersection of, 517, 528, 530
Point of division formula, 489, 490
Point of inflection, 132, 134, 137
Point-slope form, 44
Polar coordinates, 452
  acceleration in, 479
  area in, 472
  curvature in, 478
  direction of a curve in, 469
  double integral in, 637
  element of arc in, 477
  intersection of curves in, 466
  velocity in, 479
Polar equation, graph of a, 457
Polar form, equations in, 455, 462
Polynomial, derivative of a, 86, 87
Position function, 141
Position vector, 437, 441, 507, 556
Postage function, 55
Power series, 698
  computation with, 716
  continuity of a, 703
  differentiability of a, 705
  division of, 720
  integrability of a, 704
  multiplication of, 720
  sum of a, 702
Projection, 14, 494, 519
Pseudo-rectangle, 638

Quadratic equation in $x, y$, 175, 214, 218
Quadric surface, 542, 547

Radial component, 480
Radical axis, 189
Range, 54, 97, 570
Rate of change, 77, 139, 141
Ratio test, 690
Rationalize a numerator, 62
Real numbers, 1–4, 676
Rectangular coordinates, 12, 483
Reduction formulas, 398, 399, 400
Reference point, 489
Reflected ray, 229
Reflection, 150
Region, 253
  area of a, 268, 270, 641
  simple, 630
  simply-connected, 628
  under a curve, 269
Related rates, 143
Relation, 40, 285
Relative maximum (minimum), 120, 607
Remainder in Taylor's formula, 714, 726
Removable discontinuity, 71, 572
Riemann sums, 257, 259
Right-handed system, 483, 501, 502
Rolle's Theorem, 236
Rotation, angle of, 219
  equations of, 215

of a curve, 460
of axes, 214
Ruling, 538, 549

Saddle point, 607, 610
Scalar product, 29, 440, 491
Second-derivative test, 123, 610
Second moment, 368, 650
Second-order partial derivative, 580
Sequence, 256, 671, 672
   convergent (divergent), 674
   limit of a, 256, 674
   of Riemann sums, 259
Series, alternating, 693
   binomial, 711
   comparison tests for, 686
   conditional convergence of, 696
   convergent (divergent), 678, 680
   geometric, 680
   harmonic, 684
   hyperharmonic, 684
   infinite, 671, 677
   integral test for, 682
   Maclaurin, 708
   of positive terms, 692
   power, 698
   ratio test for, 690
   sum of a, 678
   Taylor, 707
   term of a, 678
   value of a, 678
Slant height, 339
Slope, 33, 83, 424, 425
Slope-intercept form, 45
Smooth arc, 630
Solid, volume of a, 644
Solid of revolution, 327, 332, 365
Speed, 442, 526
Sphere, 487, 542
Spherical coordinates, 566, 567, 660, 662
Standard forms, 213
Straight line, 43, 514
   normal form of equation, 462
   parametric equations, 514
   slope, 33
   symmetric equations, 514
Straight lines, intersection of, 47
   parallel, 34, 506
   perpendicular, 34, 35, 506
Substitution in an integral, 246
Summation, 249
Surface, 534
   area, 665
   cylindrical, 537
   element of area, 666
   quadric, 542, 547
   ruled, 549
   section of a, 535
   trace of a, 535
Surface integral, 668
Surface of revolution, 339, 539
   area of, 339, 433, 476
   centroid of a, 366
Symmetry, 150, 151, 457, 536

Tangent, 81, 163, 557
Tangent line, 82
Tangent plane, 590, 591

Tangent vector, 442, 557, 591
Tangential component, 444, 562
Taylor formula, 725
   function of two variables, 724
   remainder in, 714, 726
   with remainder, 708
Taylor polynomial, 710
Taylor series, 707
Telescoping sum, 250, 681
Torsion, 564
Torus, 334
Trace, 511, 535
Tractrix, 427
Transfer of Axis theorem, 653
Tranformation, equations of, 454
Translation of axes, 194
Transverse component, 480
Triangle inequality, 8
Triangle law of addition, 20, 487
Trigonometric functions, 277, 279, 282
Trigonometric limit, 276, 278
Trigonometric substitution, 381
Triple integral, 654, 655, 660
Triple product, 501
Two-intercept form, 45


Upper bound, 3


Variable, 40, 246, 251, 266
Vector, 15, 16
   binormal, 562
   components of a, 18, 505
   direction of a, 30
   geometric representation, 20, 485
   magnitude of a, 16, 485
   negative of a, 16, 485
   normal, 445, 560
   normalize a, 17, 485
   position, 437, 441, 507, 556
   tangent, 442, 445, 557, 559, 591
   three-dimensional, 484
   unit, 16, 485
   zero, 16
Vector derivatives, properties, 439
Vector function, 421, 437, 438, 554, 555
Vector product, 497, 498
Vectors, angle between, 30, 492, 493, 496, 499
   difference of, 16, 485
   distributive property of, 16
   in three dimensions, 483, 484
   parallel, 500
   perpendicular, 31
   sum of, 16, 484
Velocity, 141, 441, 561, 562
   in polar coordinates, 479
   radial component of, 480
   rate of change of, 141
   transverse component of, 480
Volume, by double integrals, 643
   of a solid, 644
   of a solid of revolution, 327, 332
   of a solid with known cross section, 325
Volume integral, 654, 655


Witch, 420
Work, 345, 621